SIGNAL INTEGRITY AND RADIATED EMISSION
OF HIGH-SPEED DIGITAL SYSTEMS

SIGNAL INTEGRITY AND RADIATED EMISSION
OF HIGH-SPEED DIGITAL SYSTEMS

Spartaco Caniggia
Italtel (retired) and Expert of Comitato Elettrotecnico Italiano (CEI), Italy

Francescaromana Maradei
Sapienza University of Rome, Italy

WILEY
A John Wiley and Sons, Ltd, Publication

This edition first published 2008

Registered office
John Wiley & Sons Ltd, The Atrium, Southern Gate, Chichester, West Sussex, PO19 8SQ, United Kingdom

For details of our global editorial offices, for customer services and for information about how to apply for permission to reuse the copyright material in this book please see our website at www.wiley.com.

Library of Congress Cataloging-in-Publication Data

Caniggia, Spartaco.
Signal integrity and radiated emissions of high-speed digital systems /
Spartaco Caniggia, Francescaromana Maradei.
p. cm.
Includes bibliographical references and index.
ISBN 978-0-470-51166-4 (cloth)
1. Electromagnetic interference. 2. Digital electronics. 3. Very high speed integrated circuits. 4. Crosstalk. 5. Signal processing.
I. Maradei, Francescaromana. II. Title.
TK7867.2.C36 2008
621.382′24–dc22 2008013159

A catalogue record for this book is available from the British Library

ISBN 978-0-470-51166-4 (HB)

Typeset in 10/12pt Times by Aptara Inc., New Delhi, India.
Printed and bound in Singapore by Markono Print Media Pte Ltd, Singapore.

Contents

List of Examples

Foreword

The widespread growth of high-speed and broadband systems poses increasing challenges to the designers of modern information and communication equipment. Effective *signal integrity* (SI) and *electromagnetic compatibility* (EMC) solutions are fundamental for the marketing of reliable devices able to ensure perfect functionality and compliance with legal standards.

In the area of high-speed design, it is necessary to analyze power and signal integrity issues at an early stage of the design, before the prototype board is fabricated. Typical signal integrity issues are reflections and crosstalk. Typical power integrity issues are power supply system input impedance, simultaneous switching noise, *printed circuit board* (PCB) resonance, decoupling capacitor placement, and edge radiations. Power distribution systems play an important role in power and signal integrity and *electromagnetic interference* (EMI). It is a common experience that the EMC of digital systems improves significantly when boards are subjected to careful power integrity and signal integrity analysis. To this end, it is necessary to analyze the input impedance between power and ground and further provide an equivalent circuit model for signal integrity analysis. Also, the prediction of simultaneous switching noise must be performed in the time domain, employing non-linear models for drivers and receivers.

This book provides a good overview of the above-mentioned SI and EMC issues, and discusses how to design boards with a careful consideration of signal and power integrity. Using realistic case studies and downloadable software examples, a leading expert from industry and his academic coauthor demonstrate today's best practices for designing and modeling interconnects in order to distribute power and minimize noise efficiently.

The reader will enjoy a review of the most important phenomena determining the SI and EMC of PCB systems equipped with digital circuits. The aim of this book is to highlight the effects of variation in design parameters on system performance, and to provide criteria for design. For all of the phenomena mentioned, mathematical models are specified. Many of these models are ready to implement in high-level formula-evaluation programs and/or circuit simulators, and the authors provide useful examples of such implementation.

The approach and practical examples of this book make it a valuable tool for learners and professionals concerned with signal and power integrity and electromagnetic interference, including electrical engineers, system designers, and signal integrity engineers.

In conclusion, the efforts made by the authors to produce this quality contribution should be highly praised. I am sure that the publication of this book represents a significant step forward in promoting the awareness of SI and EMC problems among the designers of electrical and electronic systems. With this in mind, I recommend this book to the reader, and I wish every success to this work and to its authors.

Flavio Canavero
Politecnico di Torino,
Turin, 2008

Preface

Following the foreword by Prof. Canavero, we would like to give more details concerning the content of this book. The book is designed to meet the needs of high-speed digital designers and as support for electrical engineering and physics students who desire to gain knowledge of *signal integrity* (SI), *electromagnetic interference* (EMI) and *radiated emission* (RE) topics. One of the two authors has had industrial experience over a period of more than 30 years in designing telecommunication equipment, and the other author is a professor of the University of Rome, with research and academic educational experience. Therefore, the intention of the authors is to provide the reader with the fundamental principles of SI, EMI, and RE by using rigorous theory, appropriate models for prediction, and experimental results for investigation and validation of the models. The book's main objective is to study the *electromagnetic* (EM) phenomena concerning *printed circuit boards* (PCBs) and their attached cables, considering the interconnect driven and loaded by digital devices. This is a very important aspect to consider, as the non-linear behavior of digital devices greatly influences the performance of the interconnect in terms of reflection, crosstalk, and switching noise. Modeling of PCBs and cables is a basic topic of the book, as this helps significantly in reducing the cost of a product and in saving design time. This is a very important issue, bearing in mind the new European EMC Directive of December 2004. One of the main novelties of the new Directive is to allow the use of appropriate models for computations/simulations in order to demonstrate that a product meets the regulatory limits. Therefore, costly and time-consuming radiated emission measurements in very large semi-anechoic chambers or open-area test sites can be avoided.

Many of the models outlined in the book are new (not presented in other books) and experimentally validated. Most of them come from past and recent works presented by the authors and/or other researchers at international symposia and published in specialized magazines such as the *IEEE Transactions*. Other models are presented for the first time in this book. The models are based on analytical approaches or on equivalent circuits suitable for implementation in popular mathematical commercial codes (MathCad, MATLAB®) or SPICE-like circuit simulators respectively. One of the most important feature of the book is the use of commercial full-wave numerical codes to investigate phenomena related to SI and RE. How to simulate PCBs in 3D structures or set-ups for measurements is outlined, with explanation of the meaning of the basic parameters used. How to build up an appropriate model taking into account the significant parts of the structure considered is also discussed. The very dangerous resonant effects occurring when a PCB structure under study is electrically large for the frequencies of interest can be evidenced by these tools.

Many of the experiments reported come from direct industrial experience. Measurements and simulations are used to investigate basic mechanisms concerning SI and RE, to validate models, and to provide design rules for practicing engineers as well. And last but not least, an important purpose of this book is that the reader may be helped in understanding and in using professional industrial software for SI and EMI predictions that can be purchased from software houses for pre- and post-layout simulations, such as Mentor (Hyperlynx), CST (CST STUDIO SUITE 2008), ANSOFT (HFFS), etc. It is worth pointing out that these modeling tools can provide the design team with reliable numerical results, taking guesswork out of the design and providing SI and EMC engineers with the credibility to get their design recommendations seriously considered by the team.

The book is organized as follows:

Chapter 1 is an introductory chapter in which the reasons for SI and RE investigations are explained, basic definitions of the physical quantities used are given, and some examples are reported to support the basic concepts. Introductory material on analytical formulation, circuit models for SPICE-like simulators and full-wave codes based on numerical methods such as *Method of Moments* (MOM) and the *Finite Integration Technique* (FIT) is also included.

Chapter 2 is devoted to a description of the main characteristics of the basic families of digital devices such as TTL, CMOS, and ECL. Some material seems not to have been updated, but the intention of the authors is to provide a background to understand the performance of the latest developments of components for high-speed applications based on CMOS, ECL, and LVDS technologies. The chapter ends with an introduction to the standard IBIS models, which are useful as a fast and accurate behavioral method for modeling I/O buffers based on I/V curve data derived from measurements or full-circuit simulations. An example of the use of this type of model by SPICE is given.

Chapter 3 is devoted entirely to the inductance concept, which greatly influences SI and RE performance. Self and mutual loop inductances are introduced with rigorous theory as the basic background for introducing the concept of partial inductance associated with a part of a loop. The concept of partial inductance is extensively used in the book for modeling interconnects and discontinuities in PCBs. Formulae for typical structures are provided in *Appendix A*.

Chapter 4 is complementary to *Chapter 3* and is devoted to the capacitance which is fundamental with the inductance for building an interconnect model. The definition of common- and differential-mode inductance and capacitance, useful for implementing EMI filters and differential signaling, is provided at the end of *Chapters 3* and *4*.

Chapter 5 provides methods for predicting reflections in interconnects with digital devices. The novelty of this chapter is how to use an exact lossless transmission line model for mathematical codes and/or SPICE in order to compute the signal launched onto the line when the interconnect has powered terminations to enhance the driver capability of a buffer.

Chapter 6 examines the very important topic of crosstalk among traces in PCBs. After the basic concept of inductive and capacitive coupling, an exact circuit model for SPICE, based on odd and even modes of propagation, is presented for two coupled lines. This model is useful for crosstalk and differential signaling simulations. For more than two coupled lines, a model based on n-decoupled modes of propagation is presented, and experimental validations with IBIS-like models for TTL and CMOS devices are provided.

Chapter 7 introduces the lossy line fundamental parameters concerning skin, proximity, and dielectric effects. The problem of modeling lossy lines in the transient domain, considering the frequency dependence of the line parameters, is for the time being not completely solved. Therefore, the main aim of this chapter is to provide methods suitable for the simulation in the time domain of SI in traces and cables. Two methods are described and validated experimentally: one is based on the Vector Fitting technique, which makes it possible to model a segment of line by a net of lumped circuit elements (resistance, inductance and capacitance), the other one is based on convolution integral of S-parameters of the line. Both methods allow simulations directly in the time domain with non-linear loads. *Appendix B* provides closed-form expressions for calculating the characteristic impedance, delay time, and attenuation of traces having the structure of a microstrip or a stripline.

Chapter 8 investigates the noise in the *power distribution network* (PDN) of a PCB that is caused by the switching of digital devices. Fixes are provided to mitigate this type of interference. The limiting effects of the effective partial inductances associated with the decoupling capacitors and leads used for lowering the PDN impedance are discussed in detail. An example of how to simulate by SPICE some types of PCB with different stack-ups and equipped with digital devices and decoupling capacitors is presented and validated experimentally in the transient domain. Three models based on the analytical formulation, circuit simulation and full-wave numerical computation, for predicting in the frequency domain the PDN impedance and resonance frequencies in a pair of power and ground planes populated by decoupling capacitors are given in *Appendix C*. Ground and power bounce phenomena due to parasitic effects associated with the packaging of digital devices is studied by circuit simulations and measurements.

Chapter 9 is devoted to the problem of *radiated emission* (RE) from PCBs and attached cables. After a short introduction concerning how to model the spectrum of typical periodic digital signals and noises, the problem of how to model common and differential emissions in the far-field zone is discussed in order to meet the RE limits of the standards. The task of separating these two modes of emission is very important for setting design rules to mitigate radiated emission by using fixes such as power and ground planes in PCBs, EMI filters, and shielded cables. Radiation mechanisms are investigated by equivalent circuits of the structures under study, considering the parasitic elements in terms of inductances and capacitances. In doing so, the concept of partial inductance introduced in *Chapter 3* becomes fundamental, as does the concept of capacitance associated with the displacement current between the structure and its environment. The concept of transfer impedance of shielded cables is also introduced, and related models for RE prediction are provided. One of the merits of this chapter is that many experimental results are reported as validation of the models presented. The chapter ends with a discussion on how to simulate complex systems in order to obtain radiation patterns by numerical techniques. *Appendix D* provides closed-form expressions for calculating radiated fields for typical structures in PCBs and set-ups for measurements according to the RE standards.

Chapter 10 investigates several topics concerning grounding in and among PCBs. The first topic deals with the *ground loop coupling* (GLC) that occurs when the return signal path is in common with other signals and subjected to interference from the environment. The concept of transfer impedance as a measure of the level of interference is introduced, and circuit models for GLC computation are given. Ground strategy for mitigating GLC is discussed, and typical connector structures for PCBs are simulated for an appropriate pin assignment. A second very important topic is the determination of the return path of the signal current in typical multilayer PCB structures. The importance of ensuring a low-impedance path for the return current by using appropriate routing and locations of decoupling and stitching capacitors is shown. A third topic deals with an investigation of the benefits offered by several fixes: mitigation of *simultaneous switching noise* (SSN) in PDN with decoupling capacitors located over the entire PCB; splitting planes to block the propagation of unwanted electromagnetic waves; realization of power islands; shorting vias; using EMI filters. These fixes are discussed and quantified by simulations for a complex structure of a PCB with an attached cable inserted in a shielded box. *Appendix E* introduces the nodal method, which is useful for computing the effective partial inductance associated with a return signal path of finite size for GLC computation and return current distribution determination.

Chapter 11 presents two key instruments useful for characterizing PCB structures and for extracting circuit parameters regarding interconnects and discontinuities occurring in high-speed digital systems. The first instrument is the *time domain reflectometer* (TDR), which performs measurements in the time domain. This is very useful for characterizing interconnects in terms of characteristic impedance, delay, and losses. The second instrument is the *vector network analyzer* (VNA), which performs measurements of S-parameters in the frequency domain. These parameters become increasingly important as the working frequency of high-speed digital devices increases, and they make it possible: to determine the resonance frequencies of PCB structures; to extract the circuit parameters of components and discontinuities in PCBs; to characterize losses in traces, as outlined in *Chapter 7*. SPICE and numerical codes are used to highlight the errors a user may encounter if the parasitic elements of the connections between the *device under test* (DUT) and instruments for measurements are not correctly accounted for. The chapter ends with an important discussion regarding the validation of the numerical models used for radiated emission predictions by comparison with measurements. It is shown that the level of agreement between computed and measured data depends on the uncertainty of the set-up used for carrying out measurements, the instruments, and the environment.

Chapter 12, the final chapter of the book, examines two important topics: differential signaling and how to model discontinuities in PCBs. Differential-mode transmission is the most effective manner for designing a very high-speed digital system for best performance in terms of SI and EMI. The main advantages offered by differential-mode signals over singled-ended signals are discussed in detail. A standard technique for implementing differential signal transmission in a system such as Advanced Telecommunications Computed Architecture (ATCA) is introduced, and an example of the realization of a motherboard according to this technique is outlined. The main characteristics of LVDS devices are investigated by measurements (SI and EMI) and by circuit simulations (crosstalk). The chapter ends by describing how to model discontinuities in PCBs such as trace bends, connectors, and ground slots with simple equivalent circuits. An example of the extraction of a parasitic inductance parameter to be associated with a ground slot is provided by using a commercial numerical code. Finally, package-type connections for integrated circuits are presented and discussed.

The basic models used for computations/simulations in the book can be downloaded from the Wiley website. A list of these files, written in MathCad and MicroCap format, is given in *Appendix F*. To run the MicroCap files, the reader must download the demo version of MC9 from www.soft-spectrum.com, while for readers who do not use MathCad the analytical models are also given in files readable by Word.

We would like to acknowledge the invaluable suggestions and constructive criticisms of Prof. F. Canavero, Prof. A. Orlandi, Prof. T. Hubing, Dr Elya Joffe, Dr M. Borsero,

Dr V. Ricchiuti, Dr B. Ross, Dr I. Novak, Prof. F. Leferink, Prof. R. DuBroff, Dr J. Knighten, and Dr E. Klink. In particular, we are grateful to Dr P. Santi of Italtel for reviewing the entire book and all the technical EMC and SI staff of Italtel who helped in performing several simulations and carried out all the measurements reported in the book.

Spartaco Caniggia (ex-Italtel and EMC consultant) and
Francescaromana Maradei (Sapienza University of Rome)
2008

1

Introduction to Signal Integrity and Radiated Emission in a Digital System

This is an introductory chapter in which the motivations for studying the subjects of *Signal Integrity* (SI) and *Radiated Emission* (RE) are discussed.

Signal integrity is a very important task and deals with the need to ensure that electrical signals are of sufficient quality for proper operation. Signal integrity affects all levels of electronics packaging, including, but not limited to, the *Integrated Circuit* (IC). For high-speed digital products, at the level of an IC package or *Printed Circuit Board* (PCB), the main issues of concern for SI are reflections occurring because of interconnect discontinuities, noise induced by neighbouring connections (crosstalk), and noise on power distribution, produced by switching of the digital devices. All these noises can cause functional problems if they are not mitigated by controlling parameters such as the characteristic impedance and spacing of interconnects, which, owing to fast switching of the actual digital devices, should be considered as transmission lines. An overview of the noises affecting SI is given in this chapter, leaving a detailed discussion to the following chapters where the different noises are introduced and investigated separately.

The interest in radiated emission is due to the fact that an apparatus or system must be electromagnetically compatible with its environment. Electronic devices generate electromagnetic fields that unintentionally propagate away from the device's structure, and they may interfere with their normal operation or the normal operation of other devices in close proximity. For this reason, the allowable radiated emissions from electronic modules are regulated by mandatory standards which must be complied with before marketing the apparatus or system. In this chapter, FCC part 15 and CISPR 22, relating to emission from digital systems, are highlighted, and the sites for measurements are discussed. Particular emphasis is given to the new EMC European Directive 2004/108/EC which makes it possible to demonstrate conformity of a product to the essential requirements of emission and immunity by using calculations and therefore computer simulations instead of measurements. The three main sources of emissions of a complex digital system (traces, integrated circuits, and cables) are investigated.

Signal Integrity and Radiated Emission of High-Speed Digital Systems Spartaco Caniggia and Francescaromana Maradei

An example of a complex system that complies with the RE requirements is reported, and its emission spectrum with and without shielding is discussed. The difficulties in mitigating radiated emission are shown by using simple radiating structures.

In the third part of this chapter, signaling parameters significant for SI are defined. Some examples of data errors when the voltage and current specifications of the devices are not met owing to reflections on the interconnects are provided. An example of an eye diagram for jitter signal evaluation is provided.

Finally, the last part of the chapter offers an overview of the methodologies suitable for developing prediction models of SI and RE problems. Advantages and drawbacks regarding mathematical, circuit, and numerical codes for simulation are discussed. A list of problems solved by simulation and reported in the book is provided.

1.1 Power and Signal Integrity

Power and signal integrity addresses two concerns in electrical design aspects: the timing and the quality of the signal. The goal of power and signal integrity analysis is to ensure reliable high-speed data transmission. This can mainly be done by setting up design rules in order to mitigate the delays and distortions of digital signals due to reflections, *crosstalk*, and switching noise (ΔI*-noise*):

- Reflection refers to signal waveform distortion caused by discontinuities along the interconnects of the digital devices, such as impedance mismatch, stubs, vias, and other line discontinuities.
- *Crosstalk* refers to the noise produced in a signal line by other lines as inductive and capacitive coupling.
- Switching noise refers to the disturbances induced in a signal line by the voltage drop along the inductive path of the power supply network for the IC and its packaging. This noise is also called *ground bounce*, ΔI*-noise* or *Simultaneous Switching Noise* (SSN).

Power and signal integrity are not regulated by standards because the associated disturbances are considered as internal noises of the system and therefore they do not interfere with the environment or other nearby equipment or systems. It is the task of the PCB designer to prepare a set of design rules to limit these types of noise which affect both timing and quality of the signal. To accomplish this goal, circuit and numerical simulations are used.

The first step to evaluate these types of problem consists in modeling by an equivalent electrical circuit the physical structure of the PCB where the digital devices are located. The physical parameters of the PCB to be considered are: the width, thickness, and spacing of the interconnects (traces); the dielectric constant of the substrate; the via or hole diameter and spacing. The modeling is usually performed by means of closed-form expressions when available, or by using field-solver programs to calculate the desired inductances, capacitances, and resistances. Once these linear network parameters are known, any required quantity, such as the characteristic impedance of the line Z_0 (ohm), the line propagation delay time T_D (seconds), and the line coupling coefficients, can be calculated. The first two parameters are defined as

$$Z_0 = \sqrt{\frac{L}{C}} \tag{1.1a}$$

$$T_D = \sqrt{LC}\,l = t_{pd}l \tag{1.1b}$$

where L is the per-unit-length inductance of the line (H/m), C is the per-unit-length capacitance of the line (F/m), l is the length of the line (m), and t_{pd} is the per-unit-length propagation delay time of the line (s/m).

The key parameters defined by Equations (1.1) are nominal and frequency independent in the frequency range of interest. They refer to a lossless interconnect and depend on the interconnect geometry. For typical stripline and microstrip trace structures used in multilayer PCBs, Z_0 and T_D can be computed with closed-form expressions as reported in *Appendix B*. These two parameters greatly affect the performance, the net design, and the noise limits of power and signal distribution, as will be shown in the following sections [1].

1.1.1 Power Distribution Network

The *Power Distribution Network* (PDN) for a typical PCB is depicted in Figure 1.1a. A *Voltage Regulator Module* (VRM) (i.e. DC/DC converter) provides the required power supply to the digital device by a pair of bus bars or solid copper planes indicated as *Power* and *Ground*. At points P and G, a digital device (i.e. IC) is connected. In this representation, the device has a gate switching from low to high level, and a step voltage ΔV_S with a rise time t_r is launched onto the line (trace) towards a receiver placed somewhere in the PCB. The traveling signal is given by

$$\Delta V_S = Z_0 \Delta I_S \tag{1.2}$$

where Z_0 is the characteristic impedance of the line (trace) and ΔI_S is the variation of current in line before and after the switching. This happens every time the double delay of the trace $2T_D$ is much higher than the switching rise time t_r or fall time t_f of the output voltage. Considering a typical per-unit-length delay time of about 6 ns/m, and rise and fall times (i.e. t_r and t_f) of about 1 ns or less, which is a common situation in a PCB, it is easy to deduce that traces must be modeled as transmission lines. When this situation does not occur, the line is said electrically *short*, and the load of the driver can be modeled by a lumped capacitance which is the sum of the trace capacitance and the receiver input capacitance. For short lines, the inductive effect can be neglected. In any case, at the output of the driver there is a current variation ΔI_S that must be provided by the PDN. When the gate switches, another impulsive current, denoted by ΔI_t, could be sunk by the gate. This current is caused by the momentary simultaneous switch-on of the two output transistors in the typical totem-pole configuration characteristic of TTL and CMOS devices that, with their complementary condition on or off, determine one of the two (high or low) logic levels (see *Chapter 2*). Therefore, the total switching current that the PDN must provide to the IC is given by

$$\Delta I = \Delta I_t + \Delta I_S \tag{1.3}$$

Denoting by Z_{PDN} the characteristic impedance of the PDN or the impedance looking back from the points P and G where an IC is connected, the voltage drop between these two points is given by

$$\Delta V_{PG} = Z_{PDN} \Delta I \tag{1.4}$$

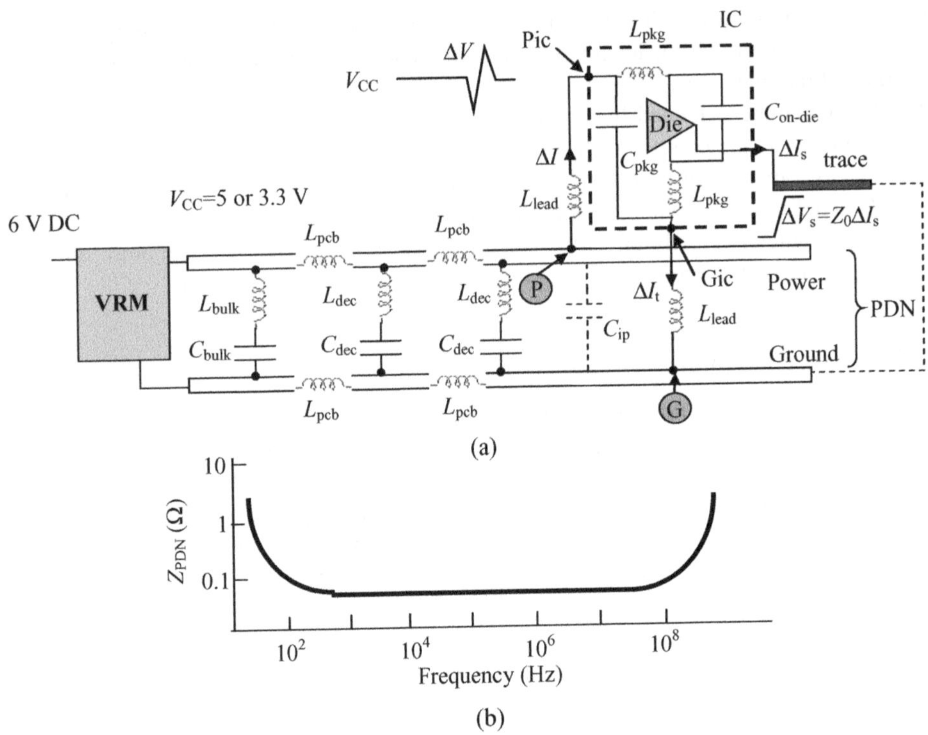

Figure 1.1 Power distribution network in PCBs: (a) equivalent circuit; (b) impedance versus frequency at points P and G

This is an impulsive disturbance, indicated as ΔI-*noise*, that sums to the DC power supply V_{CC} of the device. As many gates can switch simultaneously, this noise, known also as *Simultaneous Switching Noise* (SSN), could rise to dangerous values for the functionality of the system. Therefore, the main task of an electrical designer is to make the parameter Z_{PDN} as low as possible. This goal can be pursued by increasing the capacitance term in Equation (1.1a) by means of decoupling capacitors, and by using power and ground planes instead of bus bars in order to have a higher interplane capacitance C_{ip} and a lower PDN inductance L_{pcb}. This is usually accomplished as shown in Figure 1.1a.

At the VRM output, a large decoupling capacitor, indicated as bulk capacitance C_{bulk}, is inserted for filtering the lower-frequency components of low- and high-level changes caused by circuit switching throughout the PCB. For filtering the higher-frequency components, a number of decoupling capacitors with capacitance C_{dec} are distributed at regular intervals along the PDN and located near the devices. Between each pair of capacitors there is a power/ground effective inductance L_{pcb}. The problem with decoupling capacitors is that their action as capacitance is affected by the inductance associated with the component itself, plus the inductance associated with the component connections to the power and ground conductors, denoted by L_{bulk} and L_{dec}. The effect of these parasitic inductances is shown in Figure 1.1b, where a typical impedance Z_{PDN} is plotted versus frequency. At very low frequencies the network appears to be capacitive, while at very high frequencies the network appears to be inductive. In the mid-range, capacitances compensate for inductances, yielding a very small impedance for the PDN. The goal is to design the PDN so that the curve is flat and resistive throughout the frequency range required by the speed of the circuits.

To achieve this goal, two strategies can be applied: (1) choose an appropriate number of decoupling capacitors, located in order to minimize their parasitic inductances; (2) make the interplane capacitance large by increasing the dielectric constant ε_r and, above all, by minimizing the distance between the two power and ground planes. All this will be discussed in *Chapter* 8.

Actually, ensuring a low Z_{PDN} curve could not be sufficient for preserving the IC from malfunctions. In fact, the connection of the power and ground pins to the PDN must be realized with care taken to minimize the loop inductance associated with the connections. Looking at points Pic and Gic in Figure 1.1a, which correspond to the power and ground pins of the IC respectively, a further voltage drop on the path of the power supply between points P and Pic must be considered. This is indicated as *power bounce noise* and is given by

$$V_{\text{lead}} = L_{\text{lead}} \frac{\Delta I}{\Delta t} \tag{1.5}$$

where L_{lead} is the effective inductance associated with connection between points Pic and P. This concept of effective inductance associated with a segment of the loop, known as *partial inductance*, is very useful for package modeling, and it will be defined in *Chapter* 3 starting from the loop inductance definition.

Up to this point, it is a task of the PCB designer to minimize all these inductive effects. However, looking within the IC towards the die where the circuitries are allocated, it is important to consider the package inductance L_{pkg} associated with the pins–die connection (between points Pic and die). This inductance also produces a voltage drop on the power supply. It is a task of the device manufacturer to minimize L_{pkg} and to provide a die capacitor $C_{\text{on-die}}$ in order to have an on-die filtering that permits less impulsive current to be required from the PDN. This will be considered in depth in *Chapter* 8 by circuit simulations.

1.1.2 Signal Distribution Network

The *Signal Distribution Network* (SDN) for a high-speed digital system accounts for a considerable part of the total path delay. To minimize this delay, it is very important to examine the role that the characteristic impedance Z_0 plays in designing a SDN. The choice of an appropriate characteristic impedance Z_0 is important to all aspects of the SDN, as it affects net design, net performance, and disturbances such as reflection, crosstalk, and ΔI*-noise*. By superimposing these effects, a design space can be generated for selecting Z_0, as will be shown in the next section.

Consider the interconnect shown in Figure 1.2 where a driver, represented by its Thévenin equivalent circuit, sends a signal onto a line with a receiver R1 along the line (trace), and a cluster of receivers, R2–R4, at the end of the line. The reference plane could be the power plane or the ground plane of the PCB in accordance with the fact that the PDN impedance Z_{PDN} is extremely low in the frequency range typical of signals, as shown in Figure 1.1b, and the AC return current tends to flow along a path that ensures less loop impedance. Each

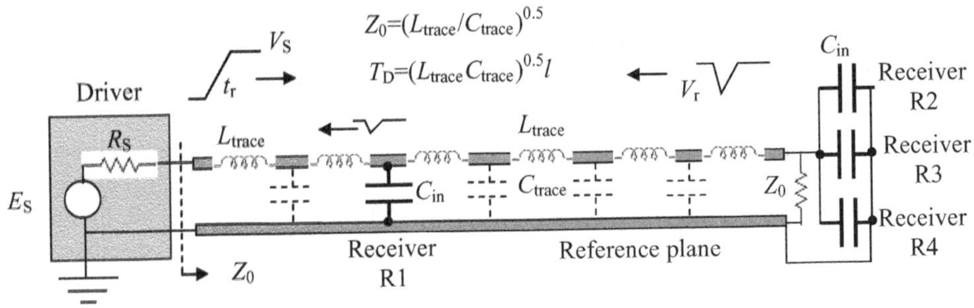

Figure 1.2 Interconnect (trace in PCB) of length l, with receivers concentrated at the end of the line causing large reflections

receiver is represented by its input capacitance C_{in} of some pF, and the line is terminated with its characteristic impedance Z_0 in order to avoid resistive mismatching which could generate large reflections. The inductance L_{trace} and the capacitance C_{trace} are parameters associated with a segment of the trace and determine the line characteristic impedance Z_0 and the delay time T_{D} according to Equations (1.1). Dangerous reflection peaks are generated by excessive capacitive loads. The mechanism can be explained in this way. When a signal V_{S} is sent out by the driver, a fraction of the unloaded voltage swing E_{S} enters the line because of the voltage divider consisting of R_{S} and Z_0. The signal V_{S} is given by

$$V_{\mathrm{S}} = \frac{Z_0}{R_{\mathrm{S}} + Z_0} E_{\mathrm{S}} \tag{1.6}$$

From Equation (1.6), two important facts must be observed. To maximize the sending signal in order to ensure switching of the receivers at the first step with a suitable margin, the driver resistance R_{S} should be minimized and the characteristic impedance Z_0 should be maximized. The upper bound of Z_0 is dictated by the fact that, above certain values, undesired capacitive reflections and excessive coupling effects between traces could occur. While the signal propagates along the line, reflections are generated at each capacitive discontinuity. When the lines are long and the losses can be neglected, the signal travels with unchanged rise time t_{r}. For a line to be considered long, each segment between two loads should have a propagation delay time T_{D} that exceeds one-half of the rise time t_{r}. At each capacitive discontinuity a negative reflection is generated that has maximum value and width given by [2]

$$V_{\mathrm{r}} = -\frac{C_{\mathrm{D}} Z_0 V_{\mathrm{i}}}{2t_{\mathrm{r}}} \tag{1.7a}$$

$$t_{\mathrm{w50}} = t_{\mathrm{r}} \tag{1.7b}$$

$$t_{\mathrm{w0}} = t_{\mathrm{r}} + 1.5 C_{\mathrm{D}} Z_0 \tag{1.7c}$$

$$T_{\Delta} = \frac{C_{\mathrm{D}} Z_0}{2} \tag{1.7d}$$

where C_{D} is the capacitance of the discontinuity, V_{r} is the peak voltage of the reflection, $V_{\mathrm{i}} = V_{\mathrm{S}}$ is the incident voltage magnitude, t_{w50} is the width of the reflection at the 50 %

points, t_{w0} is the width of the reflection at the baseline, and T_Δ is the delay added to the main line incident signal because of the discontinuity.

Thus, as C_D and Z_0 increase, the reflections V_r defined by Equation (1.7a) become larger and the delay time T_Δ given by Equation (1.7d) becomes longer. Limits should be set on these parameters. If they are not, the reflection from the load at the end of the line, where a large capacitance is formed by the cluster of the receivers, could be so large when it hits the receiver R_1 that it transiently switches into its down state, causing a logical error in a downstream latch. The discontinuity capacitance C_D associated with the load at the end of the line is the sum of the receiver input capacitances C_{in} and the capacitances associated with the connection of the receivers to the line.

A way to avoid this problem is to distribute the receivers along the main line at regular electrically short intervals, or, in other words, so that the delay of the line between two loads is less than one-half of the rise time t_r. In this case, the reflections merge together and the load capacitance is now combined with the line capacitance and treated as if it were uniformly distributed along the line. The additional line capacitance acts to lower Z_0 and increases the propagation delay time T_D. The new line parameters are expressed as

$$Z_{0eq} = \sqrt{\frac{L_{trace}}{C_{trace} + (C_{in} + C_{stub})n/l}} \tag{1.8a}$$

$$T_{Deq} = \sqrt{L_{trace}(C_{trace} + (C_{in} + C_{stub})n/l)}\,l \tag{1.8b}$$

$$T_{\Delta eq} = T_D(Z_0/Z_{0eq} - 1) \tag{1.8c}$$

where n is the number of receivers distributed for the line length l, C_{stub} is the capacitance associated with the trace connecting the receiver to the main line, and $T_{\Delta eq}$ is the added delay due to loading.

Treating each discontinuity as a lumped capacitance is helpful for understanding fundamental dependencies. However, for actual design work, the effective partial inductance associated with the leads and IC package must be accounted for as done for PDN in Figure 1.1a. Once inductances are introduced into the model, the analytical approach becomes unwieldy and circuit simulators based on SPICE must be used to predict the desired signal waveforms. Reflections will be investigated in detail in *Chapter 5*, where circuit models for their predictions will also be presented.

1.1.3 Noise Limitations and Design for Characteristic Impedance

Three types of noise generally concern the electrical PCB designer: reflection, switching noise, and *crosstalk*. Very often, reflections may be treated separately, while the other two noises can interact. An example of PCB where ΔI*-noise* and crosstalk generated by the digital devices switching can sum, causing a false switching, is shown in Figure 1.3. The upper gate of chip 1 switches and sends onto line 1 a voltage step $\Delta V_1 = Z_0 \Delta I_1$, where Z_0 is the characteristic impedance of line 1, and ΔI_1 is the current difference at the driver output before and after the switching. Recall that Z_0 depends on the trace geometry and on the dielectric constant of the PCB substrate. If line 1 is not matched or terminated with a resistance equal to Z_0, the voltage step ΔV_1 can return in part as reflection towards the driver, causing waveform

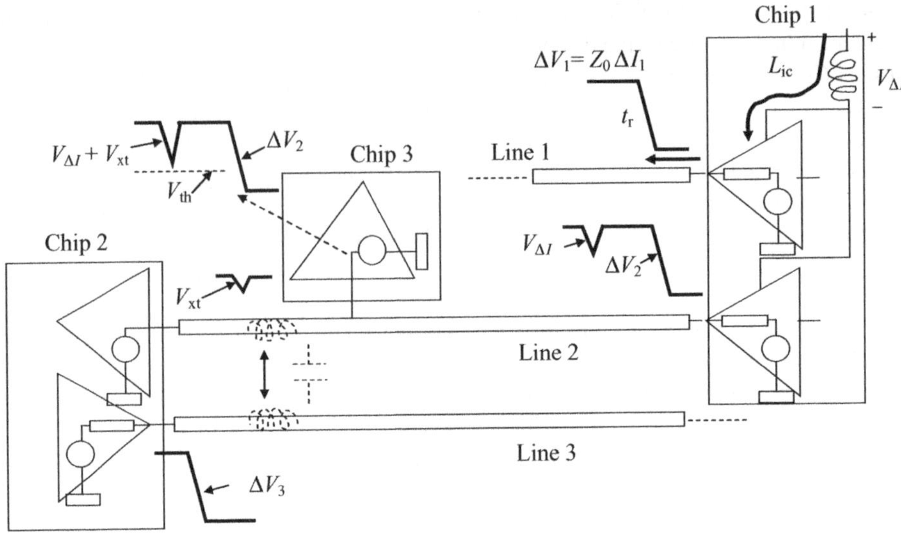

Figure 1.3 Illustration of a PCB where reflection, crosstalk, and switching noises sum

distortions. The current ΔI_1 causes the voltage drop (ΔI-*noise*) given by $V_{\Delta 1} = L_{ic}\Delta I_1/\Delta t_1$, where $L_{ic} = L_{lead} + L_{pkg}$ is the effective power supply inductance of chip 1, and $\Delta t_1 = t_r$. In this example it is assumed that the voltage drop caused by the PDN of Figure 1.1a can be neglected, and only the voltage drop on the effective inductances associated with the lead and package conductors of the IC are significant (*power bounce*). This voltage drop becomes a disturbance for the signal sent by the lower gate of chip 1 just some nanoseconds before, and superposes on the step voltage ΔV_2.

The lower gate of chip 2 switches from a high to a low state with a swing ΔV_3 and, owing to inductive and capacitive coupling between line 2 and line 3 (*crosstalk*), induces a disturbance V_{xt} on line 2. The total disturbance $V_{xt} + V_{\Delta I}$ can cause a false switching at the input of chip 3 if the voltage of the signal plus the total disturbance is lower than the threshold voltage V_{th} of the receiver. The threshold voltage is the nominal level where the receiver changes state. It is important to point out that all three kinds of noise (reflection, *crosstalk*, and ΔI-*noise*) depend on the parameter Z_0. The dependency of crosstalk on Z_0 will be investigated with suitable modeling and measurements in *Chapter 6*. Superimposition of crosstalk and disturbance in line, produced by ΔI-*noise*, happens frequently when a large number of simultaneous switchings occur in the same IC. This will be investigated by modeling and measurements in *Chapter 8*.

The aim of the designer is to find a range of values for Z_0 where the immunity of a generic receiver is maximized, as illustrated in Figure 1.4. The receiver noise immunity V_{NI} is the margin that a designer must preserve to ensure functionality of the system in the presence of other internal and external disturbances, and is defined as

$$V_{NI} = (\Delta V - V_{th}) - (V_{xt} + V_{\Delta I}) > 0 \tag{1.9}$$

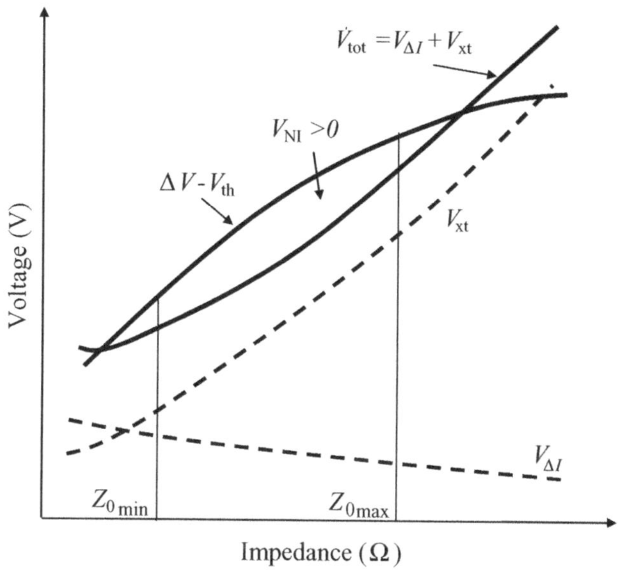

Figure 1.4 Total noise versus the characteristic impedance Z_0 of a PCB trace

where ΔV is the step signal, V_{th} is the threshold voltage of the receiver, V_{xt} is the crosstalk disturbance, and $V_{\Delta I}$ is the ΔI*-noise*.

The condition defined by Equation (1.9) must be verified for both low-to-high and high-to-low switching. As the characteristic impedance Z_0 increases, the step signal ΔV increases and the disturbance $V_{\Delta I}$ decreases because the driver requires less switching current, but, unfortunately, the disturbance V_{xt} increases more than the decrease in $V_{\Delta I}$. This is due to the fact that, to have higher Z_0 values, the traces must be positioned more distant from the reference return plane, and therefore the inductive and capacitive coupling parameters are more significant. To quantify signals and noises, it is very important to have circuit models of the digital devices and their interconnects. How to build up these models will be one of the main purposes of the following chapters.

1.2 Radiated Emission

In this book, radiated emission is considered together with signal integrity because they are strictly correlated. Reflections of signals have the effect of increasing the radiated emission from PCBs, while the switching noise produced by the digital devices generates strong radiations from cables attached to PCBs.

1.2.1 Definition of Radiated Emission Sources

Radiated Emission (RE) regards the unwanted electromagnetic field produced by PCBs and cables of an equipment or system. Radiated emission is regulated by standards because the associated disturbance is considered as an external noise that can interfere with the environment or other nearby equipment or systems. Experience, measurements, and computer simulations are the tools for preparing a set of design guidelines that take into account the technologies used. Several possible sources and different types of emission can be distinguished. This task is generally very difficult because high values of emission are often due to the unwanted *common-mode* (CM) currents on PCBs and cables. *Common-mode* currents are produced by voltage drops in power and ground planes, caused by impulsive noises flowing through parasitic inductances. The noise voltages across the parasitic inductances feed the cables attached to the PCB which act like antennas. Another source of *common-mode* current on cables is the dissymmetric structure of I/O devices which is difficult or impossible to predict. Although the *common-mode* current is much lower than the signal or *differential-mode* (DM) current (μA versus mA), it produces very high levels of emission, as it returns to the source in the form of electrical and displacement currents making large loops and often uncontrolled paths.

In a PCB, three types of emission source can be identified, as illustrated in Figure 1.5: *Integrated Circuits* (ICs), traces, and cables attached to the PCB.

- Emission from ICs is due to the switching current that flows within the device and forms a small loop. The radiated field can be calculated as radiation by a small loop antenna once the current spectrum is known.
- Emission from traces is due to the signal current. A trace and its reference conductor, generally a plane in a high-speed digital system, form a *Transmission-Line* (TL) structure.

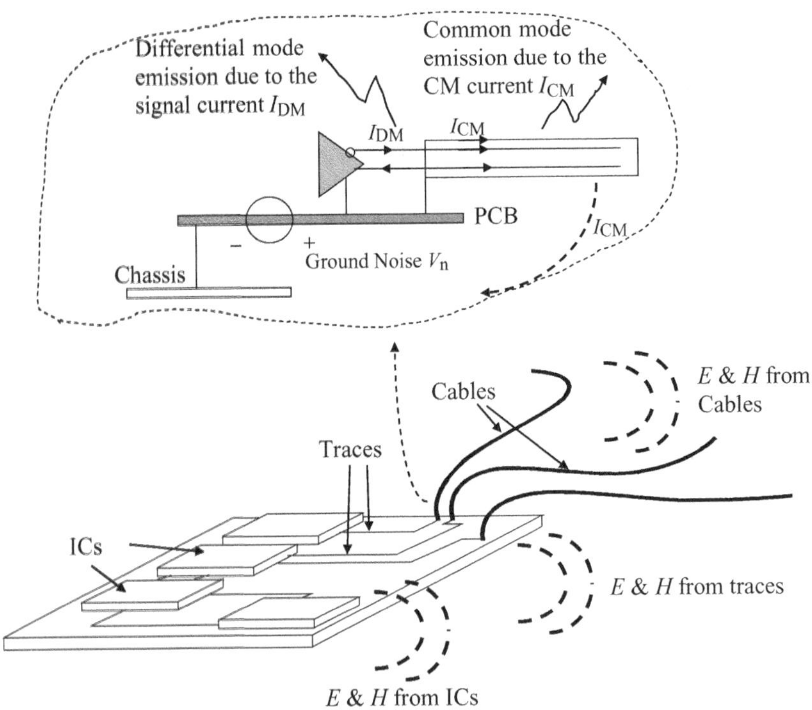

Figure 1.5 Illustration of emission sources from a PCB

Removing the plane and using as the return path a conductor with a distance twice the height of the trace from the plane (image theory), the radiated field can be calculated by segmenting each conductor in electrically small dipoles (length much smaller than the minimum wavelength of interest) and calculating the current in each segment by the TL model, and by accounting for the input/output characteristics of the drivers/receivers.

- Emission from cables is due to the wanted *differential-mode* current I_{DM} used for signaling (often this type of emission can be neglected owing to the cancellation effect that occurs in the pair of signal wires), or due to the unwanted *common-mode* current I_{CM} on the cable which can be caused by the noise V_n in the power and/or ground planes of the PCB, as well as by the unbalance of the driver. The radiated emission from cables can be calculated by combining the transmission-line model (i.e. *differential-mode* emission) and the long monopole or dipole antenna models (depending on the position of the cable), fed by the noise in the PCB, and having the cables as branches (i.e. *common-mode* emission). With attached cable, it is intended that an I/O cable, because of the low output impedance of its line driver device, behaves like a wire connected to the ground of the PCB. Even if the driver does not transmit any signal, the cable emits like an antenna fed by the voltage noise V_n occurring in the PCB.

The mechanisms of emission and the relative models will be described in detail in *Chapter 9* and throughout this book, starting from signal integrity considerations.

1.2.2 Radiated Emission Standards

There are three classes of radiated emission requirements that are imposed on digital systems:

1. Those mandatory for selling a product.
2. Those imposed by some organizations as proof of quality.
3. Those imposed by the product manufacturer.

The mandatory requirements cannot be avoided for the products to be marketed. An example is given by the European Community which obliges manufacturer to demonstrate the conformity of their products to the limits of emission imposed by the relevant standard. An example of the second type are the requirements imposed by Telcordia Technologies to the manufacturers of network telecommunications equipment [3], in which the emission limits that must be complied with are extended up to 10 GHz instead of the usual upper limit of 1 GHz. On the other hand, the emission requirements that manufacturers voluntarily impose on their products are intended to result in customer satisfaction. This section will be devoted to a brief illustration of requirements of the first type for commercial products. For more details regarding commercial and military standards, and the measurement sites and instrumentations, the reader is referred to Paul's textbook [4].

1.2.2.1 FCC and CISPR Standards

The most popular emission standards for commercial products are the FCC and CISPR standards. The *Federal Communications Commission* (FCC) published, under Part 15 of its Rules and Regulations, a requirement that has had, and will continue to have, an impact for digital products to be marketed in the United States [5]. The FCC standard sets limits for the radiated and conducted emissions of a digital device which is defined as '*any unintentional radiator (device or system) that generates and uses timing pulses at a rate in excess of 9000 pulses (cycles) per seconds and uses digital techniques* …'. Therefore, the range of frequency to be considered starts from 9 kHz. Any product that does not meet the limits imposed by this standard is illegal in the USA. The FCC classify digital device products into Class A and Class B. Class A devices are those that are marketed for use in a commercial, industrial, or business environment, while Class B are those that are marketed for use in a residential environment. The Class B limits are more stringent than those of Class A, as the susceptible devices are likely to be in closer proximity to the product seen as a source of emission. Another reason is that owners of sensitive devices do not have knowledge of how to protect their products from the interference of other products. Examples of Class B products are personal computers and their peripherals. Examples of Class A products are items of telecommunication equipment to be installed in telecommunication centers.

The FCC limits are presented in the remainder of this section, while the measurement procedures to verify compliance will be discussed in the following subsection. The frequency range considered by FCC for conducted emissions extends from 150 kHz to 30 MHz. The frequency range for radiated emissions begins at 30 MHz and extends up to 40 GHz. Radiated emissions concern the electric and magnetic fields radiated by a digital system that may be received by other electronic devices which would be victims of interference. The FCC, as well as other regulatory agencies such as the European Community, requires the radiated electric field to be measured in terms of field strength in dBμV/m (i.e. $20\log_{10}(E \times 10^6)$, with E in V/m). This enables very low and very high levels of electric fields to be plotted in the same graph. Compliance is verified by measuring the radiated electric fields from the product either in a *Semi-Anechoic Chamber* (SAC) or at an *Open Area Test Site* (OATS). The radiated emissions must be measured with the antenna in both vertical and horizontal polarizations with respect to the test site ground plane, and the product must be compliant for both the polarizations.

The upper frequencies of applicability for radiated emissions are given in Table 1.1 and are based on the highest frequency of use in the product. For example, for a personal computer having a clock frequency of 3.4 GHz, its radiated emissions will be measured up to 17 GHz.

Table 1.1 Upper limit of measurement frequency

Highest frequency generated or used in the system or on which the system operates or tunes (MHz)	Upper frequency of measurement range (MHz)
<1.705	30
1.705–108	1000
108–500	2000
500–1000	5000
>1000	5th harmonic of highest frequency or 40 GHz, whichever is lower

Table 1.2 FCC radiated emission limits

Frequency (MHz)	Class A measured at 10 m (dBμV/m)	Class B measured at 3 m (dBμV/m)
30–88	39	40
88–216	43.5	43.5
216–960	46.4	46
>960	49.5	54
>1 GHz	49.5 (AV)	54 (AV)
	69.5 (PK)	74 (PK)

The limits of radiated emissions for both FCC Class A and B products are given in Table 1.2. Up to 960 MHz, the level refers to a quasi-peak detector in the measurement receiver. For measurements above 1 GHz, the limits are referred to an average (AV) or peak (PK) detector. The distances for radiated emission measurements are 3 m for Class B and 10 m for Class A products.

A common practical method for comparing the limits for Class A systems with those for Class B systems is to add about 10 dB to the Class A limits according to the assumption that the emissions fall off linearly with increasing distance of the measurement antenna. Thus, the emissions at 3 m are assumed to be reduced by a factor of 3/10 if the measurement distance is moved to a farther distance of 10 m, and vice versa, and therefore $20\log_{10}(10/3) = 10.46 \cong 10$ dB. According to this extrapolation, it can be observed in Table 1.2 that the Class A limits are some 10 dB less stringent than the Class B limits. This assumption of 10 dB is affected by two errors. The first error is that the emissions from antennas fall off inversely with distance only if the measurement points are in the farfield zone where there are no components of the fields along the direction of propagation, and the ratio between the orthogonal electric and magnetic fields is constant and equal to 377 Ω. An approximate criterion for evaluating the farfield boundary is $d = 3\lambda_m = 3 \times 300/f_{MHz}$, where d is the distance between the *Equipment Under Test* (EUT) and the antenna (in meters) [4]. Therefore, the near-to-farfield boundary at the lowest measurement frequency of 30 MHz is 30 m, but it is 90 cm at 1 GHz. The second error is the presence of a metallic reference plane in both possible test sites, SAC and OATS. This plane causes electromagnetic reflections that algebraically sum with the direct emissions. This will be discussed in *Section 11.3*. In conclusion, the comparison appears somewhat approximate.

The majority of the governmental emission requirements for markets outside the USA are based on the work carried out by the *International Special Committee on Radio Interference* (CISPR), which is a committee of the *International Electrotechnical Commission* (IEC). Although CISPR writes standards, they are not mandatory. However, most countries adopt the CISPR recommendations. The most widely used standard is CISPR 22 [6]. This sets limits for the radiated and conducted emissions of *Information Technology Equipment* (ITE), which basically includes digital systems as described for FCC. By analogy with FCC, limits are provided for Class A and Class B equipment. CISPR 22 has been adopted by the *European Economic Area* (EEA). This includes the members of the *European Union* (EU), which was formerly known as the *European Community* (EC) or the *European Economic Community* (EEC). The new European EMC Directive 2004/108/EC, published on 31 December 2004 (the former directive was 89/336/EEC) took effect on 20 July 2007 [7] and applies to members

Table 1.3 CISPR 22 radiated emission limits for ITE equipment

Frequency (MHz)	Class A measured at 10 m (dBμV/m)	Class B measured at 10 m (dBμV/m)
30–230	40	30
230–1000	47	37

of the EEA. Although the directive refers to a large number of electromagnetic compatibility standards, the primary one is the European Norm EN 55022 [8] often mentioned in this book. This is essentially the CISPR 22 standard published by the IEC.

The radiated emission limits of CISPR 22 (EN 55022) are tabulated in Table 1.3 for both classes of ITE equipment. Note that, in this case, both Class A and Class B emissions refer to a distance of 10 m from the EUT. Moreover, similarly to FCC, the emissions are to be measured with a CISPR 16 receiver having a quasi-peak detector (QP) [4]. Whereas it is straightforward to compare FCC and CISPR emission limits for Class A equipment or systems, it is not as simple to perform comparisons for Class B, as the measurements for FCC compliance are to be carried out at a distance of 3 m, while a distance of 10 m is adopted in CISPR compliance. With the limitations discussed above in using the approximation of about 10 dB according to the inverse distance rule, the FCC limits at 3 m are scaled at -10.45 dB for comparison with CISPR limits, and the deviations between CISPR 22 and FCC limits are reported in Table 1.4. From this comparison it can be observed that CISPR 22 limits are slightly less restrictive up to 88 MHz, more restrictive in the range 88–230 MHz (up to 6.4 dB for Class A in the range 216–230 MHz), and again slightly less restrictive in the range 230–960 MHz. Above 960 MHz the CISPR 22 limits revert to being more restrictive.

CISPR standards are in continuous evolution. The basic standard for instruments and measurement procedures is CISPR 16, which is summarized in Table 1.5. For instance, CISPR-16-1-4 describes the test sites for measurement of radio disturbance field strength not only in the frequency range 30 MHz–1 GHz but also for the range 1 GHz–18 GHz. Alternative test sites such as reverberating chambers for total radiated power measurement are also considered. An example of evolution for products is CISPR 32 on multimedia equipment, which is still in preparation and will replace CISPR 13 (*Sound and television broadcast receiver and associated equipment – limits and method of measurement*) and the CISPR 22 [6] standards.

As this book is mainly focused on providing suitable models for signal integrity and radiated emission predictions, attention is directed towards the set of CISPR 16 documents that concern specifications for radio disturbance and immunity measuring apparatus, including the

Table 1.4 Deviation of CISPR 22 radiated emission limits from FCC radiated emission limits. To allow Class B comparison, a scale factor of -10.5 dBμV/m has been applied to the FCC limits of Table 1.2

Frequency (MHz)	Class A measured at 10 m (dBμV/m)	Class B measured at 10 m (dBμV/m)
30–88	−1	−0.5
88–216	+3.5	+3
216–230	+6.4	+5.5
230–960	−0.6	−1.5
960–1000	+2.5	+6.5

Table 1.5 Some CISPR publications

CISPR 16-1-1	Measuring apparatus
CISPR 16-1-2	Ancillary equipment – Conducted disturbances
CISPR 16-1-3	Ancillary equipment – Disturbance power
CISPR 16-1-4	Ancillary equipment – Radiated disturbances
CISPR 16-1-5	Antenna calibration test sites for 30 MHz to 1 GHz
CISPR 16-2-1	Conducted disturbance measurements
CISPR 16-2-2	Measurement of disturbance power
CISPR 16-2-3	Radiated disturbance measurements
CISPR 16-2-4	Immunity measurements
CISPR 16-3	CISPR technical reports
CISPR 16-4-1	Uncertainties in standardized EMC tests
CISPR 16-4-2	Measurement instrumentation uncertainty
CISPR 16-4-3	Statistical consideration in the determination of EMC compliance of mass-produced products
CISPR 16-4-4	Statistics of complaints and model for the calculation of limits

uncertainties associated with the measurement instrumentation and test site. This last item is very important in order to validate the models experimentally (this will be discussed in *Section 11.3*).

Modeling has recently become very important not only for designing but also for demonstrating conformity with essential requirements of the new European EMC Directive 2004/108/EC. *Essential requirements* means conformity to the emission and immunity limits for a specific product. One important novelty of the new directive is that the conformity can be demonstrated by technical documentation that includes the following information (see Annex IV – Technical documentation and EC declaration of conformity [7]):

> *'The technical documentation must enable the conformity of the apparatus with the essential requirements to be assessed. It must cover the design and manufacture of the apparatus, in particular:*
> *– a general description of the apparatus;*
> *– evidence of compliance with the harmonized standards, if any, applied full or in part;*
> *– where the manufacturer has not applied harmonized standards, or has applied in part, a description and explanation of the steps taken to meet the essential requirements of the Directive, including a description of the electromagnetic compatibility assessment set out in Annex II, point 1,* ***results of design calculation made****, examination carried out, test reports, etc.;*
> *– . . .'*

This means that, contrary to the previous EMC directive, it is no longer mandatory to carry out all the measurements required by the standards applicable to the product, but conformity can also be demonstrated, always referring to the standards, by calculations such as simulations performed with suitable and validated models.

1.2.2.2 Radiated Emission Set-Up for Verification of Compliance

FCC and CISPR 22 require that the radiated emission measurements for compliance should be carried out at an *Open Area Test Site* (OATS) or in a *Semi-Anechoic Chamber* (SAC). In

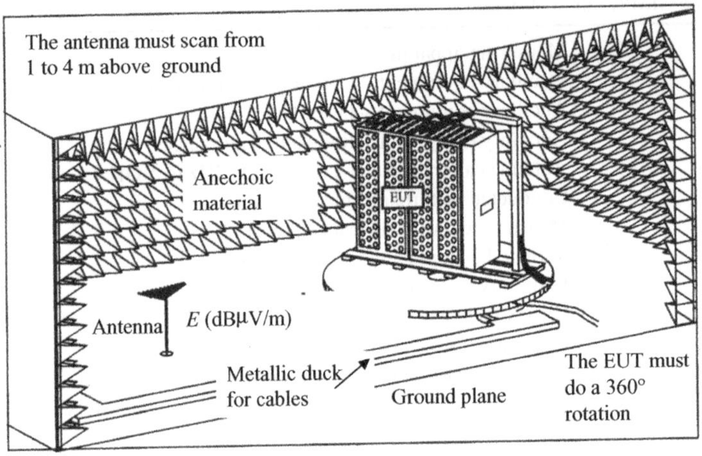

Figure 1.6 Set-up for radiated measurement in a semi-anechoic chamber

CISPR, alternative test sites such as reverberation chambers (CISPR 16-1-4) are also considered. While the OATS should be preferred, especially for FCC, SAC provides an all-weather measurement capability as well as security. An SAC consists of a shielded room lined with radio-frequency absorber material on the sides and at the top of the room to prevent reflections and simulate free space. A schematic representation is shown in Figure 1.6 where the *Equipment Under Test* (EUT) is a standing-floor system. The equipment is positioned on a turntable. The ongoing signal and power cables are to be arranged in order to maximize emissions. The radiated emissions must be measured with the measurement antenna in both horizontal and vertical polarizations with respect to the ground plane of the test site. The antenna must be elevated at a distance above the ground plane in the range 1–4 m, and the maximum emission must be recorded for each frequency. Portable products such as computers are to be placed 1 m above the floor of the chamber. The floor of the room constitutes a ground plane without an absorber, and this causes reflections that must be accounted for when performing simulations by models.

For an accurate measurement, the preferred antenna should be a tuned, half-wave dipole. A half-wave dipole is a linear antenna whose length is 0.5λ at the measuring frequency. If the frequency is changed, the dipole physical length must also be changed in order to maintain an electrical length of 0.5λ. Since this procedure is very time consuming, antennas having large bandwidth covering the whole range from 30 to 1000 MHz are used [4]. This fact must be taken into account when making comparisons with results obtained by simulations because some uncertainty should be associated with the measurements. This will be discussed in *Section 11.3*.

The main reason for choosing an SAC is to prevent external electromagnetic emission which could interfere with the measurements. An example is given in Figure 1.7, which shows the measured electric field from 30 MHz to 1000 MHz in an industrial area. It can be noted that the maximum measured electric field is due to radios and TV broadcast transmitters: they are much higher than the CISPR 22 limit for Class B equipment scaled at 3 m by the +10 dB factor, so that measurements with the EUT powered are unpredictable. In fact, as

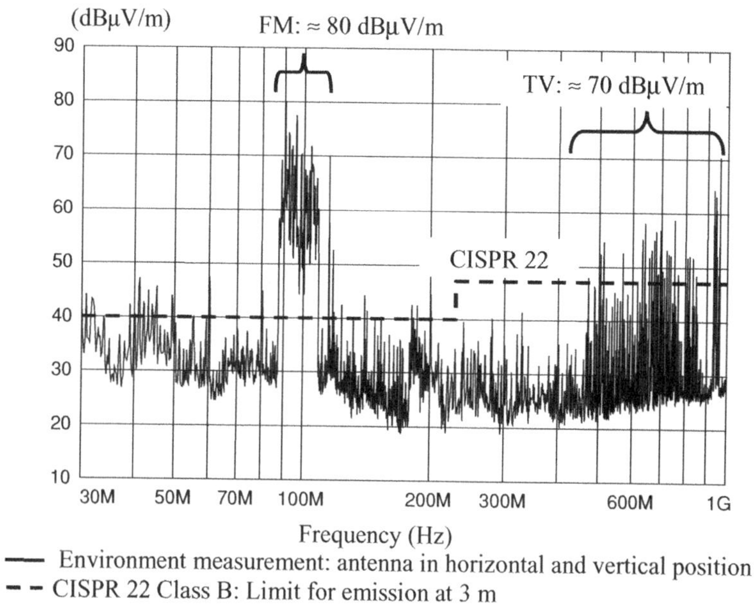

Figure 1.7 Radiated emission measurements of environment outside the semi-anechoic chamber in an industrial area. The CISPR 22 limit is indicated by the dashed line

will be shown later on, the emission profile of a typical digital system without fixes such as shielding is around 70 dBμV/m, that is, approximately of the same order of magnitude as the electromagnetic environment.

1.2.3 Radiated Emission from a Real System

As an example of a typical digital equipment emission profile, the radiated emission measured in the case of a switching telecommunication rack is shown in Figure 1.8. The system consists of several parts: a power voltage regulator (−48 V DC/logic power supply) located on a board, distributed microprocessors, memories, nets for telecommunication switching, and I/O devices for data transmission. The PCBs have several logic families with a maximum clock frequency of 155 MHz.

The equipment was designed to comply with the CISPR 22 Class A limits at 10 m. It was tested for precompliance verification in an SAC for 3 m measurements, taking CISPR 22 Class B as the design goal, with the E-field limit reported to 3 m by the scale factor of 10 dB, as previously discussed. Although design rules to minimize the levels of emission were applied, the equipment was too complex, and shielded racks and cables were required to meet the limits. In fact, with open doors and unclamped cables (which means that the shield of the cables is not connected to the metal frame of the rack), the emission profile is well above the limits of up to 25 dB and comparable with the E-fields present in the environment outside the semi-anechoic chamber. With closed doors and clamped cables (which means that the shield of the ongoing cables is well connected at 360° to the metallic frame of the rack),

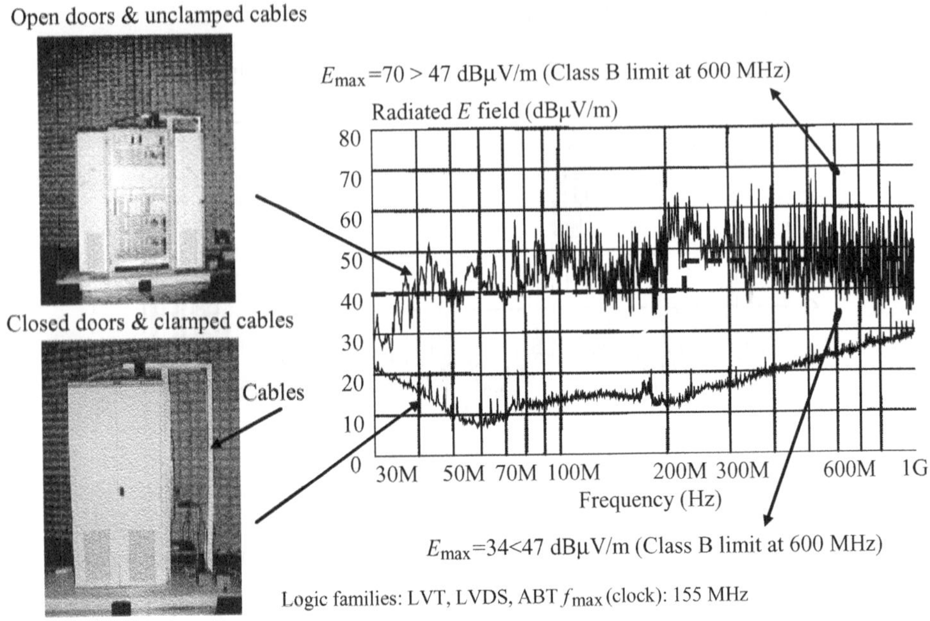

Figure 1.8 Radiated emission measurements at 3 m in a semi-anechoic chamber of switching equipment for telecommunication. The CISPR 22 limit is indicated by the dashed line

the emission profile is much less than the CISPR limit. A maximum measured field of 34 dBμV/m was measured, with 13 dB of margin with respect to the limit. From Figure 1.8 it can be noted that the emission profile has maximum values in the upper frequency range owing to the very fast switching time of the logic devices. Therefore, the apertures and the contacts of the cable shield with the metallic part of the rack must be designed with care. The numerous peaks of emission measured without shielding are due to the harmonics of the clocks and are present in the whole frequency range. Lower, random, continuous levels of emission are caused by data signaling. To meet the limit, as required by FCC and CISPR, it is necessary to check whether all the peaks are under the required level using a peak detector. If some peaks are above the limit, the measurement must be repeated using a quasi-peak detector in order to verify whether the peak is persistent. When this occurs, very often the emission is due to a harmonic of the clock, as will be discussed in *Section 9.1*. To achieve a trade-off between performance and cost of the shielding, it is very important to have design rules to mitigate emission from PCBs. One of the main tasks of this book is to outline methods and models to achieve this goal.

The emission profile of a complex system is not generally due to the sum of the contribution of many sources of emission, it could be mainly the result of one source only if this source has not been properly designed. In order to demonstrate this important issue, two simple experiments were performed with the same PCB consisting of two parallel wires of length $l = 20$ cm and diameter $d = 1$ mm, and separated by a distance $s = 2$ cm, as schematically shown in Figure 1.9a. An 8 MHz oscillator drove an inverter ACMOS device that, by an output resistance of 50 Ω, sent a periodic digital signal to a load of 100 Ω attached to the other end

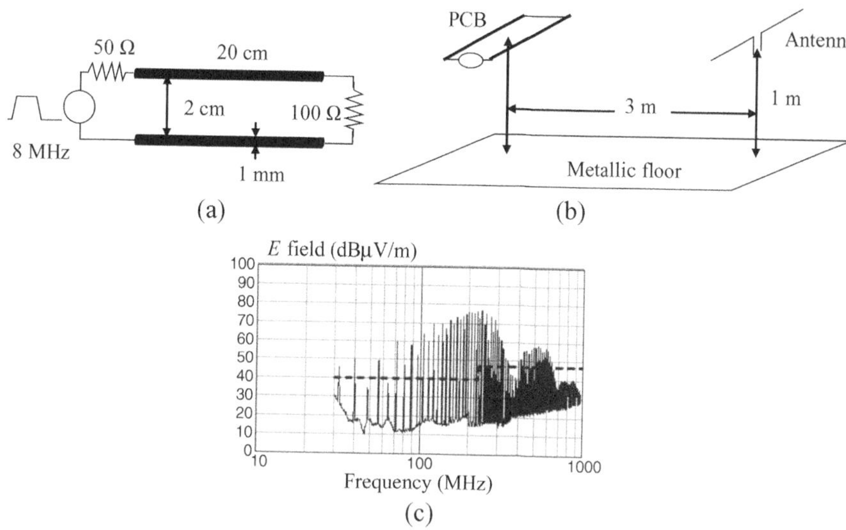

Figure 1.9 A simple experiment to demonstrate the difficulty in meeting the radiated emission limits: (a) schematic and dimensions of the tested device; (b) set-up for measurements; (c) measured radiated emission. The CISPR 22 limit is indicated by the dashed line

of the pair of wires. In order to measure the contribution of the wires only, the active device was placed within a small shielded box that also contained a voltage regulator driven by a 9 V battery. The power supply was very compact and had no connection to a commercial power system. The set-up for measurement is shown in Figure 1.9b, and the measured horizontal radiated emission is shown in Figure 1.9c. Observe that the horizontal emission exceeds the CISPR 22 Class B limit by as much as 30 dB, as in a complex system!

The second experiment was carried out with the same PCB within a shielded rack and with a cable attached to the return wire. The cable goes out by a small hole, as shown in Figure 1.10a. Without the cable, no emission was measured. With the cable, the emission rose to the levels shown in Figure 1.10b. Observe that, in this case also, the horizontal emissions exceed the CISPR 22 Class B limit by as much as 35 dB, and the peaks are mainly located in the low-frequency range, as previously found for a complex system with power and I/O signal cables. These two experiments will be considered in more detail in *Sections 9.2* and *9.6*. Moreover, the radiation mechanism will be investigated and models to predict the emission profiles will be provided.

1.3 Signaling and Logic Devices

In this subsection, the fundamental parameters concerning signaling with digital devices are defined. Digital devices belonging to different logic families will be presented in detail in *Chapter 2*. Anyway, they are all characterized by the following static parameters:

- V_{OHmin} – minimum output high (OH) voltage of the driver for a defined sourced current to ensure a high level;

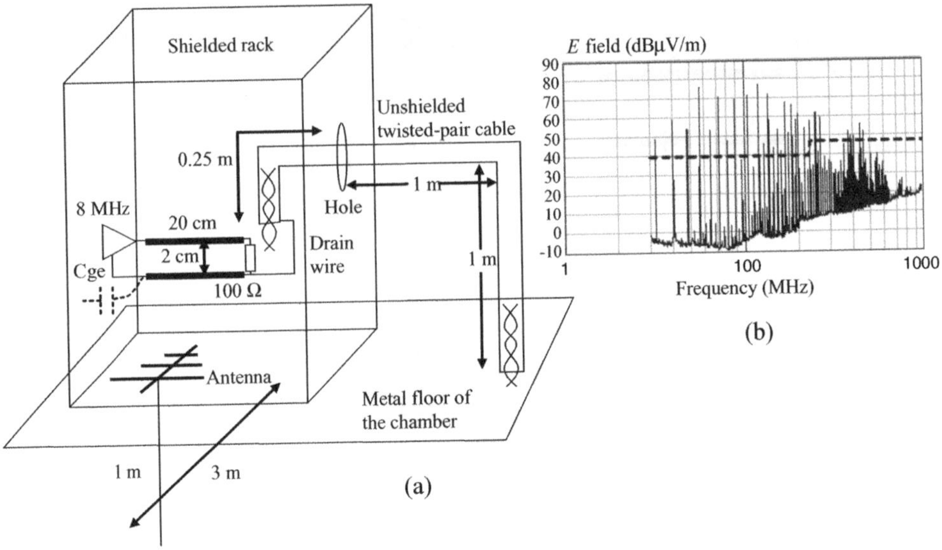

Figure 1.10 PCB with an attached cable: (a) schematic and set-up for measurements; (b) measured radiated emission with the antenna in the horizontal position. The CISPR 22 limit is indicated by the dashed line

- V_{IHmin} – minimum input high (IH) voltage of the receiver to recognize a high level;
- V_{th} – threshold switching voltage of the receiver;
- V_{ILmax} – maximum input low (IL) voltage of the receiver to recognize a low level;
- V_{OLmax} – maximum output low (OL) voltage of the driver for a defined sunk current to ensure a low level;
- $\mathrm{NM}_{\mathrm{Lmin}} = V_{\mathrm{ILmax}} - V_{\mathrm{OLmax}}$ – minimum noise margin (NM) at low level;
- $\mathrm{NM}_{\mathrm{Hmin}} = V_{\mathrm{OHmin}} - V_{\mathrm{IHmin}}$ – minimum noise margin (NM) at high level.

These parameters are guaranteed by the component manufacturer in order to ensure functionality of the device under defined conditions regarding power supply, temperature, and loading. The data sheet provides these parameters. Other parameters significant for *Signal Integrity* (SI) are:

- overshoot, undershoot, and plateau;
- noise immunity;
- set-up and hold time;
- data jitter and clock skew.

These parameters are defined and discussed in detail in the following.

1.3.1 Overshoot, Undershoot and Plateau

Overshoot and undershoot are positive and negative ringing with respect to the steady-state voltage levels. The plateau is a constant step voltage on the signal waveform that lasts twice the time delay T_{D} of the interconnect. Some examples are provided to clarify these definitions.

Consider the equivalent circuit of an interconnect with CMOS devices, as shown in Figure 1.11a. The driver and receiver devices are three inverter gates in cascade (see *Section 8.1* for more details). The driver is excited at its input by a voltage source of trapezoidal waveform with steady-state low and high levels of 0 and 5 V respectively, a rise and fall time $t_r = t_f =$ 0.2 ns, and a period $T_p = 50$ ns with a duty cycle $D = 50$ % (where D is the portion of time during which the device is operated at a high level with respect to the period). At the output of the driver there is a series resistance $R_S = 30\ \Omega$ to mitigate reflections. The device and package capacitances are represented in the equivalent circuit by $C_{in} = 3$ pF and $C_{out} =$ 10 pF. The receiver threshold $V_{th} = 2.4$ V. The interconnect is represented by a lossless *Transmission Line* (TL) of characteristic impedance $Z_0 = 60\ \Omega$ and time delay $T_D = 2$ ns. This last value corresponds approximately to a microstrip trace of length $l = 30$ cm. The simulated voltage waveforms at the points $D1_{out}$ (driver output after R_S) and $R2_{in}$ (receiver input) are shown in Figure 1.11b, where the overshoot and undershoot due to mismatching at both ends of the line can be observed. For a logic high level, the oscillations should be above the guaranteed V_{OHmin} voltage level with specified load and converge rapidly to the associated steady-state value. If this happens, the noise margin $NM_{Hmin} = V_{OHmin} - V_{IHmin}$ is preserved, as V_{IHmin} is the specified minimum voltage level for the receiver to recognize a logic high level. The same considerations hold for the logic low level, where the noise margin to preserve is

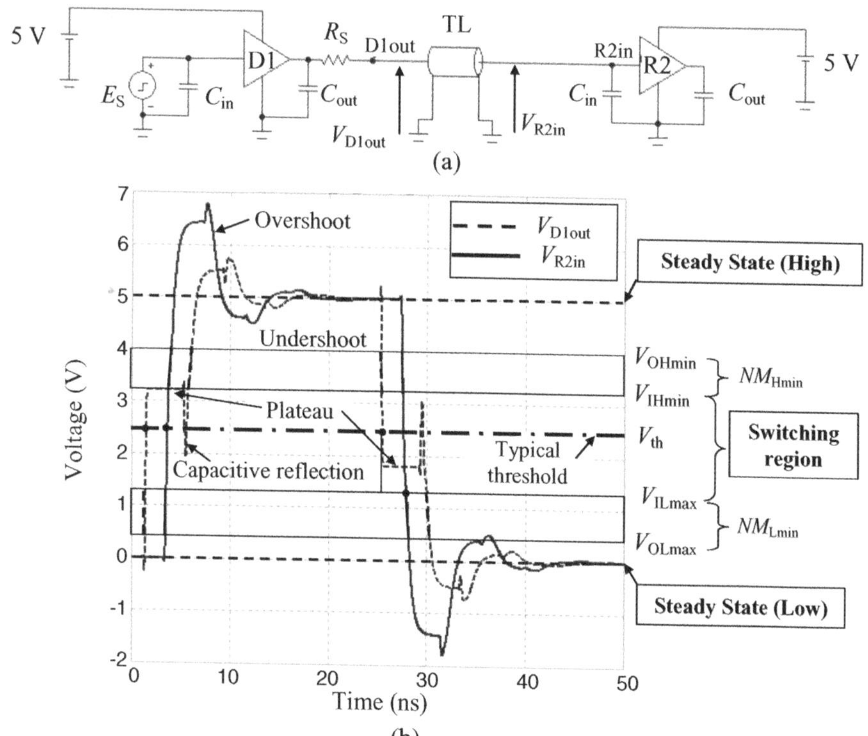

Figure 1.11 Interconnect with CMOS devices with series termination R_S: (a) equivalent circuit; (b) simulated waveforms with definitions of fundamental signaling parameters

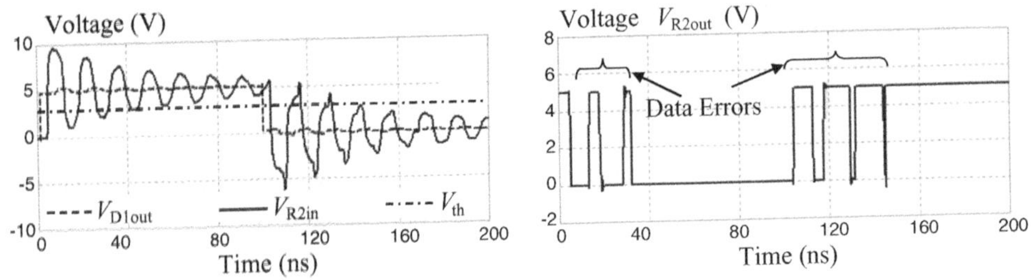

Figure 1.12 Interconnect with CMOS devices without series termination: simulated waveforms in line and data errors at the receiver

$NM_{Lmin} = V_{ILmax} - V_{OLmax}$. The values of these parameters for some popular logic families will be given in *Section 2.1*.

A very important signal distortion to consider is the plateau that can occur when the driver does not provide sufficient current to drive lines with low Z_0. This plateau lasts twice the line delay time and, if it stays in the region V_{IHmin}–V_{ILmax}, can cause data error. This concept will be clarified by the following examples.

Figure 1.12 shows what happens when the series resistance termination R_S is omitted. The overshoots and undershoots are so high that several data errors occur at the receiver output, as the oscillations cross the voltage threshold $V_{th} = 2.4$ V several times. For this example and the others that follow, the line has a delay time $T_D = 3$ ns, corresponding to a trace length $l =$ 50 cm.

In many cases, especially for CMOS devices, two clamping diodes are used to mitigate reflections, as shown in Figure 1.13a: one is connected between the receiver input and the ground, the other between the receiver input and the power supply. The simulated waveforms

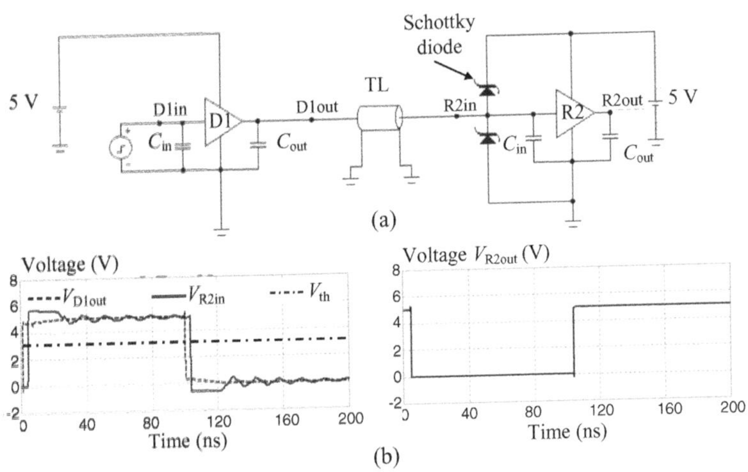

Figure 1.13 Interconnect with CMOS devices with clamping diodes: (a) equivalent circuit; (b) simulated waveforms in line and at the receiver

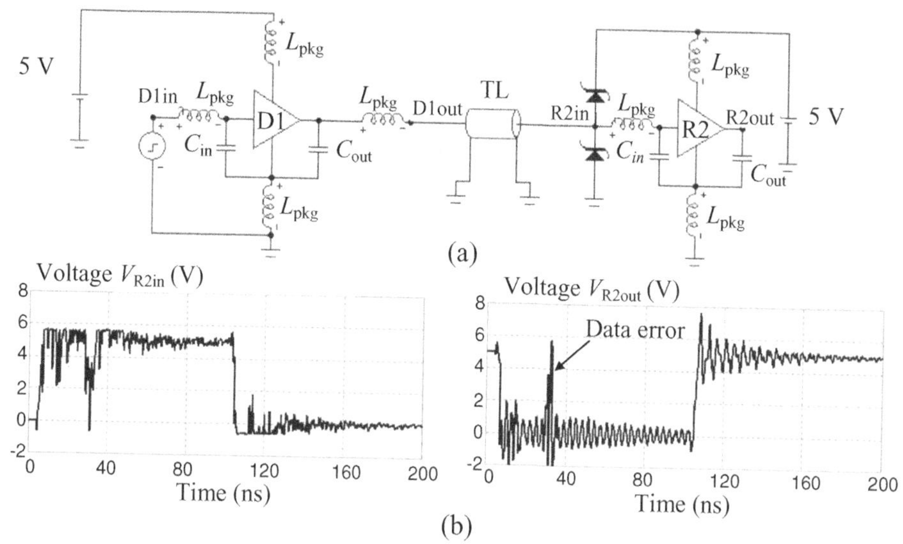

Figure 1.14 Interconnect with CMOS devices where parasitic package effects are considered: (a) equivalent circuit; (b) simulated waveforms in line and data error at the receiver

are shown in Figure 1.13b. In this case, errors do not occur, as the overshoot and the undershoot are far away from the threshold voltage $V_{th} = 2.4$ V.

A high-speed circuit is defined as 'a circuit for which the parasitic elements (resistance, inductance, and capacitance) of the PCB and its components play a significant role in performance'. The following classification is also used:

- low speed: frequency <10 MHz, edge rates >5 ns;
- high speed: frequency >10 MHz, edge rates <5 ns.

To see the importance of parasitic elements in SI performance, simulation of the structure with clamping diodes was repeated, introducing the inductance L_{pkg} associated with the package of each pin of the device, as shown in Figure 1.14a. It was verified that, with increase in the parasitic inductances to a value of 13 nH, data failure occurs. These device inductances must be minimized to avoid data errors.

Last but not least, to see the plateau effect on the line delay, consider the interconnect structure shown in Figure 1.15a, where a second receiver was added just after the series resistance $R_S = 30\ \Omega$. Since the sending waveform depends on the characteristic impedance Z_0, the simulations were repeated, adopting different values of Z_0 in the range 30–60 Ω with a step of 10 Ω in order to assess its effect. The simulated waveforms in line and at receiver output R1 are shown in Figure 1.15b. Note that for lower Z_0 an extra delay occurs for the data owing to the plateau effect. This does not occur at the input of receiver R2 because the arriving signal step doubles. Of course, if receiver R1 is positioned before the series resistance, just at the output of the driver, the extra delay does not occur because there is no partitioning

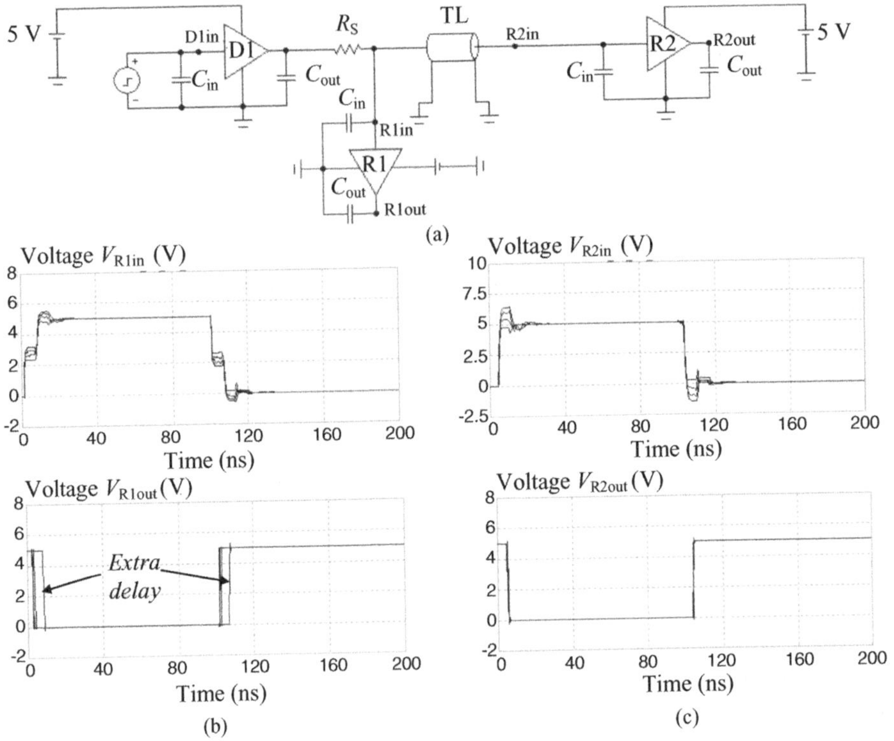

Figure 1.15 Interconnect with CMOS devices with a second receiver located at the driver end: (a) equivalent circuit; (b) simulated waveforms in line and extra delay at receiver R1 for variable Z_0; (c) simulated waveforms in line and correct data at receiver R2. Z_0 varies from 30 to 60 Ω with a step of 10 Ω

effect between the series resistance and the line characteristic impedance. Recall that, when a high-speed digital device switches, it sees at its output the characteristic impedance of the line and not the receiver located at the interconnect end. In *Chapter 5*, methods for coping with reflection problems will be provided.

1.3.2 Noise Immunity

Any signal applied to a receiver must have sufficient width t_w and amplitude V_p to be recognized from the receiver. It is possible to distinguish a '*static noise immunity*' when the receiver recognizes the input signal from its level only, regardless of the time width t_w. Generally, when the width of the spike t_w is smaller than the delay of the receiving device, the capability of the receiver to recognize the input signal depends on its width and level. In particular, the ability of a receiver to ignore very narrow signals (i.e. spikes) is called '*receiver dynamic noise immunity*' [9]. In general, very narrow pulses require more amplitude to trip the receiver, and faster logic families are more sensitive to a given spike than slower families are.

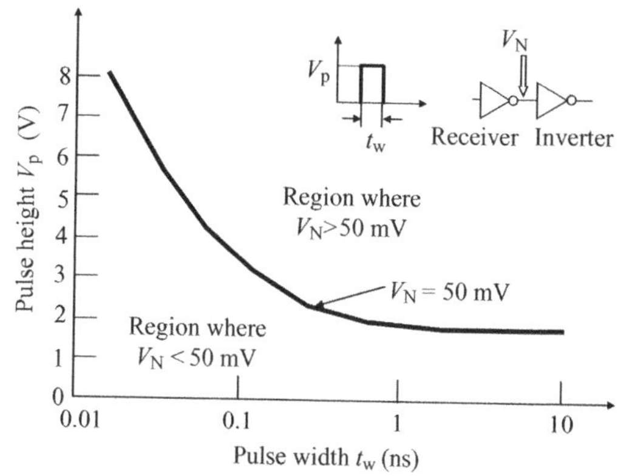

Figure 1.16 Example of dynamic noise immunity, showing the dependence of the dynamic immunity noise, defined as $V_N < 50$ mV, on pulse width t_w and height V_p

To avoid unwanted switching, the noise V_N passing through a receiver could be required not to exceed some value such as 50 mV. A simple example is shown in Figure 1.16 [9] for a 3.3 V inverter-based receiver with feedback. In the region of static noise immunity defined for $t_w > 1$ ns, the maximum pulse height $V_p = 1.65$ V ensures a noise level within the fixed 50 mV. When the width of the spike t_w is less than 0.1 ns, a pulse height V_p of more than the supply voltage 3.3 V is required to cause $V_N > 50$ mV.

1.3.3 Timing Parameters

The reflection effects and other interferences such as crosstalk and Δ*I-noise* can affect the following timing parameters essential for defining the performance of a digital system in terms of speed.

Timing parameters are defined according to the common-clock timing scheme shown in Figure 1.17, where a single clock is distributed to a driver and to a receiver by traces having

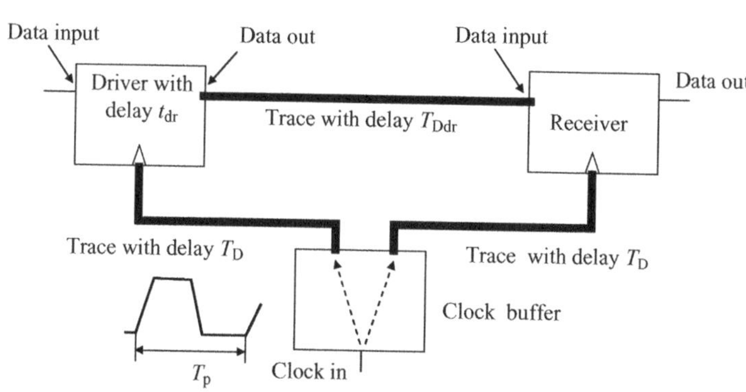

Figure 1.17 Block diagram of a common-clock bus

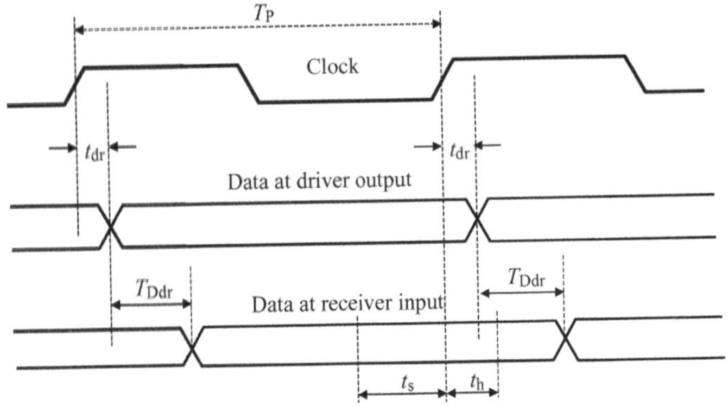

Figure 1.18 Timing diagram of a common-clock bus

the same delay time T_D [10]. The driver sends bits of data to the receiver by a trace of delay time T_{Ddr}. A complete data transfer requires two clock pulses, one to latch the data into the driver flip-flop and one to latch the data into the receiving flip-flop. Data transfer takes place in the following sequence:

1. Data are provided by the circuit core to the input of the flip-flop driver.
2. The clock to the driver is provided by the clock buffer through a trace of delay time T_D, and data transfer from the driver input to its output occurs with delay t_{dr}.
3. The bit propagates down the trace with a delay time T_{Ddr} and is latched by the clock edge coming from the clock buffer to the receiver by a trace of delay time T_D.

This process is shown as a timing diagram in Figure 1.18, where:

- The set-up time t_s is the time for which the input waveform at the receiver is settled, in other words, the bit must meet the input voltage specifications previously defined as static parameters before the clock edge acts on the receiver.
- The hold time t_h is the time after the clock edge during which the waveform must still meet input voltage specifications previously defined as static parameters.

Assuming that the data are sampled on the rising clock edge, it must be true that

$$T_p > t_{dr} + T_{Ddr} + t_s \tag{1.10}$$

Therefore, under the condition that $t_h < t_{dr} + T_{Ddr}$, the maximum clock rate is

$$f_{max} < 1/(t_{dr} + T_{Ddr} + t_s) \tag{1.11}$$

The timing diagram in Figure 1.18 implies that the clock and data edges fall at precise times. In a real system this does not occur, and many factors, such as reflection, crosstalk, and simultaneous switching noise, influence the times. The uncertainty in the arrival time of a signal edge is the '*signal jitter*', while '*clock skew*' refers specifically to skew from all sources on the clock line in a synchronous system. Clock generators inherently produce some variation in the timing of clock edges at their output, and this variation is called '*clock jitter*'. In a system driven by a single clock generator, clock jitter is included in the timing budget as uncertainty in the clock period.

The maximum clock rate is given by

$$f_{\max} < 1/(t_{\mathrm{dr}} + T_{\mathrm{Ddr}} + t_{\mathrm{s}} + \Delta T_{\mathrm{Ddr}} + \Delta t_{\mathrm{clock}} + \Delta t_{\mathrm{jitter}} + \Delta t_{\mathrm{margin}}) \tag{1.12}$$

where ΔT_{Ddr} is the signal jitter, $\Delta t_{\mathrm{clock}}$ is the clock skew, $\Delta t_{\mathrm{jitter}}$ is the clock jitter, and $\Delta t_{\mathrm{margin}}$ is the margin chosen by the designer.

Eye diagram simulations can help in calculating the signal jitter precisely, as we will show in *Chapter 7* for lossy lines. Further consideration of digital timing analysis can be found in work by Hall *et al.* [10].

1.3.4 Eye Diagram

For a long stream of bits at high rate launched onto a lossy line, it could be difficult to tell if the signal meets the design specifications by monitoring the data waveforms. An example is shown in Figure 1.19, where the distortion on the edge of the signal is due to the losses in the line (see *Chapter 7* for details). To overcome this problem, the technique is to translate rise and fall waveforms of each bit in just one time window. Superposing all the bits builds an eye diagram [11]. A good plot results when the waveform is plotted for one clock period before and after the edge. In this case, the full data bit plus a half of the one before and a half of the one after are captured. In doing so, it is assumed that the data are sampled on the rising edge of the clock.

The eye diagram is a very useful method for accurate drawing of the timing diagram for determining the maximum clock frequency not only in the case of lossy lines but also for all types of signaling where reflection, crosstalk, and switching noise can cause jitter. The eye diagram gives an indication of the signal quality: the larger the eye opening, the better is the signal quality. The data sequence can be generated by a pseudorandom sequence generator, which is a digital shift register with feedback connected to produce a maximum length sequence. Ideally, the entire bit edge should cross the threshold voltage at the same time. On

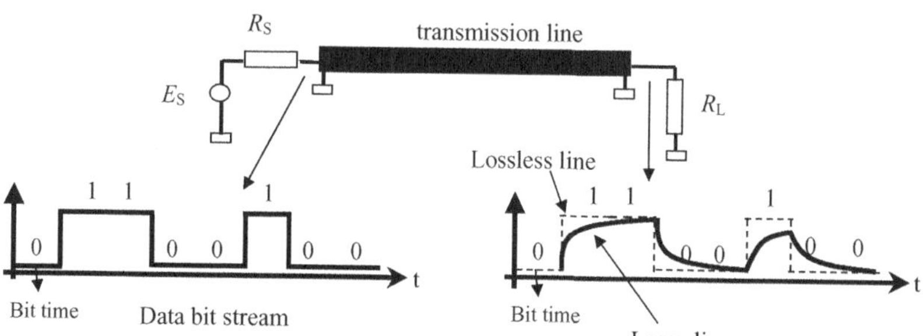

Figure 1.19 Data bit stream transmission with a lossy line

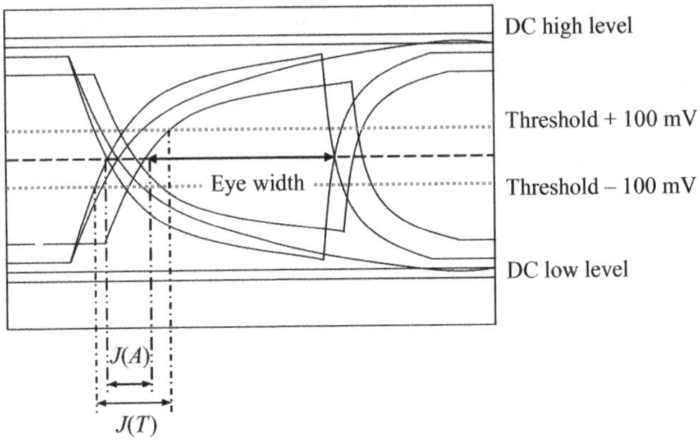

Figure 1.20 Fundamental parameters of an eye diagram

account of lossy lines, noises, and spread in threshold, this does not happen. The parameter that quantifies this fact is called 'signal jitter', whereas the vertical thickness of the line in the eye diagram is indicative of the AC voltage noise.

Figure 1.20 shows a typical eye diagram with differential signaling, where a '*zero crossing*' jitter $J(A)$ is defined together with a '*worst-case jitter*' $J(T)$. It is assumed, as an example, that the uncertainty of the threshold is between 100 mV and −100 mV.

An example of a measured eye diagram is given in Figure 1.21. A data stream at 311 Mb/s is injected onto an AWG28 twinax cable having LVPECL devices as the driver/receiver. Two measurements are shown: one with a cable length of 6 m and the other one with a cable

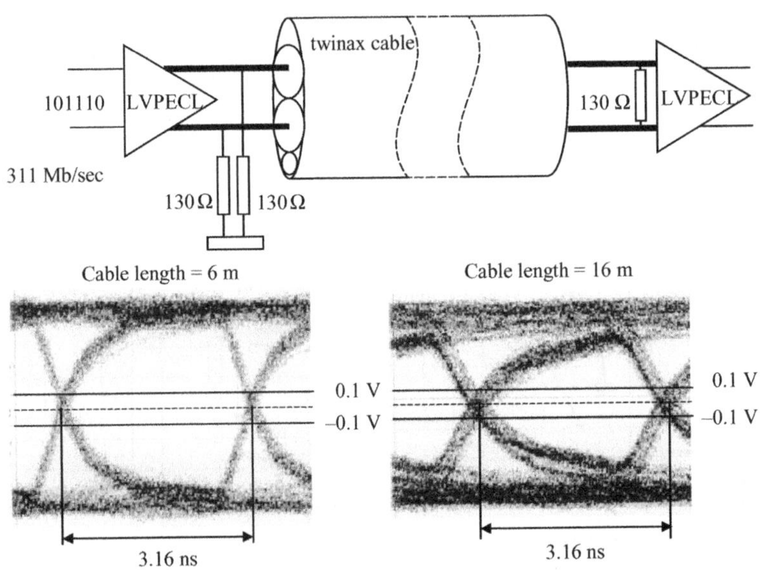

Figure 1.21 Example of eye diagram measurements with LVPECL devices and a twinax cable

length of 16 m. For the longer cable, the eye tends to close, as there are more losses in line. Another fact that determines eye closure is increase in the frequency of the data rate. Models for predicting the step response and eye diagrams directly in the time domain with lossy lines will be presented in *Section 7.2*.

1.4 Modeling Digital Systems

Modeling is very important for the design of complex digital systems. As mentioned in *Section 1.2*, it is also becoming a useful method for demonstrating conformity to the EMC standards, instead of measurements, as stated by the new EMC European Directive [7].

A digital system is usually very complex and consists of several components that need to be simulated to predict *Signal Integrity* (SI) and electromagnetic interference (EMI). An example is shown in Figure 1.22, where a multilayer PCB with its backplane is considered. Some components, such as bypass capacitors, sockets, package chips, vias, and connectors, can be modeled by lumped elements, such as resistances, inductances, and capacitances, as their maximum dimension is usually less than the minimum wavelength of interest. Other components, such as traces, cables attached to a PCB, and power and ground planes, have to be simulated by distributed models to take into account delays and points of resonance.

The available commercial tools that allow simulation can be classified depending on the models as:

- tools for mathematical model implementation;
- SPICE-like circuit simulators;
- full-wave numerical tools;
- professional simulators based on mixed formulations.

1.4.1 Mathematical Tools

Commercial mathematical tools such as *MathCad* and *MatLab* may be useful in computing reflection, crosstalk between parallel coupled lines, and radiated emission when the interconnect to be simulated is a simple point-to-point structure that consists of a driver modeled by

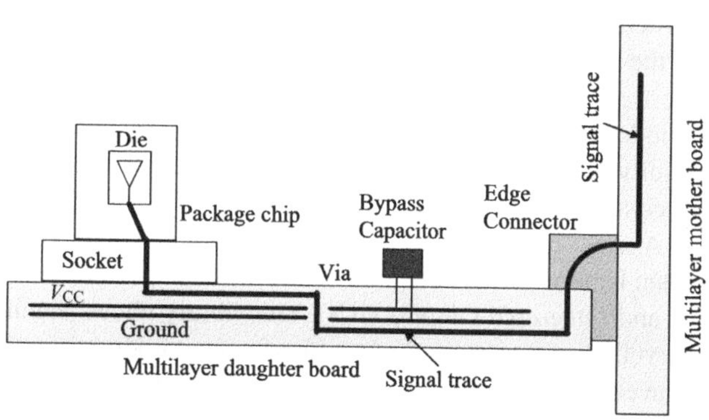

Figure 1.22 Example of parts of a digital system with components needing modeling

Table 1.6 Performances of mathematical codes

Advantages	Disadvantages
• Complex mathematical expressions for interference sources and propagation mechanisms. • Multiconductor lines analyzed as transmission lines with losses. • Fast time-domain analysis. • Matrix computation.	• Canonical structures that consist of a source, line, and load. • Simple sources and loads. • Linear load.

an equivalent Thévenin circuit and receivers represented by simple *RLC* nets. The line can be modeled by using the closed-form TL expression to compute currents and voltages along the line (see *Chapter 5*). For a frequency-domain (i.e. AC) computation as required in radiated emission problems, losses produced by skin, proximity, and dielectric effects can also be accounted for (see *Section 7.1*). Numerous examples concerning the computation of the radiated field at a certain distance from the source once the current distribution has been calculated will be provided in *Chapter 9*. When the simulation of a lossy line must be performed in the time domain, the convolution integrals based on known line scattering parameters can be numerically performed by exploiting the mathematical functions present in the code library, as will be shown in *Section 7.2*.

A powerful characteristic of these codes is the simplicity of managing matrix computation. An example of calculation of return current density in a 2D microstrip and stripline structures is provided in *Section 10.2*. Another important feature offered by matrix computation is that, applying node network theory, more topological complex circuits than point-to-point structure can be analyzed (see *Appendix E*). Some of the main advantages and disadvantages of analytical models are summarized in Table 1.6.

1.4.2 Spice-Like Circuit Simulators

For 2D problems, the best way to perform simulations in DC, AC, and the time domain is to use any circuit simulator based on SPICE. SPICE is an acronym for *Simulation Program with Integrated-Circuit Emphasis*. The original SPICE tool was developed to analyze complex electric circuits, in particular integrated circuits at diode and transistor level. It was developed in the early 1970s at the University of California at Berkeley, which is not the owner. The most widely available free-of-charge version is PSPICE SV, version 9.1, and MICROCAP Evaluation V.9 for Windows. To make the code more user friendly and to improve the performance, numerous software houses have developed and marketed SPICE-like circuit codes. Popular medium-cost commercial codes are PSPICE (*Cadence Design System*) and MicroCap (*Spectrum Software*). A more powerful and high-cost professional code is HSPICE (*Synopsys*). The most important feature of these commercial codes is the powerful component library of diodes, transistors, and *Integrated Circuits* (ICs). This library allows simulation of digital devices at transistor level, as often required when interference produced by the switching of the devices must be investigated (see the examples reported in *Chapter 8*). In particular, in *Section 8.2.3* and *Appendix C* it is shown that a multilayer PCB with ICs populated by decoupling capacitors can be simulated in the time domain to compute the noises between power

Table 1.7 Performances of SPICE-like circuit simulators

Advantages	Disadvantages
• It is based on the description of the structure by circuit elements.	
• Complex topology easily simulated.	• 2D structures.
• Complex sources and loads described at transistor level.	• Limited use of mathematical expressions.
• Time-domain analysis with non-linear loads.	• No user matrix computation.
• Short computational time.	

and ground planes modeled as a grid of electrically short transmission lines for several situations of decoupling capacitors in terms of values and allocations. The circuit model considered also makes it possible to account for resonance effects of the PCB when excited at particular frequencies. Some of the main performances of the SPICE-like circuit models are reported in Table 1.7.

Unfortunately, the model library of low-cost SPICE codes does not include coupled lines or time-domain analysis of lines with frequency-dependent losses. The models available in the library are the lossless transmission-line model and the lossy-line model with DC resistance. Therefore, users are requested to implement their own models to perform signal integrity simulation for general cases with lossy and coupled interconnects. One of the main tasks of this book is to describe these types of circuit model. In *Chapter 6*, crosstalk models based on lumped and distributed line parameters for two and n coupled lines are outlined. In *Chapter 7*, new lossy-line models for analysis directly in the time domain with non-linear loads are presented.

SPICE is also useful for radiated emission problems. For example, in *Section 9.7* it is shown how to model unshielded and shielded cables driven by a differential digital device in order to calculate radiated fields for EMC compliance. In *Chapter 10* it is shown how to use SPICE to compute grounding noise produced by return signal currents for several structures of PCBs and connectors, applying the concept of partial inductance introduced and defined in *Section 3.2*.

The need for suitable modeling of digital devices with their non-linearity for an accurate and fast prediction of signal integrity is also discussed. The behavior models that are presented in *Chapter 2* and verified experimentally in *Chapter 6* are the basis of the IBIS models, a standard for the digital behavior device modeling described in *Section 2.4*. The SPICE-based commercial software used in this book is MicroCap [12], and all the models proposed are validated experimentally.

1.4.3 Full-Wave Numerical Tools

For 3D problems, the simulations should be performed by numerical codes based on the solutions of the Maxwell equations. For this reason, these numerical codes are also called *full-wave* tools. The main advantage offered by this type of programs is the possibility to simulate 3D objects considering their metallic and dielectric parts. The main disadvantage is that only simple sources (i.e. voltage and current sources) and simple loads composed of a simple

RLC net are allowed. To overcome this limitation, some software houses are working for an integration of their 3D code with SPICE. There are a variety of *full-wave* electromagnetic modeling techniques. However, six techniques are typically used for EMI/EMC problems: the *Finite-Difference Time Domain* (FDTD) method, the *Finite Integration Technique* (FIT), the *Method of Moments* (MOM), the *Finite Element Method* (FEM), the *Partial Element Equivalent Circuit* (PEEC), and the *Transmission-Line Matrix* (TLM) [13, 14].

FDTD and FIT are volume-based solutions of time-domain Maxwell equations in differential and integral forms respectively. The entire volume, which consists of the object to be modeled and the surrounding, is represented by square and/or rectangular grids, the cell dimension of which is small compared with the shortest wavelength of interest. Commercial codes usually determine the grid automatically once the maximum frequency of interest is set. The broadband frequency response of the model is determined by performing a Fourier transform of the time-domain results at the specified monitor points. The boundary conditions of the volume-based solution (the edges of the grid) must be specially controlled to avoid reflection of the radiated field. The technique that allows this is called the *Absorbing Boundary Condition* (ABC). It usually provides an effective reflection of less than −60 dB. FDTD and FIT are not well suited for modelling wires or thin structures, as the number of grid cells increases dramatically.

MOM is a surface current technique. The metal objects to be modeled are converted into a series of plates and wires, or all wires electrically short. A set of linear equations is created to find the RF currents on each wire segment and surface patch. Once these currents are known, the E- and H-fields at any point in space can be determined by considering the radiation from each segment/patch and performing the vector summation. This makes the technique particularly suitable in solving problems with a long thin structure, such as cables attached to a PCB. The MOM is a frequency-domain solution technique, and therefore the simulation must be run for each frequency. As a digital signal with very fast edge times has a large spectrum, numerous frequencies must be computed. This is the main disadvantage of this technique, together with the fact that the dielectric substrate of a PCB is difficult to model. The first developed code based on MOM was the *Numerical Electromagnetic Code* (NEC) written in FORTRAN and available free of charge. Codes developed by software houses based on NEC and improved in graphic representation and computation techniques can be purchased [15].

FEM is another volume-based solution technique where the space is split into small elements usually having a triangular or tetrahedral shape. The field in each element is approximated by low-order polynomials with unknown coefficients. The Galerkin method is used to determine the coefficients. Once these coefficients are computed, the fields are known within each volume element. The computation is performed in the frequency domain, and results in the time domain are obtained by *Fast Fourier Transform* (FFT).

TLM and PEEC techniques are based on a representation of the volume (TLM) and surface (PEEC) elements used to decompose the computational domain by electrically short transmission lines and lumped-circuit elements, respectively, interfering with all the others.

Some of the main performances of the numerical models are summarized in Table 1.8.

The full-wave solution of several SI and RE problems by the software tool MWS based on FIT [16] and by NEC [15] will be presented in the following chapters.

MWS is used in *Section 6.5* for the analysis of SI. The crosstalk between two couple traces in a PCB is investigated considering a finite ground plane as the return for signal currents

Table 1.8 Performances of numerical codes

Advantages	Disadvantages
• FDTD, FIT, FEM and TLM are based on the differential form of the Maxwell equations and consider a volume of calculation.	• Simple sources and loads.
• MOM and PEEC methods are based on the integral formulation of the Maxwell equations and consider a surface of calculation.	• Structure with shielded cables cannot be simulated straightforwardly.
• 3D structures with dielectric material and calculation of resonance points.	• Coupled field-to-circuit simulation for shielded cable.
• Scattering parameters.	• Time-consuming.
• Radiation pattern.	• High cost of the code in terms of computer time and memory storage.

with and without cuts. The results are compared with those obtained by using transmission-line theory, which assumes an infinite continuous ground plane.

One of the most important features of the *full-wave* code is the possibility of computing the scattering S-parameters in matrix form, which can be used to extract equivalent circuits for SPICE simulations. *Section 7.2* shows an example of S-parameter computation for an electrically short segment of twisted-pair cable by MWS. This type of structure is particularly interesting because the losses due to the proximity effect cannot be computed by closed-form expressions. Another important feature of numerical software tools is the possibility of computing the resonance frequencies in cavity structures such as a multilayer PCB populated by decoupling capacitors. An example of this application is provided in *Appendix C*.

Some examples of RE problems will be presented in *Chapter 9*. In particular, two examples of using numerical codes for radiated emission problems are provided. The first example concerns a cable that links two shielded boxes. It highlights the importance of considering both antenna polarizations regardless of whether the radiating cable is in the horizontal position (see *Section 9.7*). The second example concerns how to model a simple PCB, a wire above a finite ground plane, with an attached cable in order to compute radiation patterns (see *Section 9.9*). Comparisons between results at different frequencies obtained by NEC and MWS models are given as reciprocity validation.

The great advantage of a numerical code is the feature that makes it possible to compute radiated fields for more complex structures such as a PCB with cuts in the ground plane, equipped with EMI filters and inserted in a shielded box with an attached cable outgoing from an aperture. An example of this application is described in *Section 10.3*, where the numerical results for the basic structure of PCB, a wire above a ground plane with a long wire attached to the ground plane, are compared with those obtained by closed-form expressions implemented in a mathematical code.

When using numerical codes to compute S-parameters to extract equivalent circuits of electrically short discontinuities in a PCB, such as connectors or vias, it is very important to consider the small inductance associated with a discrete port used for excitation, which can introduce significant errors in performing the simulations. This aspect is investigated in detail in *Section 11.2*, comparing actual and ideal results of S-parameters computed in both the frequency and the time domain.

An important aspect of a numerical model is its validation by measurements. However, in going through this procedure, it is fundamental that the model reproduces exactly the set-up used for the test, and all metallic parts of the radiating object must be accounted for. Another important point to consider is that the measurements are affected by uncertainties. This is demonstrated in *Section 11.3*, where the radiated fields of a shielded rack with an outgoing cable are computed by NEC and MWS, and the results are compared with the measurements carried out in two different semi-anechoic shielded rooms used for EMC compliance. The last examples of using numerical codes for signal integrity are given in *Chapter 12*, where design rules for routing single-ended and differential traces in PCB are investigated.

1.4.4 Professional Simulators

At the end of this section, it is important to mention professional simulation tools where the integration of mathematical and numerical tools with a circuit simulator is realized in order to allow a designer to perform automatic simulations of PCBs from an industrial point of view.

The information provided in this book can be used as the background for a better understanding of the performance offered by these professional simulators which, for solving signal integrity and EMC problems, enable a partial or full simulation of a PCB, starting from its layout. These tools contain: 2D field solvers for extracting *RLGC* matrices of single/coupled transmission lines; a single/coupled lossy transmission line simulator; a 3D field solver for wirebonds, vias, and metal planes; and behavior modeling of drivers and receivers such as IBIS. They are also called upon to take physical layout files as input data and to post-process simulation results in the time domain (timing and waveform measurement) and the frequency domain (impedance parameter and *S*-parameters). For more information, the reader should visit the website of software houses such as Ansoft, CST, Applied Simulation Technology, Cadence, INCASES, Mentor Graphics, Sigrity, Quantic EMC, etc.

References

[1] Davidson, E., 'Electrical Design of a High Speed Computer Package', *IBM Journal of Research and Development*, **26**(3), May 1982, 349–361.

[2] Blood, W., '*MECL System Design Handbook*', 4th edition, Motorola Semiconductor Products, Inc., Phoenix, AZ, 1983.

[3] GR-1089-CORE, '*Electromagnetic Compatibility and Electrical Safety – Generic Criteria for Network Telecommunications Equipment*', Telcordia Technologies, Morristown, NJ, October 2002.

[4] Paul, C., '*Introduction to Electromagnetic Compatibility*', John Wiley & Sons, Inc., Hoboken, NJ, 2006.

[5] Code of Federal Regulations, Title 47 (47CFR), Part 15, Subpart B, '*Unintentional Radiators*'. Available online: www.hottconsultants.com, www.gpoaccess.gov/cfr/index.html, www.wireless.fcc.gov/rules.html.

[6] International Electrotechnical Commission, CISPR 22, '*Information Technology Equipment – Radio Disturbance Characteristics – Limits and Methods of Measurement*', 4th edition, 2003–2004. Available online: www.iec.ch.

[7] 2004/108/EC, European EMC Directive, 31-12-2004. Available online: http://europa.eu.int/comm/enterprise/electr_equipment/emc/.

[8] EN 55022:1988 (CISPR 22:1997) delayed, 1995 (CISPR 22:1993), '*Information Technology Equipment – Radio Disturbance Characteristics – Limits and Methods of Measurement*'. Available online: http://europa.eu.int/comm/enterprise/newapproach/standardization/harmstds/reflist/emc/ and http://www.cenelec.org.

[9] Young, B., '*Digital Signal Integrity: Modeling and Simulation with Interconnects and Packages*', Prentice Hall PTR, Upper Saddle River, NJ, 2001.

[10] Hall, S., Hall, G., and McCall, J., *'High-Speed Digital System Design – A Handbook of Interconnect Theory and Design Practice'*, John Wiley & Sons, Inc., New York, NY, 2000.
[11] True, K.M., *'Long Transmission Lines and Data Signal Quality'*, AN 808, National Semiconductor, March 1992. Available online: www.national.com.
[12] www.spectrum-software.com.
[13] Sadiku, M., *'Numerical Techniques in Electromagnetics'*, CRC Press, Boca Raton, FL, 1992.
[14] Archambeault, B., Brench, C., and Ramahi, O., *'EMI/EMC Computational Modeling Handbook'*, 2nd edition, Kluwer Academic Publishers, Norwell, MA, 2001.
[15] www.nittany-scientific.com.
[16] www.cst.com.

2

High-Speed Digital Devices

For many calculations or simulations of *Signal Integrity* (SI) and *Radiated Emission* (RE) effects, it is very important to know the *input/output* (I/O) static and dynamic characteristics of digital devices. The aim of this chapter is to highlight the basic characteristics of the main logic families: *Transistor–Transistor Logic* (TTL), *Complementary Metal Oxide Semiconductor* (CMOS) logic, and *Emitter-Coupled Logic* (ECL). The intention is to provide a background to understand the latest development of components for high-speed applications. How to build up a linear or a simple behavioral model that takes into account the non-linear effects of the driver and receiver is outlined. This must also be considered a starting point for building up more accurate behavioral models. The chapter ends with an introduction of the IBIS models. IBIS is a standard for a fast and accurate behavioral method for modeling I/O buffers based on I/V curve data derived from measurements or full-circuit simulations. An example is given of the use of this type of model by SPICE.

2.1 Input/Output Static Characteristic

In this section the main parameters of the digital technologies are presented to find out the I/O static characteristics of a device. The objectives of the component manufacturers are to improve speed by minimizing the transistor size and to decrease the power consumption by lowering the value of the voltage supply. Other modifications to improve the performance of the components are introduced at the circuit level, but the general behavior of the static characteristics remains the same. The reader can find more detailed information by consulting textbooks [1–3] and the numerous Application Notes (ANs) prepared by device manufacturers that are available on the web [4–18].

2.1.1 Current and Voltage Specifications

For good signal transmission in digital communication, the high-level current I_{OH} sourced and the low-level current I_{OL} sunk by the driver must in absolute value be less than the fixed values I_{OHmax} and I_{OLmax} given by the data sheet in a specified range of power supply and temperature (see Figure 2.1). As will be explained in *Section 2.1.4*, in the case of ECL devices, the current

Signal Integrity and Radiated Emission of High-Speed Digital Systems Spartaco Caniggia and Francescaromana Maradei

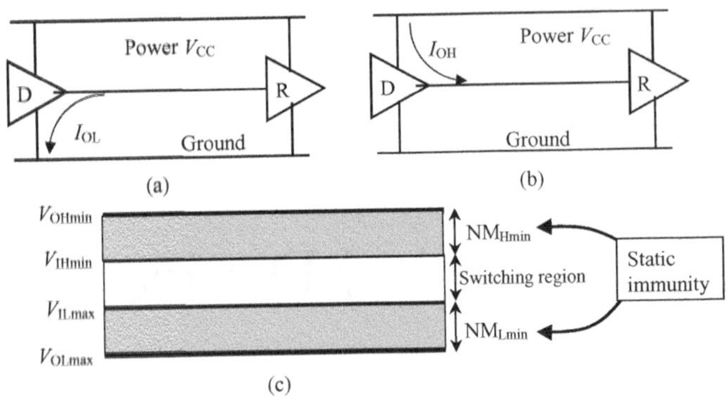

Figure 2.1 Static noise immunity of an IC device: (a) driver D at low level and (b) at high level; (c) noise margin definition

is sourced by the device for both logic levels. Under these conditions, a high-level voltage $V_{OH} \geq V_{OHmin}$ or a low-level voltage $V_{OL} \leq V_{OLmin}$ is guaranteed at the output of the driver. In this case, the worst-case static immunity at the high and low levels (i.e. $NM_{Hmin} = V_{OHmin} - V_{INmin}$ and $NM_{Lmin} = V_{ILmax} - V_{OLmax}$ respectively) is preserved. On the other hand, for good driver capability, I_{OHmax} and I_{OLmax} should be much higher than the values specified for typical logic gates used for short interconnects in order to ensure static noise immunity and therefore switching of the receivers at the first step. This is a very important requirement, especially in the case of interconnects with distributed loads such as chain and bus structures with Thévenin termination (see *Section 5.4*).

The non-linear I/O static characteristics of IC digital devices (see Figure 2.2) are very useful for calculating reflections and crosstalk in interconnects simulated as transmission

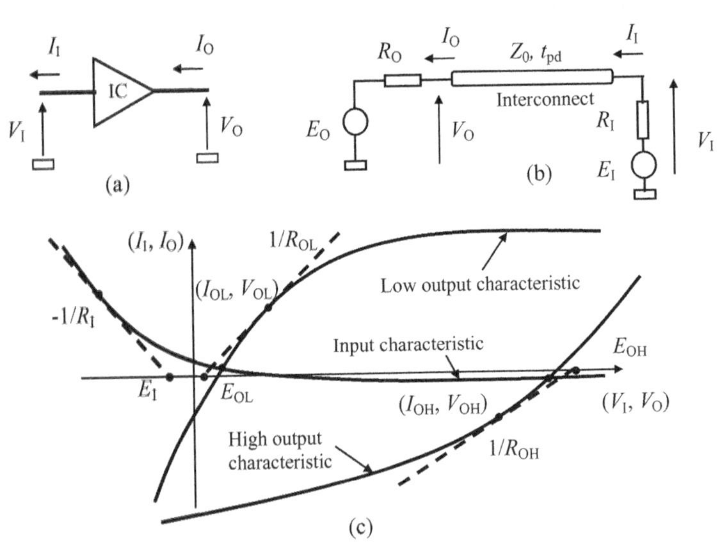

Figure 2.2 Voltage and current convention for (a) IC device and (b) interconnect. (c) Input and output static characteristics

lines. The key line parameters are: the per-unit-length propagation delay time, t_{pd}, and the characteristic impedance, Z_0. Once the static output characteristics of the driver associated with the low and high levels are known by measurements, by SPICE simulations, or by data sheet, the output impedances of the device, R_{OL} and R_{OH} for low and high levels respectively, can be determined as the slope of the line passing through the current–voltage points of interest (I_{OL}, V_{OL}) and (I_{OH}, V_{OH}) as $1/R_O = \Delta I_O/\Delta V_O$. The voltage source E_O associated with a determined R_O is the intersection between the line passing from the point (I_O, V_O) with slope $1/R_O$ and the voltage axis (i.e. line $I_O = 0$). Therefore, the two values E_O and R_O change according to the location of the point (I_O, V_O), and to the output low or high level. The same considerations apply to the input parameters of the receiver, E_I and R_I. Usually, three or four regions in the driver output static characteristic can be recognized where E_O and R_O can be considered constant, as will be shown in the next sections. This is an advantage in building appropriate simple models of IC devices, as will be illustrated in *Chapters 5* and *6*. As a designer generally wishes to maximize the signal launched onto the line to cause switching of the receivers at the first step, the regions useful for data transmission are those with low R_O. The current and voltage couple in the static condition is defined by the intersection between the receiver input and the driver output characteristics for both logic levels.

Although there are many families of digital devices based on different technologies, they can be divided roughly into three broad categories:

- *Transistor–Transistor Logic* (TTL);
- *Complementary Metal Oxide Semiconductor* (CMOS) logic;
- *Emitter-Coupled Logic* (ECL).

TTL and ECL are bipolar technologies differing in implementation techniques, while CMOS (MOS technology) differs in fundamental transistor structure and operation. In the next sections, the static output characteristics of these categories of digital devices will be introduced and discussed with the purpose of building suitable simple circuit models for reflection and crosstalk predictions by the graphical method (see *Chapter 5*) or by SPICE-like circuit simulators.

2.1.2 Transistor–Transistor Logic (TTL) Devices

The designation 'bipolar' essentially refers to the basic component used to build this family of integrated circuits, the bipolar transistor [1]. Employing a bipolar transistor in the output driver of a logic function as well as the input buffer results in a *Transistor-to-Transistor Logic* (TTL) direct connection. Older technologies were interconnected via passive components such as resistors or diodes. Since the original TTL design, several enhancements have been employed to reduce power and to increase speed [17, 18]. Common to these has been the use of Schottky diodes, which, ironically, no longer strictly result in TTL connections. Consequently, the two names, Schottky and TTL, are used in combination: *Low-power Schottky* (LS), *Advanced Low-power Schottky* (ALS), and *Advanced Schottky FAST TTL*. Typical input and output stages of a TTL device are shown in Figure 2.3.

For a long time the superior characteristics of TTL compared with CMOS have been its relatively high speed and high output capability to drive long interconnects; these advantages

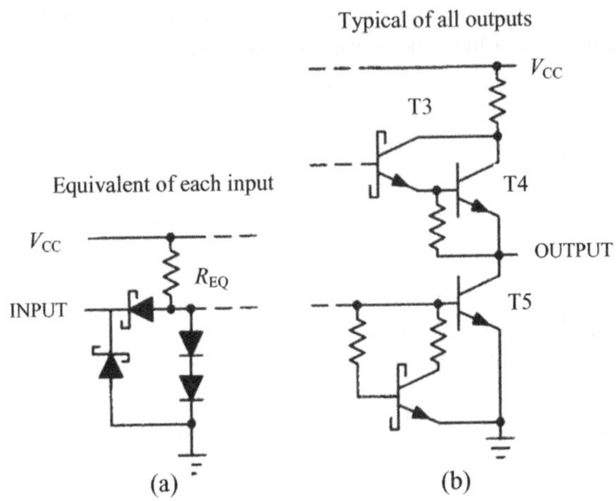

Figure 2.3 Transistor–transistor logic (TTL) device: (a) input stage; (b) output stage

are rapidly diminishing, as described in the next section. One family of devices, *Advanced BiCMOS Technology* (ABT), utilizes TTL circuitry at the inputs and outputs, and CMOS technology in between, attempting to combine the advantages of both bipolar and CMOS. Recall that a very important feature of CMOS devices are their low power dissipation.

A typical output characteristic of a TTL gate is shown in Figure 2.4, where the absence of linearity can be seen. At a high (H) logic output level, where transistors T2 and T5 of Figure 2.4a are cut off, three output resistance values R_{OH} can be distinguished for $V_O > 0$:

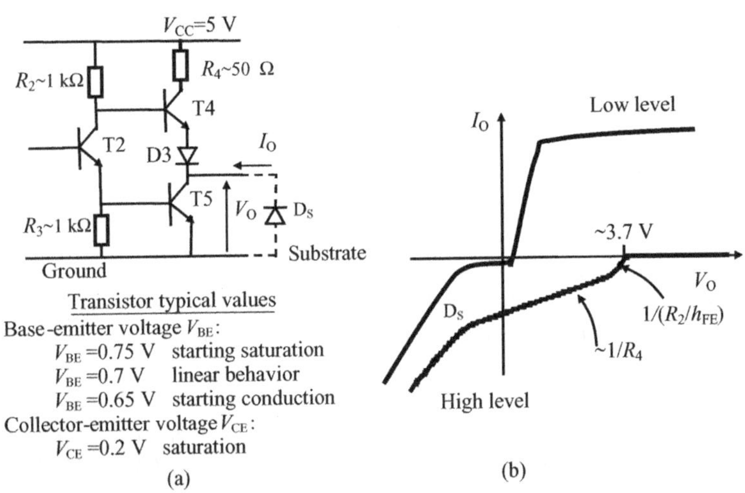

Figure 2.4 Basic TTL device: (a) output stage; (b) output static characteristics at low (L) and high (H) levels

- For values of V_O up to about 3 V, transistor T4 is in saturation, the voltage drop between the base and emitter is $V_{BE4} = 0.75$ V, and the voltage drop between the collector and emitter is $V_{CE4} = 0.2$ V. The voltage in diode D3 is $V_{D3} = 0.75$ V. The base current of transistor T4 is $I_{B4} = (3.5 - V_O)/R_2$, and the collector current is $I_{C4} = (4.05 - V_O)/R_4$. This means that $I_{C4} \gg I_{B4}$, and therefore $I_{C4} \approx I_{E4} = -I_O$. The output characteristic is then represented by the line $-I_O = (4.05 - V_O)/R_4$, and hence $R_{OH} = R_4 = 50\ \Omega$.
- When V_O approaches 3.7 V, it yields $R_{OH} = R_2/(h_{FE} + 1)$, where, in general, $h_{FE} = I_C/I_B$ is the gain of a transistor, and the current sourced by the emitter is $I_E = I_B + I_C$. This new value of output resistance is due to the fact that transistor T4 is in the active region, its base current is $I_{B4} = -I_O/(h_{FE} + 1)$, and with the adopted current convention $I_{E4} = -I_O$. Starting from the supply $V_{CC} = 5$ V, and taking into account the voltage drop across R_2, and the voltages 0.65 V across the base–emitter junction of T4 and 0.65 V across diode D3, this yields $V_O = 3.7 + R_2 I_O/(h_{FE} + 1)$. Thus, the output has a Thévenin representation, which consists of a voltage of 3.7 V in series with an output impedance $R_2/(h_{FE} + 1)$. With $R_2 = 1\ \text{k}\Omega$ and $h_{FE} + 1 = 50$, it yields $V_O = 3.7 + 20 I_O$, so that in this case $R_{OH} = 20\ \Omega$.
- For $I_O = 0$ there is no voltage drop across R_4, nor any voltage drop across the base–emitter junction of T4 or diode D3, and, in principle, V_O should become equal to $V_{CC} = 5$ V. Hence, R_{OH} has a very high value.

At low logic output level, T4 is cut off while T2 and T5 are on. The output looks directly across transistor T5, which now sinks a current I_O. The volt–ampere characteristic at the output terminals is now precisely the common emitter–collector characteristic of the transistor corresponding to the base current I_{B5} of transistor T5. Therefore, at low logic output level for $V_O > 0$, two R_{OL} output resistance values can be distinguished:

- when T5 is out of the saturation region, R_{OL} is low (a few ohm) and the voltage drop V_O across the transistor has a low value and increases very slowly with large change in current I_O;
- when T5 is in saturation, I_O and R_{OL} are high, which means that the voltage drop V_O across the saturated transistor increases rapidly with small change in current.

For $V_O < 0$, the output characteristic is determined by the substrate diode D_S.

A number of types of TTL gate are available, differing principally in the compromise made between speed and power dissipation. Thus, the basic gate is considered a medium-speed gate. A typical high-speed TTL gate is shown in Figure 2.3b. The differences with a basic gate are that another transistor T3 is introduced and diode D3 is not present. The function of D3, included in basic TTL to ensure that T4 will be cut off when T2 and T5 are saturated, is made by the voltage drop across the base–emitter junction of T3. The important characteristic of the improved gate is that, when T3 and T4 are conducting, the gate output resistance is substantially lower than in the basic TTL circuit. Repeating the calculation made for basic TTL, the output resistance is $R_{OH} \approx R_2/h_{FE}^2$ when the transistor resistance is neglected. This lowered output impedance and the use of Schottky junction technology increase the speed of operation at the gate. Any capacitance shunting the gate output will be able to charge more rapidly. It is interesting to note that, in this configuration, T4 can never saturate.

The basic input static characteristic of a TTL is shown in Figure 2.5. When $V_I = 0$, T2 is cut off and the collector current of transistor T1 is $I_{C1} = 0$. In this case, T1 can be roughly

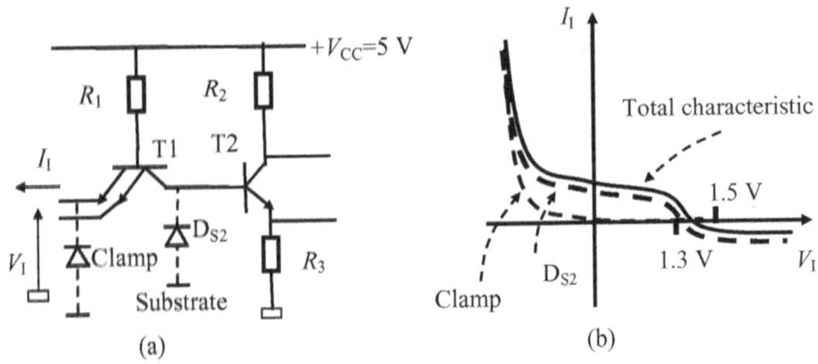

Figure 2.5 Basic TTL device: (a) input circuitry; (b) input static characteristic (not to scale)

represented as a diode. Assuming a 0.75 V junction voltage and $R_1 = 4\ \text{k}\Omega$, it yields $I_I = (5 - 0.75)/4000 = 1.06$ mA.

As V_I increases, the I_I–V_I plot departs from the straight line as T2 begins to turn on. When $I_I = 0$, all the current through the 4 kΩ resistor flows into the collector of T1 and into the base of T2. At this point, both T2 and T5 will be in saturation, and the voltage V_{B2} will be about 1.5 V. The characteristic will begin to depart from the straight line at about 1.3 V.

2.1.3 Complementary Metal Oxide Semiconductor (CMOS) Devices

Complementary Metal Oxide Semiconductor (CMOS) field-effect transistors differ from bipolar both in structure and operation [1]. The primary advantages of the CMOS are its low power dissipation and small physical geometry. Advances in design and fabrication have brought CMOS devices into the same speed and output drive capability as TTL. Again, enhancements have resulted in the evolvement of additional classifications: MG (*Metal-Gate* CMOS), HC (High-speed silicon-gate CMOS) [17], and FACT (Advanced CMOS) [8].

The most recent evolution in CMOS logic has been in reducing supply voltage without sacrificing performance. The new LCX family is one outgrowth of this trend. This family results from the joint efforts of a triumvirate of companies including Motorola, National, and Toshiba. Although each company has done its own design and fabrication, they have mutually agreed to provide identical performance specifications. In addition to the 3 V operating voltage, LCX inputs and outputs are tolerant of interfacing with 5 V devices. The CMOS inverter is shown in Figure 2.6a. The drains of p-channel and n-channel transistors are joined, and a supply voltage V_{CC} is applied from source to source. The output is taken at the common drain. The input V_I swings nominally through the range of V_{CC} that is positive. Because of the complete symmetry of the circuit, the two transistors are chosen to be reasonably alike. A typical I/O static characteristic is shown in Figure 2.6b [1].

When $V_I = V_{CC}$ and V_O is forced to move from V_{CC} to 0 V, the n-type MOSFET is on, and the p-type MOSFET is off. The output static point moves from the saturation region, where the output impedance R_{OL} is very high, to the triode region, where R_{OL} is low. In the triode region it is found that for an n-channel device

$$I_{DS} = k\left[2(V_{GS} - V_{th})V_{DS} - V_{DS}^2\right], \quad \text{with } 0 \le V_{DS} \le V_{GS} - V_{th} \tag{2.1}$$

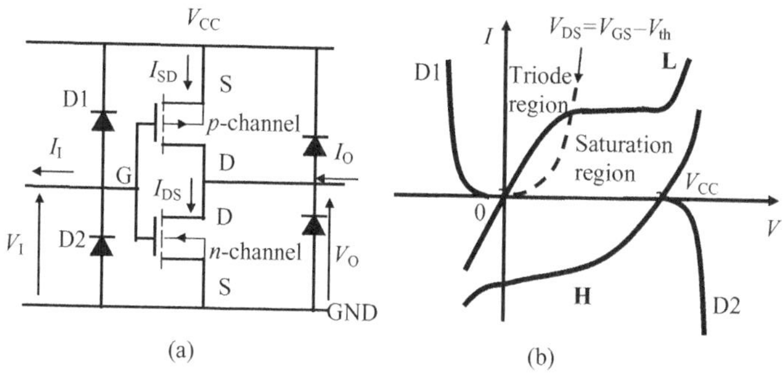

Figure 2.6 CMOS gate: (a) input and output stage; (b) output static characteristics at low (L) and high (H) levels and input characteristic

while in the saturation region

$$I_{DS} = k(V_{GS} - V_{th})^2, \quad \text{with } 0 \leq V_{GS} - V_{th} \leq V_{DS} \tag{2.2}$$

where V_{th} is the threshold voltage, $k = \chi \varepsilon w/(2tl)$ is a constant, χ is the mobility of carriers in the channel (electrons in n-channel devices), ε is the dielectric constant of oxide under the gate, t is the thickness of oxide under the gate, w is the channel width, and l is the channel length.

The impedance in the saturation region is very high; thus, operation in this region should be avoided when driving transmission lines, and matched impedance is required. If the buffer is designed for operation primarily within the triode region, the impedance will vary much less and the output impedance can be considered almost linear and equal to a few ohms.

When $V_I = 0$ and V_O is forced to move from 0 to V_{CC}, the n-type MOSFET is off, and the p-type MOSFET is on. The output static point moves from the saturation region, where the output impedance R_{OH} is very high, to the triode region, where R_{OH} is low. In a p-channel transistor, operating in the triode region the equation for device current is more conveniently written in the form

$$I_{SD} = k\left[2(V_{SG} - V_{th})V_{SD} - V_{SD}^2\right], \quad \text{with } 0 \leq V_{SD} \leq V_{SG} - V_{th} \tag{2.3}$$

and when operating in the saturation region as

$$I_{SD} = k(V_{SG} - V_{th})^2, \quad \text{with } 0 \leq V_{SG} - V_{th} \leq V_{SD} \tag{2.4}$$

Input diodes D1 and D2 are for protection and determine the input characteristic. Output diodes determine the output characteristic when $V_O < 0$ and $V_O > V_{CC}$.

In the triode area, the equivalent output Thévenin circuit is a voltage source E_{out} with a swing between 0 and V_{CC} and R_{out} of a few ohms.

2.1.4 Emitter-Coupled Logic (ECL) Devices

Emitter-Coupled Logic (ECL) derives its name from the differential-amplifier configuration in which one side of the diff-amp consists of multiple-input bipolar transistors with their emitters tied together [1]. An input bias on the opposite side of the diff-amp causes the amplifier to operate continuously in the active mode. Consequently, ECL consumes a relatively substantial amount of power in both states (one or zero), but also results in the fastest switching speeds of all logic families. An inherent benefit of ECL is the narrow switching level swing between devices (approximately 800 mV), which helps to reduce noise generation.

There have also been many evolutionary advancements in ECL: 100K (1975), 10KH (1981), and ECLinPS^{TM} (1987) [6, 15]. Of most recent vintage is the ECLinPS Lite^{TM} family of single-function devices. By focusing on simplicity, this family achieves very high performance while reducing the package size.

The basic circuit configuration employed in the logic ECL with positive power supply (PECL) is shown in Figure 2.7, where the difference amplifier is used to avoid saturation region operation of the transistors [1], giving rise to a high-speed digital device. It is possible to establish a transistor in its active region, with stability, by introducing negative feedback through the simple expedient of using a large emitter resistor. In the present application as a logic gate, the base of transistor T2 is held at a fixed reference voltage V_R while an input voltage V_I is applied to the base of transistor T1. The bias voltage V_R is set at the midpoint of the signal logic swing. When V_I is sufficiently lower than V_R, T1 will be cut off and current will flow through T2. Voltage V_R, the resistance at the collector, and the circuit at the emitter of T2 are selected to ensure that T2 operates in its active region and is not saturated. The emitter current I_E, corresponding to input level '0', is $I_E = (V_R - V_{BE})/R_E = 4$ mA, with $V_R = 3.65$ V, $V_{BE} = 0.8$ V, and $R_E = 713\ \Omega$. The voltage drop across the collector resistance $R_{C2} = 245\ \Omega$ of T2 may be calculated as $V_{RC2} = (I_E - I_{B2})R_{C2} \approx I_E R_{C2} = 0.004 \times 245 = 0.98$ V. Assuming a voltage $V_{BE} = 0.8$ V of the emitter follower transistor T3 in its active region as the loading effect of V_T and R_T, at OR output, it yields $V_{OL} = 5 - (0.98 + 0.8) = 3.2$ V. This current–voltage point is indicated in Figure 2.7b by Q_L.

When V_I rises to V_R, the current in the two transistors will be nominally equal. Finally, as V_I continues to increase, the emitter voltage V_E on resistance R_E increases, as $V_E = V_I - V_{BE1}$, with V_{BE1} approximately constant, and eventually T2 will cut off. We now have T1 operating

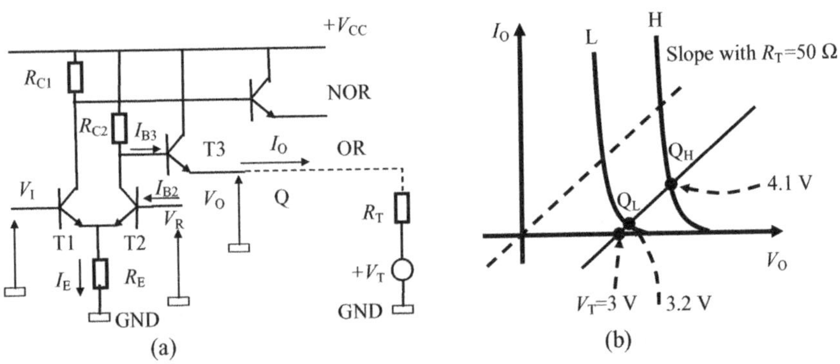

Figure 2.7 ECL device: (a) output stage; (b) output static characteristics at low (L) and high (H) states

in its active region. In this case the voltage at OR output is $V_{OH} = 5 - (0.8 + 0.1) = 4.1$ V. This current–voltage point is indicated in Figure 2.7b by Q_H, assuming that T2 is near the cut-off region with a collector current that causes a voltage drop on the collector resistance of 0.1 V.

The two outputs, OR and NOR, have complementary voltages that can be used for *differential mode* transmission. Thus, the mechanism of operation consists in switching a nominally fixed emitter current from one transistor to the other with very little change in emitter current. Outputs are taken at the collectors through emitter followers, which drops the collector voltage level one base-emitter drop. The emitter follower provides buffering and low impedance at the output terminals.

Considering the positive direction of I_O when the current is sourced by the output for convenience of representation, the *I–V* output static characteristic is shown in Figure 2.7.

Taking into account that the base current on the output transistor is $I_{B3} \approx I_O/h_{FE}$, where h_{FE} is the transistor gain, and that the current on the collector of transistor T2 is about I_E, the point QL is the interception between the line $V_{CC} - (I_E + I_O/h_{FE})\, R_{C2} - V_{BE} = V_O$ (or, adopting the convention of Figure 2.7b, $I_O = -h_{FE}/R_{C2}V_O + h_{FE}/R_{C2}(V_{CC} - I_E R_{C2} - V_{BE})$) and the line $I_O = V_O/R_T - V_T$. With $h_{FE} = 50$, the output resistance of the ECL gate is $R_{out} = R_{C2}/h_{FE} \cong 5\ \Omega$. Note that $V_O = 3.2$ V for $I_O = 0$. The point Q_H of Figure 2.7b is calculated with the $I_E R_{C2}$ term missing because T2 is cut off. In this case R_{out} is again 5 Ω and $V_O =$ 4.2 V for $I_O = 0$. The output equivalent circuit of an ECL is therefore a voltage generator with a swing from 3.2 to 4.2 V and a series resistance of 5 Ω. This means that ECL is suitable for a driving transmission line with low Z_0, e.g. 50 Ω, and in matching condition when $R_T = 50\ \Omega$.

2.1.5 Low-Voltage Differential Signal (LVDS) Devices

Low-Voltage Differential Signal (LVDS) is defined in the TIA/EIA-644 standard and can be designed using CMOS processes [7, 13]. It is a low-voltage, low-power, differential technology used primarily for point-to-point and multidrop cable driving applications. The standard was developed under the *Data Transmission Interface* committee TR30.2. It specifies a maximum data rate of 655 Mbps, although some of today's applications are pushing well above this specification for a serial data stream.

Compared with other differential cable driving standards like RS422 and RS485, LVDS has the lowest differential voltage swing of 700 mV, and a typical offset voltage of 1.25 V above ground. The output stage of an LVDS device is shown in Figure 2.8.

According to the direction of the output current I_O, the voltage V_O referred to ground (see output Q+) can have values of 0.9–1.1 or 1.3–1.5 V considering the low and high level outputs of a CMOS gate. LVDS features a low swing differential constant current source configuration, which supports fast switching speeds and low power consumption. LVDS is a valid alternative to PECL devices for differential signal transmission. The characteristic of this technology used for differential signaling will be investigated in more detail in *Section 12.1*.

2.1.6 Logic Devices Powered and the Logic Level

The average power dissipation P_{avg} of a digital device that must charge and then discharge its capacitive load C through a series resistance R every cycle with period T and swing voltage V can be calculated considering that the voltage across the capacitor during the charging time

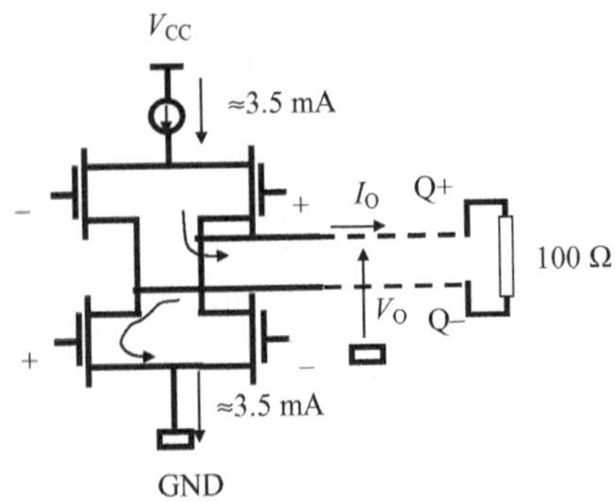

Figure 2.8 LVDS device output stage

is $V(t) = V(1 - e^{-t/RC})$. The instantaneous power drawn from the supply is

$$P(t) = V(t)I(t) = V(t)C\mathrm{d}V(t)/\mathrm{d}t = (V^2/R)e^{-t/RC} \tag{2.5}$$

Integrating the power (2.5) over a half cycle and dividing by the full period yields

$$P_{avg} = \frac{1}{T}\int_0^{T/2} \frac{V^2}{R} e^{-t/RC}\,\mathrm{d}t = CV^2 f \quad \text{for } T \gg RC \text{ and } f = 1/T \tag{2.6}$$

Therefore, for a data line with transitions on the rising or falling edge of the clock, the average power dissipation is $P_{avg} = \frac{1}{2}CV^2 f$.

With the square dependence on voltage, the reduction in voltage is the most effective way of reducing power dissipation. The trend in developing new logic devices is therefore to lower power supply and increase speed. Logic families with a power supply at 5 V include TTL, ACMOS, and PECL, and those at 3.3 V include LVT, LVC, LCX, GTL, and LVDS. The values of voltages V_{OHmin}, V_{IHmin}, V_{OLmax}, and V_{ILmax} for several logic families are shown in Figure 2.9, so that their immunity noise can be compared. These values are very important, as a successful communication requires that driver and receiver agree on what voltage levels constitute logic low and logic high. For voltage-mode signaling such as TTL, the driver directly sets the output voltage.

For current-mode signaling such as ECL in differential mode and LVDS, the driver sets the current level, and the voltage drop across a termination resistor at the receiver is detected. In either case, it is sufficient to discuss signaling in terms of the voltages set by the driver and the voltages detected by the receiver.

2.2 Dynamic Characteristics: Gate Delay and Rise and Fall Times

The main dynamic parameters of a logic gate are:

- gate or propagation delay time T_{GD};
- rise time t_r and fall time t_f.

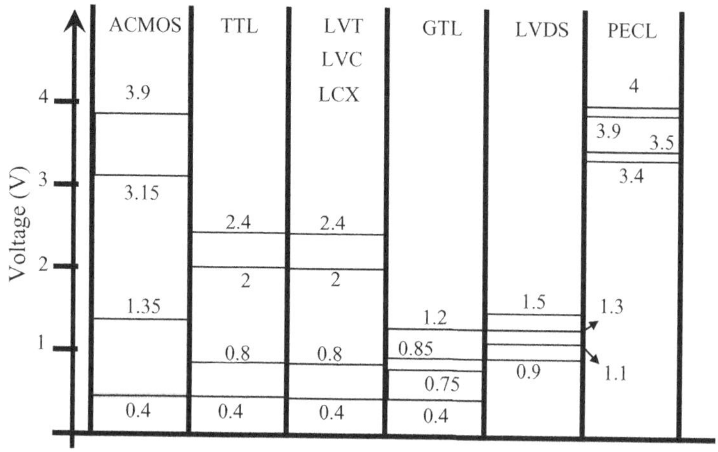

Figure 2.9 Logic voltage levels for some logic families

The gate delay time T_{GD} is defined in Figure 2.10. It is the delay between the input and output voltages of the device taken at the threshold voltage level V_{th} where the device changes state. The threshold voltage V_{th} depends on the logic device used. For example, with the TTL family, $V_{th} = 1.5$ V. The gate delay time T_{GD} for high-to-low ($T_{GD\text{-}HL}$) and low-to-high ($T_{GD\text{-}LH}$) logic state transitions is very important because it affects the jitter and skew defined in *Chapter 1*. Rise time t_r and fall time t_f parameters are defined in Figure 2.11. The rise time t_r is the time required for V_O to go from 10 to 90 % of the voltage swing V_{sw}. The fall time t_f is the time required to go from 90 to 10 % of V_{sw}. The times t_r and t_f could be different for the same device, for instance, FAST has $t_r = 3$ ns and $t_f = 1$ ns; on the other hand, FACT has $t_r = t_f = 2$ ns.

These parameters are very important when a system designer has to decide whether an interconnect must be considered as a 'short line' or as a 'long line'. In *Chapter 5* it will be shown that, if $t_r/t_f < 2t_{pd}l$, where t_{pd} is the per-unit-length propagation delay time of the

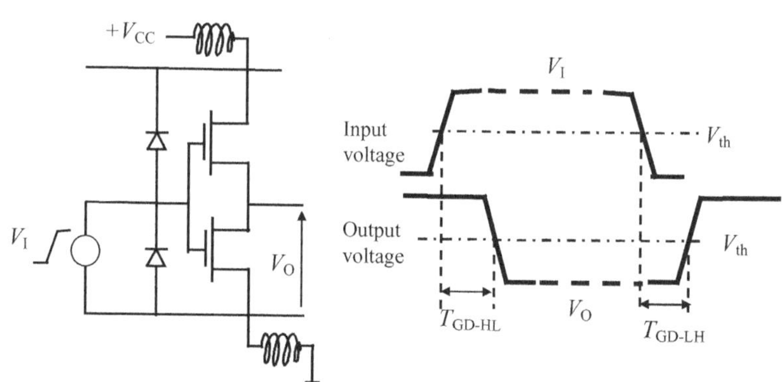

Figure 2.10 Delay parameters: (a) final stage of a CMOS gate; (b) gate delay parameters

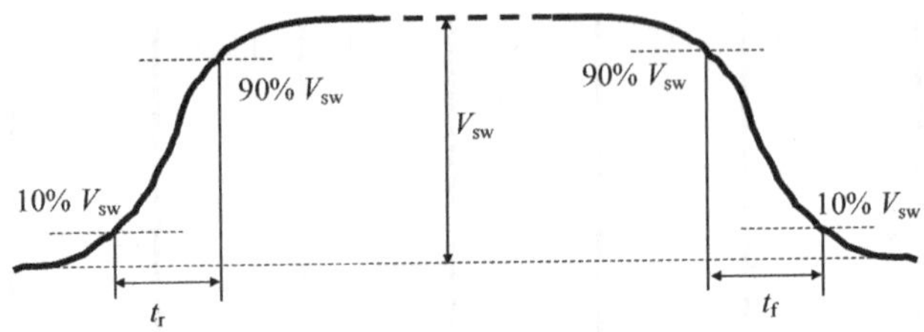

Figure 2.11 Definition of rise t_r and fall t_f times

interconnect between two gates and l is the length of this link, the line must be considered as 'long line' and simulated as a transmission line instead of a lumped line capacitance.

Dynamic parameters t_r and t_f depend mainly on the number of simultaneously switching gates, loads, decoupling capacitors, package, and power distribution. Consider that measuring t_r and t_f with only one gate switching without load leads to an overdesign. It is important to find out typical cases or an average between two extreme cases.

Typical dynamic parameters of some traditional logic families are summarized in Table 2.1. The dynamic performance of the up-to-date devices used for very high-speed systems such as LVDS and PECL will be discussed in *Chapter 12*.

2.3 Driver and Receiver Modeling

The problem of how to model the driver and the receivers of an interconnect correctly is a very important topic in signal and power integrity predictions. There are simple models suitable for predicting reflections and crosstalk, and other more complex models for predicting ΔI*-noise*. The purpose of this section is to help the reader in choosing the most appropriate model for the problem of interest.

2.3.1 Types of Driver Model

There are three general types of driver model that can be used in the simulation of digital systems:

1. Linear models.
2. Behavioral models.
3. Full transistor models.

In many practical cases of interconnects, the first two types of model are suitable for reflection and crosstalk predictions. These models are similar because they adopt a Thévenin equivalent circuit to simulate the output stage. This circuit is constructed from curves describing the I/O static current–voltage characteristic, and considering the dynamic performance during the switching of the output gate. Numerous examples of application of these models are provided in *Chapter 5* and *Chapter 6*. These types of model are the basis of the IBIS models provided by the manufacturers of devices in a standard format. The IBIS models are introduced and described in *Section 2.4*. A description of up-to-date behavioral models that consider

Table 2.1 Dynamic parameters of some logic families

Typical Commercial	TTL				CMOS				ECL		
Parameter (0–70 °C)	LS	ALS	ABT	FAST	HC	FACT	LVC	LCX	10H	100K	ECL in PS
Gate propagation delay (ns)	9	7	2.7	3	8	5	3.3	3.5	1	0.758	0.33
Flip-flop toggle rate (MHz)	33	45	200	125	45	160	200	200	330	400	1000
Output edge rate (ns)	6	3	3	2	4	2	3.7	3.6	1	0.7	0.5
Supply voltage (V)	4.5–5.5	4.5–5.5	4.5–5.5	4.5–5.5	2–6	1.2–3.6	1.2–3.6	2–6	−4.5−5.5	−4.5−5.5	−4.5−5.5

several phenomena within digital devices, such as non-linear behavior and switching noise, can be found elsewhere [19–22]. Models of the third type are used for modeling the input, intermediate, and output stages of an IC device at the transistor and diode level. They may be necessary when crosstalk and simultaneous switching noise in IC devices must be simulated (see the example given in *Section 8.4*).

2.3.2 Driver Switching Currents Path

Drivers fall into two basic categories: push-pull and current steering. The push-pull drivers use transistors to connect the output to either of two voltage levels to indicate the logic state (TTL, CMOS). The current-steering drivers use two different levels of current to indicate logic state, where the current is converted to voltage at a termination resistor to enable detection (*differential-mode* ECL, LVDS).

The main disadvantage of a push-pull driver is the extra current required from the power supply by the simultaneous conduction of the two transistors when they are switching. This current is indicated as *crowbar* or *shoot-through current* in Figure 2.12. This does not happen with current-steering drivers. Other switching currents are required to charge and discharge line and load capacitors, as illustrated in Figure 2.12.

If the effects of crowbar current can be neglected, as occurring in the presence of a quasi-ideal bypass capacitor between power supply leads and ground (associated lead inductance very low), the transmitted and received waveforms on the signal line can be simulated with the point-to-point structure shown in Figure 2.12c. In fact, ground and power supply conductors are at the same potential in the high-frequency range, and the return signal current can refer to both.

Any non-linear circuit can be made linear and represented by a Thévenin equivalent circuit. Modeling a driver with equal pull-up and pull-down output impedance consists in assigning the driver's output voltage swing and slew rate to the Thévenin voltage source, and the driver's output impedance to the Thévenin equivalent impedance. For example, if the driver is a CMOS that can swing from 0 to 5 V in 2 ns, then the Thévenin source voltage is a linear independent

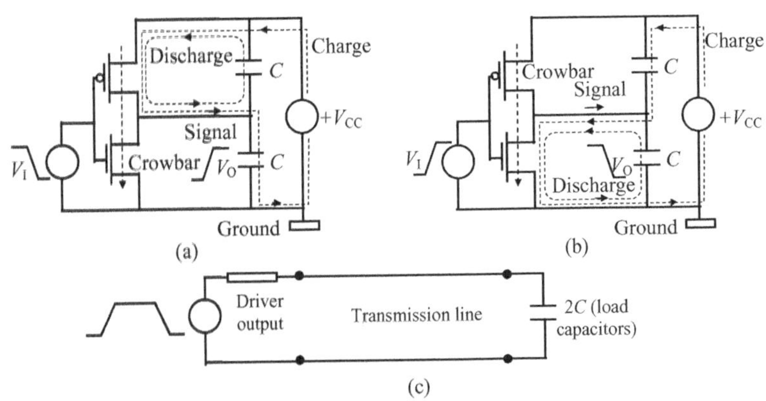

Figure 2.12 CMOS driver: (a) switching path for output changing from low to high state; (b) switching path for output changing from high to low state; (c) point-to-point interconnect with capacitive load

voltage source swinging from 0 to 5 V in 2 ns without load. The Thévenin impedance is a resistance equal to the output impedance of the driver calculated as the average value during the transition. A critical point is the choice of the edge rate in terms of rise and fall times. In fact, these times are different if the gate drives large capacitive loads or long transmission lines.

Logic families suitable for being modeled by a linear circuit are *differential-mode* ECL and LVDS because they are used with controlled interconnects (a fixed characteristic impedance is required), and a termination is used to match the line. For other logics, a non-linear model is required, especially when terminations are not used

2.3.3 Driver Non-Linear Behavioral Model

The construction of a behavioral model of a non-linear driver by using a Thévenin equivalent circuit is described considering typical CMOS static output characteristics (see Figure 2.13). The first step is to divide the low and high static curves into three segments, as shown in Figure 2.13b: L_1, L_2, L_3 for the low state and H_1, H_2, H_3 for the high state. As each of these segments for both low (L) and high (H) states in the I–V plane are described by the line equation $I_O = (V_O - E_O)/R_O$, they can be simply identified by two parameters E_{OL} or E_{OH}, as the interception of the line with the voltage axis V_O, and the line slope $1/R_{OL}$ or $1/R_{OH}$. The voltage V_{OL} or V_{OH}, as the interception of two adjacent segments, characterizes the point of change from one segment to another. For example, V_{OL2} is the interception of segments L_2 and L_3, and V_{OH1} is the interception of segments H_1 and H_2.

The second step is to assign dynamic values to the Thévenin equivalent circuit parameters, which are the voltage source E_O and the resistance R_O, considering the switching performance of the type of driver (see Figure 2.13a). For example, with a CMOS buffer that drives an

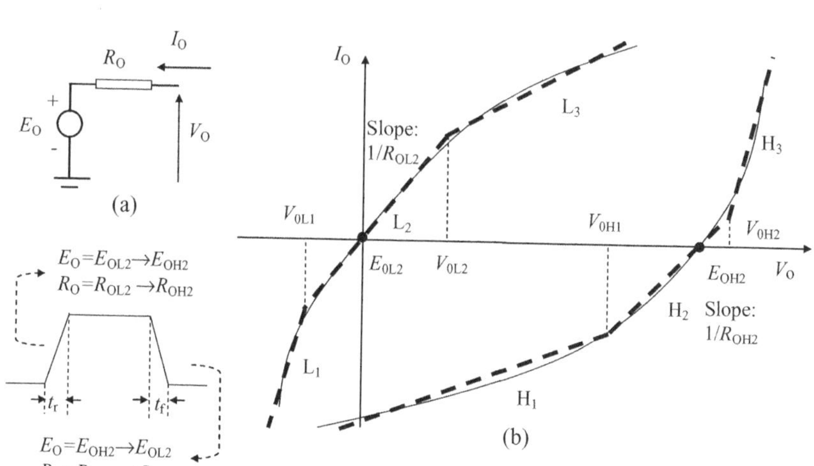

Figure 2.13 Behavioral model for CMOS: (a) output equivalent circuit; (b) output static characteristic (solid line) and its segmentation (dashed line); (c) dynamic characteristic

interconnect of typical characteristic impedance in the range 50–150 Ω, the starting point in the I–V characteristic lies on segment L_2 before the low-to-high switching (E_{OL2} in Figure 2.13b), and on line H_2 after the rise time t_r (E_{OH2} in Figure 2.13b). In the case of high-to-low switching, the starting point is on segment H_2, and after the fall time t_f it is on segment L_2. In fact, as will be explained in *Chapter 5*, the values of I_O and V_O after low-to-high switching can be found as the interception of the line with slope $-1/Z_0$ passing from the static point on segment L_2 with segment H_2. The values of I_O and V_O after the high-to-low switching can be found as the interception of the line with slope $-1/Z_0$ passing from the static point on segment H_2 with line L_2. This means that the equivalent circuit of the driver, in the interval time t_r, can be simulated by a voltage source E_O that changes from E_{OL2} to E_{OH2}, and R_O that changes from R_{OL2} to R_{OH2}. In the interval t_f, E_O changes from E_{OH2} to E_{OL2}, and R_O from R_{OH2} to R_{OL2}, as shown in Figure 2.13c. Outside this short interval, as happens with high-speed logic, E_O and R_O are a function of the output voltage V_O and have the task of reproducing, according to the value of V_O, the segment L_i or H_i, where $i = 1, \ldots, 3$, of the static output characteristic of the driver. All this can be implemented in a SPICE-like circuit simulator by using the 'if' operator and tables in the time domain.

The procedure adopted for CMOS can be repeated for TTL with some differences due to the technology. The static characteristic at the low state is divided into four segments, L_1, L_2, L_3, and L_4, and the static characteristic at the high state is divided into three segments, H_1, H_2, and H_3, as shown in Figure 2.14.

As the TTL buffer also drives interconnects with a typical characteristic impedance in the range 50–150 Ω, the starting point in the I–V characteristic lies on segment L_3 before the low-to-high switching (E_{OL3} in Figure 2.14b) and on segment H_1 after time t_r (E_{OH1} in Figure 2.14b). In the case of high-to-low switching, the starting point is on segment H_3 (E_{OH3} in Figure 2.14b) and after time t_f on segment L_3 (E_{OL3} in Figure 2.14b). In fact, the values of I_O

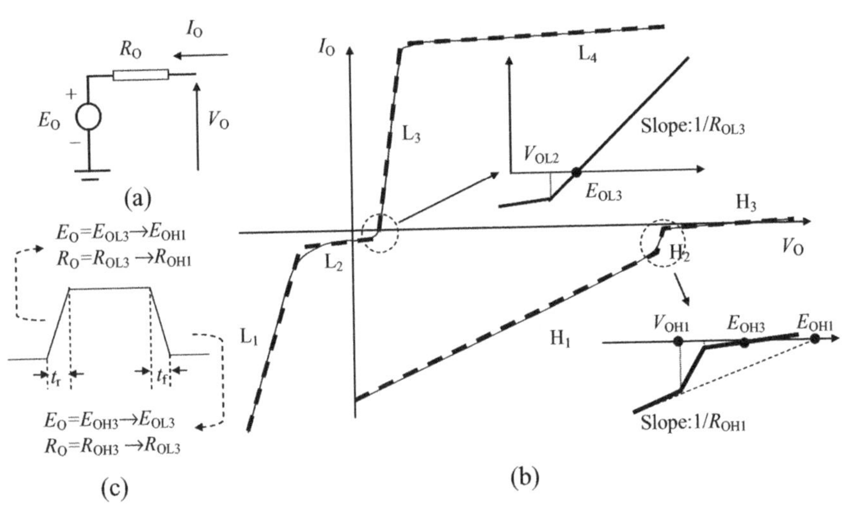

Figure 2.14 Behavioral model for TTL: (a) output equivalent circuit; (b) output static characteristic (solid line) and its segmentation (dashed line); (c) dynamic characteristic

and V_O after the low-to-high switching can be found as the interception of the line with slope $-1/Z_0$ passing from the static point on segment L_3 with segment H_1. The values of I_O and V_O after the high-to-low switching can be found as the interception of the line with slope $-1/Z_0$ passing from the static point in segment H_3 with segment L_3. This means that the Thévenin equivalent circuit of the driver, in the time interval t_r, can be simulated by a voltage source E_O that changes from E_{OL3} to E_{OH1} and R_O that changes from R_{OL3} to R_{OH1}. In the time interval t_f, E_O changes from E_{OH3} to E_{OL3} and R_O from R_{OH3} to R_{OL3}, as shown in Figure 2.14c. Outside this short interval, as happens with high-speed logic, E_O and R_O are functions of the output voltage V_O and have the task of reproducing, according to the value of V_O, the segment L_i, where $i = 1, \ldots, 4$, or H_j, where $j = 1, \ldots, 3$, of the static output characteristic of the driver.

Examples of the accuracy of these models for calculating reflections and crosstalk will be given in *Section 6.4* with TTL and CMOS gates. The package effects within the driver are also taken into account with the addition of resistance, inductance, and capacitance circuit elements.

2.3.4 Receiver Non-Linear Behavioral Modeling

The primary function of the receiver is to detect the logic level of an analog waveform in the presence of noise. To accomplish this task, the receiver offers very high impedance in the interval of the voltage logic swing and the protection of diodes outside this region. The input equivalent circuit of the receiver behavioral model is shown in Figure 2.15, as well as the input static characteristics of TTL and CMOS. In the case of TTL receivers, a diode is used when the input voltage $V_I < 0$ V, and this is fundamental in mitigating reflections. With CMOS receivers there are two diodes, one for $V_I < 0$ V and the other for $V_I > V_{CC}$, where V_{CC} is the power supply value. These two diodes act in mitigating reflection and protect the device from ESD.

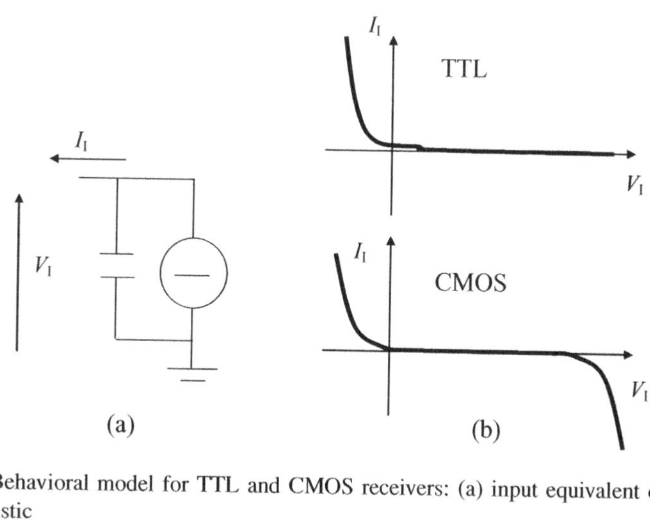

Figure 2.15 Behavioral model for TTL and CMOS receivers: (a) input equivalent circuit; (b) input static characteristic

In any case, a receiver can be represented as a circuit element by a voltage-dependent current generator using tables. Also, it is very important to include in the model the input capacitance of the receiver, which is usually some pF. This capacitance has a great effect on signal integrity, especially for lines with distributed loads or loads concentrated at the opposite end of the driver, as will be shown in *Chapter 5*.

2.4 I/O Buffer Information Specification (IBIS) Models

I/O Buffer Information Specification (IBIS) is a fast and accurate behavioral method of modeling I/O buffers on the basis of *I/V* curve data derived from measurements or full-circuit simulations. IBIS uses a standard software parsable format to create the behavioral information needed to model analog characteristics of ICs in order to simulate reflections and crosstalk in interconnects of digital devices [23, 24]. IBIS was originally started by Intel, and is presently driven by the IBIS open forum consisting of software vendors, computer manufacturers, semiconductor vendors, and universities. The first version was released in April 1993, focusing on bipolar TTL and CMOS logic components. Other versions were ratified in subsequent years, with many new capabilities that have increased its accuracy and the number of device types that are supported, such as ECL, differential, open-drain I/O devices, and expanded package model definitions to include coupling between pins and other features. The advantages of IBIS are:

- Protection of proprietary information.
- Accurate model of non-linear I/O characteristics, package structure, and devices for ESD protection.
- Signal integrity simulation on system board.
- Models available from semiconductor vendors for free.
- Faster simulation time compared with structural methods.
- Compatibility with all industrial simulation platforms.
- No additional resources required for customer support.

2.4.1 Structure of an IBIS Model

An example of a block diagram for an IBIS behavioral model is shown in Figure 2.16, while the basic elements that must be included for IBIS modeling of an I/O structure are shown in Figure 2.17. The elements of the IBIS model correspond to the keywords in the IBIS format specification. The pull-down element contains the *I/V* pull-down information, including the typical, minimum, and maximum currents for the given voltages of the pull-down. The pull-up element contains the *I/V* pull-up information, modeling the characteristics of the buffer when driven high. Note that the voltages in the pull-up and power-clamp tables are V_{CC} relative and are derived from the equation

$$V_{\text{table}} = V_{CC} - V_{\text{output}} \tag{2.7}$$

The IBIS table lists voltages from $-V_{CC}$ to $2V_{CC}$. The wide voltage range is provided to improve the accuracy of certain simulators. Many simulators benefit from including these

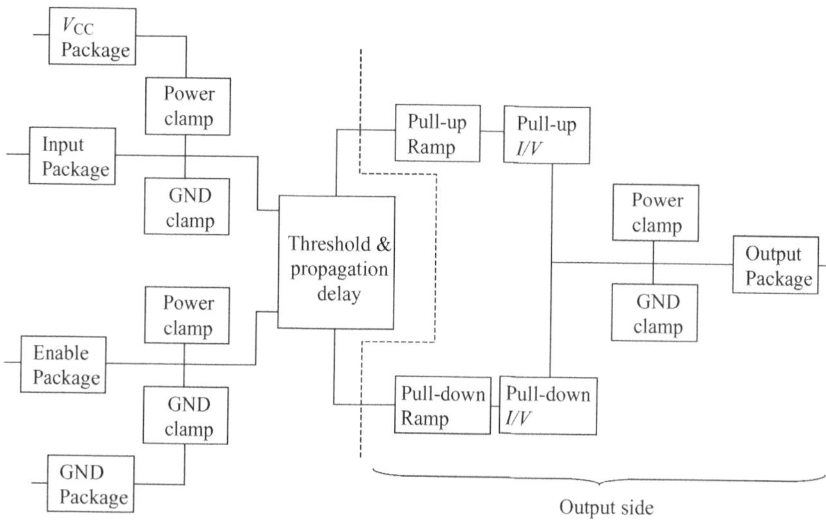

Figure 2.16 IBIS behavioral block diagram

characteristics in the model. For simulators that do not extrapolate unspecified voltages, the ranges given are more than adequate.

The clamp diode characteristics are meant to be modeled in parallel with the driver information as pull-down and pull-up, ensuring that the diode characteristics are present even when the output buffer is in a high-impedance state (off). The currents listed in the table can be large and are provided only to enable simulators to construct the proper diode curve.

The element containing the ramp time for the pull-up and pull-down structures ensures the correct time operation of the model. There is a 'typ' column together with 'min' and 'max' columns. The 'min' column represents the longest rise/fall times, and the 'max' column represents the shortest times. These values often appear very small because they are intrinsic values for transistors with all packaging and external loads removed. The packaging characteristics are added outside the transistor model. The packaging element in the model includes the inherent capacitance of the silicon portion of the die, C_{comp}, and not the package. The package is

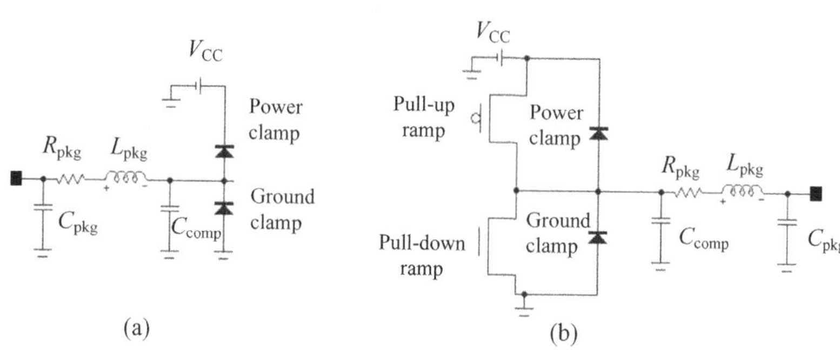

Figure 2.17 Circuit element of an IBIS model: (a) input model; (b) output model

modeled by the parameters R_{pkg}, L_{pkg}, and C_{pkg}, schematically organized as shown in Figure 2.17. The table supplies the range (minimum-to-maximum) for each parameter.

With the IBIS data, signal integrity designers can model the device characteristics for both fast and slow switching edges. The slow model can be used to determine the propagation time. The fast model can be used to simulate overshoot, undershoot, and crosstalk. A slow model can be created by combining the minimum currents with the maximum ramp time and maximum package element values. A fast model can be created with the largest currents, the fastest ramp, and the minimum package element values. The minimum and maximum data include both temperature and process variations. Voltage variation is normally adjustable within simulation tools, or can be approximated by shifting the *I*/*V* data by the desired voltage tolerance. The input model includes *I*/*V* curves for diodes only.

2.4.2 *IBIS Models and Spice*

To use IBIS models in SPICE, an IBIS-to-SPICE translator is required, as some simulators cannot use the IBIS file directly. It must be translated into a usable model language. Typically, the conversion leads to a SPICE-compatible syntax. On the other hand, codes for professional users, such as HSPICE and Eldo, handle IBIS models directly.

An example of using an IBIS model with SPICE will be provided here by using Micro-Cap version 9, which has the required translator [25]. The input file should have an IBIS extension (.ibs), and a .lib extension is usually assigned to the output file to indicate its use as a library file, although it also contains SPICE code for plotting the translated buffer models.

In the example considered the component 74AC244SC (file name *ac244sc 450.ibs.*) is adopted. This is not an up-to-date device – it refers to version 2.1. However, our intention is to give basic information concerning a typical IBIS format to help understand the more recent versions and how to use the information provided for signal integrity simulations by SPICE. Other files regarding the components of interest can be downloaded from the web as ‘xxx.ibs’, provided by the component manufacturers. The static *I*/*V* and dynamic output characteristics extracted by the IBIS file, using the ASCII data contained in the IBIS file, are shown in Figure 2.18. The input model uses the same clamp to power and ground characteristics as the output model. Voltage and current sign conventions are those used in *Section 2.3*: positive current sunk by the driver and voltages referred to ground.

The simulated waveforms shown in Figure 2.19 concern the case of the driver and receiver connected by a transmission line of characteristic impedance $Z_0 = 50\ \Omega$ and delay time $T_{\text{D}} = 3$ ns. Observe that maximum values produce overshoot and undershoot, while minimum values produce extra delay.

When a user has a low-cost circuit simulator that does not have the feature to translate automatically the IBIS format to SPICE, the information contained in the IBIS file can be used to obtain similar curves and waveforms to those in Figure 2.19. Then the component can be modeled as outlined in *Section 2.3*.

To give an idea of the type of accuracy achievable with these types of model, many examples of signal integrity investigations using models similar to IBIS obtained by measurement of the I/O characteristics will be provided in *Chapter 6*. Simple behavioral I/O models to implement into SPICE will also be described for the case where the user does not have a IBIS-to-SPICE translator available.

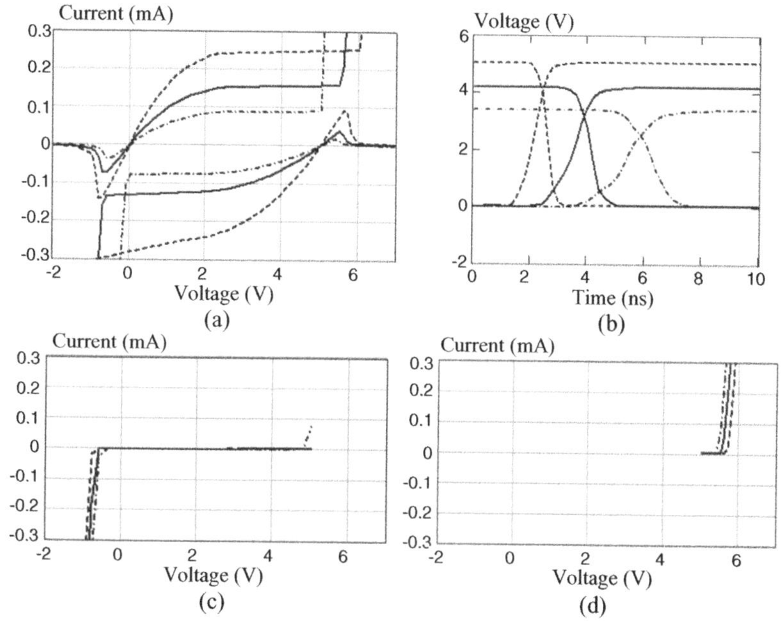

Figure 2.18 Static and dynamic characteristics of the device 74CA244 buffer as obtained by its IBIS model: (a) pull-up and pull-down *I*/*V* output; (b) rise and fall times; (c) clamp ground; (d) clamp power. Typical (solid line), minimum (dotted-dashed line), and maximum (dashed line)

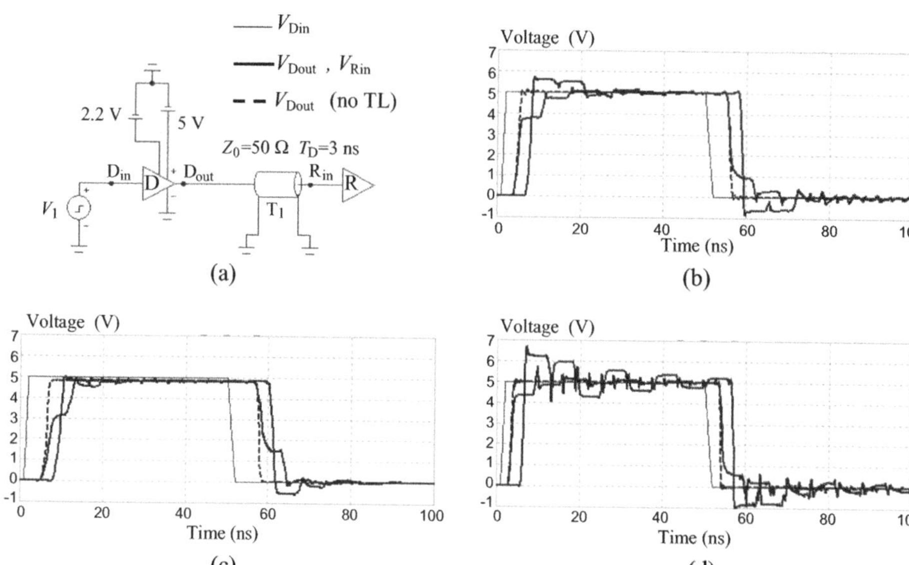

Figure 2.19 Example of signal integrity investigation by using 74AC244 IBIS models for driver (D) and receiver (R) translated into SPICE: (a) interconnect structure; (b) simulated waveforms with typical values; (c) with minimum values; (d) with maximum values

References

[1] Taub, H. and Schilling, D., *'Digital Integrated Electronics'*, McGraw-Hill Education, New York, NY, 1977.

[2] Blood, W., *'MECL System Design Handbook'*, 4th edition, Motorola Semiconductor Products, Inc., Phoenix, AZ, 1983.

[3] Morris, R.L. and Miller, J.R., *'Designing with TTL Integrated Circuits'*, McGraw-Hill, New York, NY, 1971.

[4] *'LVT Family Characteristics'*, AN-SCEA002A, Texas Instruments, 1998. Available online: www.ti.com.

[5] *'LVT (Low Voltage Technology) and ALVT (Advanced LVT)'*, AN-243, Philips Semiconductors, 1998. Available online: www.nxp.com.

[6] *'Designing with PECL (ECL at +5 V)'*, AN-1406, Motorola Semiconductor, Rev. 2, August 2001. Available online: www.freescale.com.

[7] *'LVDS Description and Family Characteristics'*, MS500530, Fairchild Semiconductor, 2000. Available online: www.fairchildsemi.com.

[8] *'FACT™ Descriptions and Family Characteristics'*, MS010158, Fairchild Semiconductor, 2000. Available online: www.fairchildsemi.com.

[9] *'ABT: Description and Family Characteristics'*, MS011558, Fairchild Semiconductor, 1999. Available online: www.fairchildsemi.com.

[10] *'LVT Family Introduction'*, MS000227, Fairchild Semiconductor, 1999. Available online: www.fairchildsemi.com.

[11] *'LVX Family Introduction'*, MS000217, Fairchild Semiconductor, 1999. Available online: www.fairchildsemi.com.

[12] *'AC/DC Family Characteristic Comparison'*, MS500120, Fairchild Semiconductor, 1999. Available online: www.fairchildsemi.com.

[13] *'LVDS Fundamentals'*, AN-5017, Fairchild Semiconductor, 2000. Available online: www.fairchildsemi.com.

[14] *'GTLP: an Interface Technology for Bus and Backplane Applications'*, AN-1065, Fairchild Semiconductor, 2001. Available online: www.fairchildsemi.com.

[15] *'ECL Backplane Design'*, AN-768, Fairchild Semiconductor, 2000. Available online: www.fairchildsemi.com.

[16] *'Transmission-line Effects Influence High-Speed CMOS'*, AN-393, Fairchild Semiconductor, 1985. Available online: www.fairchildsemi.com.

[17] *'Comparison of MM74HC to 74LS, 74S and 74ALS Logic'*, AN-319, Fairchild Semiconductor, 1983. Available online: www.fairchildsemi.com.

[18] *'FAST/FASTr Design Considerations'*, AN-661, Fairchild Semiconductor, 1990. Available online: www.fairchildsemi.com.

[19] Mutnury, B., Swaminthan, M., Cases, M., Pham, N., de Araujo, D., and Matoglu, E., 'Macromodeling of Nonlinear Transistor-level Receiver Circuits', *IEEE Trans. on Advanced Packaging*, **29**(1), February 2006, 55–66.

[20] Mutnury, B., Swaminthan, M., Cases, M., and Libous, J., 'Macromodeling of Nonlinear Digital I/O Drivers',*IEEE Trans. on Advanced Packaging*, **29**(1), February 2006 102–113.

[21] Stievano, I., Maio, I., Canavero, F., and Siviero, C., 'Reliable eye-diagram analysis of data links via device macromodels'. *IEEE Trans. on Advanced Packaging*, **29**(1), February 2006, 31–38.

[22] Stievano, I., Maio, I., Canavero, F., and Siviero, C., 'Parametric macromodels of differential drivers and receivers', *IEEE Trans. on Advanced Packaging*, **28**(2), May 2005, 189–196.

[23] *'IBIS (I/O Buffer Information Specification), Version 4.1'*, Ratified 30 January 2004.

[24] IEC 62014-1: 'Electronic Design Automation Libraries – Input/Output Buffer Information Specification (IBIS version 3.2)'.

[25] www.spectrum-software.com.

3

Inductance

Inductance is a very important line and device parameter for predicting *Signal Integrity* (SI) and *Radiated Emission* (RE). In this chapter, the self and mutual inductances of coupled loops are theoretically introduced considering two generic loops, and the associated equivalent circuits are derived. The **L** matrix concept for two coupled wires having a reference return conductor is provided as background to build an inductance matrix **L** for n-multiconductor transmission lines. As an example, closed-form expressions for **L** matrix calculation for round wires are given. The dependence of inductance on frequency is discussed.

The concept of partial inductance as a consequence of the segmentation of a loop is introduced in *Section 3.2*. Partial inductance is very useful for building up lumped-circuit models of electrically short parts of a PCB, such as the package in IC devices, vias, and connectors, in the case of SI predictions, and of a PCB with a finite ground plane in the case of RE predictions. Simple closed-form expressions of self and mutual partial inductance for round and rectangular conductors are provided. An example of an equivalent circuit with partial inductances is described for the case of a decoupling capacitor and a switching device located in a multilayer PCB.

The definitions of *differential mode* and *common mode* inductances are outlined at the end of the chapter. These definitions are very useful for crosstalk modeling, differential signaling, and filtering to mitigate radiated emission. A collection of working formulae for partial inductance calculation is given in *Appendix A*.

3.1 Loop Inductance

The definition of inductance can be rigorously based on the Maxwell equation in integral form. Two different approaches can be used to define the inductance:

- the field-based approach for formula development and theoretical treatment;
- the energy-based approach for computer computation.

The first definition is normally used to obtain formulae for traces, connectors, capacitor leads, vias, etc., as they are considered to be lumped elements.

Signal Integrity and Radiated Emission of High-Speed Digital Systems Spartaco Caniggia and Francescaromana Maradei

Simple and approximate closed-form expressions of inductance can be derived for some canonical configurations of multiconductor transmission lines in a homogeneous medium. Generally, numerical methods may be used in the case of an inhomogeneous medium. In these cases, inductance is calculated through the energy-based definition which is suitable for general-purpose electromagnetic simulators [1].

3.1.1 Inductances of Coupled Loops

To define inductances between two loops C_1 and C_2 in quasi-static condition (i.e. small loop dimensions compared with the wavelength of the electromagnetic field), the configuration shown in Figure 3.1 is considered. The currents I_1 and I_2 flow into the two loops defined by the closed paths C_1 and C_2 and placed in close proximity so that the magnetic flux generated from one loop penetrates the area enclosed by the other.

Faraday's law in integral form applied to the contour C_i, where $i = 1, 2$, is

$$\oint_{C_i} \vec{E} \cdot \mathrm{d}\vec{l}_i = -\frac{\partial}{\partial t} \int_{S_i} \vec{B} \cdot \mathrm{d}\vec{S}_i, \quad i = 1, 2 \tag{3.1}$$

where $\vec{E}$ is the electric field vector, $\vec{B}$ is the magnetic flux density vector, $\mathrm{d}\vec{l}_i$ is the elemental vector tangent to the contour path C_i and directed in accordance with the current I_i, and $\mathrm{d}\vec{S}_i$ is the elemental area on the surface S_i, oriented normally to S_i according to the right-hand rule. Equation (3.1) can be rewritten as

$$V_i(t) = \frac{\partial \Psi_i}{\partial t}, \quad i = 1, 2 \tag{3.2}$$

where V_i is the voltage induced in the ith loop and Ψ_i is the magnetic flux in the ith circuit owing to the currents in the different loops, which are respectively given by

$$V_i = -\oint_{C_i} \vec{E} \cdot \mathrm{d}\vec{l}_i, \quad i = 1, 2 \tag{3.3a}$$

$$\Psi_i = \int_{S_i} \vec{B} \cdot \mathrm{d}\vec{S}_i, \quad i = 1, 2 \tag{3.3b}$$

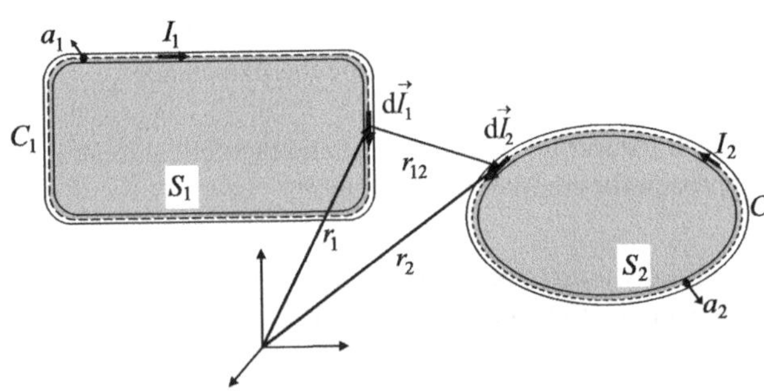

Figure 3.1 Currents and geometries of two coupled loops for inductance calculation

The magnetic flux density produced by the currents I_1 and I_2 flowing into the circuits C_1 and C_2 can be expressed as

$$\vec{B} = \vec{B}_1 + \vec{B}_2 \tag{3.4}$$

Introducing Equation (3.4) into Equation (3.3b) yields

$$\Psi_i = \sum_{j=1}^{2} \int_{S_i} \vec{B}_j \cdot \mathrm{d}\vec{S}_i = \sum_{j=1}^{2} \Psi_{ij}, \quad i = 1, 2 \tag{3.5}$$

where Ψ_{ij} is the magnetic flux in the ith loop ($i = 1, 2$) owing to the current I_j in the jth loop ($j = 1, 2$) when $I_k = 0$, where $k = 1, 2$ and $k \neq j$. This magnetic flux Ψ_{ij} can be expressed as

$$\Psi_{ij} = \int_{S_i} \vec{B}_j \cdot \mathrm{d}\vec{S}_i = L_{ij} I_j, \quad \text{for} \quad I_k = 0 \quad \text{with} \quad k = 1, 2 \quad \text{and} \quad k \neq j \tag{3.6}$$

where the coefficients L_{ij} are the self inductances for $i = j$ and the mutual inductances for $i \neq j$. The inductance L_{ij} is expressed in Henrys (H) and is defined through Equation (3.6) as

$$L_{ij} \equiv \frac{\Psi_{ij}}{I_j} = \frac{\int_{S_i} \vec{B}_j \cdot \mathrm{d}\vec{S}_i}{I_j}, \quad \text{for} \quad I_k = 0 \quad \text{with} \quad k = 1, 2 \quad \text{and} \quad k \neq j \tag{3.7}$$

Definition (3.7) is general and can be used in the case of n coupled loops. The inductance can be related to the loop geometry by the magnetic vector potential $\vec{A}$ defined by $\vec{B} = \nabla \times \vec{A}$. In fact, by introducing this expression into (3.6), and by using Stokes' theorem, the magnetic flux Ψ_{ij} in loop i averaged over the conductor cross-section a_i is [2]

$$\Psi_{ij} = \frac{1}{a_i} \oint_{C_i} \int_{a_i} \vec{A}_{ij} \cdot \mathrm{d}\vec{l}_i \, \mathrm{d}a_i \tag{3.8}$$

where $\vec{A}_{ij}$ is the magnetic vector potential along the ith path C_i produced by the current I_j in the jth loop C_j when $I_k = 0$ with $k = 1, 2$ and $k \neq j$. Assuming a uniform distribution of the current I_j over the cross-section a_j which is constant along the whole loop C_j, the magnetic vector potential $\vec{A}_{ij}$ is given by [2]

$$\vec{A}_{ij} = \frac{\mu}{4\pi} \frac{I_j}{a_j} \oint_{C_j} \int_{a_j} \frac{\mathrm{d}\vec{l}_j \, \mathrm{d}a_j}{r_{ij}} \tag{3.9}$$

where $r_{ij} = |\vec{r}_i - \vec{r}_j|$, as shown in Figure 3.1, and μ is the permeability which in vacuum is $\mu_0 = 4\pi \times 10^{-7}$ H/m. Introducing Equation (3.9) into Equation (3.8) and the resulting expression for Ψ_{ij} into definition (3.7), the following inductance L_{ij} is obtained:

$$L_{ij} = \frac{1}{a_i a_j} \frac{\mu}{4\pi} \oint_{C_i} \int_{a_i} \oint_{C_j} \int_{a_j} \frac{\mathrm{d}\vec{l}_i \cdot \mathrm{d}\vec{l}_j}{r_{ij}} \mathrm{d}a_i \, \mathrm{d}a_j \tag{3.10}$$

This formulation shows that averages are taken over the conductor cross-sections for both the flux (3.8) and the vector potential (3.9).

3.1.2 *Inductances of Thin Filamentary Circuits*

When the loops are made of thin filamentary wires (i.e. are of negligible cross-section), Equation (3.10) reduces to the Neumann formula

$$L_{fij} = \frac{\mu}{4\pi} \oint_{C_i} \oint_{C_j} \frac{d\vec{l}_i \cdot d\vec{l}_j}{r_{ij}} \tag{3.11}$$

where the subscript 'f' denotes that the current filament assumption is considered. By using Equation (3.11), Equation (3.10) can be written in the simple form

$$L_{ij} = \frac{1}{a_i a_j} \int_{a_i} \int_{a_j} L_{fij}\, da_i\, da_j \tag{3.12}$$

The idea of the average is apparent in Equation (3.12). For most geometries, closed-form solutions of the multiple integrals are hard to find or are unduly complicated. However, this approach for calculating L_{ij}, instead of using the energy-based definition as done by numerical codes, is very useful for introducing the partial inductance concept which is essential in modeling digital systems for signal integrity simulations (see *Section 3.2*).

3.1.3 *Equivalent Circuit of Two Coupled Loops*

The voltage V_i induced in the ith loop is also the voltage across the loop terminal, in principle an infinitesimal gap, where the connection with external devices (i.e. driver and receiver) occurs.

The introduction of Equations (3.5) and (3.6) into Equation (3.2) yields a system of equations that relate the voltages induced in the loops to the currents flowing through them. For the considered two-loop configuration, the system in explicit form is given by

$$\begin{cases} V_1(t) = L_{11}\dfrac{dI_1(t)}{dt} + L_{12}\dfrac{dI_2(t)}{dt} \\[2ex] V_2(t) = L_{21}\dfrac{dI_1(t)}{dt} + L_{22}\dfrac{dI_2(t)}{dt} \end{cases} \tag{3.13}$$

Therefore, with two coupled loops there are four inductances: two self inductances L_{11} and L_{22} and two mutual inductances L_{12} and L_{21}, with $L_{12} = L_{21}$.

An equivalent circuit for the two coupled loops is shown in Figure 3.2, where L_{11} and L_{22} are the loop self inductances, while the effect of coupling is modeled by two voltage sources depending on the mutual inductances L_{12} and L_{21} and the time derivative of the current in the other branch, i.e. I_2 and I_1 respectively.

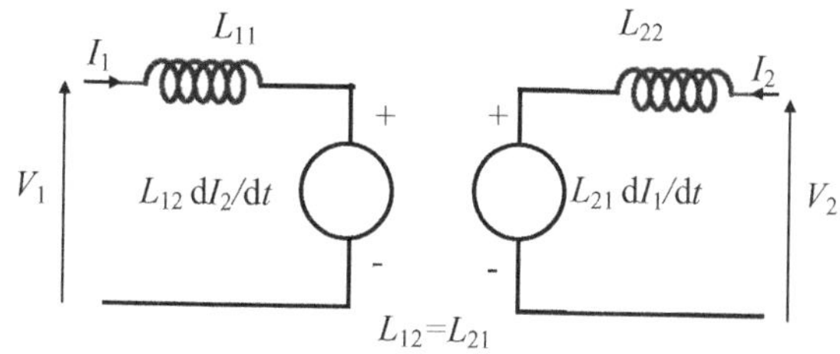

Figure 3.2 Equivalent circuit of two coupled loops

When the coupled loops are n, it is convenient to express the inductance in matrix form as

$$\mathbf{L} \equiv \begin{bmatrix} L_{11} & & \cdots & & L_{1n} \\ & \ddots & & & \\ \vdots & & L_{ii} & L_{ij} & \\ & & L_{ji} & L_{jj} & \\ L_{n1} & & & & L_{nn} \end{bmatrix} \tag{3.14}$$

where the elements L_{ij} are defined as in Equation (3.7) and can be evaluated, at least in principle, from Equation (3.12). The off-diagonal terms of the $\mathbf{L}$ matrix are the mutual inductances, while the diagonal terms are the self inductances. The flux–current relationship (3.6) for a general system composed of n loops is

$$\boldsymbol{\Psi} = \mathbf{LI} \tag{3.15}$$

where $\boldsymbol{\Psi} = [\Psi_1 \ldots \Psi_i \ldots \Psi_n]^{\mathrm{T}}$ is the flux vector, whose coefficient Ψ_i is the total flux through the ith loop generated by all n currents, and $\mathbf{I} = [I_1 \ldots I_i \ldots I_n]^{\mathrm{T}}$ is the current vector.

In relation to network analysis, Equation (3.2) in matrix form becomes

$$\mathbf{V} = \frac{\mathrm{d}\boldsymbol{\Psi}}{\mathrm{d}t} = \mathbf{L}\frac{\mathrm{d}\mathbf{I}}{\mathrm{d}t} \tag{3.16}$$

From Equation (3.16), the equivalent circuit of n coupled loops comprises n branches, each with a self inductance in series with $n-1$ voltage sources, depending on the mutual inductances and the time derivative of currents in other branches. For example, with $n = 3$, branch 1 is formed by an inductance L_{11} with in-series voltage sources $\mathrm{L}_{12}\,\mathrm{d}I_2/\mathrm{d}t$ and $\mathrm{L}_{13}\,\mathrm{d}I_3/\mathrm{d}t$.

*3.1.4 **L** Matrix of Two Coupled Conductors Having a Reference Return Conductor*

Two parallel conductors having a third conductor as reference or return for the signaling currents constitute a simple practical case of signal integrity affected by inductive coupling that can be modeled by the equivalent circuit previously introduced. This structure can be defined by two parallel traces in a PCB, as well as two parallel wires in a cable, both under the assumption that they can be considered electrically short and, hence, modeled by lumped elements. One loop is formed by conductor 1 and the return, and the other loop by conductor 2 and the return.

It is important to note that the inductance is only defined for a closed loop formed by a wire, while the wire itself does not have any inductance. However, interconnect modeling requires open loops so that circuits can be connected. Breaking the loop creates a two-wire interconnect, but the loop still has only one inductance describing it. The full inductance can be assigned to the top wire or to the bottom wire, or it can be split between the two with a mutual inductance by using the partial inductance concept, which will be introduced in the next section.

When there is no interest in the voltage drop along the return conductor, as in the crosstalk computation, the loop inductances L_{11} and L_{22} can be associated with the two conductors, respectively, and the return is used to refer voltages only. This means that, in a SPICE-like circuit simulator, the return conductor can be the reference '0' ground used for all the elements of the circuit (see Figure 3.3b). The coupling between the two inductors is represented in Figure 3.3a by the black dots placed on the two polarized inductors to indicate where the mutual inductance formula assumes that the current enters. The polarity of one can be reversed if the sign of the mutual inductance is also reversed. For SPICE and several SPICE-like circuit simulators, it is not necessary to use the equivalent circuit of Figure 3.3b to model mutual inductance, and the equivalent circuit of Figure 3.3a can be adopted by defining the coupling factor K for mutual inductance:

$$K = \frac{M}{\sqrt{L_{11}L_{22}}} \tag{3.17}$$

where K is bounded between ± 1. The K-factor is a measure of the strength of the mutual inductance, with magnitudes below about 0.1 being weak, and those above about 0.4 being strong.

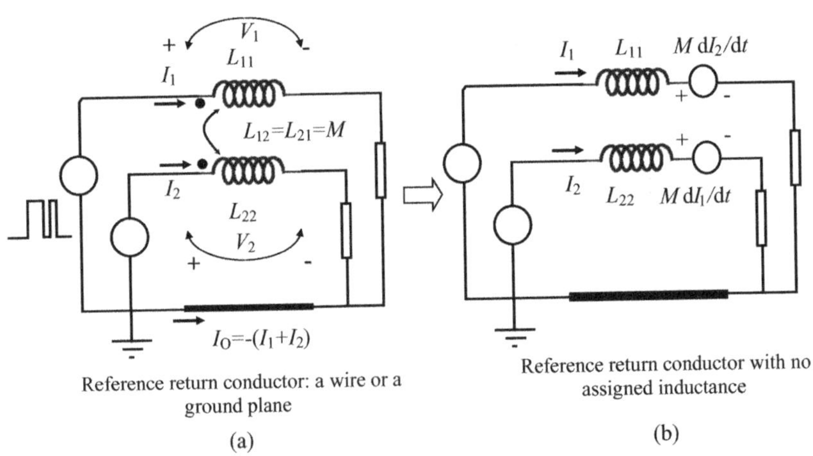

Figure 3.3 Two coupled conductors: (a) loop inductances; (b) equivalent circuit with dependent voltage sources

3.1.5 L Calculation of a Three-Conductor Wire-Type Line

Two typical examples of wire-type MTLs consisting of three conductors are shown in Figure 3.4 [3]: two parallel wires having a third wire as reference (Figure 3.4a) and two parallel wires above an infinite ground plane (Figure 3.4b).

Closed-form expressions of the per-unit-length inductances can be calculated by Equation (3.7) in terms of flux, on the assumption that the wires are separated sufficiently so that the current distributions along the peripheries of the wires are essentially uniform (i.e. wide separation) [3].

It should be noted that, in the calculation of self inductance by Equation (3.7), the magnetic flux Ψ_{ii} accounts for the portion of the magnetic field external to the wire, which contributes to the external inductance, and for the portion of the magnetic field internal to the wire, which contributes to the internal inductance. The total per-unit-length inductance is the sum of the external and internal inductances. However, for typical line dimensions, the external inductance is much larger than the internal inductance so that the per-unit-length inductance can be reasonably approximated to the external inductance. In the remainder of this section the internal inductance will be neglected. Note that the external inductance is a property of the conductor's geometry and relationship to the return path.

By applying a current I_1 on wire 1 and returning it on the reference conductor while setting the other current $I_2 = 0$ (see Figure 3.4), the resulting fluxes Ψ_{11} and Ψ_{21} through the appropriate surfaces are determined, and the inductances L_{11} and L_{12} are obtained as

$$L_{11} = \frac{\Psi_{11}}{I_1} |I_2 = 0 \tag{3.18a}$$

$$L_{21} = \frac{\Psi_{21}}{I_1} |I_2 = 0 \tag{3.18b}$$

The calculation of inductances through Equations (3.18) can be easily performed by using the fundamental relation linking the current I on a wire and the magnetic flux through a parallel surface having radial distances r_1 and r_2 from the wire, where $r_2 \geq r_1$:

$$\Psi = \frac{\mu I}{2\pi} \ln(r_2/r_1) \tag{3.19}$$

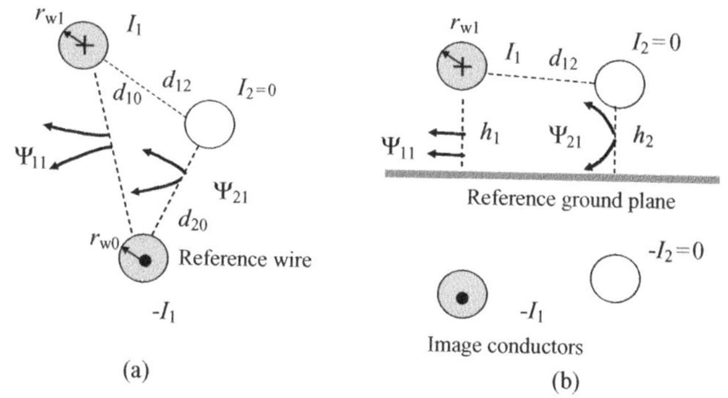

Figure 3.4 Currents and generated flux in the case of (a) a wire–wire structure and (b) a wire–plane structure

For calculation of the self inductance L_{11}, r_1 is the wire 1 radius. Observe that, in the wire–wire configuration of Figure 3.4a, the flux Ψ_{i1} $(i = 1, 2)$ is the sum of the flux generated by current I_1 in conductor 1 and the flux generated by the return current $-I_1$ in reference conductor 0 with current in conductor 2 $I_2 = 0$ according to Equation (3.18a). The computation of the flux Ψ_{i1} $(i = 1, 2)$ is performed on the per-unit-length surface between the ith conductor and the reference wire 0. In the wire–plane configuration of Figure 3.4b, the flux Ψ_{i1} $(i = 1, 2)$ is the sum of the flux generated by current I_1 in conductor 1 and the flux generated by return current in the image of conductor 1 (i.e. $-I_1$) with current $I_2 = 0$. The computation of flux Ψ_{i1} $(i = 1, 2)$ is performed on the per-unit-length surface between the ith conductor and the reference ground plane.

The results for self inductance L_{11} for wire–wire and for wire–plane structures are

$$L_{11} = \frac{\mu}{2\pi} \ln\left(\frac{d_{10}^2}{r_{w0} r_{w1}}\right) \quad \text{for wire–wire} \tag{3.20a}$$

$$L_{11} = \frac{\mu}{2\pi} \ln\left(\frac{2h_1}{r_{w1}}\right) \quad \text{for wire–plane} \tag{3.20b}$$

where r_{w0} and r_{w1} are the radius of conductors 0 and 1 respectively, d_{10} is the wire-to-wire distance between conductors 0 and 1, and h_1 is the height of conductor 1 above the ground plane. The inductance L_{22} can be calculated by similar expressions.

The per-unit-length mutual inductance L_{12} for wire–wire and wire–plane structures are

$$L_{12} = \frac{\mu}{2\pi} \ln\left(\frac{d_{10} d_{20}}{d_{12} r_{w0}}\right) \quad \text{for wire–wire} \tag{3.21a}$$

$$L_{12} = \frac{\mu}{2\pi} \ln\left(1 + \frac{4h_1 h_2}{d_{12}^2}\right) \quad \text{for wire–plane} \tag{3.21b}$$

where d_{12} is the wire-to-wire distance between conductors 1 and 2 [3]. Note that, in the case of a wire–plane configuration, self and mutual inductances are calculated by replacing the ground plane with the image wires, as shown in Figure 3.4b. This is a common practice for signal integrity and radiated emission computation when the ground plane is sufficiently large to be considered as infinite.

3.1.6 *Frequency-Dependent Internal Inductance*

The internal inductance is related to the magnetic flux inside the conductor. Under quasi-static approximation, and assuming that the current is uniformly distributed inside the conductor (i.e. low-frequency approximation), the per-unit-length internal inductance of a straight round wire is [1, 3]

$$L_{\text{int}} = \frac{\mu}{8\pi} \tag{3.22}$$

Note that the internal inductance L_{int} does not depend on the radius of the wire; all round wires of a given length have the same internal inductance. Moreover, in the case of MTLs, the

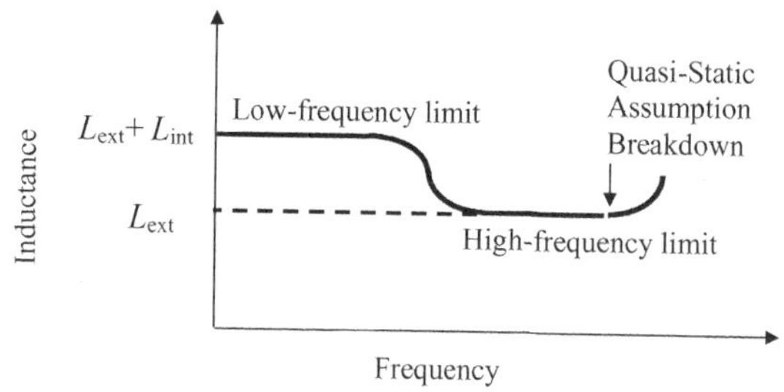

Figure 3.5 Self inductance of a conductor as a function of frequency

internal inductance of wires is not dependent on the line configuration as far as the conductors are widely separated and the proximity effect is negligible. The total self inductance L of a round wire is the sum of external L_{ext} and internal L_{int} inductances.

Whenever the quasi-static approximation holds, the field external to a wire is weakly affected by the distribution of the current inside the wire; hence, the external inductance L_{ext} is slightly dependent on frequency. On the other hand, the internal inductance depends strongly on frequency. When the skin effect is well developed, the current resides on the surface of the wire and the internal inductance L_{int} tends towards zero. The trend of the internal inductance versus frequency is sketched in Figure 3.5. A closed-form analytical expression of L_{int} for a round wire is provided and discussed in *Section 7.1*.

It is important to underline that many field solvers compute the inductance matrix from the capacitance matrix ($\mathbf{L} = \mu_0\varepsilon_0\mathbf{C}^{-1}$, where $\mathbf{C}$ is the capacitance matrix calculated for $\varepsilon_{\text{r}} = 1$). In this case, numerical calculation provides the external inductance and does not take into account the internal inductance. Therefore, the inductance at an infinity frequency is obtained, and represents the inductance lower bound.

3.2 Partial Inductance

The definition of inductance for a particular set of loops is given by Equation (3.7), where Ψ_{ij} is the flux induced in the ith closed loop, the surface of which is bounded by the loop itself, and due to the current I_j in the jth loop. As previously pointed out, inductance is a physical parameter associated with closed loops and not with wires themselves. However, any complex shaped loop may be segmented into many smaller regular pieces, each with its own inductance (i.e. partial inductance), so that the loop inductance is obtained by the sum of the partial inductances. The inductance for a piece of the loop can be defined starting from both Equation (3.10) and Equation (3.11) [2]. Although measurements must always be performed on closed loops to determine inductances, calculations can be performed on sections of closed loops to predict partial inductances. The total inductance of a loop is the algebraic sum of partial inductances.

3.2.1 Partial Inductances of Coupled Loops

The integrations over the loop lengths C_1 and C_2 in Equation (3.10) can be rewritten as summations over the straight K and M segments C'_k and C'_m used to decompose loops 1 and 2

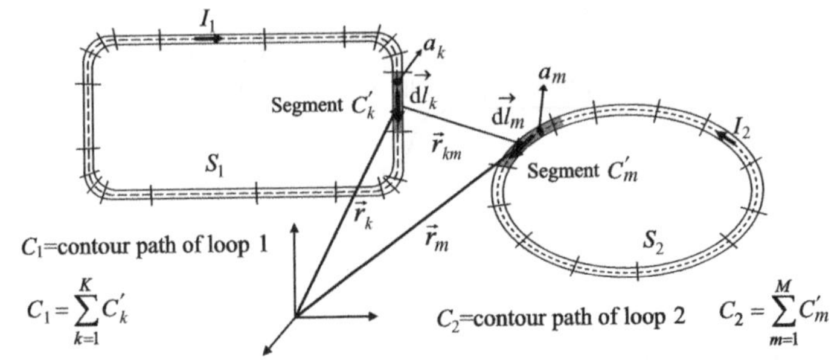

Figure 3.6 Currents and geometries of two coupled loops for inductance calculation

respectively (see Figure 3.6). All segments are oriented in accord with the current flowing through them and are allowed to have a different cross-section, i.e a_k and a_m are the cross-sections of the kth and mth segments along loops 1 and 2 respectively. With this decomposition of the integration paths, Equation (3.10) becomes

$$L_{ij} = \sum_{k=1}^{K}\sum_{m=1}^{M} \frac{\mu}{4\pi}\frac{1}{a_k a_m}\int_{a_k}\int_{a_m}\int_{C'_k}\int_{C'_m} \frac{\vec{dl}_k \cdot \vec{dl}_m}{r_{km}} da_k\, da_m \tag{3.23}$$

Partial inductances $L_{p\,km}$ are defined in general as the argument of the double summation (3.23) for the conductor segments:

$$L_{p\,km} \equiv \frac{\mu}{4\pi}\frac{1}{a_k a_m}\int_{a_k}\int_{a_m}\int_{C'_k}\int_{C'_m} \frac{\left|\vec{dl}_k \cdot \vec{dl}_m\right|}{r_{km}} da_k\, da_m \tag{3.24}$$

Partial inductances are denoted by $L_{p\,km}$ in order to distinguish them from the loop inductances L_{ij}. Then, Equation (3.23) is written in general as

$$L_{ij} = \sum_{k=1}^{K}\sum_{m=1}^{M} s_{km}\, L_{p\,km} \tag{3.25}$$

where the term s_{km} determines the sign and is given by $s_{km} = \vec{dl}_k \cdot \vec{dl}_m / |\vec{dl}_k \cdot \vec{dl}_m| = \pm 1$, with the vectors $\vec{dl}_k$ and $\vec{dl}_m$ oriented in accordance with the currents flowing through them. $L_{p\,km}$ is zero for the special case where the scalar product is identically zero, i.e. for orthogonal currents. For $k = m$, Equation (3.24) gives the self partial inductance, while for $k \neq m$ the mutual partial inductance is obtained.

3.2.2 Flux Area of Partial Inductance of Thin Filamentary Segments

Consider two thin and straight wire segments C'_k and C'_m that are not necessarily coplanar, as shown in Figure 3.7. Under the adopted thin-wire assumption, the partial inductance $L_{pf\,km}$ can be derived from Equation (3.11), and in accordance with the segmentation procedure described in the previous section, as

$$L_{\mathrm{pf}\,km} = \frac{\mu}{4\pi}\int_{C'_k}\int_{C'_m}\frac{\left|\mathrm{d}\vec{l}_k \cdot \mathrm{d}\vec{l}_m\right|}{r_{km}} \tag{3.26a}$$

The partial inductance $L_{\mathrm{pf}\,km}$ is also given by [2]

$$L_{\mathrm{pf}\,km} = \frac{1}{I_m}\int_{S_k}\vec{B}_{km}\cdot \mathrm{d}\vec{S}_k \tag{3.26b}$$

where S_k is the area bounded by the conductor segment C'_k and ∞ (infinity), and by straight lines located at the ends of segment C'_k and perpendicular to segment C'_m, as shown in Figure 3.7. The equivalence between the two Equations (3.26) can be easily derived by Stokes' theorem, which relates the surface integral over S_k to a line integral over C'_k. In fact, the vector potential $\vec{A}_{km}$ produced by the current I_m that flows on segment C'_m can be expressed by Equation (3.9), where the line integral is over segment length C'_m and the surface integral is omitted owing to the filament assumption. Therefore, Equation (3.26a) can be expressed in terms of $\vec{A}_{km}$ as

$$L_{\mathrm{pf}\,km} = \frac{1}{I_m}\int_{C'_k}\vec{A}_{km}\cdot \mathrm{d}\vec{l}_k = \frac{1}{I_m}\oint_{l_k}\vec{A}_{km}\cdot \mathrm{d}\vec{l}_k \tag{3.27}$$

where l_k is the contour path of the surface S_k shown in Figure 3.7 and is defined by the straight lines located at the ends of segment C'_k, perpendicular to segment C'_m, and extending to infinity. The two integrals in Equation (3.27) are equal, as no contribution results from the portion of the loop other than C'_k. In fact, $\vec{A}_{km}$ around the current-carrying wire C'_m is parallel to the wire and therefore normal to the two paths perpendicular to the conductor C'_m, and it goes to zero at infinity. Stokes' theorem applied to the closed contour path of Equation (3.27) leads to Equation (3.26b), in accordance with *Section 3.1.1*.

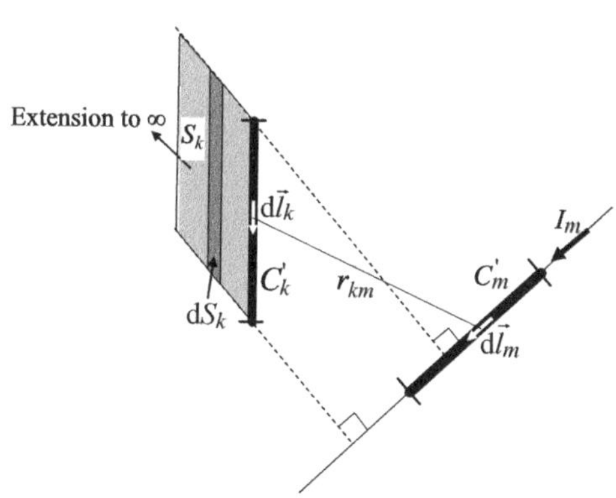

Figure 3.7 Flux area of partial inductance associated with two thin-wire segments C'_k and C'_m

3.2.3 Loop Inductance Decomposed into Partial Inductances

The concept expressed by Equation (3.27) is herein applied to a thin loop. Consider the thin-wire loop of rectangular shape shown in Figure 3.8a. Equation (3.7) defines the loop inductance as the ratio between the flux Ψ through the loop surface S and the current I, and it can be expressed equivalently either in terms of magnetic flux $\vec{B}$ or in terms of the magnetic vector potential $\vec{A}$ as

$$L = \frac{\Psi}{I} = \frac{\int_S \vec{B} \cdot \mathrm{d}\vec{S}}{I} = \frac{\oint_C \vec{A} \cdot \mathrm{d}\vec{l}}{I} \tag{3.28}$$

The rectangular loop C can be decomposed into four segments C_i, each corresponding to one edge of the loop and having an associated current $I_i = I$. The line integral of the magnetic vector potential $\vec{A}$ around the loop can be broken into the sum of the integrations along each segment as

$$L = \frac{\sum_{i=1}^{4} \int_{C_i} \vec{A} \cdot \mathrm{d}\vec{l}}{I} = \sum_{i=1}^{4} L_i \tag{3.29}$$

where L_i is the inductance that can be attributed to segment C_i of the loop and is defined by

$$L_i = \frac{1}{I} \int_{C_i} \vec{A}_i \cdot \mathrm{d}\vec{l}_i \tag{3.30}$$

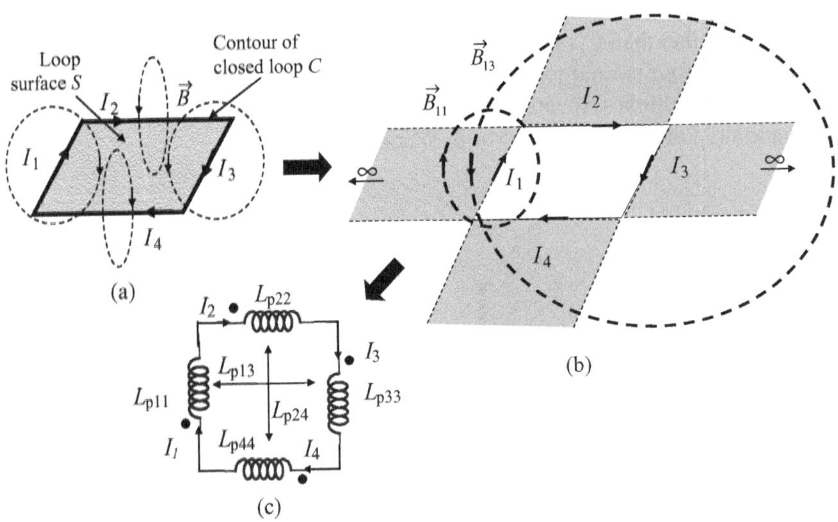

Figure 3.8 Loop inductance decomposed into partial inductances: (a) flux computation for loop inductance; (b) flux computation for partial inductances associated with a segment of the loop; (c) equivalent circuit in terms of partial inductances

where $\vec{A}_i$ is the magnetic vector potential along C_i, produced by the current in the loop (i.e. $I_1 = I_2 = I_3 = I_4 = I$).

Expressing $\vec{A}_i$ as the sum of the four potentials $\vec{A}_{ij}$ ($j = 1, 4$), each produced by the current I_j in the loop segment C_j, yields

$$L_i = \sum_{j=1}^{4} \frac{1}{I_j} \int_{C_i} \vec{A}_{ij} \cdot d\vec{l}_i = \sum_{j=1}^{4} L_{pij} \tag{3.31}$$

where $L_{p\,ij}$ is the partial inductance of the equivalent circuit in Figure 3.8c, which is defined as

$$L_{pij} = \frac{1}{I_j} \int_{C_i} \vec{A}_{ij} \cdot d\vec{l}_i \tag{3.32}$$

In this way, the self partial inductance L_{pij} of a segment C_i of a closed current loop can be uniquely defined by Equation (3.32) as the ratio between the line integral of $\vec{A}_{ii}$ along the segment C_i and the I_i current on that segment (which is the current of the ith loop). Note that the magnetic vector potential $\vec{A}_{ii}$ along segment C_i is directly proportional to the loop current $I = I_i$ that flows in this segment (see Equation (3.9)). Hence, even if this segment is a part of several different current loops (as is often the case), the value of the self partial inductance of the segment is unique and does not depend on the loops of which the segment is a part. This is a very important consideration for practical device and package modeling, such as for power and ground bounce computation, as will be described in *Chapter 8*.

Each partial inductance L_{pij} can also be calculated by the alternative expression [2]

$$L_{pij} = \frac{1}{I_j} \int_{S_i} \vec{B}_{ij} \cdot d\vec{S}_i \tag{3.33}$$

where $\vec{B}_{ij}$ is the magnetic induction due to current I_j, and the integration is computed on the area S_i between the conductor C_i and infinity, and by straight lines located at the ends of segment C_i and perpendicular to segment C_j. The equivalence between Equations (3.32) and (3.33) can be demonstrated as shown in the previous subsection. For $i = j$, Equations (3.32) and (3.33) are referred to as the self partial inductances, while for $i \neq j$ these are referred to as the mutual partial inductances. It is important to point out that the magnetic vector potential $\vec{A}_{ij}$ is parallel to the current I_j producing it. Hence, for any segment C_i orthogonal to C_j, the mutual partial inductances are zero. For this reason, in the equivalent circuit of Figure 3.8c, only the mutual partial inductances L_{p13} and L_{p24} are present, while $L_{p12} = L_{p14} = L_{p23} = L_{p34} = 0$. With reference to Figure 3.8c, the voltage drop across a segment of the loop can be uniquely and meaningfully obtained. For example, the voltage across segment 4 is given by

$$V_4 = L_{p44} \frac{dI_4}{dt} + L_{p42} \frac{dI_2}{dt} \tag{3.34}$$

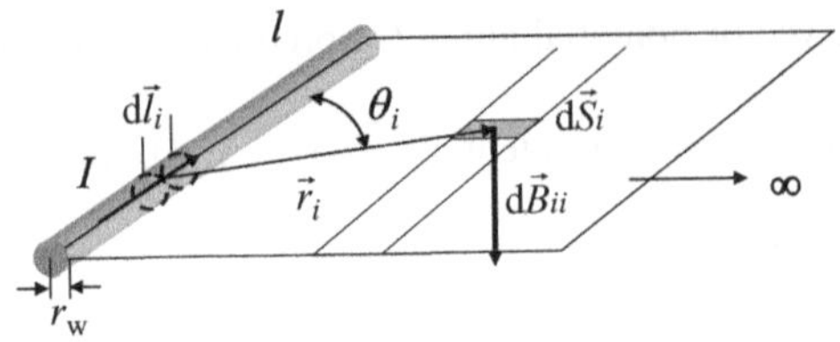

Figure 3.9 Round wire self partial inductance

3.2.4 *Self and Mutual Partial Inductance*

3.2.4.1 Self Partial Inductance of a Round Thin Wire

The self partial inductance L_p of an isolated wire of radius r_w and length l can be determined by Equation (3.33), assuming the integration area S_i and the elemental area dS_i shown in Figure 3.9 [4]. The vector $\vec{B}_{ii}$ in Equation (3.33) is orthogonal to the integration surface (i.e. the scalar product in Equation (3.33) reduces to a simple product) and represents the magnetic induction on the elemental area dS_i owing to the current I in the considered wire. It is given by the integration along the wire of length l of the magnetic flux density dB_{ii} produced by the current I in the wire segment dl_i and given by

$$dB_{ii} = \frac{\mu I \, dl_i}{4\pi r_i^2} \sin\theta_i \tag{3.35}$$

where r_i is the distance between the filament of length dl_i located in the center of the wire and the generic element of surface dS_i. It should be pointed out that dB_{ii}/μ is the magnetic field produced by an electric (Hertzian) dipole, considering the static field only [3].

The result of the integration is the simplified self partial external inductance $L_p = L_{pii}$ for a round wire

$$L_p = L_{p\,ext} \approx \frac{\mu}{2\pi} l \left[\ln\left(\frac{2l}{r_w}\right) - 1 \right] \quad \text{valid for } r_w/l \ll 1 \tag{3.36}$$

Equation (3.36) does not include the internal inductance of the wire. The total self partial inductance for a round wire $L_{p\,tot}$ is the sum of external and internal inductance. Therefore, summing to Equation (3.36) the value of the internal inductance of a conductor at very low frequencies (3.22) yields

$$L_{p\,tot} = L_{int} + L_{p\,ext} \approx \frac{\mu}{2\pi} l \left[\ln\left(\frac{2l}{r_w}\right) - \frac{3}{4} \right] \quad \text{valid for } r_w/l \ll 1 \tag{3.37}$$

Observe that the self partial inductance depends on the length l of the wire.

3.2.4.2 Mutual Partial Inductance of Parallel Thin Wires

The mutual partial inductance $M_p = L_{p12} = L_{p21}$ between two parallel conductors C_1 and C_2 both of length l can again be obtained by Equation (3.33). In this case, the surface of

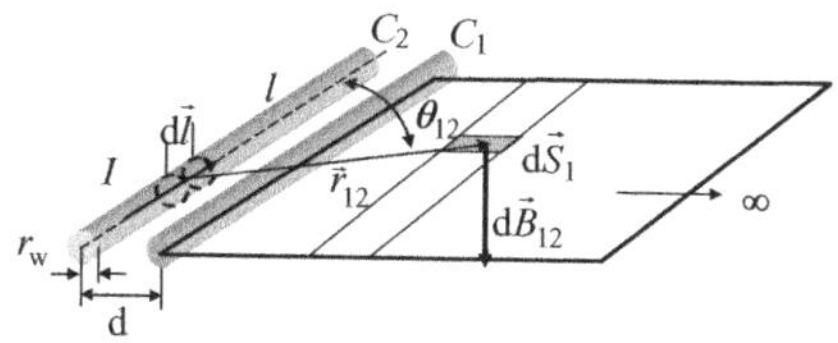

Figure 3.10 Mutual partial inductance of parallel round wire

integration S_1 is the area between the line centered on the wire C_1 and infinity, as shown in Figure 3.10. In the calculation, the current I_1 is assumed to be zero, while I_2 is the filament current associated with the wire C_2 separated from the other by a center-to-center distance d. The result of this calculation is

$$M_p = \frac{\mu}{2\pi} l \left[\ln \left(\frac{l}{d} + \sqrt{1 + \frac{l^2}{d^2}} \right) - \sqrt{1 + \frac{d^2}{l^2}} + \frac{d}{l} \right] \tag{3.38}$$

This is the mutual partial inductance between two filaments because the radius r_w is absent in the formula. When the length of the filament, l, is much larger than the separation, d, Equation (3.38) is approximated to

$$M_p \approx \frac{\mu}{2\pi} l \left[\ln \left(\frac{2l}{d} \right) - 1 \right] \quad \text{for } d \ll l \tag{3.39}$$

This is a simplified result useful for practical cases of signal integrity. It can be noted that, in the case of conductors of small cross-section, partial inductances are independent of the cross-sectional shape. Therefore, Equation (3.38) can be used with good approximation for both round and rectangular wires. All the inductances are expressed in Henry.

An alternative way to compute mutual partial inductances is by means of Equation (3.26a) [1]. This equation cannot be used to compute self partial inductance because the denominator vanishes when segments of the same loop superimpose. However, the self partial inductance can be found by computing the mutual partial inductance between a filament at the center of the round wire and the other one along its edge. In fact, using this trick, and putting r_w in place of d, Equation (3.39) coincides with Equation (3.36). In general, in the case of curved wires, to find the self partial inductance correctly, the mutual partial inductance between different short wires that are not aligned must be included.

3.2.4.3 Partial Inductance of PCB Traces

The meaning and interpretation of the partial inductances for conductors of rectangular cross-section, such as PCB traces, is the same as for wires. The calculation of them is much more complicated, and exact formulae can be found in a paper by Ruehli [2]. However, for calculation of the self partial inductance of a flat etched conductor, busbar, or ground plane of width w, thickness t, and length l, when isolated from a return path at low frequency, the following simple approximated formula can be used [4]:

$$L_p = \frac{\mu}{2\pi} l \left[\ln\left(\frac{2l}{w+t}\right) + \frac{1}{2} + \frac{2}{9}\left(\frac{w+t}{l}\right)\right] \tag{3.40}$$

For mutual partial inductance, calculation of structures such as a busbar with an adjacent return bus, or a flat conductor with an adjacent return path conductor, Equation (3.38) found for two coupled round wires can be used. In fact, this formula is independent of the cross-sectional shape and holds whenever the two conductors satisfy the wide separation assumption (i.e. $d > w$) and the width condition $w > t$.

3.2.5 *Inductance Between Two Parallel Conductors*

Extending the procedure used for the closed loop in *Section 3.1.4* to the common case of two parallel conductors, the equivalent circuits shown in Figure 3.11 can be derived. If the loop represents an interconnect, for example between a driver and a receiver, and assuming that $l \gg d$, the total inductance of the loop L_{tot} can be computed without the contribution of vertical segments d and can be associated with the upper horizontal conductor while using the lower as reference for voltages. Since the currents in the conductors are equal and opposite, the total loop inductance is given by

$$L_{tot} = L_{p1} + L_{p2} - 2M_p = (L_{p1} - M_p) + (L_{p2} - M_p) = L_{e1} + L_{e2} \tag{3.41}$$

by means of which an effective inductance L_e, given by the difference of the self partial and mutual partial inductance, can be associated with each conductor. This means that the mutual partial inductance can be included in the partial inductance associated with the conductor.

It is necessary to point out that, from a signal integrity point of view, the first circuit with the total loop inductance L_{tot} is not always convenient. In fact, the voltage drop on the return conductor of a point-to-point interconnect cannot be calculated by this representation. Knowledge of the ground noise is often very important for signal integrity and radiated emission prediction (see *Chapter 9* and *Chapter 10*).

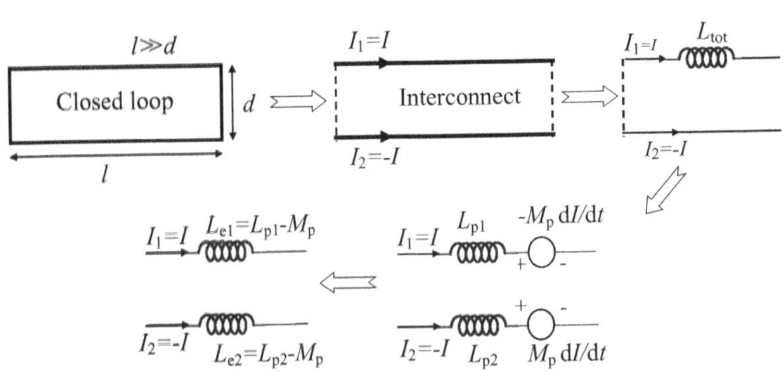

Figure 3.11 A closed loop seen as an interconnect and its equivalent circuits

In the case of round wires, both L_{p1} and L_{p2} are given by Equation (3.36) with $r_w = r_{w1}$ and $r_w = r_{w2}$ respectively, and M_p is given by Equation (3.39). By introducing these expressions into Equation (3.41), and by using the logarithm properties, the obtained expression of the total inductance coincides with Equation (3.20a).

3.2.6 Loop Inductance Matrix Calculation by Partial Inductances

The simple model applied in the case of two parallel conductors can be extended to multiconductor structures as those often occurring in signal integrity analysis. Consider, for example, a three-conductor TL. To calculate reflections and crosstalk, it is convenient to assume a conductor as reference and assign to it zero inductance. The equivalent model of three-conductor coupled lines by using the loop inductances L_{ij} is shown in Figure 3.12a.

When the voltage drop across the reference conductor 0 is required, the equivalent circuit of Figure 3.12b, based on the partial inductances L_{pij}, should be used. The link between loop inductances L_{ij} and partial inductances L_{pij} for three-conductor coupled lines can be found by considering the loop inductance definition and the properties $L_{ij} = L_{ji}$ and $L_{pij} = L_{pji}$, and is given by

$$L_{11} = \left.\frac{V_1}{dI_1/dt}\right|_{I_2=0} = L_{p11} - L_{p10} + L_{p00} - L_{p01} = L_{p11} + L_{p00} - 2L_{p10} \quad (3.42a)$$

$$L_{22} = \left.\frac{V_2}{dI_2/dt}\right|_{I_1=0} = L_{p22} - L_{p20} + L_{p00} - L_{p02} = L_{p22} + L_{p00} - 2L_{p20} \quad (3.42b)$$

$$L_{12} = \left.\frac{V_1}{dI_2/dt}\right|_{I_1=0} = L_{p00} + L_{p12} - L_{p10} - L_{p20} \quad (3.42c)$$

In the case of $n + 1$ coupled lines, the relation between the loop and partial inductances is given by [2]

$$L_{ii} = L_{pii} + L_{p00} - 2L_{pi0} \quad (3.43a)$$

$$L_{ij} = L_{p00} + L_{pij} - L_{pi0} - L_{pj0} \quad (3.43b)$$

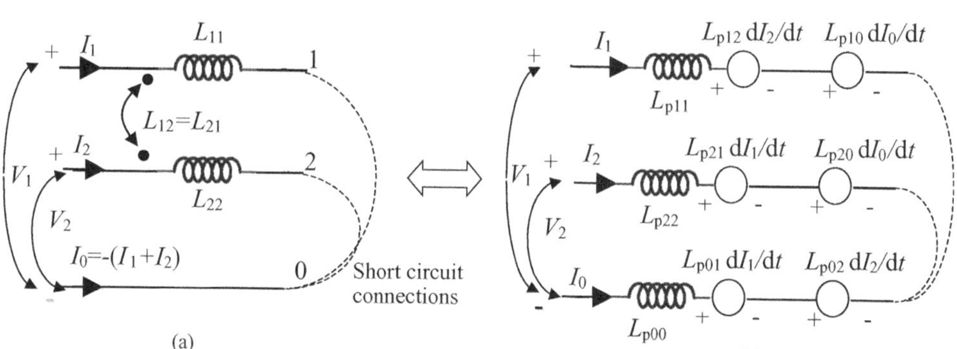

Figure 3.12 Equivalent circuits of three conductors by (a) loop inductances and (b) partial inductances

The application of the partial inductance concept leads to values of the inductance that correspond to the low-frequency limit shown in Figure 3.5. The variation in inductance with frequency is bounded between the minimum value (i.e. high-frequency limit) obtainable by the capacitance as $\mathbf{L} = \mu_0\varepsilon_0\mathbf{C}^{-1}$ and the maximum value (i.e. low-frequency limit) which can be derived by the application of the partial inductance procedure. Comparison between the results provided by the two approaches for the case of six conductors above a planar surface has revealed that the two bounds differ by less than 10 % [2].

When the equivalent circuits of Figure 3.12 are implemented in a circuit simulator, it is important to verify whether the inductance matrix **L** is passive, to avoid instabilities during the simulation [1]. In fact, active inductance matrices can cause no convergence of the circuit simulation, with computation of ever-growing voltage levels, and, most dangerously, reasonable but incorrect results. A dense matrix completely filled out through consistent electromagnetic simulation or measurements is not likely to be active; however, matrices constructed from partial results or by combinations of results generated from several techniques can be active.

It can be shown that the inductance matrix is passive only when all of its eigenvalues are greater than or equal to zero [1].

3.2.7 Partial Inductance Associated with a Finite Ground Plane

In high-speed digital systems, the return conductor for a single-end interconnect is usually a ground plane of finite dimension (see Fig. 3.13a). At the ground plane, very often there is a cable attached and the interest is focused on the voltage drop across the ground plane produced by the signal current to calculate the radiated emission of the PCB plus cable. The voltage drop depends on the inductance L_{gnd}, that is, the inductance associated with the ground plane, and has the significance of the effective inductance L_{e} of Figure 3.11. This associated inductance can be computed by modeling the ground plane as n small-cross-section conductors (filaments) and assuming the signal conductor to be the $n + 1$th filament, as shown in Figure 3.13b. In the equivalent circuit of Figure 3.13c, each small conductor is represented by its effective inductance. Self partial inductance can be computed by Equation (3.36) if each small conductor is assumed to be a round wire, or by Equation (3.40) if a rectangular conductor cross-section is adopted. The calculation of mutual partial inductances can be performed

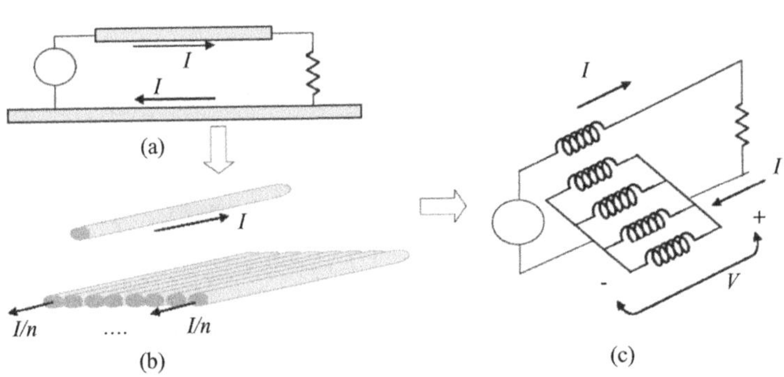

Figure 3.13 Partial inductance associated with a finite ground plane: (a) conductor above a finite ground plane; (b) n filaments of the ground plane; (c) equivalent circuit

using Equation (3.38). Then, the ground inductance is computed as $L_{gnd} = \text{Im}[V/I]/\omega$ by using a circuit simulator such as SPICE. An example of this procedure is given in *Section 10.1*. For practical PCBs with the trace close to the ground plane, L_{gnd} is usually of the order of a few nH/m.

Several closed-form expressions for L_{gnd} have been found in the past using different computation techniques. A simplified and accurate expression is given by [5]

$$L_{gnd} = \frac{\mu_0}{2\pi} l \ln\left(\frac{\pi h}{w_{gnd}} + 1\right) \tag{3.44}$$

where h is the height of the trace above the ground plane, w_{gnd} is the width of the ground plane, and l is the length of the ground plane (see *Appendix A*). A validation of the accuracy of Equation (3.44) is shown in *Appendix E* and in *Section 9.6*, where a comparison with experimental results is provided.

3.2.8 Solving Inductance Problems in PCBs

The procedure for solving inductance problems by using the partial inductance concept is based on the following steps:

- Choose the closed loop.
- Break the closed loop into several segments in order to facilitate the computation of partial inductances.
- Compute the self partial inductance for each section by closed-form expressions.
- Compute the mutual partial inductance between each pair of sections by closed-form expressions.
- In the case of coupled multiconductor lines, choose whether the return signal conductor must have associated inductance or not.
- Incorporate the equivalent circuit into a SPICE-like circuit simulation program.

Example 3.1. Calculation of Inductances Associated with the Decoupling Capacitor and IC Device in a PCB

As an example of solving inductance problems in a PCB, consider the situation depicted in Figure 3.14 where an IC device and a decoupling capacitor are connected to the ground (G) and power (P) planes of a four-layer PCB by vias. The path formed by the IC drawing current I, the planes, and the capacitor has three areas of magnetic flux that define the total loop inductance between the IC and the capacitor. These three regions are shown in Figure 3.14 as loop 1, 2 and 3. The total inductance L_{tot} encountered by the current flowing from the IC to the capacitor is the sum of the three loop inductances of Figure 3.14:

$$L_{tot} = L_{loop\,1} + L_{loop\,2} + L_{loop\,3} \tag{3.45}$$

where $L_{loop\,1}$ and $L_{loop\,3}$ represent the inductances of the connection between capacitor and ground plane and between IC and ground plane respectively, and $L_{loop\,2}$ is the inductance of the portion of the current loop that exists between the power and ground planes and the two vias where the switching current flows. In detail, $L_{loop\,1}$ is the inductance associated with the

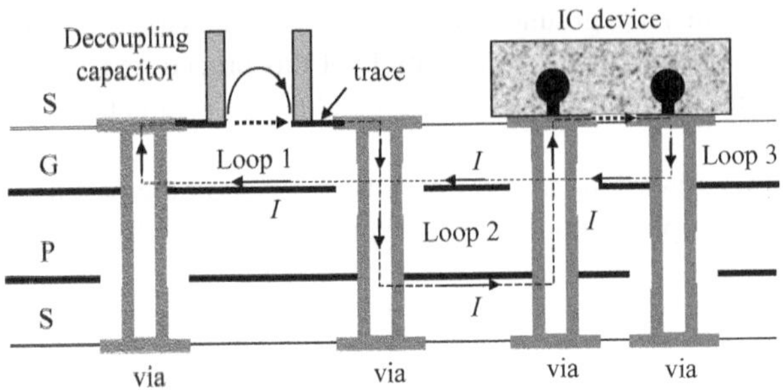

Figure 3.14 Inductance and current path between IC and decoupling capacitor in a four-layer PCB: S = signal; G = ground; P = power

rectangular loop formed by the two vias, the G-plane, and the capacitor shorted by a trace that is directly on top of the PCB, as indicated by the dashed arrow. This inductance should be increased by the portion of flux caused by the real current path through the capacitor (see the solid arrow in Figure 3.14). Since this portion of inductance should be measured using an impedance analyzer, the attention here is focused on the calculation of the loop inductance by a closed-form expression. Similar consideration can be applied to loop 3. Note that each loop has a couple of vias as vertical bounding lines that can be considered as two parallel conductors. Therefore, the two vias have a self partial inductance $L_{\mathrm{p\,via}}$ and a mutual partial inductance $M_{\mathrm{p\,via}}$ that can be calculated by the simplified formulae provided in *Sections 3.2.4.1* and *3.2.4.2* for round wires, or by the exact expressions given by Ruehli [2]. The effective inductance of each via is given by

$$L_{\mathrm{e\,via}} = L_{\mathrm{p\,via}} - M_{\mathrm{p\,via}} \tag{3.46}$$

As the currents are equal and opposite, the inductance contribution of the vias to the loop is

$$L_{\mathrm{via}} = 2L_{\mathrm{e\,via}} \tag{3.47}$$

For loop 2, $L_{\mathrm{loop\,2}} = L_{\mathrm{via}}$ because the two parallel planes do not make a significant contribution to the flux.

For loops 1 and 3 there is also a contribution from the flux owing to the horizontal geometry encompassed by the trace and the ground plane forming a microstrip structure. The inductance L_{trace} associated with the microstrip can be calculated using image theory, which consists in removing the plane and considering a second conductor parallel to the trace, with equal and opposite current, and with twice the distance between trace and plane. The resulting closed-form expression can again be the simplified formulae provided in Table A2 of *Appendix A* for rectangular conductors or the more accurate formulae given by Ruehli [2]. The inductance associated with the trace is therefore

$$L_{\mathrm{trace}} = L_{\mathrm{e\,trace}} = L_{\mathrm{p\,trace}} - M_{\mathrm{p\,trace}} \tag{3.48}$$

The total inductances $L_{\mathrm{loop}\,1}$ and $L_{\mathrm{loop}\,3}$ are for the respective loops 1 and 3:

$$L_{\mathrm{loop}} = L_{\mathrm{via}} + L_{\mathrm{trace}} \tag{3.49}$$

Therefore, Equation (3.45) becomes $L_{\mathrm{tot}} = (L_{\mathrm{via}\,1} + L_{\mathrm{trace}\,1}) + L_{\mathrm{via}\,2} + (L_{\mathrm{via}\,3} + L_{\mathrm{trace}\,3})$. Observe that, if the capacitor is moved closer to the IC, the vias of loop 2 become closer, the mutual partial inductance $M_{\mathrm{p\,via}}$ between these vias increases, L_{via} decreases according to Equations (3.46) and (3.47), and consequently the total inductance L_{tot} decreases.

The SPICE model can be built using the inductance L_{via} for each via path and L_{trace} for each trace path. The interaction between the loops can be neglected, as interest is focused on approximated values for engineering estimations. Validation of this procedure with experimental data and detailed discussion are reported elsewhere [6, 7]. The concept of partial inductance will be used intensively in *Chapter 8* to investigate the electromagnetic interference (EMI) in PCBs caused by the switching currents of the digital devices that flow through the component connections.

3.3 Differential Mode and Common Mode Inductance

The concept of *differential mode* (DM) and *common mode* (CM) inductance is widely used in high-speed digital systems for crosstalk modeling, for differential signaling investigation, and for choosing the appropriate filters to mitigate conducted and radiated emission. The effective inductance associated with each conductor for *differential* and *common modes* will be used in *Section 6.2* to introduce a transmission-line model for two symmetric lines based on two decoupled modes of propagation: even and odd modes, characterized by proper characteristic impedance and delay time. Even and odd modes are directly related to *differential* and *common modes*.

3.3.1 Differential Mode Inductance

In *Section 3.1.4* the concept of loop inductance was introduced for a structure of two conductors having a third conductor used as reference (see the equivalent circuits in Figure 3.3). When the currents I_1 and I_2 are equal in magnitude and opposite in sign, the current does not cross the reference conductor and a loop inductance regarding only the two conductors 1 and 2 can be defined. This loop inductance is defined as *differential mode* inductance, L_{DM}, and can be calculated by the following system of equations referred to the equivalent circuits of Figures 3.15a and b:

$$\begin{cases} I = I_1 = -I_2 \\ V_{1\mathrm{b}} = L_1 \dfrac{\mathrm{d}I}{\mathrm{d}t} - L_m \dfrac{\mathrm{d}I}{\mathrm{d}t} = (L_1 - L_m)\dfrac{\mathrm{d}I}{\mathrm{d}t} = L_{\mathrm{e}1}\dfrac{\mathrm{d}I}{\mathrm{d}t} \\ V_{2\mathrm{b}} = -L_2 \dfrac{\mathrm{d}I}{\mathrm{d}t} + L_m \dfrac{\mathrm{d}I}{\mathrm{d}t} = (L_m - L_2)\dfrac{\mathrm{d}I}{\mathrm{d}t} = -L_{\mathrm{e}2}\dfrac{\mathrm{d}I}{\mathrm{d}t} \end{cases} \tag{3.50}$$

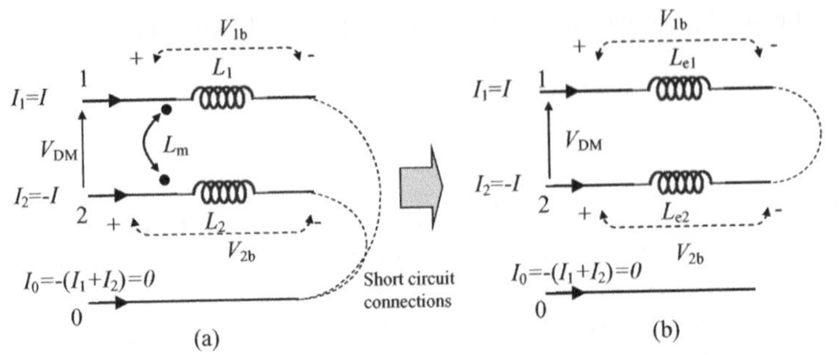

Figure 3.15 Differential mode: (a) two loop inductances and currents; (b) equivalent circuit in terms of effective inductances

where $L_{e1} = L_1 - L_m$ and $L_{e2} = L_2 - L_m$. The *differential mode* voltage V_{DM} between nodes 1 and 2 is

$$V_{DM} = (V_{1b} - V_{2b}) = (L_{e1} + L_{e2})\frac{dI}{dt} = L_{DM}\frac{dI}{dt} \tag{3.51}$$

with

$$L_{DM} = L_1 + L_2 - 2L_m \tag{3.52}$$

When the structure is symmetric, $L_1 = L_2 = L_w$, the following effective inductance can be associated with each conductor:

$$L_{eDM} = L_w - L_m \tag{3.53}$$

It is important to point out that the inductance L_{eDM} refers to one conductor and assumes the meaning of an inductance associated with the odd mode, as will be explained in *Section 6.2*. The true *differential mode* inductance is given by Equation (3.52) which yields $2L_{eDM}$, as it refers to the inductance of the loop formed by the two parallel conductors.

3.3.2 Common Mode Inductance

When the two conductors are connected at both ends and a current $2I$ is forced, the loop inductance between these conductors and the reference conductor is called the *common mode* inductance, L_{CM}, and is calculated by solving the following system of equations (see Figure 3.16):

$$\begin{cases} 2I = I_1 + I_2 \\ V_{CM} = L_1\dfrac{dI_1}{dt} + L_m\dfrac{dI_2}{dt} = L_{e1}\dfrac{dI_1}{dt} \\ V_{CM} = L_2\dfrac{dI_2}{dt} + L_m\dfrac{dI_1}{dt} = L_{e2}\dfrac{dI_2}{dt} \end{cases} \tag{3.54}$$

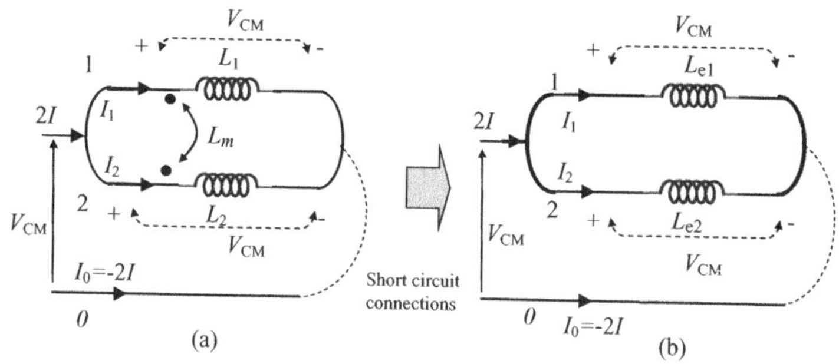

Figure 3.16 Common mode: (a) two loop inductances and currents; (b) equivalent circuit in terms of effective inductances

Substituting dI_1/dt of the second equation into the third equation and dI_2/dt of the third equation into the second equation yields

$$L_{e1} = \frac{L_1 L_2 - L_m^2}{L_2 - L_m} \tag{3.55a}$$

$$L_{e2} = \frac{L_1 L_2 - L_m^2}{L_1 - L_m} \tag{3.55b}$$

and the *common mode* inductance is

$$L_{CM} = \frac{L_{e1} L_{e2}}{L_{e1} + L_{e2}} = \frac{L_1 L_2 - L_m^2}{L_1 + L_2 - 2L_m} \tag{3.56}$$

When the structure is symmetric $L_1 = L_2 = L_w$, the following effective inductance can be associated with each conductor:

$$L_{eCM} = L_w + L_m \tag{3.57}$$

This inductance coincides with the inductance of the even mode, as will be explained in *Section 6.2*.

Observe that, if L_m increases, the conductors are more coupled, L_{DM} decreases, and L_{CM} increases. To strongly increase L_m in order to have an EMI filter working to stop the *common mode* currents, a magnetic material is used to enhance the flux interfering between the two loops. These practical filters are called chokes [3].

References

[1] Young, B., '*Digital Signal Integrity: Modeling and Simulation with Interconnects and Packages*', Prentice Hall PTR, Upper Saddle River, NJ, 2001.

[2] Ruehli, A.E., 'Inductance calculations in a complex integrated circuit environment', *IBM Journal of Research and Development*, **16**, September 1972, 470–481.

[3] Paul, C., '*Introduction to Electromagnetic Compatibility*', John Wiley & Sons, Inc., Hoboken, NJ, 2006.

[4] Grover, F.W., '*Inductance Calculations*', Dover, New York, NY, 1946.

[5] Leferink, F., 'Inductance Calculations; Methods and Equations', *IEEE International Symposium on EMC*, Atlanta, GA, 14–18 August 1995, 16–22.
[6] Knighten, J., Archambeault, B., Fan, J., Selli, G., Xue, L., Connor, S., and Drewniak, J., 'PDN Design Strategy: II. Ceramic SMT Decoupling Capacitors – Does Location Matter?', *IEEE EMC Society Newsletter*, Issue No. 208, Winter 2006, 56–67.
[7] Wang, C., Mao, J., Selli, G., Luan, S., Zhang, L., Fan, J., Pommerenke, D., DuBroff, R., and Drewniak, J., 'An Efficient Approach for Power Delivery Network Design with Closed-form Expressions for Parasitic Interconnect Inductances', *IEEE Trans. on Advanced Packaging*, **29**(2), May 2006, 320–334.

4

Capacitance

Capacitance is another very important line, filtering, and IC device parameter for *Signal Integrity* (SI) and *Radiated Emission* (RE). In this chapter, self and mutual capacitances are theoretically introduced, considering the coupling between two generic conductors. The capacitance **C** matrix concept for two coupled wires having a reference return conductor is introduced as background to build a **C** matrix for n-multiconductor transmission lines. Calculation of the inductance **L** matrix for multiconductor lines, starting with the **C** matrix, is considered. Finally, the definitions of *Differential Mode* (DM) and *Common Mode* (CM) capacitances are given.

4.1 Capacitance Between Conductors

The capacitance is a very important parameter in modeling interconnects. Starting with the basic definition of capacitance, the concept is then extended to a general multiconductor formulation which is essential, as many interconnects consist of several coupled lines. Particular emphasis is given to the capacitance **C** matrix and its negative off-diagonal terms, as this concept is fundamental to understanding the output of a static field solver for building transmission-line models.

4.1.1 Definition of Capacitance

Capacitance is a property of a geometric configuration defined by two conducting objects surrounded by a homogeneous dielectric. It describes the ability of the given configuration to store electrostatic energy [1–3]. Consider, for example, the geometry shown in Figure 4.1, where two conductors charged with a total charge Q of opposite sign are surrounded by a homogeneous dielectric medium of dielectric constant $\varepsilon = \varepsilon_0 \varepsilon_r$, where $\varepsilon_0 = 8.854 \times 10^{-12}$ F/m is the vacuum permittivity and ε_r is the relative permittivity. As a result of this charge distribution, there is an electric flux emanating from the positive charge and terminating at the negative one. Gauss' law relates the electric field $\vec{E}$ or the electric flux density $\vec{D} = \varepsilon \vec{E}$ to

Signal Integrity and Radiated Emission of High-Speed Digital Systems Spartaco Caniggia and Francescaromana Maradei

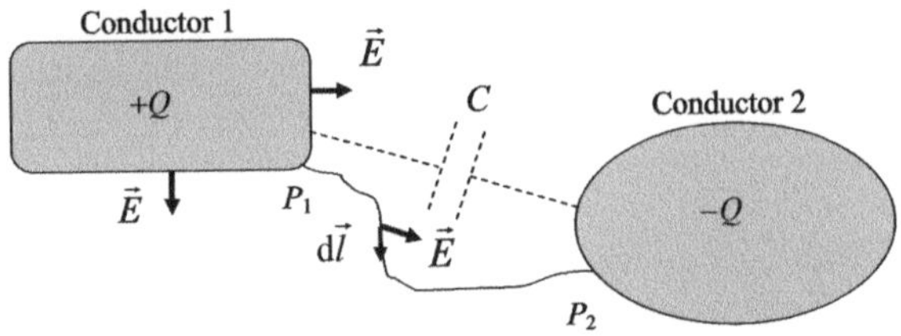

Figure 4.1 Capacitance between two conductors

the charge Q according to

$$Q = \int_V \rho \, \mathrm{d}V = \int_S \varepsilon \vec{E} \cdot \mathrm{d}\vec{S} \tag{4.1}$$

where ρ is the charge density in the volume V, and S is a closed surface around conductor 1 only. Gauss' law (4.1) simply indicates that the total electric flux density emanating from the closed surface S is equal to the total charge enclosed by the surface. Equation (4.1) exhibits a linear charge–electric field relation; this means that a doubling of Q is accompanied with a doubling of E. The voltage between the two conductors is defined by

$$V = -\int_{\mathrm{P}_1}^{\mathrm{P}_2} \vec{E} \cdot \mathrm{d}\vec{l} \tag{4.2}$$

where $\mathrm{d}\vec{l}$ is the unit vector tangent to any line between points P_1 and P_2. A linear relation can be observed in Equation (4.2) between the voltage V and the electric field. Therefore, the voltage is linearly dependent on the total charge by a coefficient c called the capacitance. The capacitance of this two-conductor system is defined as the ratio of the positive charge Q to the resulting potential difference V between the two conductors:

$$c = \frac{Q}{V} \tag{4.3}$$

Hence, the capacitance is independent of the specific amount of charge on the conductors and of the specific value of the potential difference between them.

The current I flowing through the capacitance c in the time domain is given by

$$I = \frac{\mathrm{d}Q}{\mathrm{d}t} = c\frac{\mathrm{d}V}{\mathrm{d}t} \tag{4.4}$$

For a given geometry of the two-conductor system, there are two procedures for calculating c. In the first one, a charge Q of opposite sign is assigned to the conductors, and the resulting electric field is calculated by Gauss' law or by other means. Then the potential difference between the two conductors is found by Equation 4.2, and the capacitance is obtained through Equation (4.3). In the second procedure, a potential difference between the two conductors is assigned, and the total charge on the conductors is calculated. This latter procedure requires the solution of the Laplace equation.

4.1.2 Partial Capacitance and Capacitance Matrix of Two Coupled Conductors Having a Reference Return Conductor

Capacitance is present between any two charged metallic surfaces at different potentials. When more than two conductors are present, Equation (4.3) can be repeatedly applied to obtain the partial capacitances between each couple of conductors of the whole group.

Consider, for simplicity, the three-wire configuration shown in Figure 4.2a, where conductor 0 is assumed to be grounded (i.e. $V_0 = 0$) and represents the reference conductor. In practical applications, the reference conductor is often given by a ground plane. The wires are assumed to be electrically short and charged with Q_0, Q_1, and Q_2 respectively. The potential difference between the conductors and the grounded reference are indicated as V_1 and V_2. In the considered configuration of Figure 4.2a, c_{10} and c_{20} represent the capacitive coupling between conductors 1 and 2 and the reference, while the mutual capacitance c_{12} expresses the capacitive coupling between conductor 1 and conductor 2.

The voltage V_1 between conductors 1 and 0 is supported by a charge $Q_{10} = c_{10}V_1$ on conductor 1 and by $-Q_{10}$ on conductor 0. Analogously, the voltage $(V_1 - V_2)$ between conductors 1 and 2 is supported by a charge $Q_{12} = c_{12}(V_1 - V_2)$ on conductor 1 and by $-Q_{12}$ on conductor 2. The total charge on conductor 1 is the sum of these two charges $Q_1 = Q_{10} + Q_{12}$.

The wire currents I_1 and I_2 shown in Figure 4.2b as functions of voltages and capacitances can then be expressed as

$$I_1 = c_{10}\frac{\mathrm{d}V_1}{\mathrm{d}t} + c_{12}\frac{\mathrm{d}(V_1 - V_2)}{\mathrm{d}t} = (c_{10} + c_{12})\frac{\mathrm{d}V_1}{\mathrm{d}t} - c_{12}\frac{\mathrm{d}V_2}{\mathrm{d}t} \tag{4.5a}$$

$$I_2 = c_{12}\frac{\mathrm{d}(V_2 - V_1)}{\mathrm{d}t} + c_{20}\frac{\mathrm{d}V_2}{\mathrm{d}t} = -c_{12}\frac{\mathrm{d}V_1}{\mathrm{d}t} + (c_{20} + c_{12})\frac{\mathrm{d}V_2}{\mathrm{d}t} \tag{4.5b}$$

or in matrix form as

$$\begin{bmatrix} I_1 \\ I_2 \end{bmatrix} = \mathbf{C}\frac{\mathrm{d}}{\mathrm{d}t}\begin{bmatrix} V_1 \\ V_2 \end{bmatrix} \tag{4.6}$$

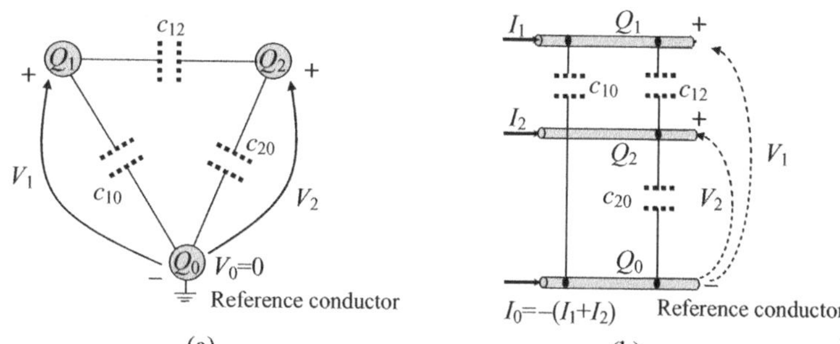

Figure 4.2 Capacitances between two conductors and their reference: (a) transversal representation; (b) longitudinal representation

where **C** is the capacitance matrix given by

$$\mathbf{C} = \begin{bmatrix} C_{11} & C_{12} \\ C_{21} & C_{22} \end{bmatrix} = \begin{bmatrix} c_{10} + c_{12} & -c_{12} \\ -c_{12} & c_{20} + c_{12} \end{bmatrix} \tag{4.7}$$

The diagonal terms in matrix (4.7) are the self capacitances and represent the displacement currents flowing from a given conductor to ground when the other conductor is grounded. It should be noted that the self capacitance is not the wire-to-ground capacitance. The off-diagonal terms in matrix (4.7) are always negative and are the mutual capacitances with changed sign.

4.1.3 Capacitance Matrix of n *Coupled Conductors Having a Reference Return Conductor*

The capacitance matrix **C** introduced in the previous section for the two-conductor configuration can be easily extended to the more general configuration of n coupled lines having a reference return conductor [4]. In this case, the diagonal and off-diagonal terms are given by

$$C_{ii} = c_{i0} + \sum_{j=1, j\neq i}^{n} c_{ij} \tag{4.8a}$$

$$C_{ij} = -c_{ij} \tag{4.8b}$$

The diagonal term C_{ii} in Equation (4.8a) is given by the sum of the ith wire-to-ground capacitance c_{i0} and the mutual capacitances c_{ij} between the ith and the other wires. The off-diagonal term C_{ij} in Equation (4.8b) is always negative and is given by the mutual capacitance between the ith and jth conductors with changed sign. It is important to point out that the coefficients of the capacitance matrix (4.7) have no physical meaning, but they can be used to calculate the physical partial capacitances between conductors.

Because of the reciprocity of capacitance, capacitance matrices are also reciprocal, so $C_{ij} = C_{ji}$, and the matrix is symmetric.

If the medium surrounding the conductors is homogeneous and characterized by permittivity ε_r and relative permeability μ_r, the following relationship between capacitance and inductance matrices holds:

$$\mathbf{LC} = \mu_0\mu_r\varepsilon_0\varepsilon_r\mathbf{I} \tag{4.9}$$

where **I** is the $n \times n$ identity matrix. Therefore, only one of the parameter matrices needs to be calculated. In general, the static field solvers firstly calculate the matrix **C** while deriving the matrix **L** by Equation (4.9).

Usually, in practical configurations, the medium surrounding the conductors is characterized by an inhomogeneous dielectric constant and permeability $\mu_r = 1$. As the inductance matrix is not affected by the dielectric inhomogeneity, it can be derived from the knowledge of the capacitance matrix $\mathbf{C}_0$ obtained by replacing the dielectric with free space. Thus, for inhomogeneous media, the computation of the capacitance matrix is performed twice, with and without the presence of the dielectric.

4.2 Differential Mode and Common Mode Capacitance

In high-speed digital systems the concept of *differential mode* (DM) and *common mode* (CM) capacitance is very useful for crosstalk modeling, for differential signaling investigation, and for choosing the appropriate filters to mitigate conducted and radiated emission. The effective capacitance associated with each conductor for *differential mode* and *common mode* will be used in *Section 6.2* to introduce a transmission-line model for two symmetric lines based on two decoupled modes of propagation (i.e. even and odd modes) characterized by proper characteristic impedance and delay. It will be shown that even and odd modes are directly related to *differential mode* and *common mode*.

4.2.1 Differential Mode Capacitance

When, in the configuration of Figure 4.2, the two voltages V_1 and V_2 are forced to be equal in magnitude and opposite in sign (see Figure 4.3), Equations (4.5) become

$$I_1 = c_{10}\frac{dV}{dt} + 2c_{12}\frac{dV}{dt} = c_{e1}\frac{dV}{dt} \quad (4.10a)$$

$$I_2 = -2c_{12}\frac{dV}{dt} - c_{20}\frac{dV}{dt} = -c_{e2}\frac{dV}{dt} \quad (4.10b)$$

where c_{e1} and c_{e2} are the effective capacitances associated with the two conductors, given by

$$c_{e1} = c_{10} + 2c_{12} \quad (4.11a)$$

$$c_{e2} = c_{20} + 2c_{12} \quad (4.11b)$$

Based on Equation (4.10), the *differential mode* circuit shown on the righthand side of Figure 4.3 is obtained, and the *differential mode* capacitance between conductor 1 and conductor 2 is given by

$$C_{DM} = \frac{c_{e1}c_{e2}}{c_{e1} + c_{e2}} \quad (4.11c)$$

If the structure is symmetric, then $c_{10} = c_{20} = c_0$. In this case, the two effective capacitances are equal to $c_{e1} = c_{e2} = c_{eDM} = c_0 + 2c_m$, with $c_{12} = c_m$, and the *differential mode* capacitance C_{DM} and the effective *differential mode* capacitance C_{eDM} are given by

$$C_{DM} = c_{eDM}/2 \quad (4.12a)$$

$$c_{eDM} = c_0 + 2c_m \quad (4.12b)$$

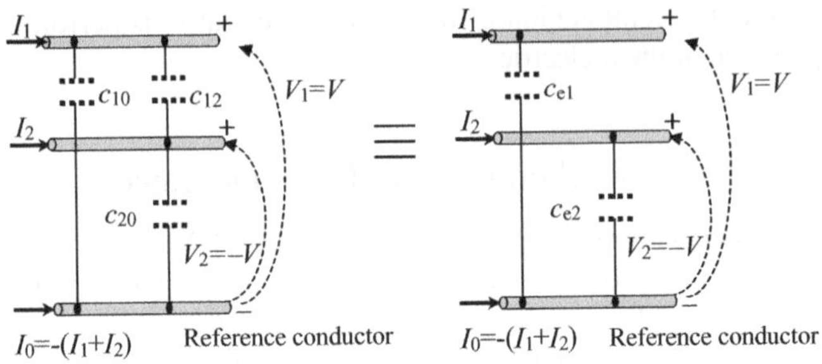

Figure 4.3 Equivalent circuits for differential mode capacitance

Equation (4.12a) provides the true *differential mode* capacitance between the two conductors. Equation (4.12b) will be used in *Section 6.2* to calculate the transmission-line parameters associated with the odd mode.

4.2.2 Common Mode Capacitance

When the two voltages V_1 and V_2 are forced to be equal in magnitude, the mutual capacitance c_{12} is not involved by any current circulation (see Figure 4.4). In this case, equations (4.5) can be rewritten as

$$I_1 = c_{10}\frac{\mathrm{d}V}{\mathrm{d}t} = c_{e1}\frac{\mathrm{d}V}{\mathrm{d}t} \tag{4.13a}$$

$$I_2 = c_{20}\frac{\mathrm{d}V}{\mathrm{d}t} = c_{e2}\frac{\mathrm{d}V}{\mathrm{d}t} \tag{4.13b}$$

where

$$c_{e1} = c_{10} \tag{4.14a}$$

$$c_{e2} = c_{20} \tag{4.14b}$$

Based on Equations (4.13), the *common mode* circuit shown on the right-side of Figure 4.4 is obtained. The *common mode* capacitance between conductors 1 and 2, which are in parallel,

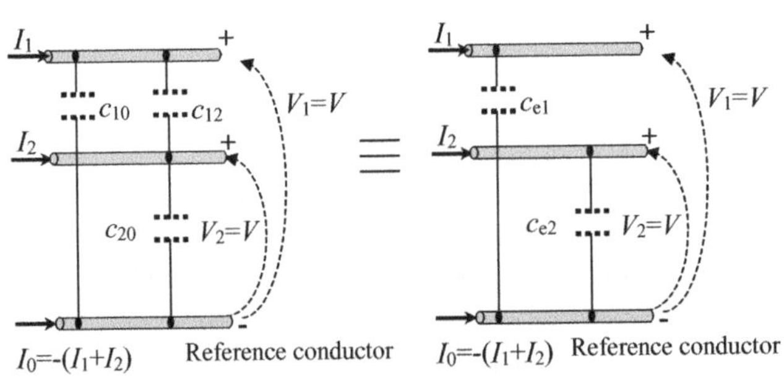

Figure 4.4 Equivalent circuits for common mode capacitance

and the reference conductor is then defined as

$$C_{\rm CM} = \frac{c_{\rm e1} + c_{\rm e2}}{2} \tag{4.15}$$

If the structure is symmetric, then $c_{10} = c_{20} = c_0$ and the *common mode* capacitance is

$$C_{\rm CM} = c_{\rm eCM} = c_0 \tag{4.16}$$

This means that the *common mode* capacitance coincides with the even-mode capacitance as defined in *Section 6.2*. Equation (4.16) will be used to calculate the line parameters associated with the even mode.

References

[1] Ramo, S., Whinnery, J., and Van Duzer, T., *'Fields and Waves in Communication Electronics'*, 3rd edition, John Wiley & Sons, Inc., New York, NY, 1994.
[2] Young, B., *'Digital Signal Integrity: Modeling and Simulation with Interconnects and Packages'*, Prentice Hall PTR, Upper Saddle River, NJ, 2001.
[3] Paul, C., *'Introduction to Electromagnetic Compatibility'*, John Wiley & Sons, Inc., Hoboken, NJ, 2006.
[4] Paul, C., *'Analysis of Multiconductor Transmission Lines'*, John Wiley & Sons, Inc., New York, NY, 1994.

5

Reflection on Signal Lines

Typical interconnects such as microstrip or stripline in PCBs, coaxial cables, and parallel or twisted-pair wire cables must be modeled as transmission lines (TLs) owing to the high speed of the digital devices used nowadays. Among the main undesired effects are reflections caused by discontinuities along the line and mismatching at the ends of the interconnects. The aim of this chapter is to provide methods to predict reflections. More detailed information on TLs can be found in many of the textbooks listed in the references. The equivalent circuit of an interconnect as a cascade of lumped elements is first discussed. An analytical solution, lattice diagram, exact SPICE model for a lossless line, and a graphical approach are outlined in order to compute incident and reflected waves. An example of the graphical method applied to *Transistor–Transistor Logic* (TTL) devices is discussed as background for other logic families. Signal distribution architectures to avoid uncontrolled situations are presented and defined. The chapter ends with a discussion about different types of line termination to enhance *Signal Integrity* (SI) and mitigate reflections. The performance of different terminations is shown by circuit simulations.

5.1 Electrical Parameters of Interconnects

An interconnect can be modelled as a transmission line in several ways. In any case, each model is based on the electric parameters of the line in terms of per-unit-length (p.u.l.) resistance, inductance, and capacitance, which depend on the geometry of the line structure. When the resistance is not significant for signal integrity investigation, the line is considered lossless. For a lossless line there are two key parameters for signal integrity investigation: the characteristic impedance and the propagation delay time of the line. In this section, these two parameters are defined, and how to model a transmission line considering its applications is outlined.

5.1.1 Typical Interconnects

Typical interconnects used in high-speed digital systems are shown in Figure 5.1. Microstrips or striplines are realized in multilayer PCBs. A microstrip consists of a conductive trace buried in or attached to a dielectric substrate. The trace has one reference plane as return conductor for the current. A stripline consists of a conductive trace between two planes separated by

Signal Integrity and Radiated Emission of High-Speed Digital Systems Spartaco Caniggia and Francescaromana Maradei

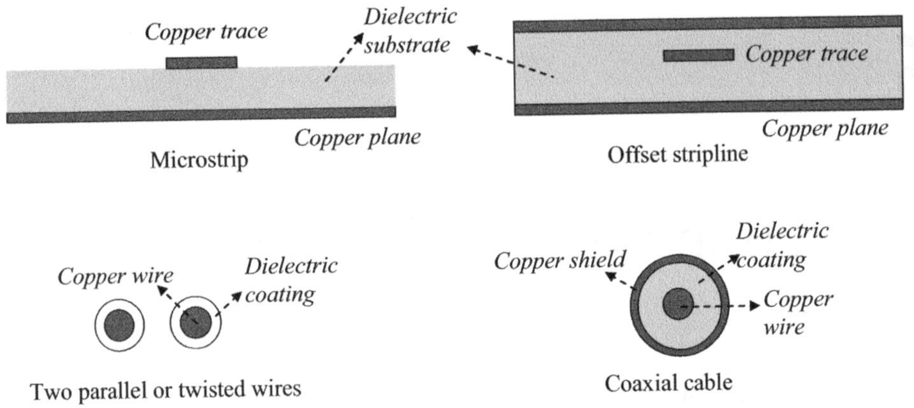

Figure 5.1 Typical interconnects

a dielectric substrate. The dielectric is usually made of FR4, which is a type of fiberglass. Two parallel or twisted-pair wires with or without a shield and coaxial cables with one or two shields are used to connect PCBs.

Owing to the fast rise and fall times of the digital devices, which are much shorter than the propagation delay time of the interconnects, these structures must be modeled as transmission lines [1–7]. The simplest way to build up a circuit model of a transmission line is to chain together sections of line electrically short.

5.1.2 Equivalent Circuit of a Short Interconnect

An electrically short transmission line, or any line segment of length $\Delta x \leq \lambda/10$, where λ is the wavelength related to the maximum frequency of interest, can be represented by an equivalent circuit composed of lumped-circuit elements, as shown in Figure 5.2, where R, L, C, G are the per-unit-length (p.u.l.) resistance, inductance, capacitance, and conductance parameters. This equivalent circuit can be obtained under the assumption that there are no components of electric and magnetic fields along the x direction of propagation, i.e. transverse electromagnetic (TEM) mode. Actually, the pure TEM mode cannot exist in practice owing to the conductor and dielectric losses and/or to the dielectric inhomogeneity of the surrounding medium. However, the deviation from a TEM field structure is typically small for good conductors and typical line cross-sectional dimensions. This is referred to as the quasi-TEM mode assumption and is accounted for in Figure 5.2 by the presence of the elements $R\,\Delta x$ (resistance) and $G\,\Delta x$

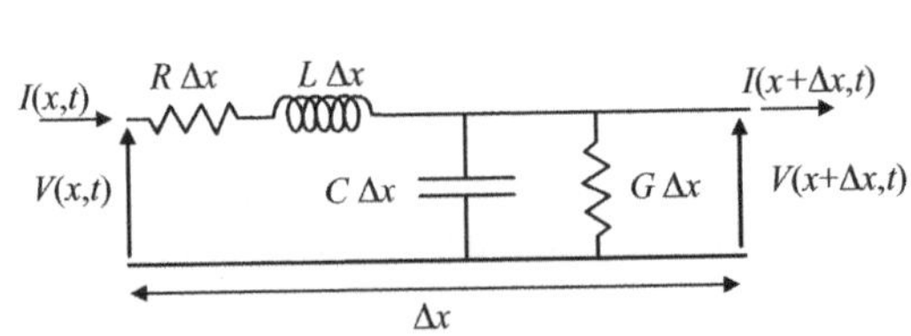

Figure 5.2 Lumped equivalent circuit of an electrically short section of interconnect

(conductance) which represent the losses due to the finite conductivity of the two conductors and the finite resistance of the dielectric respectively. The series inductor $L\ \Delta x$ represents the magnetic field, and $C\ \Delta x$ represents the electric fields between the two conductors.

By applying Kirchhoff's laws to the circuit of Figure 5.2 [5], the following equations relating voltage $V(x, t)$ and current $I(x, t)$ along the line of length l are obtained:

$$\frac{\partial V(x,t)}{\partial x} = -RI(x,t) - L\frac{\partial I(x,t)}{\partial t} \tag{5.1a}$$

$$\frac{\partial I(x,t)}{\partial x} = -GV(x,t) - C\frac{\partial V(x,t)}{\partial t} \tag{5.1b}$$

These equations, known as telegrapher's equations in the time domain, are valid in the interval $0 \le x \le l$.

The corresponding equations in the frequency domain are given by

$$\frac{\partial \hat{V}(x)}{\partial x} = -\hat{Z}(\omega)\hat{I}(x) \tag{5.2a}$$

$$\frac{\partial \hat{I}(x)}{\partial x} = -\hat{Y}(\omega)\hat{V}(x) \tag{5.2b}$$

where $\hat{V}(x)$ and $\hat{I}(x)$ are the voltage and current phasors at the frequency considered, $\hat{Z}(\omega) = R(\omega) + \mathrm{j}\omega L$ is the p.u.l. line impedance, $\hat{Y}(\omega) = G(\omega) + \mathrm{j}\omega C$ is the p.u.l. line admittance, $\omega = 2\pi f$ is the angular frequency, and j is an imaginary unit.

The basic electrical parameter that defines a transmission line is the characteristic impedance $\hat{Z}_0(\omega)$. This impedance corresponds to the input impedance of a uniform TL of infinite length, or equivalently, of a TL of finite length that is terminated on its own characteristic impedance. In other words, $\hat{Z}_0(\omega)$ is the impedance offered by the line when a signal is launched at its input. To derive the dependence of the characteristic impedance from the p.u.l. parameters, let us consider the circuit of Figure 5.3, where an electrically short line section of length Δx is loaded by $\hat{Z}_0(\omega)$. Applying the characteristic impedance definition to the circuit of Figure 5.3, and omitting for simplicity the dependence on frequency, yields

$$\hat{Z}_0 = \hat{Z}\Delta x + \frac{\left(1/\hat{Y}\Delta x\right)\hat{Z}_0}{\left(1/\hat{Y}\Delta x\right) + \hat{Z}_0} \tag{5.3}$$

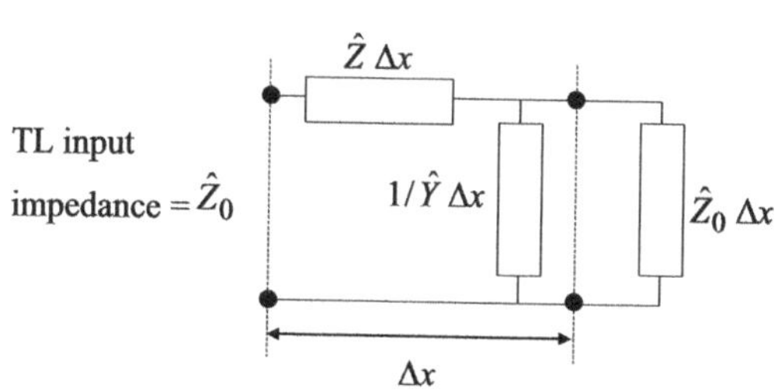

Figure 5.3 Circuit for characteristic impedance definition

After simple manipulation of Equation (5.3), and neglecting the term with Δx^2, which for $\Delta x \to 0$ is second-order infinitesimal, the characteristic impedance becomes

$$\hat{Z}_0 = \sqrt{\frac{\hat{Z}}{\hat{Y}}} = \sqrt{\frac{R + j\omega L}{G + j\omega C}} \tag{5.4}$$

In general, the characteristic impedance is a complex number with a resistive and reactive component. It is a function of the frequencies of the applied signal, and it does not depend on the line length. At very high frequency (i.e. $f \to \infty$), the characteristic impedance asymptotes to a fixed real nominal value Z_0 which is resistive.

Characteristic impedance is of primary importance for good transmission, as introduced and discussed in *Section 1.1*. Maximum power transfer occurs when the source has the same impedance as the load. Thus, for sending signals over a line, the driver must have the same characteristic impedance as the line to get the maximum signal into the line. At the other end of the line, the receiver must also have the same impedance as the line to be able to get the maximum signal. When impedances do not match, some of the signal is reflected back towards the driver, which could cause problems.

5.1.3 Lossless Transmission Lines

The lossless line model is suitable for interconnects with negligible losses. In this case, the parameters of interest are the nominal characteristic impedance Z_0 and the per unit length (p.u.l.) propagation delay time t_{pd}.

When losses can be neglected (i.e. $R = 0$, $G = 0$), the telegrapher's equations become

$$\frac{\partial V(x,t)}{\partial x} = -L\frac{\partial I(x,t)}{\partial t} \tag{5.5a}$$

$$\frac{\partial I(x,t)}{\partial x} = -C\frac{\partial V(x,t)}{\partial t} \tag{5.5b}$$

and the characteristic impedance is given by

$$\hat{Z}_0 = Z_0 = \sqrt{\frac{L}{C}} \tag{5.6}$$

which in this case is a real number. Differentiating Equation (5.5a) with respect to x and Equation (5.5b) with respect to t, and combining the results, the voltage and current wave equations are obtained as

$$\frac{\partial^2 V(x,t)}{\partial x^2} = LC\frac{\partial^2 V(x,t)}{\partial t^2} \tag{5.7a}$$

$$\frac{\partial^2 I(x,t)}{\partial x^2} = LC\frac{\partial^2 I(x,t)}{\partial t^2} \tag{5.7b}$$

The constant $1/LC$ has the dimension of (length/time)^2, so that $v_e = 1/\sqrt{LC}$ has the dimension of a velocity, and it is the propagation velocity of the line. As previously anticipated, the second fundamental parameter of a line is the p.u.l. propagation delay time t_{pd} (s/m) given by

$$t_{pd} = \frac{1}{v_e} = \sqrt{LC} \tag{5.8}$$

This is the time required by a waveform launched onto a 1 m long line by the driver to reach a receiver positioned at the end of the line. In the case of a line of length l, the delay time required by the signal to reach the end is $T_D = t_{pd}l$.

Both Z_0 and t_{pd} depend on the capacitance C, and therefore they depend on the surrounding medium and its relative dielectric constant ε_r. In the case of a homogeneous medium with relative dielectric constant ε_r, the propagation velocity is $v_e = c/\sqrt{\varepsilon_r}$, where $c = 3 \times 10^8$ m/s is the speed of light in vacuum. Z_0 and t_{pd} can be calculated analytically for simple structures, as shown in *Appendix B* for microstrip and stripline structures, while they can be derived numerically or experimentally by the time-domain reflectometric technique (see *Section 11.1*). In accordance with Equation (5.8), the characteristic impedance can be rewritten as

$$Z_0 = Lv_e \tag{5.9}$$

The solution of wave equation (5.7) is given by [1, 5],

$$V(x,t) = V^+(t - x/v_e) + V^-(t + x/v_e) \tag{5.10a}$$

$$I(x,t) = I^+(t - x/v_e) - I^-(t + x/v_e) = \frac{V^+(t - x/v_e)}{Z_0} - \frac{V^-(t + x/v_e)}{Z_0} \tag{5.10b}$$

where $V^+(t - x/v_e)$ represents the forward wave traveling in the $+x$ direction with velocity v_e, $V^-(t + x/v_e)$ represents the backward wave moving in the $-x$ direction with velocity v_e, and $I^+(t - x/v_e)$ and $I^-(t + x/v_e)$ are the corresponding forward and backward current waves. Note that Z_0 is the ratio of voltage to current for a single wave traveling in any direction at any given point and given time instant.

The equivalent circuit of a lossless line obtained by chaining together line sections that are electrically short is shown in Figure 5.4. This circuit is the basis of our discussions, although an exact TL model for a lossless line is available in any circuit simulator. The exact model will be presented in *Section 5.2*.

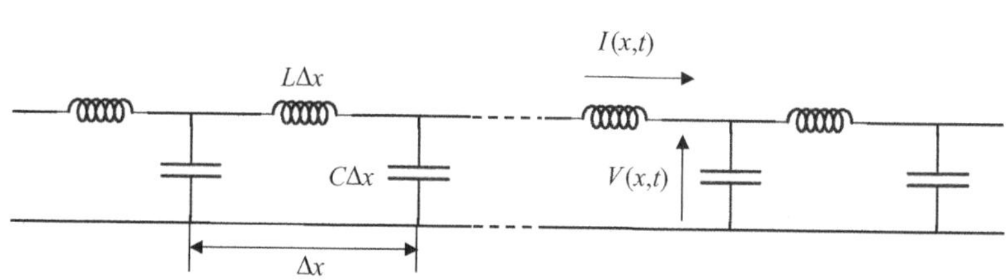

Figure 5.4 Representation of a lossless transmission line by lumped parameters consisting of cells of length Δx having p.u.l. inductance L and p.u.l. capacitance C

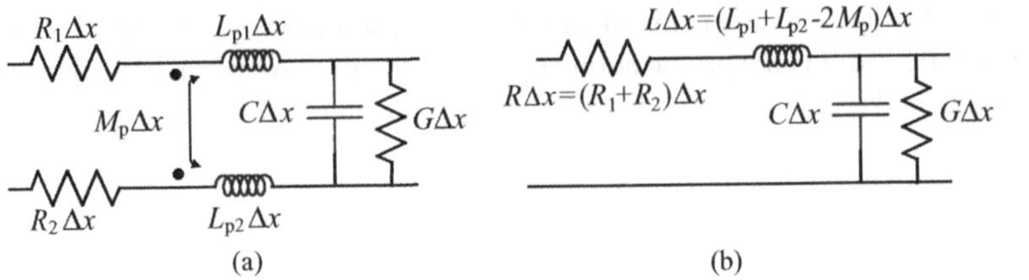

Figure 5.5 A transmission-line segment modeled (a) by partial inductances to compute the signal propagations and voltage drop along the return conductor and (b) by total inductance to compute signal propagation only

5.1.4 Transmission-Line Modeling by Using Partial Inductances

An electrically short segment of a transmission line can be modeled by two different circuits:

- The first one, shown in Figure 5.5a, is based on the partial inductance concept introduced in *Section 3.2*: the partial inductances L_{p1} and L_{p2} are associated with the signal and return conductors respectively, and the mutual partial inductance M_p between the two conductors is considered.
- The second circuit, shown in Figure 5.5b, is based on the total inductance L which can be calculated by the partial inductances defined above and is assigned to the signal conductor.

The two circuits are equivalent for current and voltage calculations. Although the first one has more circuit elements, it is more general, as it enables the voltage drop along both signal and return conductors to be calculated. Any time, in digital circuits, the interest is focused on the ground noise due to switching of a driver for electromagnetic interference in a PCB (see *Chapter* 8 and *Chapter 10*) and/or emission from cables attached to a PCB (see *Chapter 9* and *Chapter 10*), the model of Figure 5.5a must be used.

5.2 Incident and Reflected Waves in Lossless Transmission Lines

Among the most dangerous phenomena concerning signal integrity are the reflections caused by discontinuities in the interconnects. A digital signal can be seen as an incident wave traveling along a lossless line unchanged until meeting a discontinuity (i.e. resistance different from the characteristic impedance of the line or a capacitive load). The discontinuity generates a reflected wave that sums with the incident wave at the discontinuity and travels along the line towards the source of the incident wave. This mechanism occurs for each discontinuity and produces distortions of the digital signal. The purpose of this section is to investigate the phenomenon, and to provide methods and models for predicting reflected waves, considering the delay between the discontinuities and the non-linearity of the digital devices.

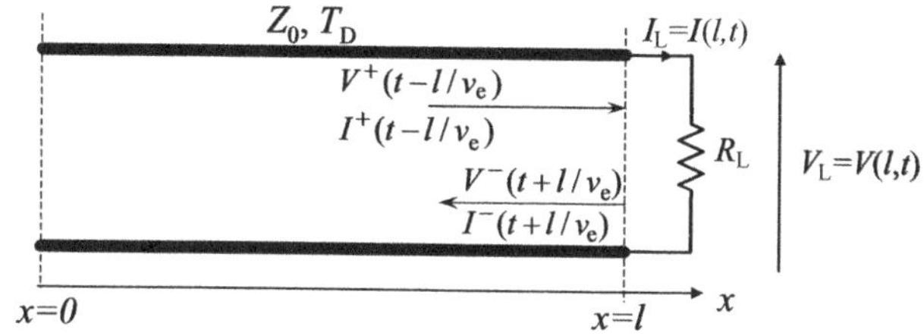

Figure 5.6 Incident and reflected waves at load location

5.2.1 *Resistive Discontinuity*

Most transmission-line problems are related to the presence of discontinuities such as resistive loads or lines with different characteristic impedance. According to Kirchhoff's laws, the total voltage and current must be continuous across the discontinuity [8–11]. According to Equations (5.10), the voltage V_L and the current I_L on the load at $x = l$ are given by (see Figure 5.6)

$$V_L = V(l,t) = V^+(t - l/v_e) + V^-(t + l/v_e) \tag{5.11a}$$

$$I_L = I(l,t) = \frac{1}{Z_0}V^+(t - l/v_e) - \frac{1}{Z_0}V^-(t + l/v_e) = \frac{V_L}{R_L} \tag{5.11b}$$

The load reflection coefficient ρ_L, defined as the ratio between the backward- and forward-traveling waves, can be derived by Equations (5.11) as

$$\rho_L \equiv \frac{V^-(t + l/v_e)}{V^+(t - l/v_e)} = \frac{R_L - Z_0}{R_L + Z_0} \tag{5.12}$$

This coefficient is limited to the range $-1 \leq \rho_L \leq 1$ and applies to voltage only [5]. The following three cases are of practical interest:

1. $R_L = 0$ (short-circuited TL), $\rho_L = -1$, the incident wave is reflected totally with a minus sign.
2. $R_L = Z_0$ (matched TL), $\rho_L = 0$, there is no reflected wave.
3. $R_L = \infty$ (opened TL), $\rho_L = 1$, the incident wave is reflected totally with a plus sign.

Of course, the case of a matched line is the goal of a good designer.

5.2.2 *Capacitive Discontinuity*

An interesting practical case is when the discontinuity is caused by a capacitive load due to the receivers distributed or concentrated along the line, as introduced in *Section 1.1*. In order to focus the attention on the reflection produced by a capacitive load (see Figure 5.7), let us assume at time $t = 0$ a forward-traveling wave arriving at the end of the line where there is a capacitance C_L and defined as [1]

$$V^+(t) = \begin{cases} 0 & \text{for} \quad t < 0 \\ V_O & \text{for} \quad t \geq 0 \end{cases} \tag{5.13}$$

According to Equations (5.10), the voltage V_L and the current I_L on the load at $x = l$ are given by

$$V_L = V(t) = V^+(t) + V^-(t) \tag{5.14a}$$

$$I_L = I(t) = \frac{1}{Z_0}V^+(t) - \frac{1}{Z_0}V^-(t) = C_L\frac{dV_L}{dt} \tag{5.14b}$$

By introducing Equation (5.14a) into (5.14b), and on the basis of the forward-traveling wave definition (5.13), the following first-order differential wave equation for the backward wave is obtained:

$$\frac{dV^-(t)}{dt} + \frac{V^-(t)}{\tau_1} = \frac{V_O}{\tau_1} \tag{5.15}$$

where $\tau_1 = C_L Z_0$. The general solution of Equation (5.15) is given by

$$V^-(t) = V_O + A\,e^{-t/\tau_1} \tag{5.16}$$

where the constant A is defined by the initial condition. At time $t = 0$, the capacitor is uncharged and, according to Equations (5.13) and (5.14a), yields $V_O + V^-(0^+) = 0$, and hence $V^-(0^+) = -V_0$. Introducing the initial condition into Equation (5.16), the constant $A = -2V_O$ is derived. The backward-traveling wave is then given by

$$V^-(t) = V_O(1 - 2\,e^{-t/\tau_1}) \tag{5.17}$$

and the voltage V_L on the capacitor by

$$V_L(t) = 2V_O\left(1 - e^{-t/\tau_1}\right) \tag{5.18}$$

The x variation of a reflected wave is obtained by replacing t with $t + x/v_e$ in Equation (5.17).

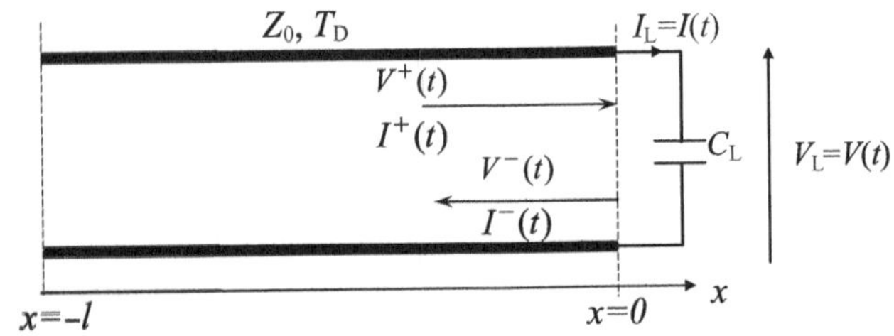

Figure 5.7 Incident and reflected waves in a line with a capacitive termination

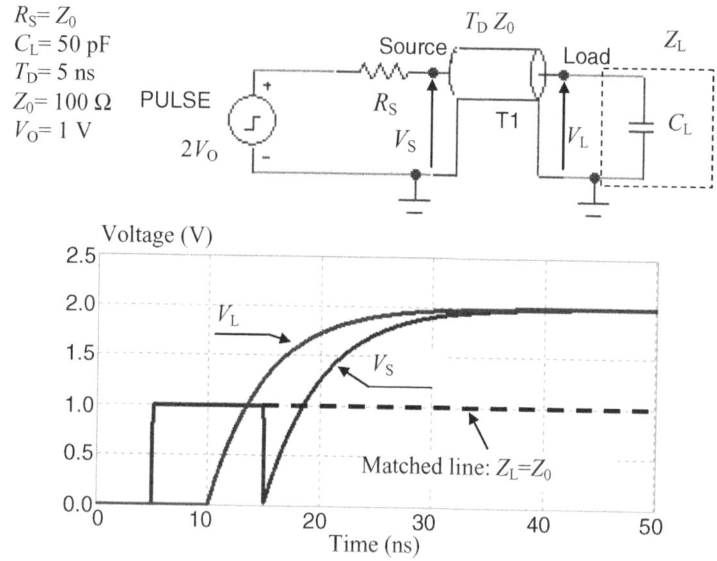

Figure 5.8 Equivalent circuit of a transmission line loaded with a capacitance and waveforms simulated by SPICE

As an example, an ideal line of characteristic impedance $Z_0 = 100\ \Omega$ and delay time $T_D = 5$ ns, matched at source $R_S = Z_0$ and terminated on a capacitive load $C_L = 50$ pF, is considered. The TL is excited by a unit step voltage source activated at time $t_0 = 5$ ns. The voltages V_S and V_L at the line end-points $x = 0$ and $x = l$, computed by SPICE, are shown in Figure 5.8. The load voltage V_L and the reflected wave voltage V^- can also be computed by Equations (5.18) and (5.17) respectively. Comparing the obtained waveform of V_S with the one obtained in the case of a matched line (dashed lines in Figure 5.8), it is interesting to note that the capacitance produces a negative reflected wave that, at the source end and at time $t = 2T_D$, causes a rapid return to zero of the voltage. After this instant, V_S rises to the maximum value $2V_O$ exponentially with time constant $C_L Z_0$. From a practical point of view, a reflected wave caused by capacitive loads could be very dangerous if there are receivers located at the source end.

5.2.3 Reflections in Interconnects Terminated with Resistive Loads

In *Section 5.2.1* the reflection generated by a resistive load has been studied and characterized through the load reflection coefficient ρ_L defined by Equation (5.12). In practical situations, the interconnect is fed by drivers characterized by the source resistance R_S. If $R_S \neq Z_0$, new reflections occur every time a backward-traveling wave is propagating towards the driver. The reflection at the source can be characterized by the source reflection coefficient ρ_S given by

$$\rho_S = \frac{R_S - Z_0}{R_S + Z_0} \tag{5.19}$$

The load and source reflection coefficients given by Equations (5.12) and (5.19) are the key parameters for calculating reflections in interconnects.

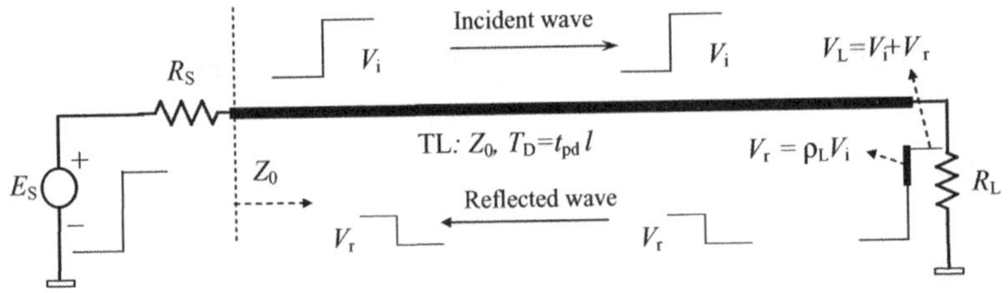

Figure 5.9 Illustration of incident and reflected wave calculation along an interconnect (TL) loaded with resistances

A typical interconnect is shown in Figure 5.9, where the line is terminated with a resistive load R_L. When the rise or fall time of the step voltage source E_S is much shorter than the line propagation delay time $T_D = t_{pd}l$, the incident waveform V_i (forward-traveling wave V^+) initially launched from the source onto the line is related to the source voltage E_S by

$$V_i = \frac{Z_0}{R_S + Z_0} E_S \tag{5.20}$$

After a time equal to the line delay time T_D, V_i reaches unchanged the load at $x = l$ and causes a load voltage V_L given by the sum of the incident wave V_i and the reflected wave towards the source $V_r = \rho_L V_i$ (backward-traveling wave V^-). When, after the time $2T_D$, the reflected wave V_r reaches the source end, a new reflected wave $\rho_S \rho_L V_i$ towards the load end is generated. This process of repeated reflections continues at regular time intervals T_D as re-reflections at the source and load ends. For a good quality of transmission (i.e. a signal on the load equal to that launched by the source), the optimal condition should be $\rho_S \ll 1$ and $\rho_L = 0$. In this case, E_S is quasi-totally transmitted and there is no reflection at the load.

5.2.4 Critical Length of Interconnects

Not all the interconnects must be studied as transmission lines. There is a limit, called the 'critical length' of the interconnect, for which the propagation must be considered. The length of a line l is said to be critical, and indicated as l_c, when the rise time of the propagating signal satisfies the relation

$$t_r = 2T_D = 2t_{pd} l_c \tag{5.21}$$

When $t_r/2T_D \ll 1$ (i.e. $l \gg l_c$) the interconnect is defined as a long line, whereas when $t_r/4T_D > 1$ (i.e. $l \ll l_c$) it is defined as a short line. For long lines (see Figure 5.10a), the reflections reach their maximum value and stay for a time equal to twice the line delay time T_D. When $l = l_c$ (see Figure 5.10b), the reflections reach their maximum value for a very short time and soon change owing to the other incoming reflection. Finally, in the case of a short line (see Figure 5.10c), the interconnect behaves as a concentrated capacitance and the inductive effect can be neglected. It is important to point out that a line can be assumed to be of critical length, short, or long depending on the rise/fall time of the driver.

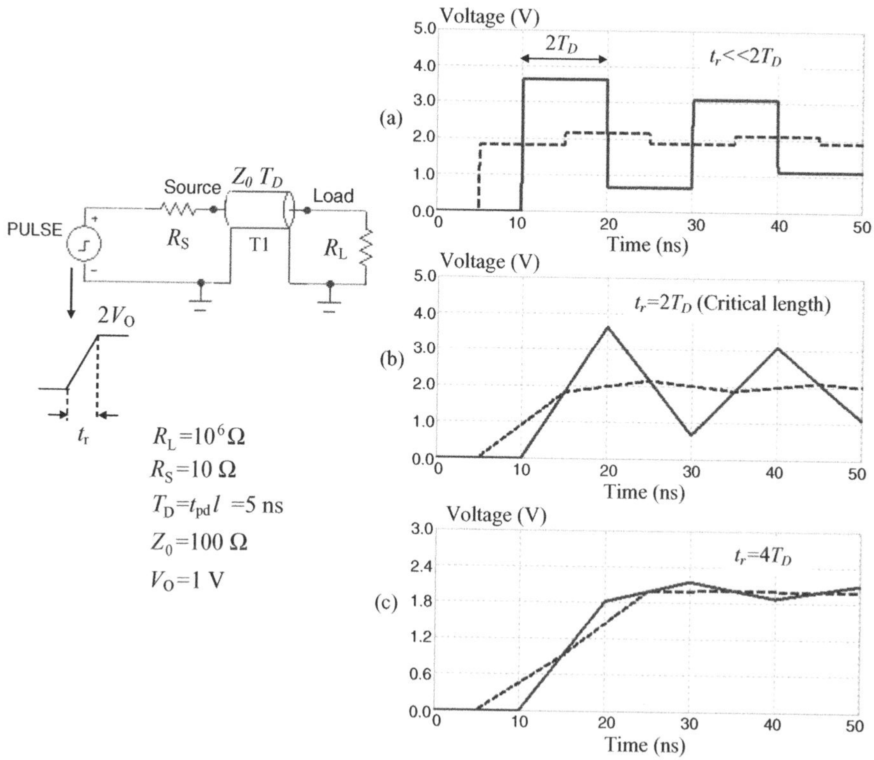

Figure 5.10 Source (dashed line) and load (solid line) simulated voltages of an interconnect with different rise times t_r: (a) long TL; (b) TL of critical length; (c) short TL

5.2.5 Lattice Diagram for Reflection Calculation

The waveforms of Figure 5.10a can be obtained by applying a graphical method for long lines called the lattice diagram, which is illustrated in Figure 5.11b. Basically, the method consists in the graphical visualization of multiple reflections at different time instants while calculating incident and reflected voltages at the two line ends by using the source and load reflection coefficients [5]. For the structure under study, shown in Figure 5.11a, we have $\rho_S = (10-100)/(10+100) = -0.818$ and $\rho_L = (10^6-100)/(10^6+100) \approx 1$. At time $t = 0^+$, immediately after switching of the driver, an incident voltage $V_i = 2V_O Z_0/(R_S + Z_0) = 1.82$ V starts from the source (point S) and travels towards the other end of the line (point L). At time $t = T_D$, V_i reaches the load, producing a load voltage $V_{L1} = V_i + \rho_L V_i = 3.64$ V and a reflected wave $\rho_L V_i = 1.82$V towards the source. At time $t = 2T_D$, the load reflection reaches the source (point S), where the voltage $(1+\rho_S)\rho_L V_i$ sums with the existing voltage V_i so that $V_{S2} = 2.15$ V. The reflected wave at the source end, equal to $\rho_S \rho_L V_i = -1.49$ V, starts to travel towards the load. After a time $t = 3T_D$, the load voltage is given by $V_{L3} = V_{L1} + \rho_S \rho_L (1+\rho_L) V_i = 0.66$ V. The process continues with the same mechanism up to the time the reflections die down. It is important to remember that voltages at both ends change after a time equal to twice the line delay time.

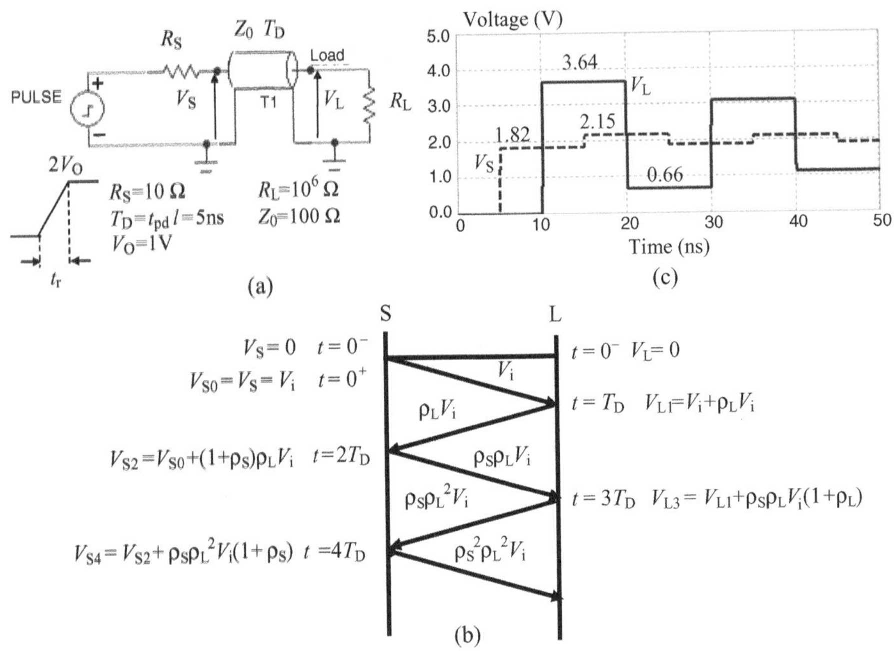

Figure 5.11 Illustration of reflection computation by the lattice diagram: (a) equivalent circuit of the interconnect; (b) lattice diagram procedure; (c) simulated waveforms at source (dashed line) and at load (solid line)

The lattice diagram can be very useful for finding reflections in a cascade connection of lines with different characteristic impedance Z_0. As an example, the cascade connection of three TLs of the same length and with characteristic impedances $Z_{01} = 50\ \Omega$, $Z_{02} = 75\ \Omega$, and $Z_{03} = 100\ \Omega$ is considered, as shown in Figure 5.12a. The configuration is matched at both ends (i.e. $R_S = Z_{01}$ and $R_L = Z_{03}$), so that at the source and load ends no reflections occur. However, owing to the different TL characteristic impedances, reflections arise at junctions J_1 and J_2 of the lines (see Figure 5.12). Let ρ_1^+ denote the reflection coefficient at junction J_1 seen by a forward-traveling wave coming from the left-hand side, ρ_1^- the reflection coefficient at junction J_1 seen by a backward-traveling wave coming from the right-hand side, and ρ_2^+ the reflection coefficient at junction J_2 seen by a forward-traveling wave from the left-hand side. The reflection coefficient ρ_2^- is not important because no reflected waves come from the load. Applying the basic mechanism at discontinuity points, voltages at source and at load can be calculated as the sum of incident, reflected, and transmitted waves arriving at different times. Figure 5.12c shows a comparison between simulated waveforms by SPICE and maximum flat values computed with the lattice diagram of Figure 5.12b.

5.2.6 Exact Model of a Lossless Transmission Line

In the previous sections, the propagation of incident and reflected waves was theoretically and graphically studied by using the reflection coefficients. The results of SPICE simulations

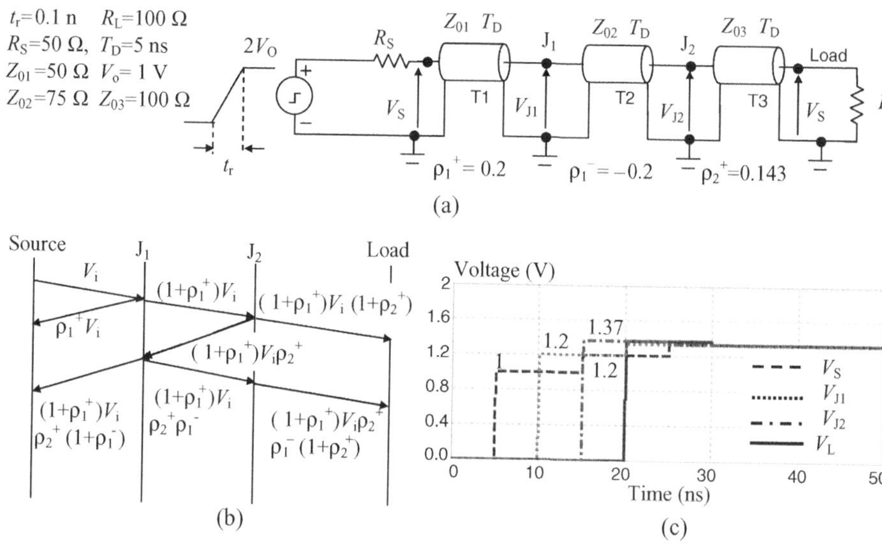

Figure 5.12 Illustration of reflection computation by lattice diagram for the case of the cascade connection of three lines: (a) interconnect equivalent circuit; (b) lattice diagram procedure; (c) simulated waveforms at source, junction J_1, junction J_2, and at load

were shown to validate the theoretical and graphical procedures. Lossless transmission lines in SPICE-based circuit simulators are described by the model shown in Figure 5.13. This equivalent circuit, attributed to Branin [12], is an exact solution of transmission-line equations in the case of uniform lossless lines. The demonstration of the model can be found in references [5] and [12], and it will be generalized in *Chapter 7* to simulate lossy lines. Basically, the model has two resistances equal to the characteristic impedance of the line Z_0 and two dependent-voltage sources $e_i(t - T_D)$ and $e_o(t - T_D)$ at the source and load ends, which are given by

$$e_i(t - T_D) = V(l, t - T_D) - Z_0 I(l, t - T_D) \tag{5.22a}$$

$$e_o(t - T_D) = V(0, t - T_D) + Z_0 I(0, t - T_D) \tag{5.22b}$$

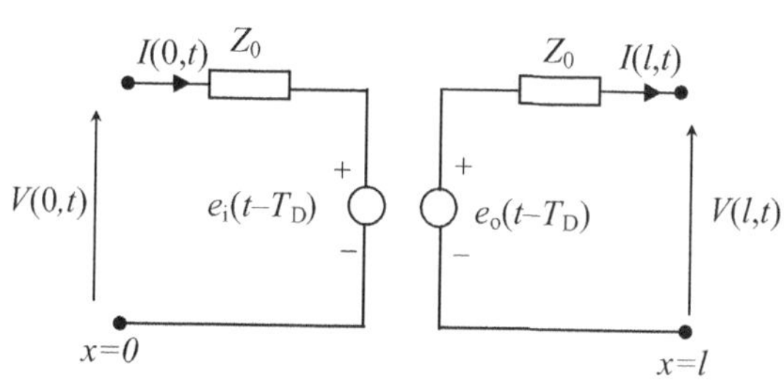

Figure 5.13 Exact distributed model for a lossless line

The controlled source $e_i(t - T_D)$ is produced by the voltage and current at the load end of the line at a time equal to the line delay time T_D earlier than the present time. Similarly, the controlled source $e_o(t - T_D)$ is produced by the voltage and current at the line input at a time equal to the line delay time T_D earlier than the present time. It is important to consider that the circuit of Figure 5.13 is absolutely general and is suitable for the handling of both non-linear and dynamic loads. In the libraries of any SPICE-based circuit simulator, the transmission-line component T_{name} is available, based on the equivalent circuit of Figure 5.13. To define the T_{name} model fully, the values of the characteristic impedance Z_0 and the line delay time T_D are required.

It must be pointed out that the circuit of Figure 5.13 does not take into account steady-state initial conditions different from zero. In practical situations, the interconnect is terminated on real driver and receiver components characterized by non-linear static characteristics and non-zero initial steady-state conditions. In this case, to account for non-zero initial conditions, the input voltage $V_S(t)$ and output voltage $V_L(t)$ of the interconnect can be expressed as

$$V_S(t) = V(0, t) = V_{S0} + \Delta V_S \tag{5.23a}$$

$$V_L(t) = V(l, t) = V_{L0} + \Delta V_L \tag{5.23b}$$

where $V_{S0} = V(0, t_0)$ and $V_{L0} = V(l, t_0)$ are the initial source and load voltages before the switching time t_0, and $\Delta V_S = \Delta V(0, t)$ and $\Delta V_L = \Delta V(l, t)$ are the voltage variations at the source and load ends after the first switch of the driver accounting for the multiple reflection process. The initial conditions V_{S0} and V_{L0} can be obtained by the direct current solution of the circuit, in which the lossless interconnect is modeled by a short circuit, which means $V_{S0} = V_{L0}$. The voltage variations ΔV_S and ΔV_L can be obtained by the Branin circuit of Figure 5.13, suitably excited.

As an example, consider a driver whose output characteristics before and after the switching can be linearly approximated, as well as the input characteristic of the receiver. In this case, both the driver and the receiver can be modeled by the Thévenin equivalent circuit, as shown in Figure 5.14a, where E_{OBS} and E_{OAS} are the output voltages of the driver, without loads, associated with the two levels before (BS) and after (AS) switching, R_{OBS} and R_{OAS} are the corresponding driver output resistances, E_L is the input DC receiver voltage due to the polarization, and R_L represents the input resistance of the receiver with possible termination. The adopted notation is extremely general and includes both low-to-high and high-to-low switching. In fact, the level before switching, as well as that after switching, can be either low (L) or high (H). The calculation of the interconnect input and output voltages is performed by Equations (5.23). The initial conditions $V_{S0} = V_{L0}$ are obtained by solving the DC circuit of Figure 5.14b. The voltage variations ΔV_S and ΔV_L are calculated by solving the dynamic circuit shown in Figure 5.14c, which is excited by the voltage source:

$$\Delta E_S = (E_{OAS} - E_{OBS}) - (R_{OAS} - R_{OBS})I_{S0} \tag{5.24}$$

This expression comes by manipulation with the following conditions which must be satisfied: $\Delta V_S = V_{SAS} - V_{S0} = Z_0 \Delta I_S = Z_0(I_{SAS} - I_{S0})$ and $E_{OAS} - R_{OAS} I_{SAS} = V_{SAS}$, where V_{SAS} and I_{SAS} are respectively the total (DC plus variation value) voltage and current in line at source S after switching of the driver.

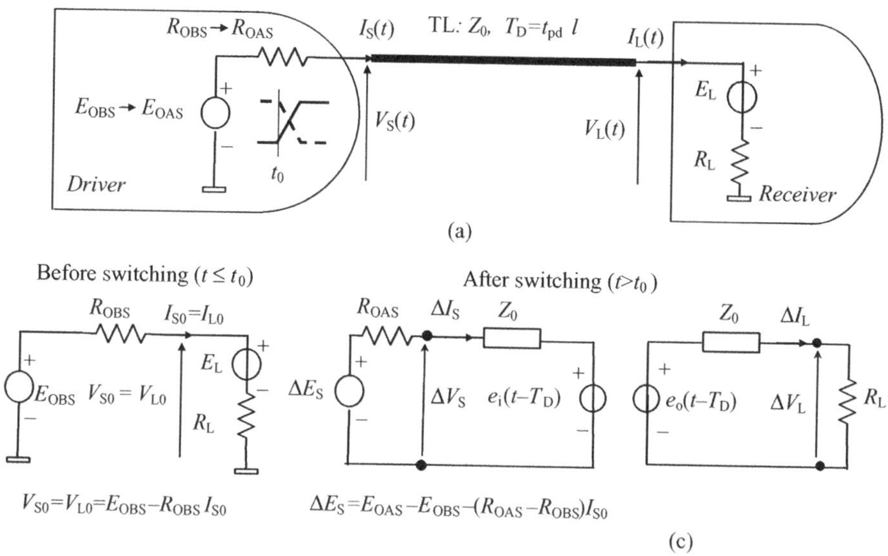

Figure 5.14 Interconnect excited by a source with a non-zero initial steady-state condition: (a) DC circuit for calculation of the initial conditions; (b) equivalent circuit to calculate the multiple reflections occurring after the first switch

In the case of non-linear drivers, the schematization shown in Figure 5.14 is still valid, but it is necessary to take into account the variability of the driver output resistance R_{OAS} and of the voltage level without loads E_{OAS} after the switching, which are functions of the total voltage at its terminal, as discussed in *Chapter 2*.

5.2.7 Graphical Solution for Line Voltages

The exact circuit model of Figure 5.14 is useful for explaining the graphical method which makes it possible to calculate reflected waves when the loads are non-linear, as often happens with digital devices. In order easily to explain the graphical procedure, let us first consider the simple driver and receiver configuration shown in Figure 5.15. According to the current and voltage notation shown in Figure 5.15a, the source output and load input characteristics are described by

$$V(0,t) = E_S - R_S I(0,t) \tag{5.25a}$$

$$V(l,t) = R_L I(l,t) \tag{5.25b}$$

The method consists in drawing in the $V - I$ plane the static characteristics (5.25). In particular, two different static characteristics describe the driver, depending on the low or high level of the voltage source. The low (L) and high (H) output characteristics of the source are given by Equation (5.25a) for $E_S = E_{OL} = 0$ and $E_S = E_{OH}$ respectively. The characteristics are the two lines with negative slope $-1/R_S$ and passing through points (0, 0) and (0, E_{OH}), as shown in Figure 5.15b. According to Equation (5.25b), the load input characteristic is a line passing

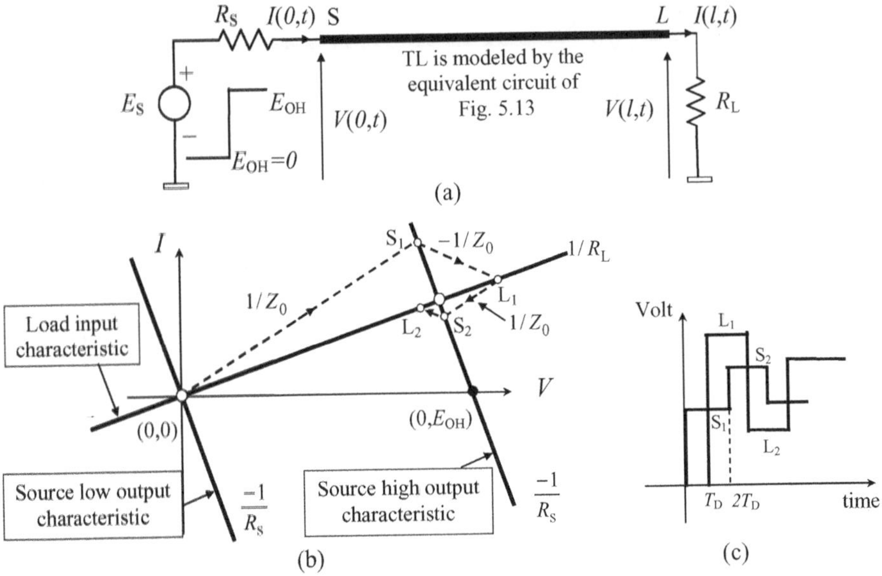

Figure 5.15 Graphical method to compute reflections for an interconnect with characteristic impedance Z_0 and delay time T_D: (a) equivalent circuit of the interconnect; (b) graphical voltage calculations at both line ends; (c) calculated waveforms for a rise/fall time equal to zero

through point (0, 0) and with positive slope $1/R_L$. The intersections of the driver output lines with the receiver input characteristic determine the steady-state low and high voltages.

When the driver switches from low to high state with a sufficiently small rise time for the interconnect to be considered a long line, the evolution in time of the voltage at the interconnect end-points can be graphically derived. Immediately after the driver low-to-high switching, the voltage $V(0, t)$ is given by

$$V(0,t) = E_{OH} - R_S I(0,t) \tag{5.26a}$$

$$V(0,t) = Z_0 I(0,t) + e_i(t - T_D) \tag{5.26b}$$

where the first equation is the source high output characteristic, and the second equation is obtained by the interconnect model of Figure 5.13. According to Equations (5.26), $V(0, t)$ in the time interval $0 \le t < 2T_D$ can be graphically obtained as the intersection between the high-level static output characteristic of the driver and the line of slope $1/Z_0$ passing through the axis origin (i.e. point S_1 in Figure 5.15b). In fact, as there is no voltage or current variation on the load for $t < T_D$, $e_i(t)$ in Equation (5.26b) is equal to zero.

After the line delay time T_D, the signal arrives at the end of the line $x = l$, and the load voltage $V(l, t)$ for $t \ge T_D$ can be calculated by

$$V(l,t) = R_L I(l,t) \tag{5.27a}$$

$$V(l,t) = -Z_0 I(l,t) + e_o(t - T_D) = -Z_0 I(l,t) + V(0, t - T_D) + Z_0 I(0, t - T_D) \tag{5.27b}$$

where the first equation is the load input characteristic, and the second equation is obtained by the interconnect model of Figure 5.13. According to Equation (5.27), $V(l, t)$ in the time interval $T_D \leq t < 2T_D$ can be graphically obtained as the intersection between the load input characteristic with slope $1/R_L$ and the line of slope $-1/Z_0$ passing through the point S_1 having coordinates $(V(0, t - T_D), I(0, t - T_D))$ (see point L_1 in Figure 5.15).

Applying the same procedure, points S_2, L_2, ... corresponding to the line input and output voltage at multiples of time T_D can be calculated up to the steady high-voltage state, as shown in Figures 5.15b and c.

In a similar way, the graphical procedure can be applied to analyze the high-to-low switching.

The main advantages of the graphical method are:

- Reflection coefficients are not required, and hence they do not need to be calculated.
- Non-linear characteristics of the driver and receiver can be considered.

Example 5.1. Measurements and Graphical Method Applied to an Interconnection with TTL Devices

The graphical method is very useful in the case of a driver and/or receiver with non-linear characteristics [13, 14]. As an example, the point-to-point interconnect with TTL fast devices is considered as shown in Figure 5.16a. The line is a microstrip of length $l = 50$ cm, characteristic impedance $Z_0 = 84\ \Omega$, and p.u.l. propagation delay time $t_{pd} = 6.6$ ns/m. Since a TTL FAST driver has a typical rise time $t_r = 3$ ns and fall time $t_f = 1$ ns, the condition for a long line is satisfied (i.e. t_r or $t_f < 2t_{pd}l = 6.6$ ns).

The static low and high output characteristics of the driver as well as the input characteristic of the receiver were obtained by measurements and are sketched in Figure 5.16b, in accordance with the convention of current directions shown in Figure 5.16a. A comparison between calculated and measured line voltages for low-to-high and high-to-low switching is also shown in Figure 5.16b. Note that the input characteristic of the receiver, as well as the high output characteristic of the driver, is characterized by low values of current which converge to zero for $V \rightarrow \infty$ with a very slight slope, which is not visible in the figure.

The graphical method is applied here to verify the capability of the driver to ensure switching of the receiver at the arrival of the first step. This problem is particularly important when a receiver is positioned very close to the driver. To apply the graphical method easily, recall that both low and high output characteristics of the driver can be approximately described by the combination of three straight lines, as discussed in *Section 2.3.3*.

(i) Low-to-High Switching

Let us start by assuming an initial low state $V_{S0} = V_{SH0} = 0.08$ V and $I_{S0} = I_{SH0} \approx 0$, given by the intersection between the receiver input characteristic and the low output characteristic of the driver, as shown in Figure 5.16b. Note that, as $I_{S0} \approx 0$, it yields $V_{S0} = V_{SH0} \approx E_{OL}$ (see the equivalent circuit in Figure 5.14b), and Equation (5.24) reduces to $\Delta E_S = E_{OH} - E_{OL}$. To calculate the effect of the first switching step at the output of the driver, starting from a non-zero steady-state condition, Equation (5.23a) is used, where, according to the circuit in Figure 5.14c, ΔV_S is given by

$$\Delta V_S = \frac{Z_0}{R_{OAS} + Z_0}(E_{OH} - E_{OL}) \tag{5.28}$$

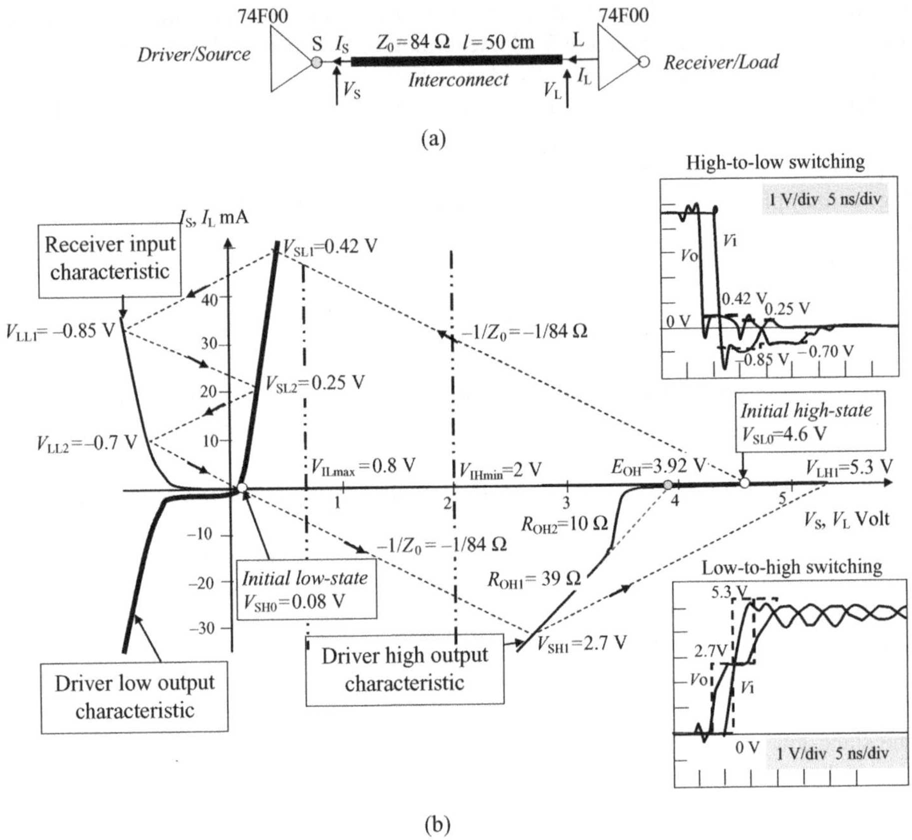

Figure 5.16 Graphical method applied to an interconnect with TTL FAST devices: (a) equivalent circuit; (b) graphical method and comparison between measured (solid line) and calculated (dashed line) voltage waveforms for high-to-low and low-to-high switching

and $R_{OAS} = R_{OH} = 39\ \Omega$, with R_{OH} being the output saturation resistance of the upper transistor (see *Chapter 2*). The high driver voltage V_{SH1} in the time interval $0 \le t \le 2T_D$ is then given by

$$V_{SH1} = Z_0 \frac{E_{OH} - E_{OL}}{R_{OH} + Z_0} + V_{SH0} = 84 \frac{3.92 - 0.08}{39 + 84} + 0.08 = 2.7\ \text{V} \tag{5.29}$$

Graphically, V_{SH1} may be calculated as the intersection between the line with slope $-1/Z_0 = -1/84$ passing through the steady-state point V_{SH0} and the line having slope $1/R_{OH1} = 1/39$ passing through the point of coordinates ($E_{OH} = 3.92$, 0). When the step voltage $\Delta V_S = V_{SH1} - V_{SH0} = 2.62$ V propagating along the line arrives at the receiver end, it doubles, giving $V_{LH1} = V_{LH0} + 2\Delta V_S = 5.3$ V, because the receiver behaves as an open circuit, and $V_{LH0} = V_{SH0}$ (see the equivalent circuit in Figure 5.14b). Graphically, the first step $V_{LH1} = 5.3$ V at the receiver may be determined as the interception of the line passing through the point with voltage V_{SH1} and having slope $1/Z_0$ with the line $I = 0$. At the same time, the receiver reflects a wave $\Delta V_S = 2.62$ V which returns towards the driver, and the driver output voltage rises to about 5 V.

It should be pointed out that, during the rise time t_r, the driver reflects three waves: two negative steps corresponding in sequence to the mismatch with the driver output impedances $R_{OH1} = 39\ \Omega$ and $R_{OH2} = 10\ \Omega$, and, after a while, a positive step due to the fact that the driver has a high output impedance for an output voltage greater than 3.6 V. These reflected waves find a high impedance at the receiver, so that other reflected waves are generated. This process lasts until voltages converge to the steady high level of 4.6 V, with the oscillations shown in Figure 5.16. These oscillations cannot be predicted by the graphical method. In any case, the capability of the graphical method to predict the first reflections is sufficient for our purposes. In fact, in practice, the interest is focused on the first step at the driver output to verify whether this step has a sufficient voltage value to cause the switching of an eventual receiver located at point S close to the driver without any delay. If $V_{SH1} < V_{IHmin} = 2$ V, the possible receiver at S would switch with a delay equal to $2t_{pd}l = 6.6$ ns (i.e. at the arrival of the wave that is reflected at the receiver end). Consequently, a slowdown in the system response would occur. If $V_{SH1} \geq V_{IHmin} = 2$ V, the receiver switches at the first step.

(ii) High-to-Low Switching

Before switching, the voltage in line is equal to the high steady-state voltage $V_{SL0} = 4.6$ V and the current $I_{SL0} \approx 0$ (see Figure 5.16b). This point is given by the intersection of the high output static characteristic of the driver and the input static characteristic of the receiver. Both characteristics have a very high equivalent resistance for voltage values greater than approximately 3.5 V. Graphically, the new voltage $V_{SL1} = 0.42$ V at the driver output after the fall time t_f is found as the intersection between the line with slope $-1/Z_0 = -1/84$ passing through the initial steady state (V_{SL0}, 0) and the output low level characteristic. The receiver voltage after the interconnect delay time T_D is given by $V_{LL1} = -0.85$ V and can be graphically obtained as the intersection between the line with slope $-1/Z_0 = -1/84$ passing through the point V_{SL1} and the input characteristic of the receiver. The other line voltages can be calculated by a similar procedure. Comparison with the measured waveforms with step values of duration $2t_{pd}l$ shows a very good agreement.

In order to have the switch at the first step of any possible receiver located at point S close to the driver, the condition $V_{SL1} < V_{ILmin} = 0.8$ V should be satisfied. If this were not the case, the switching would occur with an additional delay equal to $2t_{pd}l = 6.6$ ns. Another important application of the graphical method will be provided in *Section 5.4*, where line terminations will be discussed.

5.3 Signal Distribution Architecture

As introduced in *Section 1.1*, signal distribution is very important in designing a high-speed digital system, as the choice of interconnect structure affects the timing. Discontinuities and induced disturbances along the interconnect can enhance the total propagation delay beyond the specified limits. To avoid uncontrolled situations, it is important to choose the most suitable interconnect among defined standard structures, where signal waveforms and propagation delays can be easily evaluated.

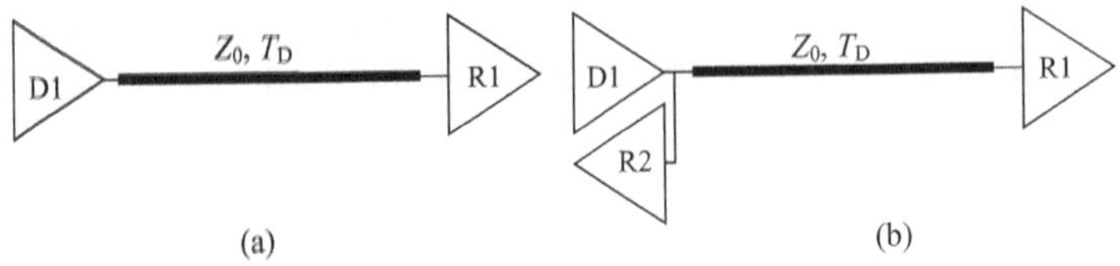

Figure 5.17 Point-to-point structure: (a) one driver and one receiver; (b) one driver and two receivers

Signal distribution in PCBs for high-speed digital systems should be realized following some basic rules in order to avoid the construction of special test boards or the need to perform detailed circuit simulations [15–22]. In general, six basic structures of interconnects should be considered: point-to-point, star, chain, bus, H-tree and comb.

5.3.1 Point-To-Point Structure

The point-to-point structure for an interconnect is the most suitable solution for ensuring the maximum speed of digital signals. It can be represented by the two configurations shown in Figure 5.17. The first consists of one driver and one receiver placed at the two ends of the line (Figure 5.17a). If the characteristic impedance Z_0 is chosen appropriately, the step signal sent by the driver doubles at the end of the line, and the receiver switches after a time delay $T_D = t_{pd}l$. The second possible point-to-point configuration consists of one driver and two receivers located as shown in Figure 5.17b. The functionality of the receiver R2 placed close to the driver D1 could be critical if the step signal sent by the driver is not sufficient in voltage to switch the receiver. When this happens, the receiver must wait for the reflected wave coming from the end of the line, and an additional delay $2T_D$ must be considered.

5.3.2 Star Structure

When a signal is to be sent to several receivers, the star structure shown in Figure 5.18 can be used. To avoid speed degradation, the length of each line must be less than the critical length l_c. In fact, in this case, the load seen by the driver is an equivalent capacitance given

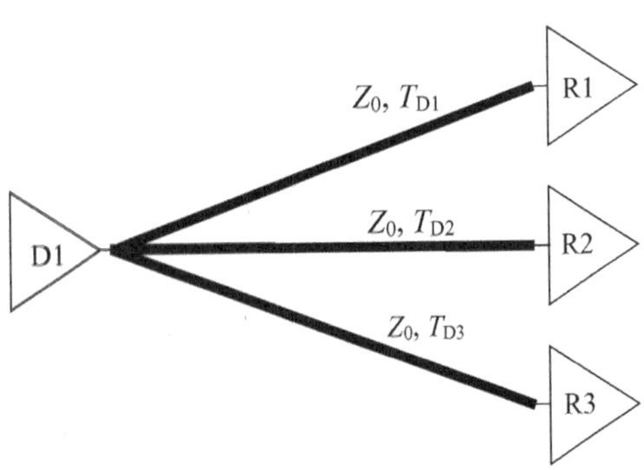

Figure 5.18 Star structure with one driver and three receivers

by the sum of the line capacitances and the receiver input capacitances. If the length of the lines were greater than the critical length, the driver would see as a load an equivalent characteristic impedance $Z_{0eq} = Z_0$/(number of branches). In some cases, Z_{0eq} could be so low that the waveform sent to the receivers is not sufficient to cause receiver switching at the first step.

5.3.3 Chain Structure

When the star configuration cannot be realized, the chain structure shown in Figure 5.19 is the most appropriate solution. All the receivers are distributed at regular intervals l_{int} and connected to the main interconnect of length l by means of stubs of length l_{stub}. To ensure good transmission performance, the length of the stub l_{stub} as well as that of the line sections l_{int} should satisfy the conditions $l_{stub} < l_c/4$ and $l_{int} < l_c/4$, where l_c is the critical length defined in *Section 5.2.4*. If these conditions are satisfied, the stub inductance can be neglected and each branch formed by the stub and the receiver has a capacitive load effect given by $C_{rec} + C_{stub}$, where C_{rec} is the receiver input capacitance of the order of 5–20 pF and C_{stub} is the capacitance of the stub. Therefore, the main interconnect can be modeled as an equivalent line with characteristic impedance Z_{0eq} and p.u.l. propagation delay time t_{pdeq} given by

$$Z_{0eq} = \sqrt{L/(C + C_d)} \tag{5.30a}$$

$$t_{pdeq} = \sqrt{L(C + C_d)} \tag{5.30b}$$

where L and C are the p.u.l. inductance and capacitance of the main line, $C_d = n(C_{rec} + C_{stub})/l$ is the distributed capacitance along the main line owing to the stub capacitance C_{stub} and the receiver capacitances C_{rec}, and n is the number of distributed receivers along the main line of length l.

Attention should be paid for avoiding values of Z_{0eq} that are too low for switching the receiver at the first step. In fact, the receivers could switch with a delay time $T_{RD} = kt_{pdeq}l$, where $k = 2, 4, \ldots$. In practical situations, t_{pdeq} could be 3 times the p.u.l. propagation delay time of the main line without loads. It will be shown in the next section that the switching of

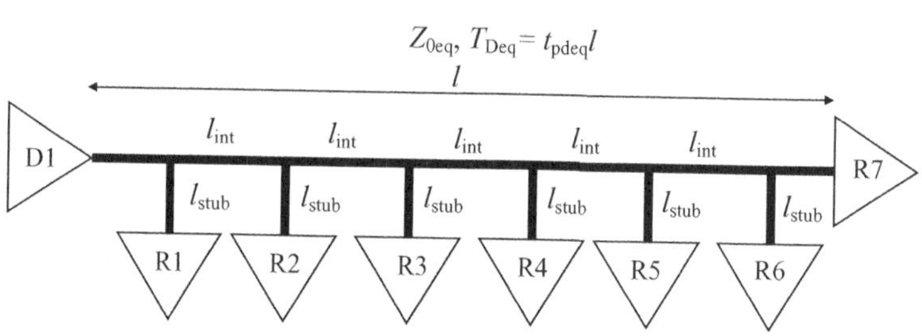

Figure 5.19 Chain structure with one driver and seven receivers distributed at regular intervals l_{int} and connected by stubs of length l_{stub} along the main line having equivalent characteristic impedance Z_{0eq} and delay $T_{Deq} = t_{pdeq}l$

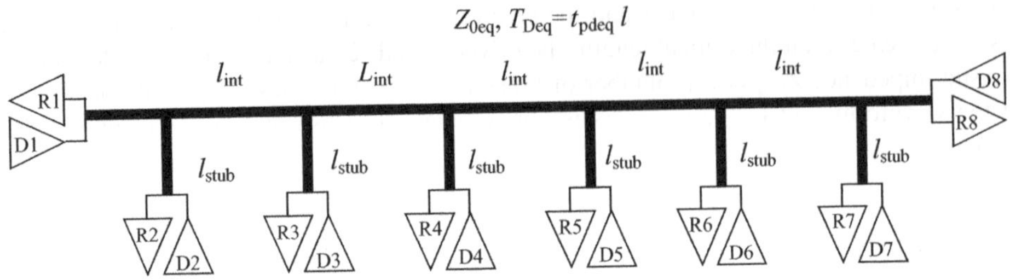

Figure 5.20 Bus structure with a couple of one driver and one receiver distributed at regular intervals l_{int} by stubs of length l_{stub} along the main line having equivalent characteristic impedance Z_{0eq} and delay $T_{Deq} = t_{pdeq} l$

the receivers at the first step can be guaranteed by the Thévenin termination as required by a clock distribution.

5.3.4 Bus Structure

Bus structures are used in data communication and are constituted by several drivers and receivers placed as shown in Figure 5.20. The loads, consisting of a driver–receiver couple, are distributed at regular intervals along the main interconnect according to the rules given for the chain structure by Equations (5.30). A critical situation of this structure is the driver in the middle of the main line, which sees a dynamic load of value $Z_{0eq}/2$. This impedance could be very low considering the capacitive load effect of the driver–receiver couple. The load is the sum of the input receiver capacitance C_{rec} and the output driver C_{dri} capacitance. The total capacitance $C_{rec} + C_{dri}$ is of the order of 5–20 pF depending on the types of device. Usually, to have a high-speed bus, a Thévenin termination is used at both ends, as will be shown in *Section 5.4*.

5.3.5 H-Tree Structure

The H-tree structure is the solution for chain and bus configurations to maintain the main interconnect with its original fundamental electric parameters Z_0 and t_{pd}. The loads (i.e. receivers or driver–receiver couples) are concentrated at one end of the main interconnect, forming a tree structure with electrically short lines of length l_{stub}, as shown in Figure 5.21. For this structure, it is very important to consider the total capacitance effect due to the concentration of the loads at one end (see *Section 5.2.2*).

5.3.6 Comb Structure

The comb structure is adopted for memory data. An example is sketched in Figure 5.22. Load separations l_{int} and stub lengths l_{stub} should follow the rules given for chain and bus structures.

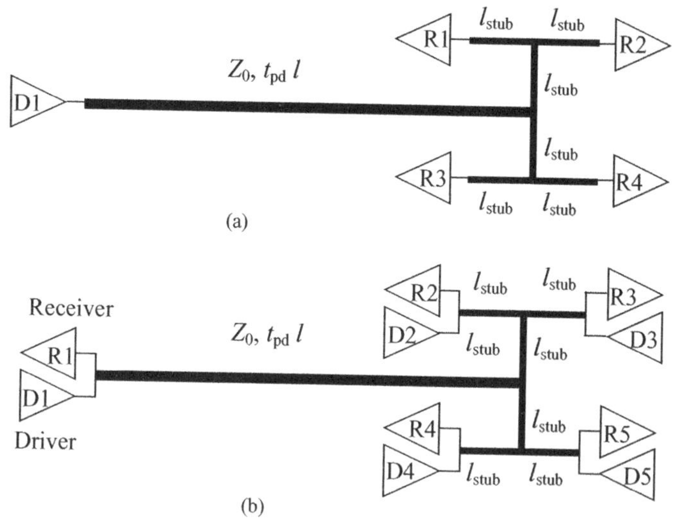

Figure 5.21 H-tree structure of a line of length l having characteristic impedance Z_0 and delay time $T_D = t_{pd}l$: (a) configuration with one driver at one end and a cluster of receivers at the opposite end; (b) configuration with a driver–receiver couple at one end and a cluster of driver–receiver couples at the opposite end

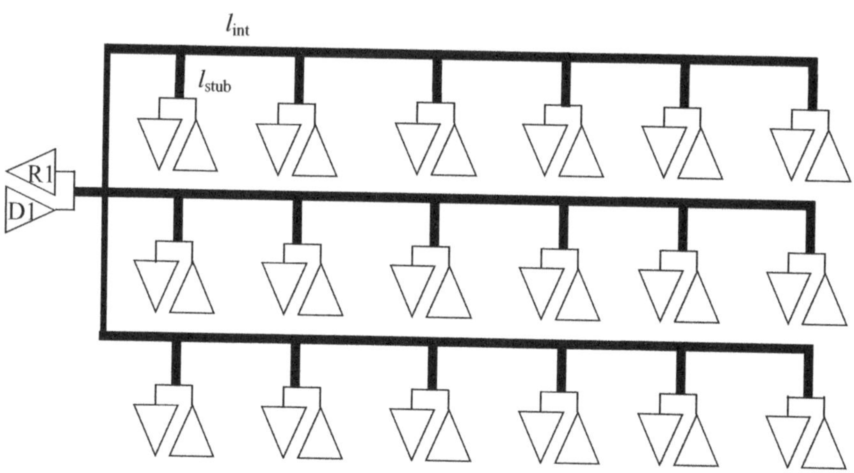

Figure 5.22 Example of a comb structure formed by three main lines of length l having characteristic impedance Z_0 and delay time $T_D = t_{pd}l$, with distributed driver–receiver couples placed at regular intervals

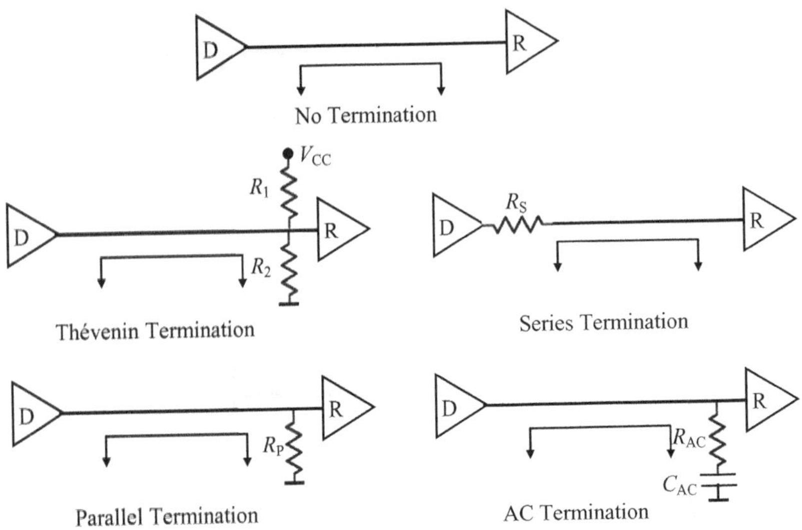

Figure 5.23 Termination schemes used to match interconnects

5.4 Line Terminations

Line terminations make it possible to realize matching conditions which are fundamental for avoiding reflections and ensuring signal integrity, as previously discussed. The four different termination schemes shown in Figure 5.23 can be used to match an interconnect [10, 15]. The choice of one termination over another depends on the type of interconnect structure (point-to-point, chain, bus, etc.) and on the power dissipation requirements.

5.4.1 Thévenin Termination

For a high-speed interconnect, the switching of a receiver at the first step voltage variation ΔV launched onto the line by the driver is a very important requirement. Thévenin termination consists of two resistances R_1 and R_2, as shown in Figure 5.23. This termination is widely used, as it enables the following goals to be achieved:

- matching of the interconnect to the characteristic impedance Z_0 in order to avoid reflections;
- enhancing the switching current of the driver, and hence guarantee of the switching of the receiver at the first step.

Low-to-high and high-to-low switching in the presence of Thévenin termination is discussed in the following by using the graphical method and by assuming an infinite input resistance of the receiver.

(i) Low-to-High Switching

Low-to-high switching can be analyzed by the equivalent circuits shown in Figure 5.24, where the Thévenin termination is modeled by a Thévenin equivalent circuit (i.e. series connection

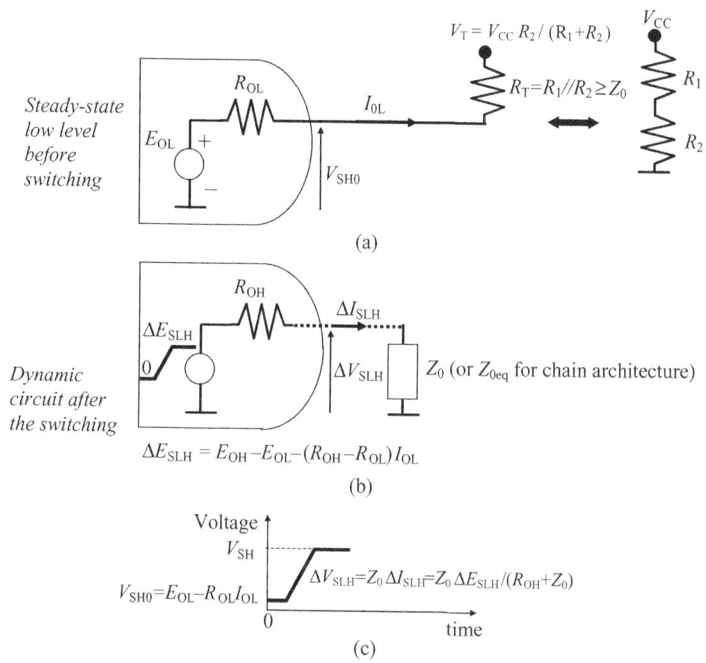

Figure 5.24 Low-to-high switching with a Thévenin termination: (a) direct current circuit for the calculation of initial steady-state low-level conditions; (b) dynamic circuit modeling the variation in voltage and current after the first switch in the time interval $0 \leq t \leq 2T_D$; (c) voltage source $V_S(t)$

of the equivalent resistance $R_T = R_1R_2/(R_1 + R_2)$ and the equivalent voltage source $V_T = V_{CC}R_2/(R_1 + R_2)$).

The steady-state initial point before the switching (i.e. V_{SH0} and I_{OL}) is obtained by the equivalent circuit of Figure 5.24a as

$$I_{OL} = (E_{OL} - V_T)\big/(R_{OL} + R_T) \tag{5.31a}$$

$$V_{SH0} = E_{OL} - R_{OL}I_{OL} \tag{5.31b}$$

where I_{OL} is the negative current in line before switching, V_{SH0} is the initial low-level steady-state output voltage, E_{OL} is the output voltage of the driver at low level without loads, and R_{OL} is the driver output resistance at low voltage in low steady-state condition.

The first voltage step can be computed by the equivalent circuit of Figure 5.24b as

$$\Delta V_{SLH} = Z_0\ \Delta I_{SLH} = \frac{Z_0}{R_{OH} + Z_0}\Delta E_{SLH} \tag{5.32}$$

where ΔI_{SLH} is the positive current variation in line after the low-to-high switching, Z_0 is the line characteristic impedance, E_{OH} is the output voltage of the driver in high steady-state condition, R_{OH} is the output resistance of the driver in high steady-state condition, and

$\Delta E_{SLH} = E_{OH} - E_{OL} - (R_{OH} - R_{OL})I_{OL}$ is the dynamic circuit excitation for the low-to-high switching according to Equation (5.24).

After switching to the high state, the line current I_{SH} can be found by simple manipulation and is given by

$$I_{SH} = \frac{Z_0 I_{OL} + E_{OH} - V_{SH0}}{Z_0 + R_{OH}} \tag{5.33}$$

The receiver will switch if the following condition is satisfied:

$$V_{SH} = V_{SH0} + \Delta V_{SLH} = E_{OL} - R_{OL} I_{OL} + \Delta V_{SLH} \geq V_{IHmin} \tag{5.34}$$

where V_{IHmin} is the threshold guaranteed by the data sheet for the receiver to recognize a high-level bit. In steady-state condition, before low-to-high switching, the relation to be satisfied is $E_{OL} - R_{OL} I_{OL} \leq V_{OLmax}$, so that the total immunity noise $V_{ILmax} - V_{OLmax}$ is preserved.

The benefit provided by Thévenin termination compared with the configuration with no termination can be easily highlighted by the graphical method. Consider the linear output characteristics of a driver in low and high state, shown in Figure 5.25. According to Equations (5.31) and the equivalent circuit in Figure 5.24a, the initial low steady-state voltage V_{SH0} and current I_{OL} are given by the intersection between the driver low output characteristic, defined by E_{OL} and R_{OL}, and the line representing the Thévenin termination defined by the Thévenin equivalent circuit parameters V_T and R_T.

The voltage and current in line, V_{SH} and I_{SH}, just after the low-to-high switching, can be computed as the intersection between the driver high output characteristic and the line with slope $-1/Z_0$ passing through the point (V_{SH0}, I_{OL}).

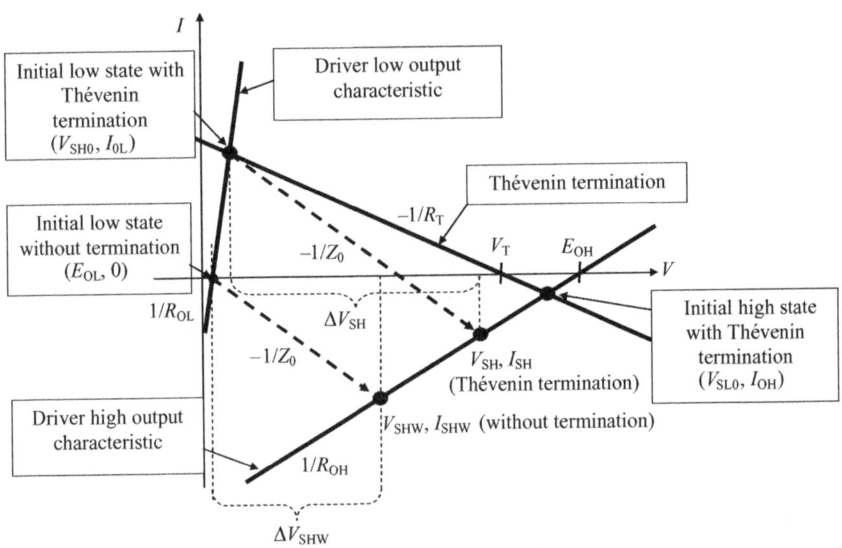

Figure 5.25 Graphical method applied to calculate reflections with Thévenin termination

It is interesting to note that, without termination, the voltage in line V_{SHW} (i.e. the intersection between the driver high voltage characteristic and the line with slope $-1/Z_0$ passing through the point $(E_{OL}, 0)$) is much lower than V_{SH} (see Figure 5.25). This shows the advantage of using a Thévenin termination to ensure switching at the arrival of the first step.

(ii) High-to-Low Switching
For high-to-low switching, similar considerations apply. In this case

$$\Delta V_{SHL} = Z_0 \Delta I_{SHL} = \frac{Z_0}{R_{OL} + Z_0} \Delta E_{SHL} \tag{5.35}$$

where $\Delta E_{SHL} = E_{OL} - E_{OH} - (R_{OL} - R_{OH})I_{OH}$ is the dynamic circuit excitation for the high-to-low switching according to (5.24).

The receiver will switch when

$$V_{SL} = V_{SL0} + \Delta V_{SHL} = E_{OH} - R_{OH} I_{OH} + \Delta V_{SHL} \leq V_{ILmax} \tag{5.36}$$

where V_{SL0} is the initial high-level steady-state output voltage before the switching, I_{OH} is the steady-state current in line that corresponds to the initial high state before switching, and V_{ILmax} is the threshold guaranteed by the data sheet for the receiver to recognize a low-level bit.

In the steady-state condition, before high-to-low switching, the equation to be satisfied is $E_{OH} - R_{OH} I_{OH} \geq V_{OHmin}$, so that the total immunity noise $V_{OHmin} - V_{IHmin}$ is preserved.

The voltage V_{SL} and current I_{SL} in line just after the switching from high to low levels may be computed in the same manner as the intersection between the low voltage driver output characteristic and the line with slope $-1/Z_0$ passing through the point (V_{SL0}, I_{OH}).

5.4.2 Series, Parallel, and AC Terminations

Series, parallel, and AC terminations can be treated by analogy with the Thévenin termination discussed in the previous section. All the formulae can be used, taking into account the following considerations:

- *Series termination* – the series resistance R_S must be added to the driver output resistances R_{OH} and R_{OL}, while $V_T = 0$ and $R_T = 0$.
- *Parallel termination* – in this case, $V_T = 0$ and $R_T = Z_0$.
- *AC termination* – in this case, $V_T = 0$ and R_T should be replaced by $Z_T = R_{AC} + 1/(j\omega C_{AC})$.

5.4.3 Series Termination and Comparison with Other Terminations by Circuit Simulations

Series termination is used to avoid power dissipation in steady-state condition. The line is matched at the driver side, choosing a resistance R_S in order to satisfy the equation $R_{out} + R_S = Z_0$, where $R_{out} = (R_{OL} + R_{OH})/2$ is the driver output average resistance.

The first step sent into the line is affected by the presence of the resistance R_S, and therefore less current I_{SH} and I_{SL} is injected into the line. Very often the first step is not able to switch

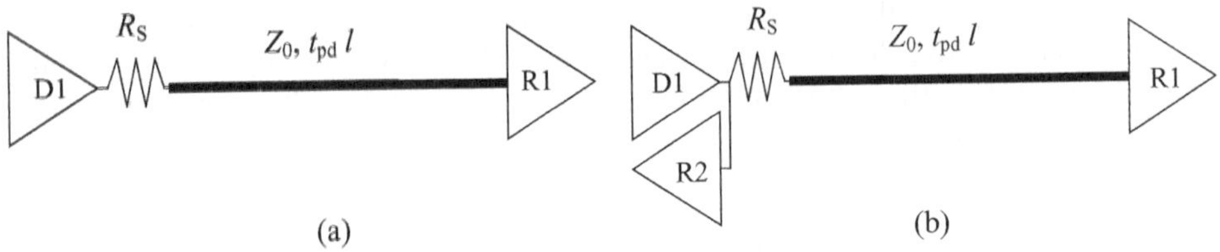

Figure 5.26 Series termination: (a) point-to-point structure with one driver D1 and one receiver R1; (b) point-to-point structure with an additional receiver R2 at the end of the driver

receivers distributed along the line, and therefore the chain structure with series termination should be avoided for very high-speed interconnects. When the step doubles at the end of the line, generally the receiver is able to switch. The reflected wave generated at the receiver stops at the driver end owing to the matching condition. As for parallel termination, static conditions before switching must be verified in order to preserve the static noise immunity requirements. Figure 5.26 shows two possible point-to-point structures with series termination. Note that for the second case the first receiver must be located very close to the driver and before the series resistance to avoid extra delay.

A comparison of different terminations (Thévenin, series, and AC) is shown in Figure 5.27. The reference is the unmatched line of Figure 5.27a. The driver considered here is the TTL point-to-point structure analyzed graphically in *Section 5.2*, adopting a linear output characteristic. Pulse source E_S switches from $E_{OL} = 0.08$ V to $E_{OH} = 4$ V in 1 ns, and at the same time the driver output resistance R_{out} switches from $R_{OL} = 6.8\ \Omega$ to $R_{OH} = 39\ \Omega$. In this way, a low-to-high logic level switching is simulated. The line has a delay time $T_D = t_{pd}l = 3$ ns and a characteristic impedance $Z_0 = 84\ \Omega$. Thévenin termination is realized with two resistances $R_1 = 140\ \Omega$ and $R_2 = 210\ \Omega$ in order to have $R_1//R_2 = Z_0$, and $V_T = V_{CC}R_2/(R_1 + R_2) = 3$ V, seeing as $V_{CC} = 5$ V. Series termination is a resistance $R_S = Z_0 - (R_{OL} + R_{OH})/2 = 61.1\ \Omega$.

From the simulated waveforms, the following observations can be derived:

- *Unmatched line* – the presence of reflections and a first step of 2.7 V at the driver output, as computed by Equation (5.34), as well as being shown by the graphical method and measurement in *Section 5.2*.
- Thévenin termination – the absence of reflections and the first step enhanced at 3.7 V, as computed by Equation (5.34).
- Series termination – no significant reflections at the receiver location and a first step of 1.9 V under the threshold of 2 V. Therefore, an eventual receiver located after R_S switches after a delay of $2T_D = 6$ ns (that is, switching at the second step).
- *RC termination* – the absence of reflections but no improvement in the first step and lower immunity at steady-state condition.

5.4.4 Thévenin Termination Applied to Chain Structures and Circuit Simulations

Thévenin termination is often used in chain structures (see Figure 5.28). This termination must be positioned very close to the last receiver, and the other receivers must be distributed

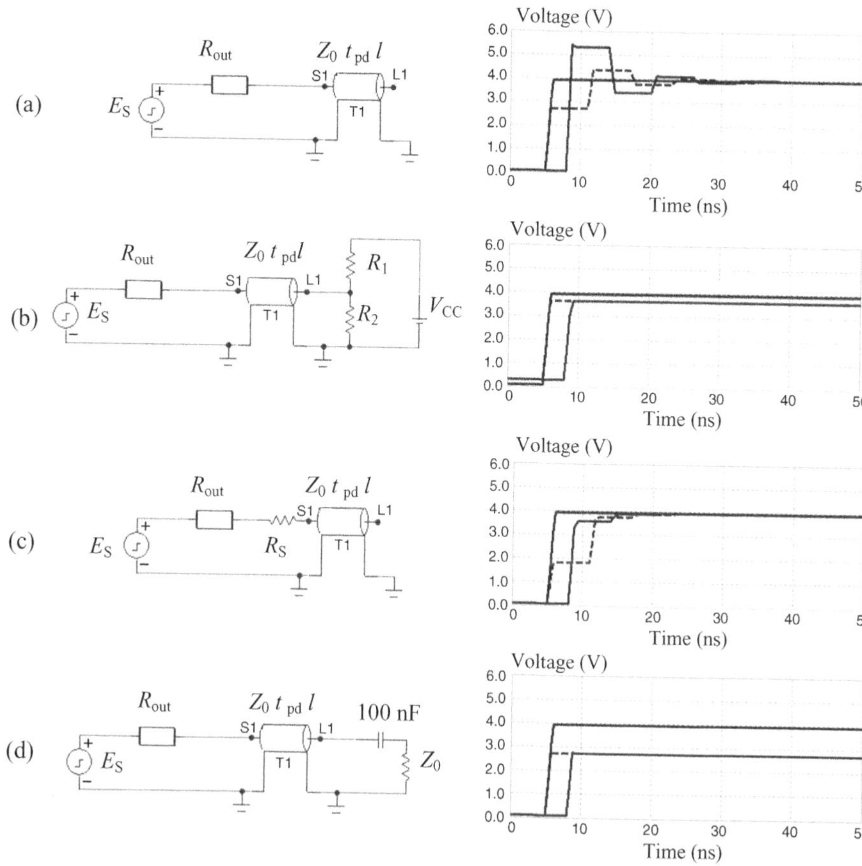

Figure 5.27 Simulated waveforms at source E_S (solid line), voltage at the input of the line (dashed line), and voltage at the end of the line (solid line) for cases of (a) an unmatched line, (b) Thévenin termination, (c) series termination, and (d) RC termination

very close to the main line. As stated in *Section 5.2.4*, the conditions l_{stub} and $l_{int} < l_{crit}/4$ should be verified in order to have a structure similar to a point-to-point one with a lower equivalent characteristic impedance Z_{0eq}. Because of the high power dissipation at steady-state condition, Thévenin termination is not used within the CMOS logic family, as the main advantage offered by this technology is lost.

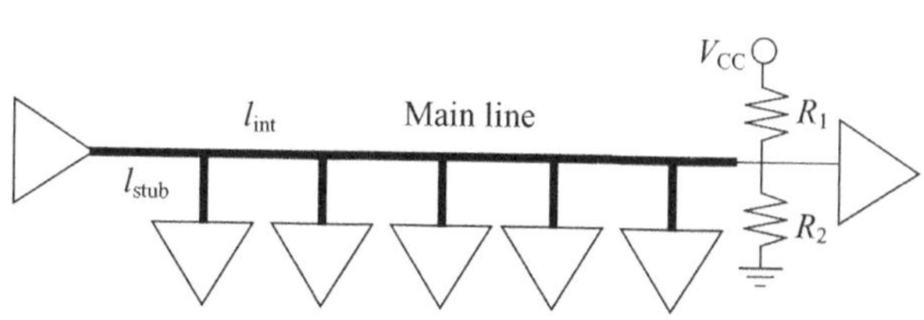

Figure 5.28 Chain structure with Thévenin termination

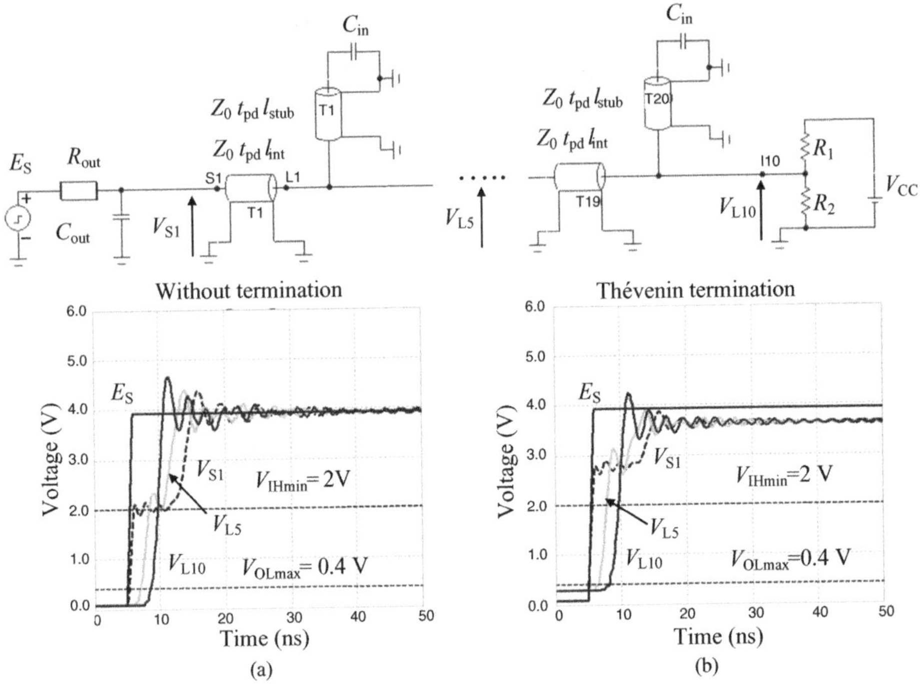

Figure 5.29 Simulated waveforms of a chain structure: (a) without termination; (b) with Thévenin termination

To verify the advantage of using Thévenin termination for chain structures, consider the example shown in Figure 5.29. The same TTL device as used before drives ten stubs with the following data: $Z_0 = 84\ \Omega$ and $t_{pd} = 6$ ns/m for the main line and stubs, $l_{int} = l_{stub} = 3$ cm, $C_{out} = 10$ pF (driver output capacitance), $C_{in} = 5$ pF (receiver input capacitance), $R_1 = 140\ \Omega$, $R_2 = 210\ \Omega$, $V_{CC} = 5$ V. This means that the main line has an equivalent characteristic impedance $Z_{0eq} = \sqrt{Ll_{int}/(Cl_{int} + Cl_{stub} + C_{in})} = 40\ \Omega$, where L and C are the p.u.l. inductance and capacitance of the main line and stubs. It should be noted that, without Thévenin termination, the switching of all the receivers at the first step is not guaranteed. With Thévenin termination, the first step V_{SH} is higher and may be computed by Equation (5.34), resulting in $V_{SH} = 2.8$ V, as given by the simulation.

5.4.5 Series Termination Applied to Chain Structures and Circuit Simulations

Series termination is not recommended for high-speed chain structures. This can be explained by the following example concerning the configuration shown in Figure 5.30 which, with the exception of the termination, is the same as that used in *Section 5.4.4* for Thévenin termination. In this case the series resistance is assumed to be $R_S = (Z_{0eq} - (R_{OH} + R_{OL})/2)$. From the simulated waveforms shown in Figure 5.31 it can be noted that series termination mitigates reflections at a high logic level but makes the interconnection slower.

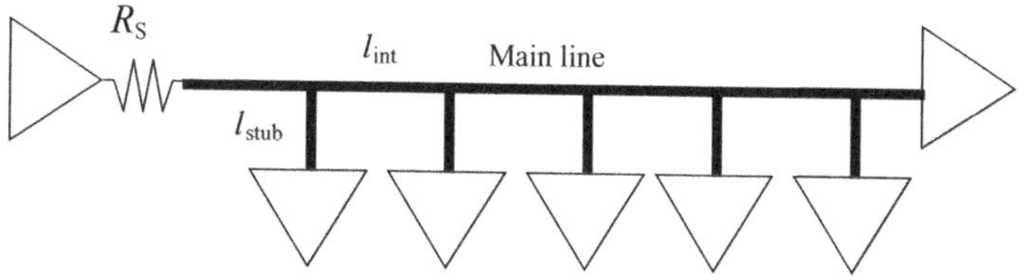

Figure 5.30 Chain structure with series termination

5.4.6 *Thévenin Termination Applied to Bus Structures and Circuit Simulations*

With a bus structure (see Figure 5.32), the Thévenin termination must be positioned at both ends. Resistance values must be doubled to have the required low and high static voltages. The distribution of the driver/receiver along the main line must be done as for chain structures. It is important to note that, for a driver switching in the middle of the structure, the first step must be computed according to the equation $\Delta V = (Z_{0eq}/2)\Delta I$.

With bus Thévenin termination, two extreme cases can occur: driver switching in the middle and driver switching at one end. Figure 5.33 shows the simulated waveforms obtained using the same geometrical and electrical parameters as those considered in the chain example.

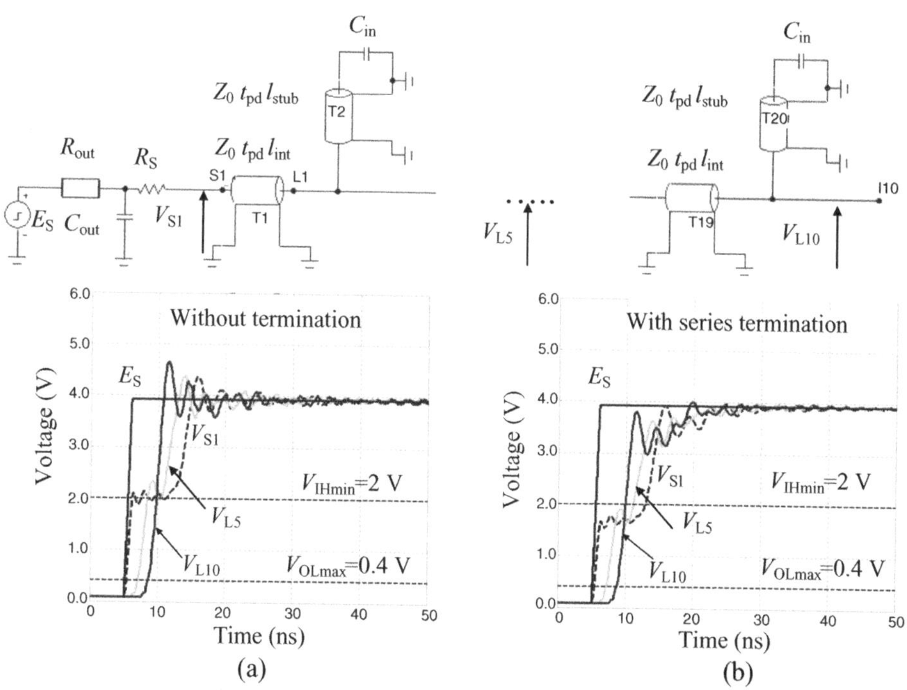

Figure 5.31 Simulated waveforms of a chain structure: (a) without termination; (b) with series termination

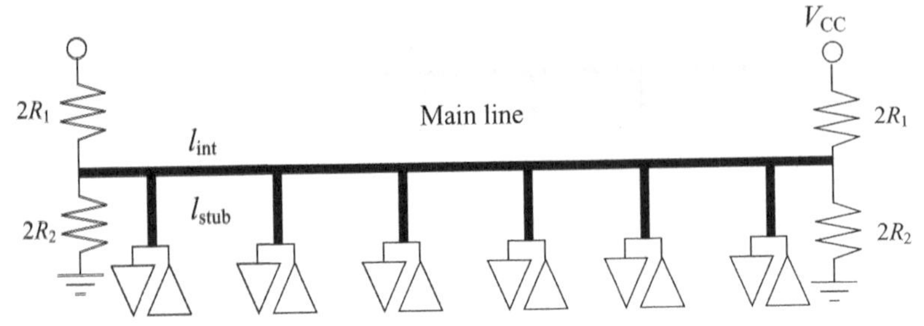

Figure 5.32 Bus structure with Thévenin termination

With the driver in the middle, the voltage in line at the driver output, V_H, may be computed by Equation (5.34) using $Z_{0eq}/2$ instead of Z_{0eq}, and it is found that $V_H = 2$ V.

With the driver at one end, V_H can be computed by Equation (5.34), taking into account that the output equivalent circuit of the driver is modified by the presence of the termination V_T and $2R_T$. For example, for low-level output it yields $E_{OLeq} = E_{OL} + R_{OL}(V_T - E_{OL})/(2R_T + R_{OL})$ and $R_{OLeq} = R_{OL}/(2R_T)$. Similar expressions can be obtained for E_{OHeq} and R_{OHeq}. Making these changes in Equation (5.34) yields $V_H = 2.4$ V, as given by the simulation.

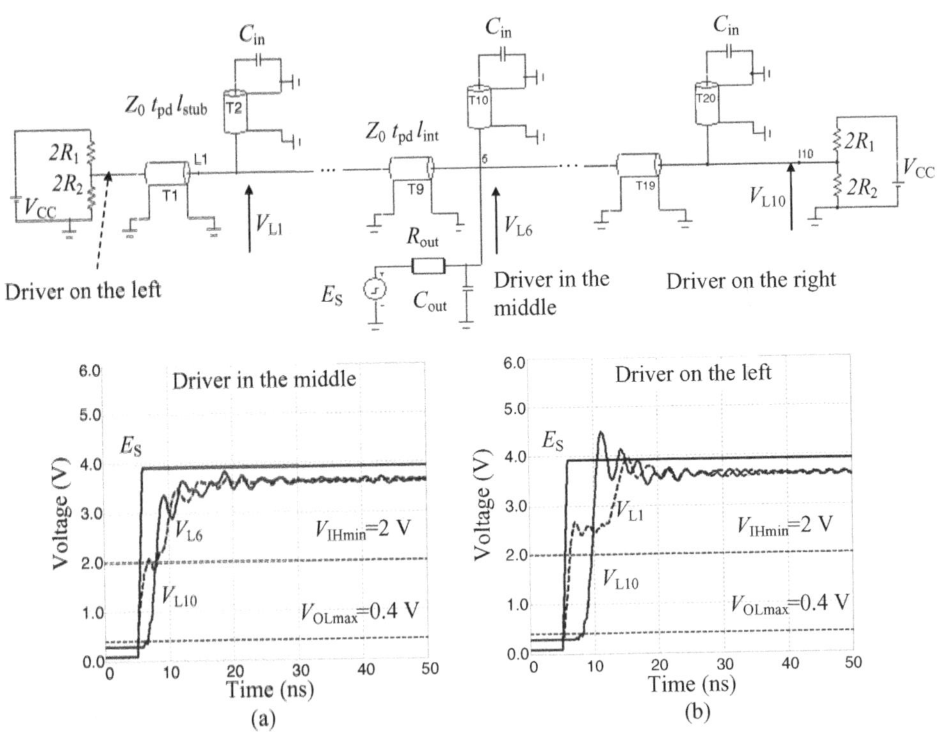

Figure 5.33 Simulated waveforms of a bus structure: (a) driver in the middle; (b) driver on the left.

5.4.7 Termination and Interconnection Structures

In conclusion, the following observations can be made about signal distribution structures and terminations:

Thévenin termination should be applied in:

- chain and bus structures with bipolar devices (Fast-ABT-ECL).

RC termination should be applied in:

- chain and bus structures with CMOS.

Series termination should be applied in:

- Point-to-point structures, with the driver having high capability.
- H-tree structures.
- Chain structures when switching of the receivers at the first step is not a requirement.
- Star structures.
- Structures in which it is desirable to minimize radiated emission.
- Structures with unknown Z_0.

5.4.8 Termination Performance

Several observations concerning the performance of terminations are summarized below:

Series termination:

- Reflections are reduced.
- There is no power dissipation at steady-state condition.
- The first step is not improved.
- The immunity of the receivers is reduced.
- The duty cycle is modified.

Thévenin termination:

- The first step is enhanced.
- Reflections are reduced.
- The duty cycle does not change.
- The static power dissipation is high.
- There is power dissipation at steady-state condition.

RC termination:

- Static dissipation is absent.
- Power is not required.
- The first step is not improved.
- The value of C must be chosen with allowance for the type of signal (clock or data).
- Dynamic dissipation is high.

References

[1] Ramo, S., Whinnery, J., and Van Duzer, T., *'Fields and Waves in Communication Electronics'*, 3rd edition, John Wiley and Sons, Inc., New York, NY, 1994.

[2] Davidson, C.W., *'Transmission Lines for Communications'*, The Macmillan Press Ltd, London, 1978.

[3] Hall, S., Hall, G., and McCall, J., *'High-Speed Digital System Design – a Handbook of Interconnect Theory and Design Practice'*, John Wiley & Sons, Inc., New York, NY, 2000.

[4] Young, B., *'Digital Signal Integrity: Modeling and Simulation with Interconnects and Packages'*, Prentice Hall PTR, Upper Saddle River, NJ, 2001.

[5] Paul, C., *'Introduction to Electromagnetic Compatibility'*, John Wiley & Sons, Inc., Hoboken, NJ, 2006.

[6] Johnson, H. and Graham, M., *'High-Speed Signal Digital Design: a Handbook of Black Magic'*, Prentice Hall PTR, Upper Saddle River, NJ, 1993.

[7] Montrose, M., *'EMC and the Printed Circuit Board: Design, Theory, and Layout Made Simple'*, IEEE, Inc., New York, NY, 1999.

[8] 'Data Transmission Lines and their Characteristics', AN-806, National Semiconductor, March 1992.

[9] Janisz, T. and Martin, R., 'Interconnection of High-speed Logic Circuits', *IEEE Trans. on Computers*, **19**(9), September 1970, 831–837.

[10] DeFalco, J., 'Reflection and Crosstalk in Logic Circuit Interconnections', *IEEE Spectrum*, **7**, July 1970, 44–50.

[11] Horna, O., 'Pulse Reflection in Transmission Lines', *IEEE Trans. on Computers*, **20**(12), December 1971, 1558–1563.

[12] Branin, F., 'Transient Analysis of Lossless Transmission Lines', *Proceedings of the IEEE*, **55**, November 1967, 2012–2013.

[13] Morris, R.L. and Miller, J.R. (eds), *'Designing with TTL Integrated Circuits'*, McGraw-Hill, New York, NY, 1971.

[14] Horna, O., 'Nonlinear Termination of Transmission Lines', *IEEE Trans. on Computers*, **21**(9), September 1972, 1011–1015.

[15] Blood, W., *'MECL System Design Handbook'*, 4th edition, Motorola Semiconductor Products, Inc., Phoenix, AZ, 1983.

[16] DeFalco, J., 'Reduce Errors in Data Transmission between Computer and Peripheral Input/Output Devices Using Properly Designed I/O Bus Circuits', *Electronic Design*, 4 February 1972, 84–86.

[17] Malaviya, S.D. and Singh, V.P., 'Transmission Lines Loaded at Regular Intervals', *IEEE Trans. on Microwave Theory and Techniques*, **27** (10), October 1979, 854–859.

[18] 'Reflections: Computations and Waveforms', National Semiconductor, AN-807, March 1992.

[19] 'GTLP: Interface Technology for Bus and Backplane Applications', AN-1065, Fairchild Semiconductor, 2001.

[20] 'ECL Backplane Design', AN-768, Fairchild Semiconductor, 2000.

[21] 'Transmission-line Effects Influence High-Speed CMOS', AN-393, Fairchild Semiconductor, 1985.

[22] 'FAST/FASTr Design Considerations', AN-661, Fairchild Semiconductor, 1990.

6

Crosstalk

Owing to the high density of traces in high-speed printed circuit boards and the fast rise and fall time of the switching devices, the electromagnetic coupling between adjacent lines, defined as *crosstalk*, is a very important topic. To cope with this problem, it is essential to have suitable circuit models to simulate complicated structures. Simplified but accurate models are of great help in understanding the *crosstalk* mechanism. This chapter starts with the description and discussion of a lumped model of two coupled lines that can be easily implemented in SPICE-like circuit simulators. By virtue of its simplicity, this is the first approach for inexperienced users when dealing with *crosstalk* and using a commercial circuit simulator that usually does not offer efficient coupled line models in its library.

In high-speed digital circuits, the analysis of symmetrical coupled lines in differential signal transmission is very often required to predict *crosstalk* or *signal integrity* (SI). The concept of even or *common mode* (CM) and of odd or *differential mode* (DM) will be introduced. This distinction will also be very useful in understanding the radiated emission mechanism from PCBs with attached cables, and EMI performance of differential signaling. An efficient distributed model based on *common mode* and *differential mode* will be presented for *crosstalk* computation. The main advantage of this model is that it is exact and provides results without the frequency limitation of a lumped model.

The model based on *common* and *differential* propagation modes is used to simulate *crosstalk* in the presence of digital devices. For accurate predictions it is fundamental to have macromodels for drivers and receivers that take into account the non-linearity of the devices and speed up the simulations. Although the procedure is presented for TTL devices, it is absolutely general and can be used for other families of digital devices. The point-to-point and bus structures are considered, and the results of simulations are compared with measurements.

A general distributed model for two or more n coupled lines, symmetrical or not, is also presented. The model is based on n decoupled modes of propagation, each characterized by its own characteristic impedance and propagation delay time. The theoretical derivation of the model is left to the referenced papers. The attention here is focused on the model implementation in SPICE simulators. Examples of simulations with five coupled lines driven and loaded with TTL and CMOS devices are given. The circuit model is validated by comparing the simulation results with measurements.

Signal Integrity and Radiated Emission of High-Speed Digital Systems Spartaco Caniggia and Francescaromana Maradei

The chapter ends with an overview of the main techniques to mitigate *crosstalk*. Some examples on the implementation of these techniques are provided and discussed by using SPICE and numerical tools.

6.1 Lumped-Circuit Model of Coupled Lines

A pair of symmetrical lines in actual PCBs and cables is a widely used configuration with which to cope. This simple structure is considered to introduce some formulations useful for *Section 6.2* where an exact model based on common and differential modes will be outlined. However, the discussion that follows concerning magnetic and electric coupling is absolutely general.

6.1.1 Equivalent Circuit of two Coupled Lines with a Reference Ground

In the case of an electrically short line (i.e. line length much shorter than the minimum wavelength of interest, $l < \lambda/10$), the interconnect can be modeled by the equivalent half-T circuit shown in Figure 6.1 [1]. This equivalent circuit will be referred to in the following as the *elementary cell*. The per-unit-length (p.u.l.) inductance **L** and capacitance **C** matrices are defined as

$$\mathbf{L} = \begin{bmatrix} L_{11} & L_{12} \\ L_{21} & L_{22} \end{bmatrix} \tag{6.1a}$$

$$\mathbf{C} = \begin{bmatrix} C_{11} & C_{12} \\ C_{21} & C_{22} \end{bmatrix} \tag{6.1b}$$

where

$$L_{11} = L_{22} = L_w, \quad L_{12} = L_{21} = L_m \tag{6.1c}$$

$$C_{11} = C_{22} = c_0 + c_m, \quad C_{12} = C_{21} = -c_m \tag{6.1d}$$

and L_w is the wire self inductance, L_m is the mutual inductance between the two wires, c_0 is the wire-to-ground capacitance, and c_m is the mutual capacitance between the two wires.

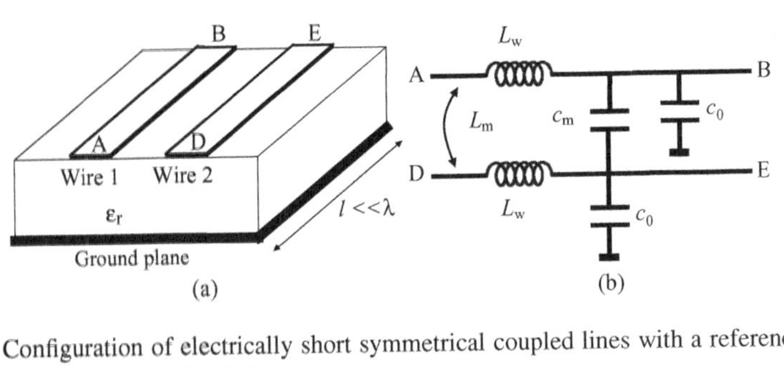

Figure 6.1 Configuration of electrically short symmetrical coupled lines with a reference ground (a) and corresponding equivalent circuit (b)

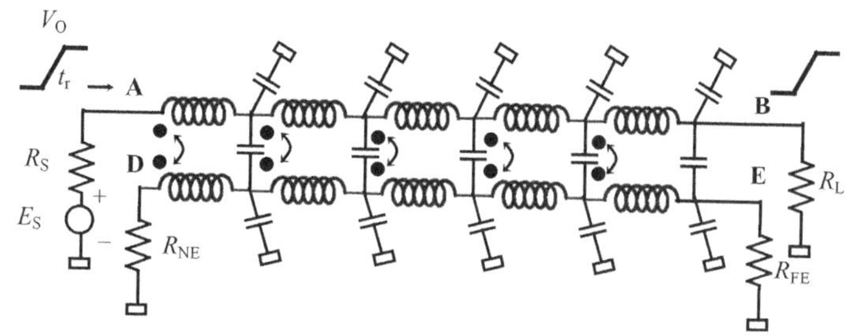

Figure 6.2 Equivalent circuit of two long coupled lines

Note that in Equations (6.1), while the inductance matrix coefficients have physical meaning, this is not the case for the capacitance matrix coefficients which are related to the physical wire-to-ground and wire-to-wire capacitances indicated by lower-case letters. This is the reason why in the circuit of Figure 6.1 the inductances appear with capital letters while the capacitances are in lower case.

When an interconnection is electrically long (i.e. l comparable with or even longer than the wavelength λ), it can be modeled as a series cascade of elementary cells, as shown in Figure 6.2 [2].

In order to illustrate the *crosstalk* phenomenon, the terminations shown in Figure 6.2 are considered. The first trace between points A and B is fed at end A by a source consisting of a voltage E_S in series with a resistance R_S and is terminated at end B on the load R_L. The second trace between points D and E is terminated at both ends on resistive loads indicated as *Near-End* (NE) resistance R_{NE} and *Far-End* (FE) resistance R_{FE} respectively. In this case, the first trace represents the signal line, while the second one represents the victim where the interference from *crosstalk* occurs. The most important parameters to be predicted in a *crosstalk* analysis are:

- *Near-End Crosstalk* (NEXT), defined as the voltage V_{NE} at point D of the victim line near to the source;
- *Far-End Crosstalk* (FEXT), defined as the voltage V_{FE} at point E of the victim line far from the source.

The step voltage launched onto the signal line by the voltage source E_S has step amplitude V_O and rise time t_r. Often, owing to the actual separation of the two coupled lines in a PCB, the weak coupling assumption can be adopted. This means that the effect of the victim line on the signal line can be neglected, and the characteristic impedance $Z_0 = (L_w/(c_0 + c_m))^{1/2}$ can be approximately associated with each line, as $c_m \ll c_0$. The amplitude of the step voltage launched onto the line is $V_O \approx E_S Z_0/(R_S + Z_0)$, where V_O is the signal step. The weak coupling assumption is used in the next sections to present an intuitive discussion of the effects of capacitive and inductive *crosstalk*.

6.1.2 Capacitive Coupling

In this subsection, the attention is focused on capacitive coupling only [2, 3]. To understand better the capacitive coupling mechanism, the equivalent circuit of an elementary cell of length

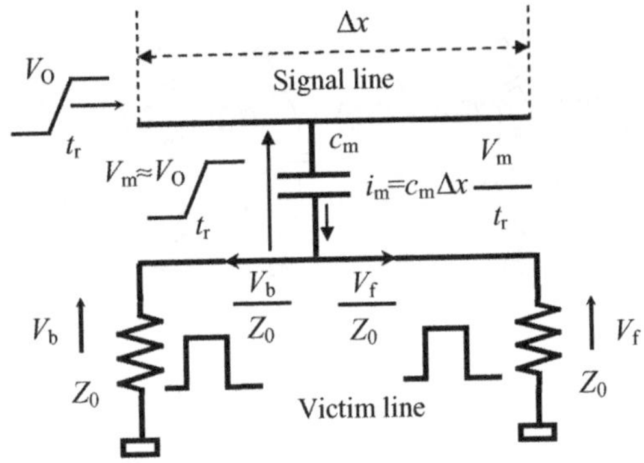

Figure 6.3 Capacitive coupling

Δx located along the two coupled lines is considered, and the contribution of the self and mutual inductances is neglected, as shown in Figure 6.3. As, under the weak coupling assumption, $c_m \ll c_O$, the voltage step across the mutual capacitance is $V_m \approx V_O$. As the voltage wave V_O launched by the source reaches the signal line segment under consideration, it injects current onto the victim line through the mutual capacitance c_m. Kirchhoff's current law applied to the victim line yields

$$\frac{V_b}{Z_0} + \frac{V_f}{Z_0} = c_m \, \Delta x \frac{V_m}{t_r} \approx c_m \Delta x \frac{V_O}{t_r} \tag{6.2}$$

where V_b and V_f are the backward and forward voltage waves on the victim line, and it is assumed that $dV/dt \approx \Delta V/\Delta t = V_m/t_r$ because V_O represents the signal voltage step with rise time t_r. The voltage is continuous, and therefore the backward V_b and forward V_f voltages are

$$V_b = V_f = \frac{Z_0 c_m \, \Delta x V_O}{2t_r} \tag{6.3}$$

The signal wave creates pulses having widths about equal to the rise time t_r and propagating in opposite directions on the victim line. Negative pulses are created in the case of high-to-low transition. This mechanism is reproduced for each segment Δx of the two coupled lines. The signal voltage V_O and the forward voltage V_f travel together towards the far end. Therefore, at the far end of the line of length l, the total noise is a single pulse of width about equal to t_r and of amplitude given by

$$V_{FEcap} = \frac{Z_0 c_m l V_O}{2t_r} \tag{6.4}$$

The signal and the backward wave on the victim line travel in opposite directions, so the overlap where the signal can inject current is only one-half of the rise time t_r. After this period, the pulse travels unchanged to the near end. These pulses are generated continuously and, if both lines are matched, the last pulse at the near end arrives after a time $2t_{pd}l$, where t_{pd} is the p.u.l. propagation delay time of both lines in the weak coupling condition. The interaction distance in Equation (6.3) is $\Delta x = \frac{1}{2} v_p \, \Delta t = t_r/(2t_{pd})$, where the phase velocity $v_p = 1/t_{pd}$.

Substituting Δx into Equation (6.3), the *near-end noise* is given by

$$V_{\text{NEcap}} = \frac{Z_0 c_{\text{m}} V_{\text{O}}}{4t_{\text{pd}}} \tag{6.5}$$

However, the following condition holds: $Z_0/t_{\text{pd}} = (L_{\text{w}}/C)^{1/2}/(L_{\text{w}}C)^{1/2} = 1/C$, where $C = c_0 + c_{\text{m}}$ is the p.u.l. capacitance of each line under the adopted weak coupling assumption. Therefore, Equation (6.5) becomes

$$V_{\text{NEcap}} = \frac{c_{\text{m}} V_{\text{O}}}{4C} \tag{6.6}$$

6.1.3 Inductive Coupling

Consider now the equivalent circuit of an elementary cell of length Δx of the two coupled line, neglecting the contribution of the line capacitances [2, 3]. The equivalent circuit to calculate the backward V_{b} and forward V_{f} voltages induced in the victim line elementary cell is shown in Figure 6.4. As the current wave launched by the source passes on the signal line segment under consideration, it injects voltages onto the victim line through the mutual inductance L_{m}. The voltage in the victim is represented in Figure 6.4 by a current-controlled voltage source. Kirchhoff's voltage law applied to the victim line yields

$$V_{\text{b}} = L_{\text{m}}\,\Delta x \frac{\text{d}I}{\text{d}t} + V_{\text{f}} \approx \frac{L_{\text{m}}\,\Delta x I_{\text{O}}}{t_{\text{r}}} + V_{\text{f}} \tag{6.7}$$

where the current time derivative is approximated as $\text{d}I/\text{d}t \approx \Delta I/\Delta t = I_{\text{O}}/t_{\text{r}}$, where I_{O} is the signal current step with rise time t_{r}. The assumptions are the same as those used for capacitive coupling, and $I_{\text{O}} = V_{\text{O}}/Z_0$. As the currents are continuous (i.e. $V_{\text{b}}/Z_0 = -V_{\text{f}}/Z_0$), the following amplitudes of backward and forward waves are obtained by eliminating V_{f} and substituting $I_{\text{O}} = V_{\text{O}}/Z_0$ in Equation (6.7):

$$V_{\text{b}} = \frac{L_{\text{m}}\,\Delta x V_{\text{O}}}{2Z_0 t_{\text{r}}} \tag{6.8a}$$

$$V_{\text{f}} = \frac{-L_{\text{m}}\,\Delta x V_{\text{O}}}{2Z_0 t_{\text{r}}} \tag{6.8b}$$

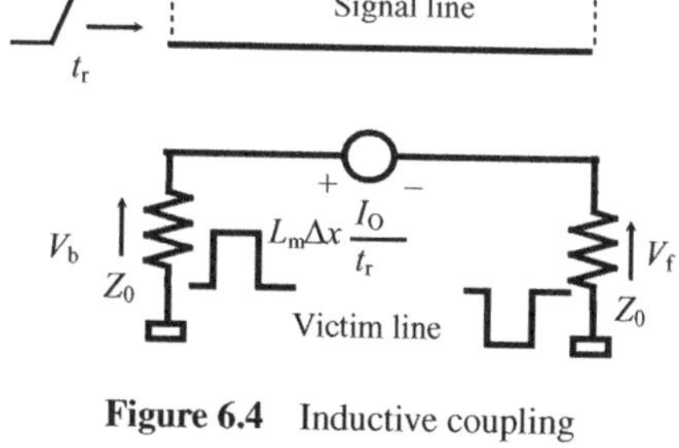

Figure 6.4 Inductive coupling

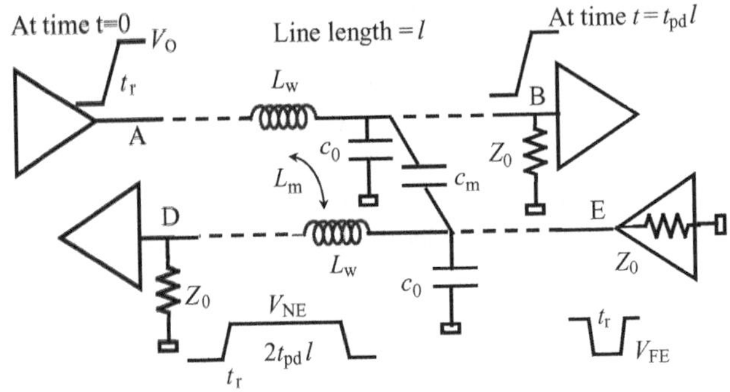

Figure 6.5 Total crosstalk

At this point, the inductive *crosstalk* derivation follows that of capacitive *crosstalk*, so that

$$V_{FEind} = -\frac{L_m l V_O}{2 Z_0 t_r} \tag{6.9}$$

$$V_{NEind} = \frac{L_m V_O}{4 L_w} \tag{6.10}$$

6.1.4 Total Coupling

In practical cases, capacitive and inductive crosstalk are simultaneously present, as shown in Figure 6.5 [2–6]. Under the weak coupling assumption, the *far-end crosstalk* is the sum of Equations (6.4) and (6.9):

$$V_{FE} = \frac{\left(Z_0 c_m - \dfrac{L_m}{Z_0}\right) l}{2t_r} V_O \tag{6.11}$$

where V_{FE} is a pulse of width t_r, and its amplitude can be zero if the numerator of Equation (6.11) is zero, as is the case for lines in a homogeneous medium (i.e. stripline structures).

The *near-end crosstalk* is obtained by summing Equations (6.6) and (6.10) and is given by

$$V_{NE} = \frac{1}{4}\left(\frac{c_m}{C} + \frac{L_m}{L_w}\right) V_O \tag{6.12}$$

where $C = c_0 + c_m$. When the lines are matched at both ends, V_{NE} is a pulse of width $2t_{pd}l$, with rise time t_r and amplitude given by Equation (6.12).

6.1.5 Simulations of Two Coupled Lines

This section concerns the prediction by simulations of crosstalk in two coupled lines considering point-to-point and chain structures.

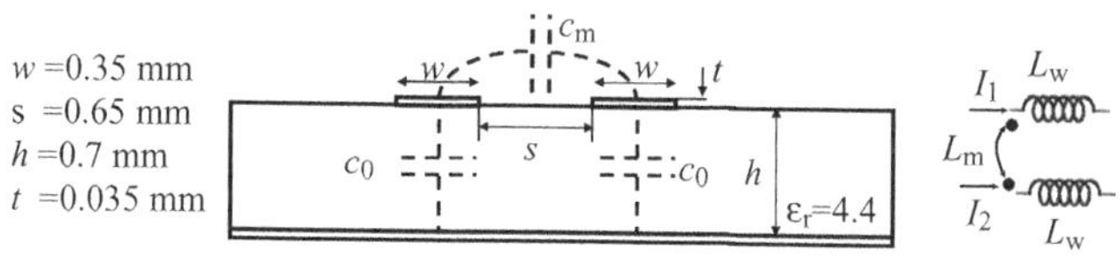

Figure 6.6 Two coupled lines with geometrical and electrical parameters

Example 6.1: Point-to-Point Structure

This example, based on the structure shown in Figure 6.6, is useful for outlining the procedure for developing a lumped-circuit model suitable for *crosstalk* simulations. To this end, the first step is to calculate the p.u.l. line parameters. A numerical program suitable for computing the p.u.l. capacitance and inductance matrices of multiconductor coupled line structures is Maxwell by Ansoft which is based on the finite element method. This code in its student version (i.e. Maxwell SV) can be downloaded from the Ansoft website for free. The calculated capacitance and inductance matrices are

$$\mathbf{C} = \begin{bmatrix} 64.67 & -8.9572 \\ -8.9572 & 64.67 \end{bmatrix} \text{pF/m}, \quad \mathbf{L} = \begin{bmatrix} 0.528 & 0.121 \\ 0.121 & 0.528 \end{bmatrix} \mu\text{H/m},$$

Using Equations (6.1b) and (6.1c) yields $L_w = 0.528\ \mu$H/m, $L_m = 0.121\ \mu$H/m, $c_0 = 55.713$ pF/m, and $c_m = 8.957$ pF/m. The signal line is driven by a voltage source V_1 of amplitude 1 V and $t_r = 1$ ns, and the source output resistance is very low. The signal line is matched at the far end, and the victim line at both ends. The two lines are 50 cm long. The most significant frequency is $f_{max} = 1/(\pi t_r) = 318$ MHz. This means that the minimum wavelength of interest is $\lambda_{min} = 300/f_{max} = 0.94$ m. The maximum length of a line segment to be simulated by lumped elements should be $\Delta x = l_{cell} = \lambda_{min}/20 = 4.7$ cm. To model the line of length $l = 50$ cm, a minimum of 10 elementary cells are necessary. To obtain more accurate results, 20 cells were used. The line parameters adopted for the circuit simulation are shown in Figure 6.7. The list is in MicroCap format [7], where '*.define X Y*' means 'assign the numerical or variable parameter *Y* to *X*'. The mutual inductance is modeled by the coupling factor $K_L = L_m/(L_w L_w)^{1/2}$.

The simulated waveforms of the two coupled lines are shown in Figure 6.8 for matched and unmatched victim lines. For a matched line it can be seen that there are no reflections after twice the line delay time $T_D = 2(L_w C)^{1/2} l = 5.84$ ns. The *near-end crosstalk* calculated with Equation (6.12) yields $V_{NE} = 0.092$ V, which is in good agreement with the value computed by SPICE. The *far-end crosstalk* calculated with Equation (6.11) yields $V_{FE} = -0.132$ V. Although this value is slightly lower than that computed by SPICE, it is in good agreement with the value computed by a distributed line model, as will be shown in the next section. With a victim line opened at the near end and short-circuited at the far end, the maximum *near-end crosstalk* doubles, and continuous oscillations with a width of about 6 ns can be observed. From these two simulations it is evident that the best condition to mitigate *crosstalk* is to match both lines.

In point-to-point structures, *far-end crosstalk* is a narrow pulse that is usually filtered by the receiver, and therefore it is not dangerous.

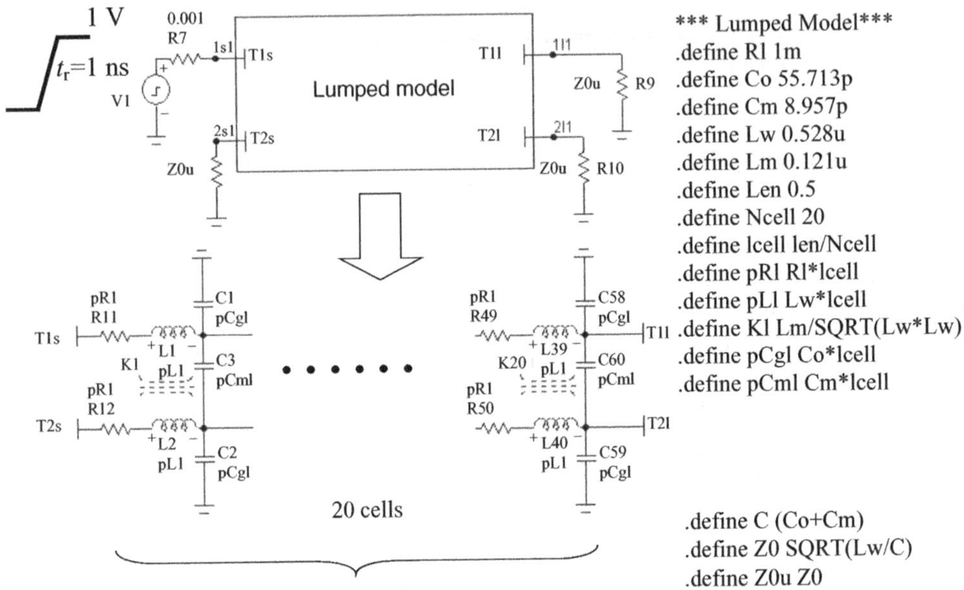

Figure 6.7 SPICE-like lumped-circuit model of two coupled lines

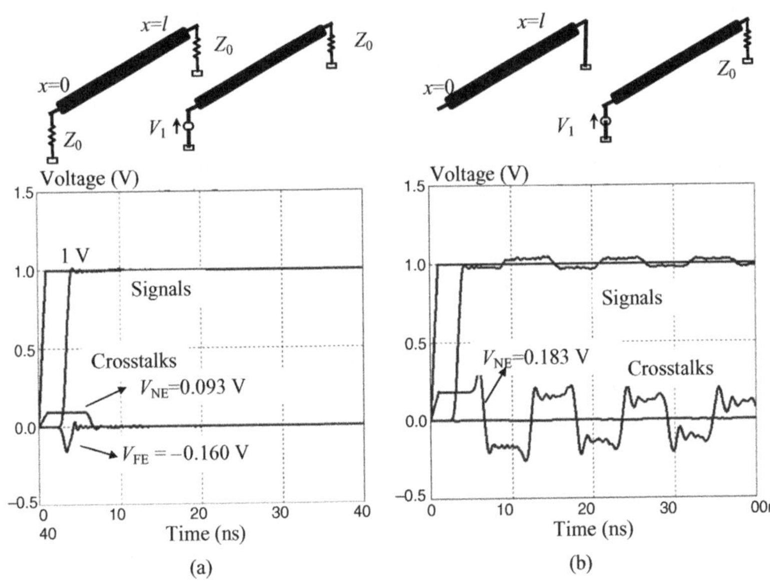

Figure 6.8 Simulated signal and crosstalk waveforms in a point-to-point structure with (a) matched lines and (b) an unmatched victim line with 10 kΩ at the near end and 1 Ω at the far end

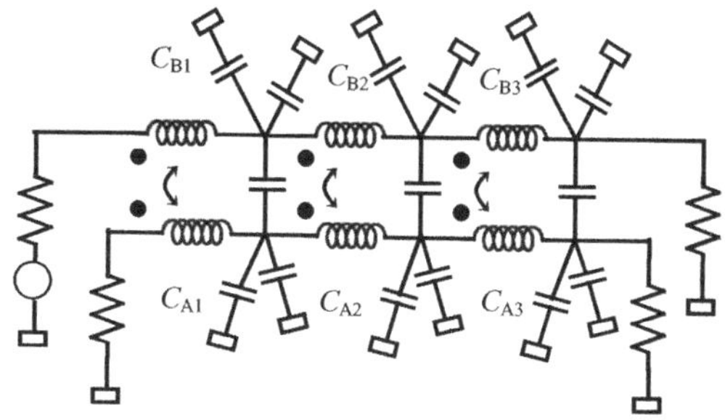

Figure 6.9 Lumped model of two coupled lines with distributed capacitive loads

Example 6.2: Chain Structure
In an interconnect with distributed loads such as receivers in chain structures, or the driver/receiver couple in bus structures, *far-end crosstalk* can be more dangerous than *near-end crosstalk*, as will be shown by the following simulations. The generic structure with elementary cells shown in Figure 6.9 is considered. If the distributed loads represented by capacitances are of equal values, $C_{Ai} = C_{Bi}$, for a generic load at position i, the line-to-ground capacitance c_0 is increased by the load capacitances $C_{Ai} = C_{Bi} = C_{Li}$. Hence, the p.u.l. equivalent capacitance between the line and the ground is obtained as

$$c_{0\text{eq}} = c_0 + \frac{1}{l}\sum_{i}^{n} C_{Li} = c_0 + c_0' \tag{6.13}$$

where l is the line length and n is the number of loads. This holds if the stub length and the spacing between two adjacent loads are electrically short at the maximum frequency of interest (see *Section 5.3*). The same structure used in *Example 6.1* and shown in Figure 6.6 is considered, assuming that for each trace there is a load $C_{Li} = 12$ pF every 3 cm. In this case, the application of Equation (6.13) gives $c_{0\text{eq}} = 55.713 + 12/3 \times 100 = 456$ pF/m.

The simulations can be carried out as in *Example 6.1*, using $c_{0\text{eq}}$ instead of c_0. The obtained results are shown in Figure 6.10. For a matched line, *near-end crosstalk* decreases and *far-end crosstalk* increases. For unmatched lines, the most dangerous situation regarding *far-end crosstalk* is when the victim line is short-circuited at the near-end point and opened at the far end, as shown in Figure 6.10b. This usually occurs in a practical situation when a driver is located at the near end. This strong increase in *far-end crosstalk* cannot be calculated exactly for every t_r by Equation (6.11) based on the approach of weakly coupled lines because the propagation velocities of the waveforms between the two lines (*differential mode*) and between the two lines and the reference plane (*common mode*) are considered to be about equal. It is the difference in speed of these two modes that causes a high level of *far-end crosstalk* when distributed loads along the interconnect are present. This will be investigated in greater detail in the next section.

6.2 Common and Differential Modes

In this section, an exact circuit model for two symmetric coupled lines based on *even* and *odd* modes of propagation is presented. The model requires knowledge of the p.u.l. parameters L and C of the lines in order to determine the characteristic impedance and delay time of the

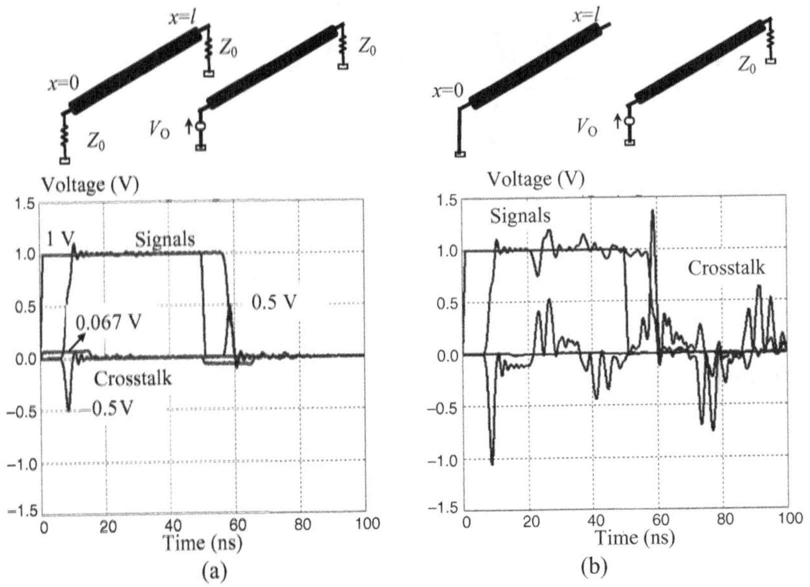

Figure 6.10 Simulated signal and crosstalk waveforms in a chain structure with (a) matched lines and (b) an unmatched victim line with 1 Ω at the near end and 10 kΩ at the far end

two decoupled modes. The utility of the model is not limited to *crosstalk* prediction, it can be extended to *common mode* to *differential mode* conversion in differential digital signaling when the loads are not symmetric and to explain the mechanism that generates high *far-end crosstalk* in structures such as a chain and bus.

6.2.1 Definition of Even and Odd Modes

Consider two coupled lines, such as two microstrip traces, and call V_1 and I_1 the voltage and current in line 1, and V_2 and I_2 the corresponding quantities in line 2 [8, 9]. These voltages and currents are the algebraic sum of two different distributions of voltages and currents, as depicted in Figure 6.11.

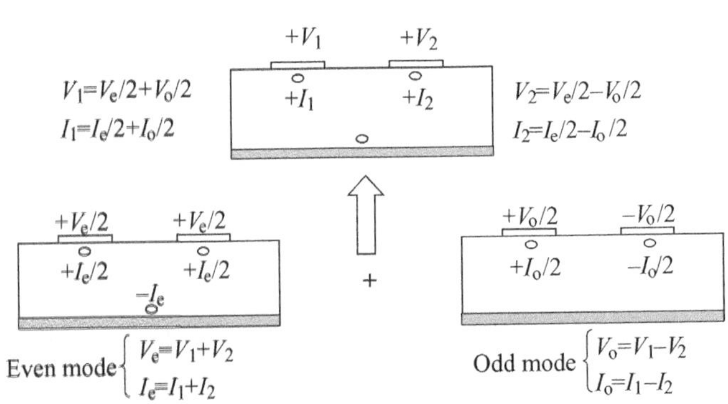

Figure 6.11 Even and odd propagation modes

Equal voltages $V_e/2$ and currents $I_e/2$, having the same sign, characterize the first distribution, indicated as the *even mode*. The total current I_e returns through the reference conductor or plane. The following equations hold: $V_e = V_1 + V_2$ and $I_e = I_1 + I_2$. Under this condition, as discussed in *Sections 3.3.2* and *4.2.2*, it is possible to associate with each line an effective inductance $L_{eCM} = L_w + L_m$ and an effective capacitance $C_{eCM} = c_0$ (see Equations (3.57) and (4.16) respectively). This means that the characteristic impedance Z_{0e} and the propagation delay time t_{pde} associated with the *even mode* are given by

$$Z_{0e} = \sqrt{\frac{L_{eCM}}{C_{eCM}}} = \sqrt{\frac{L_w + L_m}{c_0}} \tag{6.14a}$$

$$t_{pde} = \sqrt{L_{eCM}C_{eCM}} = \sqrt{(L_w + L_m)c_0} \tag{6.14b}$$

Equal voltages $V_o/2$ and currents $I_o/2$, having opposite sign, characterize the second distribution, indicated as the *odd mode*. Return currents do not interest the reference conductor or plane. The following equations hold: $V_o = V_1 - V_2$ and $I_o = I_1 - I_2$. In this case, as discussed in *Sections 3.3.1* and *4.2.1*, an inductance $L_{eDM} = L_w - L_m$ and a capacitance $C_{eDM} = c_0 + 2c_m$ can be associated with each line (see Equations (3.53) and (4.12) respectively). This means that the characteristic impedance Z_{0o} and the propagation delay time t_{pdo} of this mode are given by

$$Z_{0o} = \sqrt{\frac{L_{eDM}}{C_{eDM}}} = \sqrt{\frac{L_w - L_m}{c_0 + 2c_m}} \tag{6.15a}$$

$$t_{pdo} = \sqrt{L_{eDM}C_{eDM}} = \sqrt{(L_w - L_m)(c_0 + 2c_m)} \tag{6.15b}$$

The advantage of this approach is that the two modes are independent and can be analyzed separately to obtain the actual voltages and currents in line. Generally, the *odd mode* is faster than the *even mode*, $t_{pdo} < t_{pde}$, because t_{pdo} depends on $(L_w - L_m)$, while t_{pde} depends on $(L_w + L_m)$.

Even and *common* modes are equivalent in terms of characteristic impedance and propagation delay time, as shown in Figure 6.12a. Therefore

$$Z_{0CM} = Z_{0e} \tag{6.16a}$$

$$t_{pdCM} = t_{pde} \tag{6.16b}$$

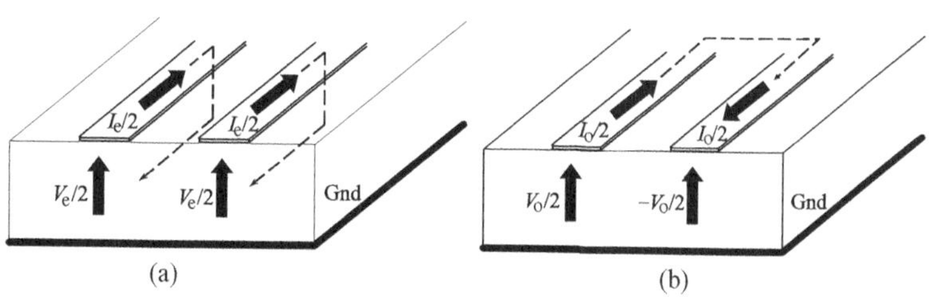

Figure 6.12 Mode representation: (a) even and common mode; (b) odd and differential mode

where Z_{0CM} is the *common-mode* characteristic impedance and t_{pdCM} is the p.u.l. propagation delay time associated with this mode.

Odd and *differential* modes have an equal p.u.l. propagation delay time, while the *differential-mode* characteristic impedance Z_{0DM} is twice the characteristic impedance of the *odd mode* (see Figure 6.12b). In fact, $Z_{0DM} = (V_o/2 - (-V_o/2))/(I_o/2) = V_o/(I_o/2) = 2Z_{0o}$, and hence

$$Z_{0DM} = 2Z_{0o} \tag{6.17a}$$

$$t_{pdDM} = t_{pdo} \tag{6.17b}$$

6.2.2 Equivalent Circuit Based on Even and Odd Modes

Since *even* and *odd* modes are independent, the propagation of both modes can be analyzed by the Branin circuit model introduced in *Section 5.2*. The total voltages and currents at the line ends can be found by coupling the two modes, and this can be easily done by the equivalent circuit shown in Figure 6.13, where the dependent voltage sources $e_o(t)$ and $e_e(t)$ make an algebraic sum with delay of the voltages and currents of respective modes at the opposite ends of the lines [10]:

At $x = 0$:

$$e_{oA}(t) = V_o(l, t - t_{pdo}l) - Z_{0o}I_o(l, t - t_{pdo}l) \tag{6.18a}$$

$$e_{eA}(t) = V_e(l, t - t_{pde}l) - Z_{0e}I_e(l, t - t_{pde}l) \tag{6.18b}$$

At $x = l$:

$$e_{oB}(t) = V_o(0, t - t_{pdo}l) + Z_{0o}I_o(0, t - t_{pdo}l) \tag{6.18c}$$

$$e_{eB}(t) = V_e(0, t - t_{pde}l) + Z_{0e}I_e(0, t - t_{pde}l) \tag{6.18d}$$

where, at the line ends A and B, the *even* and *odd* voltage and current are given by

$$V_e = V_1 + V_2, \quad I_e = I_1 + I_2, \quad V_o = V_1 - V_2, \quad I_o = I_1 - I_2$$

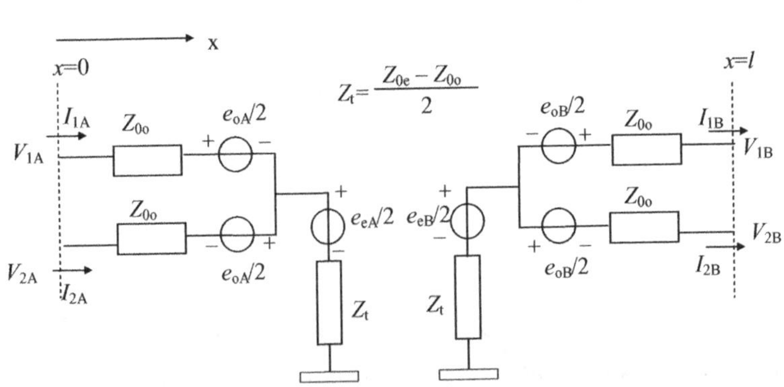

Figure 6.13 Equivalent circuit based on even and odd modes

6.2.3 *Equivalent Circuit for the Differential Transmission Mode*

The circuit in Figure 6.13 can be used to excite only one mode by introducing suitable terminations. To excite the *odd* or *differential mode*, the two lines should be driven at the line end A by two voltage sources of equal amplitude and opposite sign, and equal source impedance R_S, and should be terminated at line end B on equal load impedance R_L. On the other hand, if this condition is not realized, a *common mode* is generated, with a consequent deterioration in signal integrity and an increase in the radiated emission. To match the interconnect for the *differential mode*, the load impedance should be $R_L = Z_{0o}$ or, equivalently, $2R_L = Z_{DM}$. To match both *common* and *differential* modes, a π-resistive net should be used, as described in *Section 12.1.4*. The model can also be very useful for simulating *common-mode* disturbances converted into *differential-mode* noises on account of the non-symmetry of the sources and loads with respect to the reference ground [10].

6.2.4 *Simulations of Point-To-Point and Chain Structure by Even and Odd Modes*

The exact equivalent circuit based on even and odd propagation modes can be very useful for crosstalk calculation or simulation. This section provides the results of simulations obtained for the case of point-to-point and chain structures.

Example 6.3: Two Coupled Lines with Linear Loads

The simulations are performed adopting the same structure of two microstrip coupled lines as that considered in *Example 6.1*. The procedure for even and odd modes starts with the calculation of the mode parameters by the following closed-form expressions [6]:

$$C = c_0 + c_m, \quad Z_0 = \sqrt{L_w/C}, \quad t_{pd} = \sqrt{L_w C}$$

$$K_C = c_m/C, \quad K_L = L_m/L_w$$

$$t_{pde} = t_{pd}\sqrt{(1+K_L)(1-K_C)}, \quad t_{pdo} = t_{pd}\sqrt{(1-K_L)(1+K_C)}$$

$$Z_{0e} = Z_0\sqrt{\frac{1+K_L}{1-K_C}}, \quad Z_{0o} = Z_0\sqrt{\frac{1-K_L}{1+K_C}}$$

Adopting the p.u.l. parameters of *Example 6.1* yields

$$t_{pd} = 5.843 \text{ ns}, \quad t_{pde} = 6.01 \text{ ns}, \quad t_{pdo} = 5.477 \text{ ns},$$

$$Z_0 = 90.35\ \Omega, \quad Z_{0e} = 107.883\ \Omega, \quad Z_{0o} = 74.385\ \Omega$$

The implementation of the exact equivalent model of Figure 6.13 for the structure under consideration by using MicroCap [7] is shown in Figure 6.14. The four equivalent circuits added to the model perform the delay function. For other Spice-based simulators the procedure is similar. The results of simulations obtained with the lumped model of *Section 6.1* and with the distributed model presented in this section are compared for the case of matched and unmatched lines in Figure 6.15. Note that the distributed model provides exact results without the slight oscillation of the lumped model. For example, the maximum *far-end crosstalk* is practically equal to that calculated with Equation (6.11).

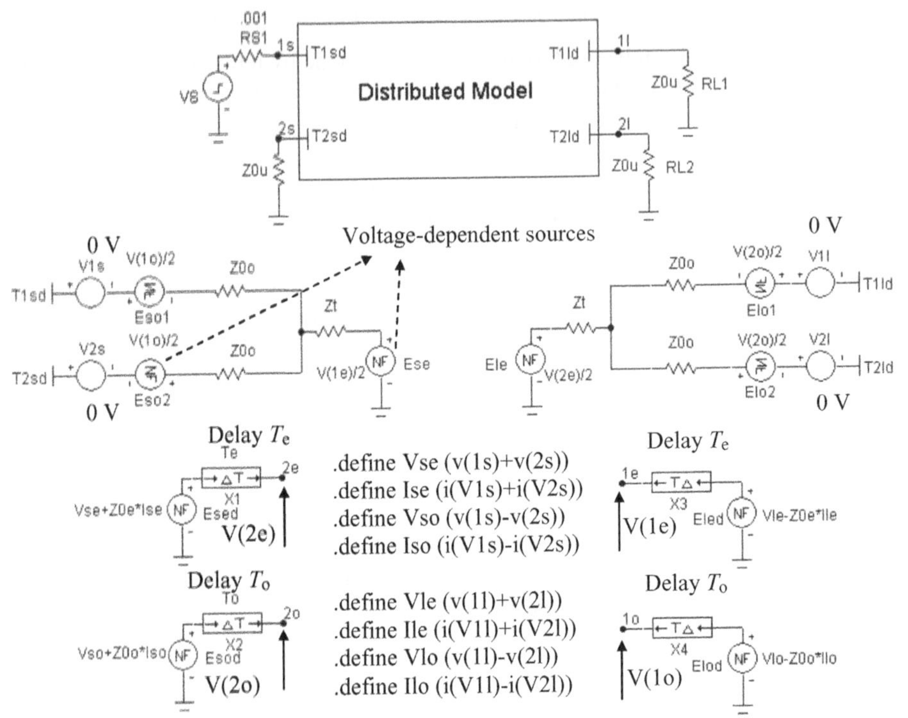

Figure 6.14 SPICE-like distributed circuit model of two symmetrical coupled lines

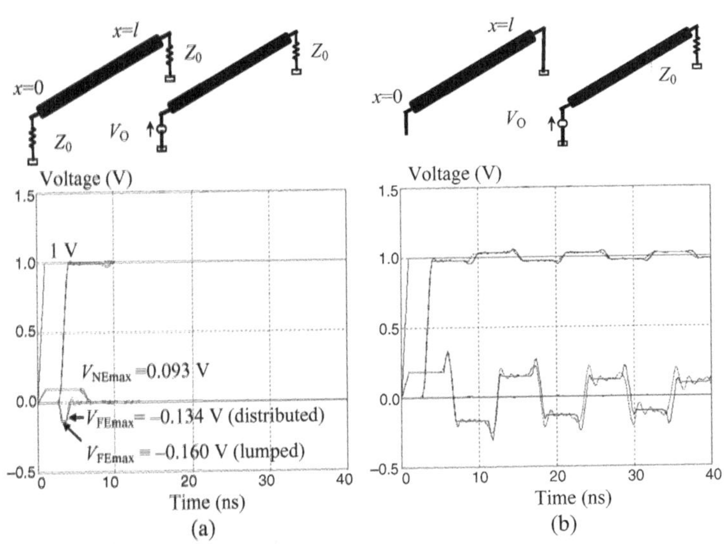

Figure 6.15 Simulated signals and crosstalk waveforms of a point-to-point structure obtained by the lumped model (dotted line) and by the distributed even–odd model (solid line) in the case of (a) matched lines and (b) an unmatched victim line with 10 kΩ at the near end and 1 Ω at the far end

Table 6.1 Line parameters for chain and point-to-point structures

	Chain structure	Point-to-point structure
t_{pd}	15.662 ns/m	5.843 ns/m
Z_0	33.7 Ω	90.35 Ω
t_{pde}	17.19 ns/m	6.01 ns/m
t_{pdo}	13.891 ns/m	5.477 ns/m
Z_{0e}	37.721 Ω	107.883 Ω
Z_{0o}	29.328 Ω	74.385 Ω

To simulate the same chain structure as that considered in *Example 6.2*, the mode line parameters need to be recalculated, replacing c_o with c_{oeq}. The new values of the line parameters are shown in Table 6.1, together with the values associated with the initial configuration. It is interesting to note that the difference between the p.u.l. propagation delay time associated with the two modes (i.e. t_{pde}-t_{pdo}) is increased from 0.53 to 3.3 ns/m going from the point-to-point structure to the chain structure. Since the *odd* mode is faster than the *even* mode and the waveform induced by the *odd* mode in the victim line is negative, the *near-end crosstalk* is negative and higher than the *crosstalk* with the line unloaded. *Near-end crosstalk* returns to zero when the positive waveform excited by the *even mode* arrives at the far-end point of the victim line. The results of simulations obtained with the lumped model of *Section 6.1* and with the distributed model presented in this section are compared for the case of matched and unmatched lines in Figure 6.16, where a good agreement between the two models can be observed.

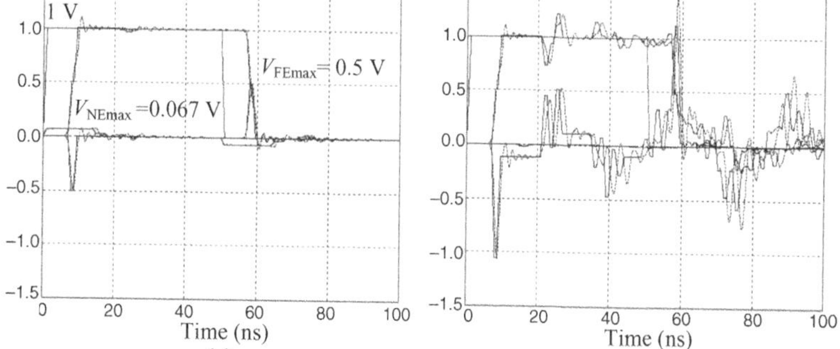

Figure 6.16 Simulated signal and crosstalk waveforms of a chain structure obtained by the lumped model (dotted line) and by the distributed even–odd model (solid line) in the case of (a) matched lines and (b) an unmatched victim line with 1 Ω at the near end and 10 kΩ at the far end

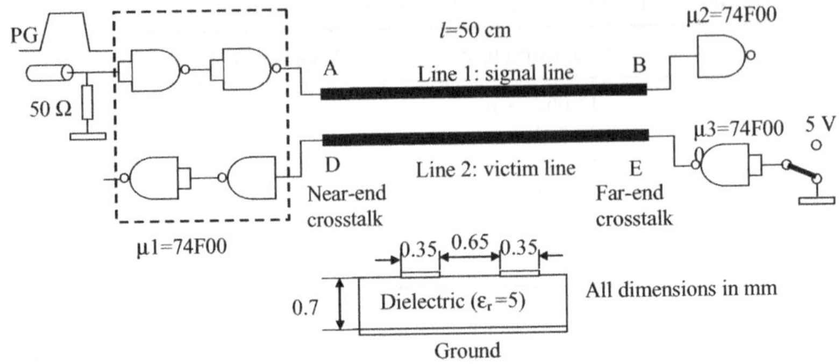

Figure 6.17 Test board for crosstalk with TTL FAST devices in a point-to-point structure

6.3 Models for Digital Devices: Simulation and Measurements

This section illustrates how to build up models for interconnects with digital devices, taking into account their non-linearity. Some examples are presented, supported by experimental data for two coupled lines.

Example 6.4: Two Coupled Traces with TTL Devices in a Point-to-Point Test Board Structure
A test board comprising two coupled microstrip lines running parallel for a length $l = 50$ cm is considered, as shown in Figure 6.17. The signal line is indicated by line 1, and the victim line is indicated by line 2. Line 2 can be at low or high static logic voltage, moving the manual switching on the right-hand side of Figure 6.17. All drivers and receivers are 74F00 TTL devices. The circuit model used in the simulation is shown in Figure 6.18. The p.u.l. parameters computed by Maxwell SV are: $c_0 = 55.713$ pF; $c_m = 8.957$ pF; $L_w = 0.528$ mH $L_m = 0.121$ mH (see Example 6.1). With these values, all the parameters describing the circuit model based on even- and odd-mode decomposition can be calculated as outlined in *Section 6.2.*

Drivers and receivers are modeled taking into account the interconnection of the package, the non-linearity of the devices, and the dynamic performance of the driver in terms of rise time t_r and fall time t_f.

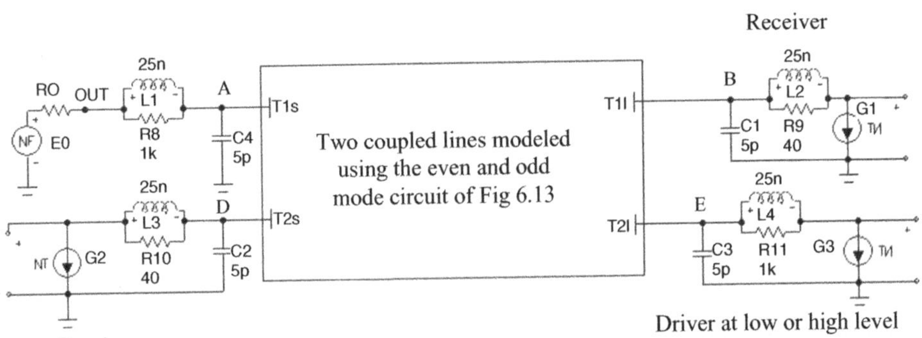

Figure 6.18 Devices and package circuit models

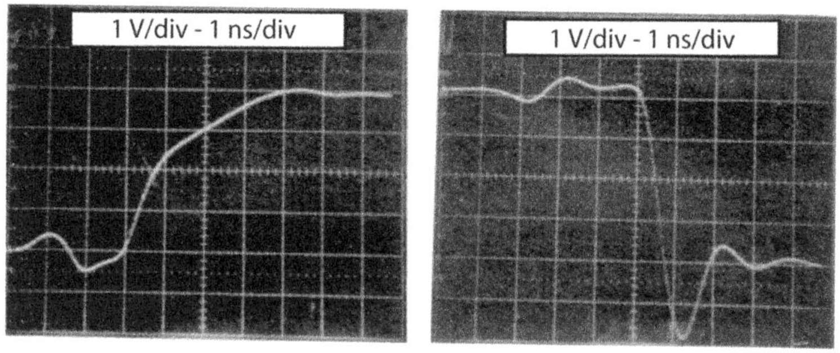

Figure 6.19 Measured rise and fall times at the 74F00 driver output without load. Scale: 1 V/div 1 ns/div

The active driver was modeled by a Thévenin equivalent circuit. Measurements of t_r and t_f were performed with the driver output unloaded, leading to the measured waveforms shown in Figure 6.19. The measured I/O static characteristics of a 74F00 device are shown in Figure 6.20. These characteristics are assigned to the output resistance R_O of the active driver in piecewise linear form, following the procedure outlined in *Section 2.3*. The low-level output characteristic is modeled by four segments, and the high-level characteristic by three segments. The measured output waveform in the time domain is assigned to the voltage source E_O in table form. The quiet driver and the receiver are simulated by a dependent current source, and the I/O characteristics are assigned in table form.

As an example, the equations used to simulate the active driver in the simulations performed by MicroCap are shown in Figure 6.21. They can be implemented in a similar manner in any other SPICE-like simulator.

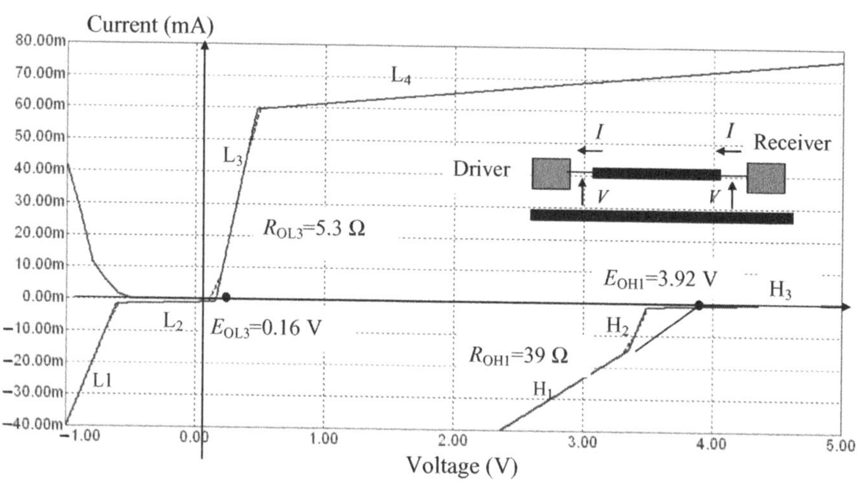

Figure 6.20 Low- and high-level state I/O static characteristics of the 74F00 device: solid line – active driver with values assigned in piecewise form; dotted line – driver at low or high level with values assigned in table form

L→H

```
.define Tsw 7n
.define Tr 3n
.define PH
TABLE(t,0,0,Tsw,0,(Tsw+1n),0.6,(Tsw+Tr),
        1,100n,1)
.define PL (1-PH)
.define EOs (PH*EOH1+PL*EOL3)
.define ROs (PH*ROH1+PL*ROL3)
.define EO IF(t<=(Tsw+Tr),EOs,EOH)
.define RO IF(t<=(Tsw+Tr),ROs,ROH)
```

H→L

```
.define Tsw 12n
.define Tf 1n
.define PH
TABLE(t,0,1,Tsw,1,Tsw+Tf,0,100n,0)
.define PL (1-PH)
.define EOs (PH*EOH3+PL*EOL3)
.define ROs (PH*ROH3+PL*ROL3)
.define EO IF(t<=(Tsw+Tf),EOs,EOL)
.define RO IF(t<=(Tsw+Tf),ROs,ROL)
```

Low state

```
.define EOL1 -0.6
.define ROL1 10
..define EOL2 1
.define ROL2 1000
.define EOL3 0.16
.define ROL3 5.3
.define EOL4 -15.7
.define ROL4 272
.define VOL1 ((EOL1/ROL1-EOL2/ROL2)/(1/ROL1-1/ROL2))
.define VOL2 ((EOL2/ROL2-EOL3/ROL3)/(1/ROL2-1/ROL3))
.define VOL3 ((EOL3/ROL3-EOL4/ROL4)/(1/ROL3-1/ROL4))
.define EOL1f IF(v(out)<=VOL1,EOL1,0)
.define EOL2f IF(v(out)>VOL1 AND v(out)<=VOL2,EOL2,0)
.define EOL3f IF(v(out)>VOL2 AND v(out)<=VOL3,EOL3,0)
.define EOL4f IF(v(out)>VOL3,EOL4,0)
.define EOL (EOL1f+EOL2f+EOL3f+EOL4f)
.define ROL1f IF(v(out)<=VOL1,ROL1,0)
.define ROL2f IF(v(out)>VOL1 AND v(out)<=VOL2,ROL2,0)
.define ROL3f IF(v(out)>VOL2 AND v(out)<=VOL3,ROL3,0)
.define ROL4f IF(v(out)>VOL3,ROL4,0)
.define ROL (ROL1f+ROL2f+ROL3f+ROL4f)
```

High state

```
.define EOH1 3.92
.define ROH1 39
.define EOH2 3.5
.define ROH2 10
.define EOH3 4.6
.define ROH3 1000
.define VOH1 ((EOH1/ROH1-
        EOH2/ROH2)/(1/ROH1-1/ROH2))
.define VOH2 ((EOH2/ROH2-
        EOH3/ROH3)/(1/ROH2-1/ROH3))
.define EOH1f IF(v(out)<=VOH1,EOH1,0)
.define EOH2f IF(v(out)>=VOH1 AND
        v(out)<=VOH2,EOH2,0)
.define EOH3f IF(v(out)>VOH2,EOH3,0)
.define EOH (EOH1f+EOH2f+EOH3f)
.define ROH1f IF(v(out)<=VOH1,ROH1,0)
.define ROH2f IF(v(out)>=VOH1 AND
        v(out)<=VOH2,ROH2,0)
.define ROH3f IF(v(out)>VOH2,ROH3,0)
.define ROH (ROH1f+ROH2f+ROH3f)
```

Figure 6.21 Equations describing low-to-high and high-to-low switching transitions and output characteristics of the 74F00 device in MicroCap format

Comparison between simulated and measured voltage waveforms at points A (signal line input) and D (victim line input) with a quiet driver at low voltage is shown in Figure 6.22. *Near-end crosstalk* is the most dangerous noise, as expected for this type of structure. A positive pulse noise of about 0.65 V appears at the receiver at point D, with a width of 6.6 ns, equal to twice the time required by the signal to go from point A to point B. After this time, the noise assumes negative values for another 6.6 ns. Then it reaches a point of rest without any other significant reflections. This can be explained by the fact that the output static characteristic of the driver at low state for negative output voltages represents a parasitic diode (see lines L1 and L2), which helps to limit the reflections.

Comparison between simulated and measured waveforms at points A (signal line input) and D (victim line input) with a quiet driver at high voltage is shown in Figure 6.23. In this case, in contrast to the low level, both *near-end* and *far-end crosstalk* damp in longer times because there is no diode action limiting the reflections, and the quiet line, for some voltage values, is opened at both ends. For *near-end* and *far-end crosstalk* there is a maximum noise of 0.7 and 0.9 V respectively, with a width equal to twice the time for the signal to travel from point A to point B. However, this is not a dangerous situation because the minimum values of V_D and V_E in the presence of crosstalk are far from $V_{IHmin} = 2$ V.

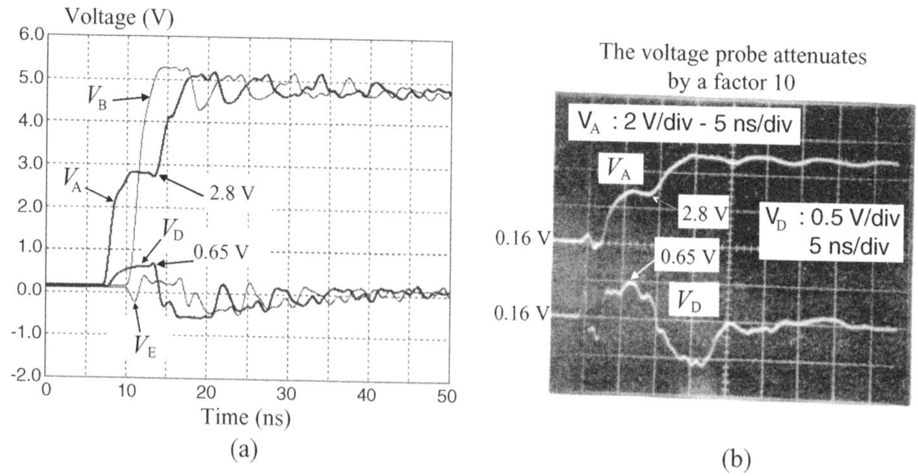

Figure 6.22 Low-to-high transition in a point-to-point structure with 74F00 devices: (a) simulated and (b) measured waveforms. Scale: 2 V/div (V_A), 0.5 V/div (V_D), 5 ns/div

If the interest is focused only on the calculation of the maximum *near-end crosstalk* V_{NE}, the equivalent circuit shown in Figure 6.24 can be used. This circuit is a simplification of the more general model of Figure 6.13. In this case, a linear output characteristic of the device is used for the part of the voltage–current range of interest. For the considered point-to-point structure, the circuit parameters of Figure 6.24 are $Z_{0e} = 108\ \Omega$, $Z_{0o} = 74.4\ \Omega$, $Z_t = 16.8\ \Omega$, and $Z_{in} = \infty$ for the receiver at point D if $V_D > 0$ V.

The first step is calculated by the circuit in Figure 6.24 and is given by

$$\begin{aligned} V_A &= (E_{OH1} - E_{OL3})(Z_{0o} + Z_t)/(R_{OH1} + Z_{0o} + Z_t) + E_{OL3} \\ &= (3.92 - 0.16)(74.4 + 16.8)/(39 + 74.4 + 16.8) + 0.16 \\ &= 2.79\ \text{V} \end{aligned}$$

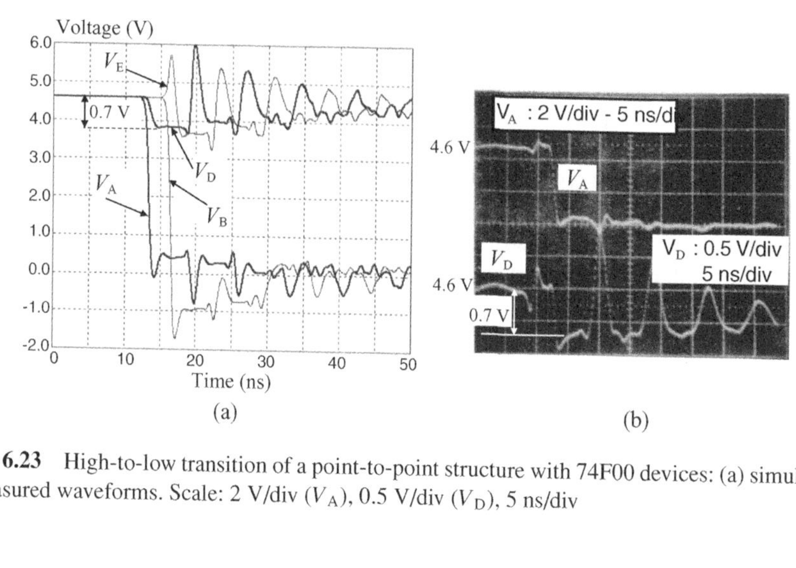

Figure 6.23 High-to-low transition of a point-to-point structure with 74F00 devices: (a) simulated and (b) measured waveforms. Scale: 2 V/div (V_A), 0.5 V/div (V_D), 5 ns/div

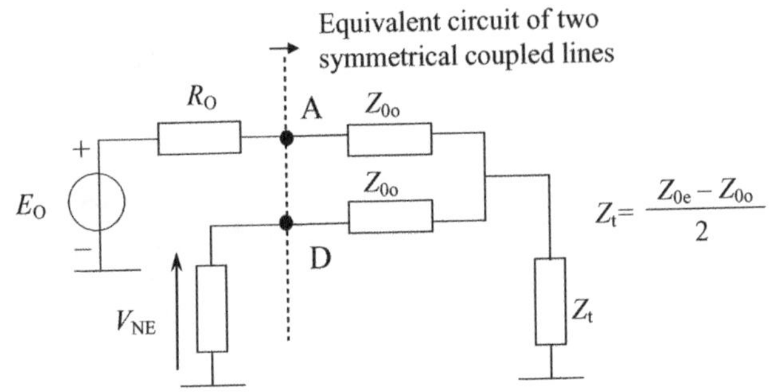

Figure 6.24 Equivalent circuit for near-end crosstalk calculation

Then, the following maximum $V_D = V_{NE}$ value in the presence of crosstalk is obtained:

$$\begin{aligned} V_D &= (E_{OH1} - E_{OL3})Z_t/(R_{OH1} + Z_{0o} + Z_t) + E_{OL3} \\ &= (3.92 - 0.16)16.8/(39 + 74.4 + 16.8) + 0.16 \\ &= 0.65 \text{ V} \end{aligned}$$

These values are in good agreement with the simulated and the measured data.

Example 6.5: Two Coupled Lines in a Bus Test Board Structure

As introduced in *Section 5.3*, a bus interconnect is characterized by a main line with distributed drivers and receivers along the line at regular small intervals compared with the wavelength associated with the maximum frequency of interest. Devices are connected to the main line by short interconnects called stubs. The stub and the driver and receiver capacitances load the main line, lowering the equivalent characteristic impedance Z_0 to such a small value that terminations are required to enhance the first step of the signal launched into the line. As discussed in *Section 5.4*, Thévenin terminations are generally used to enhance the driver capability and reduce reflections and therefore *crosstalk*.

Bus *crosstalk* is investigated by considering the test board shown in Figure 6.25, given by a motherboard with two coupled microstrips loaded every 3 cm for a length of 50 cm. Drivers are TTL 74F244, suitable for driving lines with low Z_0. For simplicity, drivers and receivers at positions 2–15 are replaced with a diode in parallel with a capacitance of 10 pF to have the same load effect. Terminations are positioned at both ends, as required in a bus structure. It will be shown that, for this type of structure, the *far-end crosstalk* is dominant owing to the loading effects of stub and device capacitances.

The test board was simulated by MicroCap according to the following assumptions:

- The active driver is modeled by its DC non-linear output characteristic using a voltage source and a resistance.
- The dynamic parameters used were $t_r = 3$ ns and $t_f = 1$ ns.
- The quiet driver output and receiver input characteristics were modeled by a current source defined in table format.

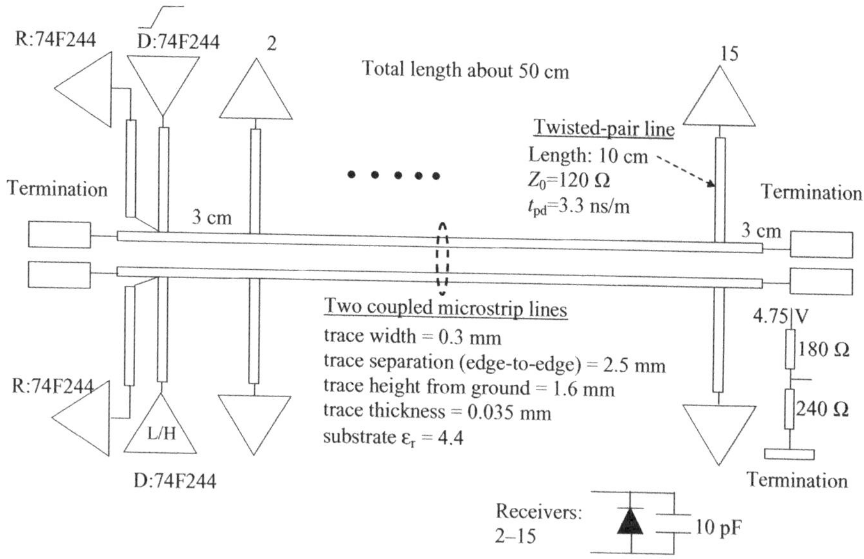

Figure 6.25 Bus test board structure

- Each section of the coupled microstrips is modeled by an R, L, and C network, the parameters of which were calculated by Ansoft's Maxwell SV 2D numerical code.
- The stubs were modeled with distributed lossless or lossy lines.
- At the end of each stub, a capacitance of 10 pF was assigned in order to take into account the possible presence of a driver and receiver.

The measured I/O static characteristic of the driver used for measurements is shown in Figure 6.26, where the parameters have the same significance as in the previous example. The simulation by Maxwell SV provided the following p.u.l capacitance and inductance matrices:

$$\mathbf{C} = \begin{bmatrix} 45.282 & -2.7213 \\ -2.7213 & 45.282 \end{bmatrix} \text{pF/m}, \quad \mathbf{L} = \begin{bmatrix} 0.709 & 0.086 \\ 0.086 & 0.709 \end{bmatrix} \mu\text{H/m},$$

Hence, by using Equations (6.1), (6.14), and (6.15), the following parameters were computed:

- For coupled lines without stubs: $L_w = 0.709\ \mu$H/m, $L_m = 0.086\ \mu$H/m, $c_0 = 42.561$ pF/m, $c_m = 2.721$ pF/m, $t_{pd} = 5.667$ ns/m, $Z_0 = 125.145\ \Omega$, $t_{pde} = 5.819$ ns/m, $t_{pdo} = 5.468$ ns/m, $Z_{0e} = 136.715\ \Omega$, $Z_{0o} = 113.908\ \Omega$.
- For coupled lines with stubs: $c_{0eq} = (42.561 + 425)$ pF/m, $t_{pdeq} = 18.262$ ns/m, $Z_{0eq} = 38.833\ \Omega$, $t_{pdeeq} = 19.286$ ns/m, $t_{pdoeq} = 17.164$ ns/m, $Z_{0eeq} = 41.248\ \Omega$, $Z_{0oeq} = 36.288\ \Omega$.

The differences $(t_{pde} - t_{pdo})$ increases when the lines are loaded with the capacitance of stubs and receivers equal to (2.75 + 10) pF/3 cm, or 425 pF/m. This means that the *far-end crosstalk* has high magnitude levels.

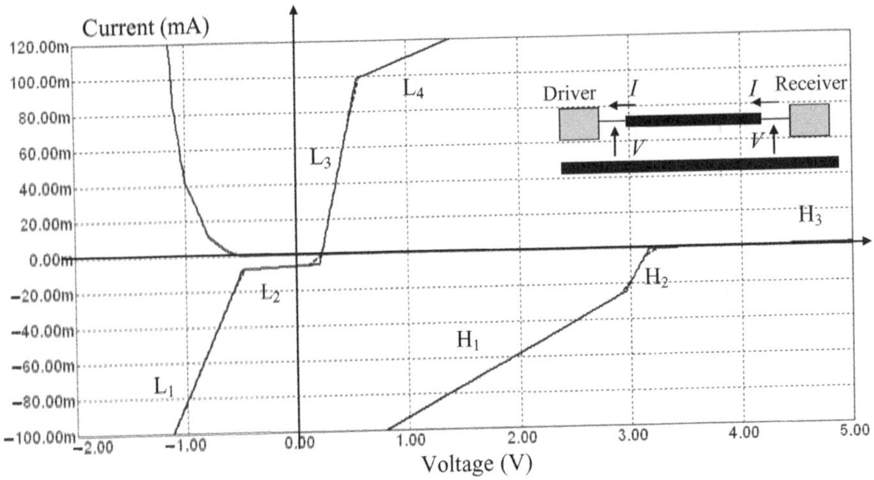

(a)

```
.define EOL1 -0.446      .define EOL3 0.217      .define EOH1 3.67      .define ROH2 8.6
.define ROL1 6.78        .define ROL3 3.33       .define ROH1 28.8      .define EOH3  EOH1
.define EOL2 1.9         .define EOL4 -3.4       .define EOH2 3.17      .define ROH3 1000
.define ROL2 300         .define ROL4 40
```

(b)

Figure 6.26 I/O DC characteristic for 74F244: (a) low- and high-level state with solid line means active driver with values assigned in piecewise form, and dotted line means driver at low or high level with values assigned in table form; (b) MicroCap listing of equivalent circuit parameters used for simulations

The circuit model used for MicroCap simulation is shown in Figure 6.27. Note that the stub was simulated as a lossy line in order to obtain more accurate results. The lossy-line model used will be explained in detail in *Section 7.2* and is based on the vector fitting technique.

The comparison between measured and simulated waveforms for the signal line is shown in Figure 6.28. Note that, although a lower bit rate was used in the simulation, the results are in good agreement. Comparison between measured and simulated *crosstalk* waveforms for the quiet line at low level is shown in Figure 6.29, and the following observations can be made:

- As expected, the far-end crosstalk V_{L2} is higher than the *near-end crosstalk* V_{S2}.
- A transient lossy model for stubs is required for better results.
- The maximum *far-end crosstalk* has a value of 1 V.

As expected, the *far-end crosstalk* V_{L2} is also higher than the *near-end crosstalk* V_{S2} in the case of a quiet driver in the victim line at high voltage, as shown in Figure 6.30.

From these comparisons, the following comments can be made:

- Although the waveforms were simulated with a limited bit rate, measured and simulated waveforms revealed a good agreement.
- Some differences are due to the fact that a simple behavioral model was used for the drivers and receivers.
- Owing to the loading effects of terminations, the same simulated waveforms can be obtained by using a linear model for the active driver setting: $E_O = E_{OL3}/E_{OH1}$, $R_O = R_{OL3}/R_{OH1}$.

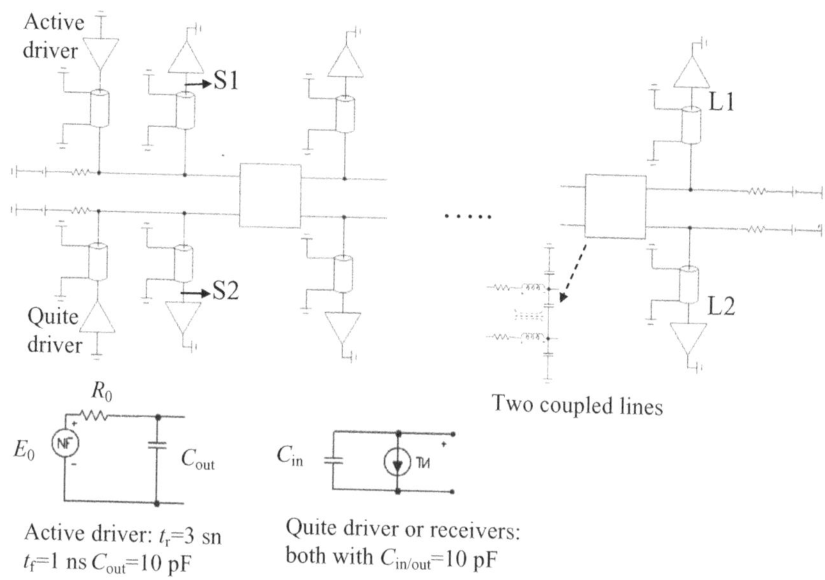

Figure 6.27 SPICE-like circuit model of the bus test board

- Better results are obtained by using a lossy-line model for stubs.
- The output driver waveform has a step of about 2.7 V > 2V (minimum guaranteed threshold at high level).
- *Far-end crosstalk* is higher than *near-end crosstalk* and has a maximum peak of about 1 V > 0.8 V (maximum guaranteed threshold at low level).

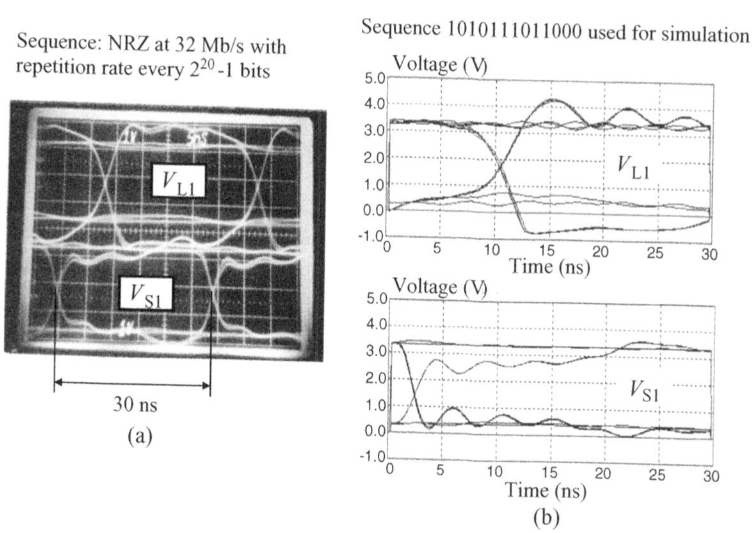

Figure 6.28 Eye diagram waveforms in an active line: (a) by measurements (scale: 1 V/div, 5 ns/div); (b) by simulations

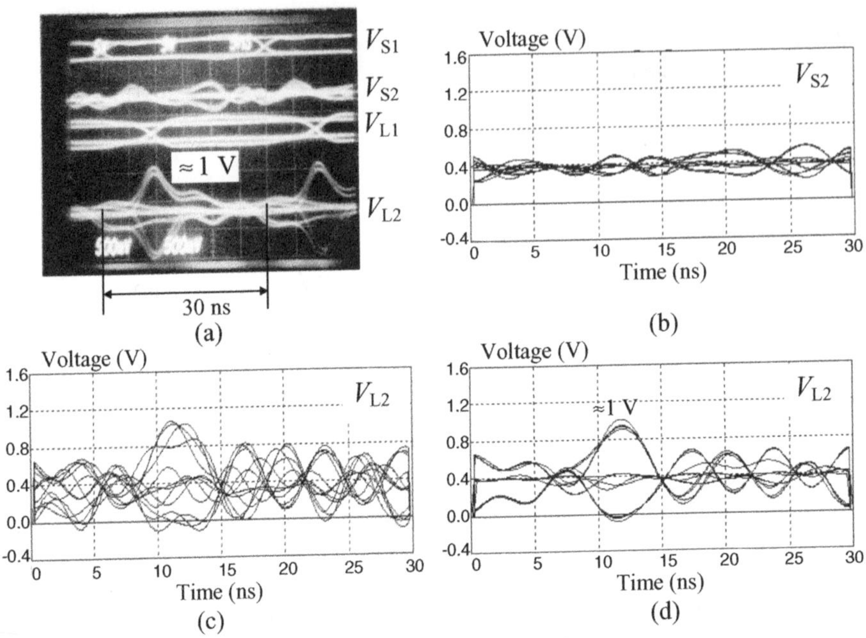

Figure 6.29 Measured and simulated eye diagram waveforms with a quiet driver in the victim line at low level: (a) measured signals (scale: 5 V/div, 5 ns/div) and crosstalk (scale: 0.5 V/div, 5 ns/div); (b) simulated near-end crosstalk; (c) simulated far-end crosstalk with lossless stubs; (d) simulated far-end crosstalk with lossy stubs

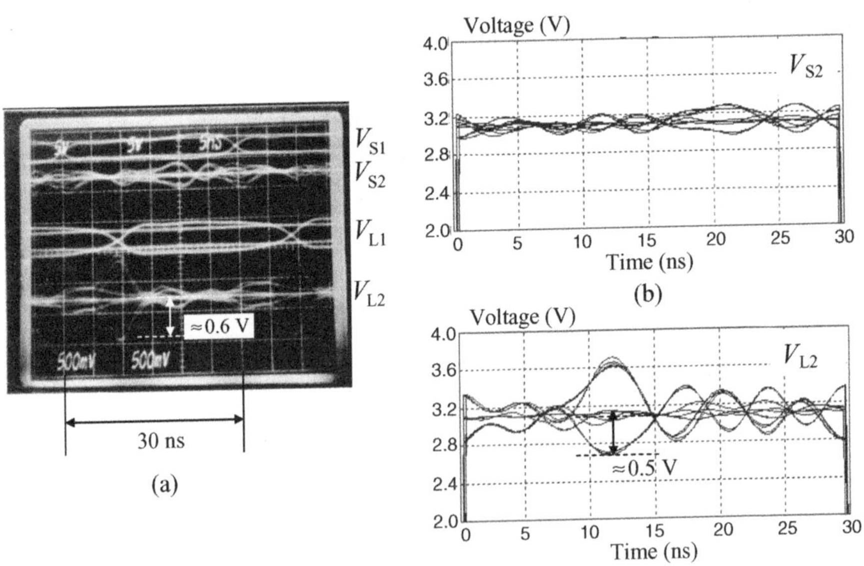

Figure 6.30 Measured and simulated eye diagram waveforms with a quiet line at high level: (a) measured signals (scale: 5 V/div, 5 ns/div) and crosstalk (scale: 0.5 V/div, 5 ns/div); (b) simulated near-end crosstalk with lossy stubs; (c) simulated far-end crosstalk with lossy stubs

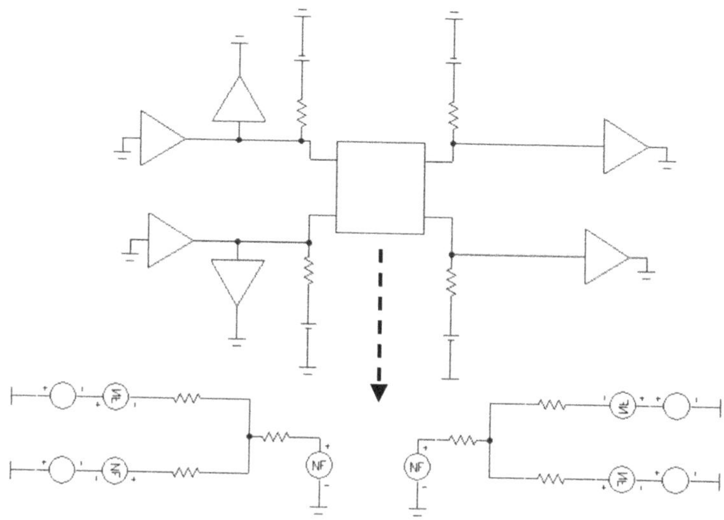

Figure 6.31 Equivalent circuit of a bus structure using the even and odd mode model of the interconnect previously shown in Figure 6.13

When interest is focused on calculation of the maximum *far-end crosstalk*, to understand better why it is so high for chain or bus structures, the following procedure can be used, referring to the equivalent circuit shown in Figure 6.31:

- *Crosstalk* estimation is performed simulating the main lines and stubs with a circuit model based on modal decomposition for two coupled lines.
- *Common-* and *differential-mode* parameters are those obtained when loading the lines with $C_{stub} + C_{in}$. The rise and fall time t_r and t_f must have the values estimated, simulated, or measured at the end of the signal line with distributed loads.

The simulated waveforms of the bus test board obtained by using the exact circuit model in Figure 6.13 are shown in Figure 6.32. If the simulations are performed with very low t_f, e.g. 0.1 ns, it is evident why the *far-end crosstalk* has an opposite sign with respect to the signal. In other words, a high-to-low transition of the signal produces a positive *far-end crosstalk*, and vice versa for a low-to-high transition. This can be explained by the fact that the capacitances of the distributed loads increase the capacitance of the *common mode* but not the *differential mode* which begins faster.

With very low t_f, the *near-end crosstalk* is a pulse of maximum amplitude 2.4 V with a width equal to the difference between the time required by the two modes, *differential* and *common*, to arrive at the end of the line. This maximum value can be lowered by t_f only. With $t_f = 5$ ns at the end of the line, the maximum *crosstalk* of 1 V is obtained, in agreement with the measured *far-end crosstalk*. This is reasonable considering that the slope of the front of the signal changes along the line owing to load effects between the lines and the reference ground plane (see Figure 6.28a). This is taken into account by the model with stubs.

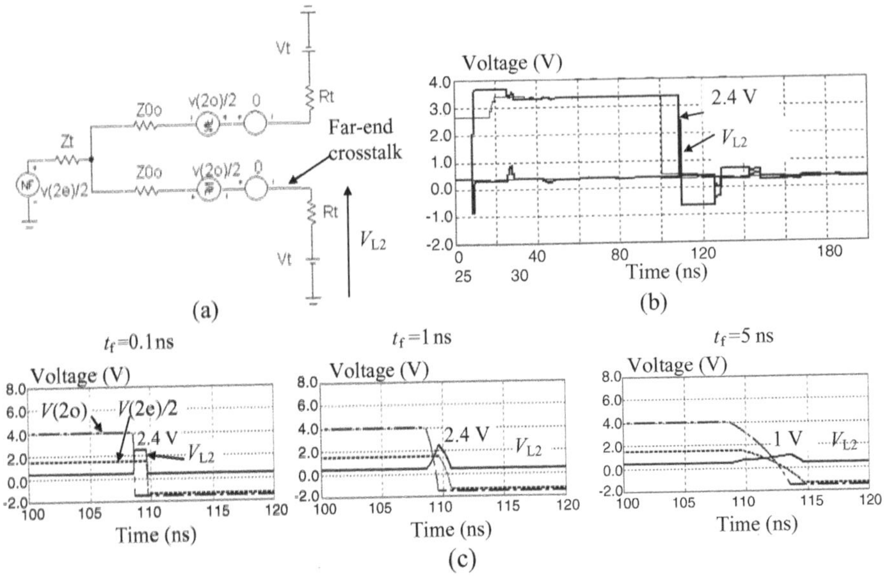

Figure 6.32 Simulated far-end crosstalk: (a) equivalent circuit; (b) simulated signals and crosstalk with $t_f = 0.1$ ns; (c) waveform of dependent voltage sources for even and odd mode signals and far-end crosstalk as function of t_f

In conclusion, for *far-end crosstalk* estimation, the following observations can be made:

- The maximum peak of *far-end crosstalk* can be calculated as superimposition effects of common e_{eB} and differential e_{oB} source voltages (Equations (6.18)).
- The propagation velocity of *common-mode* waveforms is lower than that of *differential-mode* waveforms owing to the loading effects. This is more evident when t_r and t_f are very small and *far-end crosstalk* reaches its maximum level.
- The rise t_r and fall t_f time to be used for simulation (in this case $t_r = t_f = 5$ ns) can be measured or estimated by simulating signal propagation on a single bus line.
- The shape of maximum *far-end crosstalk* depends on the shape of switching waveforms, e.g. to obtain an isosceles triangle, the t_r and t_f must be a ramp.

6.4 General Distributed Model for Lossless Multiconductor Transmission Lines

The problem of lossless *Multiconductor Transmission Lines* (MTLs) has been investigated by many researchers in the past, and several valuable contributions can be found in the literature [11–19]. In this section, essential information for implementing a general MTL model in SPICE circuit simulators only is provided. A test board consisting of five coupled traces in a microstrip structure driven by TTL and CMOS devices is used to compare simulated waveforms with measurements.

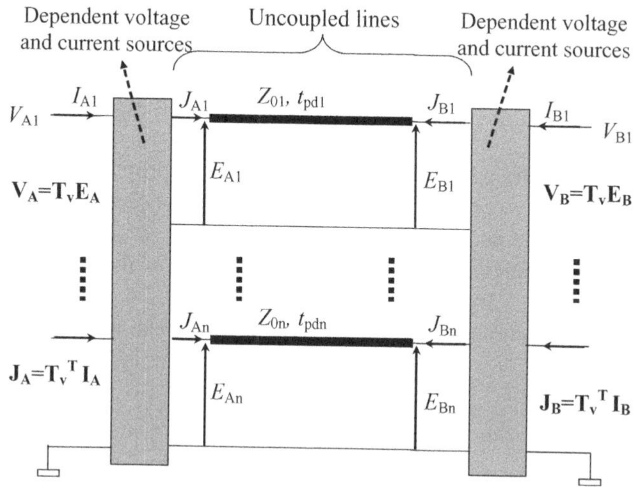

Figure 6.33 Exact circuit model of a lossless MTL derived by modal analysis

6.4.1 Equivalent Circuit of n *Coupled Lossless Lines*

A very useful equivalent circuit for modeling n coupled lossless lines can be derived by applying modal analysis [1]. In the modal domain, the MTL is decomposed into n uncoupled TLs (i.e. mode lines) that do not interact. The schematic representation of the equivalent circuit is shown in Figure 6.33. Voltages E_{Ai} and E_{Bi} and currents J_{Ai} and J_{Bi}, with $i = 1, \ldots, n$, at the mode line ends are computed by the lossless TL model available in the SPICE library. Actual voltages V_{Ai} and V_{Bi} and currents I_{Ai} and I_{Bi} at both line ends are computed by transformation matrices which can be modeled in the circuit simulator by dependent voltage and current sources that implement the relations between actual and modal voltages and currents. The key parameter characterizing the transformation from the actual to the modal domain is the transformation matrix $\mathbf{T_V}$ which can be obtained by the geometrical parameters of the line.

The derivation of the circuit model based on the modal analysis is outlined in detail by Paul [1]. In this book, only the essential steps for implementing the MTL model in SPICE circuit simulators are provided.

The mathematical procedure for obtaining the transformation matrix $\mathbf{T_V}$, the p.u.l. propagation delay times t_{pdi}, and the characteristic impedances Z_{0i} associated with the ith mode line is based on knowledge of the p.u.l. inductance $\mathbf{L}$ and capacitance $\mathbf{C}$ matrices, and is briefly summarized below. The matrices $\mathbf{L}$ and $\mathbf{C}$ can be computed by a numerical field solver such as Maxwell SV. Once these matrices are known, the following quantities are calculated:

$$\mathbf{M} = \mathbf{CL} \tag{6.19a}$$

$$\mathbf{t_{pd2}} = \text{eigenvalues } (\mathbf{M}) \tag{6.19b}$$

$$\mathbf{t_{pd}} = \mathbf{t_{pd2}}^{1/2} \tag{6.19c}$$

$$\mathbf{T} = \text{eigenvectors } (\mathbf{M}) \tag{6.19d}$$

$$\mathbf{T_V} = \text{transpose } (\mathbf{T}^{-1}) \tag{6.19e}$$

$$\mathbf{T_i} = ((\text{transpose } (\mathbf{T_V}))^{-1} \tag{6.19f}$$

$$\mathbf{L_m} = \mathbf{T_V^{-1} L T_i} \tag{6.19g}$$

$$\mathbf{C_m} = \mathbf{T_i^{-1} C T_V} \tag{6.19h}$$

where $\mathbf{t_{pd}}$ is a diagonal matrix whose coefficients are the p.u.l. propagation delay times of the decoupled mode lines: $t_{pd1}, \ldots, t_{pdn}$; $\mathbf{C_m}$ is a diagonal matrix whose diagonal coefficients are the p.u.l. capacitances of the decoupled mode lines: $C_{m\,1,1}, \ldots, C_{mn,n}$; and $\mathbf{L_m}$ is a diagonal matrix whose diagonal coefficients are the p.u.l. inductances of the decoupled mode lines: $L_{m\,1,1}, \ldots, L_{mn,n}$.

The modal characteristic impedances of the decoupled lines can then be easily obtained as

$$Z_{01} = \sqrt{\frac{L_{m1,1}}{C_{m1,1}}}, \ldots, Z_{0n} = \sqrt{\frac{L_{mn,n}}{C_{mn,n}}} \tag{6.20}$$

6.4.2 *Measurements and Simulations of Five Coupled Lines with TTL and CMOS Devices*

In this section, the performance of the model for *n*-coupled lossless lines is shown by comparing simulations with measurements for some practical cases.

Example 6.6: Five Coupled Lines with Non-linear Loads

The test board of five parallel microstrip lines of length $l = 1$ m is considered according to the geometrical parameters shown in Figure 6.34. A quiet line is located in the middle line, and TTL or CMOS devices drive the other four lines simultaneously [18, 19].

The finite element based software tool Maxwell SV was used to calculate the following capacitance **C** and inductance **L** matrices:

$$\mathbf{C} = \begin{pmatrix} 49.796 & -13.845 & -2.2026 & -0.62992 & -0.32576 \\ -13.845 & 53.615 & -13.118 & -2.0389 & -0.63149 \\ -2.2026 & -13.118 & 53.565 & -13.121 & -2.2082 \\ -0.62992 & -2.0389 & -13.121 & 53.586 & -13.852 \\ -0.32576 & -0.63149 & -2.2082 & -13.852 & 49.806 \end{pmatrix} \text{pF/m}$$

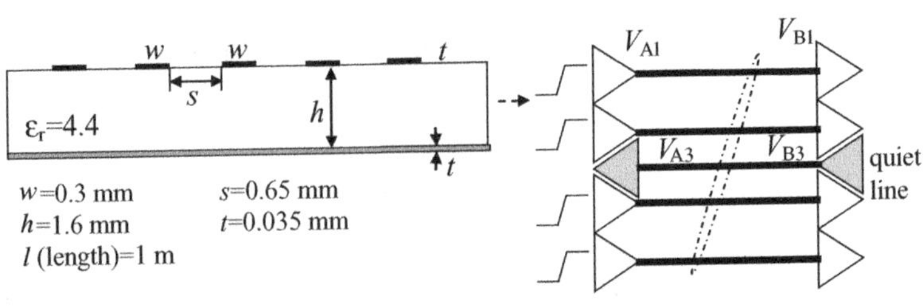

Figure 6.34 Five coupled lines with non-linear loads

Table 6.2 Coefficients of the transformation matrix $\mathbf{T_v}$ and parameters characterizing the mode lines

$Tv_{1,1} = 0.422$	$Tv_{1,2} = 0.603$	$Tv_{1,3} = 0.501$	$Tv_{1,4} = 0.107$	$Tv_{1,5} = -0.321$
$Tv_{2,1} = 0.465$	$Tv_{2,2} = 0.374$	$Tv_{2,3} = -0.274$	$Tv_{2,4} = -0.426$	$Tv_{2,5} = 0.663$
$Tv_{3,1} = 0.478$	$Tv_{3,2} = -0.01$	$Tv_{3,3} = -0.628$	$Tv_{3,4} = 0.675$	$Tv_{3,5} = -0.094$
$Tv_{4,1} = 0.463$	$Tv_{4,2} = -0.383$	$Tv_{4,3} = -0.22$	$Tv_{4,4} = -0.568$	$Tv_{4,5} = -0.585$
$Tv_{5,1} = 0.418$	$Tv_{5,2} = -0.597$	$Tv_{5,3} = 0.496$	$Tv_{5,4} = 0.177$	$Tv_{5,5} = 0.335$

Z01 = 224.624 Ω	Td1 = 6.114 ns/m	vm1 = 0.164 m/ns
Z02 = 138.914 Ω	Td2 = 5.56 ns/m	vm2 = 0.18 m/ns
Z03 = 100.519 Ω	Td3 = 5.408 ns/m	vm3 = 0.185 m/ns
Z04 = 72.615 Ω	Td4 = 5.339 ns/m	vm4 = 0.187 m/ns
Z05 = 81.487 Ω	Td5 = 5.362 ns/m	vm5 = 0.186 m/ns

$$\mathbf{L} = \begin{pmatrix} 0.712 & 0.259 & 0.143 & 0.089 & 0.06 \\ 0.259 & 0.709 & 0.257 & 0.142 & 0.089 \\ 0.143 & 0.257 & 0.708 & 0.257 & 0.143 \\ 0.089 & 0.142 & 0.257 & 0.708 & 0.259 \\ 0.06 & 0.089 & 0.143 & 0.259 & 0.71 \end{pmatrix} \mu\text{F/m}$$

Computed matrix $\mathbf{T_V}$ and decoupled line parameters required by the MTL model are shown in Table 6.2. Details about the MTL model implemented in MicroCap [7] are shown in Figure 6.35. Note that the dependent voltage sources E_i and the dependent current sources F_i use polynomial function 'Poly' to perform the associated matrix product.

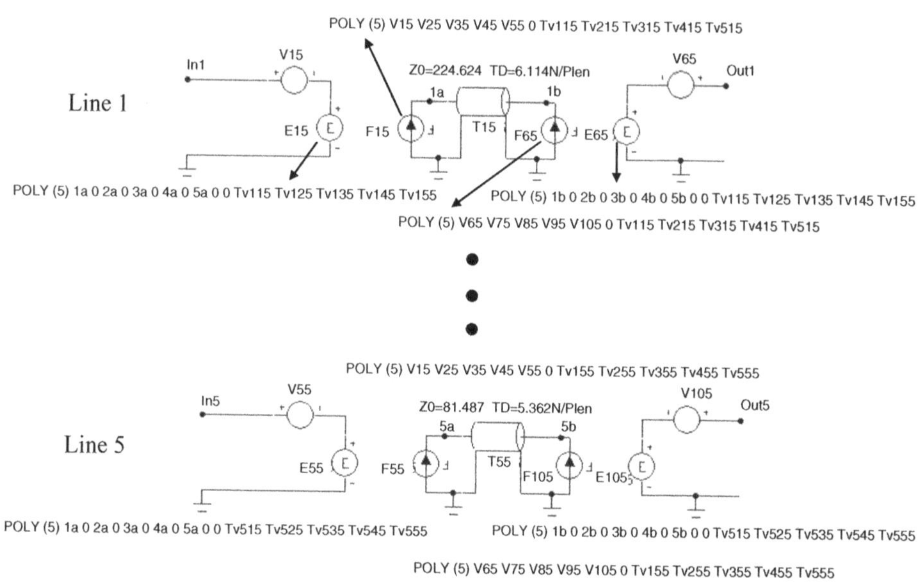

Figure 6.35 SPICE model of five coupled lines (Plen=1)

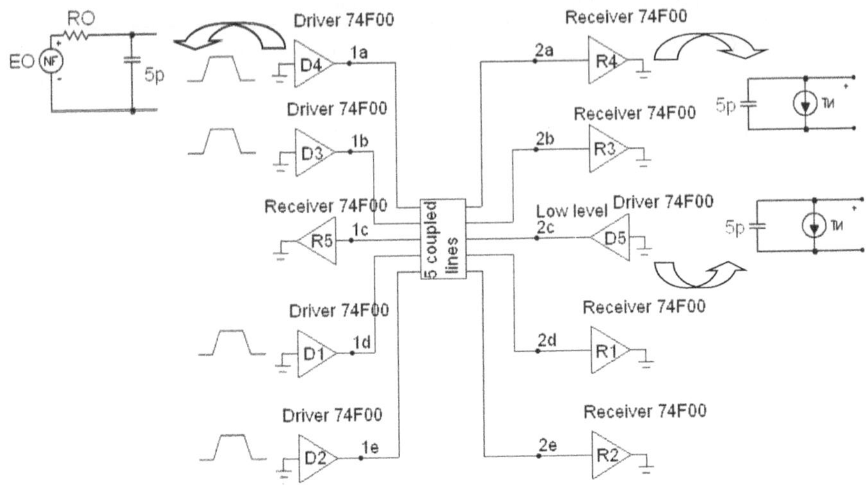

Figure 6.36 Schematic representation for circuit simulation with 74F00

Typical I/O static characteristics of TTL devices were measured on two different devices, while worst (W) and best (B) characteristics were deduced by data sheet considering I_{OLmax}, I_{OHmax}, and short-circuit current at high level. Comparing the different curves, a spread among the characteristics was observed. For this reason, in order to compare simulations with measurements, it is very important to use in the simulation the measured I/O characteristics of the same devices as those used in the experimental set-up. For the simulations, the same macromodel and parameters reported in *Section 6.3* for 74F00 were used.

The schematic representation of the test board in macromodel components for simulation is shown in Figure 6.36. The devices are modeled by the same method outlined in *Section 6.3*.

The comparison between simulated waveforms and measurements with 74F00 is shown in Figure 6.37. Note that the maximum *near-end crosstalk* peak of $V_{3A} \approx 2$ V, the signal first step of about 3.2 V, and the reflections are accurately reproduced by the simulations using the macromodeling procedure. The slight differences are due to the driver model used, which does not exactly correspond to the I/O characteristics of the driver used for measurements.

The same experiment was performed with high-speed CMOS 74AC00 devices to verify the influence on *crosstalk* of devices with different I/O characteristics. As mentioned for TTL devices, it is very important to measure the characteristics of the device used with the test board for comparison with measurements. In fact, spread of the characteristics was also observed for these devices. Although CMOS devices have different I/O characteristics to TTL devices, their characteristics can be linearized as done for TTL, and similar macromodels can be derived (see *Section 2.3*). For CMOS, three segments for high and low levels can be used for simulation, as shown in Figure 6.38. The parameters used for the simulation with 74AC00 in MicroCap format are shown in Figure 6.39.

The schematic representation for simulation of the test board by macromodel components with ACMOS devices is shown in Figure 6.40. Note that, ACMOS being faster than TTL, $t_r = t_f = 2$ ns, the package of the devices must be simulated to reproduce measurements with good accuracy.

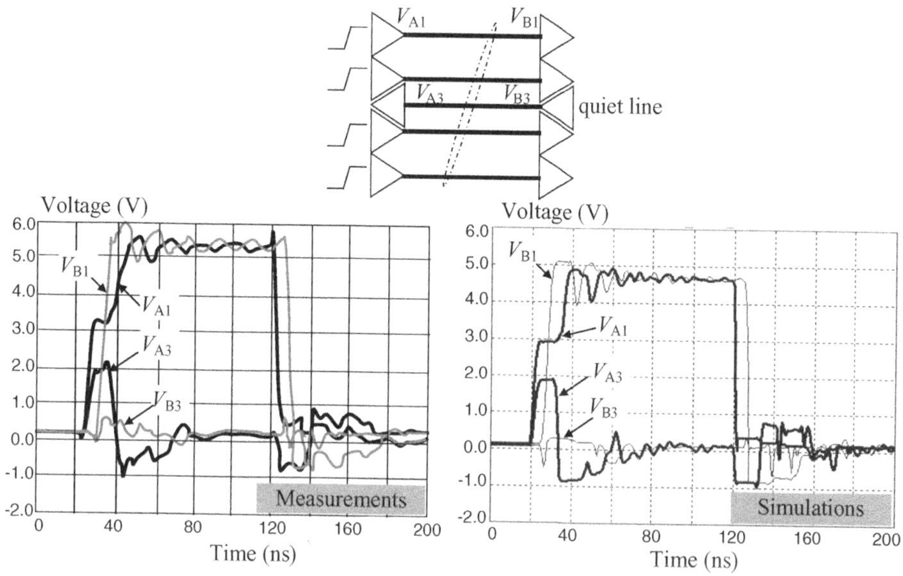

Figure 6.37 Measured and simulated waveforms of five coupled lines with 74F00 devices

Comparison between simulated and measured waveforms with 74AC00 is shown in Figure 6.41 Note that the maximum *crosstalk* peak of about 3 V and the signal first step of about 4.6 V are reproduced. Reflections are reproduced only partially. In fact, after 80 ns from the low-to-high switching, measurements show a *near-end crosstalk* peak of 1 V, as against a value of 0.5 V given by the simulation. The same difference in negative reflection can be observed on the signal at the receiver after the same time. This means that a more accurate behavioral device model should be required to reconstruct these reflections accurately.

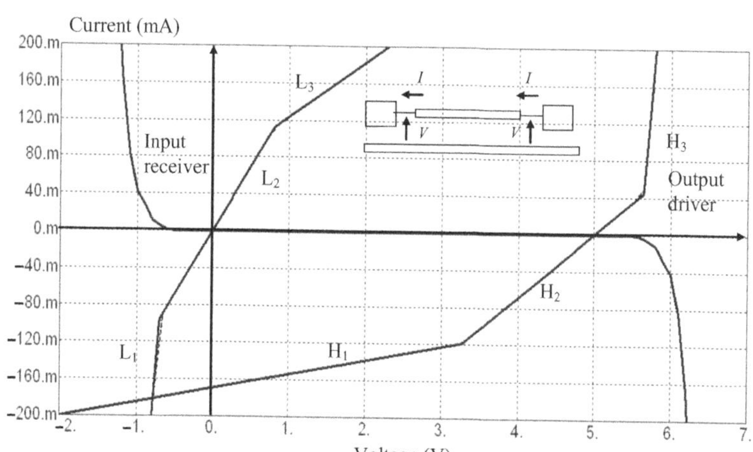

Figure 6.38 74AC00: I/O low and high static characteristics

```
define Tsw 7n
.define Tr 2n
.define Tf 2n
define Tp 60n
.define PH TABLE(t,0,0,Tsw,0,(Tsw+Tr),1,Tp,1,(Tp+Tf),0,100n,0)

.define PL (1-PH)
.define EOsLH (PH*EOH2+PL*EOL2)
.define ROsLH (PH*ROH2+PL*ROL2)
.define EOsHL (PH*EOH2+PL*EOL2)
.define ROsHL (PH*ROH2+PL*ROL2)
.define EO IF(t<=(Tsw+Tr),EOsLH,(IF(t>Tp,IF(t<(Tp+Tf),EOsHL,EOL),EOH)))
.define RO IF(t<=(Tsw+Tr),ROsLH,(IF(t>Tp,IF(t<(Tp+Tf),ROsHL,ROL),ROH)))
```

```
.define EOL1 -0.6
.define ROL1 1
.define EOL2 0
.define ROL2 7.2
.define EOL3 -1.1
.define ROL3 17
.define EOL4 -1000
.define ROL4 1K
...Same equations used
for 74F00
```

```
.define EOH1 11
.define ROH1 65
.define EOH2 5
.define ROH2 14.5
.define EOH3 5.6
.define ROH3 1
...Same equations used
for 74F00
```

Figure 6.39 Equations describing transition switching and the I/O characteristics of 74AC00 in MicroCap format

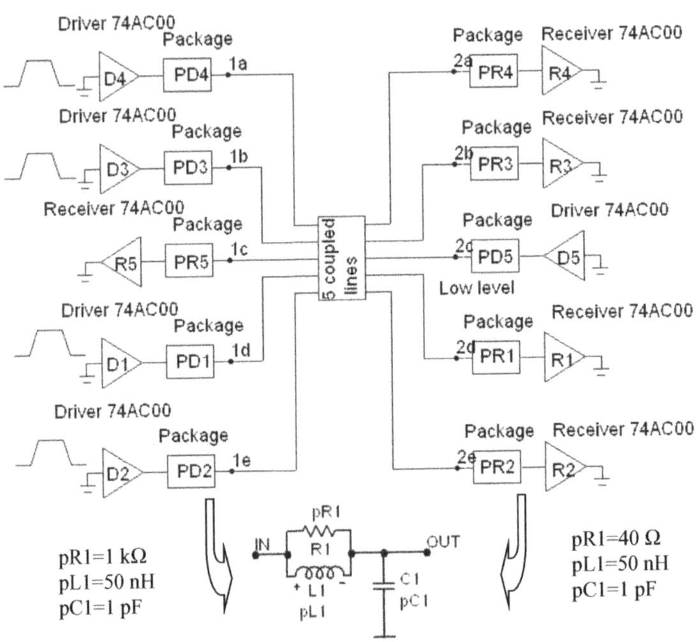

Figure 6.40 Schematic representation for SPICE (MicroCap) simulation with 74AC00 devices

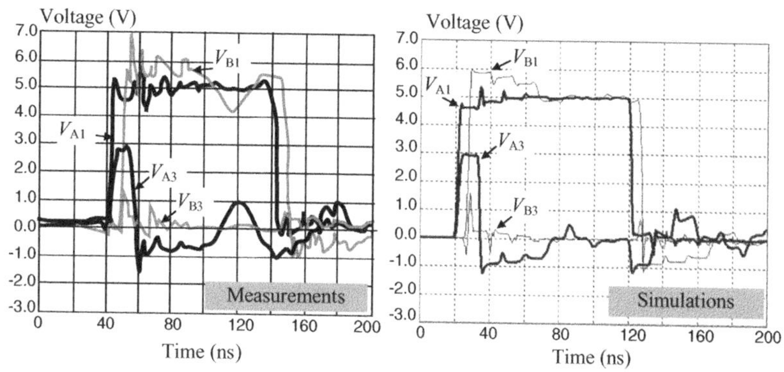

Figure 6.41 Measured and simulated waveforms of five coupled lines with 74AC00 devices

In conclusion, the following considerations can be summarized concerning crosstalk simulations:

- An equivalent circuit with few resistances and dependent sources based on modal analysis can model MTL interconnects.
- A field solver can provide accurate **L** and **C** matrices, so that the required line parameters such as delays, characteristic impedances, and coupling/decoupling $\mathbf{T}_V$ matrix can be calculated.
- A simple behavioral model based on the Thévenin equivalent circuit can simulate non-linear static and dynamic output characteristics of drivers.
- Non-linear static input characteristics of receivers can be simulated by a simple equivalent circuit consisting of a capacitance in parallel with a dependent current source.
- With the same interconnect structure, signal and *crosstalk* waveforms depend strongly on the I/O non-linear characteristics of the IC components and on their variations between worst-case and best-case values.
- IBIS models are suitable for worst-, typical-, and best-case simulations without performing measurements (see *Section 2.4*).

6.5 Techniques to Reduce Crosstalk

On the basis of what has been presented in the previous sections of this chapter, some basic rules can be deduced to mitigate the crosstalk in a PCB. Other useful information can be found in the literature [20–24]. This section ends with a discussion on how *crosstalk* investigations can be performed by using full-wave software tools.

6.5.1 Fixes to Reduce Crosstalk

Basic guidelines to mitigate crosstalk are listed below:

1. Use wide traces.
2. Locate traces near to the reference plane.
3. Increase the spacing between traces.

4. Match the lines.
5. Reduce the coupling length.
6. Provide other traces for shielding.
7. Avoid locating critical traces near to the edge of the PCB.
8. Ensure a solid ground plane without cuts.

Using wider traces or traces closer to a reference plane means higher trace-to-ground capacitance c_0 and lower self inductance L_w. Looking at the *near-end crosstalk* given by Equation (6.12), it can be observed that L_m/L_w increases while $c_m/C = c_m/(c_0 + c_m)$ decreases. However, the contribution made by the capacitive term is greater than that made by the inductive term, and therefore the crosstalk is reduced.

Increasing the spacing between traces means lower L_m and C_m and therefore less crosstalk.

Providing Thévenin termination to the line means more step signal launched onto the line, less impedance at the crosstalk point, and, above all, less reflections also on the victim line (see *Example 6.1*). The width of the *crosstalk* pulse depends linearly on the coupling length, so reducing the length means less crosstalk.

Grounding a trace at both ends between two coupled lines creates a shield effect, and therefore less *crosstalk*, as will be shown in the next subsection.

Very interesting examples of rules 2 and 3 are given by DeFalco [6], who discusses how *crosstalk* decreases with increasing spacing between traces and decreasing height of traces from the ground plane. An analytical method for *crosstalk* computation is also provided. Examples of how to investigate rules 7 and 8 will be given in *Section 6.5.3*.

6.5.2 Simulations of Coupled Lines with Grounded Traces used as a Shield

As an example of application of rule 6, the same structure as that used in *Example 6.6* with CMOS devices is simulated here using lines b and d as traces for shielding, as shown in Figure 6.42. Simulated signal and *crosstalk* waveforms for different grounding conditions of line b and d are shown in Figure 6.43. In case 1 (see Figure 6.43a), the traces are open at both ends; in case 2 (see Figure 6.43b), the traces are in open condition at the near end and grounded at the far end; in case 3 (see Figure 6.43c), the traces are grounded at both ends. Looking at the results, it can be seen that in case 1 there is no shielding effect, in case 2 the signal has less integrity and *crosstalk* is worst, and in case 3 a shielding effect is evident. Simulation verifies that traces devoted to shielding must be grounded at both ends for effective *crosstalk* mitigation.

6.5.3 Full-Wave Numerical Simulations of Two Coupled Lines

Full-wave numerical simulation codes can be used successfully to set design rules when the traces and ground plane in a PCB are considered, without the need to take into account the presence of drivers and receivers with their non-linearity. In this way, the advantages of the features offered by a 3D analysis can be exploited. Source and loads may be represented by simple circuit elements. In performing this type of analysis, some guidelines should be considered in order to avoid the need for excessive memory storage and a long computational time.

To start with, it is convenient to consider a simple structure and reproduce the computed waveforms by other tools (e.g. SPICE) or measurements. For example, consider the two

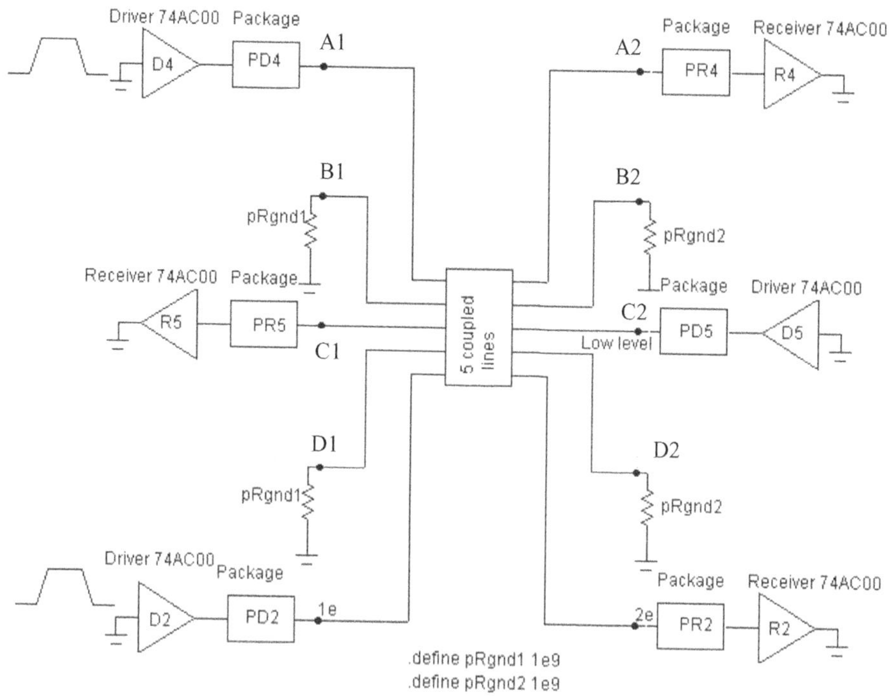

Figure 6.42 Schematic representation for circuit simulation of five coupled lines with 74AC00 devices. Traces b and d are used for shielding

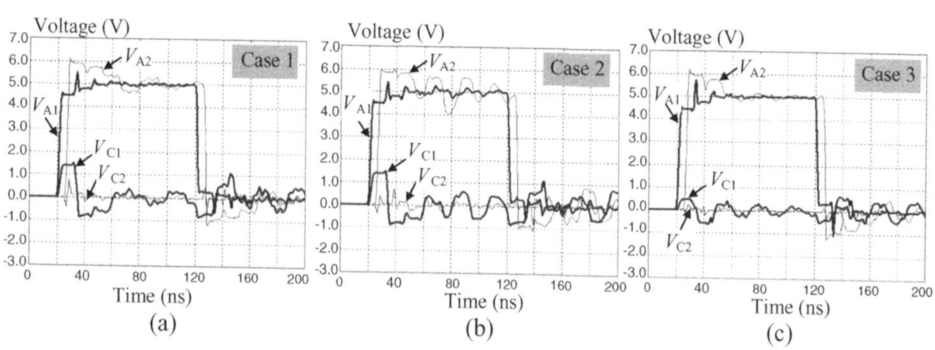

Figure 6.43 Simulations with different grounding conditions of traces for shielding: (1) shielding trace floating with pRgnd1 = 1e9, pRgnd2 = 1e9; (2) shielding trace connected to ground at the far end with pRgnd1 = 1e9, pRgnd2 = 1e-9; (3) shielding trace connected to ground at both ends with pRgnd1 = 1e-9, pRgnd2 = 1e-9

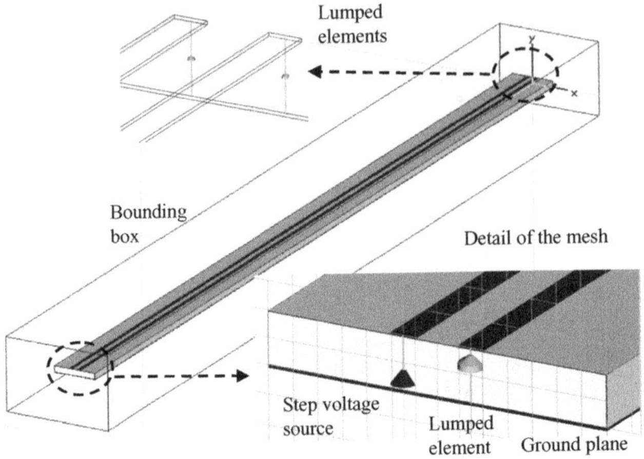

Figure 6.44 3D representation of two coupled microstrip lines of 10 cm length for crosstalk computation by the Microwave Studio software tool

coupled microstrip lines used in *Section 6.1* (*Example 6.1*). The structure under study in 3D is shown in Figure 6.44. The *MicroWave Studio* (MWS) code by CST [25] was used to perform the simulations. In preparing the model, attention should be paid to the following points:

- The ground plane should be finite with a width, w_d, roughly 6 times the height of the strips from the ground plane.
- The length of the lines, l, should be chosen in relation to the rise time t_r of the step source in order to produce maximum *crosstalk*.
- For an appropriate meshing, the frequency range of analysis should be extended to high frequency (the GHz region), correlated with a fast rise time (in the ps region) of the source.

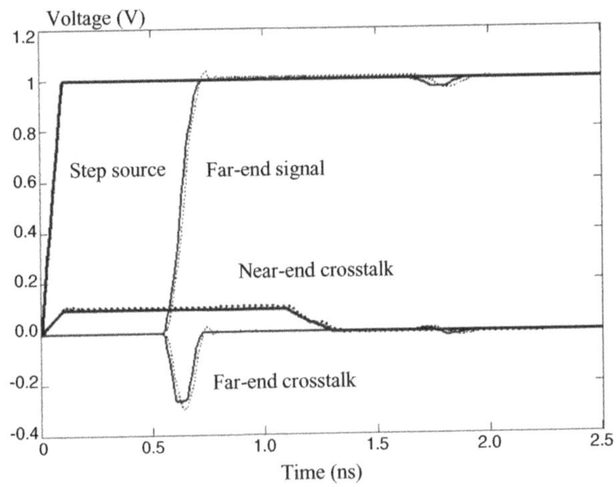

Figure 6.45 Comparison between signal and crosstalk waveforms computed by circuit model (solid lines) and by MWS (dotted lines) with the lines matched

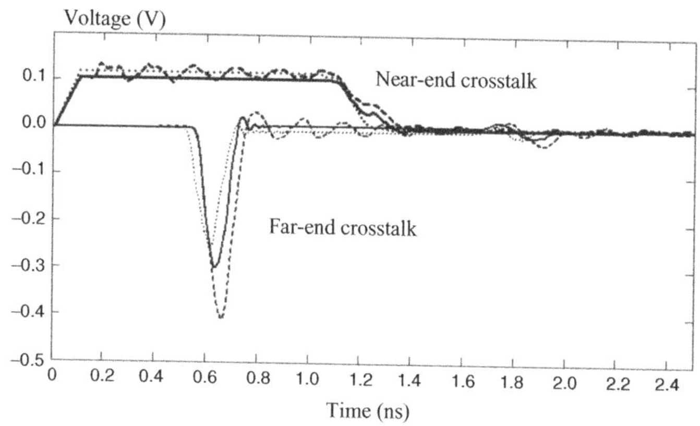

Figure 6.46 Crosstalk comparisons for three structures of two coupled microstrip lines computed by MWS: traces in the middle (solid line), traces at the edge (dotted line), and traces in the middle with cuts in the ground plane (dashed line)

For the example in Figure 6.44, the following data were used: $t_r = 10$ ps, $l = 10$ cm, $w_d = 5$ mm, $f_{min} = 0$, $f_{max} = 40$ GHz. With these values, a meshing of 114 464 cells was created within a bonding box. Figure 6.44 also shows some details of the mesh in accordance with the source location. The lumped element was assumed to be a resistance equal to the characteristic impedance of the coupled line, $Z_0 = \sqrt{L_w/(c_0 + c_m)} = 90\ \Omega$, in order to match the lines, as in *Section 6.1* and *Section 6.2*. Comparison between waveforms obtained with the exact model of *Section 6.2* based on modal decomposition and those computed by MWS is shown in Figure 6.45, where a very good agreement can be observed. Once the parameters chosen for the 3D structure have been validated, it is possible to perform other simulations considering the variations in the structure under study. For example, it could be interesting to investigate what happens to the *crosstalk* if the strips move towards the edge of the PCB or when some cuts are present in the ground plane.

Figure 6.46 shows the comparison between *crosstalk* waveforms when one of the strips is positioned exactly at the right edge of the PCB and when seven cuts are created at regular intervals perpendicular to the strips of size 4.72×0.7 mm. It can be noted an increase of the *near-end crosstalk*, that could be even worst if the spacing between the strips decreases, and the distortions introduced by the cuts.

With this simple example, the efficiency of a full-wave 3D simulation has been shown. More complicated multiconductor structures can be simulated once the user has gained confidence with this type of tool.

References

[1] Paul, C., *'Introduction to Electromagnetic Compatibility'*, John Wiley & Sons, Inc., Hoboken, NJ, 2006.

[2] Young, B., *'Digital Signal Integrity: Modeling and Simulation with Interconnects and Packages'*, Prentice Hall PTR, Upper Saddle River, NJ, 2001.

[3] Blood, W., *'MECL System Design Handbook'*, 4th edition, Motorola Semiconductor Products, Inc., Phoenix, AZ, 1983.

[4] Feller, A., Kaupp, H., and Digiacomo, J., 'Crosstalk and Reflections in High-Speed Digital Systems', Proceedings of Fall Joint Computer Conference, Las Vegas, NV, 30 November–2 December 1965.
[5] Catt, I., 'Crosstalk (Noise) in Digital Systems', *IEEE Trans. on Electronic Computers*, **16**(6), December 1967, 743–763.
[6] DeFalco, J., 'Reflection and Crosstalk in Logic Circuit Interconnections', *IEEE Spectrum*, **7**, July 1970, 44–50.
[7] www.spectrum-soft.com.
[8] Amemiya, H., 'Time-domain Analysis of Multiple Parallel Transmission Lines', *RCA Review*, June 1967, 241–276.
[9] Dvorak, V., 'Computer-aided Analysis of Pulse Signal Propagation in Digital Systems', *International Journal of Electronics*, **45**(6), 1978, 657–665.
[10] Caniggia, S., 'EMC Design of Digital System using Macromodelling Procedures for Integrated Circuits and their Interconnections', 5th Symposium and Technical Exibition on EMC, Zurich, 3–5 March 1983, 465–470.
[11] Chang, F.-Y., 'Transient Analysis of Lossless Coupled Transmission Lines in a Nonhomogeneous Dielectric Medium', *IEEE Trans. on Microwave Theory and Techniques*, **18**(9), September 1970, 616–626.
[12] Marx, K., 'Propagation Modes, Equivalent Circuits, and Characteristic Terminations for Multiconductor Transmission Lines with Inhomogeneous Dielectrics', *IEEE Trans. on Microwave Theory and Techniques*, **21**(7), July 1973, 450–457.
[13] Ho, C.W., 'Theory and Computer-aided Analysis of Lossless Transmission Lines', *IBM Journal of Research and Development*, May 1973, pp. 249–255.
[14] Ruehli, A.E., 'Survey of Computer-aided Electrical Analysis of Integrated Circuit Interconnections', *IBM Journal of Research and Development*, **23**(6), November 1979, 626–639.
[15] Agrawal, A., Fowles, H., Scott, L., and Gurbaxani, S., 'Application of Modal Analysis to the Transient Response of Multiconductor Transmission Lines with Branches', *IEEE Trans. on Electromagnetic Compatibility*, **21**(3), August 1979, 256–262.
[16] Chang, F.-Y., 'Computer-aided Characterization of Coupled TEM Transmission Lines', *IEEE Trans. on Circuits and Systems*, **27**(12), December 1980, 1194–1205.
[17] Agrawal, A. and Price, H., 'Experimental Characterization of Partially Degenerate Three-conductor Transmission Lines in the Time Domain', *IEEE Trans. on Electromagnetic Compatibility*, **23**(3), August 1981, 133–138.
[18] Caniggia, S., Catello, L., and Dorella, S., 'Noise Estimation on Signal Lines of Digital Systems using a Simple Macromodel Procedure', 7th Symposium and Technical Exhibition on Electromagnetic Compatibility, Zurich, 3–5 March 1987, 271–276.
[19] Caniggia, S., Catello, L., and Costa, V., 'Models for Circuit Simulators to Analyze Crosstalk and Radiation into Wires', IEEE International Symposium on Electromagnetic Compatibility, Seattle, 2–4 August 1988, 304–310.
[20] Johnson, H. and Graham, M., *'High-Speed Signal Digital Design: a Handbook of Black Magic'*, Prentice Hall PTR, Englewood Cliffs, NJ, 1993.
[21] Hall, S., Hall, G., and McCall, J., *'High-Speed Digital System Design – a Handbook of Interconnect Theory and Design Practice'*, John Wiley & Sons, Inc., New York, NY, 2000.
[22] Thierauf, S.C., *'High-Speed Circuit Board Signal Integrity'*, Artech House, Inc., Norwood, MA, 2004.
[23] Rainal, A., 'Transmission Properties of Various Styles of Printed Wiring Boards', *The Bell Technical Journal*, **58**(5), May–June 1979, 995–1025.
[24] Davidson, E., 'Electrical Design of a High Speed Computer Package', *IBM Journal of Research and Development*, **26**(3), May 1982, 349–361.
[25] www.cst.com.

7

Lossy Transmission Lines

At high speed of actual digital devices, interconnects behave as lossy *Transmission Lines* (TLs) in which the effects of losses can seriously degrade *Signal Integrity* (SI) quality. Accurate and efficient simulation techniques are needed during design and verification to ensure that TLs do not affect correct operation. For this reason, the problem of considering losses in simulating TLs has come to prominence.

In this chapter, lossy line fundamental parameters are introduced and their effect on signal propagation is discussed. The reflection mechanism due to losses along interconnects such as PCB traces or cables is described by the segmentation approach based on the decomposition of the line into a series cascade of elementary circuit cells. Each cell includes an impedance, representing the effects of losses, and a lossless transmission line whose parameters are the nominal characteristic impedance and delay time associated with the corresponding lossless line segment. Losses due to skin, proximity, and dielectric effects are introduced, and the frequency range where they become significant is discussed. Closed-form expressions for calculating these losses are given, including the proximity effect which cannot be directly computed. A coefficient K_{p} is introduced into the skin-effect expression to take account of the proximity effect. This coefficient can be derived by using full-wave codes or analytically considering the procedure outlined in *Appendix B* for microstrip and stripline traces. An analytical circuit approach for predicting the step response of a lossy line is also provided. This approach consists in simulating the lossy behavior of the line in the frequency domain and then using the discrete *Inverse Fourier Transform* (IFT) to obtain results in the time domain. In this way, the effect of losses in slowing down the rise time of the traveling signal can be highlighted.

A good circuit model for the interconnect is critical for accurate and efficient analysis of the transient propagation of signals. For this reason, in the second part of this chapter, modeling procedures to perform simulation of a lossy line directly in the time domain are outlined. Two main modeling procedures are presented. The first procedure is based on the segmentation approach (i.e. the TL is decomposed into a series cascade of sections) and the *Vector Fitting* (VECTFIT) technique which allows an electrically short segment of the cable with frequency-dependent losses to be represented as a network of lumped-circuit elements independent of frequency. This network reproduces the effects of frequency-dependent losses in the time domain. By this model, lossy lines with non-linear loads can be simulated directly

Signal Integrity and Radiated Emission of High-Speed Digital Systems Spartaco Caniggia and Francescaromana Maradei

in the time domain. The drawback of this modeling procedure is that the length of the lossy lines must not be too large, to avoid memory and time problems during the simulations. The second procedure for modeling lossy TLs in the time domain is based on scattering parameters which can be measured or computed in the time domain as the response of a step source for the line length of interest. Then, by a simple circuit that performs numerical derivative, convolution integral, and delay functions, the signal integrity simulation of a point-to-point structure with non-linear loads can be obtained without limitation in length. The proposed models are validated by comparing simulated waveforms with those obtained by experimental measurements.

7.1 Lossy Line Fundamental Parameters

To have suitable models for simulating lossy lines is a fundamental requirement for designing high-speed digital systems. Many approaches can be found in the literature [1–20]. The aim of this chapter is to provide simple models that make it possible, with acceptable approximation, to reproduce the eye diagrams of signaling in PCBs and cables. To build up these models, it is essential to begin with the definition and characterization of the fundamental parameters of a lossy line in different ranges of frequency.

7.1.1 Reflection Mechanism in a Lossy Line

The equivalent circuit of a lossy transmission line may be represented by a cascade connection of electrically short lumped-circuit elements, as shown in Figure 7.1 [21, 22]. This circuit allows for frequency-constant losses and is described by the following parameters:

- R_i = the p.u.l. internal resistance accounting for losses due to the signal and return conductors;
- L_0 = the p.u.l. external inductance;
- C_0 = the p.u.l. shunt capacitance;
- G_d = the p.u.l. shunt conductance depending on the conductivity of the substrate material and representing losses in the dielectric surrounding the conductors.

The nominal characteristic impedance Z_0 and the nominal velocity of propagation v_0 of a line are defined as the quantities associated with the line when losses are neglected. By this definition, if Z_0 and v_0 are known, it is possible to calculate L_0 and C_0 as follows:

$$L_0 = Z_0/v_0 \tag{7.1a}$$

$$C_0 = 1/(Z_0 v_0) \tag{7.1b}$$

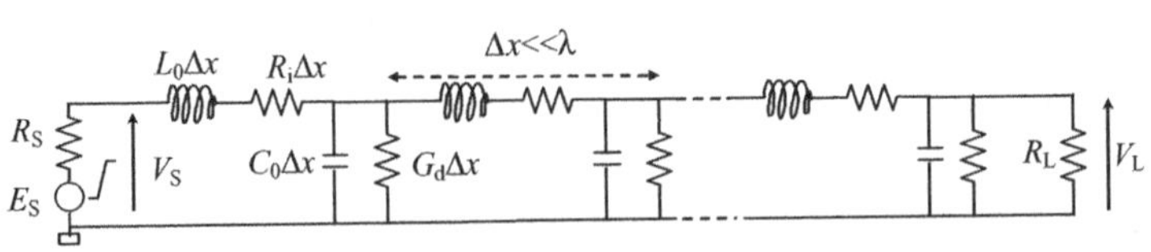

Figure 7.1 Lossy line equivalent circuit as a cascade connection of half-T cells

To understand how losses deteriorate the *signal integrity* of a transmitted signal across the line, the simple case of frequency-constant losses $R_i = R_{dc}$ is considered, and the line is modeled as a series cascade of lossless TLs of characteristic impedance Z_0 and propagation delay time $t_{pd}\Delta x$, connected to the series resistance $R_{dc}\Delta x$ representing the DC losses associated with the signal and return conductors of length Δx. Each subsection of the line of length Δx should be very short, usually at least $<\lambda/10$, where λ is the minimum wavelength of interest.

Let us consider the instant at which the voltage source E_S switches, and let us denote by V_S the voltage generated after the source resistance R_S. Owing to the partitioning effect between lumped resistances R_S and $R_{dc}\Delta x$ and Z_0, the amplitude of the voltage signal launched into the line is given by

$$V_S = E_S \frac{R_{dc}\Delta x + Z_0}{R_S + R_{dc}\Delta x + Z_0} \tag{7.2}$$

Again, owing to the partitioning effect, the voltage launched onto the first TL has the value ηV_S, where η is the transmission coefficient, as shown in Figure 7.2. When the step voltage ηV_S reaches the other end of the TL, there is a reflected voltage $\rho\eta V_S$ and a transmitted voltage $\eta V_S(1+\rho)$, where ρ is the refection coefficient, as shown in Figure 7.2. The voltage at the input of the second TL is $\eta^2 V_S(1+\rho)$, and a new reflection of value $\rho\eta V_S(1+\rho)$ is generated. This mechanism occurs with each TL, and the total waveforms at source and load ends are the algebraic sum, with suitable delays, of a large amount of reflections of this type. All this can be better represented by the lattice diagram shown in Figure 7.2. Note that the signal at the input of each TL towards the load is increasingly smaller than the starting signal V_S. This does not happen with lossless lines because the waveform that reaches the load is exactly the same as the waveform sent by the source.

In practical cases, the line p.u.l. internal resistance R_i has a more complicated expression than $R_i = R_{dc}$ and depends on frequency owing to skin and proximity effects. In order to account for frequency-dependent losses, the total p.u.l. series impedance $\hat{Z}(f)$ of any transmission line can be expressed as

$$\hat{Z}(f) = j\omega L_0 + \hat{Z}_i(f) \tag{7.3}$$

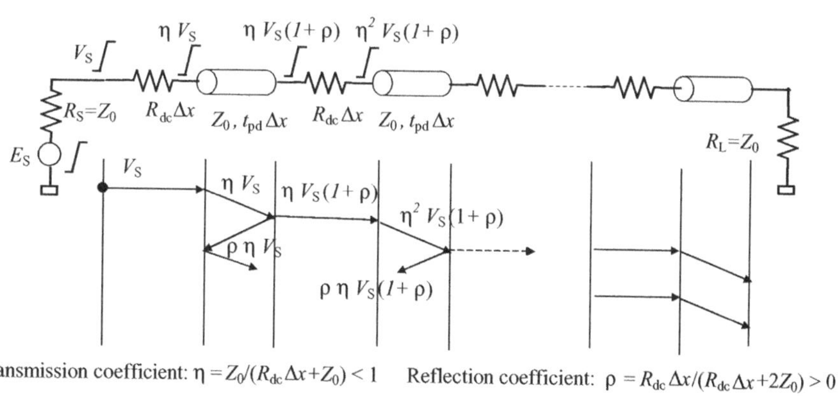

Figure 7.2 Reflection mechanism in a frequency-constant lossy line

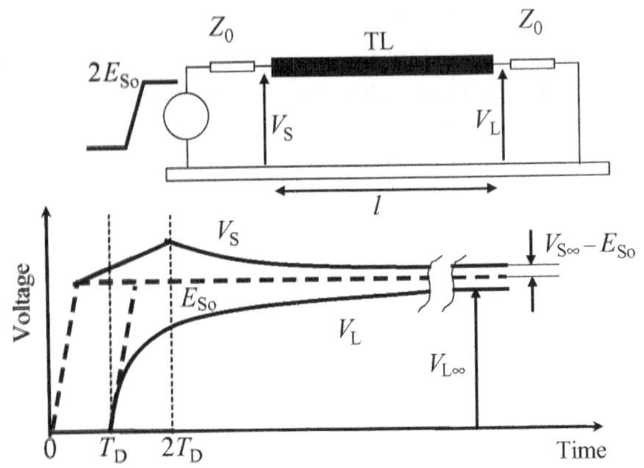

Figure 7.3 Typical source and load voltage waveforms for an interconnect matched at both ends: lossless TL (dashed line), frequency-dependent lossy TL (solid line)

where L_0 is the p.u.l. external series inductance related to the magnetic flux external to the conductor, and $\hat{Z}_i(f)$ is the line p.u.l. internal impedance.

The impedance $\hat{Z}_i(f)$ is a frequency-dependent complex parameter consisting of a real resistive part increasing with the frequency and an imaginary part representing the internal inductance of the conductor which decreases with the frequency.

In *Section 7.2* it will be shown that, by analogy with the circuit in Figure 7.2, the series cascade of $Z_i(f)\Delta x$ and a lossless TL can be used to simulate a frequency-dependent lossy line in the time domain once an appropriate network for $\hat{Z}_i(f)$ is derived. The network, which takes into account frequency-dependent losses, is extracted by the *Vector Fitting* (VF) technique and is composed of constant circuit elements R, L, and C in order to reproduce zeros and poles in the frequency domain of a small section of the interconnect.

Figure 7.3 illustrates the differences between a lossless line and a typical frequency-dependent lossy line, in terms of voltage waveforms at the input V_S and output V_L of the line of length l matched at both ends, when the source is a step voltage with a linear rise time. For a lossless line, the waveforms are equal in shape and separated by the line delay time $T_D = t_{pd}l$. For a lossy line, V_S is modified by reflections, rises to its maximum peak at time $2T_D$, and then decreases to its rest value $V_{S\infty}$ after a long period of time. Owing to the attenuation effect of the losses, V_L rises more slowly to its rest value $V_{L\infty}$. Final values depend on the total DC resistance $R_{dc}l$ of the line according to the equations

$$V_{S\infty} - E_{So} = \frac{R_{dc}l}{R_{dc}l + 2Z_0} \tag{7.4a}$$

$$V_{L\infty} = \frac{2Z_0}{R_{dc}l + 2Z_0} \tag{7.4b}$$

In *Section 7.2* it will be shown that the set-up of Figure 7.3 is very useful for obtaining parameters in order to simulate the lossy line in the transient domain by a distributed model.

In general, different types of loss characterize the line and contribute to determining the line p.u.l. internal impedance $\hat{Z}_i(f)$:

- DC losses;
- skin effect;
- proximity effect;
- radiation effect (less important and not treated in this book).

Moreover, dielectric effects should be considered to obtain a more realistic modeling of the line p.u.l. admittance $\hat{Y}(f)$ than that based on constant C_0 and G_d as shown Figure 7.1. A detailed analysis of these losses will be outlined in the following sections.

7.1.2 Skin Effect

The phenomenon of the skin effect is based on two facts: a current flowing in any real conductor produces an electric field given by Ohm's law; the current distribution and/or magnetic field distribution in a conductor is frequency dependent. For DC current in a single isolated conductor, the current density is uniform across the conductor. When alternating current is used, the current density is not uniform across the conductor. The current tends to concentrate on the conductor surface. Current density continuously increases from the conductor center to its surface, but, for practical purposes, the current penetration depth, δ, is assumed to be a dividing line for current density. The current is assumed to flow in an imaginary cylinder of thickness equal to the penetration depth δ with a constant current density throughout the cylinder thickness. The distribution of current densities for both actual and simplified models is shown in Figure 7.4 [22].

The penetration depth, expressed in meters and as function of the frequency f, is defined as

$$\delta(f) = \sqrt{\frac{1}{\pi f \mu \sigma}} \tag{7.5}$$

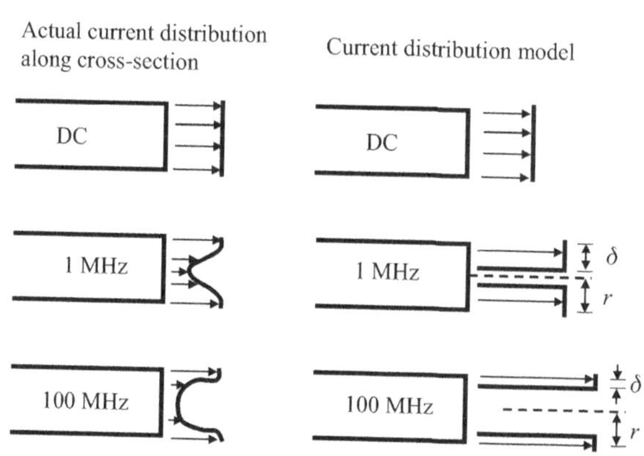

Figure 7.4 Skin effect: actual (left) and simplified (right) current distributions across a round conductor for several frequencies

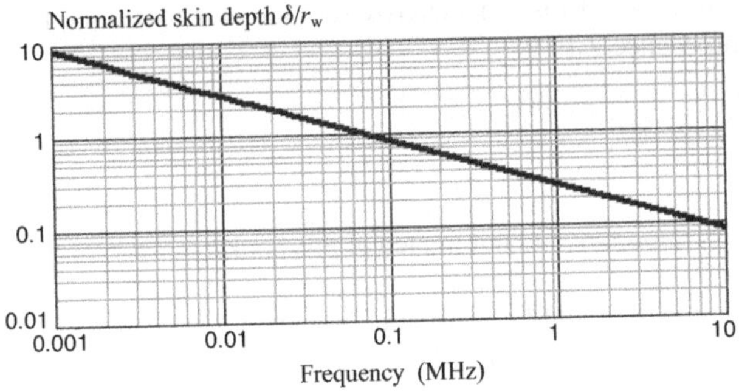

Figure 7.5 Normalized skin depth δ/r_w for a round copper wire of radius $r_w = 0.25$ mm

where μ is the magnetic permeability of the conducting material (for air $\mu = \mu_0 = 4\pi 10^{-7}$ H/m), and σ is the conductivity of the conducting material (for copper $\sigma_{Cu} = 5.8 \times 10^7$ S/m).

The variation in the penetration depth with frequency is shown in Figure 7.5 for the case of a round wire of radius $r_w = 0.25$ mm and copper material. The skin depth equals the radius at a frequency of 70 kHz and becomes 1/10 at a frequency of 8 MHz.

Because the skin effect reduces the equivalent conductor cross-sectional area, the frequency increase causes an increase in the p.u.l. effective resistance of the line. This in turn leads to an increasing attenuation with frequency. If the frequency response of a cable is plotted on log–log graph paper, log dB, or nepers versus log frequency, the curve slope will be 0.5 if the cable losses are primarily governed by classical skin effects. The slope of the attenuation curve, along with the attenuation at a particular frequency, can be used to estimate coaxial cable transient response as a function of length [22]. Losses in coaxial cable will be treated in *Section 7.2*.

7.1.2.1 Round Wires

The p.u.l. internal impedance $\hat{Z}_i(f)$ for an isolated round wire can be calculated exactly by the following equation [23]:

$$\hat{Z}_i(f) = \frac{jR_{surf}(f)}{\sqrt{2}\pi r_w}\left(\frac{Ber\,(0, q(f)) + jBei\,(0, q(f))}{dBer\,(q(f)) + jdBei\,(q(f))}\right) \tag{7.6}$$

where $Ber(0, q(f))$ and $Bei(0, q(f))$ are the real and imaginary parts of the complex Bessel function $\hat{J}_o(j^{-1/2}q)$, and

$$R_{surf}(f) = \frac{1}{\sigma\delta(f)} \tag{7.7a}$$

$$q(f) = \frac{\sqrt{2}r_w}{\delta(f)} \tag{7.7b}$$

$$dBer(q) = \frac{d}{dq}(Ber(0, q)) \tag{7.7c}$$

$$dBei(q) = \frac{d}{dq}(Bei(0, q)) \tag{7.7d}$$

These equations can be conveniently computed by mathematical programs such as MathCad™. The equations are valid on the assumption of a 'good conductor' material for which $\sigma \gg 2\pi f \varepsilon$ at all frequencies of interest. This is the case for all practical conductors.

At very low frequency or DC condition, the p.u.l. resistance R_{dc} and internal inductance $L_{i,dc} = L_{int}$ of the wire assume the values

$$R_{dc} = \frac{1}{\sigma \pi r_w^2} = \frac{1}{\sigma a} \tag{7.8a}$$

$$L_{i,dc} = \frac{\mu}{8\pi} \tag{7.8b}$$

where $a = \pi r_w^2$ is the round wire area, and R_{dc} doubles if the return conductor is equal to the signal conductor.

For *low frequencies* (LF), where the skin effect is not yet significant, $q(f)$ is small and series expansions of the Bessel functions show that $\hat{Z}_i(f)$ may be expanded as

$$\hat{Z}_{i,LF}(f) = \frac{1}{\pi r_w^2 \sigma}\left[1 + \frac{1}{48}\left(\frac{r_w}{\delta(f)}\right)^2\right] + j\frac{2\pi f \mu}{8\pi} \tag{7.9}$$

The p.u.l. resistance and internal inductance in the LF range are

$$R_{LF}(f) = \frac{1}{\pi r_w^2 \sigma}\left[1 + \frac{1}{48}\left(\frac{r_w}{\delta(f)}\right)^2\right] \tag{7.10a}$$

$$L_{i,LF}(f) = \frac{\mu}{8\pi} \tag{7.10b}$$

The first term of $R_{LF}(f)$ is the DC resistance, and the second is a correction useful for radius equal to the skin depth δ. The term $L_{i,LF}(f)$ corresponds to the low-frequency internal inductance $L_{i,dc}$ of the wire.

For *high frequencies* (HF), where $\delta < r_w$, the argument $q(f)$ is large. It may be shown that the high-frequency approximation of $\hat{Z}_i(f)$ is [23]

$$\hat{Z}_{i,HF}(f) = \frac{(1+j)R_{surf}(f)}{2\pi r_w} = \frac{(1+j)}{\sigma\delta(f)p} = \frac{(1+j)}{\sigma\delta(f_0)p}\sqrt{\frac{f}{f_0}} = (1+j)R_0\sqrt{\frac{f}{f_0}} \tag{7.11a}$$

$$R_0 = 1/(\sigma p \delta(f_0)) \tag{7.11b}$$

where f_0 is a particular frequency, chosen well above the onset frequency of the skin depth but below the non-TEM mode region, $p = 2\pi r_w$ is the perimeter of the round wire, and R_0 is the p.u.l. real part of the skin-effect impedance at f_0.

It can be noted that, in the HF range, resistance and internal reactance are equal.

The term $R_{surf}(\omega)/(2\pi r_w)$ in Equation (7.11) denotes the p.u.l. resistance of a round wire when the total current is assumed to be concentrated in an annular surface of thickness δ [24].

In fact

$$\frac{1}{\sigma\left[\pi r_{\mathrm{w}}^2 - \pi(r_{\mathrm{w}} - \delta(f))^2\right]} \approx \frac{1}{\sigma 2\pi r_{\mathrm{w}}\delta(f)} = \frac{R_{\mathrm{surf}}(f)}{p} \quad \text{for} \quad r_{\mathrm{w}} \gg \delta \tag{7.12}$$

The resistive and inductive parts at high frequencies are given by [25]

$$R_{\mathrm{HF}}(f) = \mathrm{Re}\left[\hat{Z}_{\mathrm{iHF}}(f)\right] = R_0\sqrt{\frac{f}{f_0}} \tag{7.13a}$$

$$L_{\mathrm{i,HF}}(f) = \frac{\mathrm{Im}\left[\hat{Z}_{\mathrm{iHF}}(f)\right]}{2\pi f} = \frac{R_0}{2\pi}\sqrt{\frac{1}{f f_0}} \tag{7.13b}$$

Note from Equation (7.5) that the skin depth decreases with increasing frequency as the inverse square root of the frequency. Thus, the high-frequency resistance $R_{\mathrm{HF}}(f)$ increases at a rate of 10 dB/decade. The resistance remains at the DC value up to the frequency where these two asymptotes meet, or $r_{\mathrm{w}} = 2\delta(f_\delta)$. The skin-effect frequency f_δ can be found by solving the equation $R_{\mathrm{dc}} = R_{\mathrm{HF}}(f_\delta)$ and is given by

$$f_\delta = \left(\frac{p}{a}\right)^2 \frac{1}{\mu\sigma\pi} \tag{7.14}$$

Equation (7.13b) shows that the high-frequency internal inductance $L_{\mathrm{i,HF}}$ decreases at a rate of 10 dB/decade after frequency f_δ. Below this frequency the internal inductance has a constant value (Equation (7.10b)).

For the purposes of computation by mathematical programs, $\hat{Z}_{\mathrm{i}}(f)$ can be defined by three different approximate formulae [25]:

$$\hat{Z}_{\mathrm{ia}}(f) = \begin{cases} \hat{Z}_{\mathrm{i,LF}}(f), & \text{if } f < f_\delta \\ \hat{Z}_{\mathrm{i,HF}}(f), & \text{if } f > f_\delta \end{cases} \tag{7.15}$$

$$\hat{Z}_{\mathrm{ib}}(f) = R_{\mathrm{dc}} + \hat{Z}_{\mathrm{i,HF}}(f) \tag{7.16}$$

$$\hat{Z}_{\mathrm{ic}}(f) = \sqrt{R_{\mathrm{dc}}^2 + (\hat{Z}_{\mathrm{i,HF}}(f))^2} \tag{7.17}$$

Comparison of the exact method (Equation (7.6)) for round wire and approximate methods (Equations (7.15)–(7.17)) is shown in Figure 7.6. It can be seen that $\hat{Z}_{\mathrm{ia}}(f)$ and $\hat{Z}_{\mathrm{ic}}(f)$ practically coincide with the exact impedance $\hat{Z}_{\mathrm{i}}(f)$, while $\hat{Z}_{\mathrm{ib}}(f)$ is significantly different in the transition region from DC to the high-frequency region where the skin effect dominates. The imaginary part tracks upwards as if it were an inductor of value $L_{\mathrm{i,dc}}$ under the skin-effect onset frequency f_δ. Real and imaginary parts match above f_δ; both track upwards at +10 dB/decade. From now on, the approximation $\hat{Z}_{\mathrm{ic}}(f)$ will be used for computations. The above expressions can also be used, with good approximation, for other conductors of different shape but having area a and perimeter p.

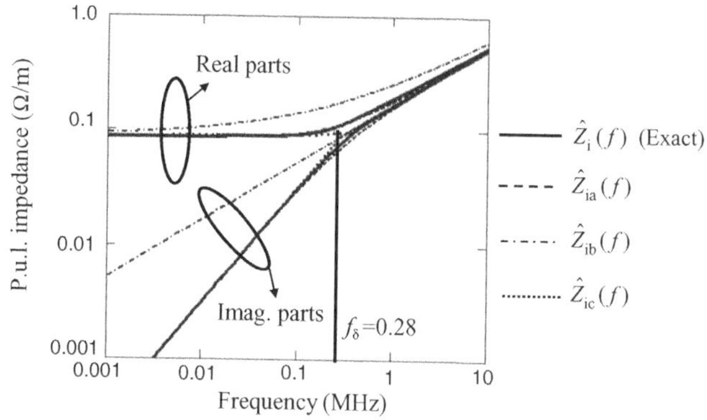

Figure 7.6 Skin-effect impedances computed by exact and approximate methods for a round copper wire of 1 m length and 0.25 mm radius

7.1.2.2 Rectangular Conductors

For a rectangular conductor such as a trace in a PCB, expression (7.17) of $\hat{Z}_{ic}(f)$ can be used, adopting in the computation the following perimeter and area:

$$p = 2(w + t) \tag{7.18a}$$

$$a = wt \tag{7.18b}$$

where w and t are the width and thickness of the rectangular conductor respectively.

This procedure can be used, as, similarly to the case of round wires, the current becomes concentrated in a thickness equal to one skin depth at the surface as the frequency increases. Actually, the current also peaks at the corners, but this fact may be practically neglected to simplify the computation. As for wires, the skin effect becomes significant when the thickness of the smaller dimensions equals two skin depths.

7.1.3 Proximity Effect

The proximity effect is the current density redistribution in a conductor owing to the mutual repulsion (or attraction) generated by currents flowing in nearby conductors [25]. The current density at those points on the conductor close to neighboring conductors is different than that occurring in isolated conductors. This current density redistribution reduces the effective cross-sectional area of the conductor, thereby increasing the p.u.l. resistance. This effect is a function of the conductor diameters, the separation of the conductors from each other, and frequency. Analytical evaluation of the proximity effect is quite complicated, and, except for certain limited cases, no general rule-of-thumb expressions have been proposed. The proximity effect is not present in coaxial cables because of their circular symmetry. The proximity effect is a significant contributor to signal losses, particularly in cases of twisted-pair cables or parallel wire lines such as in ribbon and twinax cables.

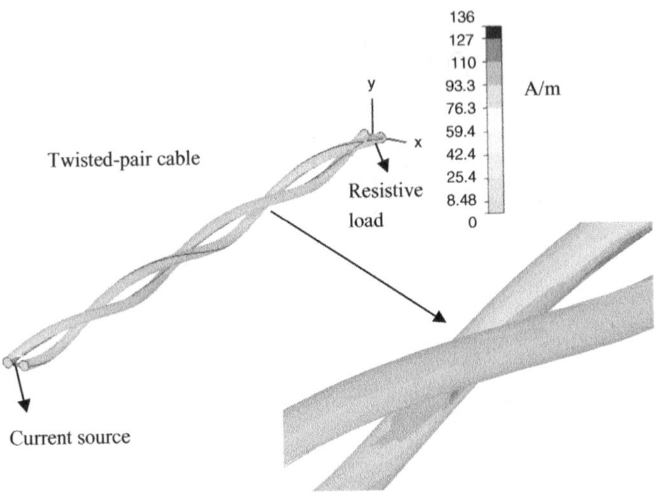

Figure 7.7 Proximity-effect example: computed tangential H-field to show the surface current (peak) at 5 GHz in an unshielded twisted-pair cable

As an example, the surface current at the frequency $f = 5$ GHz in a two-twisted-pair line computed by a full-wave numerical code is shown in Figure 7.7. The currents in each wire are equal but of opposite sign, and therefore surface currents are higher in the internal part of the cable.

To take into account analytically the proximity effect, a factor K_p is introduced into the equations used for skin-effect computations. In practice, expression (7.11a) of the high-frequency internal impedance is still valid, with the only exception that in this case the real part of skin-effect impedance R_0 is given by [25]

$$R_0 = \frac{K_p}{\sigma p \delta(f_0)} \tag{7.19}$$

where the proximity-effect factor K_p is present. Equation (7.19) is valid for operating frequency $f > f_\delta$, where f_δ is the starting frequency for skin and proximity effects, which can be derived by analogy with *Section 7.1.2.1*, and is given by

$$f_\delta = \left(\frac{p}{K_p a}\right)^2 \frac{1}{\mu \sigma \pi} \tag{7.20}$$

It is important to consider the following points:

- $K_\text{p} = 1$ for any conductor well separated from its return path.
- K_p increases as the conductor and its return path are brought closer.
- K_p may be computed by numerical codes, as analytical expressions are not available.

As will be shown by some examples, factor K_p can be obtained by TDR measurements (see *Section 11.1*), by numerical simulations (see *Section 7.2*), or by an indirect analytical approach as described in *Appendix B* for microstrip and stripline traces. Anyway, values of K_p for typical cables and traces used in PCBs are given by Johnson and Graham [25].

7.1.4 Lossy Dielectric Effect

If a time-harmonic voltage $\hat{V}(\omega)$ is applied to a material of surface area a and thickness h, the corresponding current $\hat{I}(\omega)$ is given by [25]

$$\hat{I}(\omega) = \hat{V}(\omega)\frac{a}{h}(\sigma + \mathrm{j}\omega\varepsilon) \tag{7.21}$$

where $\omega = 2\pi f$ is the angular frequency in rad/s, σ is the conductivity of the material in S/m, and ε is the permittivity of the material in F/m.

The current is therefore the sum of a term in-phase with the voltage source called the *conduction current* (it behaves like a resistor) and a term in-quadrature with the voltage source called the *displacement current* (it behaves like a capacitor).

For a good conductor, $\sigma \gg \omega\varepsilon$, which means that the current flows mostly in-phase with the voltage. Both σ and ε stay fairly constant over a wide range of frequency. For any conductor there is a critical frequency $f_c = \sigma/(2\pi\varepsilon)$ above which the displacement current dominates and the material mostly acts like a capacitor. Any material that operates at a frequency well above f_c is classified as a *good insulator*. For this material, the ratio $\sigma/\omega\varepsilon$ remains nearly constant. Equation (7.21) can also be written as

$$\hat{I}(\omega) = \hat{V}(\omega)\frac{a}{h}\mathrm{j}\omega\hat{\varepsilon}_c(\omega) \tag{7.22}$$

where $\hat{\varepsilon}_c(\omega) = \varepsilon' - \mathrm{j}\varepsilon''$ is the *complex permittivity* of a material, the real part of which $\varepsilon' = \varepsilon$ is related to the displacement current, while the imaginary part $\varepsilon'' = \sigma/\omega$ is related to the conduction current. For a good insulator, the imaginary part ε'' is much smaller than the real part ε'. In practice, for a good insulator, two parameters are specified:

- the dielectric permittivity of the material $\varepsilon' = \varepsilon$;
- the *dielectric loss tangent* $\tan\theta$, defined by

$$\tan\theta = \frac{\varepsilon''}{\varepsilon'} \tag{7.23}$$

Many textbooks tabulate the loss tangent for different materials and different frequencies, but the main properties of the dielectric loss tangent are as follows:

- It remains stable for a broad range of frequency.
- It has a positive value.

Note that in this chapter the loss tangent is referred to as $\tan\theta$ to avoid confusion with the skin depth notation, but in the literature it is widely denoted by $\tan\delta$. For a typical FR4 substrate of a PCB, $\tan\theta = 0.02$. For $\tan\theta < 0.05$, it is possible to use the approximation $\tan\theta \approx \theta$.

The permittivities of free space and air are considered practically the same and assume the value $\varepsilon_0 = 8.854 \times 10^{-12}$ F/m. Therefore, the *complex relative permittivity* $\hat{\varepsilon}_{\mathrm{rc}}(\omega)$ of the substrate of a PCB or of the insulation material in a cable is defined as

$$\hat{\varepsilon}_{\mathrm{rc}}(\omega) = \varepsilon_{\mathrm{r}}' - \mathrm{j}\varepsilon_{\mathrm{r}}'' = \frac{\varepsilon'}{\varepsilon_0} - \mathrm{j}\frac{\varepsilon''}{\varepsilon_0} \tag{7.24}$$

In the case of non-dispersive dielectric materials, the real part of the complex relative permittivity is equal to the relative dielectric permittivity ε_{r}. When the dispersion needs to be considered, as in the case of PCBs with a very high working frequency, more sophisticated models such as Debye may be used to describe the relative permittivity of the medium, and the real part of the complex relative permittivity in Equation (7.24) no longer coincides with ε_{r} [26, 27].

Looking at the line of Figure 7.1 and applying these last concepts, the shunt current for an electrically short section of length Δx can be represented by the equation

$$\hat{I}_{\mathrm{Shunt}}(x,\omega) = \hat{Y}_{\mathrm{Diel}}(\omega)\Delta \mathrm{x}\hat{V}(x,\omega) \tag{7.25}$$

where $\hat{Y}_{\mathrm{Diel}}(\omega)$ is the p.u.l. complex dielectric admittance of the transmission line, which includes the effects of the capacitance C_0 and of the conductance G_{d}. To obtain an analytical expression for $\hat{Y}_{\mathrm{Diel}}(\omega)$, the following considerations should be taken into account. The phase of the complex permittivity is $-\theta$ constant over a wide frequency range; then the log–log slope of the magnitude of the permittivity over that range must be very close to $-(2/\pi)\theta$, as stated by Bode in the general theory of phase/magnitude relations. Therefore, $\hat{\varepsilon}_{\mathrm{rc}}(\omega)$ can be represented as [25]

$$\hat{\varepsilon}_{\mathrm{rc}}(\omega) = k\,(\mathrm{j}\omega)^{-2\theta/\pi} \tag{7.26}$$

where k is an arbitrary real constant. Starting from this assumption, it can be shown that the dielectric complex admittance $\hat{Y}_{\mathrm{Diel}}(\omega)$ is given by [25]

$$\hat{Y}_{\mathrm{Diel}}(\omega) = G_{\mathrm{d}}(\omega) + \mathrm{j}\omega C(\omega) = \mathrm{j}\omega C_0 \left(\frac{\mathrm{j}\omega}{\omega_0}\right)^{-2\theta_0/\pi} \tag{7.27}$$

where $\omega_0 = 2\pi f_0$ is the angular frequency chosen in the frequency range where the lossy dielectric effect is significant, and θ_0 is the dielectric loss tangent at frequency f_0.

Another approach is that reported by Paul [24] with reference to the case of a structure in a homogeneous medium. According to this approach, the capacitance is proportional to the cross-sectional dimensions and the permittivity of the dielectric, and therefore, using ε_{r} and $\tan\theta$ definitions, the following equation holds:

$$\begin{aligned}\hat{Y}_{\mathrm{Diel}}(\omega) &\approx \mathrm{j}\omega\hat{\varepsilon}_{\mathrm{c}}k = \mathrm{j}\omega\varepsilon_0\hat{\varepsilon}_{\mathrm{rc}}k = \omega\varepsilon_0\varepsilon_{\mathrm{r}}''k + \mathrm{j}\omega\varepsilon_0\varepsilon_{\mathrm{r}}'k \\ &= \omega\varepsilon_0 \tan\theta\,\varepsilon_{\mathrm{r}}'k + \mathrm{j}\omega\varepsilon_0\varepsilon_{\mathrm{r}}'k = G_{\mathrm{d}}(\omega) + \mathrm{j}\omega C_0\end{aligned} \tag{7.28}$$

As previously outlined, numerous handbooks tabulate the loss tangent for different materials and different frequencies. As, for FR-4, which is usually used to construct PCBs, the loss tangent is fairly constant at high frequencies and is approximately 0.02, the p.u.l. conductance $G_d(\omega)$ can be written as

$$G_d(\omega) = \omega C_0 \tan\theta \approx \omega C_0 \theta \tag{7.29}$$

This result only applies to lines in a homogeneous medium, but the following consideration may be extended to lines in an inhomogeneous medium. With the loss tangent $\tan\theta$ practically constant, observe that the p.u.l. conductance increases directly with frequency, 20 dB/decade, instead of the 10 dB/decade of the skin-effect resistance.

It should be pointed out that, in model (7.27), both conductance and capacitance are frequency dependent. On the other hand, in Equation (7.28) the capacitance is constant and equal to the nominal line parameter C_0 of a lossless line, while the conductance is frequency dependent according to Equation (7.29). The two models produce the same conductance, as will be shown in *Example 7.1*.

7.1.5 Data Transmission with Lossy Lines

The main goal with lossy lines in high-speed digital links is to have methods for predicting, in the time domain, the deteriorations in the transmitted data caused by the different types of loss, which depend on frequency. To do this, the first step is to investigate the transmission-line properties in the frequency domain, which can be done by considering the transfer function of a line of length l matched at both ends [25]:

$$\hat{H}(\omega, l) = \mathrm{e}^{-l\hat{\gamma}(\omega)} \tag{7.30}$$

where $\hat{\gamma}(\omega)$ is the transmission-line propagation coefficient, given by

$$\hat{\gamma}(\omega) = \alpha(\omega) + \mathrm{j}\beta(\omega) = \sqrt{\hat{Z}(\omega)\hat{Y}(\omega)} \tag{7.31}$$

In Equation (7.31), $\hat{Z}(\omega)$ and $\hat{Y}(\omega)$ are the p.u.l. impedance and admittance of the line. Note that one of the significant effects of losses is that the amplitude of each of the sinusoidal waves that comprise the time-domain signal is attenuated as they travel down the line. This is given by the real part of $\hat{\gamma}(\omega)$. The attenuation in dB is usually defined in two ways:

$$\alpha(\omega)_{\mathrm{dB}} = -20\log\left(|\hat{H}(\omega, l)|\right) = \frac{20\,l}{\ln(10)}\mathrm{Re}[\hat{\gamma}(\omega)] = 8.6858896\,l\,\alpha(\omega)_{\mathrm{neper}},\ \mathrm{dB} \tag{7.32a}$$

$$\alpha(\omega)_{\mathrm{neper}} = \mathrm{Re}\left[\hat{\gamma}(\omega)\right],\ \mathrm{neper/m} \tag{7.32b}$$

Therefore, 1 neper = 8.6858896 dB. The velocity of propagation can be computed as

$$v(\omega) = \frac{\omega}{\beta(\omega)} \tag{7.33}$$

This means that, as digital pulses are composed of a large number of sinusoidal components that travel down the line at different speeds and reach the load with different amplitudes, the starting pulse signal will be reconstructed on the load with distortion. This is called *dispersion.* It will be shown that losses attenuate the higher-frequency components more than the lower-frequency components. Hence, the bandwidth of the pulse is reduced and the pulse rise/fall times are increased.

The other important parameter of an interconnect is the complex characteristic impedance, defined as

$$\hat{Z}_0(\omega) = \sqrt{\frac{\hat{Z}(\omega)}{\hat{Y}(\omega)}} \tag{7.34}$$

7.1.5.1 AC Analysis of Loss Influence on TL Performance

In this section an investigation is made of the influence of different types of loss in the whole frequency range starting from DC to very high frequency [25].

For traces in PCBs and cables, three regions can be distinguished where, in turn, the DC effect, the skin plus proximity effect, and the dielectric effect become significant. Adopting this subdivision of the frequency spectrum, the p.u.l. impedance and admittance can be summarized as

$$\hat{Z}(\omega) = \begin{cases} R_{dc} + j\omega L_0 & \text{DC or LC region} \\ \hat{Z}_{ic}(\omega) + j\omega L_0 & \text{skin- and proximity-effect region} \\ \hat{Z}_{ic}(\omega) + j\omega L_0 & \text{dielectric-effect region} \end{cases} \tag{7.35}$$

$$\hat{Y}(\omega) = \begin{cases} j\omega C_0 & \text{DC or LC region} \\ j\omega C_0 & \text{skin- and proximity-effect region} \\ \hat{Y}_{Diel}(\omega) & \text{dielectric-effect region} \end{cases} \tag{7.36}$$

where L_0 and C_0 are the p.u.l. line external inductance and capacitance, R_{dc} is the p.u.l. DC line resistance given by Equation (7.8a), $\hat{Z}_{ic}(\omega)$ is the p.u.l. frequency-dependent internal impedance of the line according to expression (7.17), and $\hat{Y}_{Diel}(\omega)$ is the p.u.l. line admittance accounting for frequency-dependent dielectric losses given by Equation (7.27).

(i) DC Region

In the first region, also called the LC region, the propagation coefficient and the characteristic impedance of the TL are given by

$$\hat{\gamma}_{LC}(\omega) = \sqrt{(R_{dc} + j\omega L_0)(j\omega C_0)} \tag{7.37a}$$

$$\hat{Z}_{0,LC}(\omega) = \sqrt{\frac{R_{dc} + j\omega L_0}{j\omega C_0}} \tag{7.37b}$$

For practical purposes, this is the region characterized by $f_{LC} < f < f_\delta$, with $f_{LC} = R_{dc}/(2\pi L_0)$, and f_δ is given by Equation (7.20). In this case, the characteristic impedance (Equation (7.37b)) is approximately equal to the nominal characteristic impedance of a lossless line, i.e. $\hat{Z}_{0,LC}(\omega) \approx Z_0 = \sqrt{L_0/C_0}$.

Below the frequency $f_{\rm LC}$, another region, called the RC region, should be considered, where only the line parameters $R_{\rm dc}$ and C_0 are significant. As the study of this type of region is beyond the scope of this book, the reader can find details about the general behavior within the RC region in the work by Johnson and Graham [25].

(ii) Skin- and Proximity-Effect Region

This second region is defined in the frequency range $f_\delta < f < f_\theta$, where f_δ is the onset frequency of the skin and proximity effect given by Equation (7.20), and f_θ is the onset frequency of the dielectric effect, the definition of which is given in the following frequency region. The propagation coefficient and the characteristic impedance of the TL are given by

$$\hat{\gamma}_{\rm Skin}(\omega) = \sqrt{(\hat{Z}_{\rm ic}(\omega) + {\rm j}\omega L_0)({\rm j}\omega C_0)} \tag{7.38a}$$

$$\hat{Z}_{0,{\rm Skin}}(\omega) = \sqrt{\frac{\hat{Z}_{\rm ic}(\omega) + {\rm j}\omega L_0}{{\rm j}\omega C_0}} \tag{7.38b}$$

It can be shown that the attenuation, in neper/m, is [25]

$$\alpha_{\rm Skin}(\omega) = {\rm Re}\,[\hat{\gamma}_{\rm Skin}(\omega)] = \frac{1}{2}\frac{R_0}{Z_0}\sqrt{\frac{\omega}{\omega_0}}, \quad \text{neper/m} \tag{7.39}$$

where R_0 is given by Equation (7.19). To obtain Equation (7.39), it was assumed that $|{\rm j}\omega L_0| \gg |\hat{Z}_{\rm ic}(\omega)|$, which is reasonable in the frequency range of interest.

(iii) Dielectric-Effect Region

In this region the propagation coefficient and the characteristic impedance of the TL are given by

$$\hat{\gamma}_{\rm Diel}(\omega) = \sqrt{(\hat{Z}_{\rm ic}(\omega) + {\rm j}\omega L_0)\hat{Y}_{\rm Diel}(\omega)} \tag{7.40a}$$

$$\hat{Z}_{0,{\rm Diel}}(\omega) = \sqrt{\frac{\hat{Z}_{\rm ic}(\omega) + {\rm j}\omega L_0}{\hat{Y}_{\rm Diel}(\omega)}} \tag{7.40b}$$

It can be shown that the attenuation, in neper/m, is [25]

$$\alpha_{\rm Diel}(\omega) = {\rm Re}\,[\hat{\gamma}_{\rm Diel}(\omega)] = \frac{1}{2}\frac{\theta_0\omega}{v_0}\left(\frac{\omega}{\omega_0}\right)^{-\theta_0/\pi}, \quad \text{neper/m} \tag{7.41}$$

where θ_0 is the angle corresponding to the dielectric loss tangent $\tan\theta$ at the frequency f_0. This region begins with the frequency f_θ. The definition of f_θ comes equating the attenuation of the skin and dielectric regions and neglecting the slowing varying term $(\omega/\omega_0)^{-\theta_0/\pi}$, so that

$$f_\theta = \frac{1}{2\pi\omega_0}\left[\frac{v_0}{Z_0}\frac{R_0}{\theta_0}\right]^2 \tag{7.42}$$

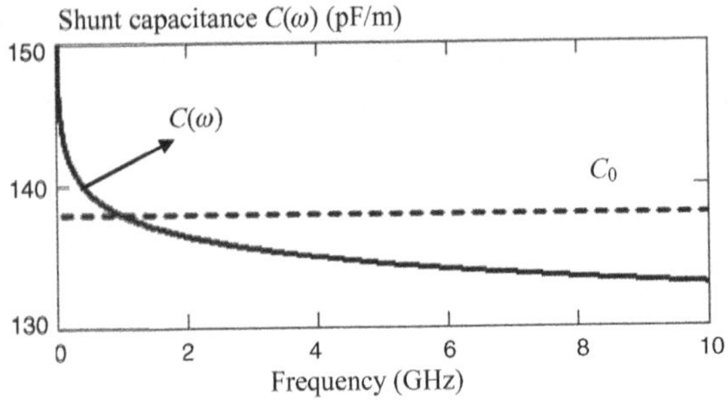

Figure 7.8 Shunt capacitance as a function of frequency

Example 7.1: Trace with 50 Ω Characteristic Impedance

For discussion purposes, consider a copper trace in a PCB with the following characteristics [25]: size 0.150 mm × 0.0174 mm; length $l = 1$ m; $L_0 = 346$ nH; $C_0 = 138$ pF; $\tan\theta_0 = 0.025$ at $f_0 = 1$ GHz; $K_p = 1$. These parameters give $Z_0 = 50\ \Omega$, $R_{dc} = 6.6\ \Omega$, $f_{LC} = 3$ MHz, $f_\delta = 71.9$ MHz, and $f_\theta = 205.6$ MHz. The response of the trace up to 10 GHz is calculated.

The frequency-dependent capacitance $C(\omega)$ obtained by Equation (7.27) is compared with the nominal capacitance C_0 in Figure 7.8. Note that the capacitance decreases slightly with increasing frequency in the frequency range where the dielectric effect becomes significant, in other words, above 205.6 MHz. The p.u.l. shunt resistance $R_d(\omega) = 1/G_d(\omega)$ obtained by the two models (7.27) and (7.29) is plotted in Figure 7.9. The two approaches provide the same results.

The real part of the complex characteristic impedance $\hat{Z}_0(\omega)$ is plotted in Figure 7.10 for the three types of region discussed above. The imaginary part (not plotted) goes to zero at 100 MHz. Note that the characteristic impedance approaches its nominal value $Z_0 = \sqrt{L_0/C_0}$ at about 10 MHz.

The propagation coefficient $\hat{\gamma}(\omega)$, with its real (attenuation) and imaginary (phase) parts, versus frequency is shown in Figure 7.11 for the PCB trace considered. The case denoted as $\hat{\gamma}_{\mathrm{Diel}}(\omega)$ considers all kinds of loss. The scale is logarithmic for both frequency and propagation coefficient. Below f_{LC}, both attenuation and phase rise together in proportion to the

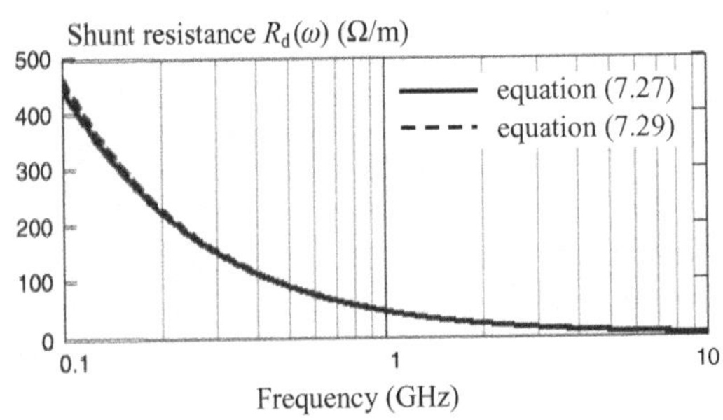

Figure 7.9 Shunt resistance R_d as a function of frequency, obtained by the two different models (7.27) and (7.29)

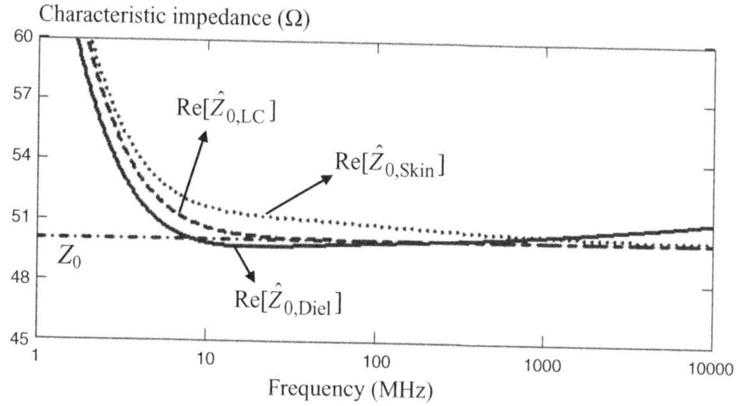

Figure 7.10 Real part of complex characteristic impedance

square root of frequency. Above f_L, phase grows linearly with increasing frequency for each type of loss, while attenuation is flat when DC lossy is considered only.

When skin-effect losses are considered above f_δ, the attenuation curve has a slope proportional to the square root of frequency (dotted line). When dielectric-effect losses are considered above f_θ, the attenuation curve tends to a slope proportional to the frequency (solid line). The plot also shows the case of $\hat{\gamma}(\omega)$ calculated without considering the internal impedance (longer dashed line), in other words, considering the dielectric losses only. The slope of this curve is proportional to the frequency. Above f_θ, dielectric losses dominate. A smooth arc represents the slope change from one region to another.

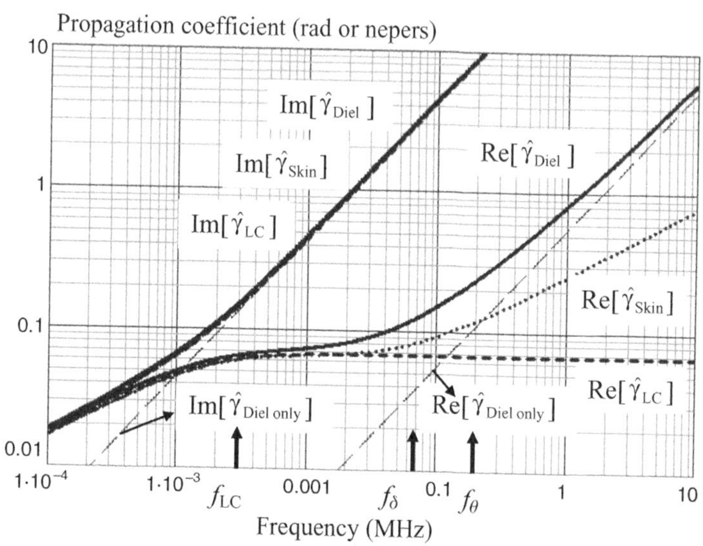

Figure 7.11 Propagation coefficients plotted as a function of frequency

7.1.5.2 Step Response of a Lossy Line

As there are no analytical time-domain formulations to compute the step response of a line including all kinds of frequency-dependent loss, the traditional way to analyze lossy interconnect is to perform the calculation in the frequency domain and then move the solution to the time domain by the *Inverse Fast Fourier Transform* (IFFT). To do this, three steps are required:

- Fourier transform representation in the frequency domain of a pulse with finite rise and fall times;
- frequency-domain solution of transmission-line equations of a lossy line loaded at both ends;
- inverse Fourier transform of the frequency-domain results to obtain time-domain solutions.

The procedure is outlined considering the same PCB trace as used in *Example 7.1*. The goal is to compute the step response at the end of the matched trace of length $l = 0.5$ m considering the different types of loss. The step is characterized by a pulse width $t_{pw} = b = 40$ ns and a rise time $t_r = 0.05$ ns. The computation of the pulse function Fourier transform and its implementation in MathCad™ are reported in Table 7.1 [25].

The following values were used in the simulation: $\Delta t = 0.01$ ns, $m = 13$, $b = 40$ ns, $\tau = 1$ ns, $t_r = 0.05$ ns. With these values, the number of points in the time domain is $N = 8192$, and therefore the pulse has a duration $N\Delta t = 81.92$ ns. The frequency step is

Table 7.1 Implementation of Fourier transform for computation of a step response of a lossy line in MathCad™

Fourier Transform	Discrete Fourier Transform (MathCad notations)
Parameters: $b = N\ \Delta t/2$ pulse width; $\Delta t =$ chosen step time; $m =$ chosen integer number; $n = 0, 1\ldots,$ N (number of points in time domain) with $N = 2^m$; $t_n = n\Delta t$ (discrete time); $N\Delta t =$ duration of the pulse; $k = 0, 1\ldots, N/2$ (number of frequencies); $\Delta f = 1/(N\Delta t)$ (frequency interval); $f_k = k\ \Delta f$ (number of frequencies); $\tau =$ delay; $t_r =$ rise time	
Pulse of width b	
$PulN(\omega) = \dfrac{1 - e^{-j\omega\cdot b}}{1 - e^{-j\omega\cdot\Delta t}}\cdot\Delta t$	$PulN_k := \text{if}\left(k = 0, b, \dfrac{1 - e^{-1j\cdot\frac{2\pi\cdot k}{N}\cdot\frac{b}{\Delta t}}}{1 - e^{-1j\cdot\frac{2\cdot\pi\cdot k}{N}}}\cdot\Delta t\right)$
Delay operator of amount τ	
$Dly(\omega) = e^{-j\omega\cdot\tau}$	$Dly_k := e^{-1j\cdot\frac{2\cdot\pi\cdot k}{N\cdot\Delta t}\cdot\tau}$
Linear-ramp with 10–90% rise-fall t_r with $q_{Lr} = 1.25\ t_r$	
$Lin(\omega) = \dfrac{\sin(\omega q_{Lr}/2)}{(\omega q_{Lr}/2)}$	$Lr_k := \text{if}\left[k = 0, 1, \left(\dfrac{\sin\left(\frac{\pi\cdot k\cdot q_{Lr}}{N\cdot\Delta t}\right)}{\frac{\pi\cdot k\cdot q_{Lr}}{N\cdot\Delta t}}\right)\right]$
In MathCad™ notations ":=" means define; "1j" means imaginary number j; the third term of "if" has the meaning of otherwise	

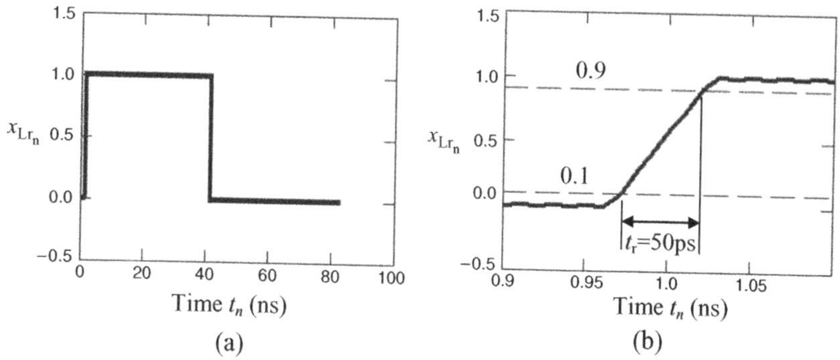

Figure 7.12 Computed impulse: (a) full view; (b) detail of the rise time

$\Delta f = 12.207$ MHz. By using MathCad™, the IFFT of the pulse DFT must be normalized to $1/(N\Delta t)$. To obtain the pulse source signal, including the starting delay and linear rise time, the computation in the frequency domain is performed by MathCad™ notations:

$$X_{\mathrm{Lr}_k} := \mathrm{PulN_k} \cdot \mathrm{Dly_k} \cdot \mathrm{Lr_k}$$

where $\boldsymbol{X}_{\mathrm{Lr}}$ is a vector of index k and ':=' means '='. The pulse source in the time domain can be obtained by the IFFT with the suitable normalization required by MathCad™ as

$$x_{\mathrm{Lr}} := \frac{1}{\mathrm{N} \cdot \Delta t} \cdot \mathrm{IFFT}\,(\mathrm{X_{Lr}})$$

where $\boldsymbol{x}_{\mathrm{Lr}}$ is a vector of index n. The computed source is shown in Figure 7.12. The step response at the load end can be computed as the IFFT of the transfer function $\hat{H}(\omega, l)$ multiplied by the impressed voltage $\hat{V}_S(\omega) = \hat{E}(\omega)$, under the assumption that $R_S = 0$. The propagation term $\hat{\gamma}(\omega)$ with the losses of interest is assigned at $\hat{H}(\omega, l)$. For example, in MathCad™ notations

$$\mathrm{H_{Diel_k}} = \mathrm{e}^{-\mathrm{Len}\cdot\gamma_{\mathrm{Diel_k}}} \quad \mathrm{X_{Diel_k}} := \mathrm{PulN_k} \cdot \mathrm{Dly_k} \cdot \mathrm{Lr_k} \cdot \mathrm{H_{Diel_k}} \quad \mathrm{x_{Diel}} := \frac{1}{\mathrm{N}\cdot\Delta \mathrm{t}} \cdot \mathrm{IFFT}\,(\mathrm{X_{Diel}})$$

where $\boldsymbol{x}_{\mathrm{Diel}}$ is a vector of index n that includes all kinds of loss.

The voltage $V_{\mathrm{L}}(t)$ computed at time t_n has the values provided by the component of index n of the vector $\boldsymbol{x}_{\mathrm{type}}$, where 'type' could be LC, Skin, Diel or Diel only, according to the definition of the propagation constant $\hat{\gamma}(\omega)$. The results are shown in Figure 7.13. Note that, for the case of DC losses only (i.e. the LC region), the load voltage is practically a translation of the source step with the rise time $t_r = 50$ ps. Taking into account the skin effect, the load voltage has a delay very close to the delay of a lossless or LC line and a sharper initial rise time owing to the larger magnitude of its high-frequency content, but a slowly evolving tail owing to its frequency response near DC. By accounting for dielectric losses, there is an anticipation of the voltage step on the load owing to the fact that at very high frequencies (i.e. above f_θ) the propagation velocity of the transmission line slightly exceeds the nominal value ν_0 and the

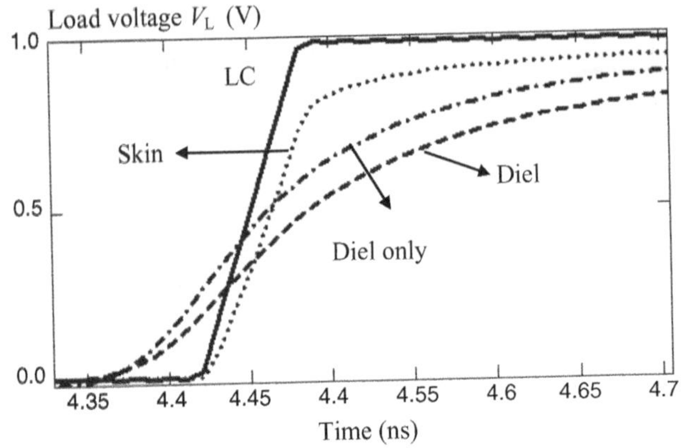

Figure 7.13 Computed time response $V_L(t)$ at the load end of the line for several types of loss

rise time increases significantly (from 10 to 90 %). This is not visible in Figure 7.11 owing to the scale chosen.

The considerations outlined are valid in the particular case of $R_S = 0$ and $R_L = Z_0$. For general cases, transmission-line equations with terminal conditions need to be solved. After simple mathematical steps [25], the load voltage of a line excited by a source with a generic impedance $\hat{Z}_S(\omega)$ and terminated on a generic load impedance $\hat{Z}_L(\omega)$ in the frequency domain is given by

$$\hat{V}_L(\omega) = \frac{\hat{E}_S(\omega)}{\left[\frac{\hat{H}(\omega)^{-1} + \hat{H}(\omega)}{2}\right]\left(1 + \frac{\hat{Z}_S(\omega)}{\hat{Z}_L(\omega)}\right) + \left[\frac{\hat{H}(\omega)^{-1} - \hat{H}(\omega)}{2}\right]\left(\frac{\hat{Z}_S(\omega)}{\hat{Z}_0(\omega)} + \frac{\hat{Z}_0(\omega)}{\hat{Z}_L(\omega)}\right)} \tag{7.43}$$

The input impedance of the line is

$$\hat{Z}_{\text{input}}(\omega) = \hat{Z}_0(\omega)\frac{\frac{\hat{H}(\omega)^{-1} + \hat{H}(\omega)}{2} + \frac{\hat{Z}_0(\omega)}{\hat{Z}_L(\omega)}\frac{\hat{H}(\omega)^{-1} - \hat{H}(\omega)}{2}}{\frac{\hat{H}(\omega)^{-1} - \hat{H}(\omega)}{2} + \frac{\hat{Z}_0(\omega)}{\hat{Z}_L(\omega)}\frac{\hat{H}(\omega)^{-1} + \hat{H}(\omega)}{2}} \tag{7.44}$$

and the voltage at the line input in the frequency domain is

$$\hat{V}_S(\omega) = \frac{\hat{E}_S(\omega)\hat{Z}_{\text{input}}(\omega)}{\hat{Z}_S(\omega) + \hat{Z}_{\text{input}}(\omega)} \tag{7.45}$$

The time-domain voltages $V_S(t)$ at the line input and $V_L(t)$ on the load can be obtained by the IFFT of the corresponding frequency-domain expressions (7.45) and (7.43).

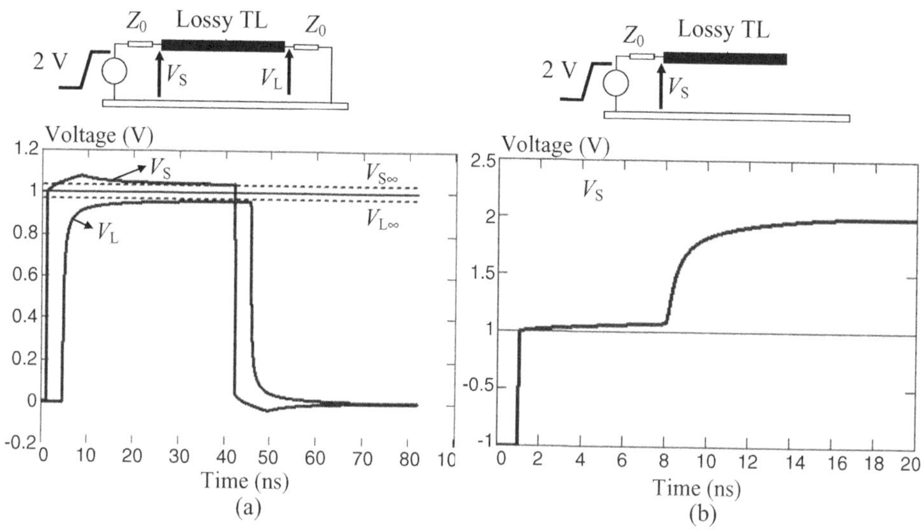

Figure 7.14 Computed voltage waveforms with (a) the line matched at both ends and (b) the line open at one end (i.e. TDR response)

Using the same PCB trace with the parameters $\Delta t = 0.005$ ns and $m = 14$ for better accuracy, $E_{\text{So}} = 1$ V, and the line matched at both ends to reproduce the typical waveforms of Figure 7.3, the results shown in Figure 7.14a are obtained. The source and load waveforms tend respectively to the limit values (see Equation (7.4) and Figure 7.3):

$$V_{\text{S}\infty} = E_{\text{So}} + \frac{R_{\text{dc}} l}{R_{\text{dc}} l + 2Z_0} = 1.032\ V, \quad V_{\text{L}\infty} = \frac{2Z_0}{R_{\text{dc}} l + 2Z_0} = 0.968\ V$$

The computed waveform of the voltage $V_{\text{S}}(t)$ when the trace is open at one end is shown in Figure 7.14b. This waveform is useful, as it can be compared with TDR measurements in order to determine the coefficient K_{p} for modeling by closed-form expressions the losses of a line including the proximity effect. A detailed description of this calculation is presented in *Section 11.1.3* for the case of microstrip and stripline trace structures. Another method is described in *Appendix B*.

7.2 Modeling Lossy Lines in the Time Domain by the Segmentation Approach and Vector Fitting Technique

To have appropriate lossy line models to simulate practical interconnects is strategic in order to verify if the data transmission specifications are met. It is also very important to have models suitable for performing a simulation directly in the transient domain without using the *Fast Fourier Transform*.

Different approaches can be found in the literature for lossy line modeling in the transient domain [1–20]. However, in this book, two methods suitable for implementation in a

Spike-like simulator are outlined:

- the *Vector Fitting* (VF) technique applied in the frequency domain to a short section of cable for equivalent circuit extraction [28, 29];
- the *Scattering* (S) parameters technique applied in the time domain to the full length of the cable [30].

The VF technique will be outlined in this section for two kinds of cable structure: coaxial cables where the skin-effect losses can be represented by closed-form analytical equations, and twisted-pair cables where the lossy line parameters must be obtained by a full-wave numerical tool. The same approach can be applied to other types of interconnect such as ribbon, twinax cables, and PCBs. The scattering parameters technique will be described in *Section 7.3*.

7.2.1 Circuit Extraction of Coaxial Cables

The configuration considered here is a coaxial cable over a ground plane, as shown in Figure 7.15. The radii of the inner wire and the internal and external shields are r_c, r_{si}, and r_{se}. The inner wire conductor and the shield have conductivities and permeabilities σ_c, μ_0, σ_s, and μ_s respectively. The relative permittivity of the dielectric is ε_{rd}.

The coaxial cable over a ground plane can be considered as a *Multiconductor Transmission Line* (MTL) of two independent conductors, and two different notations associated with the mesh and phase representations for voltages and currents can be adopted.

In particular, the most largely used *mesh representation* is based on considering two loops: the inner loop defined by the inner conductor and the internal shield, and the outer loop defined

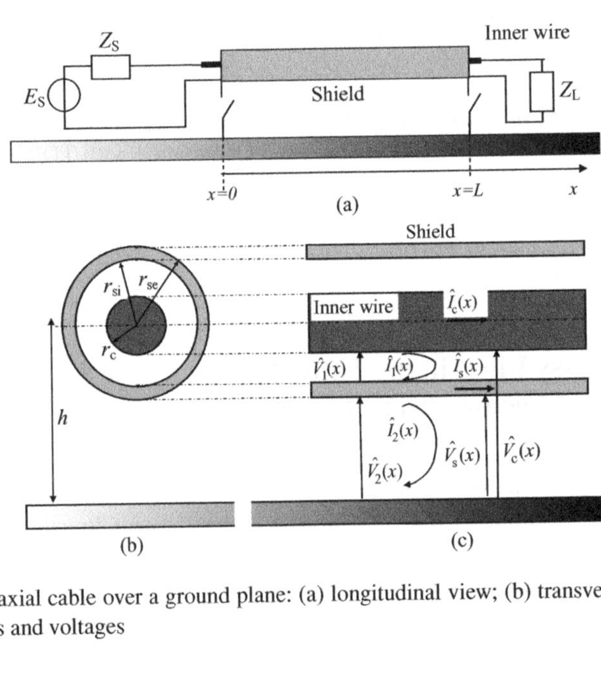

Figure 7.15 Coaxial cable over a ground plane: (a) longitudinal view; (b) transversal view; (c) phase and mesh currents and voltages

by the external shield and the ground plane. The mesh representation adopts the following notation:

$\hat{V}_1(x)$ wire-to-shield voltage;
$\hat{V}_2(x)$ shield-to-ground voltage;
$\hat{I}_1(x)$ current in the inner loop flowing in the internal wire and returning through the shield internal surface;
$\hat{I}_2(x)$ current in the outer loop flowing in the shield external surface and returning through the ground plane;
$\hat{\mathbf{Z}}(\omega)$ per-unit-length mesh impedance matrix;
$\hat{\mathbf{Y}}(\omega)$ per-unit-length mesh admittance matrix.

The second notation refers to the *phase representation* of voltage and currents and is defined by the following symbols:

$\hat{V}_c(x)$ voltage between the internal wire and the ground plane;
$\hat{V}_s(x)$ voltage between the external surface of the shield and the ground plane;
$\hat{I}_c(x)$ current on the inner wire and returning through the ground plane;
$\hat{I}_s(x)$ current on the shield and returning through the ground plane;
$\hat{\mathbf{Z}}_f(\omega)$ per-unit-length phase impedance matrix;
$\hat{\mathbf{Y}}_f(\omega)$ per-unit-length phase admittance matrix.

Note that in mesh representation the voltage of the inner wire refers to the shield, while the voltage of the shield refers to the ground plane. On the other hand, in the phase representation the ground plane represents the unique reference for the voltages.

The link between voltages and currents used in the two representations can be easily derived by Figure 7.15 and is given by

$$\hat{V}_c(x) = \hat{V}_1(x) + \hat{V}_2(x), \quad \hat{V}_s(x) = \hat{V}_2(x) \tag{7.46}$$

$$\hat{I}_c(x) = \hat{I}_1(x), \quad \hat{I}_s(x) = \hat{I}_2(x) - \hat{I}_1(x) \tag{7.47}$$

which can also be written in matrix form as

$$\hat{\mathbf{V}}_f(x) = \mathbf{T_V}\hat{\mathbf{V}}(x) \tag{7.48}$$

$$\hat{\mathbf{I}}_f(x) = \mathbf{T_I}\,\hat{\mathbf{I}}(x) \tag{7.49}$$

where

$$\hat{\mathbf{V}}_f = [\hat{V}_c(x) \quad \hat{V}_s(x)]^T \tag{7.50a}$$

$$\hat{\mathbf{I}}_f = [\hat{I}_c(x) \quad \hat{I}_s(x)]^T \tag{7.50b}$$

$$\hat{\mathbf{V}} = [\hat{V}_1(x) \quad \hat{V}_2(x)]^T \tag{7.51a}$$

$$\hat{\mathbf{I}} = [\hat{I}_1(x) \quad \hat{I}_2(x)]^T \tag{7.51b}$$

$$\mathbf{T_V} = \begin{bmatrix} 1 & 1 \\ 0 & 1 \end{bmatrix} \tag{7.52a}$$

$$\mathbf{T_I} = \begin{bmatrix} 1 & 0 \\ -1 & 1 \end{bmatrix} \tag{7.52b}$$

In the following, the mesh representation will be used. Phase representation could be required when loads between the inner wire and ground are present. However, loads of this kind can be easily taken into account by using a suitable termination network which can be derived by Equations (7.48) and (7.49) as described elsewhere [31].

The frequency-domain MTL equations describing the cable over the ground plane adopting the mesh representation are given by

$$\frac{\mathrm{d}\hat{\mathbf{V}}(x,\omega)}{\mathrm{d}x} = -\hat{\mathbf{Z}}(\omega)\hat{\mathbf{I}}(x,\omega) \tag{7.53a}$$

$$\frac{\mathrm{d}\hat{\mathbf{I}}(x,\omega)}{\mathrm{d}x} = -\hat{\mathbf{Y}}(\omega)\hat{\mathbf{V}}(x,\omega) \tag{7.53b}$$

7.2.1.1 Coaxial Cable Simulation Model

In Equation (7.53), the p.u.l. symmetric mesh representation impedance is given by

$$\hat{\mathbf{Z}}(\omega) = \begin{bmatrix} \hat{Z}_{11}(\omega) & \hat{Z}_{12}(\omega) \\ \hat{Z}_{12}(\omega) & \hat{Z}_{22}(\omega) \end{bmatrix} \tag{7.54}$$

where $\hat{Z}_{11}(\omega)$ and $\hat{Z}_{22}(\omega)$ are the self-impedances of the two meshes, and $\hat{Z}_{12}(\omega)$ is the mutual impedance between the meshes related to the cable transfer impedance.

The p.u.l. symmetric admittance matrix in Equation (7.53) is

$$\hat{\mathbf{Y}}(\omega) = \begin{bmatrix} \hat{Y}_{11}(\omega) & \hat{Y}_{12}(\omega) \\ \hat{Y}_{12}(\omega) & \hat{Y}_{22}(\omega) \end{bmatrix} \tag{7.55}$$

where $\hat{Y}_{11}(\omega)$ and $\hat{Y}_{22}(\omega)$ are the self-admittances of the two meshes, and $\hat{Y}_{12}(\omega)$ is the mutual impedance between the meshes related to the cable transfer admittance. The coefficients of the impedance and admittance matrices are explicitly given below [32, 33].

Coefficient $\hat{Z}_{11}(\omega)$ – This coefficient is the p.u.l. impedance of the inner loop formed by the inner wire and the shield. It is the sum of three terms:

$$\hat{Z}_{11}(\omega) = \hat{Z}_{\mathrm{c}}(\omega) + \hat{Z}_{\mathrm{cs}}(\omega) + \hat{Z}_{\mathrm{si}}(\omega) \tag{7.56}$$

where $\hat{Z}_{\mathrm{c}}(\omega)$ is the p.u.l. impedance of the inner wire, $\hat{Z}_{\mathrm{cs}}(\omega)$ is the p.u.l. external impedance between inner wire and shield, and $\hat{Z}_{\mathrm{si}}(\omega)$ is the p.u.l. internal impedance of the shield.

These impedances are given by

$$\hat{Z}_{\mathrm{c}}(\omega) = R_{\mathrm{c,dc}}\frac{r_{\mathrm{c}}}{\delta_{\mathrm{c}}}(1+\mathrm{j})\coth\left[\frac{r_{\mathrm{c}}}{\delta_{\mathrm{c}}}(1+\mathrm{j})\right] \tag{7.57a}$$

$$\hat{Z}_{\mathrm{cs}}(\omega) = \mathrm{j}\omega\frac{\mu_0}{2\pi}\ln\left[\frac{r_{\mathrm{si}}}{r_{\mathrm{c}}}\right] = \mathrm{j}\omega L_{\mathrm{in}} \tag{7.57b}$$

$$\hat{Z}_{\mathrm{si}}(\omega) = R_{\mathrm{si,dc}}\frac{d}{\delta_{\mathrm{s}}}(1+\mathrm{j})\coth\left[\frac{d}{\delta_{\mathrm{s}}}(1+\mathrm{j})\right] \tag{7.57c}$$

where $R_{c,dc} = 1/\left(\pi\sigma_c r_c^2\right)$ is the p.u.l. DC resistance of the inner wire, $\delta_c = 1/\sqrt{\pi f \mu_0 \sigma_c}$ is the skin depth of the internal wire, $R_{si,dc} = 1/(2\pi\sigma_s r_{si} d)$ is the p.u.l. DC resistance of the internal shield, $\delta_s = 1/\sqrt{\pi f \mu_s \sigma_s}$ is the skin depth of the shield, and $d = r_{se} - r_{si}$ is the shield thickness.

Coefficient $\hat{Z}_{22}(\omega)$ – This coefficient is the impedance of the outer loop formed by the shield and the ground plane. It is the sum of two terms:

$$\hat{Z}_{22}(\omega) = \hat{Z}_{se}(\omega) + \hat{Z}_{sp}(\omega) \tag{7.58}$$

where $\hat{Z}_{se}(\omega)$ is the p.u.l. internal impedance of the external shield and $\hat{Z}_{sp}(\omega)$ is the p.u.l. external impedance between the shield and the ground plane.

These impedances are computed as

$$\hat{Z}_{se}(\omega) = R_{se,dc}\frac{d}{\delta_s}(1+j)\coth\left[\frac{d}{\delta_s}(1+j)\right] \tag{7.59a}$$

$$\hat{Z}_{sp}(\omega) = j\omega\frac{\mu_0}{2\pi}\ln\left[\frac{2h}{r_{se}}\right] = j\omega L_{out} \tag{7.59b}$$

where $R_{se,dc} = 1/(2\pi\sigma_s r_{se} d)$ is the p.u.l. DC resistance of the external shield and $\delta_p = 1/\sqrt{\pi f \mu_0 \sigma_p}$ is the skin depth of the ground plane.

Coefficient $\hat{Z}_{12}(\omega)$ – This coefficient is the p.u.l. mutual impedance between the inner and outer meshes and is given by

$$\hat{Z}_{12}(\omega) = -\hat{Z}_t(\omega) \tag{7.60}$$

where $\hat{Z}_t(\omega)$ is the p.u.l. transfer impedance of the cable. This parameter depends on the kind of cable and defines the coupling between internal and external meshes. Low values of $\hat{Z}_t(\omega)$ mean small coupling. How to compute the transfer impedance regarding tubular and braided shields is given by several books or papers [33–35]. For modeling purposes, very often the following simplified expression can be used

$$\hat{Z}_t(\omega) = R_t + j\omega L_t \tag{7.61}$$

This means that the transfer impedance can be approximately represented by a constant term (the DC resistive term of the shield) and an inductive term. Using this simplified expression, it is possible to avoid applying the vector fitting technique to carry out an equivalent *RLC* circuit, as will be described in detail later in this section.

Coefficients $\hat{Y}_{11}(\omega)$ *and* $\hat{Y}_{22}(\omega)$ – These coefficients represent the admittance between the inner wire and the shield and between the shield and the ground plane respectively. They are given by

$$\hat{Y}_{11}(\omega) = j\omega C_{in} = j\omega \frac{2\pi\varepsilon_0\varepsilon_{rd}}{\log(r_{si}/r_c)} \tag{7.62}$$

$$\hat{Y}_{22}(\omega) = j\omega C_{out} = j\omega \frac{2\pi\varepsilon_0}{\log(2h/r_{se})} \tag{7.63}$$

Coefficient $\hat{Y}_{12}(\omega)$ – This coefficient represents the transfer admittance of the shield. For a tubular shield $\hat{Y}_{12}(\omega) = 0$. For a braided shield $\hat{Y}_{12}(\omega) \neq 0$. Details on the calculation of $\hat{Y}_{12}(\omega)$ can be found elsewhere [33, 34]. This parameter becomes important at very high frequencies. An approximate expression can be obtained by modeling this term as a simple capacitance whose value can be measured by experimental methods.

7.2.1.2 Equivalent Circuit of a Coaxial Cable by the Segmentation Approach

Once the cable simulation model is known, the lossy coaxial cable can be modeled by the segmentation approach as a cascade connection of multiport networks, each representing an electrically short segment of length Δx, and characterized by telegrapher's equations (7.53). Each cable section can be modeled by the Π-type equivalent circuit shown in Figure 7.16, where the following expressions are used:

$$\hat{Z}_{in}(\omega) = \hat{Z}_{11}(\omega) - j\omega L_{in} \tag{7.64a}$$

$$\hat{Z}_{out}(\omega) = \hat{Z}_{22}(\omega) - j\omega L_{out} \tag{7.64b}$$

These expressions are very useful, as they enable to extract the inductive terms L_{in} and L_{out} representing the p.u.l. inductances of each line to be extracted while including the frequency-dependent losses in the series impedances $\hat{Z}_{in}(\omega)$ and $\hat{Z}_{out}(\omega)$.

A different option for modeling a coaxial cable section of length Δx is the TL-type equivalent circuit shown in Figure 7.17 which is equivalent from the theoretical viewpoint to the Π-type equivalent circuit. The two lossless TLs (TL_{in} and TL_{out} in Figure 7.17) are simply

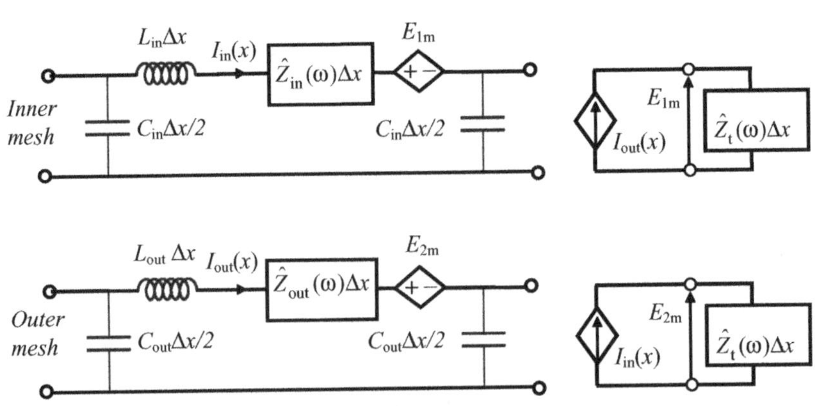

Figure 7.16 Π-type equivalent circuit of a coaxial cable section of length Δx

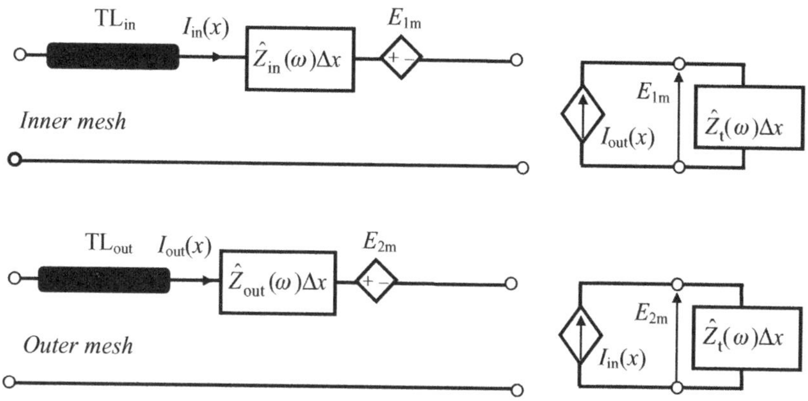

Figure 7.17 TL-type equivalent circuit of a coaxial cable section of length Δx

defined by their nominal characteristic impedance and delay time, given by

$$Z_{0\text{in}} = \sqrt{L_{\text{in}}/C_{\text{in}}} \tag{7.65a}$$

$$T_{\text{Din}} = \sqrt{L_{\text{in}}C_{\text{in}}}\ \Delta x \tag{7.65b}$$

$$Z_{0\text{out}} = \sqrt{L_{\text{out}}/C_{\text{out}}} \tag{7.66a}$$

$$T_{\text{Dout}} = \sqrt{L_{\text{out}}C_{\text{out}}}\ \Delta x \tag{7.66b}$$

This last circuit has the great advantage of reducing the computational time, and it provides more accurate results. Note that it is consistent with the approach used in Figure 7.2 for discussing the loss effects on signals traveling down a line.

In order to implement the equivalent circuits of Figures 7.16 and 7.17 in any SPICE-based circuit simulator, the frequency-dependent impedances $\hat{Z}_{\text{in}}(\omega)$, $\hat{Z}_{\text{out}}(\omega)$, and $\hat{Z}_{\text{t}}(\omega)$ need to be modeled by an equivalent *RLC* network which can be obtained by the vector fitting technique as described in the next section.

7.2.1.3 SPICE Circuit Extraction by Vector Fitting

The *Vector Fitting* (VF) technique makes it possible to approximate any complex function of frequency by a rational polynomial expression in terms of poles and residues, as described in detail by Gustavsen and Semlyen [28]. A code implementing the VF procedure in Matlab is kindly provided by these authors and can be downloaded from the website http://www.energy.sintef.no/Produkt/VECTFIT/index.asp.

The derivation of the *RLC* network by VF can be shown for the case of a generic frequency-dependent impedance $\hat{Z}_h(\omega)$, which can be $\hat{Z}_{\text{in}}(\omega)$, $\hat{Z}_{\text{out}}(\omega)$, and $\hat{Z}_{\text{t}}(\omega)$, by simply replacing the subscript '*h*' with 'in', 'out', or 't'. By the VF procedure, $Z_h(\omega)$ is expressed as

$$\hat{Z}_h(\omega) \approx A_{h0} + \text{j}\omega L_{h0} + \sum_{i=1}^{N} \frac{A_{hi}}{\text{j}\omega - p_{hi}} \tag{7.67}$$

where N is the order of the approximating function, and A_{hi} and p_{hi} are the ith residue and pole respectively. By simple manipulations and by separating the contribution of the N_r real poles from those of the N_c couples of complex conjugate poles (i.e. $N = N_r + 2N_c$), Equation (7.67) can be written as

$$\hat{Z}_h(\omega) \approx R_{h0} + \mathrm{j}\omega L_{h0} + \sum_{i=1}^{N_r} \frac{\mathrm{j}\omega \tilde{A}_{hi}}{\mathrm{j}\omega - p_{hi}} + \sum_{k=1}^{N_c} \left[\frac{\hat{A}_{hk}}{\mathrm{j}\omega - \hat{p}_{hk}} + \frac{\hat{A}^*_{hk}}{\mathrm{j}\omega - \hat{p}^*_{hk}} \right] \tag{7.68}$$

where the symbol * denotes a complex conjugate term and

$$R_{h0} = A_{h0} - \sum_{i=1}^{N_r} \frac{A_{hi}}{p_{hi}} \tag{7.69a}$$

$$\tilde{A}_{hi} = \frac{A_{hi}}{p_{hi}} \tag{7.69b}$$

Each real pole in Equation (7.68) can be modeled by an RL parallel circuit, as shown in Figure 7.18a, where the parameters are given by

$$R_{hi} = \tilde{A}_{hi} \tag{7.70a}$$

$$L_{hi} = -R_{hi}/p_{hi} \tag{7.70b}$$

Each complex conjugate pair enclosed in square brackets in Equation (7.68) can be modeled by the equivalent circuit of Figure 7.18b, where the parameters are defined as

$$C_{hk} = \frac{1}{2\mathrm{Re}[\hat{A}_{hk}]} \tag{7.71a}$$

$$R_{phk} = \frac{2\left(\mathrm{Re}[\hat{A}_{hk}]\right)^2}{\mathrm{Im}[\hat{A}_{hk}]\mathrm{Im}[\hat{p}_{hk}] - \mathrm{Re}[\hat{A}_{hk}]\mathrm{Re}[\hat{p}_{hk}]} \tag{7.71b}$$

$$L_{hk} = \left\{ \frac{(\mathrm{Re}[\hat{p}_{hk}])^2 + (\mathrm{Im}[\hat{p}_{hk}])^2}{2\mathrm{Re}[\hat{A}_{hk}]} + \frac{\left(\mathrm{Im}[\hat{A}_{hk}]\mathrm{Im}[\hat{p}_{hk}]\right)^2 - \left(\mathrm{Re}[\hat{A}_{hk}]\mathrm{Re}[\hat{p}_{hk}]\right)^2}{\left(\mathrm{Re}[\hat{A}_{hk}]\right)^3} \right\}^{-1} \tag{7.71c}$$

$$R_{shk} = -L_{hk} \frac{\mathrm{Re}[\hat{A}_{hk}]\mathrm{Re}[\hat{p}_{hk}] + \mathrm{Im}[\hat{A}_{hk}]\mathrm{Im}[\hat{p}_{hk}]}{\mathrm{Re}[\hat{A}_{hk}]} \tag{7.71d}$$

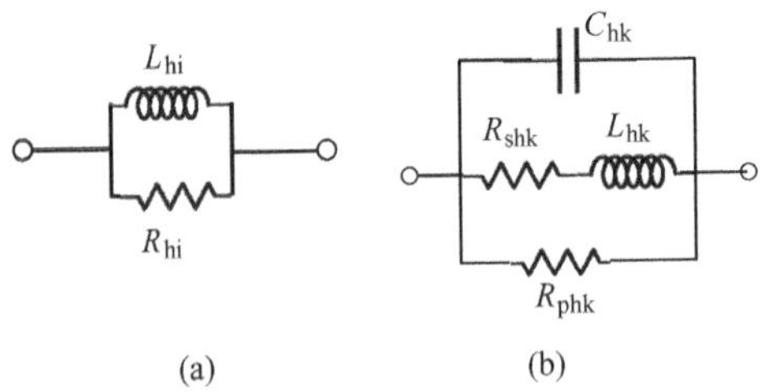

Figure 7.18 Equivalent circuits associated with (a) the ith real pole and (b) the kth complex conjugate pairs

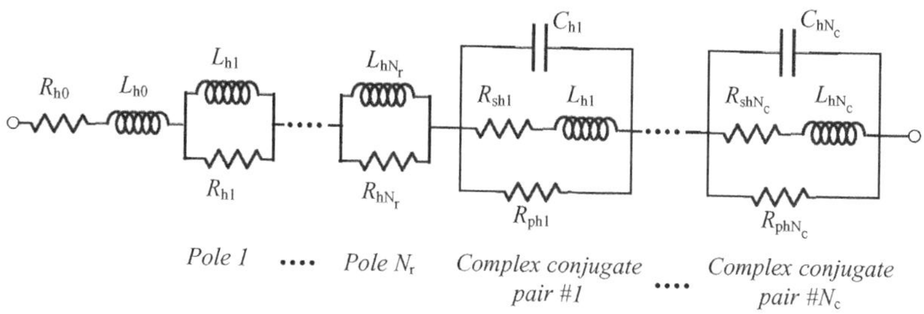

Figure 7.19 Equivalent circuit of the frequency-dependent impedance $\hat{Z}_h(\omega)$

where Re[.] and Im[.] are operators that allow the real and imaginary part of a complex number to be computed.

The equivalent circuit of impedance (7.68) is shown in Figure 7.19 and comprises the series connection of the following components:

- resistance R_{h0};
- inductance L_{h0};
- *RL* parallel circuits associated with the N_r real poles, the parameters of which are given by Equation (7.70);
- *RLC* circuits associated with the N_c conjugate pairs, the parameters of which are given by Equation (7.71).

In general, $\hat{Z}_{in}(\omega)$ and $\hat{Z}_{out}(\omega)$ are characterized by a magnitude increasing with frequency, and their polynomial approximations obtained by VF consist of real poles only. On the other hand, the transfer impedance of the shield presents a polynomial approximation according to Equation (7.68), including both real and complex conjugate pair poles.

At this point, after the VF procedure and circuit synthesis, the following observations can be made concerning the equivalent circuits obtained:

- The *RLC* parameters are constant irrespective of frequency, and are appropriate for making the circuit suitable for transient analysis.
- The circuit models of a cable section of length Δx shown in Figures 7.16 and 7.17 can be easily implemented in any circuit simulator in commerce, and the impedances $\hat{Z}_{in}(\omega)\Delta x$, $\hat{Z}_{out}(\omega)\Delta x$, and $\hat{Z}_t(\omega)\Delta x$ can be defined by suitable subcircuits added to the SPICE library.
- The equivalent circuit models are suitable for radiated field computation, as they provide the current distribution along the cable shield.
- The models can be extended to the case of multiconductor cables such as twinax or shielded twisted-pair cables. The difficulty in this case is knowledge of the cable parameters. In fact, in the case of multiconductor cables, an analytical approach accounting for the proximity effect between the inner conductors is not available. For this reason, the p.u.l. parameters, including frequency-dependent losses, can be obtained by measurements or by numerical methods.

Some examples focused on signal integrity are provided below to show the validity and efficiency of the circuits described in Figures 7.16 and 7.17. A further application for predicting interference by an ESD event in an RG58 coaxial cable can be found elsewhere [36, 37]. Note that the procedure can be extended to other types of interconnect such as other cables and PCB traces.

Example 7.2: Signal Integrity in a Lossy Coaxial Cable

As an example, consider a coaxial cable having the following parameters: inner wire radius $r_c = 0.395$ mm, internal shield radius $r_{si} = 1.397$ mm, external shield radius $r_{se} = 1.524$ mm, dielectric relative permittivity $\varepsilon_{rd} = 2.3$, shield and wire conductivity $\sigma_c = \sigma_s = 5.8 \times 10^7$ Ω^{-1}/m, vacuum permeability $\mu_0 = 4\pi 10^{-7}$ H/m, and vacuum permittivity $\varepsilon_0 = 8.854 \times 10^{-12}$ F/m. The calculation of the parameters of the equivalent circuit of the inner mesh of Figure 7.17 was performed adopting a cable segment of length $\Delta x = 5$ cm. The lossless transmission line $\mathrm{TL_{in}}$ has a nominal characteristic impedance $Z_{0\mathrm{coax}} = \sqrt{L_{\mathrm{in}}/C_{\mathrm{in}}} = 49.942\ \Omega$ and a delay time $T_{\mathrm{Dcoax}} = \sqrt{L_{\mathrm{in}}C_{\mathrm{in}}} = 252.935$ ps, with $L_{\mathrm{in}} = 12.632$ nH and $C_{\mathrm{in}} = 5.065$ pF. The equivalent circuit of the frequency-dependent impedance $\hat{Z}_{\mathrm{in}}(\omega)\Delta x$ obtained by the VF procedure is shown in Figure 7.20, where the notation is that of the circuit simulator MicroCap [38] (the notation '.define X Y' means that the number or variable Y is assigned to the variable X). This circuit was obtained by adopting 10 poles, which is an appropriate number to ensure accuracy in the frequency range of interest (0–10 GHz). Note that the series inductance $L_{10} = L_{\mathrm{in}}$ is not present in Figure 7.20 because it is included in $\mathrm{TL_{in}}$. The first resistance $R_1 = 2.244$ mΩ is the DC resistance of the inner wire plus the shield. In order to generalize the circuit for an arbitrary length of cable section, the parameter d$x = \Delta x$ is defined in Figure 7.20. The comparison between the impedance obtained by the circuit in Figure 7.20 and the impedance computed analytically using the previous equations based on skin depth for a coaxial cable is shown in Figure 7.21, where a very good agreement can be observed in the full range of interest.

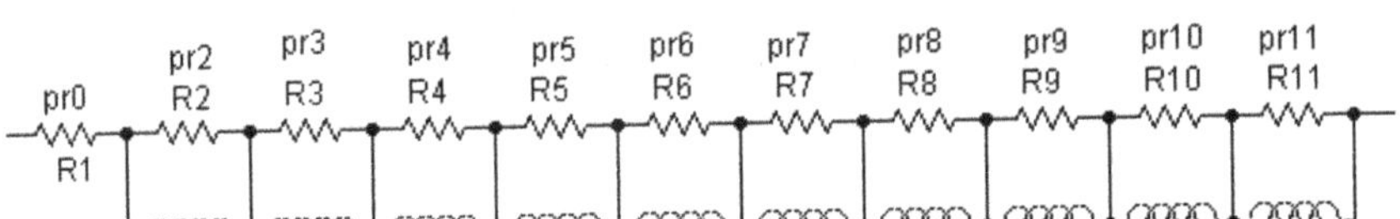


```
.define dx 0.05
.define pl2 (1.7340e-011*dx/0.05)
.define pl3 (3.1515e-011*dx/0.05)
.define pl4 (6.6198e-011*dx/0.05)
.define pl5 (1.3469e-010*dx/0.05)
.define pl6 (2.5327e-010*dx/0.05)
.define pl7 (4.5323e-010*dx/0.05)
.define pl8 (8.0524e-010*dx/0.05)
.define pl9 (1.1427e-009*dx/0.05)
.define pl10 (4.0739e-009*dx/0.05)
.define pl11 (1.8906e-007*dx/0.05)

.define pr0 (2.244m*dx/0.05)
.define pr2 (1.0527e+000*dx/0.05)
.define pr3 (3.5811e-001*dx/0.05)
.define pr4 (1.6176e-001*dx/0.05)
.define pr5 (7.3875e-002*dx/0.05)
.define pr6 (3.3826e-002*dx/0.05)
.define pr7 (1.6371e-002*dx/0.05)
.define pr8 (8.4250e-003*dx/0.05)
.define pr9 (4.1854e-003*dx/0.05)
.define pr10 (3.5315e-003*dx/0.05)
.define pr11 (1.1879e-006*dx/0.05)
```

Figure 7.20 Equivalent circuit of impedance $\hat{Z}_{\mathrm{in}}(\omega)$ for a coaxial cable of generic length dx

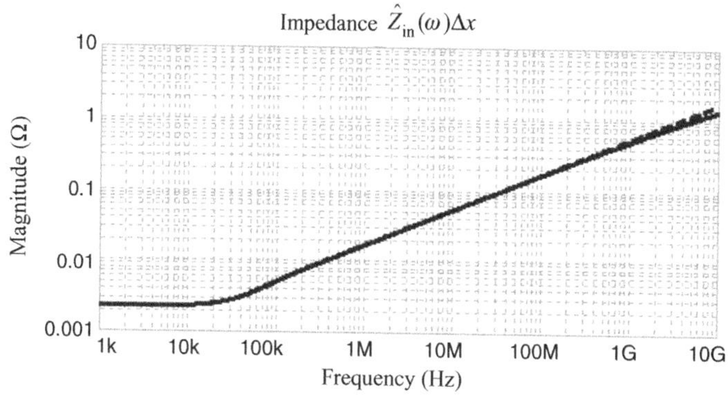

Figure 7.21 Magnitude of $\hat{Z}_{in}(\omega)\Delta x$ as a function of frequency: analytical computation (dashed line); equivalent circuit with 10 poles (solid line). $\Delta x = 5$ cm

The scattering parameters of the coaxial cable of length $l = 5$ cm are shown in Figure 7.22, where the results obtained by the following four different circuit models are compared:

1. The cable is simulated as a four-port net where the p.u.l. parameters are computed by the previous equations, and S-parameters are obtained by the exact formulation described in *Section 11.2*.

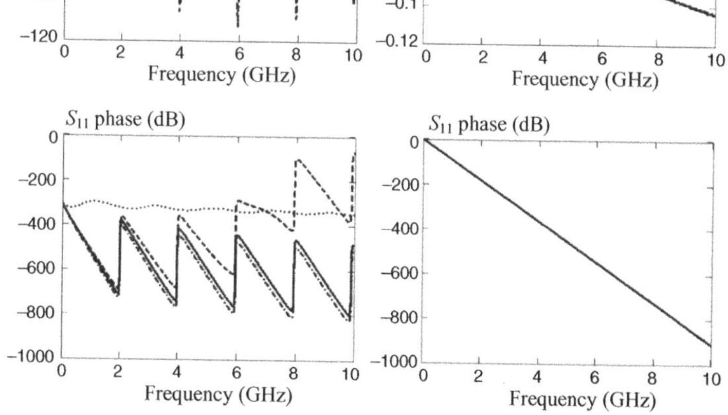

Figure 7.22 S-parameters of a 5 cm long coaxial cable: analytical approach (solid line); one-cell model (dotted line); 10-cell model (dashed line); 100-cell model (dashed-dotted line)

2. The cable is simulated with one cell (i.e. $\Delta x = l = 5$ cm) according to Figure 7.17, assuming $E_{1m} = 0$, and using the equivalent circuit of Figure 7.20 to model $\hat{Z}_{in}(\omega)\Delta x$.
3. The cable is simulated with a cascade connection of 10 cells of length $\Delta x = 0.5$ cm.
4. The cable is simulated with a cascade connection of 100 cells of length $\Delta x = 0.05$ cm.

In should be noted that only the cable model based on the cascade connection of 100 cells reveals a very good agreement in the calculation of the scattering parameter S_{11} up to 10 GHz. In fact, at 10 GHz the wavelength is $\lambda = 300/f_{MHz} = 300/10\,000 = 0.03$ m $= 3$ cm, and, to satisfy the electrically short condition on which the segmentation approach is based, the cable segment should be less than one-tenth of the wavelength, and therefore less than 0.3 cm. On the other hand, a very good agreement among the four models can be observed in the case of parameter S_{21}. This has an important practical effect, as generally the interest is focused on data transmission at the end of the cable.

As a second application, an UltraWide Band (UWB) signal transmitted along an RG58 coaxial cable of 1.83 m length, matched at both ends, is considered. The UBW signal has a bandwidth of 3.1–6.5 GHz and is therefore suitable for testing the proposed skin-effect model based on the VF technique. To account suitably for this very high frequency content, the cable was modeled by a cascade connection of 610 cells of 0.3 cm length. In a theoretical lossless line, considering the partitioning effect between $R_S = Z_0$ and Z_0, the signal $E_S/2$ shown in Figure 7.23 is launched onto the coaxial cable and transmitted to the load without attenuation. In reality, the signal is attenuated by the skin effect. A very good agreement can be observed in Figure 7.23 between the voltage on the load V_L obtained by the circuit model and the measurement performed on a LeCroy digital oscilloscope (Wavemaster 8600A).

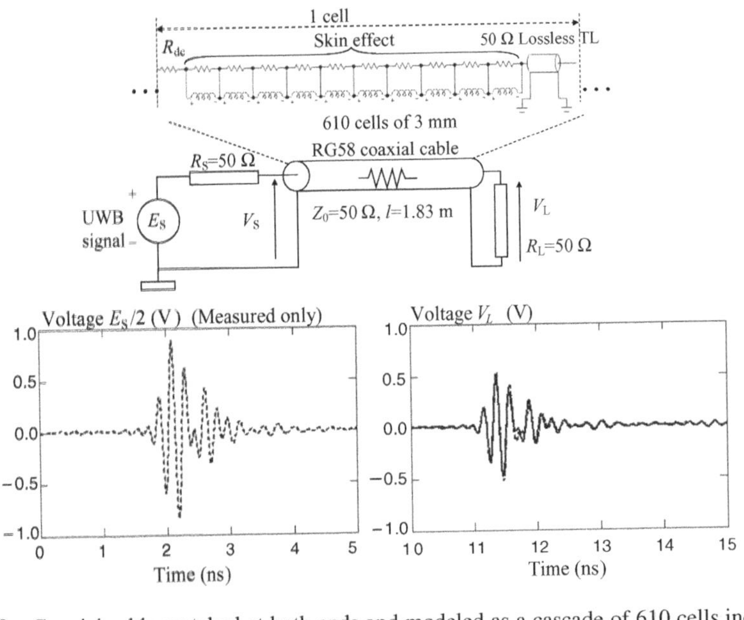

Figure 7.23 Coaxial cable matched at both ends and modeled as a cascade of 610 cells including the skin effect: comparison between measured (dashed line) and computed (solid line) waveforms

7.2.2 *Circuit Extraction of Twisted-Pair Cables*

Twisted-pair cables are widely used in several applications for their capability to reduce EMI effects induced by external magnetic fields and crosstalk produced by parallel wires. In particular, unshielded twisted-pair (UTP) cables are involved in the realization of local area networks (LANs) connecting personal computers, workstations, etc. Owing to the continuous development of high-speed communication technology, UTPs are required to support signals propagating at frequencies up to some gigahertz.

Typically, UTPs comprise two dielectrically insulated copper wires twisted together inside a dielectric sheath. This means that the proximity effect plays an important role regarding losses. An analytical approach for calculating the cable simulation model like the one used for coaxial cables cannot be used for cables of this type. However, a circuit model can be built starting from the S-parameters computed numerically by a full-wave tool for an electrically short cable segment (see *Section 11.2*). S-parameters can be measured, but this is very difficult and almost impossible for a short segment of cable.

Once the scattering matrix $\hat{\mathbf{S}}$ is obtained by the full-wave simulation, it can be used to extract two different SPICE equivalent circuits:

- Π-type equivalent circuit;
- TL-based equivalent circuit.

Both circuits are derived by a different port representation of the cable, which can be easily derived by the scattering matrix. For instance, the impedance matrix $\hat{\mathbf{Z}}$ is given by [39]

$$\hat{\mathbf{Z}} = z_{\text{rif}} \left(\mathbf{I} + \hat{\mathbf{S}}\right) \left(\mathbf{I} - \hat{\mathbf{S}}\right)^{-1} \tag{7.72}$$

where $\mathbf{I}$ is the identity matrix and z_{rif} is the port normalization impedance.

7.2.2.1 Π-type Equivalent Circuit

The first UTP macromodel is the Π-*type equivalent circuit* shown in Figure 7.24a. The transversal and longitudinal admittances of this circuit can be obtained as [32], [39]

$$\hat{Y}_{\text{long}}(\omega) = \frac{\hat{Z}_{21}}{\det \hat{\mathbf{Z}}} \tag{7.73a}$$

$$\hat{Y}_{\text{trasv}}(\omega) = \frac{\hat{Z}_{11} - \hat{Z}_{21}}{\det \hat{\mathbf{Z}}} \tag{7.73b}$$

where $\hat{Z}_{11}$ and $\hat{Z}_{21}$ are the coefficients of the impedance matrix $\hat{\mathbf{Z}}$ defined by Equation (7.72), and 'det' stands for the matrix determinant. The VF procedure is applied to carry out a rational approximation for the frequency-dependent admittances $\hat{Y}_{\text{long}}(\omega)$ and $\hat{Y}_{\text{trasv}}(\omega)$, so that the generic admittance $\hat{Y}_h(\omega)$ (with h = long or trasv) can be approximated by

$$\hat{Y}_h(\omega) = d_h + \mathrm{j}\omega e_h + \sum_{i=1}^{N} \frac{A_{hi}}{\mathrm{j}\omega - p_{hi}} \tag{7.74}$$

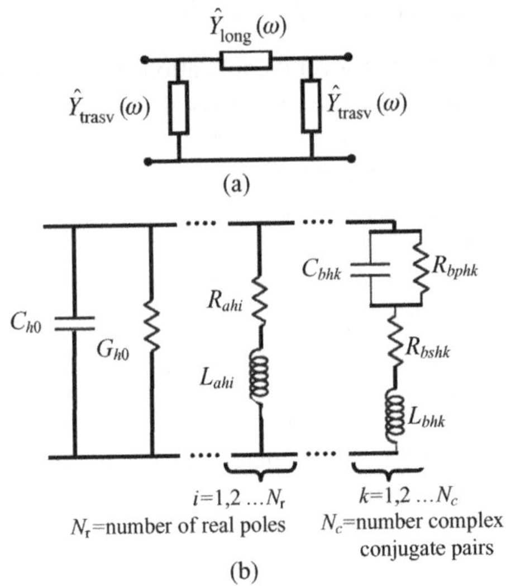

Figure 7.24 (a) Π-type equivalent circuit and (b) equivalent circuit of the frequency-dependent admittance $\hat{Y}_h(\omega)$ with h = long or trasv

where N is the order of the approximating function, and A_{hi} and p_{hi} are the ith residue and pole respectively. The rational function (7.74) can be easily modeled by an equivalent *RLC* network suitable for direct implementation in a CAD circuit simulator. To derive the equivalent circuit, let indicate N_r be the number of real poles and N_c the number of complex conjugate pairs, so that Equation (7.74) becomes

$$\hat{Y}_h(\omega) = d_h + j\omega e_h + \sum_{i=1}^{N_r} \frac{A_{hi}}{j\omega - p_{hi}} + \sum_{k=1}^{N_{pc}} \left[\frac{\hat{A}_{hk}}{j\omega - \hat{p}_{hk}} + \frac{\hat{A}^*_{hk}}{j\omega - \hat{p}^*_{hk}} \right] \tag{7.75}$$

where the asterisk denotes a complex conjugate. The equivalent circuit associated with Equation (7.75) is shown in Figure 7.24b and is given by the parallel connection of the following branches:

- a conductance $G_{h0} = d_h$;
- a capacitance $C_{h0} = e_h$;
- *RL* series circuits associated with N_r real poles, the parameters of which are

$$L_{ahi} = 1/A_{hi} \tag{7.76a}$$

$$R_{ahi} = -p_{hi}/A_{hi} \tag{7.76b}$$

- *RLC* circuits associated with the N_c complex conjugate pairs, the parameters of which are given by

$$L_{bhk} = 0.5/\text{Re}[\hat{A}_{hk}] \tag{7.77a}$$

$$R_{bshk} = \frac{\text{Im}[\hat{A}_{hk}]\text{Im}[\hat{p}_{hk}] - \text{Re}[\hat{A}_{hk}]\text{Re}[\hat{p}_{hk}]}{2\left(\text{Re}[\hat{A}_{hk}]\right)^2} \tag{7.77b}$$

$$C_{bhk} = 2\left(\mathrm{Re}[\hat{A}_{hk}]\right)^3 / \left\{ \left(\mathrm{Re}[\hat{A}_{hk}]\right)^2 \left((\mathrm{Re}[\hat{p}_{hk}])^2 + (\mathrm{Im}[\hat{p}_{hk}])^2\right) + \left(\mathrm{Im}[\hat{A}_{hk}]\mathrm{Im}[\hat{p}_{hk}]\right)^2 - \left(\mathrm{Re}[\hat{A}_{hk}]\mathrm{Re}[\hat{p}_{hk}]\right)^2 \right\} \tag{7.77c}$$

$$R_{bphk} = -\frac{\mathrm{Re}[\hat{A}_{hk}]}{C_{bhk}\left(\mathrm{Re}[\hat{A}_{hk}]\mathrm{Re}[\hat{p}_{hk}] + \mathrm{Im}[\hat{A}_{hk}]\mathrm{Im}[\hat{p}_{hk}]\right)} \tag{7.77d}$$

It should be noted that the applied VF procedure makes it possible to ensure overall passivity of the admittances but not local passivity, as some negative components can be present in the SPICE circuit. This fact can bring instability in transient analysis when cascading a large number of cells.

7.2.2.2 TL-based Equivalent Circuit

To avoid instability problems and simplify the equivalent circuit of a section of the cable in order to speed up the simulation while reducing the number of circuit elements, the TL-based model shown in Figure 7.25a can be used. This model consists of three parts:

- A frequency-dependent impedance $\hat{Z}_{\mathrm{UTP}}(\omega)$, which takes into account the skin and proximity effects and is computed using the analytical approach outlined in *Sections 7.1.2* and *7.1.3*.
- A frequency-dependent admittance $\hat{Y}_{\mathrm{UTP}}(\omega)$, which takes into account the dielectric effect and is computed using the analytical approach outlined in *Section 7.1.4*.
- A lossless transmission line $\mathrm{TL}_{\mathrm{UTP}}$ associated with the section of cable under consideration. It includes the nominal p.u.l. inductance L_0 and capacitance C_0 used to characterize a lossless line. This can be done because the propagation coefficient and therefore the delay time is slightly dependent on the type of losses (see Figure 11.1).

The frequency-dependent impedance $\hat{Z}_{\mathrm{UTP}}(\omega)$ and admittance $\hat{Y}_{\mathrm{UTP}}(\omega)$ can be modeled by *RLC* networks, which can be obtained by the VF procedure as discussed in the previous section.

A simplified TL-based model can be obtained by neglecting the effect of dielectric losses, and then by modeling the shunt admittance by the simple capacitance C_0, the contribution of which is accounted for in the lossless line, as shown in Figure 7.25b.

Example 7.3: Signal Integrity in a Lossy UTP Cable

This example illustrates how to apply the *Vector Fitting* (VF) technique and the analytical models described in *Section 7.1* in order to find a suitable equivalent circuit for a lossy UTP

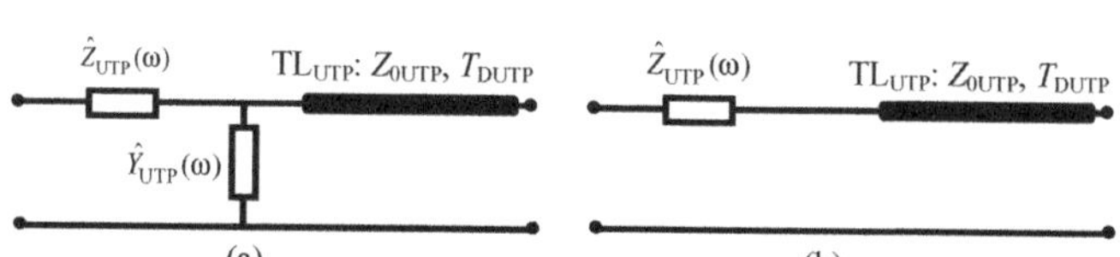

Figure 7.25 (a) TL-based equivalent circuit and (b) simplified TL-based equivalent circuit

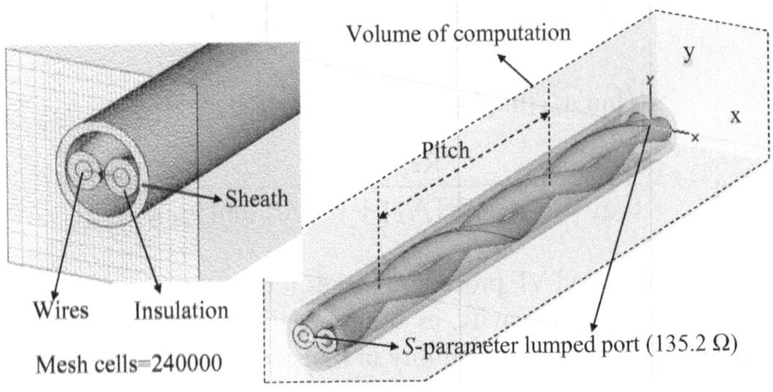

Figure 7.26 Structure of the UTP cable simulated by MWS

line. Consider the UTP cable having the structure shown in Figure 7.26 and described by the parameters reported in Table 7.2. The UTP configuration was modeled by the 3D full-wave numerical tool *MicroWave Studio* (MWS) based on the *Finite Integration Technique* [40]. The 3D model of the considered UTP cable is shown in Figure 7.26, where two pitches of the twist are considered. The simulation of a longer UTP involving several pitches would require a bigger simulation time owing to the increasing complexity. The two wires are modeled with copper material so that the internal losses are taken into account in the full-wave simulation.

Owing to the fine details of the cable, a local subgrid was used in order to obtain accurate results with a reasonable computational effort. The simulations were carried out by exciting the UTP cable at the ports (i.e. cable terminations), adopting as the port impedance the characteristic impedance $Z_0 = 135.2\ \Omega$ of the cable computed by the code with its *Time Domain Reflectometer* (TDR) macro (see *Section 11.1*). It should be noted that this choice makes it possible to increase numerical accuracy and efficiency, leaving out multiple reflections.

The VF procedure applied to the Π-type equivalent circuit of the cable under investigation leads to a good accuracy up to 5 GHz, adopting 10 poles: two real poles and four complex conjugate pairs.

The equivalent Π-type SPICE circuits as well as the equivalent circuits of the admittances $\hat{Y}_{\mathrm{long}}(\omega)$ and $\hat{Y}_{\mathrm{trasv}}(\omega)$ implemented into the MicroCap simulator are shown in Figure 7.27. The equivalent TL-based circuit is shown in Figure 7.28a, and the equivalent circuits of $\hat{Z}_{\mathrm{UTP}}(\omega)$ and $\hat{Y}_{\mathrm{UTP}}(\omega)$ fitted by eight poles are shown in Figures 7.28b and 8.28c respectively. The decoupling capacitance of 1 pF was put in series with $\hat{Y}_{\mathrm{UTP}}(\omega)$ to decouple the

Table 7.2 Geometrical parameters of the simulated UTP cable

Wire radius	Copper	$r_{\mathrm{w}} = 0.255$ mm
Insulation	Polyethylene $\varepsilon_{\mathrm{r}} = 2.25$	$r_{\mathrm{i}} = 0.535$ mm
Internal sheath radius		$r_{\mathrm{si}} = 1.12$ mm
External sheath radius	Thermoplastic copolymer $\varepsilon_{\mathrm{r}} = 3.15$	$r_{\mathrm{sc}} = 1.42$ mm
Center-to-center distance		$d = 1.17$ mm
Pitch length		$l_{\mathrm{p}} = 1.5$ cm

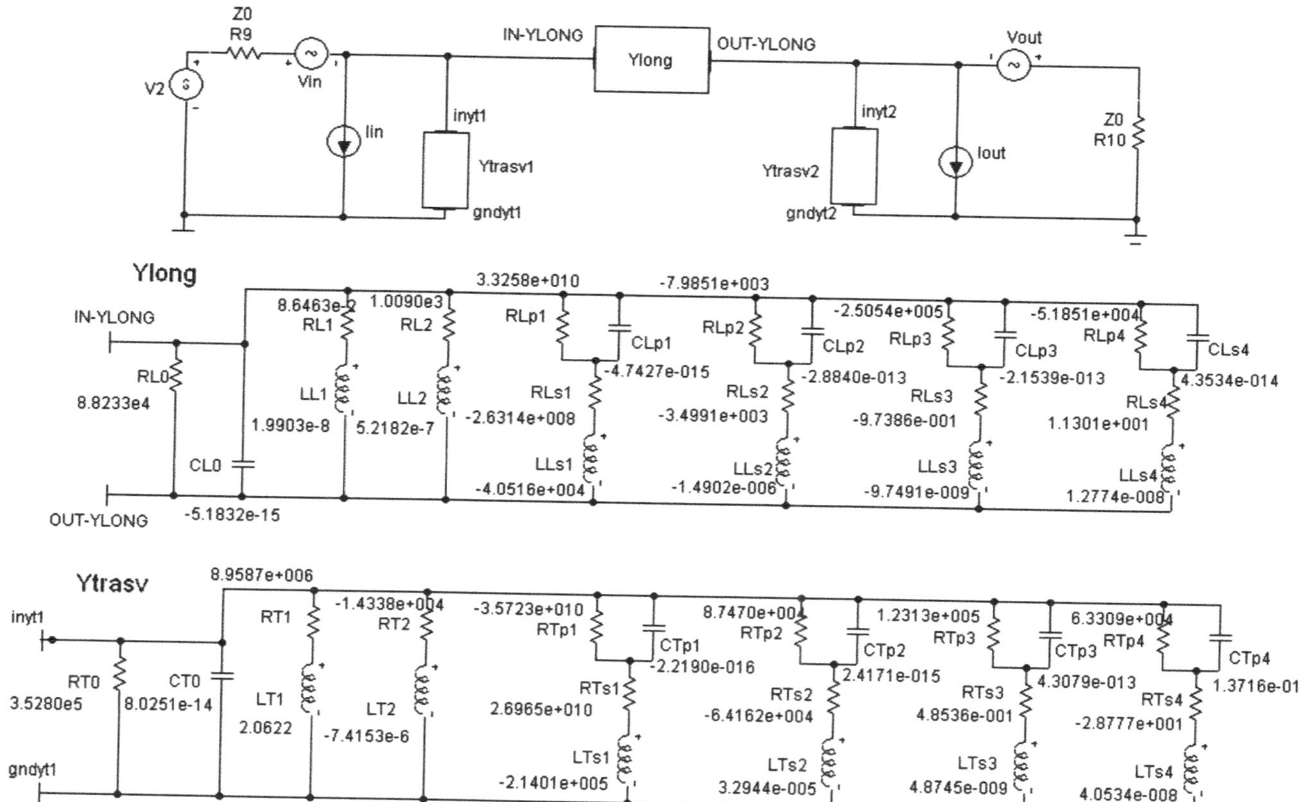

Figure 7.27 Π-type model and equivalent circuits of the admittances $\hat{Y}_{\text{long}}(\omega)$ and $\hat{Y}_{\text{trasv}}(\omega)$ of the UTP cable of Figure 7.26 implemented in the MicroCap simulator

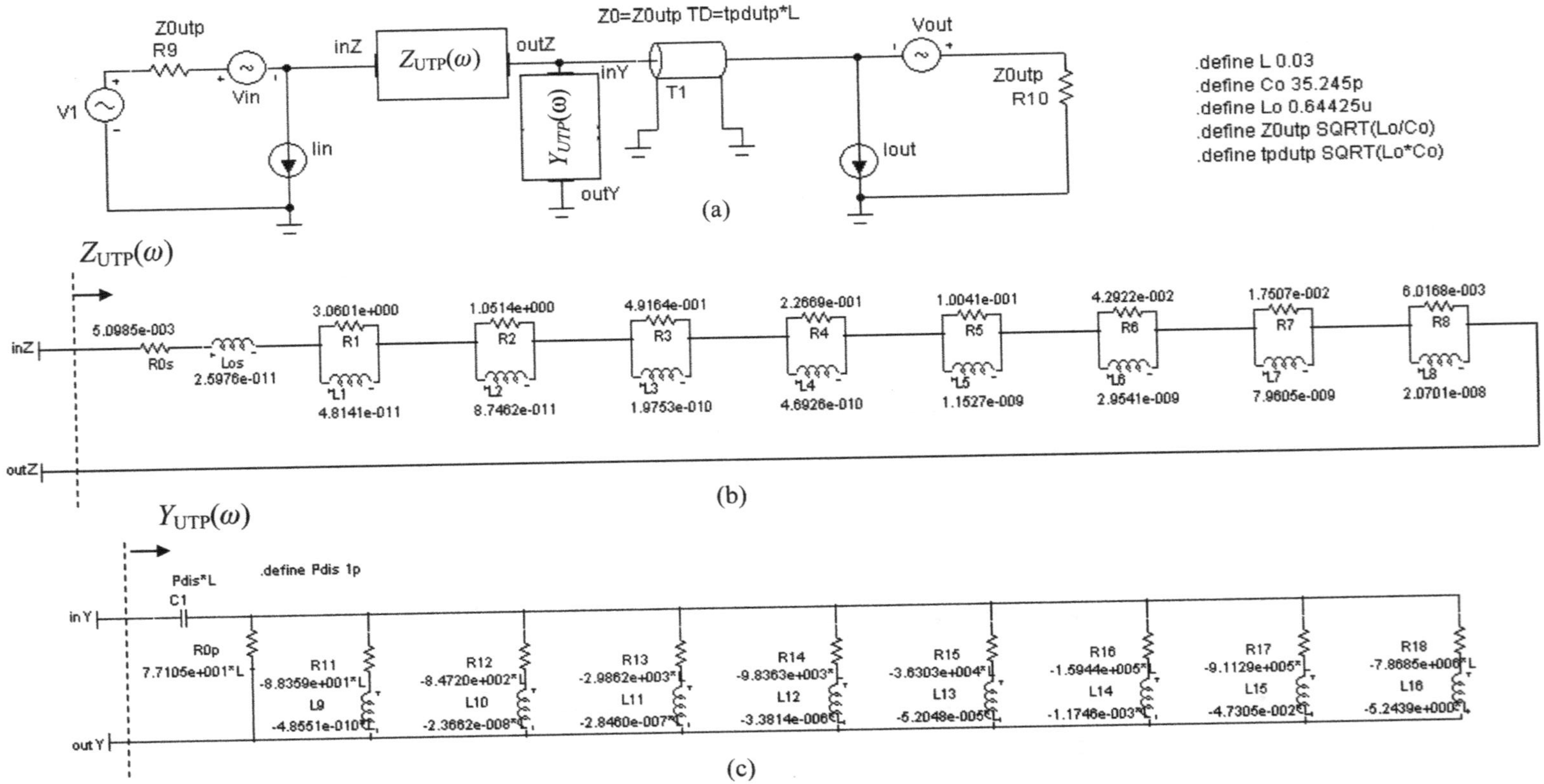

Figure 7.28 SPICE equivalent circuit of the TL-based model of the UTP cable of length $l = 3$ cm: (a) equivalent circuit to compute S-parameters; (b) equivalent circuit of $\hat{Z}_{\mathrm{UTP}}(\omega)$ accounting for skin and proximity effects; (c) equivalent circuit of conductance $\hat{Y}_{\mathrm{UTP}}(\omega) = G_{\mathrm{d}}(\omega)$ associated with the dielectric effect. The contribution of L_0 and C_0 are included in the lossless transmission line model

Table 7.3 General UTP parameters used for computation

Parameters	Value	Note
UTP segment of length l	$l = 3$ cm	Corresponds to two pitches
Characteristic impedance Z_0	$Z_0 = 135.2\ \Omega$	Computed by MWS using TDR option output
Round wire resistivity $\rho_c = 1/\sigma_c$	$\rho_c = 1.724 \times 10^{-8}\ \Omega$/m	σ_c is the conductivity of copper
Vacuum permeability μ_0	$\mu_0 = 4\pi 10^{-7}$ H/m	
Vacuum permittivity ε_0	$\varepsilon_0 = 8.854 \times 10^{-12}$ F/m	
Vacuum velocity $c_0 = 1/\sqrt{\mu_0 \varepsilon_0}$	$c_0 = 3 \times 10^8$ m/s	This is the velocity when all the objects have $\varepsilon_r = 1$
UTP velocity v_0 which takes into account the wire having an insulation with $\varepsilon_r = 2.25$ and an external sheath with $\varepsilon_r = 3.15$	$v_0 = 0.7138\ c_0$ m/s	The coefficient 0.7318 was computed by MWS considering the delay between source and load ports
p.u.l. capacitance $C_0 = 1/(v_0 Z_0)$	$C_0 = 135.245$ pF/m	Z_0 = nominal characteristic impedance
p.u.l. inductance $L_0 = Z_0/v_0$	$L_0 = 0.6442\ \mu$H/m	v_0 = nominal propagation velocity of the cable

net in DC conditions. The parameters and equations required to define the TL-based circuit model are given in Tables 7.3, 7.4, and 7.5, with comments. It must be pointed out that the equivalent circuit of $\hat{Y}_{\mathrm{UTP}}(\omega)$ was obtained by applying vector fitting to the admittance $\hat{Y}_{\mathrm{UTP}}(\omega) = \mathrm{Re}[\hat{Y}_{\mathrm{Diel}}(\omega)] = G_{\mathrm{d}}(\omega)$, as Equation (7.27) was used and the nominal capacitance C_0 (see Equation (7.28)) was extracted and included in the lossy line model. Another thing to consider is that the hypothetical UTP cable modeled as shown in Figure 7.26 had a low loss tangent $\tan\theta$. This explains why the onset frequency of the dielectric losses reported in Table 7.5 is so high. In any case, the purpose of this example is to compare the proposed equivalent circuits. In *Example 7.4* the simulation of an actual UTP cable with experimental validation will be presented.

The impedances of the TL-based model computed with the parameter values and formulae of Tables 7.3–7.5 are shown in Figure 7.29. The magnitude of the impedance $\hat{Z}_{\mathrm{UTP}}(\omega)$ representing the losses in closed-form expression by Equation (7.17) is perfectly consistent with the result produced by the equivalent circuit in Figure 7.28b. As regards the phase, there are some slight non-significant differences around the value of 45°, as expected when the amount of the real part equals the imaginary part. The results produced by three different possible representations of the admittance $\hat{Y}_{\mathrm{UTP}}(\omega)$ are also compared in Figure 7.29:

- simplified model based on C_0 only (no dielectric losses);
- analytical formulation taking into account dielectric losses;
- equivalent circuit coming from the VF technique.

The results produced by the three different models are in very good agreement both for the magnitude and the phase, which is always −90° in the whole frequency range.

Table 7.4 UTP parameters used to calculate skin and proximity effects

Parameters	Value/Expression	Note
DC correction factor K_a	$K_a = 2$	It accounts for the additional DC resistance of the return path
Correction factor determined by the proximity effect K_p	$K_p = 7$	Value obtained by matching analytical S-parameters with those computed by MWS.
Round wire surface a	$a = \pi r_w^2$	r_w is the radius of the wire in meters
Round wire perimeter p	$p = 2\pi r_w$	
Frequency f_δ Equation (7.20)	$f_\delta = \left(\frac{p}{K_p a}\right)^2 \frac{1}{\mu \sigma_c \pi} = 5.5$ kHz	It is the frequency beyond which skin- and proximity-effect losses become significant.
Frequency f_0	$f_0 = 10$ MHz	It is a particular frequency chosen well above the frequency f_δ where the skin-effect resistance equals the DC resistance.
Total DC series resistance in Ω/m	$R_{dc} = K_a \rho_c / a$	This expression accounts for signal and return wire DC resistance.
Skin depth at frequency f_0	$\delta_0 = \sqrt{\rho_c/(\pi f_0 \mu_0)}$	
Real part of the skin-effect impedance R_0 in Ω/m	$R_0 = K_p \rho_c/(p\delta_0)$	
Skin-effect impedance in Ω/m	$\hat{Z}_{i,HF}(f) = (1 + j)R_0\sqrt{f/f_0}$	This expression accounts for skin and proximity effects for frequency f well above the onset frequency f_δ.
UTP p.u.l. series internal impedance in Ω/m	$\hat{Z}_{UTP} = \sqrt{R_{dc}^2 + (\hat{Z}_{i,HF}(f))^2}$	Practical model approximating the exact formulation for an isolated round wire

It should be noted that, while many of the parameters in Tables 7.3–7.5 are set considering the geometric and electric characteristics of the UTP cable considered, generally the parameters K_p and θ_0, related to the proximity and dielectric effects respectively, are unknown a priori and can be computed by matching the S-parameters calculated using the closed-form expressions presented in *Section 7.1* with those computed by MWS. In many practical cases, such as traces in a PCB and commercial cables, θ_0 is known while K_p must be determined. The frequency-domain scattering parameters for the UTP cable under investigation are shown in Figure 7.30. While the magnitudes S_{21} are very close in the full frequency range, the magnitudes S_{11} show large differences above 1.5 GHz. However, this is not very important from a practical viewpoint because S_{21} represents the transfer function of the cable while S_{11} represents the reflections at the source end. In data transmission the interest is mainly focused on the waveforms of the signals transmitted to the load to obtain the eye diagram. Remember that in a lossless line $S_{11} = 0$ and $S_{21} = 1$. To support this statement, some validation of the models will be given below.

Table 7.5 Parameters used to account for the dielectric effect in the UTP cable

Parameters	Value or Expression	Note
Loss tangent of the dielectric material θ_0	$\theta_0 = 0.00115$	Computed at frequency f_0. Value obtained by fitting analytically computed S-parameters with those computed by MWS. This value is very low because, in the numerical model of Figure 7.26, $\tan\theta$ was set very low for the insulation materials.
Angular frequency in rad/s	$\omega = 2\pi f$	
Angular frequency ω_0 in rad/s	$2\pi f_0$	Computed at frequency f_0
Frequency f_θ Equation (7.42)	$f_\theta = \dfrac{1}{2\pi\omega_0}\left(\dfrac{v_0 R_0}{Z_0\theta_0}\right)^2 = 62.5\ \text{GHz}$	It is the frequency beyond which dielectric losses exceeds skin-effect losses. This values is very high because θ_0 is very low for this hypothetical cable.
Arbitrary reference angular frequency $\omega_{0\text{d}}$ in rad/s for computing Y_{Diel}	$\omega_{0\text{d}} = 10 \times 2\pi\, f_\theta$	In this example it is assumed that $C_0 = 1/(Z_0 v_0)$ and θ_0 does not change with increasing the frequency f_0.
Complex admittance $\hat{Y}_{\text{Diel}}(\omega)$ of the line in S/m; Equations (7.27) and (7.29)	$\hat{Y}_{\text{Diel}}(\omega) = j\omega C_0 \left(\dfrac{j\omega}{\omega_{0d}}\right)^{-2\cdot\theta_0/\pi}$ $\hat{Y}_{\text{Diel}}(\omega) = G_{\text{d}}(\omega) + j\omega C(\omega) \approx \omega C_0 \tan\theta + j\omega C_0$	The real part of $\hat{Y}_{\text{Diel}}(\omega)$ represents dielectric losses within the line. According to Equation (7.29), these losses are represented by the conductance $G_{\text{d}} = \omega C_0 \tan\theta$. The ratio between the real and the imaginary parts of $\hat{Y}_{\text{Diel}}(\omega)$ is the dielectric loss tangent $\tan\theta \approx \theta$

The scattering parameters obtained by the Π-type equivalent circuit in Figure 7.24, by the TL-based model in Figure 7.25a, and by the simplified TL-based model in Figure 7.25b are compared with those computed by MWS in Figure 7.31, and the following observations can be made:

- The Π-type model reproduces very well the parameter S_{11} computed by MWS both in magnitude and in phase in the whole frequency range of interest, while the magnitude of S_{21} is slightly different above 1.5 GHz. A point of resonance appears for both S-parameters at about 3.5 GHz. This resonance is also present in the admittances $\hat{Y}_{\text{long}}(\omega)$ and $\hat{Y}_{\text{trasv}}(\omega)$ computed by Equation (7.73) using the S-parameters obtained by MWS. The discrepancies above 1.5 GHz are mainly due to the fact that the 3 cm cable segment is not electrically short at frequencies above 1 GHz, while the simulation extends up to 5 GHz. At 1 GHz the wavelength is $\lambda = 30$ cm and the segment to be electrically short should be less than $\lambda/10$. However, in building up the circuit model, it is necessary to compromise between accuracy and the need to limit the number of cells in cascade representing the line in order to save computer memory and to speed up the simulation.

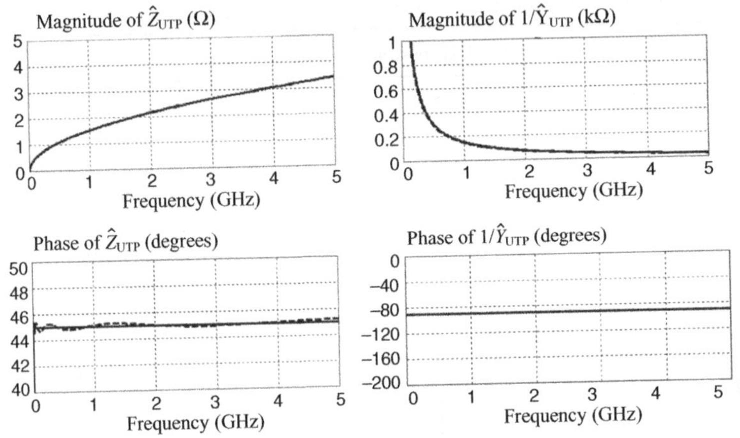

Figure 7.29 Magnitude and phase of the impedance $\hat{Z}_{UTP}(\omega)$ and $1/\hat{Y}_{UTP}(\omega)$ of the UTP cable of length $l = 3$ cm. The lines used for representing $Z_{UTP}(\omega)$ are the analytical expression (solid line) and the SPICE circuit (dashed line). Those used for representing $1/\hat{Y}_{UTP}(\omega)$ are the analytical model (solid line), the SPICE circuit (dashed line), and the simplified C_0 model (dotted line)

- The TL-based model in Figure 7.25a provides an S_{21} parameter very close to the one obtained by MWS both in magnitude and phase. The magnitude of S_{11} increases with frequency and has a profile close to the peak values of S_{11} computed by MWS. The phase of S_{11} is different from that computed by MWS. However, as will be shown later by validation tests, this fact does not compromise the time-domain results.

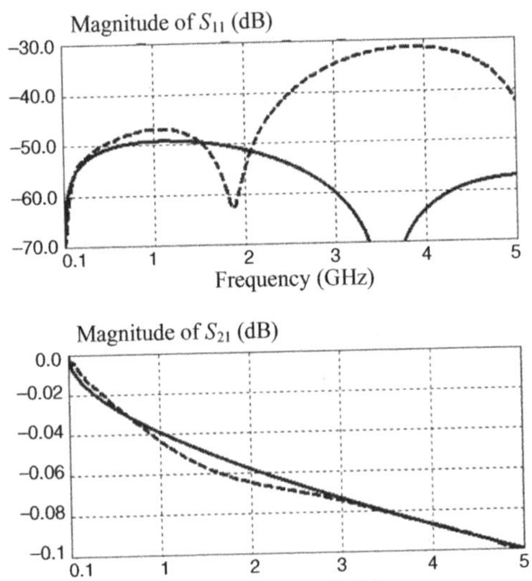

Figure 7.30 S-parameters of the UTP cable computed by MWS (solid line) and obtained by closed-form equations (dashed line)

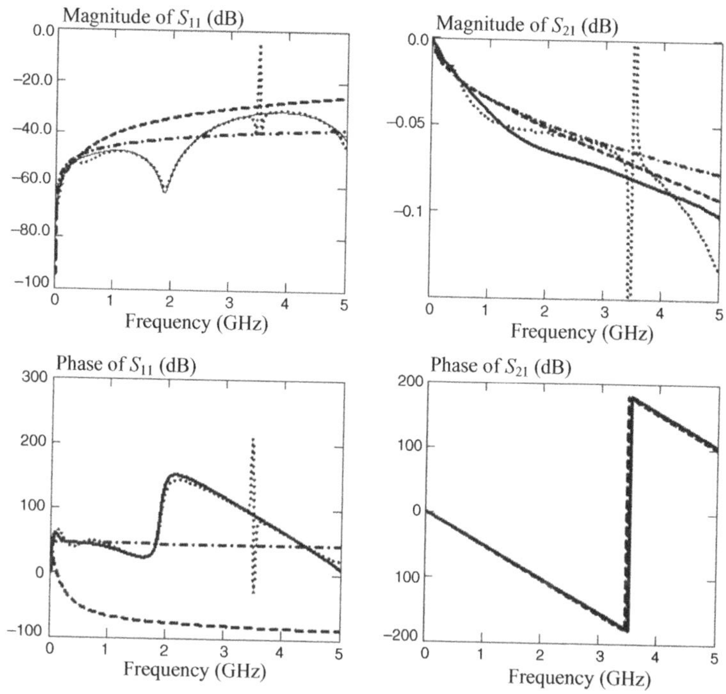

Figure 7.31 *S*-parameters of the UTP cable: MWS (solid line); Π-type circuit model (dotted line); TL-based model (dashed line); simplified TL-based model (dashed-dotted line)

The differences between the Π-type circuit and the TL-based model in the time domain are shown in Figure 7.32. Simulations were performed for a point-to-point UTP interconnect of 3 m length and having the structure of Figure 7.1. Two conditions were simulated: the line matched at both ends, and the line mismatched at the source and load ends. For both models, 100 cells of 3 cm length were used to simulate the UTP cable. The line was excited by a step source of 2 V amplitude with a rise time $t_r = 0.5$ ns. In matched conditions, both models give similar results and reproduce the typical source and voltage waveforms of a lossy line as depicted in Figure 7.3. The Π-type model presents some slight signal distortions owing to the fact that the cell of this model consists of three complex lumped-element nets. In mismatched conditions, distortions on the signal waveforms are again present, and some slight differences in delay time can be observed.

Although the Π-type model seems to provide closer *S*-parameters in magnitude and phase to the numerically computed waveforms in the full frequency range, except at the resonance frequency point, it could cause instability in transient simulation, especially when the number of cells in cascade is too large. For this reason, and considering the complexity of the nets, the Π-type circuit is not recommended. It could be used when the highest frequency of interest makes it possible to avoid resonance points and fewer cells are used. The TL-based model has the great advantage over the Π-type circuit that the simulations run in a few seconds instead of several minutes.

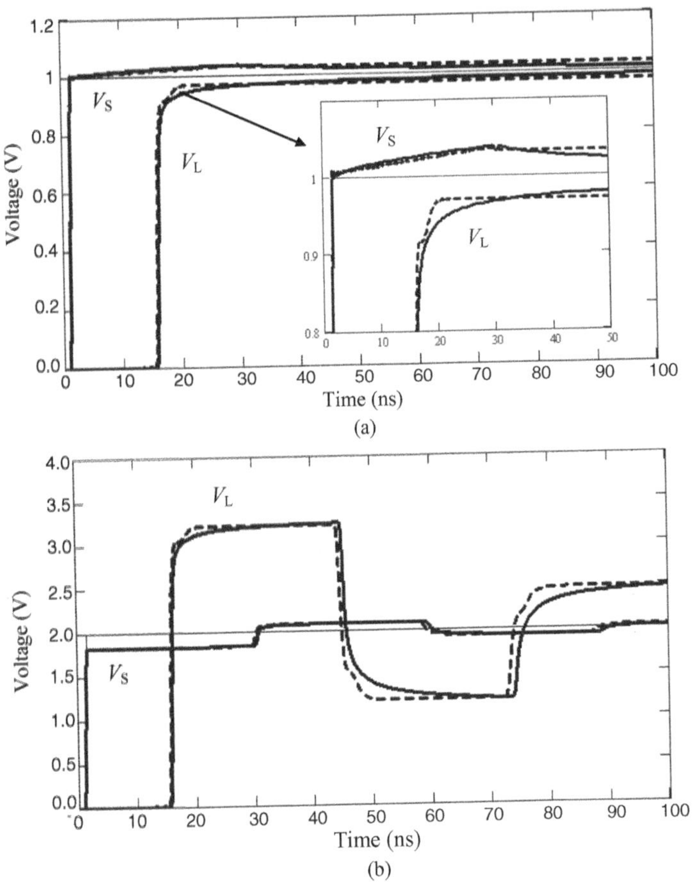

Figure 7.32 Simulated voltages at input V_S and output V_L of the UTP cable: (a) matched source and load conditions ($R_S = R_L = Z_0$); (b) mismatched conditions ($R_S = Z_0/10$ and $R_L = Z_0 10$). Π-type circuit (dashed line) and TL-based model (solid line)

The accuracy of the VF technique applied to the TL-based model is also proven by comparing the results with those obtained by the IFFT of the exact AC analytical model of lossy TLs outlined in *Section 7.1.5.2*, and indicated here as the AC-IFFT model. Referring to Table 7.1, the following parameters where used for the IFFT: $m = 14$, $\Delta t = 0.05$ ns, $\tau = 1$ ns, $t_r = 0.5$ ns. The cable parameters are those given in Tables 7.3–7.5. The results are shown in Figure 7.33, where a very good agreement can be observed.

As a further validation of the TL-based model, an UTP cable of 90 cm length (60 pitches) was manually constructed, and 30 cells were used to perform the transient simulation of the UTP cable terminated at both ends with 50 Ω resistances. The cable was excited with a unit step voltage source of 2 V amplitude with a rise time $t_r = 0.25$ ns. The transient input and output voltages are shown in Figure 7.34, where a good agreement between simulations and measurements can be observed. The slight oscillations in the measured waveforms are due to the non-uniformity of the twisted-pair wires used for the experiment.

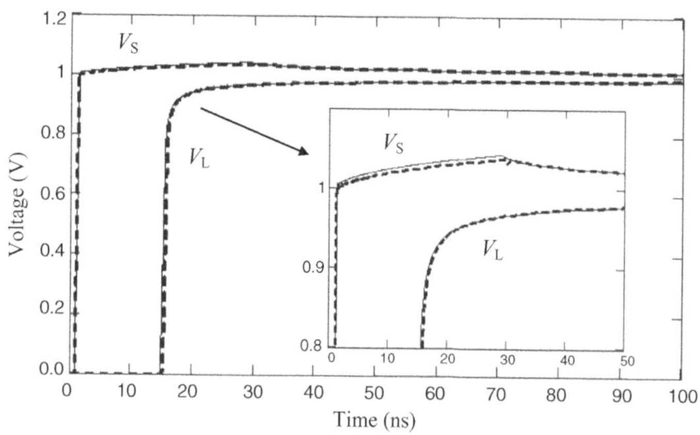

Figure 7.33 Simulated voltages at input V_S and output V_L of the UTP cable in matched source and load conditions. TL-based circuit model (dashed line) and AC-IFFT model (solid line)

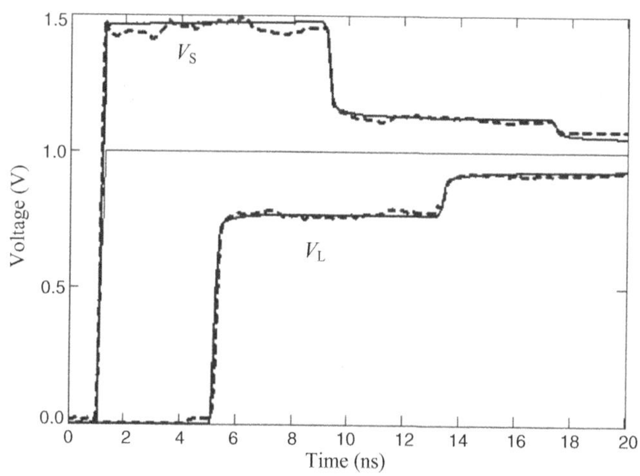

Figure 7.34 Simulated and measured voltages at input V_S and output V_L of the UTP cable in matched source and load conditions. Measurements (dashed line) and TL-based circuit model (solid line)

7.3 Modeling Lossy Lines in the Time Domain by the Scattering Parameters Technique

The technique described in this section should be applied to the case of long cables when the segmentation technique becomes too expensive in terms of the number of cells to be put in cascade. The main advantage of this approach is that the model is independent of the cable length.

Consider a cable representing a transmission line with nominal characteristic impedance $Z_0 = \sqrt{L_0/C_0}$ and time delay $T_D = t_{pd}l$, where t_{pd} is the per-unit-length propagation delay time of the cable and l is the cable length. As practical cables of interest have electrically short dimensions in the transversal section, the cable can be considered as a two-port network

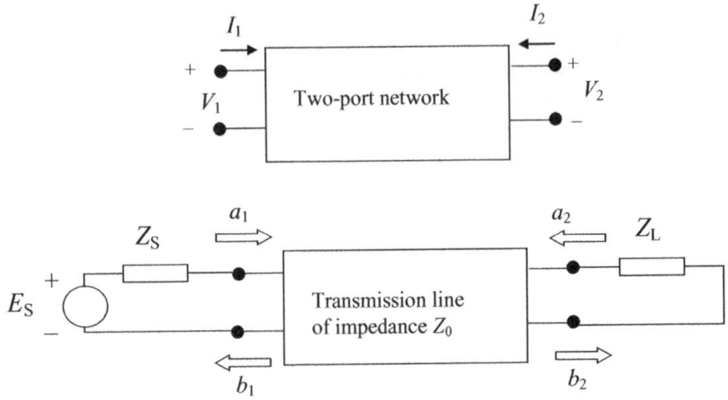

Figure 7.35 Incident and reflected waves applied to a transmission line seen as a two-port network

described by the generalized scattering parameters which are linked to the voltages and currents at the line ends as shown in Figure 7.35. The incident (a_1, a_2) and reflected (b_1, b_2) waves can be calculated from the voltages and the currents at both ends of the line using the normalizing factor $2\sqrt{Z_0}$, where Z_0 is the cable nominal characteristic impedance. The following equations hold in both the frequency and the time domains:

$$a_1 = (V_1 + I_1 Z_0)/2\sqrt{Z_0} \tag{7.78a}$$

$$a_2 = (V_2 + I_2 Z_0)/2\sqrt{Z_0} \tag{7.78b}$$

$$b_1 = (V_1 - I_1 Z_0)/2\sqrt{Z_0} \tag{7.79a}$$

$$b_2 = (V_2 - I_2 Z_0)/2\sqrt{Z_0} \tag{7.79b}$$

By a demonstration similar to that given by Paul [24] to obtain the Branin circuit presented in *Section 5.2.6*, it can be shown that the TL described by Equations (7.78) and (7.79) can be modeled by the equivalent circuit of Figure 7.36, where

$$e_i(t) = 2\sqrt{Z_0}\, b_1(t) \tag{7.80a}$$
$$e_o(t) = 2\sqrt{Z_0}\, b_2(t) \tag{7.80b}$$

In order to define the equivalent circuit of Figure 7.36, it is necessary to find suitable expressions for the voltage sources (7.80), taking into account frequency-dependent losses. To this end, let us start with the representation of a lossy TL in the frequency domain by the S-parameters:

$$\begin{bmatrix} \hat{b}_1(\omega) \\ \hat{b}_2(\omega) \end{bmatrix} = \begin{bmatrix} \hat{S}_{11}(\omega) & \hat{S}_{12}(\omega)\mathrm{e}^{-\mathrm{j}\omega T_D} \\ \hat{S}_{21}(\omega)\mathrm{e}^{-\mathrm{j}\omega T_D} & \hat{S}_{22}(\omega) \end{bmatrix} \begin{bmatrix} \hat{a}_1(\omega) \\ \hat{a}_2(\omega) \end{bmatrix} \tag{7.81}$$

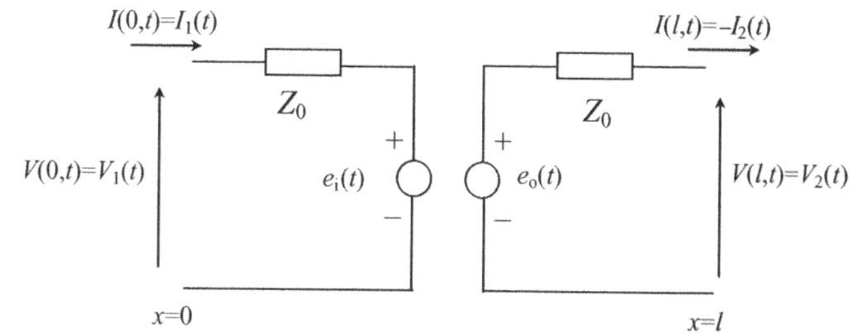

Figure 7.36 Equivalent circuit of a transmission line, using the concept of reflected waves

Note that the off-diagonal coefficient in Equation (7.81) is written by extracting the term related to the time delay, or equivalently to the phase constant. In this way, the two main contributions associated with the line propagation delay and with the frequency-dependent losses are separately highlighted, which is reasonable, as the TL propagation coefficient is slightly dependent on types of loss, as was shown in Figure 7.11.

In practice, the following relations hold: $\hat{S}_{11} = \hat{S}_{22}$ and $\hat{S}_{12} = \hat{S}_{21}$. By definition, the off-diagonal coefficient $\hat{S}_{12}(\omega)\mathrm{e}^{-\mathrm{j}\omega T_\mathrm{D}}$ in Equation (7.81) represents the TL transfer function, while the diagonal terms represent the reflections at the line ends.

Moving (7.81) to the time domain yields

$$\begin{bmatrix} b_1(t) \\ b_2(t) \end{bmatrix} = \begin{bmatrix} \int_0^t S_{11}(t-\tau)a_1(\tau)\mathrm{d}\tau + \int_0^t S_{12}(t-\tau)a_2(\tau - T_\mathrm{D})\mathrm{d}\tau \\ \int_0^t S_{22}(t-\tau)a_2(\tau)\mathrm{d}\tau + \int_0^t S_{21}(t-\tau)a_1(\tau - T_\mathrm{D})\mathrm{d}\tau \end{bmatrix} \tag{7.82}$$

The right-hand side of Equation (7.82) contains convolution integrals that must be solved numerically to compute the dependent voltage sources $e_\mathrm{i}(t)$ and $e_\mathrm{o}(t)$ defined by Equations (7.80). From the definition of incident waves, and with the current notation introduced in Figure 7.36, we have

$$\begin{bmatrix} 2\sqrt{Z_0}a_2(\tau - T_\mathrm{D}) \\ 2\sqrt{Z_0}a_1(\tau - T_\mathrm{D}) \end{bmatrix} = \begin{bmatrix} V(l, \tau - T_\mathrm{D}) - Z_0 I(l, \tau - T_\mathrm{D}) \\ V(0, \tau - T_\mathrm{D}) + Z_0 I(0, \tau - T_\mathrm{D}) \end{bmatrix} \tag{7.83}$$

Finally, by using Equations (7.82) and (7.83), the dependent voltage sources (7.80) at each time instant are calculated as the sum of two convolution integrals between the S-parameters and the algebraic combination of voltages and currents at the line ends.

In the case of a lossless TL matched at both ends, the S-parameters are

$$S_{12} = S_{21} = 1$$
$$S_{11} = S_{22} = 0$$

and $e_i(t)$ and $e_o(t)$ reduce to

$$\begin{bmatrix} e_i(t) \\ e_o(t) \end{bmatrix} = \begin{bmatrix} 2\sqrt{Z_0}b_1(t) \\ 2\sqrt{Z_0}b_2(t) \end{bmatrix} = \begin{bmatrix} 2\sqrt{Z_0}a_2(t-T_D) \\ 2\sqrt{Z_0}a_1(t-T_D) \end{bmatrix} = \begin{bmatrix} V(l,t-T_D) - Z_0 I(l,t-T_D) \\ V(0,t-T_D) + Z_0 I(0,t-T_D) \end{bmatrix} \tag{7.84}$$

In this case, Equation (7.84) coincides with Equation (5.22), and therefore the equivalent circuit of Figure 7.36 is equivalent to the Branin circuit introduced in *Chapter* 5 to model lossless TLs and present in the SPICE library.

The time-domain S-parameters necessary to calculate the convolution integrals in Equation (7.82) can be obtained as the response of a narrow Gaussian pulse for the frequency range of interest, and this can be done by full-wave numerical tools (i.e. MWS based on the FIT technique), or by measurements according to the set-up shown in Figure 7.37, where the TL is excited by a step generator. The rise time of the step generator should be fast enough to obtain the waveforms of Figure 7.3. Moreover, as in practice both the step generator and the oscilloscope have an internal resistance equal to the standard 50 Ω value, to match the TL at both ends, two resistances of value $R = Z_0 - 50\ \Omega$ are required.

The S-parameters to be used in the convolutions (7.82) are related to the voltages V_S and V_L at the line ends (see Figure 7.37) by

$$S_{11}(t) = \frac{\mathrm{d}(V_S(t) - E_{So})}{\mathrm{d}t} \tag{7.85a}$$

$$S_{21}(t) = \frac{\mathrm{d}V_L(t)}{\mathrm{d}t} \tag{7.85b}$$

In summary, the computation of $e_i(t)$ and $e_o(t)$ in the lossy TL circuit model of Figure 7.36 must be performed by the following operations:

- Numerical derivative of the voltage step responses of the matched line to calculate by Equations (7.85) the time-domain scattering parameters $S_{11}(t)$ and $S_{21}(t)$.
- Numerical convolution integral between scattering parameters $S_{11}(t)$ and $S_{21}(t)$ and incident waves $a_1(t)$ and $a_2(t)$ according to Equation (7.82). The incident waves are functions of the voltages and currents at opposite ends, as stated by Equations (7.80).
- When performing convolution with $S_{12} = S_{21}$, the incident waves must be delayed by a time T_D, as shown in Equation (7.82).

The main problem in implementing the lossy TL circuit model of Figure 7.36 is related to the convolution, as many SPICE-like simulators do not offer this feature. However, popular commercial mathematical codes such as MathCad or Matlab can be used with success, as shown by the following examples.

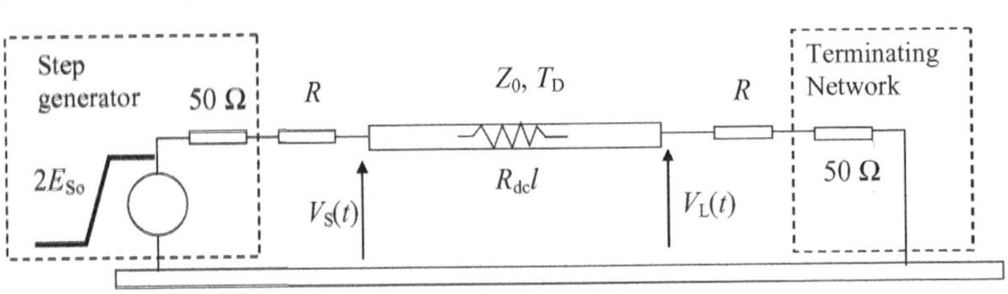

Figure 7.37 Set-up to measure the S-parameters in the time domain

Example 7.4: Eye Diagram of a 75 m Unshielded Twisted-pair Cable Driven by an RS422 Device

To validate the model with experimental results, the test set-up shown in Figure 7.38 is considered. The cable is a twisted-pair cable of Cat.5e and length $l = 75$ m. The driver is a differential CMOS 34C87 device and the receiver is a 34C86 device (RS422 standard). The data stream is an NRZ sequence at 32 Mb/s. The line is terminated with three resistances having the task of partially matching the line, and of ensuring polarization of the receiver when the cable is disconnected, as often required in practice.

As the interest is focused on the eye diagram, and the driver output characteristic is symmetric, the simulation can be performed by the equivalent circuit of the interconnect shown in Figure 7.38. The driver swings from -3 V to 3 V and has an output resistance $R_S = 10\ \Omega$. The line has a differential characteristic impedance $Z_0 = 100\ \Omega$, a p.u.l. propagation delay time $t_{pd} = 5$ ns/m, and a total DC resistance $R_{dc}l$. The load resistance R_L is assumed to be the parallel connection between 150 Ω and 2 kΩ, because in AC condition the ground and the power pins of the receiver are at the same potential.

The waveform in Figure 7.3, suitable for computing S-parameters in the time domain, can be obtained in two ways:

- by measurements according to the set-up shown in Figure 7.37 if the sample of cable is available;
- by the analytical approach outlined in *Section 7.1*, considering the typical electrical characteristic of an UTP Cat.5e cable as reported by Johnson and Graham [25] (see Table 8.2 for $\tan\theta$ and Table 8.3 for K_p).

In this second case, the main cable parameters of Tables 7.3–7.5 are set as follows: $r_w = 0.2285$ mm (wire radius), $v_0 = 0.7c = 2.099 \times 10^8$ m/s, $l = 75$ m, $Z_0 = 100\ \Omega$, $f_0 = 10$ MHz, $\theta_0 = 0.0115$, $K_a = 2$, and $K_p = 2.7$. These parameters give $f_\delta = 46$ kHz and $f_\theta = 203$ MHz. The dielectric losses become significant over 200 MHz. The input and output waveforms computed by the AC-IFFT model are shown in Figure 7.39, where the voltage on the load is shown without the delay $T_D = v_0 l$ for plot representation. The final voltages $V_{S\infty} = 1.073$ V and $V_{L\infty} = 0.927$ V are also indicated. For clarity, only the significant parts of the waveforms are shown. A very good agreement can be observed between measured and

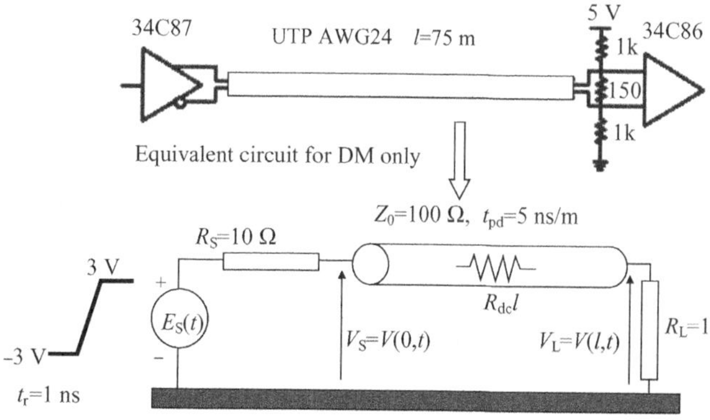

Figure 7.38 Point-to-point interconnect with differential driver/receiver devices and its equivalent circuit for differential mode signaling

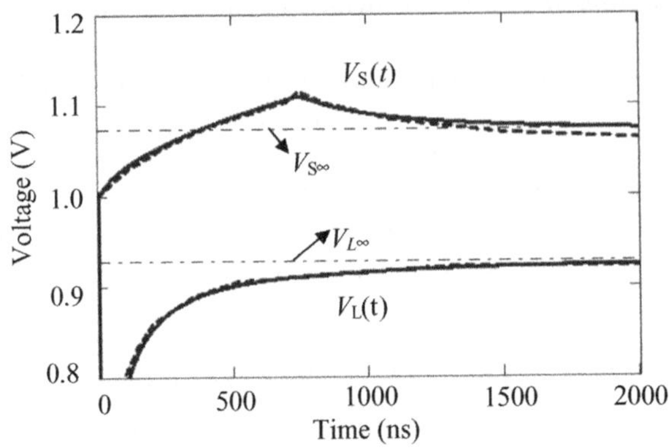

Figure 7.39 Input and output voltages of the UTP Cat-5e cable of 75 m length matched at both ends according to the set-up of Figure 7.3: measured (dashed line) and computed by the AC-IFFT (solid line)

predicted waveforms by the AC-IFFT model, which shows the efficiency of the analytical approach for long cables. Therefore, in the absence of a cable sample having the length of interest, the *S*-parameters can be estimated in the time domain by matching the frequency-domain *S*-parameters calculated by the closed-form analytical approach with those computed by a full-wave code for an electrically short segment of cable, as outlined in *Section 7.2.1*.

The simulated waveforms when the line is sourced with a *non-return to zero* (NRZ) sequence of bits with a data rate of 32 Mb/s are shown in Figure 7.40. The convolution integrals

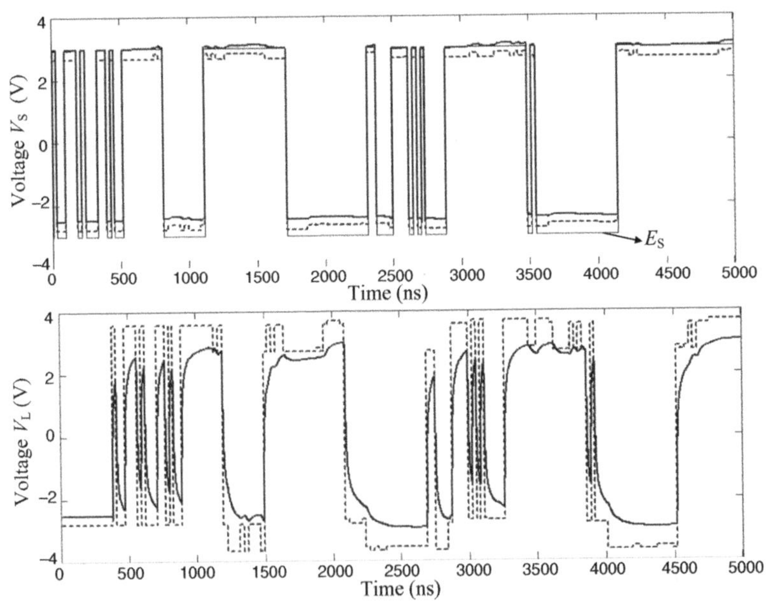

Figure 7.40 Input and output voltages of the UTP Cat-5e cable of 75 m length matched at both ends and excited by a NRZ sequence of bits at 32 Mb/s: TL lossless model (dashed line) and TL lossy model (solid line)

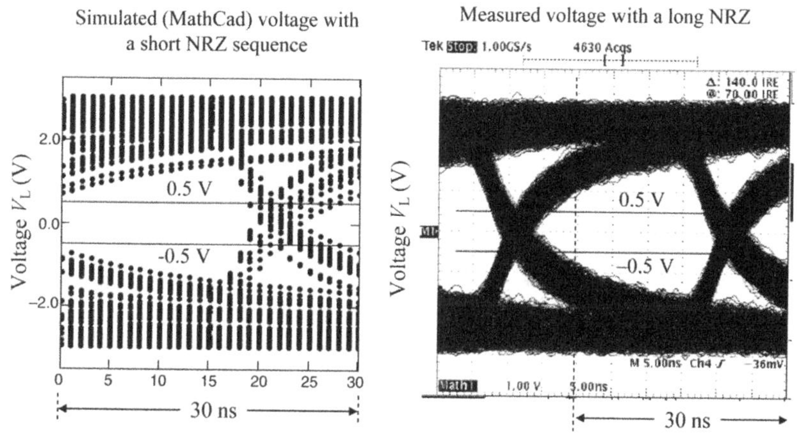

Figure 7.41 Eye diagrams of the voltage on the load for the 75 m Cat-5e UTP cable with a NRZ sequence of 32 Mb/s

were performed starting from the measured voltages in Figure 7.39. Very similar results can be obtained using the computed S-parameters. The comparison between waveforms obtained by a lossless model and those obtained by a lossy line model is quite interesting. Significant discrepancies can be observed, especially for the voltages on the load V_L. The lossy model is validated by comparison between simulated and measured waveforms using the eye diagram representation as shown in Figure 7.41 for the voltage on the load. Very good agreement can be observed, although the sequence of bits used for simulation is shorter than the measured one.

Example 7.5: Coaxial Cable Driven by an Ultrawide-band Signal Simulated by the TL Model Based on the Time-domain S-parameters

As a further validation of the lossy TL model based on the time-domain S-parameters, the same coaxial cable driven by the UWB signal as that used in *Example 7.2* is considered. To perform simulations, the following steps are required:

1. Find the equivalent circuit of an electrically short section of cable by the VF technique; in this case $\Delta x = 3$ mm is suitable, as in *Example 7.2*.
2. Run the simulation of the cable equivalent circuit obtained by cascading 610 cells, and adopting matching conditions at both the cable ends.
3. Calculate by Equation (7.85) the time-domain S-parameters necessary to implement the convolution integrals of the dependent sources $e_p(t)$ and $e_a(t)$ in Figure 7.36.

All these steps are illustrated in Figure 7.42. It is necessary to point out that $S_{11\text{step}}(t) = V_S(t) - E_{So}$ and $S_{21\text{step}}(t) = V_L(t)$ must be computed adopting a time interval small enough to avoid numerical attenuation effect. In the proposed example, $\Delta t = 0.01$ ns is used, as the rise time of the step source is $t_r = 0.1$ ns. Comparison between simulated and measured waveforms of the load voltage is shown in Figure 7.43.

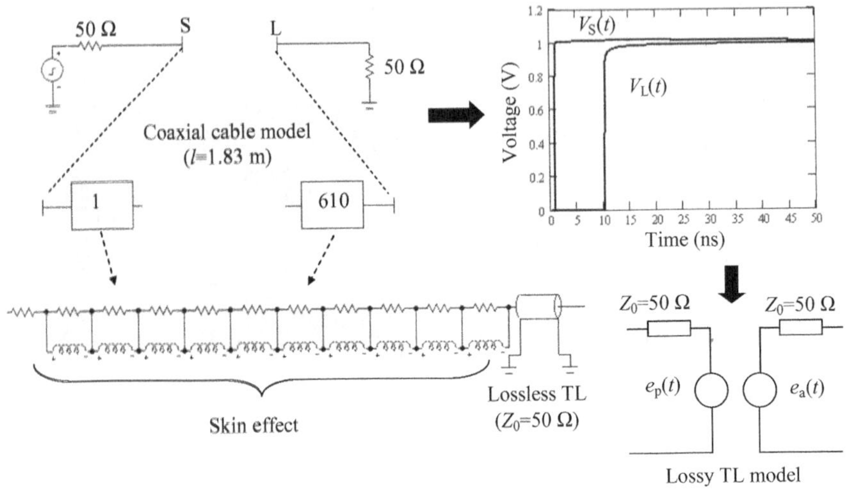

Figure 7.42 Computed input and output voltages of the coaxial cable by 610 cells of length $\Delta x = 3$ mm to be used with the lossy TL model based on the time-domain S-parameters

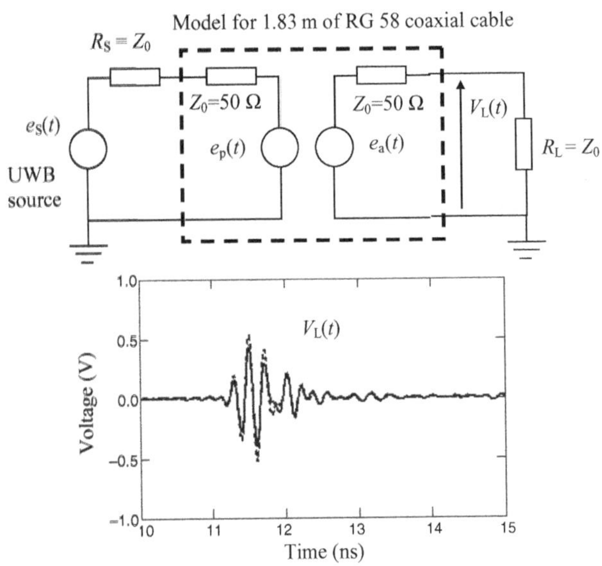

Figure 7.43 Coaxial cable matched at both ends and modeled as a simple network based on S-parameters in the time domain: comparison between measured (dashed line) and computed (solid line) waveforms

7.4 Conclusions

To conclude this section, it is possible to make the following observations:

- The two approaches outlined for simulating lossy TLs in the time domain, *Vector Fitting* (VF) and time-domain S-parameters, are suitable for modeling several structures such as traces on PCBs (microstrips, striplines) and cables (coaxial, twinax, UTP, SFTP), once the Z-, Y-, or S-parameters of an electrically short segment of cable are known.
- The VF method requires knowledge of the Z- and Y-parameters of an electrically short section of the line. These circuit parameters can be obtained by analytical expressions when available (e.g. coaxial cables), or by numerical computations of S-parameters in the frequency domain in order to consider losses such as skin, proximity, and dielectric effects. The equivalent circuit of the cable section reproducing the computed S-parameters can be easily implemented in any SPICE-like simulator and is suitable for performing simulations directly in the time domain. It has been demonstrated that each cell can be simulated by a Π-type model, or better by a TL-based model where a lossless TL is extracted and two networks for Z and Y are used to model frequency-dependent losses. This modeling procedure can be useful for short lossy TLs inserted in a general topology of interconnect structures when the required number of cells for each line is not too large.
- The S-parameters method in the time domain requires knowledge of the time-domain S-parameters for the full length of the line. These parameters can be obtained by TDR-like measurements or by a SPICE-like simulator, using the models produced by the VF technique. The great advantages of this model are its simplicity in terms of circuit elements and its independence from the length of the line. It can be easily implemented in a mathematical code for point-to-point interconnect or in a SPICE-like simulator for general interconnect structures once a numerical convolution in the time domain is performed by the mathematical or chosen circuit-based simulation code. The model can also be used to simulate even and odd modes of propagation by using the circuit model presented in *Section 6.2* [30].

When a sample of cable to be simulated with the length of interest is not available for measuring the time-domain S-parameters, the following procedure can be used:

- The S-parameters of an electrically short segment of cable are numerically calculated by full-wave simulations.
- The S-parameters calculated by the TL-based model are matched with those numerically computed, by varying the coefficients for proximity (K_{p}) and dielectric ($\tan\theta$) (when unknown) effects, until an acceptable approximation is obtained.
- The time-domain S-parameters of the line are found for the length of interest as the response of a step source performing simulation in the frequency domain and using the IFFT to obtain time-domain results (AC-IFFT model).
- Simulations of the interconnect structure are performed by the model of the lossy TL based on the computed time-domain S-parameters with a code that allows numerical derivative, convolution integral, and delay functions.

References

[1] Weeks, W., Wu, L., McAllister, M., and Singh, A., 'Resistive and Inductive Skin Effect in Rectangular Conductors', *IBM Journal of Research and Development*, **23** (6), November 1979, pp. 652–660.

[2] Gruodis, A., 'Transient Analysis of Uniform Resistive Transmission Lines in a Homogeneous Medium', *IBM Journal of Research and Development*, **23** (6), November 1979, 675–681.

[3] Gruodis, A. and Chang, C., 'Coupled Lossy Transmission Line Characterization and Simulation', *IBM Journal of Research and Development*, **25** (1), January 1981, 25–41.

[4] Yen, C.-S., Fazainc, Z., and Wheeler, R., 'Time-domain Skin-effect Model for Transient Analysis of Lossy Transmission Lines', *Proceedings of the IEEE*, **70** (7), July 1982, 750–757.

[5] Yen, C.-S., 'Distribution and Equalization of Signal on Coaxial Cables used in 10 Mbit/s Baseband Local Area Network', *IEEE Trans. on Communication*, **31** (10), October 1983, 1181–1186.

[6] Djordjevic, A., Sarkar, T., and Harrington, R., 'Time-domain Response of Multiconductor Transmission Lines', *Proceedings of the IEEE*, **75** (6), June 1987, 743–764.

[7] Chang, C.S., 'Electrical Design of Signal Lines for Multilayer Printed Circuit Boards', *IBM Journal of Research and Development*, **32** (5), September 1988, 647–657.

[8] Deutsch, A. *et al*, 'High-speed Signal Propagation on Lossy Transmission Lines', *IBM Journal of Research and Development*, **34** (4), July 1990, 601–615.

[9] Barr, A., 'Calculation of Frequency-dependent Impedance for Conductors of Rectangular Cross Section', *AMP Journal of Technology*, **1**, November 1991, 91–100.

[10] Djordjevic, A. and Sarkar, T., 'Closed-form Formulas for Frequency-dependent Resistance and Inductance per Unit Length of Microstrip and Strip Transmission Lines', *IEEE Trans. on Microwave Theory and Techniques*, **42** (2), February 1994, 241–248.

[11] Deutsch, A., 'Electrical Characteristics of Interconnections for High-performance Systems', *Proceedings of the IEEE*, **86** (2), February 1998, 315–357.

[12] Antonini, G., Orlandi, A., and Paul, C., 'Internal Impedance of Conductors of Rectangular Cross Section', *IEEE Trans. on Microwave Theory and Techniques*, **47** (7), July 1999, 979–985.

[13] Ruehli, A. and Cangellaris A., 'Progress in the Methodologies for the Electrical Modeling of Interconnects and Electronic Package', *Proceedings of the IEEE*, **89** (5), May 2001, 740–771.

[14] Achar, R. and Nakhla, M., 'Simulation of High-speed Interconnects', *Proceedings of the IEEE*, **89** (5), May 2001, 693–728.

[15] Deutsch, A. *et al.*, 'On-chip Wiring Design Challenges for Gigahertz Operation', *Proceedings of the IEEE*, **89** (4), April 2001, 529–555.

[16] Coperich, K., Morsey, J., Okhmatovski, V., Cangellaris, A., and Ruehli, A., 'Systematic Development of Transmission-line Models for Interconnects with Frequency-dependent Losses', *IEEE Trans. on Microwave Theory and Techniques*, **49** (10), October 2001, 1677–1685.

[17] Antonini, G., Drewniak, J., Orlandi, A., and Ricchiuti, V., 'Eye pattern evaluation in high-speed digital systems analysis by using MTL modeling', *IEEE Trans. on Microwave Theory and Techniques*, **50** (7), July 2002, 1807–1815.

[18] Elfadel, I., Huang, H.-M., Ruehli, A., Dounavis, A., and Nakhla, M., 'A Comparative Study of Two Transient Analysis Algorithms for Lossy Transmission Lines with Frequency-dependent Data', *IEEE Trans. on Advanced Packaging*, **25** (2), May 2002, 143–153.

[19] Coperich Branch, K., Morsey, J., Cangellaris, A., and Ruehli, A., 'Physically Consistent Transmission Line Models for High-speed Interconnects in Lossy Dielectric', *IEEE Trans. on Advanced Packaging*, **25** (2), May 2002, 129–135.

[20] Grivet-Talocia, S., Huang, H.-M., Ruehli, A., Canavero, F., and Elfadel, I., 'Transient Analysis of Lossy Transmission Lines: an Efficient Approach Based on the Method of Characteristics', *IEEE Trans. on Advanced Packaging*, **27** (1), February 2004, 45–56.

[21] *'Data Transmission Lines and Their Characteristics'*, AN-806, National Semiconductor, April 1992. Available online: www.national.com.

[22] *'Long Transmission Lines and Data Signal Quality'*, National Semiconductor, AN-808, March 1992. Available online: www.national.com.

[23] Ramo, S., Whinnery, J., and Van Duzer, T., *'Fields and Waves in Communication Electronics'*, 3rd edition, John Wiley & Sons, Inc., New York, NY, 1994.

[24] Paul, C., *'Introduction to Electromagnetic Compatibility'*, John Wiley & Sons, Inc., Hoboken, NJ, 2006.

[25] Johnson, H. and Graham, M., *'High-speed Signal Propagation: Advanced Black Magic'*, Pearson Education, Inc., Prentice Hall PTR, Upper Saddle River, NJ, 2003.
[26] Gupta, K.C., Garg, R., Bahl, I., and Bhartia, P., *'Microstrip Lines and Slotlines'*, Artech House, Norwood, MA, 1996.
[27] Araneo, R., Celozzi, S., and Maradei, F., 'Digital Signal Transmission through Differential Interconnects: Full-wave vs. Spice Modeling', *IEEE 2003 International Symposium on Electromagnetic Compatibility*, Boston, MA, 18–22 August 2003, 855–858.
[28] Gustavsen, B. and Semlyen, A., 'Rational Approximation of Frequency Domain Responses by Vector Fitting', *IEEE Trans. on Power Delivery*, **14** (3), July 1999, 1052–1061.
[29] Caniggia, S. and Maradei, F., 'Equivalent Circuit Models for the Analysis of Coaxial Cables Immunity', *IEEE 2003 International Symposium on EMC*, Boston, MA, 18–22 August 2003, 881–886.
[30] Caniggia, S., 'EMC Design of Digital Systems Using Macromodelling Procedures for Integrated Circuits and their Interconnections', *5th Symposium and Technical Exibition on EMC*, Zurich, 3–5 March 1983, 465–470.
[31] Caniggia, S. and Maradei, F., 'SPICE-like Models for the Analysis of the Conducted and Radiated Immunity of Shielded Cables', *IEEE Trans. on Electromagnetic Compatibility*, **46** (4), November 2004, pp. 606–616.
[32] D'Amore, M., *'Electromagnetic Compatibility'* (in Italian), Edizioni Scientifiche Siderea, Rome, Italy, 2003.
[33] Vance, E., *'Coupling to Shielded Cable'*, John Wiley & Sons, Inc., New York, NY, 1978.
[34] Tesche, F., Ianoz, M., and Karlsson, T., *'EMC Analysis Methods and Computational Models'*, John Wiley & Sons, Inc., New York, NY, 1997.
[35] Tsaliovich, A., *'Cable Shielding for Electromagnetic Compatibility'*, Chapman and Hall, New York, NY, 1995.
[36] Caniggia, S. and Maradei, F., 'Circuit and Numerical Modeling of ESD Coupling to Shielded Cables', *EMC Europe 2004 – International Symp. on Electromagnetic Compatibility*, Eindhoven, The Netherlands, 6–10 September 2004.
[37] Caniggia, S. and Maradei, F., 'Circuit and Numerical Modeling of Electrostatic Discharge Generators', *IEEE Trans. on Industry Applications*, **42** (6), November/December 2006, 1350–1357.
[38] www.spectrum-soft.com.
[39] Young, B., *'Digital Signal Integrity: Modeling and Simulation with Interconnects and Packages'*, Prentice Hall PTR, Upper Saddle River, NJ, 2001.
[40] www.cst.com.

8

Delta I-Noise

The noise caused by the switching of digital devices on PCB power and ground distribution networks is presented in this chapter. This impulsive voltage noise is referred to as ΔI*-noise*. The importance of using decoupling capacitors for current switching demand of ICs and for lowering the impedance of the power distribution network is introduced and discussed. The path of the impulsive switching current is analyzed for a typical PCB. The noise produced on power and ground distribution by the switching current of a typical high-speed CMOS device is simulated for stripline and microstrip power and ground distribution networks, taking into account the inductance associated with the lead of the components present in a PCB. Main design rules are given and discussed.

The contribution given by the interplane capacitance between power and ground planes in a multilayer PCB and the action of decoupling capacitors to mitigate ΔI*-noise* are investigated analytically, experimentally, and by means of circuit simulations. The resonances produced by a multilayer PCB are investigated. Three models for predicting impedances and resonance frequencies in a pair of power and ground planes populated by decoupling capacitors are outlined and validated in *Appendix C*. These three models are based on: a set of closed-form expressions for a resonant cavity, an equivalent circuit consisting of a grid of lumped elements, and a 3D structure used as input for numerical simulations. Measurements and simulations of ΔI*-noise* are presented with reference to two test boards with CMOS digital devices: one is a standard board with and without decoupling capacitors and the other is based on buried capacitance technology. The negative effects of the inductances associated with the package of components in reducing power noise are highlighted.

The *ground* and *power bounce* mechanisms caused by simultaneous switching currents in the same digital device are investigated in detail. Simulations are used to explain the ground bounce with CMOS devices, considering the inductive effects of package and lead connections. The ground and power bounce are quantified experimentally as a function of the number of simultaneous switching devices and types of load for a LVT digital device. Some unpredictable effects are also discussed, and several design rules for reducing ground bounce are given.

The chapter ends with a description of an experiment where, owing to simultaneous switching occurring within the same IC, crosstalk on signal interconnects and ground bounce on the driver sum up. In this case it is shown that the disturbances on the quiet line of the interconnect cannot be predicted by using a macromodel of the devices for reflections and crosstalk only. In fact, a model at transistor level of the driver (micromodel), as well as a transmission-line model of PCB ground and power distribution, is required for an accurate simulation.

8.1 Switching Noise

As mentioned in *Section 1.1*, the switching noise or ΔI*-noise* produced by digital devices in a PCB is a primary source of noise affecting signal integrity and radiated emission. To mitigate these unwanted effects, a suitable power distribution network (PDN) must be realized. The main goal is to have a PDN with very low input and transfer impedance. A low input impedance decreases the source of the noise; a low transfer impedance decreases the transfer of the noise across the board. This noise can cause an intense radiated emission from cables attached to the PCB. The main technique employed to obtain a PDN with low impedance is to use power and ground planes with decoupling capacitors distributed along the PCB. A capacitor for decoupling acts as an ideal capacitance up to the frequency where the parasitic inductance associated with the component and its connections to the power and ground planes become significant. For evaluation of the electromagnetic interference performance of a PDN in all the wide frequency range of interest on high-speed digital devices, it is therefore important to consider also the parasitic inductances of the on-chip package. Suitable equivalent circuits for SPICE constitute the best method for predicting switching noise effects.

8.1.1 Power Distribution Network

When an I/O buffer switches, especially if it has to drive electrically long interconnects or heavy capacitive loads (tens of pF), the following points should be considered [1]:

- I/O buffers require high-frequency current for switching.
- When drivers are switching, there will be an impulsive current ΔI, usually approximated in the time domain by a triangular shape flowing in the power and ground connections in a time Δt.
- The power and ground connections, called the *Power Distribution Network* (PDN), are not ideal, and therefore the impedance $\hat{Z}_{\text{PDN}}$ of the power–ground distribution is not zero.
- The switching current will induce a voltage drop across the PDN impedance (also called ΔI*-noise*), given in the frequency domain by

$$\Delta\hat{V} = \hat{Z}_{\text{PDN}}\Delta\hat{I} \tag{8.1}$$

  If the inductive effect only is considered (i.e. $\hat{Z}_{\text{PDN}} = j\omega L_{\text{PDN}}$), Equation (8.1) becomes

$$\Delta V = L_{\text{PDN}}\Delta I/\Delta t \tag{8.2}$$

  where Δt is the rise or fall time of the switching current.
- A *Voltage Regulator Module* (VRM), i.e. DC/DC converter, provides a stable DC voltage to the system with the aid of decoupling capacitors.

The concept of PDN has been introduced in *Section 1.1.1*. The components of a typical PDN in a PCB are shown schematically in Figure 8.1a, where a bulk capacitor immediately after the VRM and only one decoupling capacitor are present, although in a PCB populated by ICs there are several capacitors of different category (see Figure 1.1). As regards the path connecting the VRM to the chip, the following inductances can be found:

- L_{ps} = power supply inductance between the VRM and the bulk decoupling capacitor, which also includes the internal inductance of the VRM;
- L_{dec} = inductance associated with the decoupling capacitor;
- L_{bulk} = inductance associated with the bulk capacitor;
- L_{pcb} = inductance associated with the path between the pin of the IC and the nearest decoupling capacitor or between two decoupling capacitors;
- L_{pin} = inductance associated with the connection mechanisms between the device's ground/power pin and the PCB (this inductance is bigger when the device is connected to the PCB through a socket);
- L_{chip} = inductance associated with the bond wire from the device's die to its package pin, and of the pin itself.

The decoupling capacitors are required to provide the ΔI current to the chip, avoiding the inductance L_{ps} and the consequent voltage drop that would mean voltage noise occurring on the power pin of the IC when the decoupling capacitor is not present and the ΔI current being provided by the VRM [2, 3].

After the VRM, the second largest source of charges is the electrolytic bulk capacitor, the capacitance C_{bulk} of which ranges between hundred of μF to as high as a few mF. This component is able to supply charge with sufficient speed to meet the demands of systems characterized by time constants as low as a few hundred ns and even shorter. Consider that, if a printed circuit board draws 300 mA in 1 μs through an inductance $L_{ps} = 5\ \mu$H (cord interconnects), a voltage drop of 3 V occurs in the supply voltage when the bulk capacitor is absent, and this can be sufficient to cause a malfunctioning of the components on the board. A bulk decoupling capacitor reduces drastically the power supply impedance as viewed from the board, as C_{bulk} provides an impedance $\hat{Z}_{bulk} = 1/(j\omega C_{bulk})$ in parallel with the impedance $\hat{Z}_{ps} = j\omega 2L_{ps}$, under the assumption that the output impedance of the VRM is zero.

The third source of charge is represented by 'high-frequency ceramic capacitors' able to support charge demand from circuits with a time constant as low as a few tens of ns.

Finally, when the PDN is formed by parallel metallic planes (i.e. power and ground planes, see Figure 1.1), a further source of charge is the capacitance between the two planes, which is able to delivery charge to circuits whose constants are shorter than a few tens of ns, i.e. a charge demand frequency above several hundred MHz.

The VRM and the bulk capacitors are usually few in number and are located in specific areas of the PDN owing to their dimensions and other constraints. High-frequency decoupling capacitors are usually large in number and are typically easily located with a great flexibility. To be effective, these decoupling capacitors must have an inductance, L_{dec}, as low as possible, and must be positioned as close as possible to the devices to reduce the inductance L_{pcb} representing the inductance of the power distribution trace on the board when power and ground planes are not used. The inductance L_{pcb} of this trace, usually a busbar, is typically of the order of 100 nH or greater. A device drawing a current of 30 mA in 1 ns through $L_{pcb} = 100$ nH can

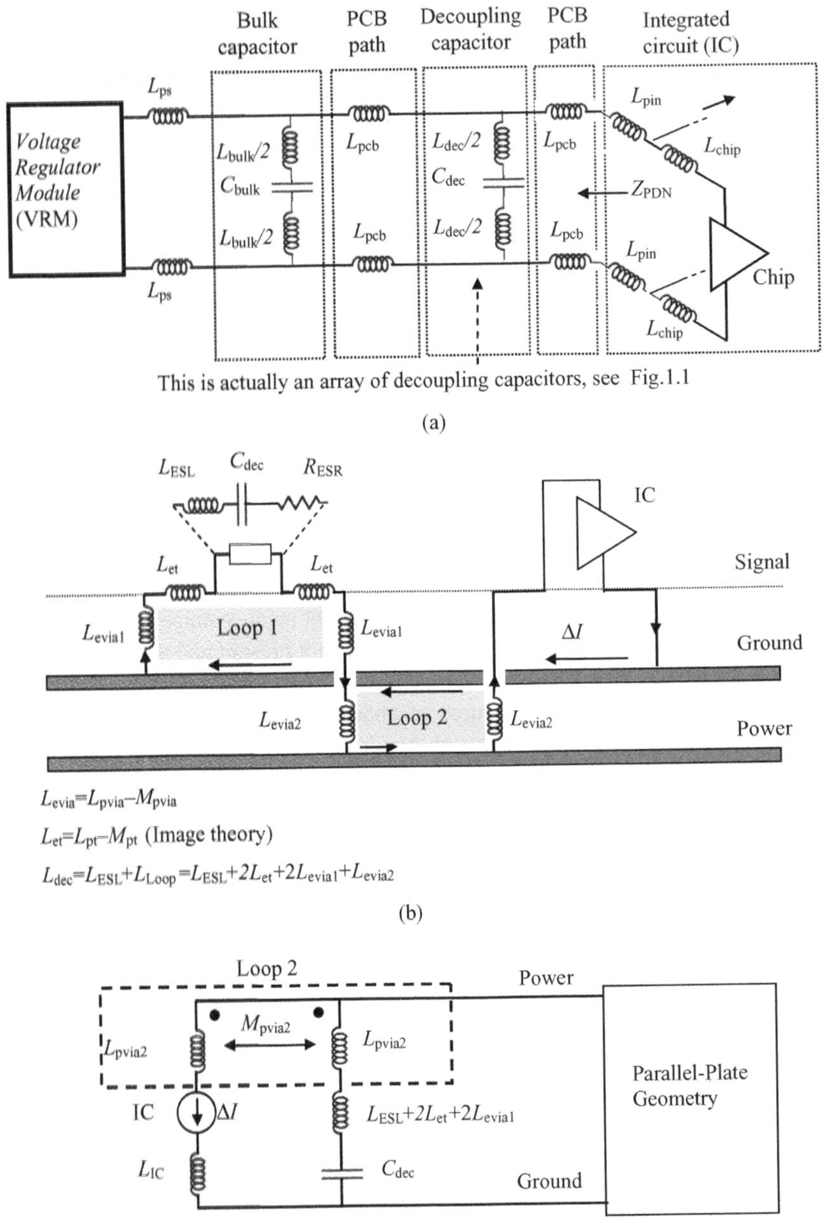

Figure 8.1 Power distribution network (PDN): (a) typical equivalent circuit of a PDN in a PCB; (b) example of parasitic inductances associated with a decoupling capacitor in a multilayer PCB; (c) equivalent circuit showing mutual inductance in loop 2

produce a 3 V transient at its own power pins and at the pins of every component connected to the power traces beyond the switching device. A decoupling capacitor between the power and ground pins of each switching component helps to mitigate this problem. On the other hand, in a multilayer board the inductance L_{pcb} is typically of the order of 0.05 nH/cm, and therefore it can be neglected in the equivalent circuit of Figure 8.1a.

In addition to its capacitance, the decoupling capacitor exhibits parasitic inductance and resistance, as shown in Figure 8.1b. The parasitic resistance is usually referred to as *equivalent series resistance* R_{ESR}. The parasitic inductance consists of two inductances: the first one L_{ESL} is an *equivalent series inductance* associated with the capacitor itself; the second one L_{Loop} is associated with the connecting path of the capacitor between the power and ground planes. This connecting path is built up by the solder pads used to secure the capacitor to the PCB and any traces and/or vias used to make the electrical connections. An example of the inductance calculation associated with traces and vias was given in *Section 3.2.8*, and reported again in Figure 8.1b for convenience [4]. The switching current provided to the digital device by the decoupling capacitor is denoted by ΔI. The loop inductance $L_{Loop} = L_{Loop\,1} + L_{Loop\,2}$ is the sum of the effective inductances associated with two loops (see Figure 8.1b): loop 1 is formed by the pad and the trace of the capacitors, the ground plane, and the two via paths between the signal layer and ground plane; loop 2 is defined by ground and power planes and two vias. Note that the effective inductance L_{evia2} associated with the via path depends on the distance between the two vias of the loop. Considering that, in the parallel coupled-pair conductors sharing the same loop, the current ΔI flows in opposite directions, the effective inductance associated with trace and via is computed as the difference between the self partial inductance of the conductor L_p and the mutual partial inductance M_p, as indicated in Figures 8.1b and c, where the coupling between the vias of loop 2 can be seen. Because the trace has the ground plane as the return conductor, image theory must be used for L_{et} inductance calculation. In conclusion, the total parasitic inductance L_{dec} of the decoupling capacitor (i.e. the equivalent inductance between ground and power planes looking towards the decoupling capacitor) is given by $L_{dec} = L_{ESL} + 2L_{et} + 2L_{evia1} + L_{evia2}$. The expressions for calculating the inductances can be found in *Appendix A*. To minimize the effective inductance $L_{evia2} = L_{pvia2} - M_{pvia2}$ (see Figure 8.1c), the via connecting the IC power pin with the power plane should be very close to the via of the decoupling capacitor. In this way the partial mutual inductance M_{pvia2} is maximized, while the self partial inductance L_{pvia2} remains constant. This means that the position of the decoupling capacitor with respect to the IC is important for minimizing the inductance associated with the connections. The inductance $L_{IC} + L_{evia2}$ in Figure 8.1c is the total inductance associated with the IC connection to the PDN, and L_{IC} can be calculated by analogy with the decoupling capacitor inductance. Moreover, for an exact analysis, all the mutual inductances of the nearby vias should be considered and not only those of the vias belonging to loop 1 and loop 2.

The inductance L_{dec} is also referred to as L_{SMT} for high-frequency *Surface-Mounted Technology* (SMT) capacitors, and as L_{bulk} for bulk capacitors.

When combined with the capacitance of the capacitor, the parasitic inductance and resistance form a series resonant circuit whose impedance dips to a minimum at the frequency where the inductive and capacitive reactance cancel. At frequencies higher than this resonant frequency, the capacitor behaves inductively and is ineffective in decoupling. If the resonant frequency is shifted higher in frequency by lowering the parasitic inductance, decoupling can be made more effective at higher frequencies. Using power and ground planes instead of busbars for PDN, with appropriate design rules for capacitor location and connection, the parasitic inductance can be lowered to less than 1 nH instead of the usual typical values of 2.5–10 nH or greater.

The equivalent series inductance L_{ESL} of the decoupling capacitors is a function of the length, width, and height of the capacitor itself. The size of the SMT decoupling capacitors has been reduced to obtain L_{ESL} of less than 1 nH. The equivalent series resistance R_{ESR} and inductance L_{ESL} are usually provided by the capacitor manufacturers who measure the values by impedance analyzers and/or network analyzers. In both cases, special fixtures are utilized along with calibration procedures and measurement techniques in order to minimize the parasitic elements associated with the measurement set-up itself [5] (see also *Section 11.2*).

A very large selection of decoupling capacitors is available to designers. The distinction is between package sizes, materials (electrolytic, tantalum, or ceramic X7R, X5R, Y5V, etc.), and manufacturing technologies (*MultiLayer Configuration* (MLC), *Low-Inductance Chip Capacitor* (LICC), *InterDigitated Configuration* (IDC), or *Low-Inductance Chip Array Configuration* (LICA)). Ceramic capacitors are characterized by lower L_{ESL} than electrolytic and tantalum capacitors, even if the latter are usually available in the same package size [6].

One of the main tasks of a designer is to find rules to reduce as much as possible the voltage drop on the ground and power supply pins caused by all the inductances considered above. The first step to realize a PDN for medium-to-high-speed digital PCBs is to choose PCBs formed by one or more pairs of conducting planes used as power and ground (power return). The low inductance associated with the charge delivery path from the planes to active elements allows a better decoupling. Often the term *power bus* is used to identify an individual plane pair, whereas the term PDN is used for the entire supply power system for active circuits placed on the PCB. A typical power bus is connected to a variety of devices found on digital PCBs, i.e. DC/DC converters, ICs, and decoupling capacitors. Noise generated in the PDN can be easily propagated throughout the board. Propagation noise can affect the operation of other active devices (signal/power integrity), as well as producing radiation (EMI) that can cause violation of the regulatory requirements. A key element to mitigate these two phenomena is the appropriate use of decoupling capacitors. To reduce the PDN inductances and decouple different parts of the board, several remedies can be adopted: board stack-up design, power/ground plane pair, usage of dielectric losses, power islands, board edge termination, etc. These techniques will be illustrated in *Section 8.1.4*.

The design of a PDN must satisfy two primary requirements:

1. The first requirement is to ensure the functionality of the PCB. The PDN can be seen as a simple circuit where the DC voltage is provided to the IC device by a PDN impedance in series with the variable impedance of the IC device. Changes between low and high states cause a changing current demand by the IC, which causes the DC voltage across the IC to fluctuate. This voltage fluctuation represents an AC voltage ripple on the DC voltage level, which is a source of AC noise on the PDN. The magnitude of this ripple is related to the magnitude of the current required by the switching devices and the PDN impedance. The requested current is unavoidable, so it is essential to lower the PDN impedance. This will be discussed in the following sections.
2. The second requirement is to minimize the noise injected into the power and ground-reference plane pair and thus to reduce the possibility of noise propagation in the board and EMI emission from the circuit board. Several mechanisms for EMI emission can be

identified. One is the emission caused by the edge voltage between the two power and ground planes [7]. Assuming that the distance h between the two plates is electrically very short, the radiation is produced by the electric field distribution, which is proportional to the voltage between the two plates by the parameter h. The plates do not radiate since it can be shown that the electric current on the outer side of the metallic surface is much smaller than the current on the inner side, so that it can be neglected [8]. Alternatively, PDN noise may couple onto input/output (I/O) connector pins or onto a ground cable shield, and be directly coupled out of the metal enclosure through any of the cables attached to the PCB. This will be discussed in *Chapter 9*. To minimize radiated emission, the impedance of the PDN should be low over a wide frequency range including the spectrum of the critical signals and their harmonics.

8.1.2 Switching Current Path

The previous section discussed the importance of lowering the impedance of the PDN for a correct functionality of the PCB (signal/power integrity) and to mitigate EMI as well as radiated emission. The purpose of this section is to investigate the switching current path. The DC voltage source V_{CC} has the task of powering active devices on the PCB, and typical values are 5 V, 3.3 V, 1.8 V, etc. The voltage across the power/ground pins V_d of the IC device is $V_d = V_{CC} - V_n$, where V_n is the voltage drop across the impedance Z_{PDN}, sometimes referred to as the AC ripple voltage, which is superimposed on the DC rail voltage V_{CC} powering the power/ground plane pair. The DC resistance effect is neglected in this discussion, as the attention is focused on investigating AC effects.

There are other mechanisms that can contribute to power bus noise [1, 9]. The IC itself contains its own PDN, as can be seen in Figure 8.1. The inductances associated with this internal PDN produce additional noise (*ground* and *power voltage bounce*), as will be explained for the switching current path when the IC switches from low-to-high or high-to-low output state.

Consider by way of example the simplified driver/receiver schematic shown in Figure 8.2. The two CMOS IC devices are powered by the voltage source V_{CC} with the aid of a

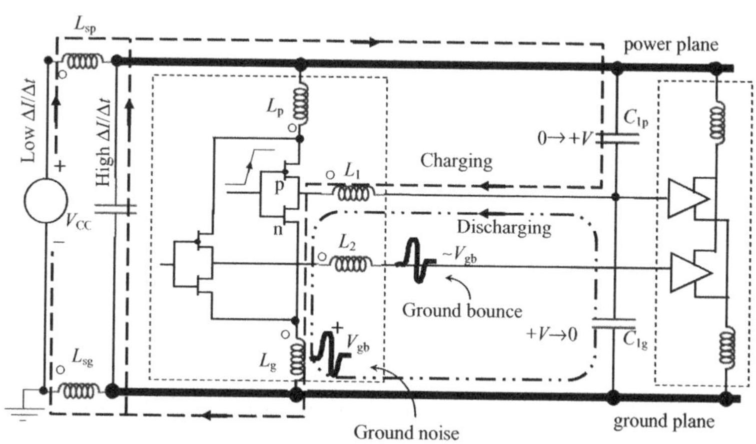

Figure 8.2 Switching current paths for high-to-low transition

decoupling capacitor. The parasitic inductance of the decoupling capacitor is not indicated in Figure 8.2. As this section is focused on I/O current, Figure 8.2 shows a typical *totem pole* output driver configuration of an I/O driver for a CMOS IC, as described in *Section 2.1.3*. When one of the two-output buffers of the first IC switches from high to low state, the upper p-MOS transistor exhibits high impedance and the lower n-MOS transistor exhibits low impedance. This means that the capacitance C_{1p} between the signal trace and the power plane passes from 0 voltage to $+V$ voltage, where V is the voltage swing of the driver. The dashed line in Figure 8.2 indicates the charging current. The path of this current involves in part the VRM represented here by V_{CC} with its low-frequency components, and in part the decoupling capacitor with its high-frequency components. As discussed in the previous section, the voltage drop V_n associated with the VRM and decoupling capacitor inductances involves the power and ground pins of the digital devices. In any case, all the current flows through the output pin, which is inductively coupled with an adjacent pin of a quiet buffer (see L_1 and L_2 in Figure 8.2), and contributes to a voltage drop V_{gb} on the ground pin inductance L_g. The second contribution to this voltage drop is made by the discharge current involving capacitance C_{1g} between signal trace and ground, which passes from voltage V to zero. The third and last contribution is made by the *shoot-through current*. As the two transistors switch states, there is a moment when both transistors exhibit relatively low impedance, allowing the current to flow from the IC power supply directly to the ground reference. This current occurs during both low-to-high and high-to-low transitions. The voltage drop V_{gb}, also called the *ground bounce*, and the mutual inductance between L_1 and L_2 are responsible for noise in the quiet trace. This noise should be added to V_n caused by the external PDN impedance. The total noise is referred to as *ΔI-noise* or *Switching Noise* (SN).

When one of the output buffers switches from low to high state, the paths of the switching currents are those shown in Figure 8.3. In this case a *power bounce* produced by the voltage drop V_{pb} on the power pin inductance L_p occurs.

Therefore, there are two important categories of current demanded from the power bus during switching: a load current demand, which occurs once during each clock cycle, and

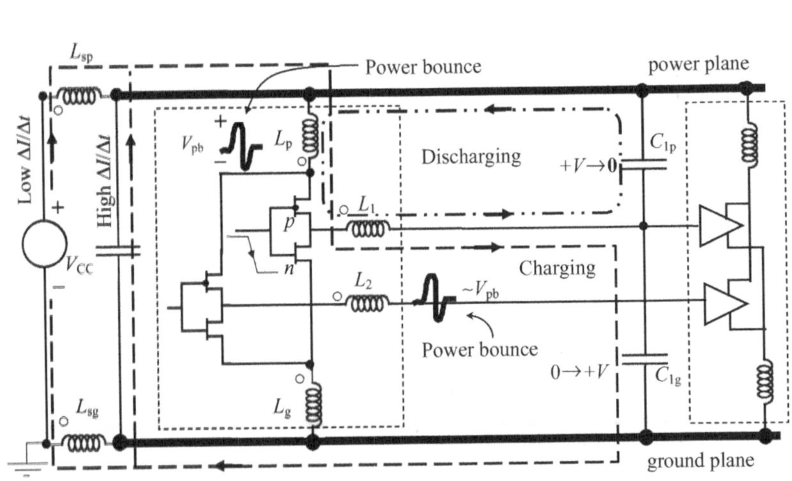

Figure 8.3 Switching current paths for low-to-high transition

a shoot-through current, which occurs twice during each clock cycle. These currents cause noises on power supply and ground pins and noise on output pins owing to mutual inductances. The functionality of the devices can be compromised if design rules are not applied to mitigate these noises.

Example 8.1: Switching Noise Simulations of Two Stripline and Microstrip Structures

A useful method for investigating design rules and for understanding the mechanisms generating the *Switching Noise* (SN) is to use a SPICE-like circuit simulator such as MicroCap [10]. With this approach, both PDN and ICs can be modeled at transistor level as sources of noise, and their interactions can be simulated considering also the resonant phenomena of the full structure. An example of this procedure will be given for the simple structures shown in Figure 8.4.

The investigation is based on the following assumptions:

- High-speed CMOS gates are used as switching devices.
- The PDN is realized by using busbars 50 times larger than the trace. In fact, as shown in *Section 10.2* using the method of moment, and in *Appendix E* using the circuit nodal approach, the return current on the bars is crowded just underneath the trace.
- The busbar is characterized by a cross-sectional dimension much smaller than its length. Therefore, resonance phenomena are expected in one direction only, and the PDN can be modeled as a transmission line. The observations that will be made can be repeated with a PCB having power and ground planes where resonances occur in both orthogonal directions of the planes, as will be outlined in *Section 8.2*.
- Lossless lines are used for the sake of simplicity.
- A cell of the PCB (power–trace–ground) of 1 cm length is considered to be electrically short for the maximum frequency of interest, and it is modeled with a lumped equivalent circuit considering mutual inductances and capacitances.

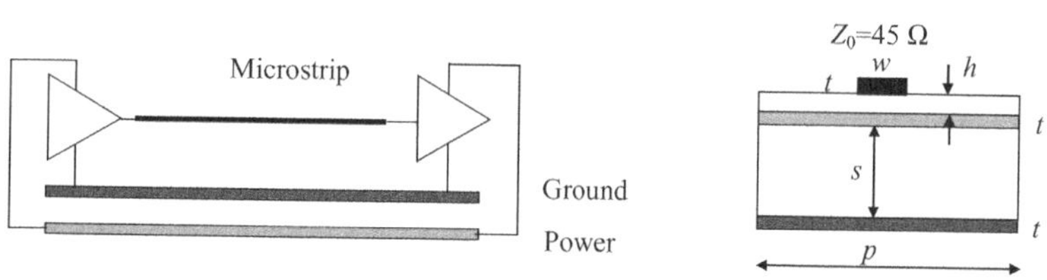

p=10 mm, t=0.04 mm, w=0.2 mm, h=0.14 mm, s=0.4 mm, ε_r=4.4, l=10 cm

Figure 8.4 Point-to-point structures used for simulations of ΔI-*noise*

- The VMR is considered ideal, and the power is distributed to the first device (the driver) by a 10 cm long trace. This means that there is a significant inductance between the power source and the driver.
- Decoupling capacitors are connected between the power and ground pins of the driver and receiver to mitigate the effects of the inductances associated with the busbars.
- The trace between the output of the driver and the input of the receiver is 10 cm long. Other geometrical parameters useful for computing equivalent circuits are shown in Figure 8.4.

The two structures are modeled as shown in Figure 8.5a, where it is possible to distinguish:

- at the center, the 10 cm stripline or microstrip structure modeled as a cascade of 10 cells;
- to the left, the 5 V power source of 0 Ω impedance (ideal source) and the effective inductance associated with the power and ground busbars for the 10 cm path between the voltage source and the first driver;
- the decoupling capacitor located between the power and ground pins of the devices;
- the internal PDN of the devices.
- the AC line termination located between the input pin of the receiver and its ground pin.

Details of the equivalent circuit of a stripline or microstrip cell are shown in Figure 8.5b, where self and mutual partial inductances, as well as capacitances between the three power–trace–ground conductors, computed by a 2D field solver are present. The numerical values shown in the figure are in nH/m (inductances) and in pF/m (capacitances). In theory, the self-inductance values of the conductors of the two structures should be equal because the conductors have the same dimensions. Actually, the slight differences are due to the fact that the 2D field solver computed the inductance matrix from the capacitance matrix with $\varepsilon_r = 1$ and taking as a reference conductor a plane sufficiently distant to consider the three conductors practically isolated. This procedure makes it possible to compute with good approximation the required self and mutual partial inductances with infinity as the reference conductor (see *Section 3.2*).

The values of the parameters of the listed components are shown in Figure 8.5c. Regarding the devices, note that the resistance R_{ic} with the 1 μF capacitance in series represents the loading effect of the IC. The internal decoupling capacitor C_{die} is normally designed to provide charges for switching at chip level. The chosen values shown in Figure 8.5c are those typically found in practice.

The driver was simulated at transistor level with a cascade of three stages, as shown in Figure 8.6, where the output static and dynamic characteristics are also reported. Observe that:

- The output static characteristic is that of a typical CMOS device, as described in *Section 2.1.3*.
- The output voltage has rise and fall times $t_r = t_f = 0.4$ ns, making the interconnect between driver and receiver electrically long and hence making it necessary to match the line using AC termination.
- The switching current I_{CC} required by the power has a triangular shape and, in accordance with an output change from low to high state, has a maximum peak value because it is the sum of two contributions: the required current to charge the load capacitance of 15 pF and the shoot-through current.

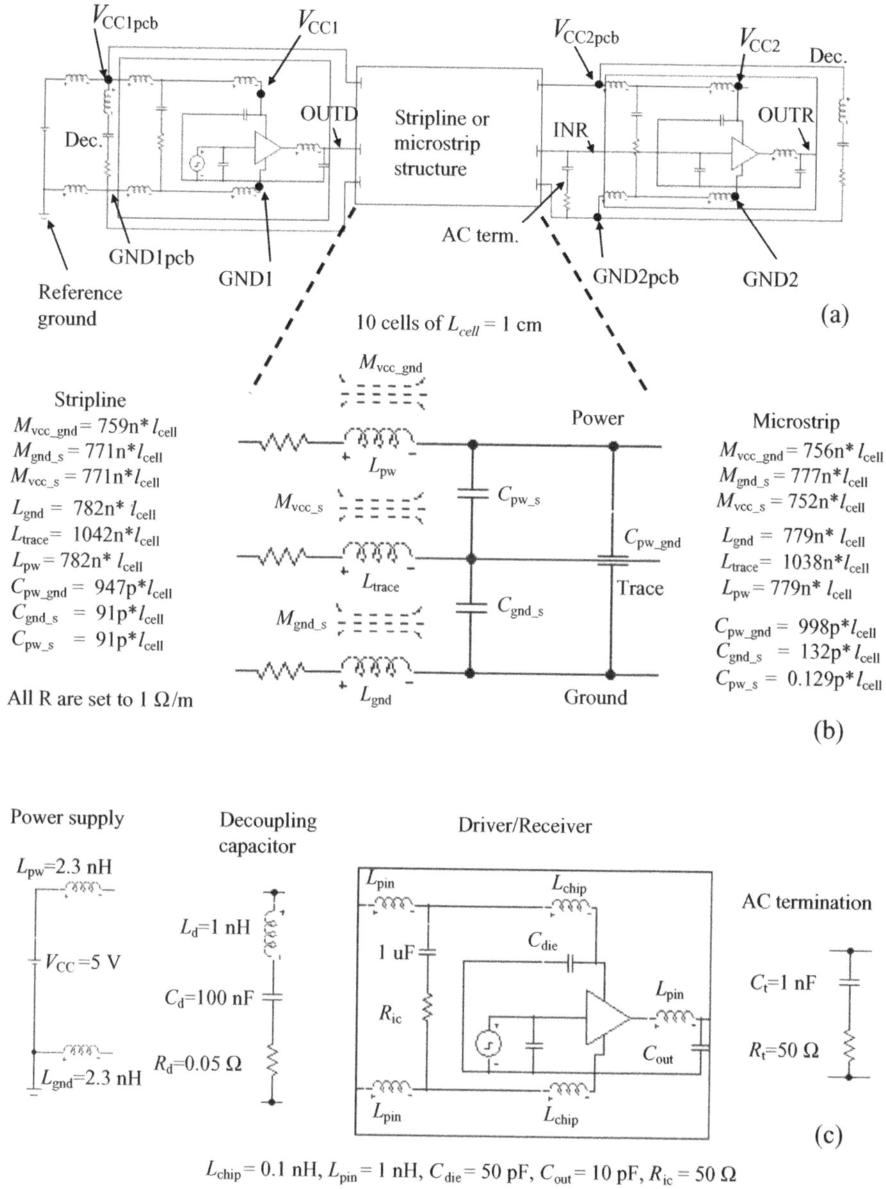

Figure 8.5 Equivalent circuit used to simulate ΔI-*noise* of the structures of Figure 8.4 by the MicroCap simulator: (a) main circuit; (b) stripline or microstrip equivalent circuit; (c) equivalent circuits of power supply, decoupling capacitor, driver/receiver, and termination

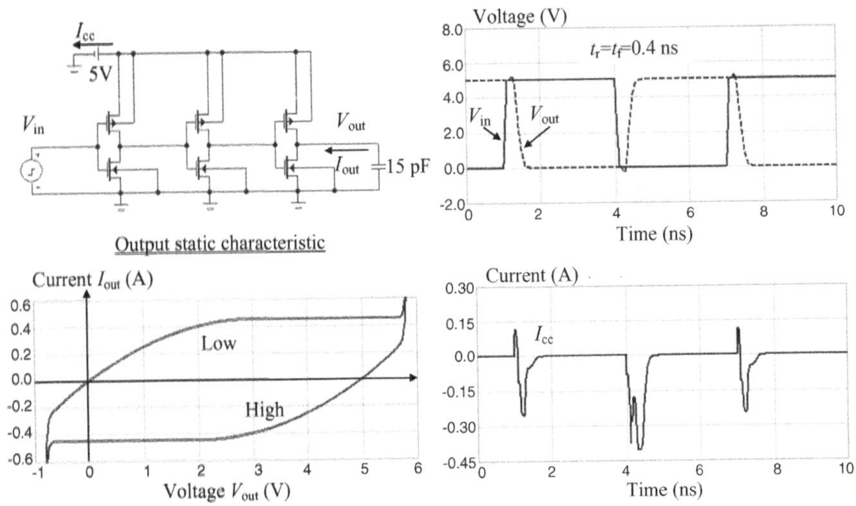

Figure 8.6 Static and dynamic characteristic of the CMOS gates

Before showing the results of the simulations that were performed by MicroCap [10], it is interesting to observe the external PDN impedance at driver location versus frequency, $\hat{Z}_{\text{PDN}}(\omega) = \hat{V}(\omega)/\hat{I}(\omega)$ shown in Figure 8.7c. A comparison is made with and without considering the contribution of the lead parasitic inductance L_{d} associated with the decoupling capacitor. The equivalent circuit used to carry out this investigation is shown in Figure 8.7a. Power and ground busbars were simulated in two ways: with a lossless transmission line model having characteristic impedance Z_0 and delay time T_{D} computed from the per-unit-length parameter inductance and capacitance of the busbars, as shown in Figure 8.7a, and with a cascade of 10 cells having the equivalent circuit of Figure 8.7b. The presence of the trace was omitted, as it is not significant for the purposes of these simulations, while the 50 Ω loading effects of the devices were taken into account. The VRM was simulated with a resistance of 1 $\mu\Omega$. The results of the simulations are shown in Figure 8.7c, where the PDN impedance magnitude Z_{PDN} computed in the frequency range 10 MHz–10 GHz is shown.

Concerning the results, it can be observed that:

- Lumped and distributed models of power and ground busbars provide the same results up to 4 GHz. After this frequency, the lumped model loses its validity.
- A lumped model can be used for simulations with active devices because the second break angular frequency, as defined in *Section 9.1*, is $1/\pi t_{\text{r}} = 0.8$ GHz.
- Z_{PDN} is about 50 mΩ above 100 MHz with $L_{\text{d}} = 0$.
- When $L_{\text{d}} \neq 0$, Z_{PDN} rises from low values up to about 10 Ω in the range 10 MHz–300 MHz. After this frequency there are peaks and valleys of value between 20 Ω and 1 Ω. Without the loading effect of the devices, the range of variation in Z_{PDN} should be much larger without changing the points of resonances.

This simulation is very important, as it makes it possible to deduce the following first design rule: L_{d} must be very low to achieve low voltage drops along the power supply distribution.

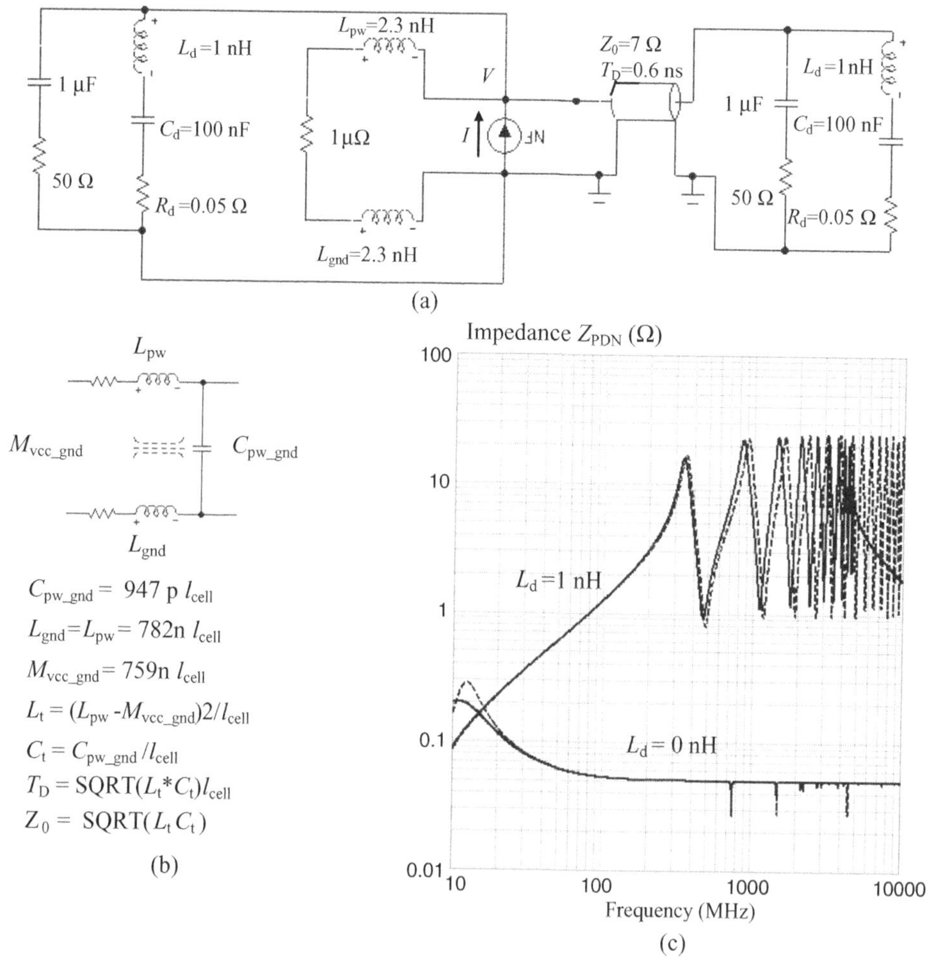

Figure 8.7 External PDN impedance computations: (a) equivalent circuit; (b) equivalent circuit of one cell of the power–ground line; (c) simulations with the distributed model (dashed line) and the lumped model (solid line)

The simulations of the full structure in Figure 8.5a were performed with the power distribution in 'ideal' (i.e. $L_{chip} = L_{pin} = L_d = 0$ nH) and in 'actual' (i.e. $L_{chip} = 0.1$ nH, $L_{pin} = L_d = 1$ nH) conditions. Remember that $L_d = 0$ would be the ultimate goal, but it is impossible to obtain, as the loop inductance of the bypass capacitors has a practical limit dictated by internal geometry and lead connections to the PDN. However, this ideal structure is chosen as reference to show how the filtering performance of the decoupling capacitors strongly depends on the total loop inductance which must be minimized.

In the simulations shown in Figures 8.8, 8.9, and 8.10, the following voltages were monitored:

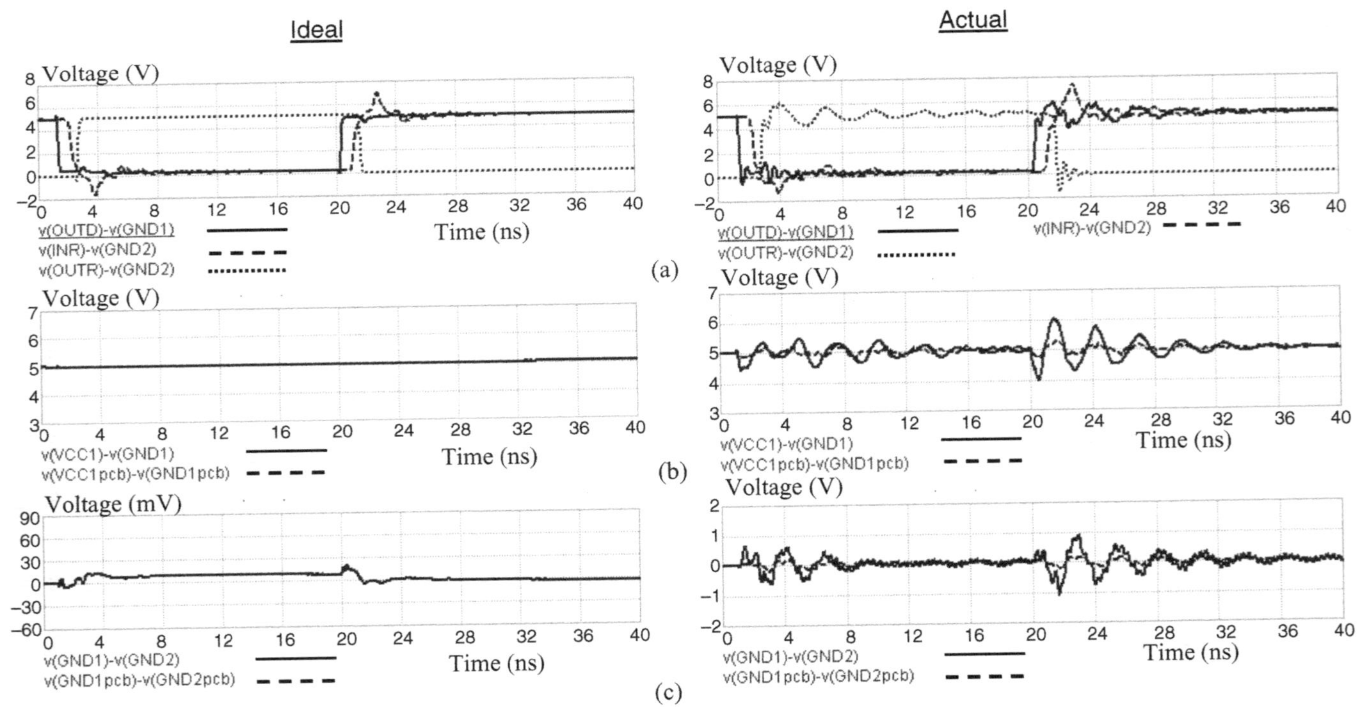

Figure 8.8 Simulated waveforms of a stripline structure at 25 MHz: (a) signals at driver output, receiver input, and receiver output; (b) noise between power and ground points; (c) noise between ground points

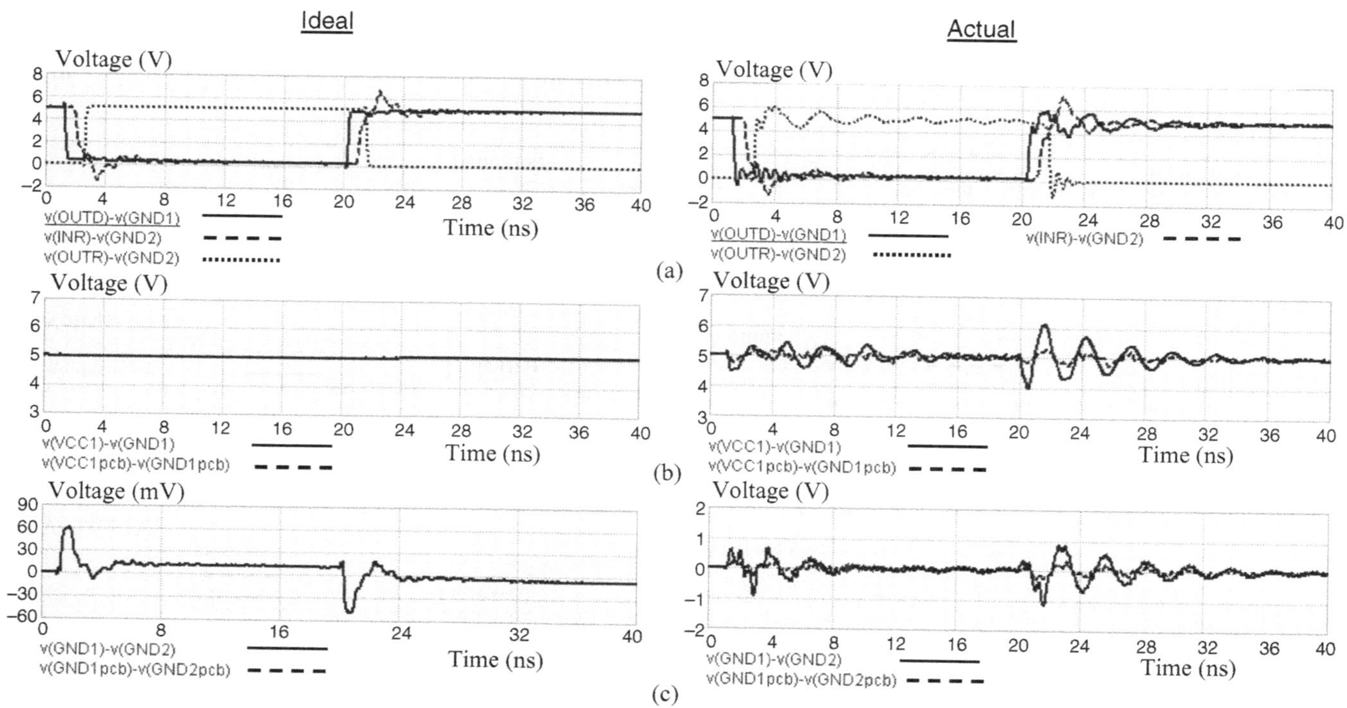

Figure 8.9 Simulated waveforms of microstrip structure at 25 MHz: (a) signals at driver output, receiver input, and receiver output; (b) noise between power and ground points; (c) noise between ground points

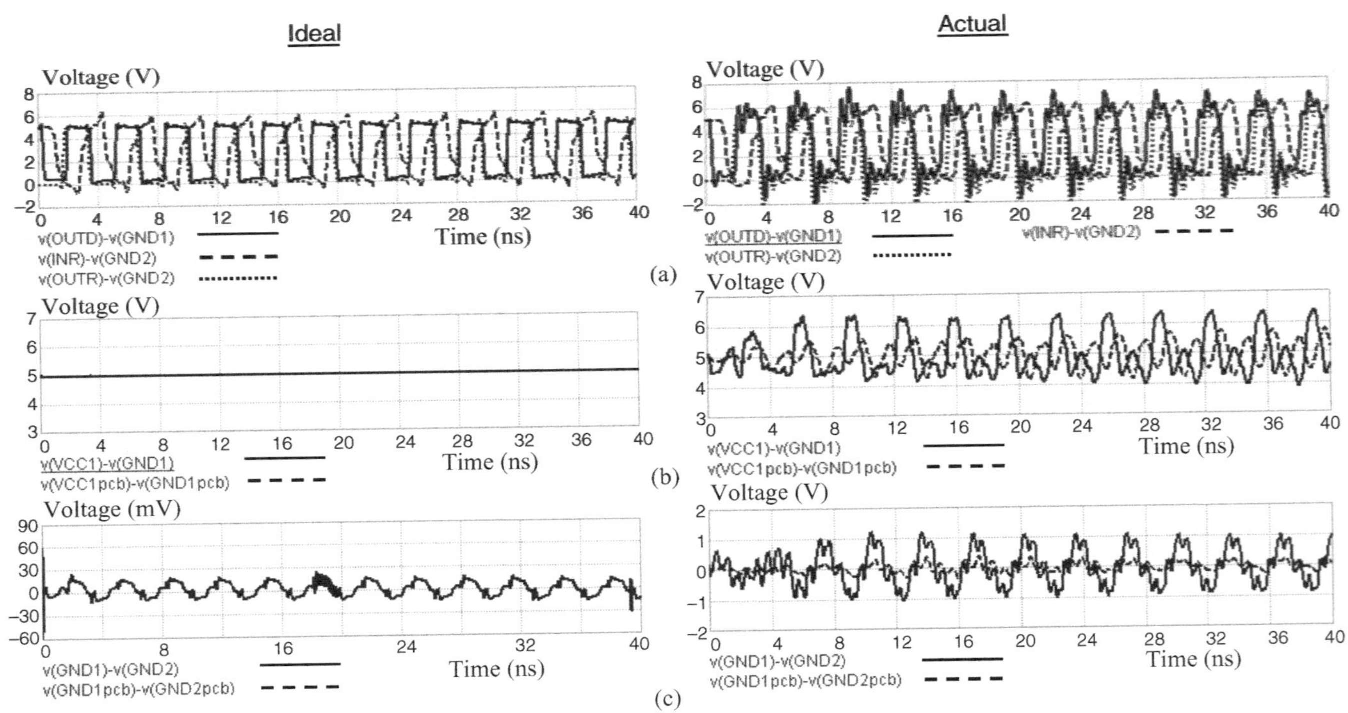

Figure 8.10 Simulated waveforms of stripline structure at 305 MHz: (a) signals at driver output, receiver input, and receiver output; (b) noise between power and ground points; (c) noise between ground points

First plot:

- the output voltage of the driver, indicated as the voltage difference between the points OUTD and GND1: *v*(*OUTD*) – *v*(*GND1*);
- the input voltage of the receiver, indicated as the voltage difference between the points INR and GND2: *v*(*INR*) – *v*(*GND2*);
- the output voltage of the receiver, indicated as the voltage difference between the points OUTR and GND2: *v*(*OUTR*) – *v*(*GND2*).

Second plot:

- the internal power voltage of the driver, indicated as the voltage difference between the points VCC1 and GND1: *v*(*VCC1*) – *v*(*GND1*);
- the external power voltage of the driver, indicated as the voltage difference between the points VCC1pcb and GND1pcb: *v*(*VCC1pcb*) – *v*(*GND1pcb*).

Third plot:

- the internal ground voltage of the driver and receiver, indicated as the voltage difference between the points GND1 and GND2: *v*(*GND1*) – *v*(*GND2*);
- the external ground voltage of the driver and receiver, indicated as the voltage difference between the points GND1pcb and GND2pcb: *v*(*GND1pcb*) – *v*(*GND2pcb*).

The comparison between simulated waveforms of the full stripline structure obtained in the case of ideal and actual structures is shown in Figure 8.8. The frequency of the signal at the driver input is 25 MHz. Looking at the results, it can be noted that:

- The signal integrity of the waveforms at the driver output, at the receiver input, and at the receiver output are significantly modified by the presence of the inductances associated with the leads (as in actual conditions).
- The noise between power and ground points of the chip has a peak-to-peak value of up to 2 V in actual conditions, and it is greater than the noise between the power and ground points in PDN. The internal noise is worst, as the voltage noise on PDN sums with the power bounce. In ideal conditions the noise is absent.
- Noise between ground points of driver and receiver is much larger in actual condition ($L_d = 1$ nH) than in ideal condition ($L_d = 0$).

The comparison between simulated waveforms for the full microstrip structure obtained in the case of ideal and actual structures is shown in Figure 8.9. Similar considerations to those for the stripline structure can apply. Note that in this case the noise between the ground points is increased because the structure is not symmetric. This means that the switching current returns mainly across the ground busbar.

The noise situation of the signal becomes even more critical when the frequency is increased up to 305 MHz, as shown in Figure 8.10. At this frequency, the power bus impedance presents a higher impedance value than that at 25 MHz, as shown in Figure 8.7c. Comparing the two ideal and actual situations, it is evidently necessary to minimize the inductance effects. All these noises must be quantified because they have a great influence in determining the signal skew when a random data stream is present at the driver input.

For the case under investigation, the performed simulations lead to the following conclusions:

- Stripline and microstrip structures have similar performance. Stripline is slightly better, as the signal return current divides equally between the power and ground busbars.
- The power distribution impedance Z_{PDN} and therefore the switching noise depend strongly on the lead inductance L_d of the decoupling capacitor.
- The inductance L_d should be minimized to extend the range of low Z_{PDN} values to higher frequencies and to avoid very high values of external Z_{PDN} at resonance frequencies.
- The power distribution inductances internal to the device, L_{pin} and L_{chip}, should also be minimized to mitigate device internal or power/ground bounce noise.
- Noise does not change significantly if the trace has the power or the ground plane as reference.

8.1.3 Design Rules

At the end of this first section devoted to switching noise problems in PCB power and ground distribution, sufficient knowledge has been gained to provide a set of design rules and recommendations. Some fixes have been discussed above and others merely mentioned in passing. The measurements and simulations presented in the following sections of this chapter will address the latter in greater depth.

(i) Decoupling Capacitors

- These serve two purposes:
  - to meet the demands for charge from switching ICs;
  - to reduce noise in power/ground-reference busbars or planes.
- Decoupling capacitors should be used to design low-impedance PDNs for good signal integrity and EMI behavior such as radiated emission.
- A hierarchical array of capacitors should be used to satisfy charge delivery demands. Large (bulk) capacitors provide large amounts of charge slowly. Small capacitors provide lesser amounts of charge rapidly.
- Either a logarithmic array of ceramic decoupling capacitor values, or just a few values (larger value in a package size) can keep the PDN impedance satisfactory low over a wide frequency range.
- Positioning should be done at regular intervals to reduce impedance uniformly between power and ground planes. In particular:
  - Ceramic capacitors should be located near the ICs with connection to the power and ground planes in order to maximize the mutual inductance between the via of the decoupling capacitor and the via of the IC. In other words, these two vias should be as close as possible and with opposite directed switching currents.
  - The use of traces to connect solder pad to via should be avoided.
  - Low-inductance solder pads should be used for SMT decoupling capacitors.
  - Capacitors having low equivalent series inductance may enhance decoupling effectiveness at high frequencies.

(ii) Power and Ground Distribution

- Stack-ups in PCBs with adjacent power and ground planes should be used to increase intrinsic capacitance between the two planes. This is a decoupling capacitance with very low inductance.
- Attention should be paid to the resonances of power/ground distribution planes. The behavior is like that of a resonant cavity, and the input impedance at any point presents high peak values which depend on the size and electric characteristics of the PCB.
- Three common design techniques for reducing PDN impedance are:
  - To reduce the spacing between the power and ground planes. This is the most effective design technique.
  - To increase the dielectric constant of the material between the power bus planes. Increasing the dielectric constant by a factor of 4 results in a lowering of the PDN impedance by approximately 12 dB for frequencies lower than the resonant frequency. An undesirable side effect of raising the dielectric constant is a downward frequency shift of the power bus resonances.
  - To increase the dielectric loss of the material between the power bus planes. The use of this fix is of limited utility because the resulting reductions in power bus impedance and voltage ripple are small, except for specific resonant frequencies.

(iii) Digital Devices

- An IC package with low inductances should be used. SMT is better than *Plate Through Hole* (PTH).
- The use of sockets should be avoided, as they greatly increase the parasitic inductance effects of the package.
- Packages with a great number of power and ground pins should be used.
- Different power and ground rings between the core and I/O cells for an *Application-Specific Integrated Circuit* (ASIC) package should be used.
- Simultaneous switching of all the internal functions of a chip should be avoided; this could cause severe power and ground bounce noise. The suggested number is 50 %.

8.2 Filtering Power Distribution

Section 1.1 and *Section 8.1* have shown the importance of an appropriate filtering operated by decoupling capacitors in order to have a low impedance value for the *Power Distribution Network* (PDN). In this section, the results of simulations and measurements are outlined and compared, highlighting the performance of decoupling capacitors in a multilayer PCB.

8.2.1 Filtering Multilayer PCBs

In *Section 8.1*, the importance of the decoupling capacitor location on the PDN was discussed, and the noise produced by the switching current when it flows through the inductance associated with the component leads along the path of the PDN was investigated. To mitigate this noise, appropriate filtering is required when designing for digital circuits with medium-to-high switching speed. To this end, the following key points should be considered [11]:

- PCBs for high-speed digital circuits use power and ground plane structures for power distribution.
- These planes are normally solid planes, often on adjacent layers.
- The power and ground planes constitute the power bus which is actually a parallel-plate capacitor providing a PDN impedance much lower than the impedance of a power distribution net comprising traces and not planes.
- Usually, the impedance offered by the bare board (board without capacitor) is not sufficiently low for appropriate filtering, and therefore bulk and high-speed decoupling capacitors are inserted between the two planes.
- At middle frequencies, parasitic parameters associated with these capacitors introduce resonances and, as a consequence, peaks of impedances.
- At higher frequencies, the power bus behaves as a cavity resonator. This introduces additional resonant peaks of the power bus impedance which can be measured by a *Vector Network Analyzer* (VNA) or predicted by several approaches based on analytical, circuit, and numerical methods.

The use of decoupling capacitors enables two important requirements to be met. The first concerns the power bus impedance which preferably should remain low over a wide range of frequency. This ensures that the noise voltages that are created by the impulsive currents produced by the switching devices remain small. The second requirement is to provide adequate charge in a timely manner to ICs that are switching to ensure proper operations. To satisfy these two goals, analyses in the frequency and time domains are required. In the following, results obtained both in the frequency and in the time domain with two test boards are presented and discussed.

The first test board is used to discuss the validity of the low-frequency lumped equivalent circuit model of the PDN by comparison with the cavity model which makes it possible to compute the resonance effects of the PDN impedance owing to the electrical dimensions of the PCB. In this context, the power and ground planes constitute a parallel-plate waveguide [12]. The impulsive current that is due to the switching devices travels along the vias connecting the PDN to the devices and causes voltage waves to propagate between the PDN planes, in a radial manner away from the exciting point. These voltage waves cause changes in the V_{CC} on the board. In particular, two different effects can be distinguished. Firstly, the switching gives rise to a voltage drop at the switching chip itself, with possible malfunctions. Secondly, the propagated voltage wave can lead to false switching of other ICs at some distance from the exciting point. In the frequency domain, these voltage drops are quantified by the transfer impedance $\hat{Z}_{12} = \hat{V}_2/\hat{I}_1$, where $\hat{V}_2$ is the voltage drop between the planes at position 2 owing to a current excitation $\hat{I}_1$ at position 1 on the PCB. When the two points are coincident, the input impedance is defined as $\hat{Z}_{11} = \hat{V}_1/\hat{I}_1$, where $\hat{V}_1$ is the voltage drop at position 1. For practical PCBs, the waves are of the TM_z type because the vias are along the vertical z axis and because the thickness of the dielectric is much smaller than the other dimensions and than the minimum wavelength associated with the frequency spectrum of the excitation current. On account of the finiteness of the board, the waves will reflect at the sides of the board. It can be shown that the reflection coefficient at the edges of the board is very close to 1. Therefore, the PCB can be considered as a perfect resonator with four perfect magnetic conducting walls and two perfect electric conductors (the power planes). This means that the quality factor of this structure will be high, and, as a direct consequence, the switching of an IC can cause long-lasting ringing effects in the voltages of the board. The resonant frequencies of the resonator will become apparent as a sequence of alternating peaks and valleys in the frequency spectrum of the impedance magnitude. In a real PCB populated by decoupling capacitors, these effects are significantly attenuated owing to the Q-factor of the cavity, which is reduced by dielectric losses and by the large number of vias. Therefore, the PDN impedance appears almost flat in the range of frequencies where the capacitors are effective.

The value of the impedances can be computed by three different methods which are briefly outlined in *Appendix C*: the analytical cavity resonant model, the equivalent SPICE circuit model, and three-dimensional full-wave numerical simulation [12, 13]. In the first example, the results obtained by the lumped model will be compared with those obtained by the cavity resonant model, and the effects of decoupling capacitors will be discussed.

In the second test, two boards are used to perform measurements and simulation in the time domain to carry out ΔI*-noise* in different conditions of switching, filtering, and distance between the power and the ground planes. The test boards are populated by digital devices with and without decoupling capacitors distributed uniformly or clustered in a portion of the area of the PCB. It will be demonstrated that by an equivalent SPICE circuit model it is possible to reproduce with great accuracy the measured ΔI*-noise* between power and ground planes. The switching devices are modeled at transistor level, and the power and ground planes are modeled as radial transmission lines able to reproduce the resonance effects. The benefit of reducing the ΔI*-noise* by increasing the intrinsic capacitance between the power and ground planes by the technique called *Buried Capacitance* (BC) is also highlighted.

Example 8.2: Test Board for Prediction of Power Bus Impedance by Several Methods
Consider the PCB shown in Figure 8.11 having two planes: one for power supply and one for power reference or ground. The switching device is simulated by a current source of unit amplitude $I = 1$ A for all the frequencies of interest. The DC load (typically 50 Ω) of the active

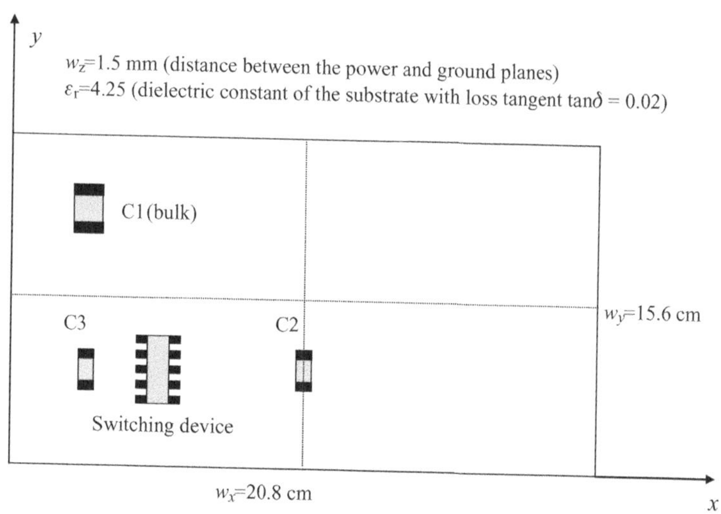

Figure 8.11 PCB used as the test board for analytical investigation and location of the devices: switching device ($w_x/4$, $w_y/4$); C1 ($w_x/8$, $3w_y/4$); C2 ($w_x/2$, $w_y/4$); C3 ($w_x/8$, $w_y/4$)

devices on the board can be neglected because it has very little effect on the overall board response, as demonstrated in *Section 8.1*. Here, C1 is the bulk capacitor, and it has the task of providing the low-frequency components of the switching current. It is characterized by a high capacitance value and an associated high lead inductance. It is generally positioned near the voltage power supply. Usually at this location a parallel combination of large-value and small-value capacitors is used to increase the frequency coverage. However, Paul [14] shows that, above the highest self-resonant frequencies of both capacitors, the impedance of the parallel combination is reduced almost by 6 dB. It is suggested that this high-frequency reduction of about 6 dB may not be worth the expense of an additional capacitor or its installation, and could be attained by using only a larger-value capacitor while simply cutting its lead length in half. C2 and C3 are high-speed decoupling capacitors having the task of providing the high-frequency components of the switching current. They are characterized by medium capacitance value and low lead inductance. They are generally distributed on the PCB in order to be as close as possible to the switching devices. For this example, they are located as indicated in Figure 8.11.

A pair of planes in a PCB stack-up forms a parallel-plate capacitor, often indicated as an interplane capacitor. The value of this interplane capacitance can be approximately estimated by

$$C_{\text{pcb}} = \varepsilon_{\text{r}} \frac{w_x w_y}{w_z} \tag{8.3}$$

where w_x, w_y, and w_z are the board dimensions (see Figure 8.11).

The equivalent circuit of the test board at frequencies lower than the first resonance frequency of the PCB is shown in Figure 8.12 [15]. The first resonance of the PCB occurs when the frequency of the exciting source rises to a value where the PCB acts for the first time as a resonance cavity. In this case, if the VRM is placed on the board, the inductances L_{ps} and L_{pcb} representing the component-to-component power distribution impedance, see Figure 8.1, can be neglected, being of the order of 0.05 nH/cm. This means that the switching current is supplied not only by the nearest decoupling capacitor but by all the capacitors connected between power and ground. Conversely, the inductance associated with every decoupling capacitor, accounting for the parasitic equivalent series inductance as well as the inductance of the connecting means to the planes, such as traces and vias, is typically 1–10 nH and needs

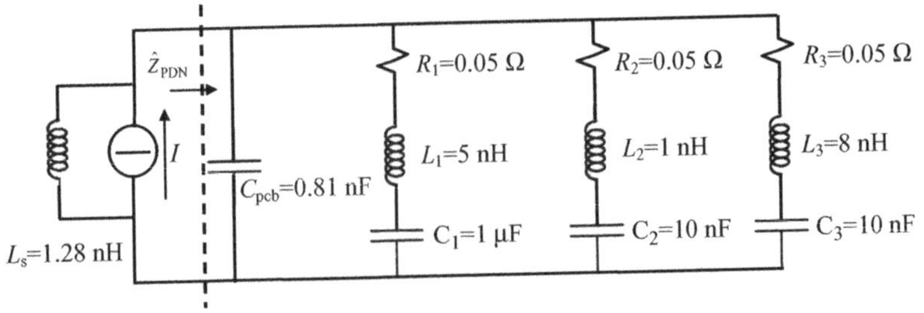

Figure 8.12 Lumped-element model of the test board with the exciting current source having its impedance defined by the inductance L_s

to be considered. Different values of total inductance were assigned at C2 and C3 to simulate possible different connections to the power and ground planes. Capacitor C3 has the worst leading inductance. The effectiveness of a particular decoupling capacitor at a given frequency depends on its impedance compared with the impedance of the other decoupling capacitors and the board capacitance. At the frequencies at which the model in Figure 8.12 is valid (typically below 200–300 MHz, as will be shown by this example), it is evident that the position of a decoupling capacitor is not nearly as important as its interconnect inductance.

The capacitance C_{pcb} is calculated by Equation (8.3), adopting the electrical and geometrical parameters of the PCB given in Figure 8.11. At low frequencies, charge stored in the planes is proportional to the value of the capacitance. The larger the capacitance, the greater is the charge stored and available to support logic transitions. At the same time, the impedance of the plane pair is inversely proportional to the capacitance value. A large value of this capacitance can be obtained with small values of w_z or large values of ε_r.

The equivalent circuit of Figure 8.12 is valid only at frequencies below the lowest resonant frequency of the PCB, given by the following expression:

$$f_{res}(m, n) = \frac{1}{2\pi\sqrt{\varepsilon_0\varepsilon_r\mu_0}}\sqrt{\left(\frac{m\pi}{w_x}\right)^2 + \left(\frac{n\pi}{w_y}\right)^2} \tag{8.4}$$

where m and n are integer numbers and never both equal to zero, ε_0 and ε_r are respectively the vacuum and relative dielectric constant, and μ_0 is the permeability of the dielectric material.

At low frequencies, the wavelength is long by comparison with the PCB dimensions, and the displacement current in the dielectric material is nearly uniform over the plane surface, except near the PCB edges. The PDN impedance magnitude Z_{PDN} seen at the point where the current source is placed, and obtained with a bare board, or, in other words, without the three decoupling capacitors, is shown in Figure 8.13a. In this figure, the results calculated by the lumped model of Figure 8.12 are compared with those obtained by the distributed model

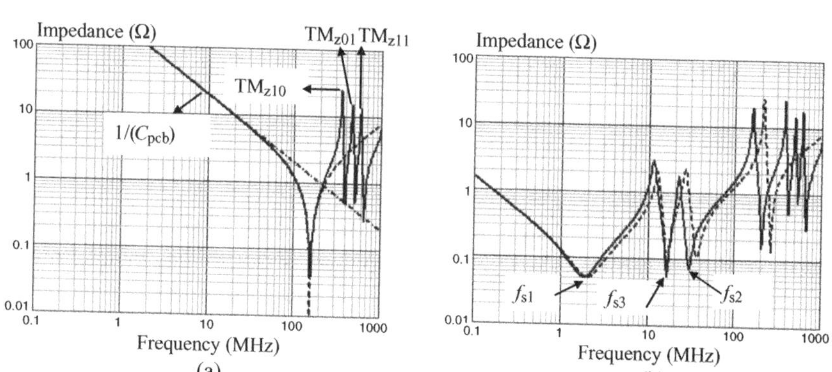

Figure 8.13 PDN impedance magnitude Z_{PDN} of the test board, computed using the cavity model (solid line) and the lumped model (dashed line and dashed-dotted line without L_s): (a) bare board; (b) with decoupling capacitors

based on the mathematical equations of the resonant cavity [13], as reported in *Appendix C*. The dashed-dotted line represents the impedance of an ideal capacitance decreasing with a slope of 20 dB/dec as the frequency increases. The other two curves, the solid line and the dashed line, represent the impedance of the bare board seen as a resonant cavity and as a simple capacitance respectively. The resonance between 100 and 200 MHz is due to the combination of interplane capacitance and the inductance of the feeding port. The closed-form expression of a resonant cavity includes this inductance [13], and this was considered in the lumped equivalent circuit of Figure 8.12 (see inductance L_s) in order to establish a comparison. At higher frequencies, the inductive behavior associated with the port is superimposed by the distributed resonance frequencies of the power and ground planes, and the impedance generally rises with frequency, with resonant peaks and valleys as shown in Figure 8.13a. Therefore, when standing waves are established between planes, the input impedance and the transfer impedance between two points of the PCB exhibit peaks and valleys (poles and zeros) in an alternating pattern. These impedances can attain high values at the frequencies corresponding to the resonant peaks, which are related to the board physical dimension and dielectric constant by Equation (8.4). The modes associated with those resonance frequencies are TM_z, i.e. transverse magnetic [13]. The three resonance frequencies with the bare boards correspond to the first modal resonance frequencies: $f_{res}(1, 0) = 350$ MHz, $f_{res}(0, 1) = 467$ MHz, $f_{res}(1, 1) = 583$ MHz. Observe that the lumped model is valid up to about 200 MHz.

The PDN impedance magnitude Z_{PDN} modifies when the three decoupling capacitors are present as shown in Figure 8.13b. The slight difference in resonance frequencies exhibited by the two approaches is due to the fact that the cavity model takes into account the separation between the components. A drastic reduction of the impedance under 100 MHz can be observed; above this frequency the impedance rises at high values, as in the bare board. This is due to the inductance effect associated with the decoupling capacitors. The three zeros correspond to the series resonant frequency of each decoupling capacitor, computed as

$$f_{si} = \frac{1}{2\pi\sqrt{L_i C_i}} \tag{8.5}$$

where $i = 1, 2, 3$. For the case considered in Figure 8.13b, the following theoretical values of the resonance frequencies are found: $f_{s1} = 2.25$ MHz, $f_{s2} = 50.3$ MHz, $f_{s3} = 18$ MHz. The second resonance frequency f_{s2} is better positioned towards high frequencies because the second decoupling capacitor has the lowest associated inductance. At frequencies lower than the first series resonance frequency, the impedance is approximately that of an ideal capacitor of value given by the sum of the four present capacitances. Above the first series resonance frequency f_{s1}, the poles and zeros must alternate, so there is exactly one parallel resonance between each pair of series resonance frequencies. An exact expression for parallel resonance frequencies is difficult to obtain for an arbitrary number of decoupling capacitors. In the frequency range 2–70 MHz, the decoupling capacitors provide most of the decoupling and the impedance fluctuates owing to the closely spaced series and parallel resonance frequencies. The average impedance in this band is still below that of bare board. In the frequency range 70–200 MHz, the last parallel resonance causes the decoupled board impedance to be higher than the bare board impedance. Above 200 MHz, the decoupling capacitors have little effect on the board impedance, where the resonance modes of the board with or without decoupling capacitors dominate. This also occurs if the number of decoupling capacitors increases.

One disadvantage of discrete decoupling capacitors is their limited effective frequency range due to interconnect inductance. This is demonstrated in Figure 8.13 where the power bus impedance of the board with and without discrete decoupling capacitors is compared. At low frequencies, the local decoupling capacitors do a good job in lowering the impedance Z_{PDN}. Above 70 MHz, the decoupling capacitors have too much connection inductance to be effective. There is no significant difference between these two curves above 300 MHz other than minor shifts in the power bus resonance frequencies. In a PCB populated by decoupling capacitors, this limit can be elevated to 500 MHz but no more owing to the limitation imposed by the lead inductance of the decoupling capacitors. A practical way to improve filtering is to increase the interplate capacitance by a technique called embedded capacitance or buried capacitance, as will be shown in *Example 8.3*.

From the results obtained in this example, the following observations can be made:

- At frequencies less than the first series resonance f_s, the impedance of the board is approximately that of an ideal capacitor.
- The built-in capacitance of the two power and ground planes is a more effective source of current than decoupling capacitors at high frequencies.
- At frequencies higher than the series resonance f_s, the decoupling capacitor branch begins to behave inductively.
- Parallel resonances correspond to poles in the board impedance and occur between two series resonance frequencies.
- Decoupling capacitors significantly reduce the PDN impedance below f_{sbulk} (f_{s1} in this example) owing primarily to the large bulk decoupling capacitor.
- In the frequency range between f_{sbulk} and the highest f_s (f_{s2} in this case), the 'local' 10 nF capacitors provide most of the decoupling. The impedance fluctuates but is still below that of the bare board because the associated interconnect inductances are low.
- At frequencies where the board cannot be considered short compared with the wavelength of interest, a distributed model should be used to consider cavity resonance effects. At these frequencies the decoupling capacitors are not effective owing to the lead inductance, which normally ranges between 2.5 and 10 nH and, with particular connections, can be decreased to about 1 nH.
- In conclusion, to reduce the impedance Z_{PDN} in a large range of frequencies, the most effective way is to reduce the distance between the power and ground planes and/or to increase the dielectric constant of the substrate in order to bring the interplate capacitance from some nF to several tens or hundreds of nF with practically zero associated inductance.

In general, to improve the Z_{PDN} impedance, each decoupling capacitor should have the lowest associated parasitic inductance. Of course, to have low and flat Z_{PDN} impedance, other decoupling capacitors are necessary. Two equivalent approaches are presented and discussed by Knighten *et al.* [2]:

- *Approach A* is based on a logarithmic array of ceramic decoupling capacitors of values 2.2 nF, 4.7 nF, 10 nF, . . . , 47 000 nF, 100 000 nF, in a quantity of four for each value;

- *Approach B* is based on the choice of a few capacitance values in greater quantity: 20 of 47 nH, 24 of 4700 nH, and 16 of 100 000 nH, or, as a subset of approach B, 44 of 47 nH and 16 of 10 000 nH.

For both cases, the ceramic capacitors are 60, and, with the aid of an electrolytic capacitor of 3.3 mF, it is shown that the target of $Z_{PDN} < -20$ dBΩ (i.e. 0.10 Ω) up to 100 MHz is obtained if the equivalent series resistance R_{ESR} is less than 200 mΩ, the equivalent series inductance L_{ESR} is less than 1.4 nH, and the interconnect inductance of the decoupling capacitors is less than 2 nH.

In *Appendix C*, a set of closed-form expressions of the cavity model are given. The computed analytical results of the input impedance and of some transfer impedances of the PCB shown in Figure 8.11 are compared with those obtained by a SPICE equivalent circuit and by full-wave software.

8.2.2 Measurement of Power Distribution Network Impedance

An alternative way to quantify the power distribution network impedance is to perform measurements [3]. A *Vector Network Analyzer* (VNA) with an *S*-parameter test set can be used (see *Section 11.2*). A couple of semi-rigid coaxial cables can be connected to the printed circuit board power plane to measure the frequency-domain scattering parameters $\hat{S}_{11}$ and $\hat{S}_{21}$. In an actual PCB, these probes can be soldered directly to decoupling capacitor-bonding pads. The center conductor of the coaxial cable can be connected to the ground plane and the outer shield to the power plane, or vice versa. The printed circuit board impedance $\hat{Z}_{PDN}$, in the frequency range where the lumped model holds, can be directly related to the measured $\hat{S}_{21}$ between the attached coaxial cable probes as [1]

$$\hat{S}_{21} = \frac{\hat{Z}_{PDN}}{\hat{Z}_{PDN} + Z_0/2} \tag{8.6}$$

where $Z_0 = 50$ Ω is the nominal characteristic impedance of the measurement system. In theory, the board impedance could also be derived from a one-port measurement as

$$\hat{Z}_{PDN} = Z_0 \frac{1 + \hat{S}_{11}}{1 - \hat{S}_{11}} \tag{8.7}$$

This procedure is not very accurate though. In fact, as the board impedance is primarily reactive, the S_{11} magnitude is nearly 1 and an accurate determination of the phase of S_{11} is critical.

In practical cases, except possibly at the parallel resonance frequencies, $Z_{PDN} \ll Z_0$, and Equation (8.6) simplifies so that the impedance Z_{PDN} is given by

$$\hat{Z}_{PDN} = 25\hat{S}_{21} \tag{8.8}$$

Using Equation (8.8), the Z_{PDN} magnitude can be easily extracted from a measurement of S_{21} without requiring phase information. It should be noted that the inductance associated with the connection of the test ports was accounted for in the derivation of Equation (8.6). However, Hubing *et al.* [3] showed that considering this inductance in the analysis leads to a resonance frequency error of less than 1 %. It was also shown that, up to 60 MHz, the results provided by Equation (8.6) agree with the measured S_{21}, with the discrepancy within 1 dB. This technique can also be very useful for measuring the interconnect inductance associated with each capacitor. In this case the measurement must be done by shorting the capacitor mounting pads and measuring the frequency at which the interconnect inductance resonates with the plane capacitance. For better accuracy, the short should be implemented using a wide strap soldered across the narrow gap of the interconnect. The strap should be the full width of the bonding pad or approximately the width of a capacitor body. Thus, the contribution of this inductance is small relative to the via and trace inductance. For normal PCBs the measured values are in the range 2.5–10 nH [3].

When a *SubMiniature version A* (SMA) connector and a low-loss precision coaxial cable are used for connecting the test board with the network analyzer port to measure Z_{PDN}, the measurement should be carried out by the following steps (see *Section 11.2*):

- the calibration plane should be at the SMA test connectors;
- a 12-term error correction model should be used, considering open, short, and load conditions (or termination) in the calibration;
- a port extension should be performed to move the measurement planes to the coaxial cable feed terminals looking into the power bus.

Numerous comparisons between measurements and full-wave simulations of S_{11} and S_{21} in a DC power bus can be found in the literature; good examples are the studies by Wang *et al.* [13] and Xu *et al.* [16]. An example is also provided in *Section 10.4* of this book, where the solution of partitioning the power plane to protect sensitivity devices is investigated.

8.2.3 PCB Circuit Model Based on Radial Transmission Line Theory

In this section a prediction model of power and ground planes suitable for SPICE-like circuit simulators is proposed. This circuit model, as well as the one reported in *Appendix C*, is based on the decomposition of the PCB into cells, and the TL modeling of each cell. However, while the circuit model reported in *Appendix C* adopts orthogonal TLs, here the theory of radial transmission lines is applied as an alternative but equivalent approach. The method will be validated by comparing the results with the measurements carried out on realistic PCBs populated by ICs in *Example 8.3*.

The proposed circuit model is derived under the following assumptions:

- The power and ground planes have the impedance of a parallel plate capacitor and can be modeled by using the theory of radial transmission lines.
- The power and ground planes act as a resonator with modes characterized by the E-field perpendicular to the planes, as the distance between planes is very small compared with the wavelength of the highest frequency of interest.
- The edges of the board reflect the waves with a reflection coefficient of almost 1, making them a perfect magnetic wall.

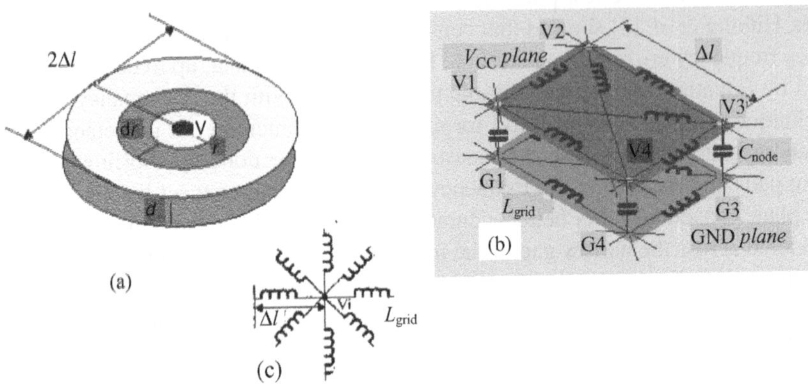

Figure 8.14 Equivalent circuit of the power distribution in a PCB seen as a radial transmission line: (a) geometrical representation as non-uniform transmission (V = center of the radial transmission line, d = distance between the planes); (b) lumped inductances and capacitances regarding four nodes; (c) inductances at one node

Starting from the assumption that the field perturbation due to the switching devices travels outwards from the component with cylindrical symmetry, the power planes can be modeled by a grid as shown in Figure 8.14. The elementary cell is modeled with lumped elements, considering that initially the waves generated by the switching device move as in a radial transmission line. Inductance and capacitance values of the grid can be determined considering that a radial transmission line has the same behavior as a non-uniform transmission line with line inductance and line capacitance parameters dependent on the radial distance r from the origin, and therefore given by [17, 18]

$$C_{\text{rad}}(r) = \varepsilon 2\pi r/d \tag{8.9a}$$

$$L_{\text{rad}}(r) = \mu d/(2\pi r) \tag{8.9b}$$

where d is the distance between the two planes, ε and μ are the dielectric constant and permeability of the dielectric material respectively, r is the radius, $C_{\text{rad}}(r)$ is the radial capacitance at location r for radius variation dr of the two planes, and $L_{\text{rad}}(r)$ is the radial inductance at location r, obtained using the relationship between the per-unit-line inductance and capacitance for a homogeneous material: $LC = \mu\varepsilon$.

As shown in Figure 8.14, the average characteristic impedance $Z_{0,\Delta l}$ for the segment Δl of the radial line can be defined as

$$Z_{0,\Delta l} = \frac{1}{\Delta l}\int_{r_{\min}}^{\Delta l}\sqrt{\frac{L_{\text{rad}}(r)}{C_{\text{rad}}(r)}}\,\mathrm{d}r \tag{8.10}$$

where $r_{\min}$ is the via radius. Since the capacitance C_{node} associated with each node (see Figure 8.14b) can be calculated as the ratio between the interplane capacitance and the number of nodes n_{node}, the grid inductance associated with each of the n_{br} branches incident at the node V_i, with $i = 1, 2, \ldots, n_{\text{node}}$ (see Figure 8.14c), can be obtained by the definition of the characteristic impedance of a lossless line as

$$L_{\text{grid}} = n_{\text{br}} Z_{0,\Delta l}^2 C_{\text{node}}/2 \tag{8.11}$$

where the factor 2 is introduced because the inductance is split on both planes, as shown in Figure 8.21b. In this way the power and ground pins of every component on the board see a *locally* radial transmission line. The field perturbations travel outwards from a node in all directions, with waves very similar to those in a radial waveguide. The model is valid up to the frequencies where $\Delta l < \lambda_{\text{min}}/10$ holds, where λ_{min} is the wavelength corresponding to the maximum frequency of interest.

The proposed circuit model is suitable for performing simulations directly in the time domain of PCBs while accounting for the presence of components. To this end, models for ICs, loads, and decoupling capacitors with their associated parasitic equivalent series and connecting lead inductances are required. Note that, for voltage noise simulation, or ΔI*-noise*, IBIS-like models of components, like those used in *Chapter 6* for crosstalk predictions, are not adequate. In fact, they do not simulate the shoot-through current occurring when the output stage of the switching device has both totem pole transistors conducting for a short time. A micromodel at transistor level is therefore required.

Example 8.3: Measurements and Simulations with Standard Buried Capacitance (SBC) Test Boards

As demonstrated in *Example 8.1*, in a PDN a very important parameter is its impedance Z_{PDN}. Smaller impedance enables the power system to provide more step current ΔI, owing to switching devices, for a given voltage drop ΔV called ΔI*-noise* (see Equation (8.1)). This is a key parameter in designing a board, as it affects signal quality and radiated emission. In fact, the noise voltage ΔV can be considered as a source of *common-mode* current for cables attached to the board, making these cables act as very good antennas, as will be shown in *Section 9.6*.

The use of multilayer boards with an adequate layer arrangement and planes for power and ground distribution reduces drastically the PDN impedance Z_{PDN} owing to the intrinsic capacitance of the power/ground pair of planes and the distributed decoupling capacitors with minimized parasitic equivalent series and connection inductances. The purpose of this experiment is to demonstrate:

- the partial benefit provided by distributed decoupling capacitors on a PCB populated by ICs in reducing voltage noise on power supply;
- the complete benefit offered by the technique called *Buried Capacitance* (BC), which allows a great increase in the intrinsic capacitance between adjacent power and ground planes;
- the level of accuracy that can be obtained by SPICE-like circuit simulators when the two planes are modeled as transmission lines and the switching ICs are modeled at transistor level (micromodel).

To fulfill these tasks, two kinds of PCB were constructed utilizing two different structures [19, 20]. The first, shown in Figure 8.15a, was made using a *standard* stack-up (indicated as STD). The second board, shown in Figure 8.15b, uses the *buried capacitance technology*

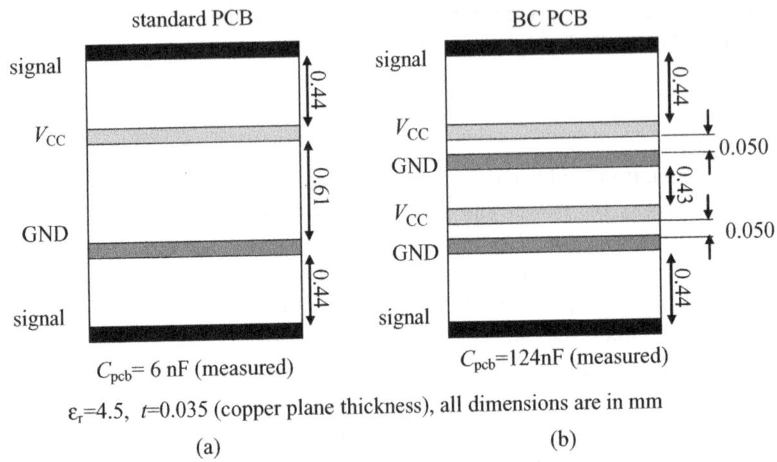

Figure 8.15 Layer arrangement of SBC test boards: (a) standard technology for the PCB; (b) *Buried Capacitance* (BC) technology for the PCB

(ZBC-2000™ technology, and denoted by BC) or *embedded capacitance*, which maximizes the intrinsic capacitance of the power planes of the board. Embedded capacitance is an alternative technique to discrete decoupling capacitors. Boards with embedded capacitance exploit the natural capacitance between the power and return planes to provide power bus decoupling. By minimizing the spacing between the two solid planes up to 0.050 mm, and filling this space with a material with high relative permittivity, the interplane capacitance can be greatly enhanced. Consequently, it is possible to eliminate most or all of the mounted decoupling capacitors [16, 21].

On account of the dimensions of the PCB and its dielectric constant, the intrinsic capacitance of the STD board was measured to be 6 nF, instead of 4.9 nF calculated by using Equation (8.3), whereas it was 124 nF for the BC board. Both boards had a dielectric constant $\varepsilon_r = 4.5$. The measurements of the intrinsic capacitances were performed by the TDR technique (see *Section 11.1*). On both boards the same functions were implemented with an identical layout to that shown in Figure 8.16. In particular, the 15 74AC244 ICs, each having eight output buffers, were loaded at each output pin with a load formed by a 56 pF capacitance in parallel with a 255 Ω resistance. The layout was carefully designed to avoid any transmission-line effects in order to consider the load assigned to each output buffer as a simple lumped element. The pulse generators attached to the BNC connectors drive the components and determine the number of outputs switching simultaneously. The boards were fed by a 5 V power supply. At the power input of the boards, a bulk decoupling capacitor was realized with two electrolytic capacitors of 47 μF and two ceramic capacitors of 100 nF.

The full PCB with components was simulated by an equivalent circuit in which the model presented in *Section 8.2.3* was used for power planes, and a simplified SPICE micromodel provided by the manufacturer was used to model the ICs. To model the power planes, 12×12 segments were used with $\Delta l = 2$ cm, the dimensions of the boards being about 24×25 cm. For the frequencies of interest in this experiment, as first-order approximation the losses can be neglected. Consider again that the interplane capacitance of the standard

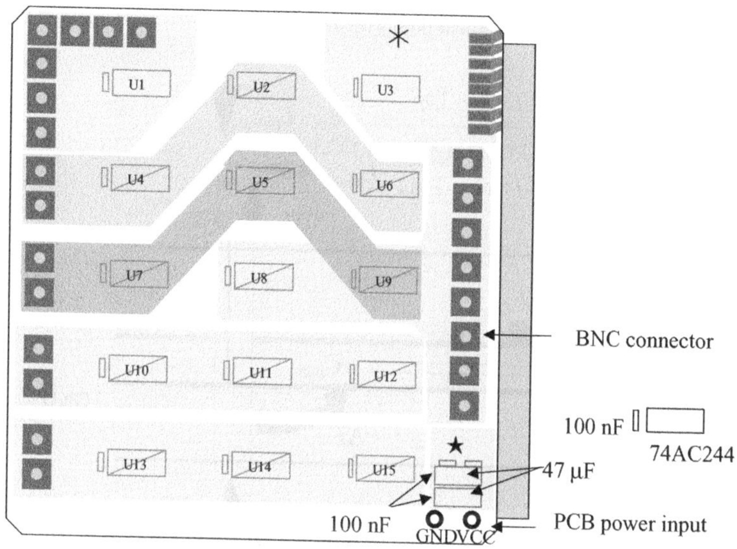

Figure 8.16 View of the component placement

board is 6 nF, while for the board in BC technology it is 124 nF (see Figure 8.15). This makes a substantial difference, as will be shown by results, between the performance of the STD and BC technologies. It is important to point out that this circuit model based on the radial TL theory is not the only way to simulate a pair of power planes by SPICE. Another possible approach is based on modeling the planes with a grid of orthogonal segments of lossy transmission lines along the major sides of the PCB. This other approach is shown and validated in *Appendix C*.

Actually, the circuit model used for the simulations was modified to take into account the TDR response of the power pins of the component. A resistance and a parallel capacitance were added between the power and ground pads to simulate respectively the loading effect of the device and its internal decoupling capacitor (see *Example 8.1*). The output loads of the ICs were modeled as simple RC elements, as the layout was designed with care in order to avoid transmission-line effects. Decoupling and electrolytic capacitors were modeled with their *RLC* equivalent circuits. The parasitic equivalent series inductance and resistance were provided by the data sheets. The following values of parasitic inductances were estimated from PCB geometries and data sheets (see Figure 8.17 for the devices and Figure 8.1 for the capacitors): $L_{\text{via}} = 0.9$ nH, $L_{\text{socket}} = 3$ nH, $L_{\text{pin}} = 2$ nH, L_{bond} (power) $= 15$ nH (high value, as the devices used were dual in-line ceramic packages), L_{bond} (signal) $= 5$ nH, $L_{\text{dec}} = 20$ nH for the decoupling capacitor of 47 μF, and $L_{\text{dec}} = 6$ nH for the decoupling capacitor of 100 nF. For all the capacitors, the equivalent series resistance was $R_{\text{ESR}} = 200$ mΩ. The full PCB model with components for different solutions of filtering, switching gates, and clock frequency is validated by comparing the results with measurements.

For each switching device, the current path, the parasitic inductance involved, and the point of measurements are illustrated in Figure 8.17. Note that the segment of via where the current flows also has an associated inductance which is not indicated in Figure 8.17. The

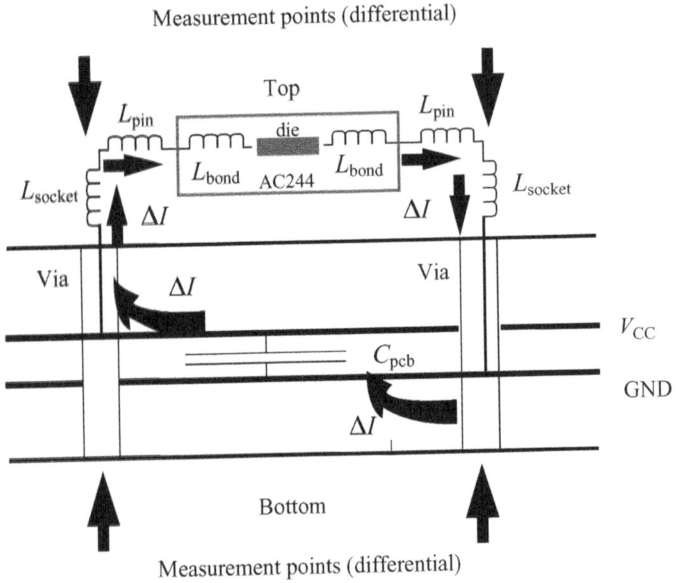

Figure 8.17 Points of measurement of switching noises and switching current path

measurement points for differential oscilloscope probes, indicated as *bottom*, provide the voltage noise ΔV between the two parallel power and ground planes as a product of Z_{PDN} and the switching current ΔI. It will be shown that this noise propagates along the board almost unchanged when the decoupling capacitors are uniformly distributed. The measurement points indicated as *top* provide the noise voltage on the power and ground pins of the ICs, which is the sum of the voltage noise ΔV and the voltage drop caused by the via and socket inductances. This is a local noise associated with the switching device. The measurements of voltage noise between the power and ground terminals of the die are not practically allowed. Only circuit simulation can provide this type of information.

Three different cases were considered, and for each of them the following voltage noise measurements were carried out:

- Case 1 – three simultaneous switchings (SS) in U8 at 5 MHz with:
  - STD board without decoupling;
  - STD board with 100 nF for every IC (STDF);
  - STD board with 15 × 100 nF in U10–U15 (STDFC);
  - BC board.
- Case 2 – 24 simultaneous switchings (SS) in U5, U7, and U9 at 5 MHz with:
  - STD board without decoupling;
  - STD board with 100 nF for every IC (STDF);
  - BC board.
- Case 3 – three simultaneous switchings (SS) in U8 for a clock frequency from 1 to 100 MHz.

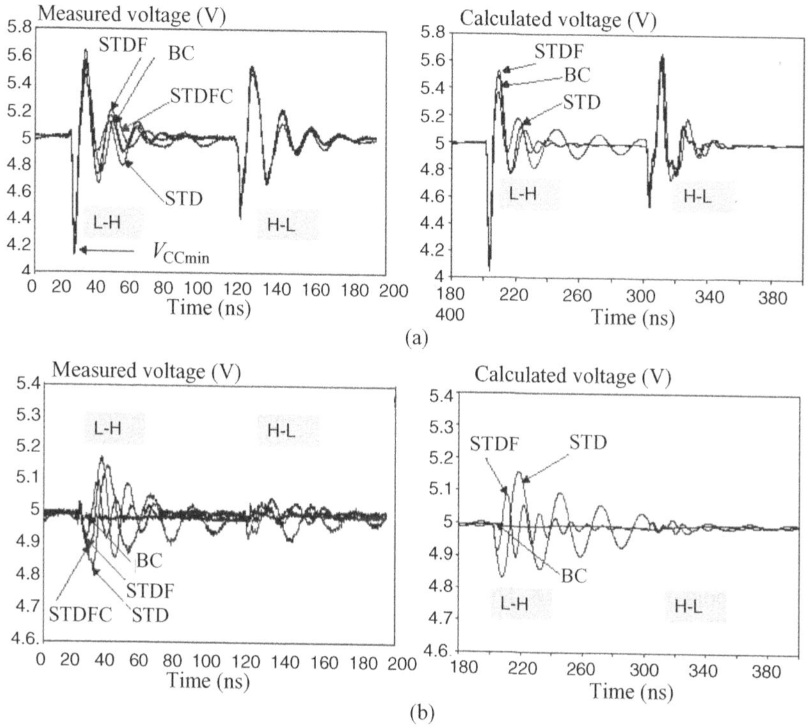

Figure 8.18 Measurements (left) and simulations (right) of V_{CC} plus ΔI-*noise* on U8 with three simultaneous switchings in U8: (a) top position (IC); (b) bottom position (PDN)

(i) Case 1 – Three Simultaneous Switchings in U8 at 5 MHz

Measured waveforms on U8 (top and bottom side) and simulated waveforms are shown in Figure 8.18 on the left- and on the right-hand side respectively, and a very good agreement between simulations and measurements can be observed.

Let us focus on discussing the results obtained with different technologies and filtering. The top-side V_{CC} noise on the active component U8, at IC pins, occurring for low-to-high state switching (L-H), is slightly different for BC and STDF. STD and STDFC have a smaller V_{CC} noise but a larger settling time. For the high-to-low state switching (H-L), the V_{CC} noise is similar for every configuration. This is because in the H-L event the current path mainly involves the load and the ground pin.

The bottom-side V_{CC} noise, at the PDN connection, is less than the top-side V_{CC} noise, as expected, because the components are socket mounted, and the inductance associated with the via and socket pin do not affect the V_{CC} noise at these points, as can be observed from the current path in Figure 8.17. Considering the difference between the nominal power supply of 5 V and the minimum peaks, the noise of the different solutions is negligible for BC, less than 100 mV for STDF, about 100 mV for STDFC, and less than 200 mV for STD. The settling time of the noise exhibits the same trend as the top results. Measurements of the power noise on the bottom side on U5 showed a very similar settling time and noise amplitude.

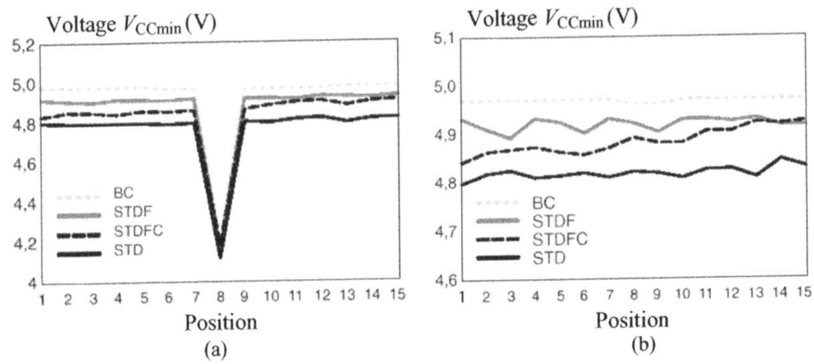

Figure 8.19 V_{CC} noise on all ICs with three simultaneous switchings in U8: (a) top position (IC); (b) bottom position (PDN). The position of the components U1–15 is shown in Figure 8.16

The minimum peak value of V_{CC} noise measured in all positions regarding the four solutions for the top and bottom measurement points is summarized in Figure 8.19. For BC, STDF, and STD the noise shown in Figure 8.19a is practically equal for all positions except for the active component in position 8. This occurs because the capacitances are uniformly distributed on the PCB. The amplitude of the noise depends on the total decoupling capacitance and their associated inductances. STDFC shows a non-uniform noise because the decoupling capacitors are clustered from U10 to U15. In fact, the noise in these positions is less than that in U1–U9.

The bottom-side V_{CC} noise is shown in Figure 8.19b, where the same information as for the top-side noise is provided, with the exception that the noise in position 8 is now practically aligned with the values measured in all the other positions of the board. This is due to the fact that the two planes can be considered to be practically at the same potential, the inductance of the planes between two positions being negligible. Again, STDFC shows a non-uniform voltage drop, as the decoupling capacitors are non-uniformly distributed.

(ii) Case 2 – 24 Simultaneous Switchings in U5, U7, and U9 at 5 MHz

The measured and simulated V_{CC} noise waveforms for STD, STDF, and BC cases are shown Figure 8.20. In this case a very good agreement can again be observed between measurements and simulations. The STDFC configuration is not considered any further because the results obtained in the previous case are enough to understand the behavior of the clustered decoupling solution. The results shown in Figure 8.20 show the importance of a high interplane capacitance, as offered by the BC solution. As can be seen from the top waveforms, STDF and BC have the same settling time but different minimum amplitudes: $V_{CCmin} = 3.7$ V for STDF and $V_{CCmin} = 4.0$ V for BC. STD exhibits a smaller amplitude but a larger settling time.

The amplitude differences for bottom measurements are $V_{CCmin} = 4.1$ V for STD, $V_{CCmin} = 4.7$ V for STDF, while V_{CCmin} is very close to 5 V for BC. Hence, the interplane capacitance for a BC board (124 nF) is suitable for feeding the full switching of the three components and for keeping the voltage gradient very small. On the other hand, in the STDF configuration, in spite of its high global capacitance (6 nF + 15 × 100 nF), the V_{CC} noise is high. This is due to the inductances of the decoupling capacitor, which reduce the capacitor performance for frequencies beyond its series resonance.

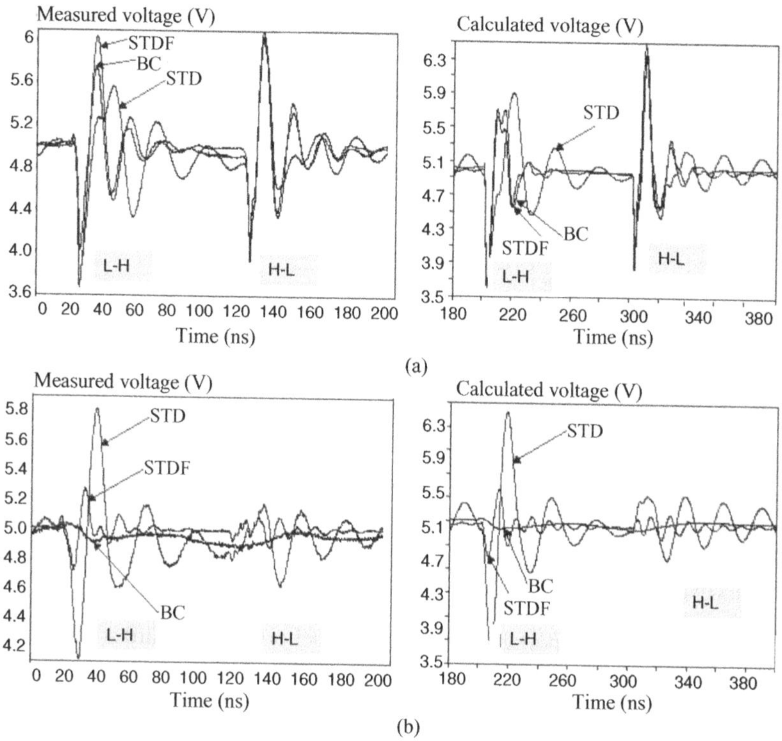

Figure 8.20 Measurements (left) and simulations (right) of V_{CC} noise on U7 with 24 simultaneous switchings: (a) top position; (b) bottom position

As for case 1, it was verified that the noise amplitudes at the top in active positions 5, 7, and 9 have a maximum negative peak, while the noise is uniform in all other positions. For bottom measurements it was verified that the BC solution stabilizes V_{CC} over the whole board at 5 V. For low-to-high (L-H) switching, the minimum V_{CC} is 4.8 V for STDF and 4.3 V for STD [19].

(iii) Case 3 – Three Simultaneous Switchings in U8 for a Clock Frequency from 1 to 100 MHz
The comparison between measurements and simulations is shown in Figure 8.21a. V_{CCmin} for top noise is reported in the ordinate of the graphic. In this case, only the STD PCB was considered because the behavior of the other two, STDF and BC, was similar. In fact, the noise depends mainly on ICs and socket parameters. The maximum absolute noise was measured at 35 MHz. Simulations were performed only for some frequencies and were in good agreement with measurements, especially in the range where the variation in V_{CCmin} is significant. The comparison for bottom noise is shown in Figure 8.21b, where the simulations confirm minimum and maximum noise values.

Further measurements carried out on the bottom side of the PCBs and not reported here showed markedly different filtering performances for the boards under study. STD exhibits maximum noise at 20 MHz and 40 MHz up to a minimum of 4.65 V. With decoupling STDF, V_{CCmin} is no less than 4.9 V up to 75 MHz, after which it goes to 4.6 V at 90 MHz with a

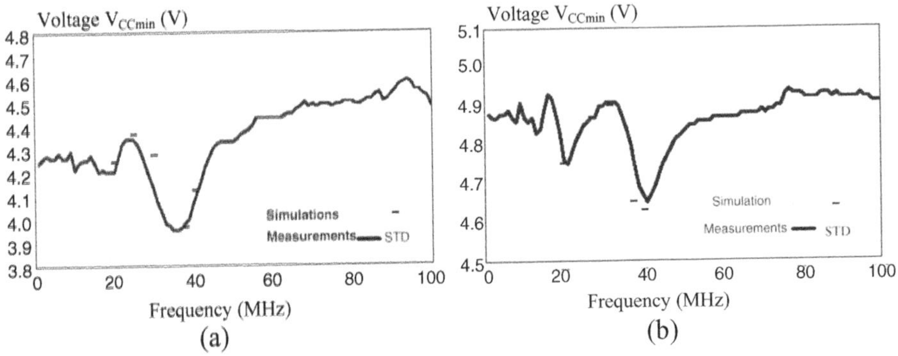

Figure 8.21 Measurements and simulations of V_{CC} noise on U8 with three simultaneous switchings for a standard board (STD): (a) top; (b) bottom

constant slope. This is due to the fact that with STDF PCB the devices do not switch correctly over 75 MHz. With BC, V_{CCmin} is very close to nominal $V_{CC} = 5$ V, and 74AC244 devices switch correctly up to 100 MHz.

To conclude this experiment, the following points can be made:

- For effective decoupling, it is important to place the capacitors on the board so that they are distributed in a regular grid pattern and near to the ICs to obtain low and uniform V_{CC} noise along the board. In fact, it has been shown that V_{CC} noise is not uniform when clustering of decoupling capacitors located on one-third of the board is adopted.
- Comparing V_{CC} noise of a standard board with that of decoupling capacitors and a buried capacitance board shows the importance of raising the interplane capacitance value because it has a very small associated parasitic inductance.
- Analyzing V_{CC} noise between the power and ground plane pins of the ICs (bottom-side measurements), and varying the clock frequency from 1 MHz to 100 MHz, it turns out that the best performances are obtained when parasitic inductances are minimized, as with BC technology.
- Power planes can be modeled as a grid of L and C elements whose values can be computed by using the radial transmission line theory or other similar approaches based on segmentation of the boards by transmission lines. Simulated waveforms for the analyzed cases are in good agreement with measurements if a SPICE model at transistor level provided by the manufacturer of the digital device is used.

8.3 Ground Bounce

Ground and *power bounce* are the voltage drops across the IC pin and on-chip package inductances caused by simultaneous switching of gates sharing the same IC. It is particularly important to investigate this phenomenon when large-scale integration digital devices are involved. The aim of this section is to investigate the bounce mechanism by using SPICE simulations and measurements. It is also shown by experiments that the bounce phenomenon very often is unpredictable owing to the complexity of the circuit and layout of digital devices. The only effective methods for mitigating the problems produced by ground bounce are to limit the number of gates that switch at the same time and to minimize the effective inductance associated with the package of the device.

8.3.1 Ground Bounce Mechanism

In *Section 8.1*, the concept of *ground bounce* due to the simultaneous switching of chips within the same IC has been introduced. In this section, the ground bounce is investigated in more detail [22, 23]. Consider the configuration shown in Figure 8.22 where a simple circuit model for a CMOS device in a lead frame is driving a standard test load through a transmission line. The inductors L_{GND}, L_{PW}, and L_{OUT} represent the effective inductances of the ground, power, and output leads of the package respectively. The capacitor C_L and resistor R_L comprise the standard equivalent circuit of the input stage of a receiver that is connected to the output of the driver by a lossless transmission line of characteristic impedance Z_0 and delay time T_D. The investigation is conducted assuming that the line is matched (i.e. $R_L = Z_0$).

The three waveforms shown in Figures 8.22b, c, and d depict how ground bounce is generated. The first waveform (Figure 8.22b) shows the voltage variation ΔV_L in the time interval Δt across the load as the device output switches from a logic high state to a logic low one. The output slew rate is dependent upon the characteristics of the output transistor, the inductors L_{OUT} and L_{GND}, and the load capacitance C_L.

The second waveform (Figure 8.22c) shows the variation in current ΔI that is generated as the capacitor C_L discharges owing to the voltage variation ΔV_L.

The third waveform (Figure 8.22d) shows the voltage drop V_{GND} that appears across the inductance in the ground lead owing to the discharging current ΔI. This voltage creates what is known as *ground bounce*. The inductor L_{GND} between the external system ground and the internal device ground causes the internal ground to be at a different potential than the external ground. The voltage drop across L_{GND} causes the device input and output to behave differently to their expected behavior because they are referenced to the internal device ground, while the

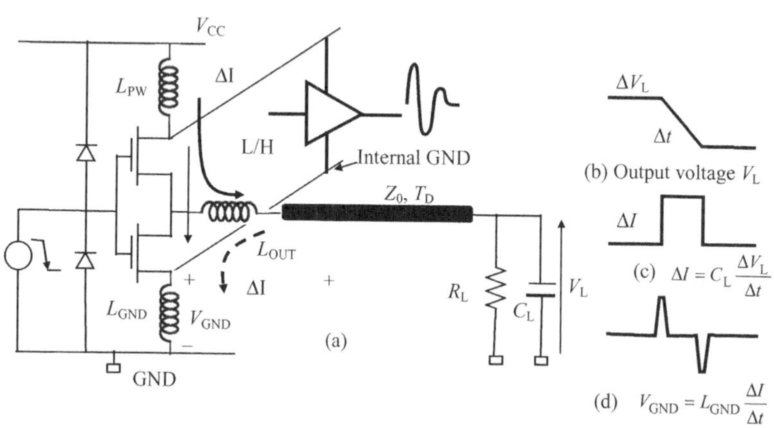

Figure 8.22 Ground bounce circuit model: (a) output circuit of a digital device with parasitic inductances; (b) high-to-low output voltage; (c) switching current on load capacitance; (d) ground bounce voltage

devices that are either driving into the input or being driven by the output are referenced to the external system ground. Outside the device, ground bounce causes input thresholds to shift and output levels to change. This situation is very similar to that of large systems where voltages can develop across extended ground networks. Note that everything discussed here concerning the ground bounce can be applied to the opposite effect, V_{CC} bounce or *power bounce*. V_{CC} bounce is the dual of ground bounce.

The following points should be considered when the *ground bounce* mechanism is investigated:

- Faster switching times of the devices cause higher transient currents at their output as they discharge load capacitances, $\Delta I = C_L \Delta V_L / \Delta t$.
- These higher currents, which are generated when multiple outputs of a device switch simultaneously from logic high to low state, can cause a board-level phenomenon known as *ground bounce*.
- Load capacitance, the effective inductances of the pin-to-die path, and the number of switching outputs are the predominant conditions that influence the magnitude of *ground bounce* in programmable digital devices. This noise generated by the ground bounce can propagate through the system, causing false switching.
- This noise is also important as it can drive cables attached to the board (*common-mode voltage driven mechanism*) and strong radiated emissions can arise.
- Since many factors contribute to *ground bounce*, no standard test methods are available to predict ground bounce magnitude for all possible PCB environments and/or configurations.
- Therefore, only experimental testing makes it possible to investigate each condition and each device's relative contributions to *ground bounce*.

8.3.2 Circuit Simulations to Understand the Ground Bounce Mechanism

For a better understanding of *ground bounce* effects in a typical situation (see Figure 8.23), circuit simulations were performed on the basis of the following assumptions:

- *Ground bounce* is produced on a quiet chip at low and high output state by two other chips switching simultaneously, and sharing the same IC internal ground.
- A high-speed CMOS chip is used as a switching device, using the same model as that employed in *Section 8.1.3*.
- The two drivers send a signal to an RC termination through a 50 Ω interconnect with delay time $T_D = 2$ ns. The line is lossless and perfectly matched.
- An ideal 5 V VRM powers the IC.
- Power distribution is realized by busbars, as in *Example 8.1*.
- The decoupling capacitor is located at a distance of 1 cm from the pins of the IC.
- The series connection of the resistance $R = 50\ \Omega$ and the decoupling capacitors of 1 μF applied between the pins of the IC takes into account the resistive load effect produced by the IC during the switching of the gates within the IC. It is important to consider this load in order to reproduce by SPICE the real noise waveforms on the power and ground pins of an IC.

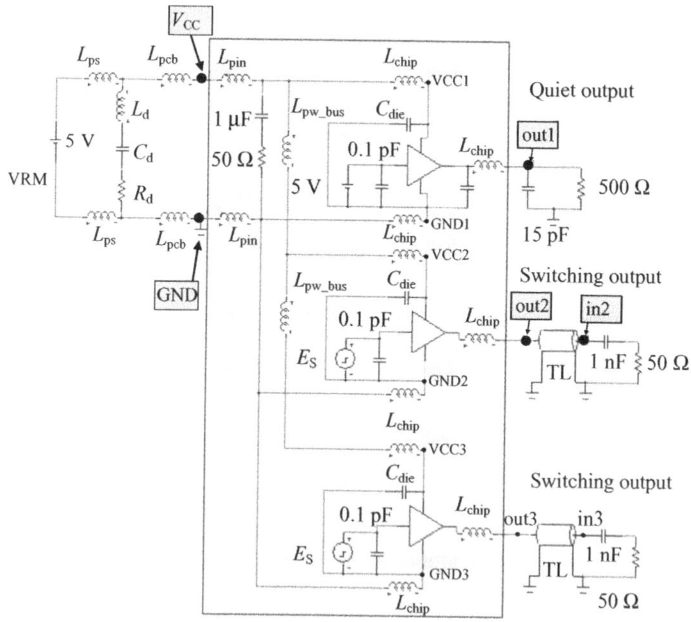

L_{ps}=2.3 nH L_{pcb}=0.23 nH L_{pin}=1 nH L_{pw_bus}=0.1 nH L_{chip}=0.1 nH

C_{die}=50 pF L_d=1 nH C_d=100 nF R_d=0.05 Ω TL: Z_0=50 Ω, T_D=2 ns

Figure 8.23 SPICE model used to evaluate ground bounce

The equivalent circuit of this configuration is shown in Figure 8.23, where the values of the circuit elements are also listed.

The simulations were addressed to calculate:

- input and output signals at the switching gate;
- the output signal at the quiet gate for both low and high voltage levels, with and without considering the lead inductance effect.

In the following, the world *ideal* means the absence of the inductances L_{chip}, L_{pin}, L_{pw_bus}, L_{pcb}, and L_d, and the world *actual* means the presence of these inductances. Simulated waveforms for *ideal* and *actual* situations are shown in Figure 8.24. The results indicate the negative effects of effective inductances associated with the leads on the integrity of the transmitted signal and the noise generated in the quiet line at low and high voltage levels.

8.3.3 Measurements of an LVT Benchmark

Simulations are very useful in understanding bounce phenomena and in investigating the influence of the intrinsic inductance of the components. Unfortunately, what happens within an IC is so complex that it is almost impossible to predict the reality. Therefore, measurements

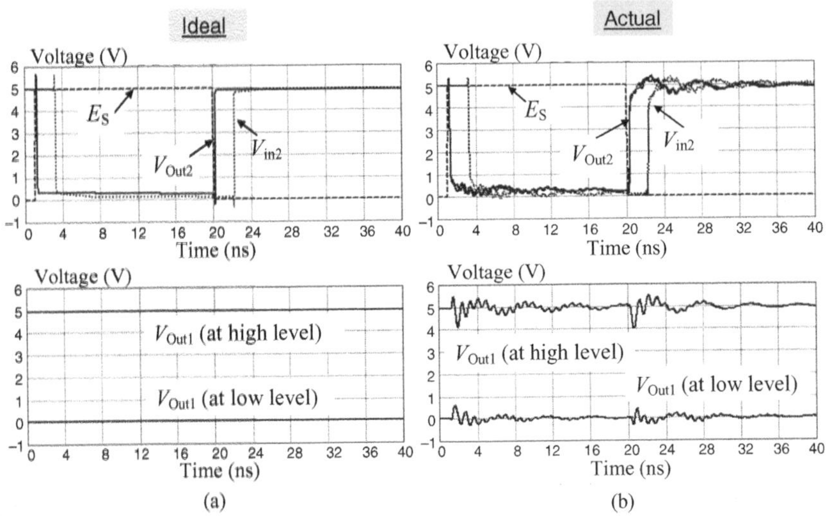

Figure 8.24 Simulated waveforms with ground bounce effects on signal pins: (a) without lead inductance (ideal); (b) with lead inductance (actual)

are the only possible way to quantify the problem [23]. With these considerations in mind, a test board was built to evaluate *ground bounce* and the factors affecting it. The test has the following purposes and characteristics (see Figure 8.25):

- The board was designed for evaluating *ground/power bounce* under different conditions.
- The *Device Under Test* (DUT) was the 74LVT16244 buffer of Texas Instruments.
- The 74LVT16244 buffer has 48 pins in total: eight for ground, four for V_{CC}, and the remainder for signal.
- Pin 2 was considered as the point of measurement for *ground bounce*.
- The test board has a multilayer structure with power and ground planes having a very small interplane distance.
- Good filtering was provided: one 47 μF electrolytic decoupling capacitor and one 10 nF ceramic decoupling capacitor positioned very close to the DUT.

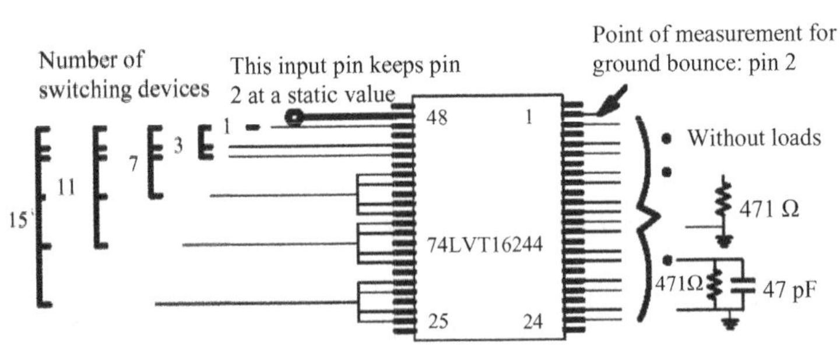

Figure 8.25 DUT and its point of measurement

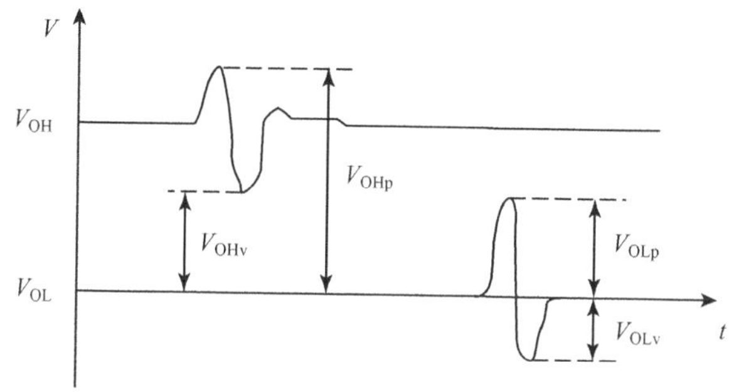

Figure 8.26 Parameter definitions: V_{OLp} and V_{OLv} for ground bounce; V_{OHp} and V_{OHv} for power bounce

- Short interconnects on the ground plane were realized to avoid reflections.
- An increasing number of simultaneously switching devices was used: 1, 3, 7, 11, and 15.
- Three different loads were considered for the output: open circuit, 471 Ω resistance, and 471 Ω resistance in parallel with a 47 pF capacitance.
- The working frequency of the device under test was 16 MHz.
- Measurements were performed for switching current and *ground/power bounce*.

The parameters measured to quantify the *ground* and *power bounce* are shown in Figure 8.26. V_{OL} is the DC value of the pin 2 when at low level; V_{OH} is the DC value of the pin 2 when at high level.

The measured currents on the ground pin as a function of the number of simultaneous switching devices are shown in Figure 8.27a for the two transitions: low-to-high (L → H) and high-to-low (H → L). The load for each output pin is the parallel between a 471 Ω resistance and a 47 pF capacitance. The measurements were performed by using a wire that tied all ground pins of the DUT together. Then the wire was passed through a high-frequency current probe for measurements (Tektronix CT1) and connected to the reference PCB ground. Observe that, by increasing the number of simultaneous switching gates, the measured current flowing through the ground increases. The worst-case current occurs for high-to-low transition because the discharge current of the capacitance sums with the shoot-through current caused by the simultaneous conducting phase of the two output totem pole transistors. The currents measured with the other two load conditions exhibited lower peak values but the same behavior: the peak current increases with the number of switching gates.

It is interesting to observe the results shown in Figure 8.28, where the parameters V_{OLp} and V_{OLv}, which characterize the *ground bounce*, are plotted for several load conditions as a function of the number of simultaneous switching devices and types of load. Note that, for low-to-high transition and with RC load, V_{OLp} and V_{OLv} are not correlated with current switching: the noise decreases instead of increasing when the number of simultaneous switching devices increases. This is considered to be anomalous or unpredictable behavior, and could be due to the particular layout inside the device.

The worst case of *power bounce* is shown in Figure 8.29 and occurs with low-to-high transition and output loaded with a capacitance. In fact, in this case the current required to charge the load capacitance and the shoot-through current flow in the power lead.

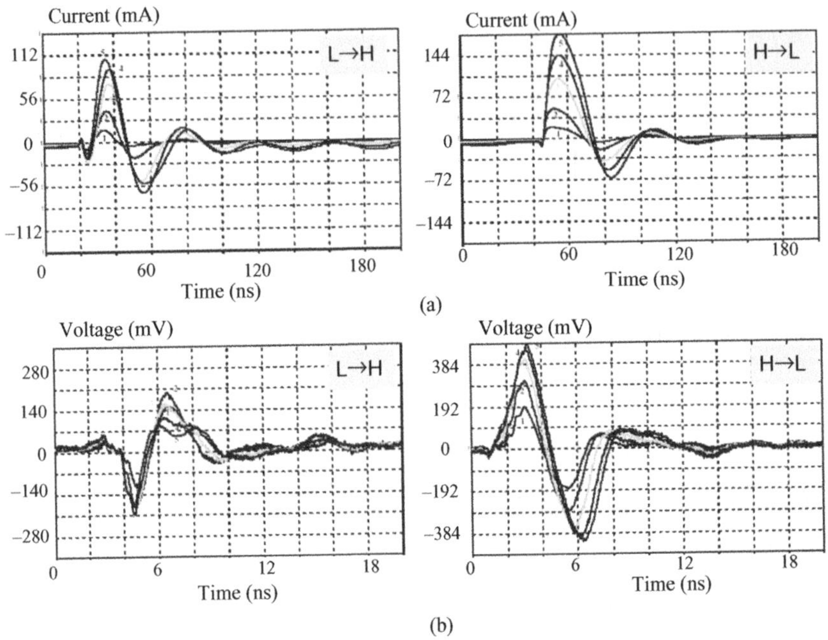

Figure 8.27 Measured waveforms: (a) switching current on GND with RC loads; (b) ground bounce with RC loads. The current and voltage peaks increase with the number of simultaneous switchings: 1, 3, 7, 11, and 15. Each number indicates how many gates switch simultaneously

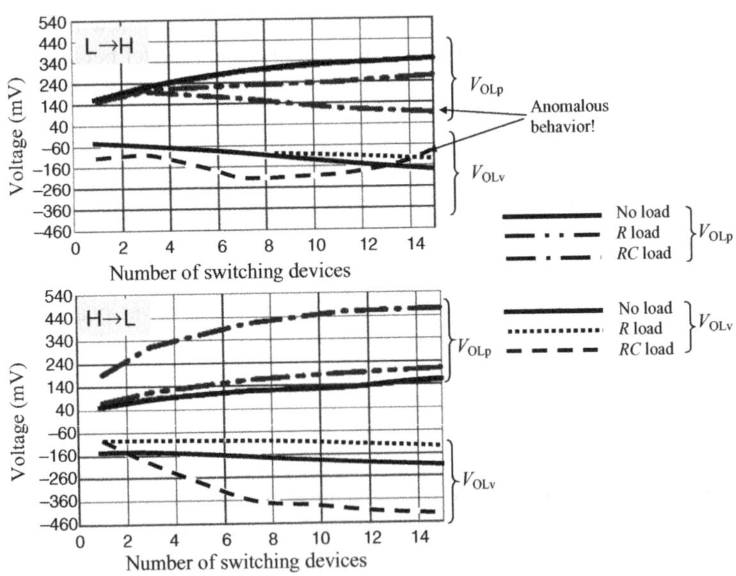

Figure 8.28 Measured ground bounce parameters (see Figure 8.25) as a function of simultaneous switching devices for low-to-high (L → H) and high-to-low (H → L) transitions

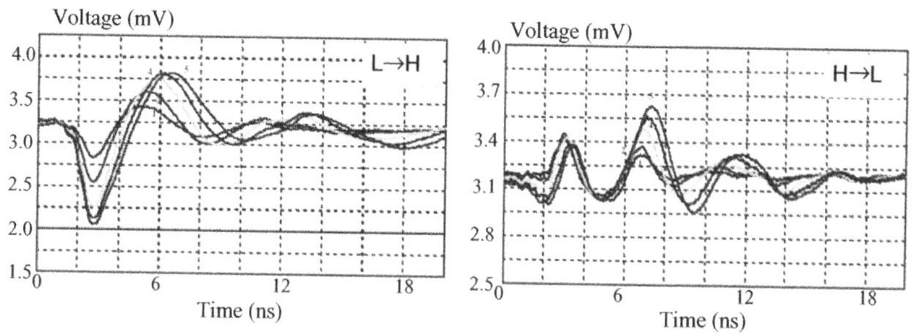

Figure 8.29 Measured power bounce with RC loads. The voltage peak increases with the number of simultaneous switchings: 1, 3, 7, 11, and 15

The parameters V_{OHp} and V_{OHv}, which characterize the power bounce for several load conditions as a function of the number of switching devices and types of load, are shown in Figure 8.30. Note that, with a maximum number of simultaneously switching devices and RC load, V_{OHv} touches the V_{IHmin} limit. This means that, according to the definition of the parameter V_{IHmin} in Section 2.1, all the available immunity margin is lost, and therefore the maximum number of simultaneous switching devices must be reduced. No anomalous behavior was observed for the *power bounce*.

To conclude this experiment, the following observations can be made:

- *Ground bounce* and *power bounce* are generally correlated with the switching currents.
- Only for *ground bounce* and RC loads is there no correlation (when current increases, noise decreases).

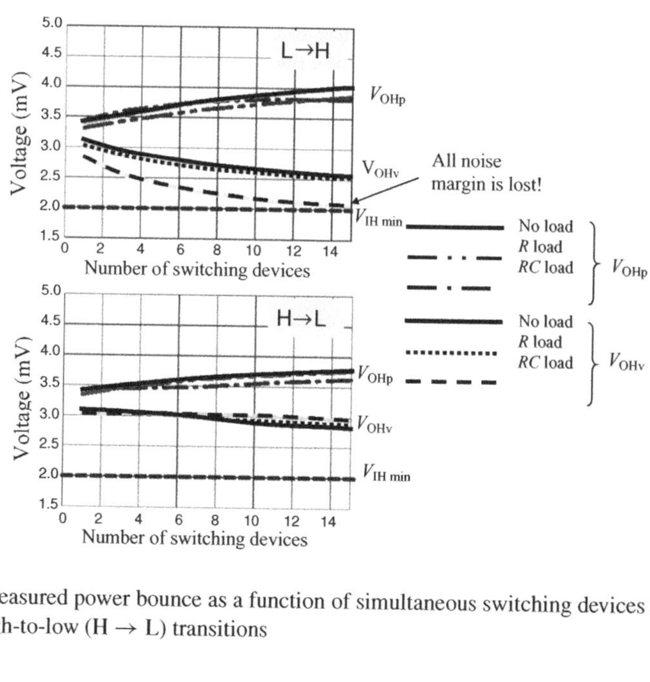

Figure 8.30 Measured power bounce as a function of simultaneous switching devices for low-to-high (L → H) and high-to-low (H → L) transitions

- *Ground bounce* is always less than the V_{ILmax} limit of 0.8 V for LVT with a margin of 320 mV.
- *Power bounce* is very close to the V_{IHmin} limit of 2 V with RC load and maximum switching devices.
- For this reason, the maximum number of simultaneously switching devices should be no more than 8.

This experiment underlines that *ground bounce* is not easily predictable by SPICE simulations because it depends on package layout and by the solutions adopted to design the IC and not provided by the manufacturer. Our opinion is that a better way to investigate the *ground bounce* phenomenon is to carry out measurements on the device of interest by preparing a test board and a set-up as just described in this section. Circuit simulations are very useful for understanding the effects produced by the *ground/power bounce* on signal integrity, but not for quantifying it in a real application on account of its complexity.

8.4 Crosstalk and Switching Noise

This last section is devoted to investigate the problem of modeling *ground bounce* and *crosstalk noise* when they occur at the same time. It is shown that, especially in a post-layout signal integrity simulation, care must be taken in choosing appropriate IC models if the noises involved need to be accurately reproduced.

With regard to the signal integrity issue in digital systems, post-layout analysis is a very important step in the design flow of *Printed Circuit Boards* (PCBs) [24]. Constraint violations can be identified and corrected even before making a prototype. To carry out post-layout analysis, several software codes are available on the market. Many commercial post-layout simulators are based on the following procedure to calculate currents on the traces: the layout interface reads the layout database, identifies the traces, and automatically runs the field solver for the appropriate geometries. To predict currents, a time-domain analysis is done by modeling the traces as coupled transmission lines (TLs), and, at the same time, the non-linear circuit behavior of the ICs is accounted for. The models used for the traces are of the same type as those presented in *Section 6.4*. The models for ICs are non-linear macromodels that reproduce with a few circuit elements the behavior of the input and output characteristics of the ICs in order drastically to reduce the computational time. They are similar to the device models presented and discussed in *Section 2.3*, *Section 2.4*, and *Section 6.3*. Unfortunately, these *IBIS-like* models do not take into account the effects of the simultaneous internal switching of the devices that produce the ΔI*-noise* and the *ground bounce*. To investigate this problem, a test board called the SQ-test was built [25].

8.4.1 Measurements and Simulations of the SQ-Test Board with Three Coupled Lines and 74AC04 Devices

The layout of a structure consisting of three coupled lines in a point-to-point configuration is shown in Figure 8.31. Three 74AC04 gates drive the three coupled lines in the same IC. The receivers are likewise three 74AC04 gates packaged in the same IC.

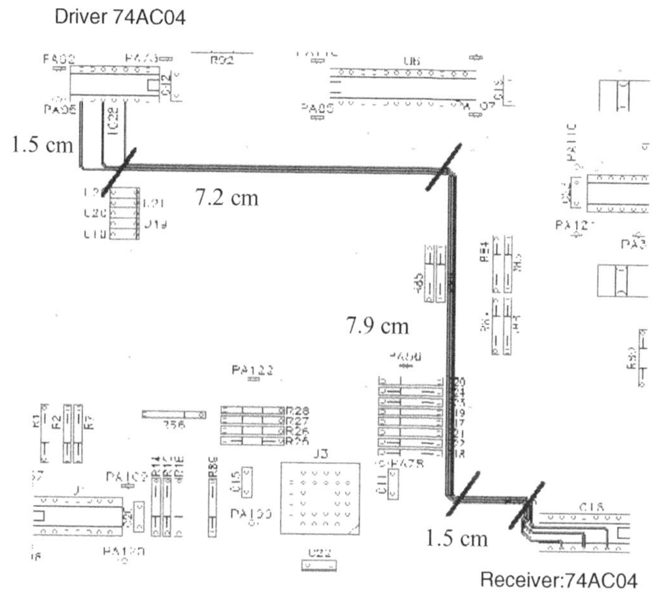

Figure 8.31 Layout of the SQ-test board with three coupled lines driven by a 74AC04 device and loaded by another 74AC04 device

The cross-section of the three coupled lines is shown Figure 8.32, with **L** and **C** matrices computed by a field solver. The substrate was a typical FR4-like dielectric used for digital PCBs. The length of the lines was about 20 cm, and their path is detailed in Figure 8.31. The simulations were carried out at the post-layout analysis stage. The topology files and the cross-sections were built up directly from layout tools, and macromodels were used in the

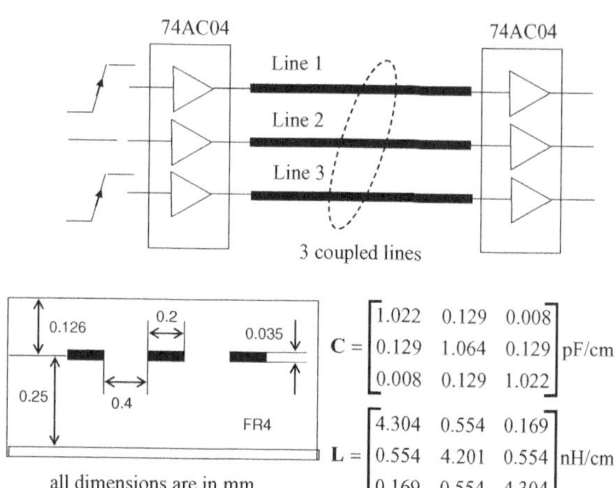

$$\mathbf{C} = \begin{bmatrix} 1.022 & 0.129 & 0.008 \\ 0.129 & 1.064 & 0.129 \\ 0.008 & 0.129 & 1.022 \end{bmatrix} \text{pF/cm}$$

$$\mathbf{L} = \begin{bmatrix} 4.304 & 0.554 & 0.169 \\ 0.554 & 4.201 & 0.554 \\ 0.169 & 0.554 & 4.304 \end{bmatrix} \text{nH/cm}$$

Figure 8.32 Three coupled lines in an embedded microstrip structure and the line parameters of the SQ test board

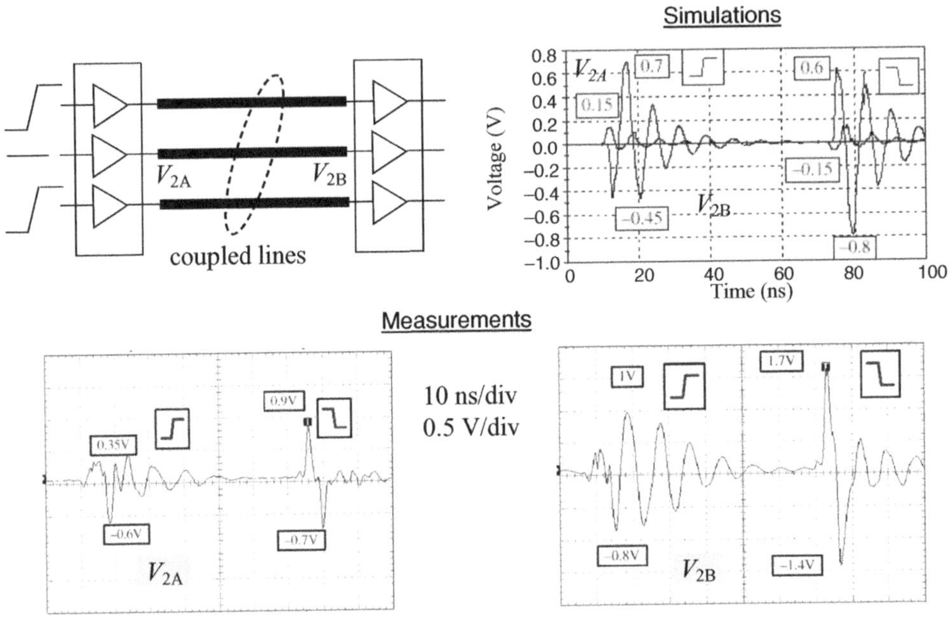

Figure 8.33 Simulated and measured waveforms for both low-to-high and high-to-low transitions with the IC devices modeled by circuit macromodels

simulation process provided by the software. Two lines were active simultaneously, and the victim was the middle line kept quiet at low state. The three traces were simulated considering each segment of the interconnect with the related type of coupling: uncoupled lines and two- or three coupled lines according to the path shown in Figure 8.31. The inductances associated with the package of the ICs were also assumed to be of about 4 nH for each power, ground, and signal pin.

The measured *near-end crosstalk* (V_{2A}) and *far-end crosstalk* (V_{2B}) with their peak values are shown in Figure 8.33. The simulated waveforms obtained by a commercial tool (different codes were used, obtaining more or less the same results) are also reported with their peak values for comparison. The differences are clear: simulations underestimate the measured values significantly!

To understand the reasons for these differences, the set-up shown in Figure 8.34 was considered. The victim line was disconnected from the components and terminated with a 15 Ω resistor at the driver end and with a 1.1 kΩ at the receiver end to simulate output and input impedance of the device. Now, with this variation, good agreement can be observed when measured and simulated waveforms are compared.

Other measurements were carried out to understand the reasons for the original discrepancy. Measurements were taken at the middle line with the active pins of the driving IC disconnected (see Figure 8.35). This means that no signals were traveling along the 'active' lines. Therefore, the measured near-end V_{2A} and far-end V_{2B} voltage waveforms are not due to crosstalk but to an internal phenomenon of the chips known as *ΔI-noise* or *ground bounce*.

At this stage of the investigation it was decided to improve the simulations using a micromodel, a model at transistor level, for the ICs, and to model power and ground planes of

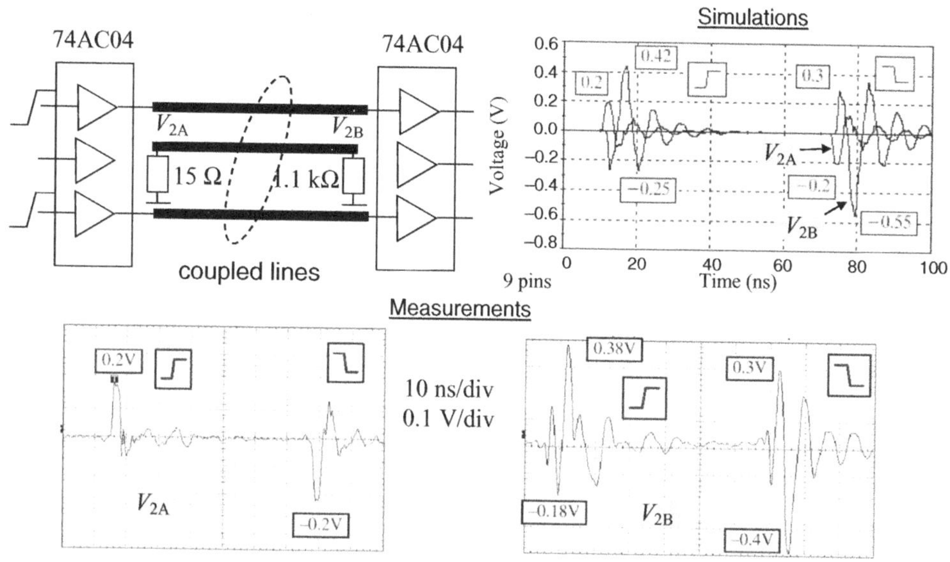

Figure 8.34 Simulated and measured waveforms for both low-to-high and high-to-low transitions with the devices modeled by macromodels and the quiet line loaded with resistances instead of the gates of the devices

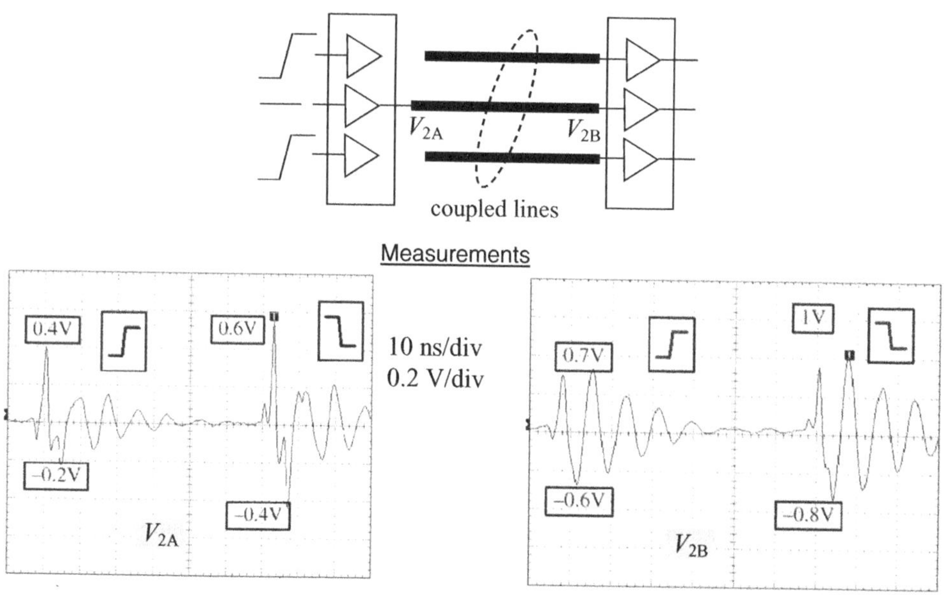

Figure 8.35 Measured waveforms for both the low-to-high and high-to-low transitions with the disturbing devices disconnected from their interconnect

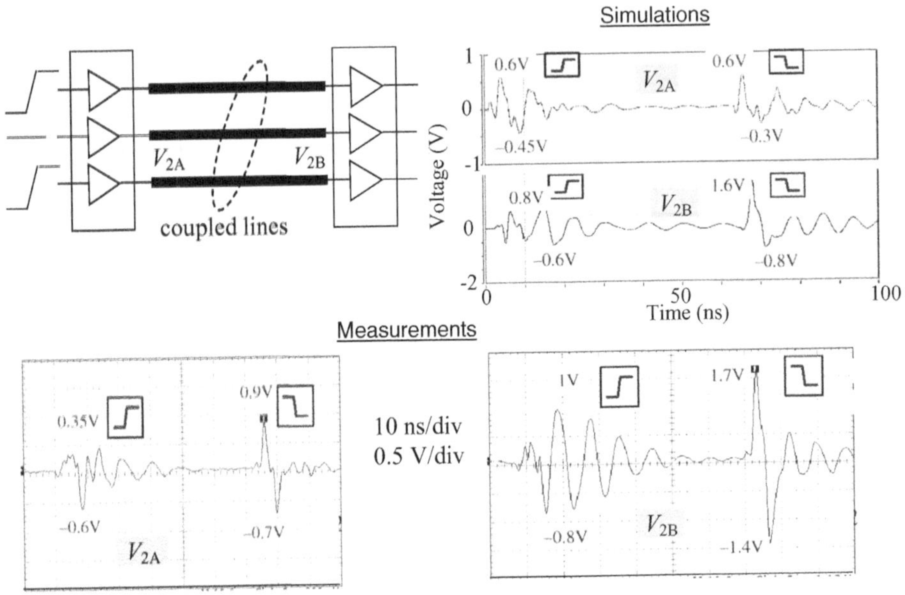

Figure 8.36 Simulated and measured waveforms for both the low-to-high transition and the high-to-low transition switching with the devices modeled by micromodels

the test board by the radial transmission line technique described in *Section 8.2*. The results obtained with this new circuit model are shown in Figure 8.36. Now, the simulated waveforms and their peaks are more similar to the measured waveforms. A very good agreement was not obtained owing to the difficulty in reproducing the complexity of the *ground bounce* phenomenon, as explained in *Section 8.3*.

From the results of this experiment, the following points can be made:

- The most appropriate way to solve ΔI*-noise* and crosstalk could be to develop reasonably approximate models with some extra features compared with those offered by an *IBIS-like* macromodel.
- There should be a close relationship between customers and manufacturers or between customers, manufacturers, and signal integrity tool vendors in order to have the necessary information for building appropriate device models.
- An evaluation should be made of whether it makes sense always to use these new models in PCB post-layout analysis with possibly prohibitive computational time.
- It could be important to consider a sort of 'mixed' simulation where both macromodels and micromodels are used. The micromodels should be used only for those nets defined as critical by the designer and with simultaneous switching gates within the same ICs.

References

[1] Hall, S., Hall, G., and McCall, J., *"High-Speed Digital System Design – a Handbook of Interconnect Theory and Design Practice"*, John Wiley & Sons, Inc., New York, NY, 2000.

[2] Knighten, J., Archambeault, B., Fan, J., Selli, G., and Connor, S., 'PDN Design Strategy: I. Ceramic SMT Decoupling Capacitors – What Value Should I Choose?', *IEEE EMC Society Newsletter*, (207), Fall 2005, 34–41.

[3] Hubing, T., Drewniak, J., Van Doren, T., and Hockanson, D., 'Power Bus Decoupling on Multilayer Printed Circuit Boards', *IEEE Trans. on Electromagnetic Compatibility*, **37**(2), May 1995, 155–167.

[4] Knighten, J., Archambeault, B., Fan, J., Selli, G., Xue, L., Connor, S., and Drewniak, J., 'PDN Design Strategy: II. Ceramic SMT Decoupling Capacitors – Does Location Matter?', *IEEE EMC Society Newsletter*, (208), Winter 2006, 56–67.

[5] http://www.agilent.com, in particular 'LCR Meters, Impedance Analyzers'.

[6] Smith, L., Anderson, R., Forehand, D., Pelc, T., and Roy, T., 'Power Distribution System Design Methodology and Capacitor Selection for Modern CMOS Technology', *IEEE Trans. on Advanced Packaging*, **22**(3), August 1999, 284–291.

[7] Leone, M., 'The Radiation of a Rectangular Power-bus Structure at Multiple Cavity-mode Resonances', *IEEE Trans. on Electromagnetic Compatibility*, **45**(3), August 2003, 486–492.

[8] Balanis, C., *"Antenna Theory"*, 2nd edition, John Wiley & Sons, Inc., New York, NY, 1997.

[9] Knighten, J., Archambeault, B., Fan, J., Selli, G., Rajagopal, A., Connor, S., and Drewniak, J., 'PDN Design Strategy: IV. Sources of PDN Noise', *IEEE EMC Society Newsletter*, (212), Winter 2007, 54–64.

[10] www.spectrum-soft.com.

[11] Knighten, J., Archambeault, B., Fan, J., Selli, G., Xue, L., Connor, S., Drewniak, J., 'PDN Design Strategy: III. Planes and Materials – Are they Important Factors in Power Bus Design?', *IEEE EMC Society Newsletter*, (210), Summer 2006, 58–69.

[12] Van Den Berghe, S., Olyslager, F., De Zutter, D., De Moerloose, J., and Temmerman, W., 'Study of Ground Bounce Caused by Plane Resonances', *IEEE Trans. on Electromagnetic Compatibility*, **40**(2), May 1998, 111–119.

[13] Wang, C., Mao, J., Selli, G., Luan, S., Zhang, L., Fan, J., Pommerenke, D., DuBroff, R., and Drewniak, J., 'An Efficient Approach for Power Delivery Network Design with Closed-form Expressions for Parasitic Interconnect Inductances', *IEEE Trans. on Advanced Packaging*, **29**(2), May 2006, 320–334.

[14] Paul, C., 'Effectiveness of Multiple Decoupling Capacitors', *IEEE Trans. on Electromagnetic Compatibility*, **34**(2), May 1992, 130–133.

[15] Alexander, M., *"Power Distribution System (PDS) Design: Using Bypass/Decoupling Capacitors"*, AN XAPP623, XILINX, v2.1, 28 February 2005.

[16] Xu, M., Hubing, T., Chen, J., Van Doren, T., Drewnial, J., and DuBroff, R., 'Power-bus Decoupling with Embedded Capacitance in Printed Circuit Board Design', *IEEE Trans. on Electromagnetic Compatibility*, **45**(1), February 2003, 22–30.

[17] Ramo, S., Winnery, J., and Van Duzer, T., *"Fields and Waves in Communication Electronics"*, 3rd edition, John Wiley & Sons, New York, NY, 1994.

[18] Schelkunoff, S., *"Electromagnetic Waves"*, D. Van Nostrand Inc., Princeton, NJ, 1943.

[19] Caniggia, S., Costa, V., and Preatoni, R. 'Investigation of EMI on Multilayer Circuit Boards: ΔI-noise and Power Supply Decoupling', IEEE 1996 International Symposium on Electromagnetic Compatibility, Santa Clara, 19–23 August 1996, 436–441.

[20] Caniggia, S., Costa, V., and Vitucci, L., 'Investigation of EMI on Multilayer Circuit Boards: Radiated Emission', *IEEE 1996 International Symposium on Electromagnetic Compatibility*, Santa Clara, 19–23 August 1996, 316–321.

[21] Ricchiuti, V., 'Power-supply decoupling on fully populated high-speed digital PCBs', *IEEE Trans. on Electromagnetic Compatibility*, **43**(4), November 2001, 671–676.

[22] *"Understanding and Minimizing Ground Bounce"*, AN-640, June 1989 Revised February 2003. Available online: www.fairchildsemi.com.

[23] Shear, D., *"EDN's Advanced CMOS Logic Ground-bounce Tests"*, EDN, 2 March 1989.

[24] Caniggia, S., Costa, V., O'Riordan, N., Preatoni, R., and Ronco, C., 'Analysis of Noise in High-speed Circuit-boards and Approaches for a Full Automatic Computer Aided Design Solution', *10th Symposium and Technical Exhibition on Electromagnetic Compatibility*, Zurich, 9–11 March 1993, 219–224.

[25] Caniggia, S., Costa, V., Meani, C., and Preatoni, R., 'Printed Circuits Boards Post-layout Analysis: from Signal Integrity to Radiated Emission', *IEEE 1998 International Symposium on Electromagnetic Compatibility*, Denver, 24–28 August 1998, 975–980.

9

PCB Radiated Emission

The essential equations for calculating the spectrum in the frequency domain of typical signal and noise waveforms in digital systems are given in this chapter. These equations will be useful for calculating the emission profile of radiating structures such as a PCB and attached cables.

The basic concepts for predicting radiated emission from a PCB to be compared with the limits of the standards are described. The distinction between *differential-mode* (DM) and *common-mode* (CM) emission is outlined. Analytical procedures are given and validated by experimental results. Closed-form expressions for calculating radiated fields from wire structures, as often found in PCBs with attached cables, are provided in *Appendix D*.

Radiated emission from typical trace structures in PCBs are investigated. Models based on *Transmission-Line* (TL) and long-dipole theory for predicting emission profiles are outlined. Experimental results obtained by some test boards are compared with simulation results. The importance of properly connecting the ground planes between them and with the circuits in order to avoid *common-mode* emissions is underlined.

It will be shown that the radiated emission from ICs can be predicted by a simple formula based on small dipole antenna formulation. Experimental results will be provided to demonstrate that radiated emission is independent of the decoupling capacitors in the absence of cables attached to the board.

Radiated emission from a PCB, including traces and ICs, will be investigated. The goals of this study are to quantify the contribution of both traces and ICs and to provide models. Experimental results are given and compared with the simulations.

A large part of this chapter is devoted to the investigation of radiated emission from the cable attached to the PCB. It is shown that the main sources of emission are *common-mode* sources generated by the PCB as ground noise. *Voltage-* and *current-driven* mechanisms are explained, and models based on transmission lines for current computation on cable are provided. These models are validated by experimental results considering simple test boards with attached cables placed outside and inside a shielded rack.

The effects of *common-mode* current on radiated emission due to imbalance between rise and fall times of the differential outputs of a line driver are investigated by experimental measurements and SPICE simulations. A reproducible set-up is defined in order to characterize

Signal Integrity and Radiated Emission of High-Speed Digital Systems Spartaco Caniggia and Francescaromana Maradei

the EMC performance of drivers, connectors, and cables. The cables used in this investigation are *Unshielded Twisted Pairs* (UTPs) and *Shielded Foil Twisted Pairs* (SFTPs) of Cat.5 in order to minimize the contribution due to the imbalance of the wires. Results with connectors such as a 5 × 2 pin Z-pack, RJ45, and 9 pin D-SUB with SFTP cable are also presented.

The main source of emission from a complex system consisting of several PCBs within a shielded rack is investigated. Procedures for predicting emission from screened cables and apertures are outlined, and the proposed models are experimentally validated.

The radiation diagram pattern concept is also introduced. The radiation pattern of a simple structure such as a PCB with an attached cable is calculated by using some numerical commercial codes based on MOM and FIT techniques. A comparison with results obtained by the analytical TL model for maximum radiated field computations are also presented. The chapter ends with a list of points to remember and design rules.

9.1 Frequency Characterization of a Digital Signal

When dealing with radiated emissions from a digital system, the investigation should be performed in the *frequency domain*, as the limits of standards are given in this domain. As shown in *Section 1.2*, the peaks of emission are due to the clock fundamental and higher-order harmonics. Therefore, the attention here is focused on the spectrum of a clock signal and periodic noises caused by the switching of the devices, and on the information that is essential for performing radiated emission predictions. A rigorous theory and the *frequency-domain* representation of non-periodic signals such as data transmission can be found in the work by Paul [1].

9.1.1 Spectrum of a Trapezoidal Waveform

The clock is ideally a periodic trapezoidal waveform $v(t)$ described by amplitude V_{g0}, pulse rise time t_r, pulse fall time t_f, pulse width t_{pw} (between the points of the waveform having an amplitude of 50 %), and period T_p, as shown in Figure 9.1a. It can be expressed in terms of its Fourier series as [1]

$$v(t) = c_0 + \sum_{n=1}^{\infty} 2\,|\hat{c}_n| \cos\left(n\omega_0 t + \angle \hat{c}_n\right) \tag{9.1}$$

where c_0 is the DC term, n is the harmonic order, with $n = 1, 2, 3$, etc., $|\hat{c}_n|$ is the magnitude of the harmonic of order n, $\angle \hat{c}_n$ is the phase value of the harmonic of order n, and $f_0 = 1/T_p = \omega_0/2\pi$ is the fundamental frequency.

To allow a simple frequency-domain characterization, the following assumptions are adopted:

- 50 % duty cycle (i.e. $t_{pw} = T_p/2$);
- equal rise and fall times (i.e. $t_r = t_f$).

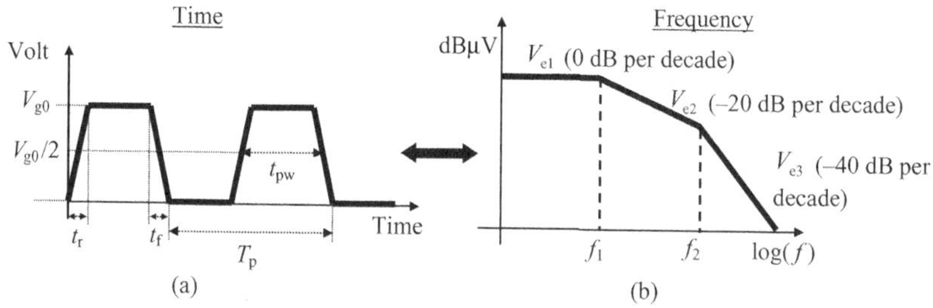

Figure 9.1 Clock waveform: (a) time-domain representation; (b) envelope of the frequency-domain representation

Under these assumptions, the coefficients in Equation (9.1) are given by

$$c_0 = \frac{t_{pw}}{T_p} V_{g0} \tag{9.2a}$$

$$\hat{c}_n = \frac{1}{2} V_{ean}(2\pi\, n f_0)\, e^{-j\pi n f_0 (t_{pw}+t_r)} \tag{9.2b}$$

where $V_{ean}(f)$ is the envelope of the spectral components:

$$V_{ean}(f) = 2c_0 \frac{\sin\left(\pi f t_{pw}\right)}{\pi f t_{pw}} \frac{\sin\left(\pi f t_r\right)}{\pi f t_r} \tag{9.3}$$

and $2\,|\hat{c}_n| = V_{ean}(2\pi n f_0)$. From a practical viewpoint, when the Fourier series representation (9.1) is used to derive the function $v(t)$ in the *time domain*, the sum is limited to an arbitrary, sufficiently large integer number, instead of the original extension to infinity.

The frequency interval of interest for radiated emission computation is usually determined by the standards, i.e. CISPR and FCC usually require $f_{start} = 30$ MHz and $f_{end} = 1000$ MHz. For signals with a 50 % duty cycle, odd harmonics only are present. In practice, the clock is not an ideal trapezoidal waveform with a 50 % duty cycle, and therefore even harmonics are present too. Since the main interest is focused on the radiated emission profile and not on the calculation of each harmonic, a convenient way to calculate this bound is to represent the spectrum of the source by its envelope. The continuous envelope of a trapezoidal waveform is shown in Figure 9.1b, and is defined by

$$V_e(f) = \begin{cases} V_{e1} & \text{for } f < f_1 \\ V_{e2}(f) & \text{for } f_1 < f < f_2 \\ V_{e3}(f) & \text{for } f > f_2 \end{cases} \tag{9.4}$$

where

$$V_{e1} = 2V_{g0}\frac{t_{pw}}{T_p} \tag{9.5a}$$

$$V_{e2}(f) = 2\frac{V_{g0}}{\pi T_p f} \tag{9.5b}$$

$$V_{e3}(f) = 2\frac{V_{g0}}{\pi^2 T_p t_r f^2} \tag{9.5c}$$

$$f_1 = \frac{1}{\pi t_{pw}} \tag{9.5d}$$

$$f_2 = \frac{1}{\pi t_r} \tag{9.5e}$$

For scale convenience, the spectrum is usually represented in dBμV and defined as

$$V_{e,dB\mu V}(f) = 20 \log \left|V_e(f)\, 10^6\right| \tag{9.6}$$

According to Equation (9.4) and the visualization shown in Figure 9.1b, three frequency intervals can be identified (log scale):

1. For $f < f_1$, the envelope is constant and equal to V_{e1}.
2. For $f_1 \le f < f_2$, the envelope is equal to $V_{e2}(f)$ and drops with a slope of −20 dB/dec.
3. For $f \ge f_2$, the envelope is equal to $V_{e3}(f)$ and drops with a slope of −40 dB/dec.

Note that the two break frequencies (9.5d) and (9.5e) are a function of the pulse width and of the rise time respectively. Looking at these spectral bounds, it is evident that the high-frequency content of a trapezoidal pulse train is primarily due to the rise/fall time of the pulse. Pulses having small rise/fall times will have larger high-frequency spectral content than pulses having larger rise/fall times. This could mean more problems in meeting the governmental regulatory requirements on radiated and conducted emissions, as will be demonstrated by the following example.

Example 9.1: Computations and Measurements of a Clock Spectrum

The analytical spectrum as well as the harmonics and the envelope of two different clocks is shown in Figure 9.2. The first clock has a frequency $f_{clock} = 8$ MHz, a rise time $t_r = 2$ ns, and an amplitude $V_{g0} = 5$ V. The second clock has a frequency 10 times higher (i.e. $f_{clock} = 80$ MHz), a rise time 10 times smaller (i.e. $t_r = 0.2$ ns), and a voltage amplitude 5 times smaller (i.e. $V_{g0} = 1$ V). It is interesting to note that, at frequencies higher than 30 MHz, the second clock has lower harmonics or envelope values and produces more emissions for the same radiating structure. Nevertheless, the second trapezoidal waveform has a smaller voltage amplitude. The shorter the rise time, the higher is the frequency at which the frequency spectrum starts to decrease in magnitude (e.g. 40 dB/dec). This means that, to avoid significant radiated emission, high-speed devices should be used only if necessary.

A 74AC244 device generating an 8 MHz clock with a series resistance of 50 Ω is considered. A comparison between the clock measured spectrum and the computed envelope is shown in Figure 9.3. Note that, to allow comparison between the analytical formula of the envelope (9.4) and the measurement, it is necessary to subtract 9 dB from the envelope:

- 6 dB owing to the resistance partition related to the point of measurement (see Figure 9.3);
- 3 dB owing to the fact that the *Spectrum Analyzer* measures the rms value.

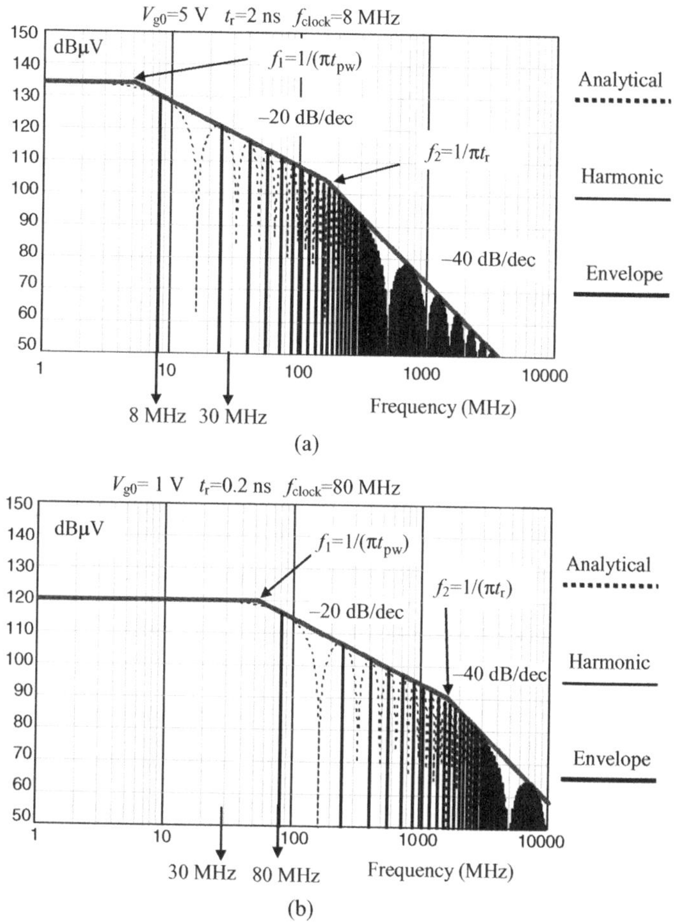

Figure 9.2 Examples of the clock spectrum for two frequencies: (a) $f_{\text{clock}} = 8$ MHz; (b) $f_{\text{clock}} = 80$ MHz

Note also that:

- Both odd and even harmonics are present.
- The first harmonic shown is the fourth at 32 MHz.

The good agreement between computed envelope and measurement is very important for radiated emission predictions. In fact, once the radiating mechanism is known, the emission profile can be estimated by using the envelope of the radiating source instead of a time-consuming computation harmonic by harmonic. This approach enables a satisfactory accuracy to be achieved, as will be shown by several examples in the following sections of this chapter. Harmonics computation will be used in special cases only.

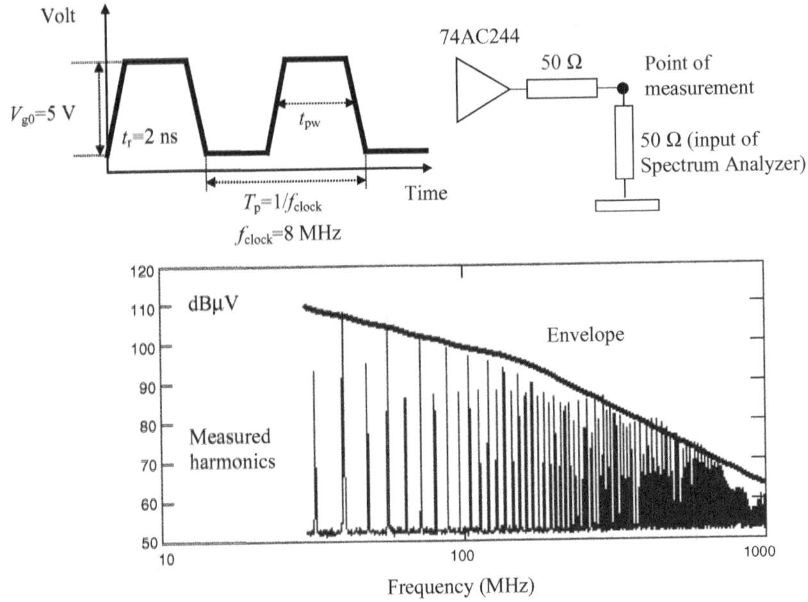

Figure 9.3 Measured and computed clock spectra

Another important point to discuss is how many of the harmonics are needed to reconstruct the trapezoidal waveform in the time domain with a reasonable approximation. Looking at the spectral bound of the signal, it can be observed that, above the second breakpoint, $f_2 = 1/(\pi t_r)$, the harmonics drop off at a rate of -40 dB/dec. Hence, if harmonics up to this frequency are used, the probability that the time waveform is reconstructed without significant distortion is quite high. To be conservative, a limit of $1/t_r$ (i.e. approximately 3 times the second breakpoint) is suggested. Hence, the *bandwidth* (BW) of a digital clock signal is chosen as

$$\mathrm{BW} = \frac{1}{t_r} \tag{9.7}$$

If the rise time is expressed in ns, the bandwidth is in GHz. For example, a signal having a rise time of 1 ns would have a bandwidth of 1 GHz. This is an acceptable criterion, as shown by Paul [1].

9.1.2 Spectrum of Typical Noises

In the previous section a Fourier series of clock periodic signals was presented, and a representation of the associated line spectra amplitudes by the envelope was proposed. A single pulse occurring only once in time is a non-periodic signal. The simplest way to approach the problem of non-periodic signals is to consider the non-periodic signal as a periodic one and to let the period go to infinity. As the interest here is focused only on periodic noises that can cause a high level of emission, a rigorous theory on the *frequency-domain* representation of

Table 9.1 Time-domain representation of periodic noise sources with period T_p (the analytical expression holds in the first period of the signal), and the corresponding magnitude spectrum envelope in the frequency domain

Time-domain noise of period T_p	Magnitude spectrum envelope in frequency domain
Triangular $V_t(t) = \begin{cases} 0 & \text{for } \lvert t\rvert > t_r \\ V_{g0}\left(1 - \frac{\lvert t\rvert}{t_r}\right) & \text{for } \lvert t\rvert < t_r \end{cases}$	$V_{ea}(f) = 2\frac{V_{g0}t_r}{T_p}\frac{\sin(\pi f t_r)^2}{(\pi f t_r)^2}$
Gaussian $V_{gauss}(t) = V_{g0}\,e^{-2(t/t_r)^2}$	$V_{egauss}(f) = 2\frac{V_{g0}t_r}{T_p}\sqrt{2\pi}\,e^{-0.5(\pi t_r f)^2}$
Damped oscillation $V_{OSC}(t) = \begin{cases} 0 & \text{for } t < 0 \\ V_{g0}\,e^{-\alpha t}\sin(\beta t) & \text{for } t \geq 0 \end{cases}$	$V_{eOSC}(f) = 2\frac{V_{g0}}{T_p}\frac{\beta}{(\alpha^2+\beta^2) - \omega(f)^2 + j4\alpha\pi f}$

Note: The factor 2 in the expression in the right-hand column makes it possible to plot the spectrum of the signal in the range of frequencies from zero up to the higher frequency of interest. Line spectra amplitudes: $f = nf_0$, where $f_0 = 1/T_p$ and $n = 1,2,\ldots$, etc.

non-periodic signals by the Fourier transform and its relationship with the Fourier series can be found in the work by Paul [1].

Periodic *time-domain* waveforms and the corresponding magnitude spectrum envelope of typical noises occurring in a digital system are reported in Table 9.1 and shown in Figure 9.4. As discussed in *Chapter 8*, triangular or Gaussian waveforms could represent the switching current of a digital device that produces the ΔI-*noise* voltage on power distribution. Dumped oscillation could represent the ringing waveform produced by resonant phenomena in a PCB, or the overshoot and undershoot on a clock signal waveform caused by reflections on the signal line as the signal level changes from one logic level to another. The waveforms in Figure 9.4a are normalized to a unit amplitude.

When the noise waveform is periodic with period T_p and maximum amplitude V_{g0}, the envelope of the line spectrum has the expressions reported in the right-hand column of Table 9.1. Periodic noises having triangular, Gaussian, and damping oscillation waveforms are represented in the frequency domain by the spectrum envelopes shown in Figure 9.4b, where a dB-log scale is adopted. It is interesting to note that:

- The damped oscillation envelope $V_{eosc,dB\mu V}(f)$, has a peak at the frequency corresponding to the oscillations. This means that, when it sums to a trapezoidal waveform representing a digital signal current along a mismatched line, a higher level of emission occurs than that produced by the same trapezoidal waveform representing a signal current in a matched line.
- $V_{e,dB\mu V}(f)$ displayed in Figure 9.4b is the envelope of the spectrum of the triangular waveform calculated from the trapezoidal clock equations (9.4) and (9.5) when $t_{pw} = t_r$ is assumed. This means that Equation (9.4) also provide the upper bound for a triangular waveform.

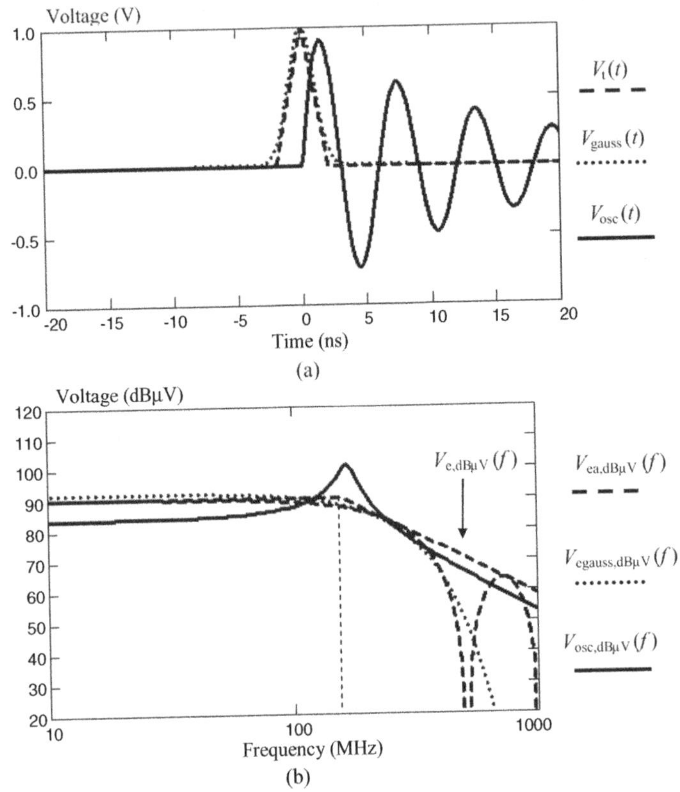

Figure 9.4 Typical noise sources with triangular, Gaussian, and damped oscillation waveforms: (a) time domain; (b) envelope of the frequency-domain magnitude spectrum in the case of periodic noise (data: $t_r = 2$ ns, $\alpha = 0.0693 \times 10^9$, $\beta = \pi/3 \times 10^9$, $V_{g0} = 1$ V, $T_p = 1/(8$ MHz))

- The Gaussian envelope $V_{egauss,dB\mu V}(f)$ has lower harmonic amplitudes than triangular noise after the second breakpoint frequency. This waveform gives more realistic results when used for ΔI-*noise* prediction, as the real waveform does not change so sharply as the triangular waveform.

9.2 The Radiated Emission Problem

In the previous section, possible sources of radiated emission were characterized in the *frequency domain*, starting from their *time-domain* representation. This approach is advantageous, as the currents generated on structures such as PCBs with attached cables can be calculated directly in the *frequency domain*. Once the currents are known, the radiated fields can be easily computed. Denoting by $\hat{I}(f)$ the spectrum of the current $I(t)$ at a generic point of the structure, the electric field $\hat{E}(f)$ at a certain distance and for a certain direction is given by

$$\hat{E}(f) = \hat{T}_E(f)\,\hat{I}(f) \tag{9.8}$$

where $\hat{T}_E(f)$ is the electric field transfer function. By definition, the transfer function $\hat{T}_E(f)$ is the radiated electric field caused by a current of unit amplitude and having a flat spectrum in the range of frequencies of interest. In the same manner, the radiated magnetic field $\hat{H}(f)$ is given by

$$\hat{H}(f) = \hat{T}_H(f)\hat{I}(f) \tag{9.9}$$

where $\hat{T}_H(f)$ is the magnetic field transfer function. When the dB representation for current and transfer functions is used, the products in Equations (9.8) and (9.9) become a sum.

The main task is therefore to find tools suitable for computing the current $\hat{I}(f)$ at each location of the radiating structure, starting from the spectrum of the voltage or current source exciting the structure, and the transfer functions $\hat{T}_E(f)$ and $\hat{T}_H(f)$ characterizing the radiation mechanism. The radiating structure can be represented as a grid of wires and/or as small surfaces of electrically short maximum dimension (size much smaller than the minimum wavelength associated with the maximum frequency of interest). At any observation point, the radiated field is given by the sum of the fields radiated by the different current contributions. In general, this task is not simple to accomplish, as the radiation at high frequencies of structures whose size is comparable with the wavelength is highly complex and directive. Full-wave codes can perform this type of calculation, taking into account the metallic and dielectric parts of the radiating structure. The standards for radiated emission measurements enable this difficulty to be overcome. In fact, to ensure repeatability of radiated emission measurements, the international standards for commercial equipment, such as FCC Part 15 and CISPR 22 (EN55022), define the minimum frequency and the distance of the antenna from the *Equipment Under Test* (EUT), with the purpose that the radiated field at the antenna location is in plane wave condition for most of the frequency range. In plane wave conditions, the electric and magnetic fields are perpendicular to the direction of propagation and orthogonal to each other. Moreover, the ratio between the amplitudes of the electric and the magnetic fields is constant (i.e $E/H = 377\ \Omega$) and depends on the inverse of the distance from the source only. When this occurs, the antenna is in the *Far-Field* (FF) region. This means that the electric field only needs to be computed and that the transfer function $\hat{T}_E(f)$ is significantly simplified. For FCC the measurement distance is 3 m for Class B products and 10 m for Class A products, while for CISPR 22 it is 10 m for both classes. Actually, a wavelength $\lambda = 10$ m corresponds to the lower frequency $f_{\min} = 30$ MHz, whereas at a frequency of 1 GHz the wavelength is $\lambda = 30$ cm. The antenna is therefore in the near field of the EUT for the lower-frequency range, and in the far field for the higher-frequency range of the regulatory limits. Anyway, as a first approximation, prediction methods based on the far-field assumption will be described in the following sections. Comparison between computed and experimental results will confirm the validity of this approach.

Full-wave codes are expensive and time consuming. Fortunately, many radiating structures of interest in digital systems can be modeled as simple wire antennas or as transmission lines. In this manner, the radiated fields are computed by a procedure based on the application of closed-form expressions, or, alternatively, by the two-step procedure in which the current distribution on the radiating structure is first obtained by SPICE-like circuit simulators, and then closed-form expressions coming from small-dipole theory are applied to derive the electric radiated field. These two approaches, as well as the utilization of full-wave codes to investigate the radiation of PCBs, cables, and apertures in shielded equipment, will be highlighted and discussed in this chapter and partially in *Chapter* 10.

It should be pointed out that knowledge of the signal current of the circuit, known as the *differential-mode* current, is not sufficient for an accurate prediction of the radiated emission. To this end, it is very important to have methods for calculating the current distribution involving the circuit and the environment. This is known as *common-mode* current or *antenna current*. In many cases the *common-mode* current is responsible for high levels of emission, although its value is much less than that of the *differential-mode* current (i.e. microamperes versus milliamperes). Important tasks of this section are to provide methods suitable for recognizing and calculating *common-mode* currents in structures such as PCBs with attached cables, and to give design rules to mitigate the *common-mode* current effects. Let us start with the computation of the radiation produced by a structure that can be represented by wires.

9.2.1 *Radiation from a Wire Antenna*

In the far-field region, the field radiated by a wire antenna at a given frequency can be calculated by the following procedure [2]:

1. The radiating structure is divided into electrically short subelements having a length $\Delta\xi$ much shorter than the minimum wavelength λ of interest (see Figure 9.5).
2. The current $\hat{I}(\xi)$ on the element $\Delta\xi$ is calculated by the *Transmission-Line* (TL) theory in the case of traces in PCBs and cables with a reference ground plane, or, when possible, by the simple antenna formulation for a small loop or long dipole.
3. The magnitude of the electric field vector $\hat{E}(r)$ radiated by the generic current $\hat{I}(\xi)$ at a distance r is obtained by using far-field approximation for a small dipole valid for $r > \lambda_0/2\pi$:

$$\hat{E}(r) = \frac{j\eta\ \beta_0(\omega)\hat{I}(\xi)\ \Delta\xi}{4\pi}\frac{e^{-j\beta_0(\omega)r(\xi)}}{r(\xi)} \tag{9.10}$$

where $\eta = 377\ \Omega$ is the far-field wave impedance, $\beta_0(\omega) = 2\pi/\lambda_0(f)$ is the phase constant, $\hat{I}(\xi)$ is the antenna current in ξ, and f is the frequency of $\hat{I}(\xi)$.

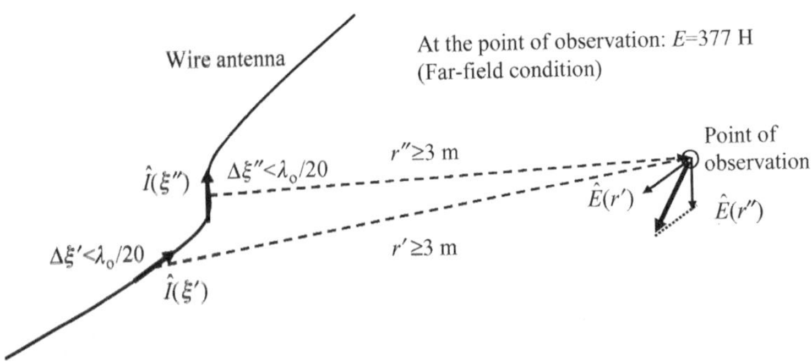

Figure 9.5 Computation of the electric field radiated by a wire antenna

4. Sum as vectors the fields $\vec{\hat{E}}(r)$ due to each current $\hat{I}(\xi)$ at the observation point, and compare with the limits required by the standards.

This procedure will be extensively used in the following sections.

9.2.2 Common- and Differential-Mode Currents and Radiations

Traces in a PCB can be represented as wire antennas. Consider, for example, the case of the PCB shown in Figure 9.6, where a signal is transmitted by two parallel conductors: in wire 1 flows the signal current which returns to the source through wire 2. The metallic plane in Figure 9.6 could be the chassis of the PCB or the floor of a semi-anechoic shielded room where the radiated emission measurements are carried out. A Thévenin equivalent circuit comprising the series connection of the voltage source $\hat{V}_S$ with the source impedance $\hat{Z}_S$ is used to represent the source device. The load is represented by the impedance $\hat{Z}_L$. If the interconnect is assumed to be electrically short, the currents indicated in Figure 9.6 are constant for the full length of the line. As a first approximation, to compute the signal current in the wires while neglecting the line inductance and capacitance, the signal current is given by the simple expression $\hat{I}_{sig} = \hat{V}_S/(\hat{Z}_S + \hat{Z}_L)$. For radiated emission computation, this approach is not appropriate. In fact, a small quantity of current returns as displacement current between the wires and the environment, in this case the metallic plane. This displacement current path is represented herein by the capacitances C_S and C_L. The displacement current assumes the characteristic of a *common-mode* current. In fact, as introduced in *Section 6.2*, *common-mode* currents in the two wires at the same cross-sectional position flow in the same direction and are equal in magnitude and phase. *Common-mode* current is mainly generated by the asymmetric position of the source with respect to the wires [3], and returns to the source by the parasitic capacitances between the circuit and the environment [4]. On the other hand, the signal or functional current I_{sig} assumes the characteristic of *differential-mode* current. In fact, in this case, the currents in the two wires at the same cross-sectional position are equal in magnitude and phase but flow in opposite directions. Note that *common-mode* currents are also present in the absence of a nearby metallic object or when the circuit is considered isolated. In this case the path of the *common-mode* current is represented approximately by a capacitance between the source and the load ends.

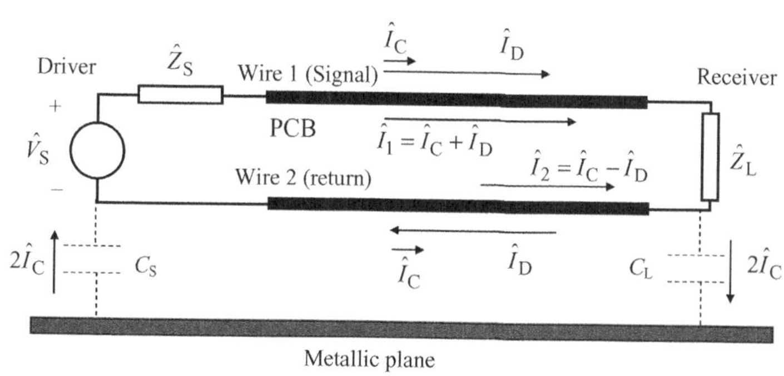

Figure 9.6 Common- and differential-mode currents in an interconnect with a reference plane

After these considerations, the two currents $\hat{I}_1$ and $\hat{I}_2$ flowing in wire 1 and wire 2 respectively can be decomposed into their *common-* and *differential-mode* components, $\hat{I}_C$ and $\hat{I}_D$ as [5, 6]

$$\hat{I}_1 = \hat{I}_C + \hat{I}_D \tag{9.11a}$$

$$\hat{I}_2 = \hat{I}_C - \hat{I}_D \tag{9.11b}$$

Adding and subtracting Equations (9.11) yields

$$\hat{I}_C = \frac{\hat{I}_1 + \hat{I}_2}{2} \tag{9.12a}$$

$$\hat{I}_D = \frac{\hat{I}_1 - \hat{I}_2}{2} \tag{9.12b}$$

Once these two currents are known, they can be used to compute the radiated fields. This decomposition is very useful because from Figure 9.6 it can be seen that the *differential-mode* current is associated with the loop defined by source-wire 1–load-wire 2. This loop in practical situations can be much smaller than the loop associated with the *common-mode* current flowing in the path wires-capacitance-metallic plane-capacitance. In other words, the dangerous effect of the *common-mode* current can be expressed as follows:

- *differential-mode* emission is the result of the current $\hat{I}_D$ (in the mA range) flowing in wires 1 and 2. The signal-wire current produces a field that tends to cancel the field produced by the return-wire current.
- *common-mode* emission is the result of the currents $\hat{I}_C$ (in the μA range) which produce the same or a higher level of radiated emission because they flow in a common direction and produce fields that tend to add.

In order to see why this occurs, consider a pair of isolated parallel wires or wires with their reference plane such a distance away that the influence of the metallic plane can be neglected. Let l be the length of the wires, and s their separation. Consider the radiated electric field in the plane of the wires, along the line passing through their central point and placed at a distance d from the line (see Table D.1 in *Appendix D*). If the wires are electrically short at the frequencies of interest, $l \ll \lambda_0$, then each wire can be approximated as a single Hertzian dipole and the resulting fields given by Equation (9.10) are superimposed. By this procedure, the maximum radiated field magnitudes in V/m at a distance d in the far-field region are given for *differential* and *common modes* by using Equation (9.10) [5–7]

$$E_{D_{max}} = 1.316 \times 10^{-14} \left|\hat{I}_D\right| f^2 l \frac{s}{d} \tag{9.13a}$$

$$E_{C_{max}} = 1.257 \times 10^{-6} \left|\hat{I}_C\right| f l \frac{1}{d} \tag{9.13b}$$

From Equations (9.13), the following observations can be made:

- Maximum *differential-mode* emission E_{Dmax} depends on the loop area $A = ls$ and on f^2. This dependence on frequency means that the transfer function rises with a slope of +40 dB/dec in a log-dB scale (i.e. increases at a rate of +40 dB/dec). By multiplying this transfer function by the input signal spectrum developed for the trapezoidal waveform in *Section 9.1*, the spectrum of the *differential-mode* radiated field is obtained. In a log-dB scale, this is equivalent to adding the transfer function of the *differential-mode* electric field (9.13a) to the envelope representation of the trapezoidal waveform (9.4). The resulting waveform is a plot where the field increases at a rate of +40 dB/dec up to the breakpoint frequency $f_1 = 1/(\pi t_{\text{pw}})$, then increases at a rate of +20 dB/dec in the frequency interval between f_1 and $f_2 = 1/(\pi t_{\text{r}})$, and finally remains constant at its maximum value above the frequency f_2.
- Maximum *common-mode* emission E_{Cmax} depends on the length l and the frequency f according to Equation (9.13b), and not on the loop area A. This dependence on frequency means that the transfer function increases at a rate of +20 dB/dec. The spectrum of the *common-mode* radiated electric field is obtained by multiplying this transfer function and the input signal spectrum developed for the trapezoidal waveform in *Section 9.1*. In a log-dB scale, this is equivalent to adding the transfer function of the *common-mode* electric field (9.13b) to the envelope representation of the trapezoidal waveform (9.4). The resulting waveform is a plot where the field increases at a rate of +20 dB/dec up to the breakpoint frequency $f_1 = 1/(\pi t_{\text{pw}})$, then remains constant at its maximum value in the frequency interval between f_1 and $f_2 = 1/(\pi t_{\text{r}})$, and finally decreases with a slope of −20 dB/dec above f_2.

These observations highlight the following important points:

1. *Differential-mode* emission dominates at high frequencies. It is generally produced by the circuits on the PCBs (ICs and traces), as their dimensions are comparable with the wavelength associated with frequencies above about 300 MHz.
2. *Common-mode* emission dominates at low frequencies. It is generally produced by the cables attached to the PCBs, as they have dimensions comparable with the wavelength associated with frequencies below about 300 MHz.
3. *Differential-mode* emission depends on rotation of the two wires, and the maximum value is obtained for points of observation in the same plane of the two wires. To mitigate *differential-mode* emission, the current level, the area between the wires, and the line length need to be reduced.
4. *Common-mode* emission does not depend on rotation of the two wires. To mitigate *common-mode* emission, the current level and the line length should be reduced.

Points 1 and 2 are also very important for more complicated structures, such as digital systems, and can be used at the diagnostic stage. In *Section 9.8*, a method for distinguishing the types of emission will be discussed in order to choose appropriate fixes to reduce radiated emission. Two electrically short parallel wires are an elementary source of emission. The computation of currents and transfer functions for actual structures occurring in digital systems will be discussed in the next sections.

9.2.3 Emission Due to Line Asymmetrical Feed

As anticipated in *Section 9.2.2*, in the case of a circuit consisting of two parallel wires of arbitrary length l, the *common-mode* current $\hat{I}_{\text{C}}$ is generated by an asymmetrical position of

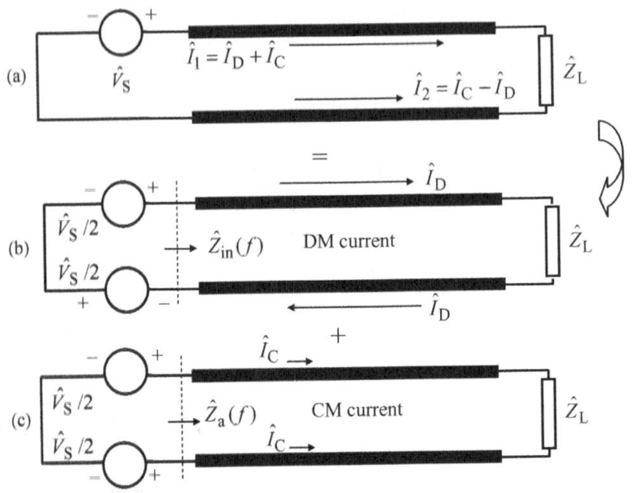

Figure 9.7 Emission due to asymmetrical feed of a line: (a) line structure; (b) differential-mode (DM) currents and common-mode (CM) current

the source, as indicated in Figure 9.7a [8]. In this case the radiated field can be computed as the superimposition of the fields associated with the two structures shown in Figures 9.7b and 9.7c respectively. The first one in Figure 9.7b has two voltage sources positioned symmetrically, equal in value $\hat{V}_S/2$ and opposite in sign, producing *differential-mode* current only. The other in Figure 9.7c has two voltage sources, positioned symmetrically, equal in value $\hat{V}_S/2$ and in sign, producing *common-mode* current only. It can be shown that:

- *Differential-mode* current along the line can be computed starting from the value of the current at the line input, given by $\hat{I}_{D,in} = \hat{V}_S/\hat{Z}_{in}$, where $\hat{Z}_{in}$ is the input impedance of a transmission line of length l terminated with the load $\hat{Z}_L$. The distribution of *differential-mode* current on the structure can be computed exactly by closed-form transmission-line expressions.
- *Common-mode* current along the line can be computed starting from the value of the current at the input of the line, given by $\hat{I}_{C,in} = \hat{V}_S/(2\hat{Z}_a)$, where $\hat{Z}_a$ is the input impedance of a thin-wire antenna of length l. The distribution of *common-mode* current can be computed approximately by closed-form expressions available for simple antennas.

9.2.4 Differential-Mode Current and Radiated Emission of a Transmission Line

Consider the circuit of Figure 9.8, where a *differential-mode* current only is considered. This equivalent circuit can model two different configurations: two parallel wires separated by a distance $2h$, sourced by a voltage generator of value $\hat{V}_S$ and loaded with an impedance $\hat{Z}_L$; one wire at a distance h above a ground plane, sourced by a voltage generator of value $\hat{V}_S/2$ and loaded with an impedance $\hat{Z}_L/2$. This second configuration has the equivalent circuit

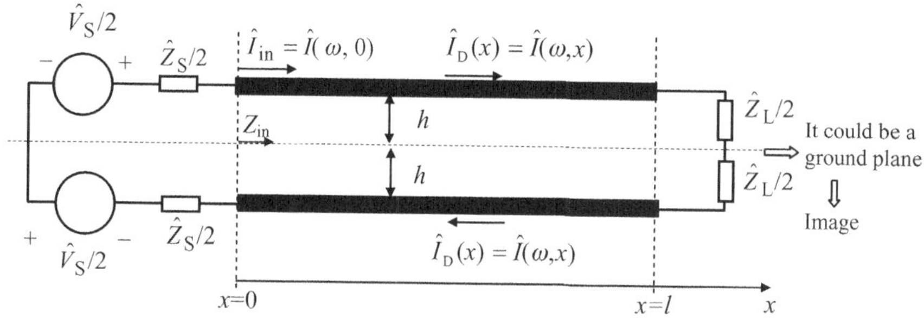

Figure 9.8 Differential-mode (DM) current calculation by the transmission-line (TL) model

of Figure 9.8 applying the method of image [2]. Both the configurations produce the same *differential-mode* current.

The current $\hat{I}(\omega, x)$ along the line can be obtained by using the transmission-line theory for lossless uniform lines [7]:

$$\hat{I}(\omega, x) = \hat{V}_S(\omega)\frac{Z_0 \cos[\beta_l(\omega)(l-x)] + j\hat{Z}_L(\omega)\sin[\beta_l(\omega)(l-x)]}{Z_0\left[\hat{Z}_S(\omega)+\hat{Z}_L(\omega)\right]\cos[\beta_l(\omega)l] + j\left[Z_0^2 + \hat{Z}_S(\omega)\hat{Z}_L(\omega)\right]\sin[\beta_l(\omega)l]} \tag{9.14}$$

where $\hat{V}_S(\omega)$ is the spectrum of the source exciting the line, Z_0 is the nominal characteristic impedance of the lossless line, $\beta_l(\omega) = 2\pi/\lambda_l(\omega)$ is the phase constant, $\lambda_l(\omega) = v_l/f$ is the wavelength at the frequency f, $v_l = 1/(\varepsilon_e\varepsilon_0\mu_0)^{-1/2}$ is the velocity of propagation in the line, ε_e is the equivalent relative dielectric constant of the line, with $\varepsilon_e = \varepsilon_r$ for a homogeneous dielectric such as in a stripline structure (see *Appendix B*), and ε_0 and μ_0 are the vacuum dielectric and permeability constants respectively.

The *differential-mode* current at the input of the line (i.e. $x = 0$) can be expressed as

$$\hat{I}_{in}(\omega, 0) = \frac{\hat{V}_S(\omega)}{\hat{Z}_S(\omega) + \hat{Z}_{in}(\omega)} \tag{9.15}$$

where $\hat{Z}_{in}(\omega)$ is the line input impedance given by

$$\hat{Z}_{in}(\omega) = Z_0\frac{\hat{Z}_L(\omega)\cos(\beta_l(\omega)l) + jZ_0\sin(\beta_l(\omega)\mathrm{l})}{Z_0\cos(\beta_l(\omega)\mathrm{l}) + j\hat{Z}_L(\omega)\sin(\beta_l(\omega)\mathrm{l})} \tag{9.16}$$

Equation (9.15) will be compared below with the results of measurements.

Once the *differential-mode* current $\hat{I}_D(x) = \hat{I}(\omega, x)$ is known for each wire, the total radiated field can be easily computed as the superimposition of the fields generated by the two wires when isolated, or as the superimposition of the fields due to the two wires and their image conductors when the PCB is above a metallic floor. The total field at any point is then given by the contribution of the currents associated with each electrically short segment used to decompose the lines, considering distance, amplitude, and phase of each term. Recall that the transfer function for field computation in the near-field region has three terms dependent on $1/r$, $1/r^2$, and $1/r^3$ and not only $1/r$ as occurs with the far-field region [9] (see also *Section 9.8.2*).

When the interest is focused on the far-field region, which is mostly the case for the commercial FCC and CISPR standard requirements, the calculation of the radiated field can be performed by the following simple two-step procedure:

1. The average current $\hat{I}_t(\omega)$ along each wire is calculated as outlined in Table D.1 of *Appendix D* for the long wire case.
2. The total radiated field $\hat{E}_t(\omega)$ is obtained as the contribution of the current $\hat{I}_t(\omega)$ of each wire, as outlined in Table D.2 of *Appendix D*. The contribution of the image current exists with the presence of a metal floor.

9.2.5 Common-Mode Current and Radiated Emission of a Transmission Line

While the *differential-mode* current distribution along the radiating structure can be computed exactly by the transmission-line theory, the *common-mode* current can be determined approximately using antenna theory for simple configurations. In the proposed approach, the *common-mode* circuit of Figure 9.7c is modeled by an equivalent dipole of branches with length l and fed by the voltage source $\hat{V}_S/2$ as shown in Figure 9.9. As *common-mode* emission dominates in the low-frequency range, the dipole structure can be modeled by a lumped-element circuit defined by the following parameters [10]:

$$L_d = \frac{\mu_0 l_a}{3\pi}\left[\ln\left(\frac{2l_{dip}}{a}\right) - \frac{11}{6}\right] \tag{9.17a}$$

$$C_d = \frac{\varepsilon_r \varepsilon_0 l_{dip} \pi}{\ln(2l_{dip}/a)} \tag{9.17b}$$

$$R_{rd}(\omega) = \begin{cases} R_r(\omega) & l \le \lambda_0(\omega)/4 \\ R_{rHF} & l > \lambda_0(\omega)/4 \end{cases} \tag{9.17c}$$

$$\hat{I}_C(\omega) = \frac{\hat{V}_S(\omega)/2}{R_{rd}(\omega) + j\left(\omega L_d - \frac{1}{\omega C_d}\right)} \tag{9.17d}$$

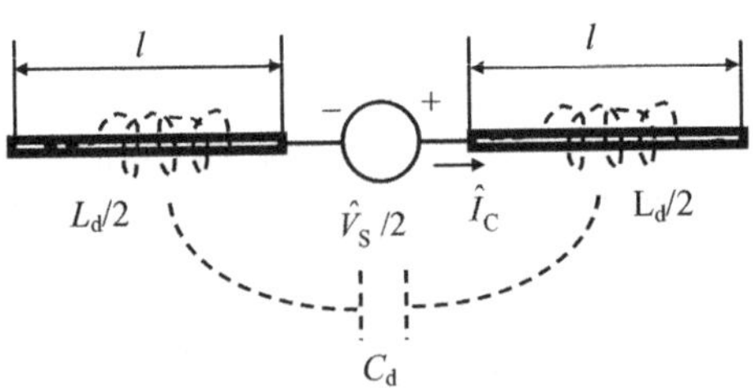

Figure 9.9 Common-mode (CM) current calculation by the dipole model

where $l_{dip} = 2l$ is the total dipole length, a is the wire radius, C_d represents a distributed parasitic capacitance between the two branches of the dipole that provides a return path for the *common-mode* or antenna current, $R_r(\omega) = 80\pi^2(l_{dip}/\lambda_0(\omega))^2$ is the radiation resistance with a linear distribution of antenna current for low frequencies, and $R_{r,HF} = 73\ \Omega$ is the high-frequency radiation resistance for $l = \lambda_0/4$.

Once the *common-mode* current $\hat{I}_C(\omega)$ is known for each wire, the radiated field can be computed following the same procedure used for *differential-mode* currents, with the difference that the currents in the two wires are equal in value and direction. Detailed formulae for calculating the radiated field can be found in Table D.2 of *Appendix D*. The contribution of image currents exists with a metal floor only.

Example 9.2: Calculations and Measurements of Emission Produced by P-Test Boards
Two test boards were built to investigate *differential-* and *common-mode* radiation mechanisms as described in previous sections. Board 1 included identical parallel wires over a wooden support, and was terminated with a 100 Ω load resistor, as shown in Figure 9.10a. Board 2 included one wire above a ground plane separated by a wooden support. The geometrical dimensions, source, and load impedance of board 2 were chosen, as shown in Figure 9.10b, to have the same signal current. The source for both boards was an 8 MHz shielded oscillator driving a 74AC244 digital device with a 50 Ω output resistance. The voltage source having a trapezoidal waveform with $V_{g0} = 5$ V, $t_r = 2$ ns, and $T_p = 125$ ns was represented in the frequency domain by the magnitude spectrum envelope described in *Section 9.1*. The size of the metallic box, where the active circuits were located to measure radiated emission from the lines only, was 7 cm × 14.5 cm × 4.5 cm. A battery power supply was chosen to avoid emission from cords of public AC/DC power supply. In principle, the two boards should have the same emission, as they have the same signal current, but it will be shown that this is not the case.

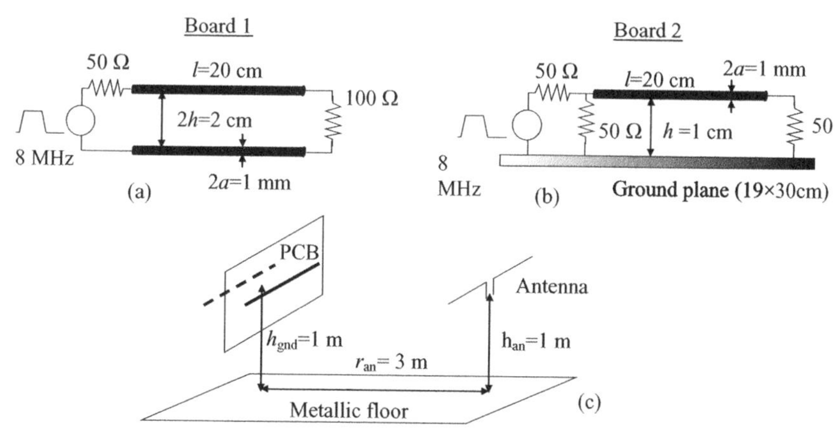

Figure 9.10 Schematic representation of test boards used to investigate differential- and common-mode emission: (a) two parallel wires; (b) one wire above a ground plane; (c) set-up for radiated emission measurements

The following measurements were performed in a semi-anechoic-shielded room for 3 m compliance with the standards:

- *common-mode* current at the input of the line for board 1 (i.e. the board with parallel wires) by using a high-frequency current probe wrapping two wires;
- *differential-mode* current at the input of the line for board 1 and board 2 (i.e. the board with a ground plane) by using a high-frequency current probe wrapping one wire;
- radiated *E*-fields with the boards oriented as shown in Figure 9.10c and the antenna in horizontal polarization.

Measured values were compared with the results of simulations based on the approaches outlined in this section, and performed according to the following steps:

- The source was modeled in the frequency domain by the magnitude spectrum envelope defined by Equations (9.4) and (9.5).
- The *differential-* and *common-mode* current distributions along the radiating structure were computed by Equations (9.14) and (9.17) respectively.
- The *differential-* and *common-mode* *E*-fields were calculated using the expressions reported in Tables D.1 and D.2 of *Appendix D*.

The comparison between measurements and simulations is shown in Figure 9.11. Note that the magnitude spectrum envelopes predicted by the simulations were obtained considering the

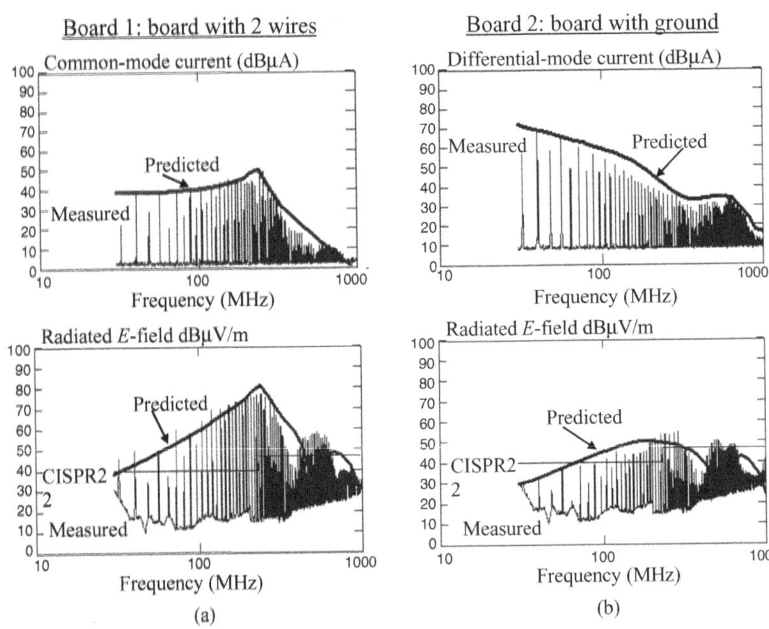

Figure 9.11 Measured and predicted values for the P-test boards: (a) board with two wires (CM prediction); b) board with a ground plane (DM prediction)

common-mode current only for board 1, and the *differential-mode* current only for board 2. The following observations can be made from the results of Figure 9.11:

- The radiated electric field from the board with two wires is much higher than that from the board with a ground plane. This is due to the fact that *common-mode* radiation dominates in the board with two wires because the structure is asymmetric.
- The radiated electric field from the board with a ground plane is quasi-totally due to *differential-mode* current and there is a reduction of emission of up to about 30 dB! This occurs because the ground plane makes the structure symmetric on account of the image theory. Some spurious *common-mode* emissions are present at high frequencies because the construction of the board is not perfect: a finite ground plane, the shielded box for the circuits is not ideally connected to the ground plane of the PCB, etc.
- Radiated fields are compared with the CISPR 22 Class B limit at 3 m: 40 dBμV/m up to 230 MHz, and 47 dBμV/m above. Note that *common-mode* emission is about 40 dB above the limit at 200 MHz, although the circuit is small and simple!

Taking these results into account, the benefit given by the ground plane used as the return path for the signal currents in order to avoid *common-mode* radiated emission is evident.

9.2.6 Image Plane

Another method for drastically reducing *common-mode* radiated emission for PCBs such as the one shown in Figure 9.10a is to place a metallic plane beneath the PCB so that each conductor is parallel to the reference plane, and the distance between each conductor and the plane is the same and indicated here as h_{gnd}. This fix was experimentally verified by German *et al.* [8]. The strong reduction is caused by the image of the dipole current that needs to be considered. The image current flows in the opposite direction to the original PCB current and produces fields that tend to cancel the fields from the original wires (see Table D.2 of *Appendix D*). This also occurs for the original *differential-mode* currents, but its effect is less significant.

When the PCB has an attached cable, represented in our investigation as a long wire connected to the PCB ground-return wire at the point where the wire is connected to the resistor, a different mechanism of *common-mode* emission arises. If the wire is stretched horizontally, it appears as an extension of the ground-return wire. Signal and ground-return wires or traces are usually placed close together to minimize the loop inductance. However, the inductance of each wire is non-zero. If the wires are electrically short, or in other words if they are shorter than the wavelength of interest, this distributed inductance can be replaced with an equivalent lumped inductance for each wire. When the source voltage changes, the resulting change in current causes a voltage drop across the inductance of the ground-return wire:

$$\hat{V}_{\mathrm{gnd}}(\omega) = \mathrm{j}\omega L_{\mathrm{gnd}} \hat{I}_{\mathrm{sig}}(\omega) \tag{9.18}$$

This voltage can drive the cable as a non-symmetric dipole or, if there is another cable attached to the source end of the ground-return wire, as a symmetric dipole antenna producing a *common-mode* current along the cable. The current $\hat{I}_{\mathrm{sig}}(\omega)$ can be easily computed by the equivalent circuit of Figure 9.8. The inductance L_{gnd} is the effective inductance associated with the ground-return wire. In the case of a PCB with a ground plane, it can be derived by the approach proposed in *Appendix E* which is based on the representation of the ground plane by wires and the application of the partial inductance concept introduced in *Section 3.2*. Since in the two wires of Figure 9.8 the same current $\hat{I}_{sig}(\omega)$ flows but in the opposite direction (for this computation the *common-mode* current contribution can be neglected), the effective inductance can be computed as

$$L_{gnd} = L_p - M_p \tag{9.19}$$

where L_p is the self partial inductance of the two wires which does not change when the separation between the wires changes, and M_p is the mutual partial inductance between the two wires which increases as the distance between the wires s is reduced.

Starting from the expressions for L_p and M_p, which are well known in the literature and reported in *Appendix A*, the effective inductance L_{gnd} has the expression given in Table A.1 for two parallel wires with opposite currents, or the expression given in Table A.2 for parallel busbars or traces with opposite currents. Reducing the separation between the two conductors, L_{gnd} decreases as well as the noise $\hat{V}_{gnd}(\omega)$, and the *common-mode* current along the cables decreases.

The presence of an image plane just beneath the PCB at distance h_{gnd} can result in another reduction in the effective inductance associated with the ground-return wire [8]. In fact, in Equation (9.19), in addition to the mutual partial inductance M_p between the signal wire and the ground-return wire, the mutual partial inductance contribution of the two image conductors must be included. The expression to be used for these partial inductance computations is given in Table A.1 of *Appendix A*. However, in the presence of an image plane, the L_{gnd} is only slightly less than the value without an image plane. This means that this lowering of L_{gnd} is not as effective for reducing *common-mode* currents along the cable as connecting the return conductor of the PCB to the image plane by a very low impedance [8]. To be effective, the connection must have an impedance much lower than the series impedance of the voltage source that drives the cable as an antenna. In fact, the equivalent circuit to be used to compute the *common-mode* current is the voltage source with its series impedance having in parallel both the shunt impedance of the connection and the input impedance of the cable varying with the frequency. Then, the current from the source has two possible paths. One path takes the current out onto the cable, resulting in radiation. The other path couples current through the image plane, bypassing the cable. Sources on PCBs tend to have source impedances with magnitudes of the order of 100 Ω. Therefore, when the shunt impedance is much less than 100 Ω, *common-mode* current along the cable will be significantly reduced [11].

In conclusion, an image plane added to a PCB without a ground plane brings two benefits:

1. The image plane dramatically reduces the radiated emissions, even though the plane is not electrically connected to the PCB.
2. The image plane dramatically reduces the emissions from a PCB with an attached cable when the image plane is correctly connected to the PCB.

Care should be taken when using an image plane with PCBs having a ground plane, as will be shown in *Section 9.3*. The radiated emission from a PCB with an attached cable will be investigated in depth in several sections of this chapter and partially in *Chapter* 10. It will be shown that the most effective way to reduce emission from cables is to use a metallic plane as the ground-return conductor, or in other words to use multilayer boards. However, this fix might not be sufficient.

9.3 Emission from Traces

In the previous section, the great advantage offered by a ground plane in a PCB for reducing drastically the radiated emission was shown. In this section, the performance offered by typical trace structures used in multilayer PCBs such as microstrips and striplines will be investigated. In the past, such investigation was performed by the transmission-line (TL) method [12] and by numerical methods such as the method of moment (MOM) [13] or the finite element method (FEM) [14]. The trace position on the board [15] and its path [16] were also investigated. Other authors focused their studies on providing design rules to comply with the radiated emission standard [17, 18]. In this section, measurements performed on some test boards will be described in order to verify the validity of simulation procedures based on the TL method and some simple antenna models. In addition, a discussion on the resonance phenomena occurring in PCBs when an image plane is present is also provided.

9.3.1 Antenna Models for Calculating the Radiation of Microstrip and Stripline Structures

A trace on a PCB with a ground plane may have the structures shown in Figure 9.12a:

- a conductor above a return metallic plane (microstrip);
- a conductor between two non-connected metallic planes (ground + image planes);
- a conductor between two connected metallic planes (stripline).

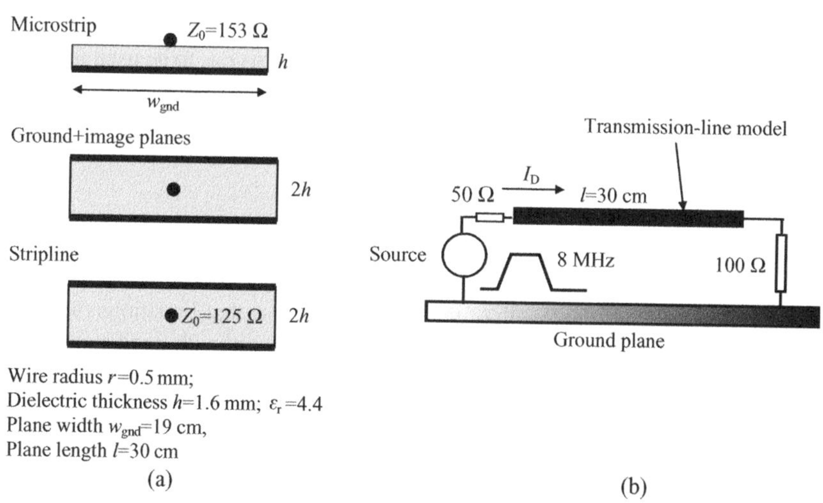

Figure 9.12 S-test board: (a) schematic representation of microstrip, ground + image planes, and stripline structures; (b) equivalent circuit

As previously discussed in *Section 9.2.3*, the electric field radiated from PCBs can be obtained as the superimposition of *differential-* and *common-mode* radiations (i.e. $\hat{E} = \hat{E}_D + \hat{E}_C$). In principle, *common-mode* current cannot flow on a PCB with an infinite ground-return plane, as the plane acts like a mirror to generate another circuit that makes the structure perfectly balanced. However, the finite ground-return plane of an actual PCB produces an imperfect image that causes the ground-return plane itself to radiate as a dipole. The trace above a finite ground plane produces a voltage drop on the ground plane directly under the trace. This voltage drop, denoted by $\hat{V}_{gnd}$, is related to the effective inductance L_{gnd} associated with the ground plane by the relation $\hat{V}_{gnd}(\omega) = j\omega L_{gnd}\hat{I}_{sig}(\omega)$, where $\hat{I}_{sig}(\omega)$ is the signal or functional current which can be derived by the equivalent circuit of Figure 9.8 based on the TL model. Recall that this concept concerning the image plane and a PCB without a ground plane was introduced in *Section 9.2.6*. Actually, this voltage drop is distributed along the board as shown in Figure 9.13a for a microstrip structure. The inductance L_{gnd} for a finite plane is usually an order of magnitude less than the inductance associated with a trace, i.e. some nH or less. It can be calculated by the approximate expression provided by Table A.2 of *Appendix A* and given here for convenience:

$$L_{gnd} = \frac{\mu_0}{2\pi} l \ln\left(\frac{\pi h}{w_{gnd}} + 1\right) \tag{9.20}$$

where w_{gnd} is the width of the ground plane, and l is the length of the line.

The total field $\hat{E}$ radiated from a microstrip PCB is defined by the superimposition of the following two radiation mechanisms, according to the scheme shown in Figure 9.13a:

- *Differential-mode* emission (i.e. $\hat{E}_D$) produced by the signal or *differential-mode* current $\hat{I}_{sig}(\omega)$ which can be calculated by the equivalent circuit model of Figure 9.8 based on the TL model.
- *Common-mode* emission (i.e. $\hat{E}_C$) produced by the *common-mode* current which can be calculated by the equivalent circuit of Figure 9.9 based on the dipole model fed by the voltage source $\hat{V}_{gnd}$. The equivalent dipole length l_{dip} and radius r_{dip} to be used in the dipole model for the microstrip structure are given by [20]

$$l_{dip} = \sqrt{\varepsilon_r}\frac{l}{2} \tag{9.21a}$$

$$r_{dip} = w_{gnd}/4 \tag{9.21b}$$

where ε_r is the relative dielectric constant, and l is the length of the PCB ground plane which, in this experiment, matches the length of the signal wire. Note that Equations (9.21) enable the fields both in the dielectric and in air to be taken into account with good approximation.

Usually, without a cable attached to the PCB and with the trace far away from the edges of the PCB or not crossing a gap in the plane (the gap effect will be discussed in *Section 11.2*), *differential-mode* emission is greater than the *common-mode* emission which can be neglected. When a trace approaches one edge of the PCB, the ground inductance L_{gnd} increases and the *common-mode* emission could become significant, especially with attached cables [21].

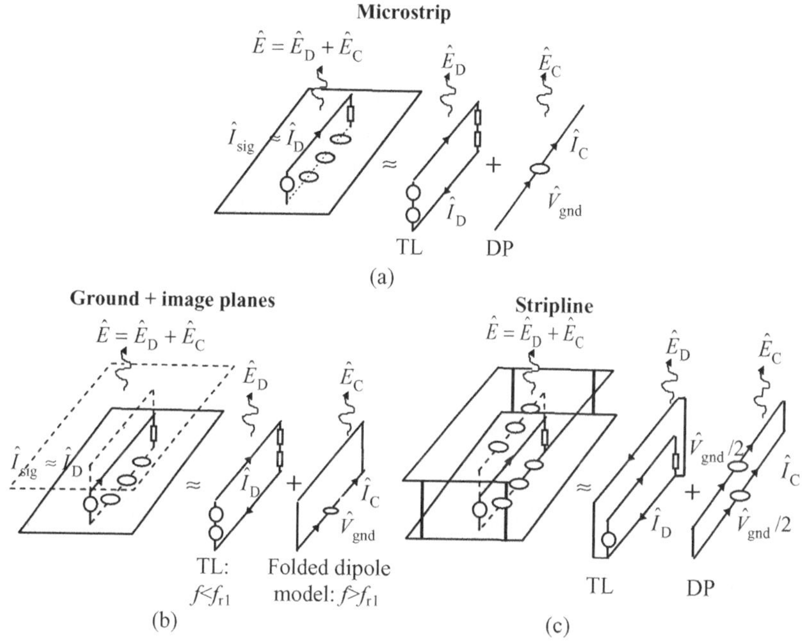

Figure 9.13 Schematic representation of radiated emission calculation as the sum of differential mode and common mode in the case of (a) microstrip, (b) ground + image plane, and (c) stripline configurations

For the stripline structure the radiation profile is strongly conditioned by the grounding of the second plane. In fact, if the second plane is not connected or badly connected to the ground plane, it acts as an image plane producing an emission profile with resonance peaks [20]. The field radiated from a ground + image plane PCB can still be defined as the superimposition of *differential-* and *common-mode* radiations. However, in this case, to describe the *common-mode* emission properly, the folded dipole model [2] is used in place of the dipole model, as schematically shown in Figure 9.13b. The *differential-mode* emission dominates up to the first resonant frequency f_{r1}, when the board has length $\lambda/2$, because below this frequency there is no current flowing in the image plane. On the other hand, for frequency $f > f_{r1}$ the *common-mode* emission dominates in the radiated field calculation. The folded dipole is a thin rectangular loop fed at the center of one wire by the voltage $\hat{V}_{gnd}$. A folded dipole basically operates as an unbalanced transmission line. The *common-mode* current $\hat{I}_C$ of Figure 9.13b, and hence the corresponding *common-mode* radiated field $\hat{E}_C$, can be obtained as the superimposition of the folded dipole (fdip) *differential* and *common modes* as

$$\hat{I}_C = \hat{I}_{fdip,C} + \hat{I}_{fdip,D} \tag{9.22a}$$

$$\hat{E}_C = \hat{E}_{fdip,C} + \hat{E}_{fdip,D} \tag{9.22b}$$

These two modes can be modeled as discussed in detail by Balanis [2] and Caniggia *et al.* [19].

If the ground and image planes are closely tied, as in the case of a stripline PCB, the structure is quasi-symmetric, the signal current is quasi-equally divided between the two planes, and the emission is only due to a small percentage difference between the currents flowing through the two planes. The emission profile can be predicted as shown in Figure 9.13c by the TL model applied to the simplified configuration consisting of a conductor modeling the signal current inserted between two conductors representing the two planes separated by a distance equal to $2h$. The model takes into account that the return current in the ground plane is not spread more than 4–5 times the distance between the strip and one of the ground planes, as will be demonstrated in *Section 10.2*. The signal current $\hat{I}_{\text{sig}}$ along the strip is computed by Equation (9.14). The radiated field is computed by the expressions in Table D.2 of *Appendix D*, considering the values and direction of the three currents: $\hat{I}_{\text{sig}}$ in the strip and $-\hat{I}_{\text{sig}}/2$ in the two planes modeled as two filaments. The *common-mode* emission is very low owing to the cancellation effect caused by the connection of the two planes (see *Figure* 9.13*c*).

By some examples it will be shown that:

- A ground plane is required to avoid strong *common-mode* emission, and, in the case of a microstrip structure, the radiated electric field can be accurately predicted using the TL model.
- An image plane added to a microstrip structure to form a sandwich with the trace in the middle produces the worst emission profile owing to the resonance frequencies generated in the PCB. The emission profile can be predicted by combining the TL and the folded dipole models.
- A stripline structure produces very low emission if the ground and image planes are closely connected.

Example 9.3: Calculations and Measurements of Emission from S-Test Boards

The models described above for the microstrip, ground + image planes, and stripline PCB structures (see Figure 9.12) are validated by comparing the results with measurements.

The three elementary S-test boards shown in Figure 9.12a were built [19]. A round wire conductor was used instead of a trace to make the construction of the boards quick and simple. The PCBs were excited by a source composed of a quartz 8 MHz oscillator driving a 74AC244 digital gate with output resistance $R_S = 50\ \Omega$ and enclosed in a shielded box of 5.5 cm × 5 cm × 2.5 cm size. The source was placed behind the plane of the PCB, and the shielded box was tightly connected to the PCB's plane in order to avoid the spurious *common-mode* emission observed in the similar experiment described in the previous section. The output connector was an SMA type. This provided a trapezoidal pulse train of frequency $f_{\text{clock}} = 8$ MHz having a 50 % duty cycle, a rise and fall time $t_r = t_f = 2$ ns and an amplitude $V_{g0} = 5$ V without load. A battery-powered supply, consisting of a 9 V battery and a 7805 regulator, provided the DC 5 V voltage for the oscillator. It is important to note that connection to the commercial power system was not used to avoid *common-mode* emissions. All the experiments were performed in a semi-anechoic chamber for 3 m tests.

The signal current and the radiated field for the three structures are shown in Figure 9.14. The PCBs and the antenna at a distance of 3 m from the EUT were positioned at 1.1 m from the metallic floor of the semi-anechoic chamber, adopting the orientation shown in Figure 9.14. A good agreement between the measured (harmonics) and calculated (envelope) current values can be observed. It can be noted that, in spite of the signal current values of

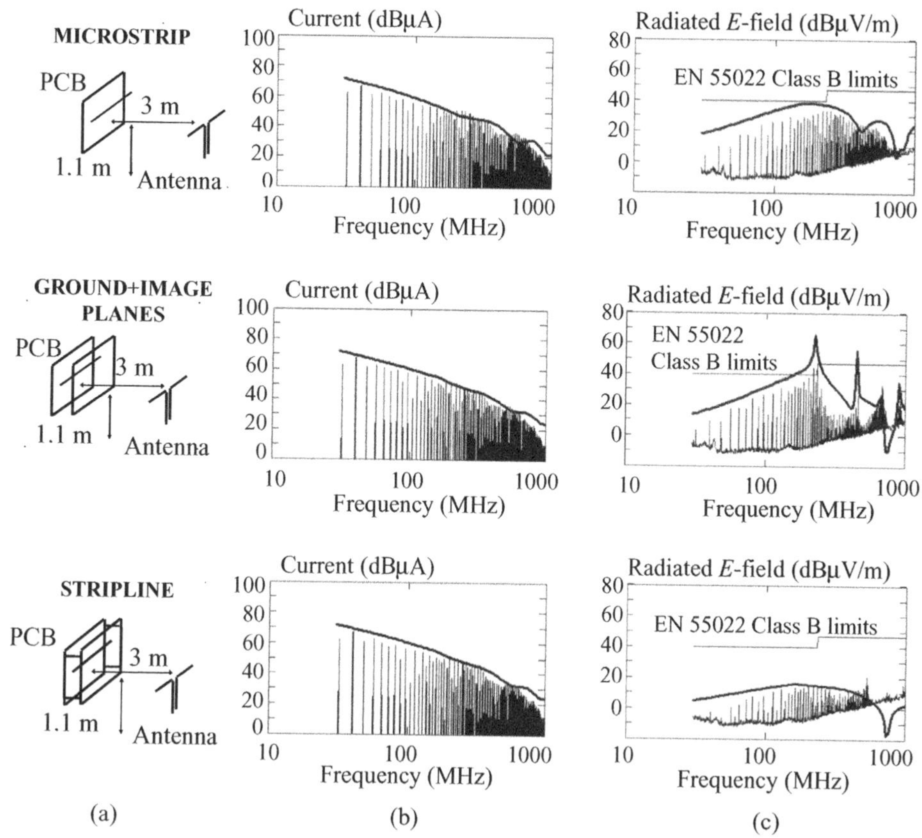

Figure 9.14 Schematic set-ups and comparison between the computed envelope and the measured harmonics for the three structures under investigation: (a) S-test board configurations; (b) signal current; (c) radiated *E*-field

the three board structures being practically equal, they generate very different radiated emissions. From the analysis of the obtained results it is possible to make the following comments:

- For the microstrip structure, the radiated field is due to *differential-mode* current only, and it is below the EN55022 Class B limit for 3 m measurement.
- For the ground + image plane PCB, at a frequency below the first resonance occurring at 240 MHz, the emission profile coincides with that measured for the microstrip structure, as the current does not flow in the image plane. This first resonance and the others occurring at 480, 720, and 960 MHz are related to the dielectric and maximum length of the PCB, and were predicted by using the folded dipole model. At the first resonance the radiated emission is some dB above the limit. This means that the image plane does not give the same benefit in lowering radiation as was shown for PCBs without a ground plane. On the contrary, it makes the emissions worse.

- For the stripline structure, the emission profile is much lower than the limit. In the calculation based on the TL model for three wires, a dissymmetry of about 20 % was adopted.

This experiment provides two important design rules:

1. Do not use floating metallic parts in multilayer PCBs.
2. Use stripline structures to reduce drastically radiated emission, with the two planes well connected by numerous vias.

Example 9.4: Calculation and Measurements of Emission from A-Test Boards

As a further validation of the models previously presented and the derived design rules, other two PCBs were built for experimental measurements.

In the case of microstrip and stripline structures, to obtain meaningful results it is necessary to reproduce as closely as possible the characteristics present in PCBs used in a real system.

Hence, some simple A-test boards were built very close to the structures met in an actual PCB. Each test board consists of five traces (either five microstrips or five striplines) connected to SMA connectors at the near end and terminated on matched loads at the far end. The layout of these test boards is illustrated in Figure 9.15.

The radiated emission measurements were carried out in the same manner as for the S-test boards with round wires of Example 9.3. Comparison between measured and predicted emission profiles by using the TL and antenna models just outlined is shown in Figure 9.16 for the centered traces. Good agreement between measured harmonics and calculated emission profile can be observed even in this new experiment. Note that for both the PCBs the emission profile is less than those measured with the S-test boards, as expected, because the traces are closer to the return ground plane.

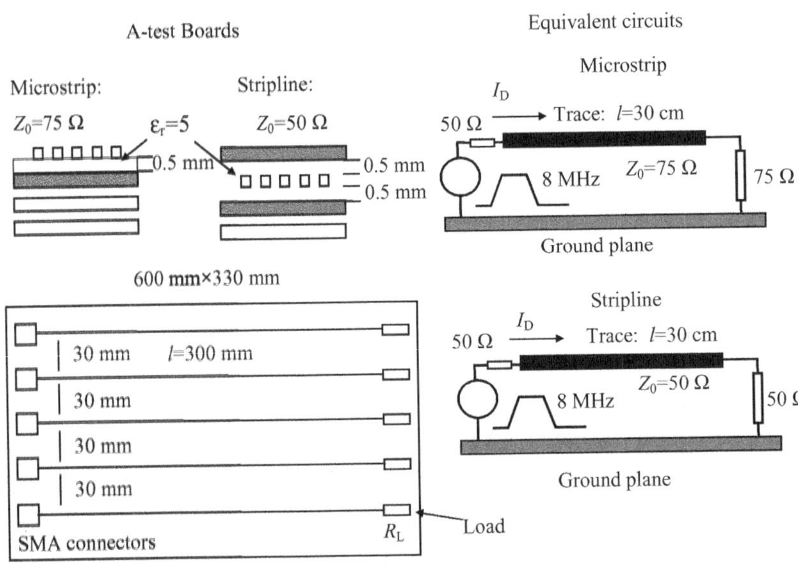

Figure 9.15 Schematic of A-test boards and their equivalent circuit for signaling

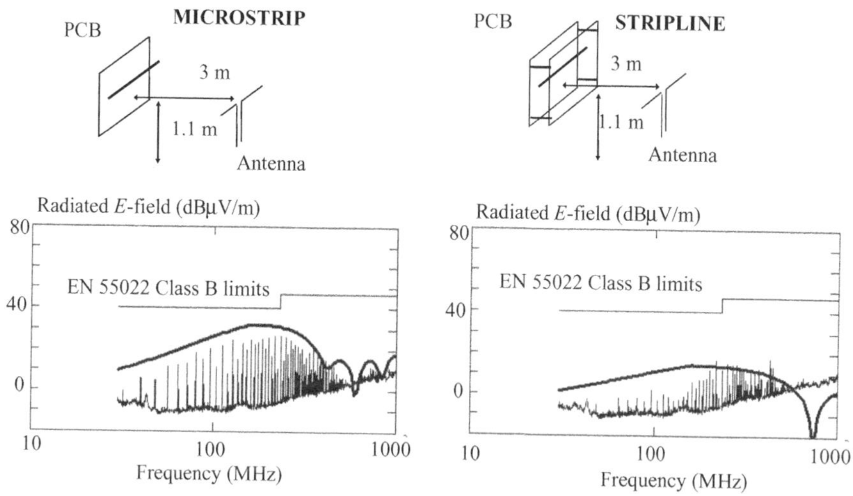

Figure 9.16 Comparison between computed (envelope) and measured (harmonics) radiated E-fields for the two structures under A-test board investigation

Moving the traces towards the edges, no significant changes in emission were observed, as verified by Daijavad *et al.* [20]. This is due to the fact that the return current concentrates beneath the signal trace, as will be shown in *Section 10.2*. A significant increase in the emission occurs only for traces very close to the edges [20, 21]. Again, it is shown that good connections of the ground planes to the circuits is the best rule for avoiding *common-mode* emissions that generate resonance peaks in the emission profile.

9.4 Emission from ICs

In *Section 9.2* and *Section 9.3* the radiation mechanisms of interconnects used in digital PCBs were analyzed, and design rules to mitigate the emission were provided in order to meet the emission limits required by the standards. To this end, it is important to make the traces short, to take care of their location, and to use multilayer boards. However, in a PCB there is another very important source of emission given by the ICs, and this needs to be considered, as it could dominate the other sources. The present section, as well as *Section 9.5* and partially *Section 9.6* and *Section 9.7*, will be devoted to this very important issue. An experiment is outlined where the emission from the ICs is emphasized by using sockets for mounting the devices in the PCB in order better to investigate the emission mechanism. Other useful information can be found in the literature [22–25], and by consulting the six documents of IEC 61967 which provide general information and definitions concerning the measurement of conducted and radiated electromagnetic disturbances from integrated circuits in the range from 150 kHz to 1 GHz. The IEC documents also provide a description of measurement conditions, test equipment, and set-up, as well as the test procedure and required test report contents.

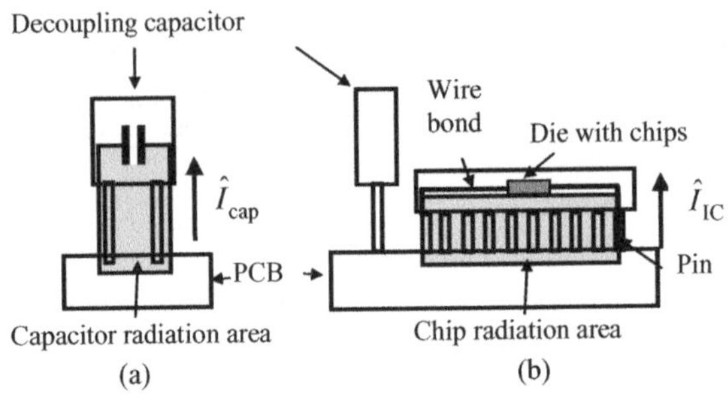

Figure 9.17 Radiation by current loops caused by a component and its associated decoupling capacitor: (a) side view; (b) front view

9.4.1 Radiated Emission Mechanism from Components in a PCB

The radiated emission mechanism from components in a PCB is mainly due to current loops, as illustrated in Figure 9.17, where it is possible to identify two loops (highlighted in gray) defined by:

- the area formed by the decoupling capacitor and the board;
- the area formed by the IC chip and the board.

The first loop (see Figure 9.17a) is crossed by the current $\hat{I}_{cap}$, and the second (see Figure 9.17b) by the switching current $\hat{I}_{IC}$ inside the IC chip. As these loops are electrically short (i.e. their maximum dimension is much smaller than the minimum wavelength of interest, which is 30 cm at 1 GHz), the radiated field can be predicted by Equation (9.13a) which defines the *differential-mode* emission generated by two electrically short dipoles. The maximum radiated field magnitude $E_{max}(f)$ is then given by

$$E_{max}(f) = 1.316 \times 10^{-14} I_{loop} f^2 \frac{S}{d} \quad (\mathrm{V/m}) \tag{9.23}$$

where I_{loop} is the magnitude of the current flowing in the loop (i.e. $I_{loop} = |\hat{I}_{cap}|$ or $I_{loop} = |\hat{I}_{IC}|$), f is the frequency, S is the loop area shown in Figure 9.17 by the highlighted gray area, and d is the distance of the observation point from the loop.

This investigation is very important because radiated emission from ICs could be significant, as well as the radiated emission from traces, especially in the case of ICs with chips switching simultaneously, as will be shown in *Section 9.5*.

As done for traces, the best way to carry on investigations is to analyze experimental results and compare them with the predictions performed by appropriate models.

Example 9.5: Calculations and Measurements of Emission from Multilayer SBC-Test Boards

The experiment presented here refers to the same test boards as those used in Example 8.3 to investigate ΔI*-noise* voltage [26], and shown in Figure 8.15. In summary, the considered test boards are of two types:

- The first was made by using a standard technology (STD).
- The second was made by using a particular technology (buried capacitance (BC)) able to maximize the intrinsic capacitance between the two pairs of power planes.

On both types of board the same functions were constructed with an identical layout (see Figure 8.16): fifteen components (74AC244 devices each having eight output buffers and distributed on a regular pattern grid) were loaded on R with C in parallel (255 Ω//56 pF). The layout was designed with care taken to avoid transmission-line effects, thus enabling the driver loads to be considered as simple lumped loads while avoiding emission from traces. This precaution is necessary in order to measure the contribution to emission of the devices only.

Three typologies of boards were tested:

(a) a standard board without decoupling (STD);
(b) a standard board with 100 nF PTH decoupling capacitors placed very close to each of the 15 components (STDF);
(c) a buried capacitor board without decoupling (BC).

For all three PCBs, a bulk filtering at the PCB power pins was present, consisting of two 100 nF ceramic and two 47 μF electrolytic decoupling capacitors. These boards were used to study the ΔI*-noise* voltage and radiated emissions.

As regards radiated emission, several measurements were made. A total of 24 ports of the components U5, U7, and U9 were simultaneously switched. In order to measure radiated emission from the board only, the PCB was powered by a battery pack. The component buffers were switched using a 8 MHz shielded oscillator, which was powered by an internal battery. The antenna was placed at a distance of 3 m. The boards were rotated 360° around their vertical axes, adopting an angular step of 45°. Little variation in the radiated field was noted.

The radiated emission from the considered test boards was investigated by simulations according to the following steps:

- The switching current in the time domain was calculated by SPICE analysis of the circuit modeling the PCB with its devices as described in *Chapter* 8. Applying this approach, the switching currents on devices U5, U7, and U9 and their decoupling capacitors were computed.
- These calculated time-domain currents were moved to the frequency domain by FFT.
- The radiated emission for each involved loop was calculated using the loop formula (9.23), and the total electric field was obtained by the superimposition effect of the three 74AC244 in positions U5, U7, and U9.

More details on the simulation model are given by Caniggia *et al.* [26].

The results obtained by simulations and measurements are shown in Figure 9.18. Comparing the radiated emission in the three cases of STD, STDF, and BC boards, it can be noted that they are quite similar. The reason for this can be easily understood by considering the time-domain currents I_{IC} (i.e. on the V_{CC} pin of the 74AC244 component) shown in Figure 9.18

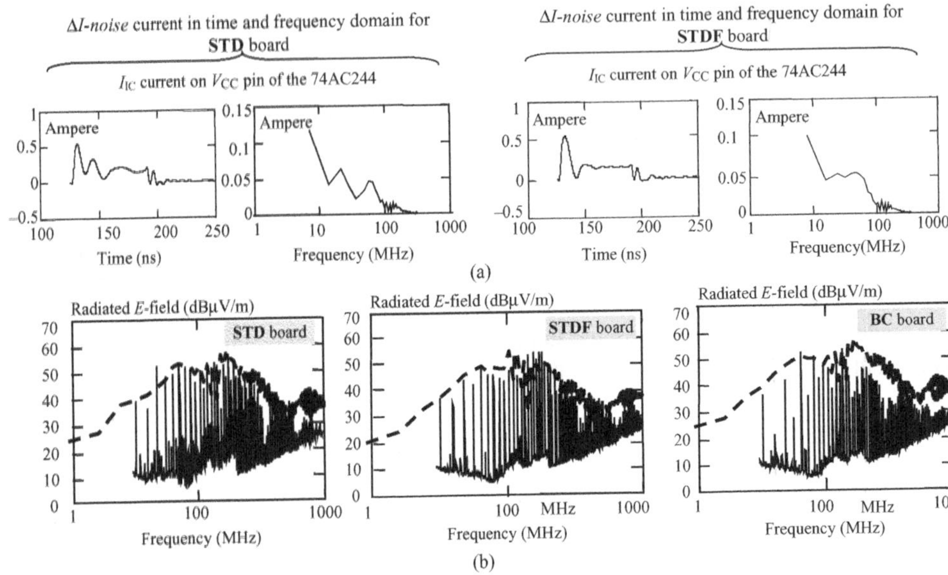

Figure 9.18 SBC test boards: (a) ΔI-*noise* current computed in the time domain and computed in the frequency domain; (b) radiated emission measured (solid line) and computed envelope (dashed line) for a Standard Test Board (STD) without a decoupling capacitor, for a Standard Test Board with a decoupling capacitor (STDF), and for a Buried Capacitor (BC) test board

for STD and STDF boards, and their frequency-domain spectrum. The current inherent to the BC board is not reported, as it was verified to be practically equal to the case of the STDF board. From these graphs, the currents I_{IC} for all of the cases studied being quite similar, it can be deduced that the radiation from the capacitor area can be neglected for this experiment. In fact, the role of the decoupling capacitors is not so important for radiated emission from ICs if they are properly connected to the pair of power planes, or rather by means of short connections. In conclusion, from this type of experiment it can be said that the emission from ICs does not depend on the type of filtering. As will be shown in *Section 9.6*, this is not the case when there is a cable attached to the PCB, where the ΔI-*noise* between the power and ground planes plays a determinant role in radiated emission from the cable.

9.5 Emission from a Real PCB

Previously it was shown that *common-mode* emission can be neglected when cables attached to the PCB are not present and when traces are not very close to the edges of a PCB with ground and power planes. Therefore, in a multilayer PCB populated by devices, the attention can be focused on two types of *differential-mode* current as responsible for producing radiated fields: ΔI-*noise* current onto digital devices and signal current on traces. The loops that radiate are therefore of two types. The first loop is defined by the path of ΔI-*noise* current that flows through the chips placed in the die of the ICs and, by pins and vias, arrives at the pair of power planes. The second loop is formed by the trace and the ground plane where the signal current returns. For radiated field computation, image theory should be applied, and the area of these loops must be doubled.

With the following experiment it will be shown that, in a PCB having devices placed on sockets and traces with geometries encountered in the real word, the radiated emission due to devices can dominate over that due to traces when the devices have a significant number of simultaneous switchings. The choice to adopt ICs placed on sockets instead of having them soldered, as occurs in practice, made it possible to perform experiments with and without active devices, emphasizing in this way the contribution made by the devices to radiated emission.

Example 9.6: Calculations and Measurements on a Test Board Equipped with CMOS Devices Interconnected by a Star Structure

Experimental measurements of radiated emission were performed on the same test board as in *Section 8.4*, referred to as the SQ-test board, but using a different net [27]. The structure under study is illustrated in Figure 9.19a, denoted as the AC star net. One 74AC00 drives a main line of 18 cm length and embedded in the dielectric, as shown in Figure 9.19b. The other

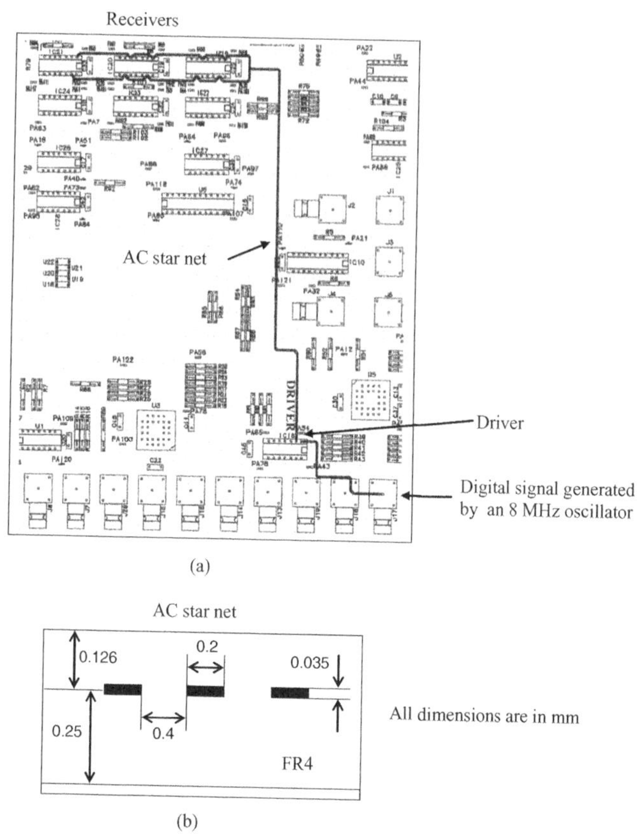

Figure 9.19 SQ test board: (a) layout of the interconnect, indicated as AC star, used for radiated emission measurements; (b) cross-view of the AC star net

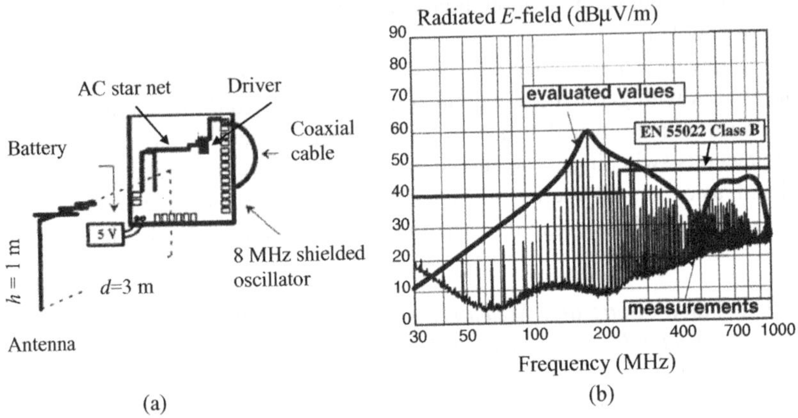

Figure 9.20 Radiated emission of the SQ test board: (a) measurement set-up; (b) computed (envelope) and measured (harmonics) radiated emission with the AC star net fully equipped with AC devices

end of the line is loaded with three 74AC00 devices. Two of them have four gates switching, while the third device has only two gates switching. Therefore, at the end of the interconnect, there are two parallel traces, connected at one end to five distributed receivers for a total of ten input gates. No loads were connected to the outputs of the receivers.

The radiated emission experiments were performed in a semi-anechoic chamber for 3 m measurements. The PCB and antenna were oriented as shown in Figure 9.20a. In order to minimize *common-mode* emission, the PCB was powered by a 5 V battery placed very close to the PCB. The signal for the driver of the AC star net was provided by a 8 MHz shielded oscillator, having an AC device as output, well connected to the ground plane in the rear of the PCB. The signal arrived at the input of the driver along a semi-rigid 50 Ω coaxial cable with minimized length. All the ICs were placed on sockets to perform experiments with and without the active devices.

The measured (harmonics) radiated field is shown in Figure 9.20b, together with the envelope of the magnitude spectrum obtained by simulations, and a good agreement can be observed.

The prediction of the radiated field was performed by superimposing the radiation from the interconnect and the radiation from devices.

The radiation from the interconnect was calculated by the following steps:

- The image theory was applied so that the return current on the ground plane was modeled by an image conductor.
- The source was modeled in the frequency domain by the magnitude spectrum envelope defined by Equations (9.4) and (9.5).
- The *differential-mode* current distribution along the traces were computed by Equation (9.14).
- The radiated field due to the interconnect was calculated taking into account the main horizontal path only. This part of the line was divided into five segments, and the field was calculated by superimposition using the equation of a small dipole with the current of each

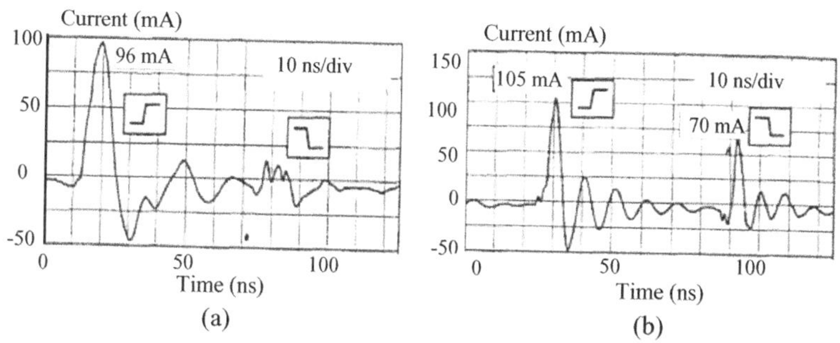

Figure 9.21 Measured Δ*I-noise* currents: (a) driver; (b) receiver

segment, as described in *Appendix D*. At one end, the main line was terminated on the linear macromodel of the driver, and at the other end on the receiver capacitances.

To calculate the radiation from devices, an IC is modeled by a rectangular loop perpendicular to the board. Before estimating the radiation caused by the Δ*I-noise* current, it is necessary to know the waveforms of the switching current in the ICs. To this end, the current on the V_{CC} pin of the driver and the receivers was measured by using a current probe, and the results are shown in Figure 9.21. The current of the driver is the sum of the current required by the interconnect to charge the distributed load capacitances and the switching current of the two output transistors. The current of the receivers is mainly of the second type because the outputs were not loaded. To estimate the radiated emission due to the ICs only, the following simple procedure was used:

- The measured waveforms shown in Figures 9.21a and 9.21b were both approximated by a triangle and by damped sinusoid functions, as reported in *Section 9.1*.
- The radiated field due to ICs was calculated by the equation of a small radiating loop (9.23) and by adopting the frequency-domain analytical expressions of the switching currents (see Table 9.1).

To validate the simulation procedure used, and to identify the contributions of traces and ICs to radiation, a further investigation was performed. To separate the contribution to the radiated field by the traces from the contribution to the radiated field by the devices, a commercial post-layout simulator taking no account of Δ*I-noise* effects was used to calculate the emission, considering the contribution of the traces only. The results provided by the post-layout simulator revealed a good agreement with the estimated emission from the line without ICs by the proposed prediction procedure. On the other hand, in the frequency range 100–300 MHz the results provided an emission profile up to 30 dB lower than that of the fully equipped net shown in Figure 9.20, which is higher than the limit of CISPR 22 or EN55022 for Class B equipment.

To investigate this discrepancy, several radiated emission measurements were carried out with the AC star net in different conditions. One measurement was performed by replacing the receivers with their input equivalent circuit realized with a 4.7 pF capacitor in parallel with

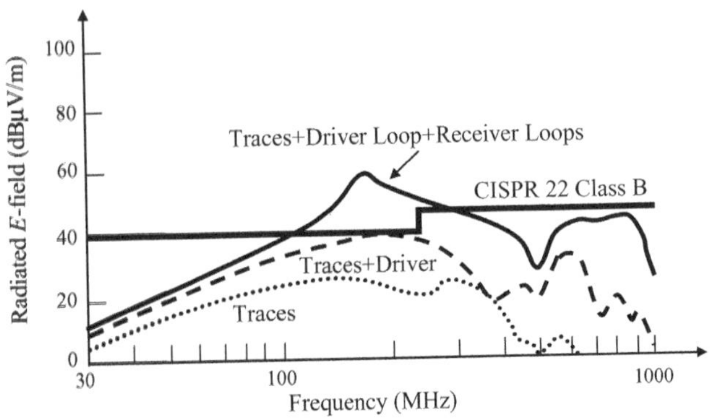

Figure 9.22 Computed emission profile with the SQ test board under several conditions: fully equipped (solid line); driver and traces (dashed line); traces only (dotted line)

a diode in order to avoid emissions from the receivers. The measured and computed results, reported by Caniggia *et al.* [27], indicated a reduction of about 20 dB in the range 100–300 MHz, and the limit of CISPR 22 was met.

Another measurement was performed, bypassing the driver and connecting the output of the shielded oscillator to the input of the AC star net in order to exclude the contribution of the driver to the radiated field. This was accomplished by a short wire which connected the input and output holes for the pins of the driver which was kept away from the PCB. In this case a reduction of 10 dB was obtained by measurement and computation. This confirms the result of the post-layout simulations that, without the contribution of the driver and receiver areas, there is a lowering of 30 dB in the range of frequencies of maximum emissions.

The emission profiles of the SQ-test board obtained by the prediction procedure considering several conditions are shown in Figure 9.22. As for the present experiment, the following considerations are necessary for a correct interpretation of the simulation results:

(a) The estimated emission from the line without ICs is similar to that obtained with a post-layout simulator where the actual geometry is taken into account in detail. Some discrepancies with the experimental results occurred around 200 MHz for the case of a transmission line only [27]. This is probably due to the *common-mode* effects of the unavoidable resonant structure consisting of battery + PCB + 50 Ω semi-rigid coaxial cable + shielded oscillator attached to the ground plane of the SQ-test PCB.
(b) The rise/fall times and the oscillation frequency of the measured waveforms of the devices switching currents are affected by the parasitic inductances introduced by the wire used to allow measurement of the current on the V_{CC} pin by a current probe. This means that the intrinsic resonance associated with the ΔI*-noise* loop, which depends on the inductance and capacitance parasitic parameters of the package, is shifted towards the lower frequency range. Therefore, for the calculation, a slightly shorter rise time was used, and a higher oscillation frequency was provided by the simulations.

In conclusion, the proposed experiment has highlighted the importance of the ICs in producing significant electromagnetic emission, especially when the device is complex and with many gates switching. Note that in the experiment the devices were mounted in sockets (not recommended for the final products), which makes the radiated emission profile worse. However, the difference between the contributions made by the ICs and the traces in multilayer PCBs is so high that the final result does not change substantially. This justifies the work by the IEC to update the international standards in respect of the emission measurements to be performed with ICs in the range from 150 kHz to 1 GHz.

9.6 Emission from a PCB with an Attached Cable

9.6.1 Sources of Emission

Cables attached to a PCB represent an important source of *radiated emission* (RE) in an electronic system. The radiated emission is mainly due to three main sources:

1. Functional current or *differential-mode* (DM) current on the cable.
2. *Common-mode* (CM) current on the cable, generated by the parasitic effects of the PCB.
3. *Common-mode* current on the cable, generated by the intrinsic unbalance of the differential drivers.

In principle, the radiated emission due to the cable functional current can be studied by TL models, as outlined in *Section 9.2*. However, its effect can often be negligible compared with the effects produced by the sources defined by points 2 and 3, as the signal is usually transmitted by shielded cables such as coaxial or twisted-pair cables in which the cancellation effect allows a very low *differential-mode* emission. The second source of radiation will be treated in this section, and the third source in *Section 9.7*. To study the second source of radiation, it is not necessary to have the driver of the cable switching to cause *common-mode* current on the cable. In fact, *common-mode* currents on the cable can arise simply because the driver connects the cable electrically to the PCB with its low output impedance, or, in the case of a shielded cable, because the shield is connected to the PCB. For these reasons, the experiments reported in this section were performed simulating the presence of an I/O cable with a simple wire attached to the ground of the test board.

9.6.2 Current- and Voltage-Driven Mechanisms with a Trace in a PCB

The mechanism by which an intentional signal induces *common-mode* currents on attached cables to a PCB can be generally divided into two categories [28, 29]:

1. A *current-driven* mechanism due to the voltage drop $\hat{V}_{gnd}$ in the return conductor of a trace produced by the signal current $\hat{I}_{sig}$ or *differential-mode* current $\hat{I}_{DM}$, and by the effective partial inductance L_{gnd} associated with the return-signal conductor, which in a multilayer PCB is a ground plane (see *Section 3.2.7*).
2. A *voltage-driven* mechanism due to the noise produced by the signal voltage $\hat{V}_{DM}$ between the trace and its return conductor and the parasitic capacitances of the structure formed by the trace, the return conductor, the cable and the environment.

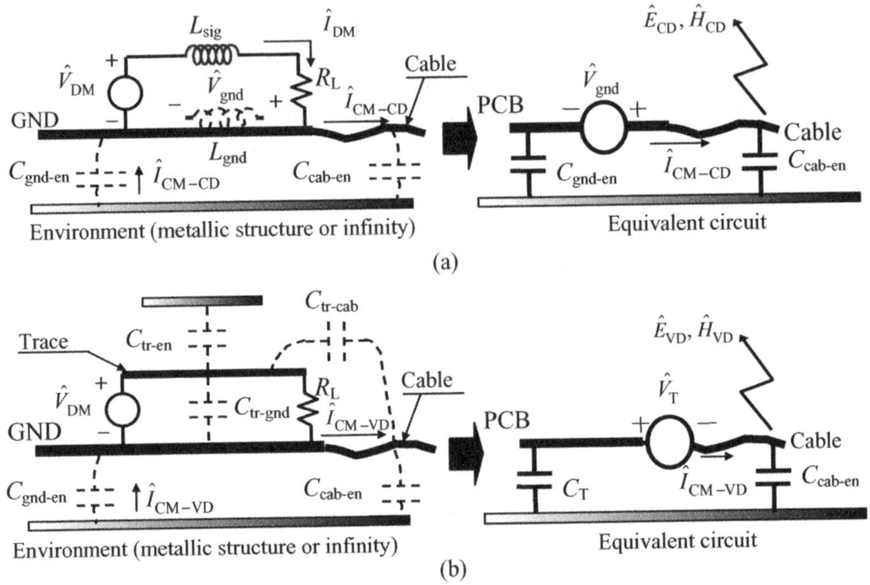

Figure 9.23 Basic coupling mechanism in a PCB for inducing common-mode current on attached cables: (a) current-driven mechanism; (b) voltage-driven mechanism

Examples of these coupling mechanisms are illustrated in Figure 9.23, where the *environment* is either a metallic surface near the PCB, such as the chassis or the metallic floor of a semi-anechoic chamber for radiated field measurements, or infinity if the PCB is considered to be isolated (i.e. with other metallic objects far way from the PCB).

9.6.2.1 Current-driven Mechanism

The configuration of a simple interconnect structure in a PCB composed of a driver, a trace with its return, and a receiver where the *current-driven* mechanism occurs is shown in Figure 9.23a. If the line is electrically short, the mechanism can be studied by a lumped-circuit model. The following notation is adopted here:

- $\hat{V}_{DM}$ = signal voltage at the output of the driver.
- L_{sig} = effective partial inductance associated with the signal trace as the difference between the self partial inductance of the trace and the mutual partial inductance between the trace and the return conductor indicated by GND.
- R_L = resistive load of the signal line, usually equal to the line characteristic impedance.
- $\hat{I}_{DM}$ = signal or *differential-mode* current on the line.
- L_{gnd} = effective partial inductance associated with the PCB return conductor as the difference between the self partial inductance of the return conductor and the mutual partial inductance between the signal and its return conductor.

- $C_{\text{gnd-en}}$ = self capacitance of the return conductor, representing the stray electric field lines that terminate at a nearby metallic reference or at infinity, referred to here as the environment.
- $C_{\text{cab-en}}$ = self capacitance between the cable and the environment.

The *current-driven* mechanism is associated with the magnetic fields that wrap around the finite width of the signal return plane and is represented in Figure 9.23a by the inductance L_{gnd}. The effective voltage drop $\hat{V}_{\text{gnd}}$ across the return conductor induces *common-mode* currents on the cables. The induced *common-mode* current due to the *current-driven* mechanism $\hat{I}_{\text{CM-CD}}$ is directly proportional to the signal current $\hat{I}_{\text{DM}}$, which justifies the name 'current-driven' mechanism. If there are two cables connected to opposite sides of the board, they are driven like a dipole antenna. With one cable attached, $\hat{I}_{\text{CM-CD}}$ can be calculated using the equivalent circuit of Figure 9.23a with

$$\hat{V}_{\text{gnd}} = \text{j}\omega L_{\text{gnd}}\hat{I}_{\text{DM}} \tag{9.24}$$

Note that in the equivalent circuit of Figure 9.23a (right-hand side) the contribution made by the self partial inductance of the ground plane in series with the source $\hat{V}_{\text{gnd}}$ has been omitted, as it is negligible compared with the cable inductance and not shown in Figure 9.23a for simplicity (see Figure 10.3 of *Section 10.1*, where an exact equivalent circuit of the PCB–environment structure is given and discussed).

The total capacitance C_{tot}, which determines the amount of *common-mode* current $I_{\text{CM-CD}}$, is given by the series connection of $C_{\text{gnd-en}}$ and $C_{\text{cab-en}}$. As the cable length is much longer than the board dimension, $C_{\text{cab-en}} \gg C_{\text{gnd-en}}$, and the total capacitance C_{tot} is approximately equal to $C_{\text{gnd-en}}$:

$$C_{\text{tot}} = \frac{C_{\text{gnd-en}}C_{\text{cab-en}}}{C_{\text{gnd-en}} + C_{\text{cab-en}}} \approx C_{\text{gnd-en}} \tag{9.25}$$

Considering that the self capacitance of a rectangular plane is similar to that of a circular disc of the same area, which is approximately $8\varepsilon_0 r$, where r is the radius of the disc [30], the capacitance C_{tot} of the wire antenna model can be approximated as

$$C_{\text{tot}} \approx C_{\text{gnd-en}} \approx 8\varepsilon_0\sqrt{\frac{A_{\text{board}}}{\pi}} \tag{9.26}$$

where A_{board} is the board area. For example, for a PCB of 20×20 cm size, $C_{\text{gnd-en}} = 8$ pF.

Assuming, for simplicity, that below the first resonance frequency $\hat{I}_{\text{DM}} \approx \hat{V}_{\text{DM}}/R_{\text{L}}$, the *common-mode* current on the cable produced by the *current-driven* mechanism is given by

$$\hat{I}_{\text{CM-CD}} \cong -\omega^2 C_{\text{gnd-en}}L_{\text{gnd}}\frac{\hat{V}_{\text{DM}}}{R_{\text{L}}} \tag{9.27}$$

where a dependence on ω^2 can be observed.

As a knowledge of the effective inductance L_{gnd} is very important for *common-mode* current calculation, many researchers have developed closed-form expressions for configurations in which the return conductor is a ground plane [21, 31–33]. A working formula is reported in

Table A.2 of *Appendix A*. It will be shown by experimental studies that the model of Figure 9.23a provides reasonable estimates of the induced *common-mode* currents and radiated fields due to the *current-driven* mechanism.

9.6.2.2 Voltage-Driven Mechanism

Not only are the inductances associated with the two conductors of an interconnect important for *common-mode* current on cables, so are the parasitic capacitances of the structure formed by the PCB, the cable, and the environment. This is shown in Figure 9.23b by a lumped-circuit model where the following capacitances are present in addition to the ones already defined in the previous section:

- $C_{\text{tr-gnd}}$ = mutual capacitance between the trace and the PCB ground plane;
- $C_{\text{tr-cab}}$ = mutual capacitance representing electric field lines that directly link the trace and cable (negligible).
- $C_{\text{tr-en}}$ = self capacitance of the trace with respect to the environment.

The capacitance $C_{\text{tr-gnd}}$ contributes to the *differential-mode* current that returns to the source through the plane. These parasitic capacitances, together with the signal voltage $\hat{V}_{\text{DM}}$, are responsible for the *voltage-driven* mechanism. In fact, by using the equivalent Thévenin representation at the point where the cable connects to the PCB, the equivalent circuit of Figure 9.23b is obtained, where

$$C_{\text{T}} = C_{\text{gnd-en}} + C_{\text{tr-en}} \tag{9.28a}$$

$$\hat{V}_{\text{T}} = -\hat{V}_{\text{DM}} \frac{C_{\text{tr-en}}}{C_{\text{tr-en}} + C_{\text{gnd-en}}} \tag{9.28b}$$

As indicated in Figure 9.23b, the amount of *common-mode* current on the cable is determined by the capacitance C_{T} which is the equivalent capacitance between the ground/power PCB and the environment. According to Equation (9.28a), this equivalent capacitance is the series combination of the self-capacitances $C_{\text{gnd-en}}$ and $C_{\text{tr-en}}$, and, as in general $C_{\text{tr-en}} \ll C_{\text{gnd-en}}$, Equations (9.28) can be approximated as

$$C_T \approx C_{\text{gnd-en}} \tag{9.29a}$$

$$\hat{V}_{\text{T}} \approx -\hat{V}_{\text{DM}} \frac{C_{\text{tr-en}}}{C_{\text{gnd-en}}} \tag{9.29b}$$

The capacitance $C_{\text{tr-en}}$ required to compute $\hat{V}_{\text{T}}$ can be determined analytically [34] or by a static field solver, while the capacitance $C_{\text{gnd-en}}$ is the board capacitance provided by approximation (9.26).

With reference to the equivalent circuit in Figure 9.23b, where C_{T} and $\hat{V}_{\text{T}}$ are given by approximations (9.29), and considering that the length of the cable is much longer than the length of the board (i.e. $C_{\text{cab-en}} \gg C_{\text{gnd-en}}$), the *common-mode* current on the cable $\hat{I}_{\text{CM-VD}}$ is then given by

$$\hat{I}_{\text{CM-VD}} \cong -j\omega C_{\text{tr-en}} \hat{V}_{\text{DM}} \tag{9.30}$$

where a linear dependence on f can be observed. Note that this source mechanism is referred to as *voltage driven* because the magnitude of the *common-mode* current induced on the cable is proportional to the signal voltage.

9.6.2.3 Emission Due to the Common-Mode Current on Cables Attached to PCBs

In PCBs with microstrip traces and with attached cables, both *current-* and *voltage-driven* mechanisms contribute to the emission profile which can be obtained by superimposing the radiations associated with the two mechanisms as

$$\hat{E} = \hat{E}_{\text{CD}} + \hat{E}_{\text{VD}} \tag{9.31}$$

where the subscripts 'CD' and 'VD' refer to the *current-driven* and *voltage-driven* mechanisms respectively.

Generally, the *voltage-driven* mechanism is likely to dominate when the sources are high-impedance circuits. On the other hand, the *current-driven* mechanism is likely to dominate when the sources are low-impedance circuits. In matched circuits the two mechanisms of emission may contribute nearly equally [30]. In CMOS circuits terminated with a capacitive load, the *voltage-driven* mechanism tends to dominate at low frequencies and the *current-driven* mechanism often becomes more important at high frequencies.

The *voltage-driven* mechanism becomes very important in the case of *ΔI-noise* voltage between ground and power planes. In this case the trace in Figure 9.23b is replaced by the power plane and $\hat{V}_{\text{DM}}$ becomes the *ΔI-noise* voltage, as will be shown by Example 9.10.

In general, when the cable is near to a metallic plane, a *transmission-line* (*TL*) model can be used to calculate *common-mode* currents along the cable that occur both by the *current-driven* and by the *voltage-driven* mechanisms of Figure 9.23, and therefore the resonance frequencies associated with the structure, as will be shown later by experiments.

Example 9.7: Calculations and Measurements of Emission from Simple Test Boards (P-Boards) with an Attached Cable

To investigate the radiated emission mechanism of a cable attached to a PCB more thoroughly, some experimental measurements were carried out by using the test boards presented in *Section 9.2* without and with a ground plane and referred to as P-test boards.

The configuration adopted in the first experiment is shown in Figure 9.24. The test board with two parallel wires is placed within a shielded rack in order to avoid measurement of emission from the circuit. An *Unshielded Twisted-Pair* (UTP) cable covered by a metallic foil is attached to the return wire of the board by its drain wire, as indicated in Figure 9.24. The cable goes out from the rack through a small hole without touching the metal, and follows

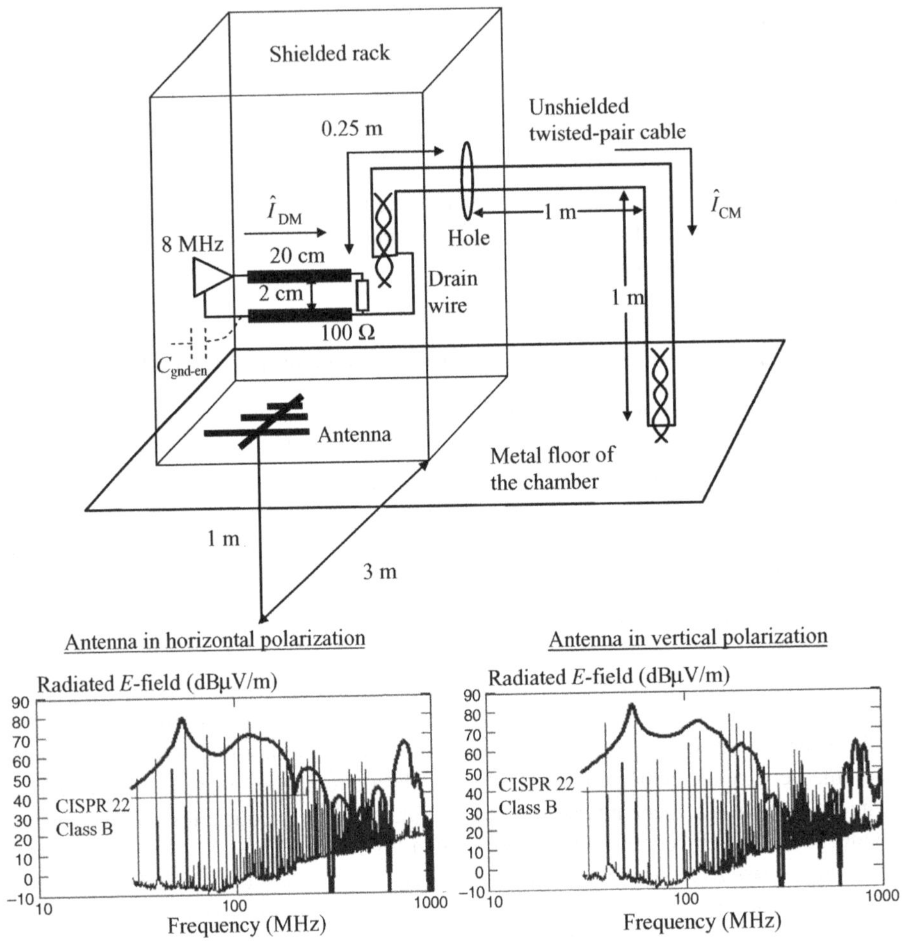

Figure 9.24 Test board set-up and comparison between computed (emission profile) and measured (harmonics) radiated E-field for a PCB composed of two wires with an attached cable outgoing from a shielded rack

a path of 1 m horizontally and a further 1 m vertically. The antenna for measurements was positioned at a distance of 3 m and at 1 m above the metal floor of the chamber. All the measurements were performed in a shielded semi-anechoic chamber for 3 m measurements. Measured (harmonics) and predicted (envelope) radiated fields for both horizontal and vertical polarization of the antenna are also shown in Figure 9.24. Observe that the emissions are up to 40 dB higher than the CISPR 22 Class B limits at low frequencies.

The predicted emission profiles were calculated considering the *current-driven* mechanism only in accordance with the frequency-domain procedure based on the following steps.

(i) Step 1 – Modeling of the Signal on the PCB
The *differential-mode* current $\hat{I}_{DM}$ in the PCB was calculated by the following approach:

- The voltage source was assumed to be described by the same envelope of the 8 MHz clock waveform as described in *Example 9.1* and used in *Example 9.2*. The source output resistance was $R_S = 50\ \Omega$.
- The signal current $\hat{I}_{DM}$ was calculated by Equation (9.15) as the input current of a TL of length $l = 20$ cm and characteristic impedance $Z_0 = 120\ \ln(2s/r) = 526\ \Omega$ (separation between the two wires $s = 2$ cm, wire radius $r = 0.5$ mm), terminated with a load $R_L = 100\ \Omega$. For simplicity, the relative dielectric of the wood support for the wires was not considered.
- The effective partial inductance L_{gnd} associated with the return-signal conductor was calculated by the equations provided in Table A.1 of *Appendix A*, and reported here for convenience:

$$L_p = \frac{\mu_0 l}{2\pi}\left[\ln\left(\frac{2l}{r}\right) - 1\right] \tag{9.32a}$$

$$M_p = \frac{\mu_0 l}{2\pi}\left[\ln\left(\frac{2l}{s}\right) - 1\right] \tag{9.32b}$$

$$L_{gnd} = L_p - M_p = 148\,\text{nH} \tag{9.32c}$$

(ii) Step 2 – Calculation of the Common-mode Current along the Attached Cable
The current along the cable generated by voltage drop $\hat{V}_{gnd} = j\omega L_{gnd}\hat{I}_{DM}$ on the return wire of the board was computed by the equivalent circuit of Figure 9.23a according to the following procedure:

- The cable is modeled as a transmission line having as the reference plane the wall of the rack for the vertical path and the metallic floor of the chamber for the horizontal path. For both paths a cable characteristic impedance $Z_{0cab} = 60\ \ln(2h_{cab}/r_{cab}) = 401\ \Omega$ was calculated, adopting $h_{cab} = 1$ m and $r_{cab} = 2.5$ mm.
- The cable input impedance $\hat{Z}_{cab,in}$ with the other end open was computed by the TL closed-form expression (9.16), considering a total length $l = 1.25$ m for the horizontal path and $l = 1$ m for the vertical path. The two lines were considered to be in series.
- The capacitance $C_{gnd\text{-}en}$ between the PCB and the environment, which in this case is the metallic walls of the rack, was estimated to be $C_{gnd\text{-}en} = 5$ pF.
- The radiation resistance for low frequencies was also considered to be $R_{cab,rad}(f) = 80\pi^2[(l_{cabh} + l_{cabv})/\lambda(f)]^2$ for linear current distribution [2], where the geometric parameters $l_{cabh} = l_{cabv} = 1$ m are the lengths of the horizontal and vertical paths of the cable outside the rack. This resistance is useful for computing the *common-mode* current at the input of the cable, taking into account the radiation from the cable.
- The cable input current $\hat{I}_{CM,in}$ is calculated by Equation (9.15) as $\hat{I}_{CM,in} = \hat{V}_{gnd}/(\hat{Z}_{cab,in} + R_{cab,rad}(f) + 1/(j\omega C_{gnd\text{-}en}))$.

- Once the cable input current is known, the current distribution along the cable is calculated by $\hat{I}_{CM}(x) = \hat{I}_{CM,in}\cos(\beta x) - j\frac{\hat{V}_{CM,in}}{Z_{0cab}}\sin(\beta x)$, where β is the phase constant and $\hat{V}_{CM,in} = \hat{I}_{CM,in}(\hat{Z}_{cab,in} + \hat{R}_{cab,rad})$ is the voltage at the input of the cable [7].
- The average current on both horizontal and vertical paths, useful for radiated field calculation, was obtained by dividing each path outside the rack into 10 segments.

(iii) Step 3 – Calculation of the Field Radiated from the Current along the Attached Cable
The radiated field was calculated by using the far-field formulation and image theory reported in Table D2 of *Appendix D*, considering the structure of one conductor to be above a ground plane. The horizontal cable path has an image with current of opposite sign; the vertical cable path has an image with the same sign. These are due to the presence of the metallic floor of the shielded room [35].

When the board with two wires was replaced with a board with a ground plane, there was a drastic reduction in radiated emission, as shown in Figure 9.25. This is due to the significant reduction in L_{gnd}. In fact, in this case, by using the expression given in Table A.2 of

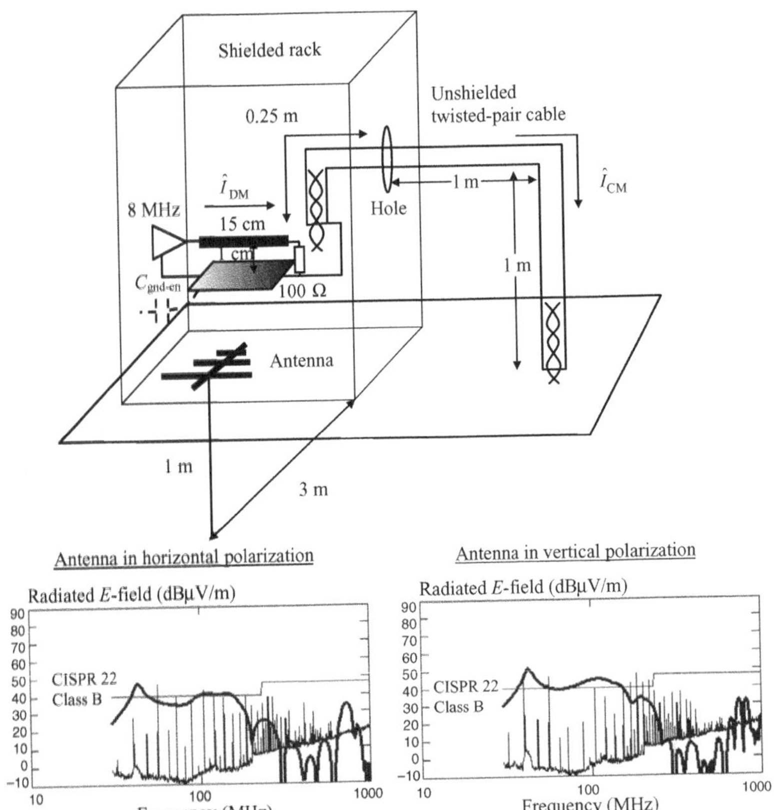

Figure 9.25 Set-up and comparison between computed (emission profile) and measured (harmonics) radiated E-field for a PCB composed of one wire above a ground plane with an attached cable outgoing from a shielded rack

Appendix A, the inductance L_{gnd} is given by

$$L_{gnd} = \frac{\mu_0 l}{2\pi} \ln\left(\frac{h\pi}{w_{gnd}} + 1\right) = 5.6 \text{ nH} \tag{9.33}$$

where $l = 20$ cm (wire plus interconnection to the driver output), $h = 1$ cm (height of the wire), and $w_{gnd} = 21$ cm (width of the ground plane). The same procedure for radiated fields as that previously described was applied, using the new L_{gnd} value and an estimated value for the capacitance $C_{gnd\text{-}en}$ of 20 pF.

From the results it is possible to make the following observations:

- A good correlation can be seen between measured (harmonics) and calculated (magnitude spectrum envelope) values up to 300 MHz. For higher frequencies, the TL model for calculating *common-mode* current along the cable is no longer valid because the distance between the cable and its reference plane is no longer electrically short.
- However, maximum emissions from cables are in the range 30–300 MHz, as discussed in *Subsection 9.2.2* for *common-mode* radiation, and in this range the TL model is valid.
- Emissions are much higher than the standard limit of CISPR 22 Class B in the low-frequency range for the board with two wires.
- Emissions are closer to the standard limit for the board with a ground plane.

The presence of a ground plane gives a substantial benefit in lowering radiated emissions from cables. Another significant reduction in emission could be obtained by cables with a braided shield well connected at 360° to the rack. In this case the *common-mode* current is diverted to the source by the metallic wall of the rack and does not go out. Another fix could be the use of EMI filters, such as a choke, to stop the *common-mode* current on the cable placed at the point where the cable leaves the rack. This means that the internal path of the cable must be minimized at very few cm of length. These fixes and others will be treated in the next sections.

Example 9.8: Calculations and Measurements of Emission from S-Test Boards with an Attached Cable

To investigate the effect of adding a second plane to a structure, similarly to a microstrip with an attached cable, the same S-test boards as those considered in *Section 9.3* were used again. A wire was connected to the ground plane to simulate a cable. The PCB + cable structures were positioned as illustrated in Figure 9.26, and the *common-mode* current was measured (harmonics) and compared with the predicted values (envelope).

The input cable current was computed using the equivalent circuit of Figure 9.23, where the circuit parameters of the different board structures were obtained as follows:

- *Microstrip structure*: $\hat{V}_{gnd}$ was computed by Equation (9.24) with $L_{gnd} = 1.57$ nH obtained using Equation (9.33).
- *Ground + image plane structure*: the equivalent circuit of Figure 9.23 was modified by adding, in parallel to the voltage source $\hat{V}_{gnd}$ given by Equation (9.24), a series circuit formed by a resistance R_{shunt}, an inductance L_{shunt}, and a capacitance C_{shunt} to take into

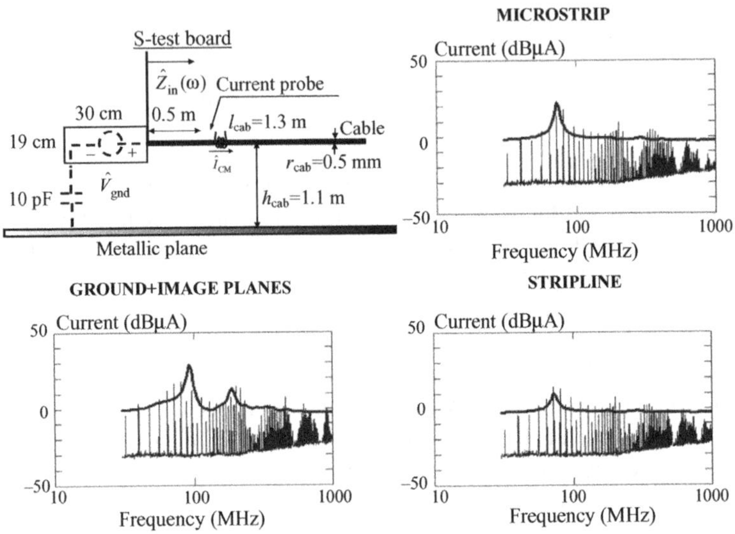

Figure 9.26 Schematic set-up of the S-test board with an attached cable for the measurement of common-mode (CM) current in the cable and comparison between computed (profile) and measured (harmonics) values

account the loading effect of the second floating ground plane. The following values were used: $R_{shunt} = 200\ \Omega$; $L_{shunt} = L_{gnd}$ equal to the effective partial inductance of the ground plane; $C_{shunt} = 789$ pF calculated as the capacitance between the two planes.

- *Stripline structure*: it was assumed that $\hat{V}_{gnd} = j\omega L_{gnd}\hat{I}_{DM}/2$; the reason for this will be explained later.

The *common-mode* currents were calculated considering the cable as a lossy transmission line with nominal characteristic impedance Z_{0cab}:

$$Z_{0_{cab}} = 60 \ln\left(\frac{2h_{cab}}{r_{cab}}\right) \tag{9.34}$$

and calculating the cable input impedance $\hat{Z}_{in}(\omega)$ as [7]

$$\hat{Z}_{in}(\omega) = Z_{0_{cab}} \frac{\hat{Z}_{cend}(\omega) + Z_{0_{cab}} \tanh(\hat{\gamma}(\omega) l_{cab})}{Z_{0_{cab}} + \hat{Z}_{cend}(\omega) \tanh(\hat{\gamma}(\omega) l_{cab})} \tag{9.35a}$$

$$\hat{\gamma}(\omega) = \sqrt{j\omega\mu_0(\sigma_{air}(\omega) + j\omega\varepsilon_0)} \tag{9.35b}$$

$$\sigma_{air}(\omega) = \omega\varepsilon_0/10 \tag{9.35c}$$

where $\hat{Z}_{cend}(\omega)$ is the impedance of the cable with respect to the metallic plane at the far end. In this case it is infinite, the cable being open. Equations (9.35) are for lossy lines and are used for the cable instead of Equation (9.16) in order to have more realistic peaks of current values in accordance with the resonance frequencies of the structure.

The *common-mode* current at each position along the cable was calculated as outlined in Step 2 of Example 9.7. Note that l_{cab} in Equation (9.35a) is the sum of the length of the wire attached to the PCB (1 m) and the length of the PCB (30 cm), because the resonant structure for current calculation is the combination of cable plus PCB. The computed *common-mode* currents are plotted in Figure 9.26, where they are compared with the measurements obtained by positioning the current probe at 50 cm from the PCB.

The calculated *common-mode* current along the cable was used to predict the radiated E-field at 3 m. In order to perform the computation considering the average values of current amplitude and phase, the cable was divided into 10 segments. The radiated E-field in the far-field region was then obtained as indicated in Table D2 of *Appendix D*. In this case the configuration is defined by one conductor and its image.

The measured and calculated radiated emissions are plotted in Figure 9.27. To highlight more clearly the contribution made by the cable, each graph shows: the emission profile of the PCB without the attached cable, as calculated in *Section 9.3*, the emission profile of the cable, and the total emission profile of the PCB with the attached cable, to be compared with the measurements. Observe that the emissions due to the cable dominate at low frequencies for all three PCB structures. In particular, there is a strong peak of emission at the first resonance frequency around 100 MHz owing to the PCB plus cable structure. The stripline structure, which, without an attached cable, has very low emission, with an attached cable has an emission profile only a few dB lower than the one produced by the microstrip. This high level of emission profile is probably due to the imperfections in building the stripline and most probably to the non-perfect placement of the finite-dimension oscillator connected to one of the two planes. These inevitable actions probably make the structure so asymmetric as to produce strong *common-mode* emission, as highlighted in *Section 9.2*. Current and electric field measurements confirmed this fact, as shown by Figures 9.26 and 9.27.

Example 9.9: Calculations and Measurements of Emission from A-Test Boards with an Attached Cable

To obtain further confirmation of the results obtained with the S-test boards, the same calculations and measurements were repeated for the two A-test boards of *Example 9.4*. As asserted in the previous example, these test boards were built to reproduce as closely as possible the typical microstrip and stripline structures found in a real PCB. The computed (harmonics) and measured (envelope) radiated fields are shown in Figure 9.28. Observe that in this case also the emission profile of the stripline PCB is comparable with the emission profile of the microstrip PCB, although the stripline PCB has a very low profile when the attached cable is not present. As already said when commenting on the results obtained with the S-test boards, the need to use a finite oscillator placed at one side of the PCB with its DC batteries negates the advantage of using the stripline structure which provides very low emission without an attached cable. However, without the asymmetries introduced by these elements, which are necessary to carry out the experimental measurements, the advantages offered by the stripline structure in lowering significantly the radiated emission are real, with and without attached cables, as confirmed by full-wave simulations.

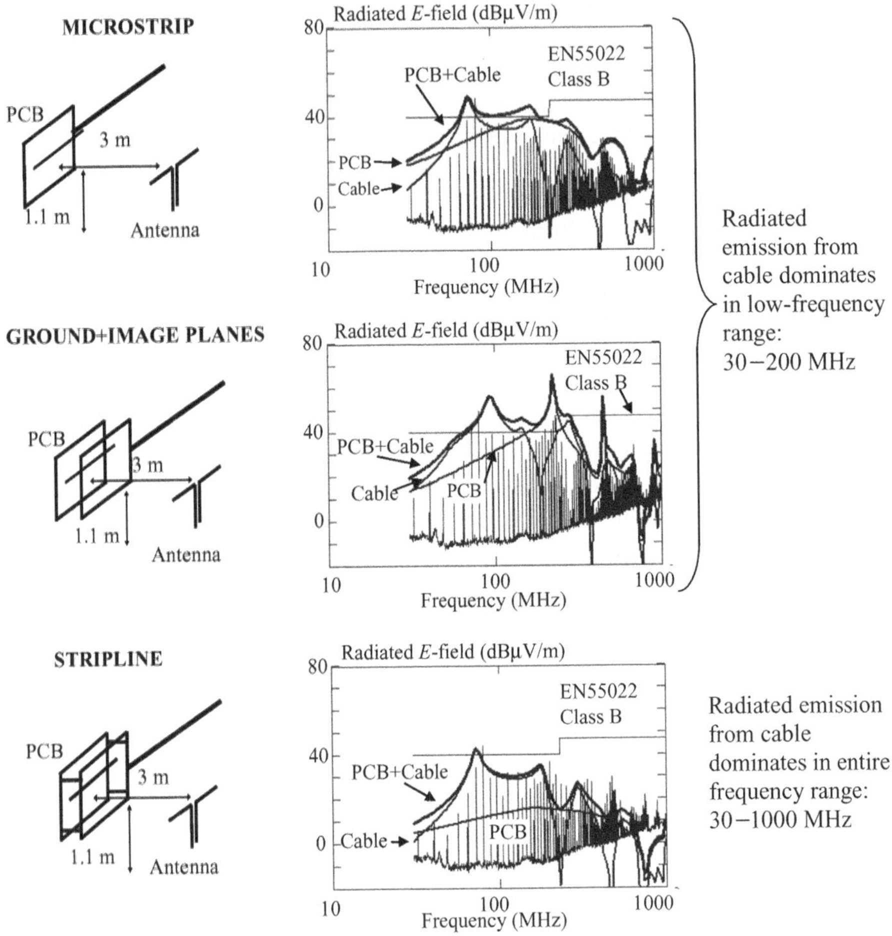

Figure 9.27 Schematic set-ups and comparison between computed (envelope) and measured (harmonics) radiated E-fields for the three S-test boards with an attached cable

Example 9.10: Calculations and Measurements of Emission from SBC-Test Boards with an Attached Cable due to the Voltage-driven Mechanism

The test boards previously considered do not have a power plane, and the switching devices were within a small shielded box well connected to one of the ground planes. Therefore, the emissions were due to the trace and cable only. As a first approximation, and to make the calculation process simpler, the *current-driven* mechanism was used to compute the emission profile of the test boards only. With the current experiment it will be shown that the *voltage-driven* mechanism becomes the dominant phenomenon when ΔI*-noise* is present in a multilayer PCB with a very short trace in order to avoid significant emission from the interconnects.

The test boards considered were the same as those used for investigating ΔI*-noise* in Example 8.3: standard boards in FR4 substrate with and without decoupling capacitors and buried

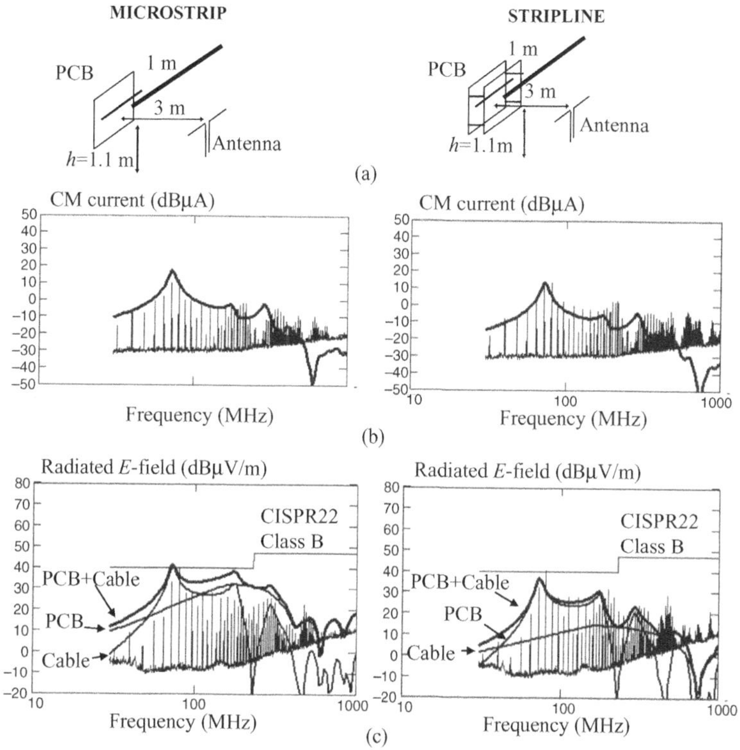

Figure 9.28 Comparison between computed (envelope) and measured (harmonics) values for two A-test boards with an attached cable: (a) schematic set-up; (b) common-mode currents; (c) radiated *E*-electric fields

capacitances capable of having a high value of intrinsic filtering offered by power and ground planes [26]. The case with devices 74AC244 in positions U5, U7, and U9 switching, for a total of 24 simultaneous switching gates, was considered.

In these PCBs, the signal traces were made so short that they did not cause significant voltage drops on the ground plane to feed the cable. We suspected that the only *current-driven mechanism* was due to the switching ΔI*-noise* current between the switching device and its decoupling capacitor for the STDF board with 100 nF decoupling capacitors. This investigation was performed by the equivalent circuit of Figure 9.23a, where now the voltage source $\hat{V}_{DM}$ is the ΔI*-noise* voltage $\Delta\hat{V}_{CC}$ between the two planes (i.e. $\hat{V}_{DM} = \Delta\hat{V}_{CC}$), L_{sig} and L_{gnd} are the effective partial inductances of the power and ground planes respectively, and the load is the nearby decoupling capacitor of 100 nF. The equivalent voltage $\hat{V}_{gnd}$ of the equivalent circuit of Figure 9.23a was computed, assigning to $\Delta\hat{V}_{CC}$ the FFT of the simulated waveforms in the time domain of Figure 8.20b for the case of STDF. The inductances were computed using the expressions of Table A.2 of *Appendix A* for busbars. The calculated spectrum of $\hat{V}_{gnd}$ has a maximum of 1 mV at about 50 MHz. This is absolutely negligible compared with the spectrum of $\hat{V}_T$, as is shown by the following calculation for the *voltage-driven* mechanism applied to power and ground planes.

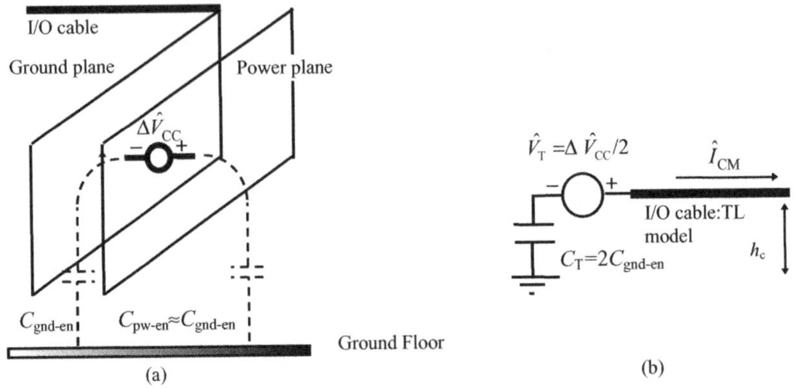

Figure 9.29 Multilayer PCB with power and ground planes: (a) schematic representation of the voltage-driven mechanism; (b) equivalent circuit for common-mode (CM) current calculations on attached cable

The equivalent circuit for the *voltage-driven* mechanism is that shown in Figure 9.23b with $\hat{V}_{DM} = \Delta\hat{V}_{CC}$ and $C_{tr\text{-}en} = C_{pw\text{-}en}$, where $C_{pw\text{-}en}$ is the stray capacitance between the power plane and the environment which in this case is the metallic floor of the shielded room where the measurements were carried out (see Figure 9.29a). The equivalent circuit is shown in Figure 9.29b and, according with Equations (9.28), has

$$\hat{V}_T = -\Delta\hat{V}_{CC}/2 \tag{9.36a}$$

$$C_T = 2C_{gnd\text{-}en} \tag{9.36b}$$

The voltage $\Delta\hat{V}_{CC}$ in the frequency domain was calculated starting from its representation in the time domain Δv_{CC} as obtained by the simulation described in *Example 8.3* (see Figure 8.20b), and shown again in Figure 9.30a with a different timescale. The spectrum for each board obtained by FFT is shown in Figure 9.30b. With these values, and estimating $C_{gnd\text{-}en} \approx$ 10 pF, the current along the cable and the corresponding radiated E-field were computed in the same manner as in *Example 9.8*.

Measured (harmonics) and computed (profile) radiated E-fields for the three considered cases are shown in Figure 9.31, where a good agreement can be observed. For comparison, measured emission profiles in the case of a PCB without an attached cable are also shown by the dashed line. Looking at these results, and adopting as reference the case of an STD board without decoupling, the radiated emission from an STD board with decoupling is about 3 dB lower, while that from a BC board is 12 dB lower. Moreover, maximum emission due to the cable corresponds to the resonance frequency of the PCB + cable structure. This occurs when the length of PCB + cable is a quarter of the wavelength, i.e. 75 MHz in the case considered. For a frequency above 200 MHz, the radiated emission is dominated by the contribution of the IC loops in the board. This consideration is very important, as the transmission-line model for the cable is valid up to about 300 MHz, as mentioned before. Note that, owing to the perfect decoupling of the buried capacitance solution (no parasitic inductance effect), the cable does not make any contribution to the radiation.

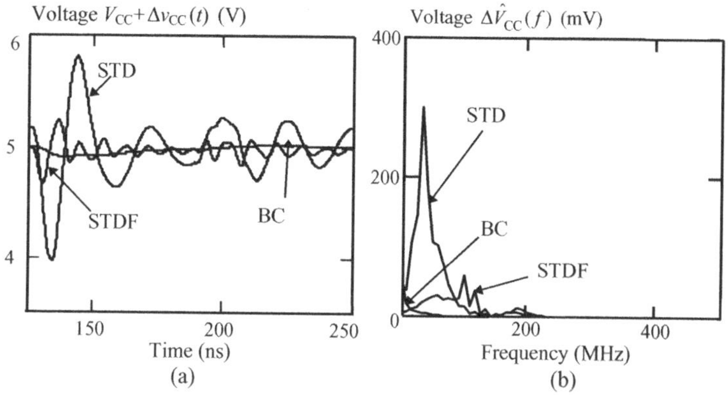

Figure 9.30 Simulated ΔI-*noise* voltage for three types of filtering: Standard Board without decoupling (STD); Standard Board with decoupling (STDF); PCB with buried decoupling (BC). (a) Time-domain waveforms; (b) frequency-domain waveforms

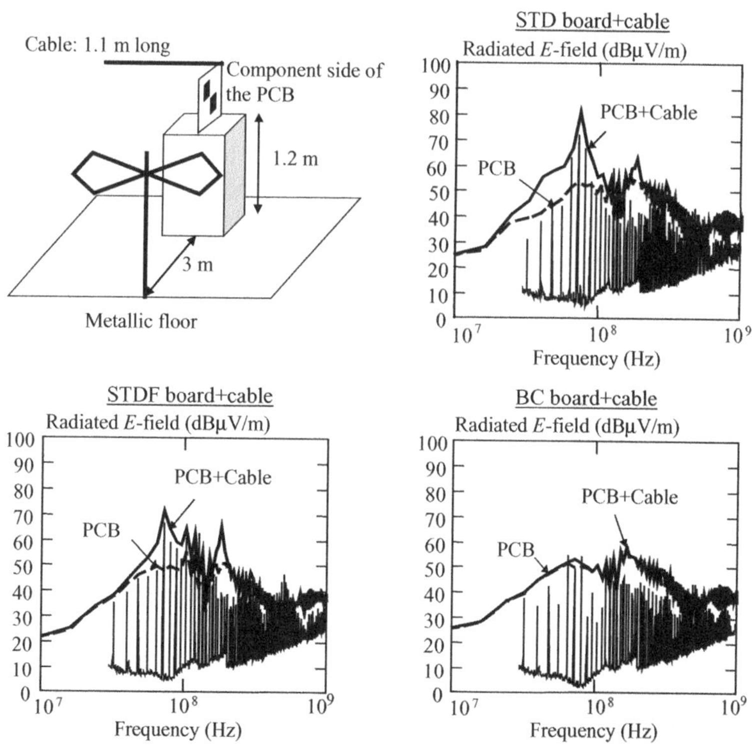

Figure 9.31 Radiated emission set-up with SBC-test boards. Measured (harmonics) and computed (profile) radiated emission with boards without cable (PCB) and with boards with attached cable (PCB + Cable)

In conclusion, the experiment presented in this example on a multilayer PCB with an attached cable has demonstrated that:

- The *common-mode* current on the cable responsible for radiated emission can be predicted by applying the equivalent circuit of the *voltage-driven* mechanism, where the source is the ΔI*-noise* voltage between the power and ground planes.
- It is very important to mitigate this noise by an appropriate filtering.
- The radiated emission from the cable dominates over that due to the circuits in the low-frequency range 30–300 MHz, as already verified for the *current-driven* mechanism.

9.7 Differential Drivers as Sources of Emission

Differential drivers and receivers are used in multilayer PCBs for very high-speed systems, and for communicating between racks and subracks by means of twinax or twisted-pair cables (this will be examined in *Chapter 12*). The choice is dictated by the high level of immunity offered by this type of transmission compared with that provided by a single-ended interconnect. The internal noise of a digital system, such as *crosstalk*, or external interference, such as radiated fields from broadcast antennas or from electrostatic discharge events, couples with the differential line as *common-mode* noise which is rejected by the receiver recognizing *differential-mode* signals only. To avoid conversion of *common-mode* noise into *differential-mode* noise, which is dangerous for signal integrity, a perfect balance of the two signal conductors with respect to the reference conductor is required. This is easily accomplished, for instance, in shielded twisted-pair cables where the shield of the cable is the reference. In theory, if the immunity is brought to an extremely high level with *differential-mode* signaling, the radiated emissions should also be far away from the regulation limits owing to the quasi-perfect cancellation of the fields generated by two conductors transporting the same current but in opposite directions. This is particularly true if the noise in the PCB which could feed the cable as an antenna is minimized. Unfortunately, this does not occur if no attention is paid to the *common-mode* current produced by the driver owing to the intrinsic unbalance of the output.

9.7.1 Common-Mode Current with Differential Drivers

Section 9.6 studied closely and systematically how the noises on ground and power planes of a PCB can act as sources of emission for cables attached to the board. Now, a third source of emission will be investigated, represented by differential drivers used for signal transmission on cables. This is schematically shown in Figure 9.32.

Ideally, a driver should have only two voltage sources $\hat{V}_{\mathrm{DM}}/2$ equally located in the output equivalent circuit. If this occurs and the interconnect is perfectly symmetric, only *differential-mode* current is present on the cable conductors, as defined in *Section 6.2*. Unfortunately, this does not occur in practice owing to the imbalance between the rise and fall times of the driver differential outputs. Therefore, at the equivalent circuit of the driver it is necessary to add the voltage source $\hat{V}_{\mathrm{CM}}$ which depends on the driver type. This aspect will be investigated for three popular types of driver in the following experiments.

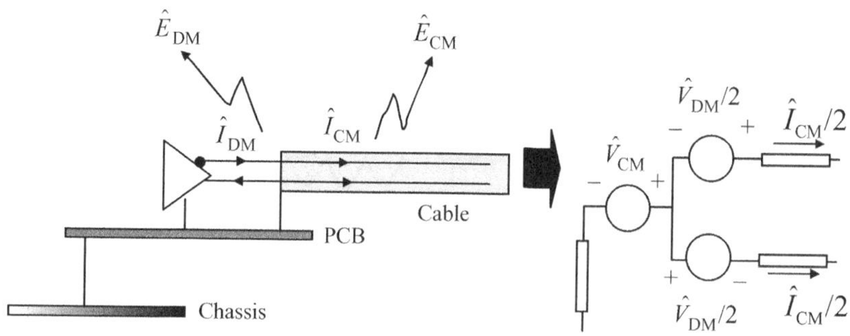

Figure 9.32 Schematic representation of common and differential currents produced by a differential driver (on the left) and its equivalent circuit (on the right)

9.7.2 *Radiated Field Mechanism of UTP and SFTP Cables*

The mechanism producing a *common-mode* radiated field with *Unshielded Twisted-Pair* (UTP) cables is illustrated in Figure 9.33, where a link between two shielded boxes is considered. Within each box there is a PCB with a differential driver and receiver respectively. The *common-mode* source $\hat{V}_{CM}$ of Figure 9.32, owing to the skew and the difference in t_r and t_f of the single-ended output waveforms, produces the *common-mode* current $\hat{I}_{CM}$ transmitted along the cable and responsible for radiated emission. A radiating loop is created between the UTP cable and the metallic ground floor of the shielded chamber for radiated emission measurements. The current $\hat{I}_{CM}$ returns to the source through the metallic ground floor of the chamber and the vertical wall of the shielded boxes. Therefore, contributions for both polarizations of the antenna used for radiated emission measurement are present.

With a *Shielded Foil Twisted-Pair* (SFTP) cable instead of the UTP in the structure of Figure 9.33, the current $\hat{I}_{CM}$ flows through each wire and returns to the driver, using as its path the internal side of the shield, as shown in Figure 9.34. Owing to the cable and connector transfer impedances, denoted by $\hat{Z}_{tc}$ and $\hat{Z}_{tg}$ respectively, the distributed voltage sources $\Delta\hat{V}_c$ and lumped $\hat{V}_g$ are generated on the external sides of the shield and of the connector respectively [36–41]. These voltage sources have the expressions indicated in Figure 9.34 and produce

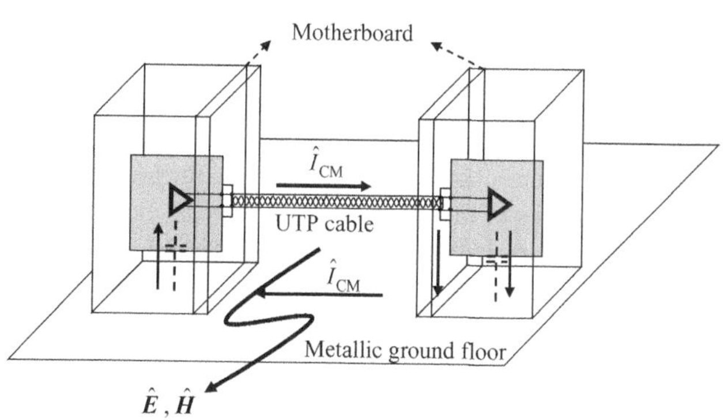

Figure 9.33 Radiated field mechanism with UTP cable

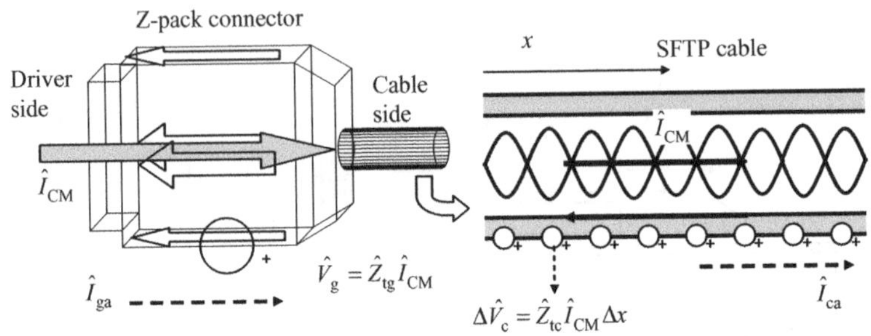

Figure 9.34 Transfer impedance of shielded twisted-pair (SFTP) cable: (a) schematization of the radiated emission mechanism; (b) measured transfer impedance of an SFTP Cat.5e cable of 30 cm length

antenna currents $\hat{I}_{ga}$ for the connector and $\hat{I}_{ca}$ for the cable that are responsible for the radiated emission. As for the UTP cable, a radiating loop is created between the cable and the metallic floor of the chamber, with the difference that in this case $|\hat{I}_a| \ll |\hat{I}_{CM}|$, where $\hat{I}_a = \hat{I}_{ga} + \hat{I}_{ca}$.

The transfer impedance $\hat{Z}_{tc}$ of shielded cables can be approximately represented as the sum of a resistive part R_{tc},which dominates up to 1 MHz, and an inductive part L_{tc}, which dominates over 10 MHz:

$$\hat{Z}_{tc} \approx R_{tc} + j\omega L_{tc} \tag{9.37}$$

The low-frequency resistance R_{tc} coincides with the DC resistance of the shield.

To investigate the performance offered by commercial devices, connectors, and cables more thoroughly, an experiment was performed.

Example 9.11: Calculations and Measurements of Emission from D-Test Boards with Differential Drivers and Receivers Interconnected by UTP and SFTP Cables

The investigation was performed with the set-up schematically shown in Figure 9.35, where the position of the current probe for conducted emission measurements on the cable and the position of the antenna for radiated emission measurements can be seen [42]. Consider that:

- The conducted (30 MHz–300 MHz) and radiated (30 MHz–1 GHz) measurements were carried out in a semi-anechoic shielded room for 3 m antenna distance.
- A Fisher-33 current probe was used to measure *common-mode* current on the cable in the range 30 MHz–300 MHz, where usually in a digital system the emission from the cable dominates over the other sources of radiation.
- UTP/SFTP Cat.5e cables of 1 m length were used to link test boards placed inside two shielded boxes (see Figure 9.36a).
- Three typologies of commercial connectors were employed for performance comparison: 5 × 2 pin Z-pack (also known as *hard metric 2 mm pitch connector*), 9 pin D-Sub, and RJ45. They were plugged in to allow connection between the two shielded boxes by SFTP cables, as shown in Figure 9.36b. For a UTP cable, only the unshielded 5 × 2 pin Z-pack connector was used.

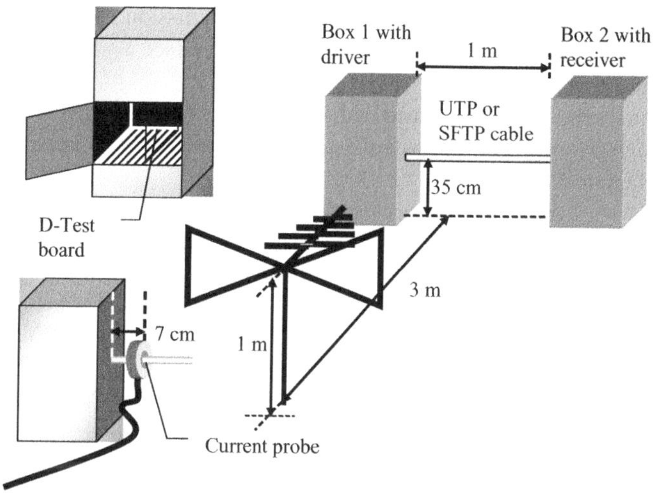

Figure 9.35 Experimental set-up with differential drivers and receivers in a semi-anechoic chamber. The shielded boxes are in contact with the metallic floor of the room

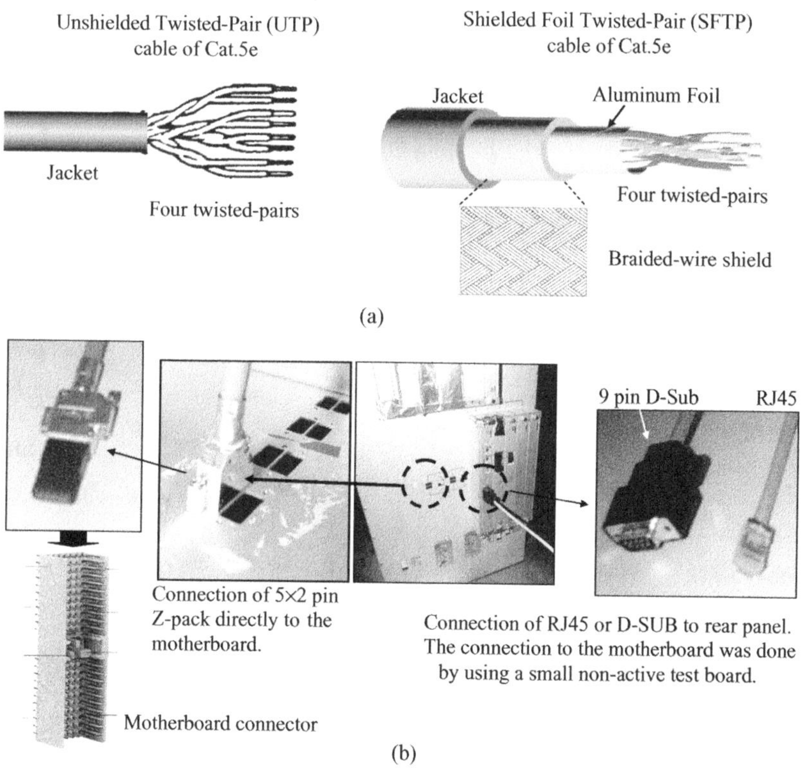

Figure 9.36 Hardware details: (a) UTP and SFTP cables; (b) shielded connectors and shielded box used for measurements

- The test boards were designed to allow differential transmission by RS422 devices, and high-speed LVDS and LVPECL drivers/receivers were used (see *Section 12.1* for more details regarding these devices).
- The layout of the test board was designed with care in order to avoid significant contribution of the *voltage-driven* and *current-driven* mechanisms to the *common-mode* current on the cable, as described in *Section 9.6*.
- The components were positioned very close to the connector of the multilayer PCB with appropriate filtering, and very short interconnects were realized.
- The working frequency was 16 MHz. The choice was dictated by the fact that the RS422 component is characterized by much slower speed (maximum 50 MHz of clock) than LVDS and ECL, but it has more *common-mode* rejection, ±7 V as against ±1 V, and is able to transmit signals at longer distances.

(i) Measurements

The structure of the links is shown in Figure 9.37, together with the measured single-ended voltages at the receiver (i.e. voltages between one wire and the reference ground) and the differential voltages at the terminations. The RS422 driver has typical rise/fall times of

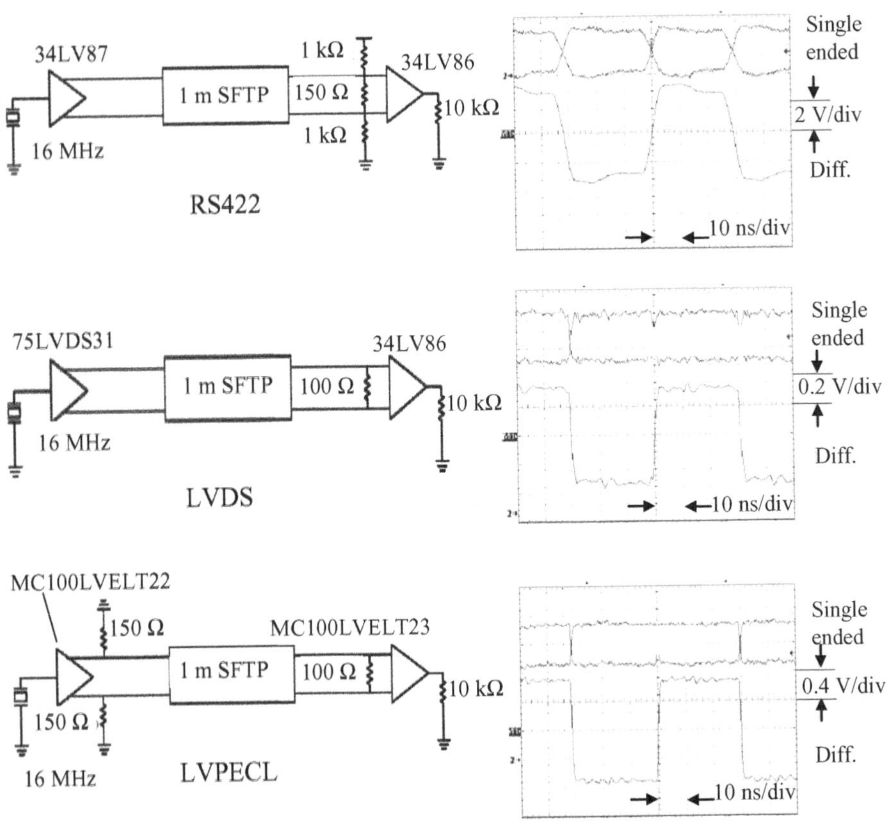

Figure 9.37 Links used for differential transmission and signal measurements at the terminations: voltage between one end and ground (single ended) and between two ends (differential)

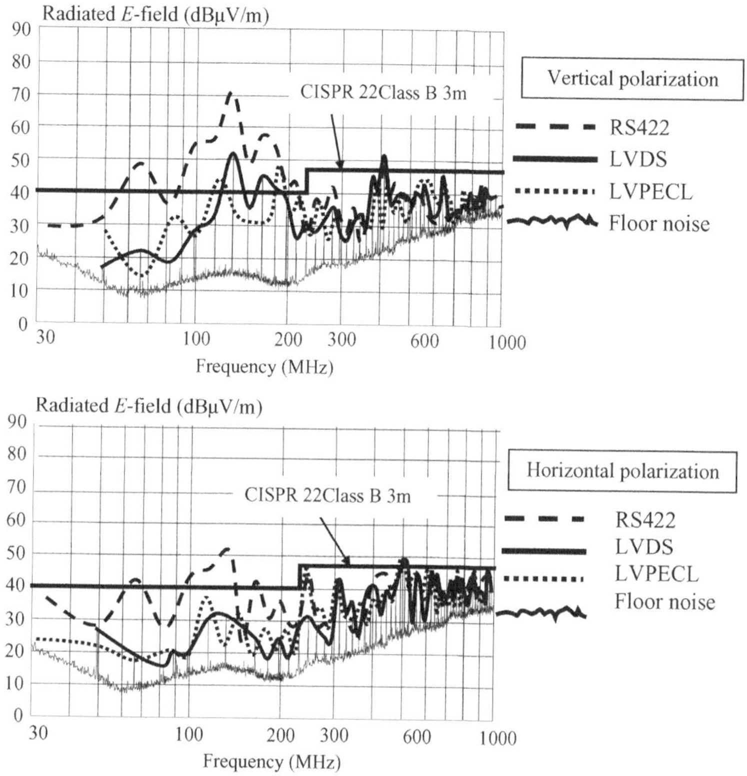

Figure 9.38 Radiated emission measurement profiles with three types of driver, where the link consists of: unshielded 5 × 2 pin Z-pack connectors and UTP cable category 5a. The profiles are the envelopes of the maximum value of the harmonics. The harmonics are shown for an LVDS driver only

4.2/4.7 ns. Observe the faster rise/fall times of the LVDS (typically 0.35 ns) and PECL (0.4 ns) devices which are matched to the characteristic impedance of the cable. The link with the RS422 was partially matched, as a polarized termination is required to maintain one logic level at the receiver even when the cable is disconnected. The other links do not have this problem.

Measurements of conducted emission current performed between 30 and 300 MHz revealed an emission profile for RS422 of around 30 dBμA, with a current peak of 45 dBμA at the eighth harmonic (i.e. 128 MHz). The other two devices exhibited an emission profile 20 dB lower [42]. This means that the RS422 device has less symmetry in the output waveforms than the other two devices.

The measurements of the radiated emission profiles with a UTP cable and the three types of driver are shown in Figure 9.38. For LVDS, harmonics and emission profile are both shown, while for the other two devices the emission profile only is shown for comparison purposes. Analysis of the results leads to the following observations:

- In vertical polarization, all the drivers have an emission profile above the CISPR 22 limit for Class B equipment.
- Between 100 and 200 MHz, the RS422 driver exhibits an emission profile about 20 dB higher than the other two drivers which have comparable values.

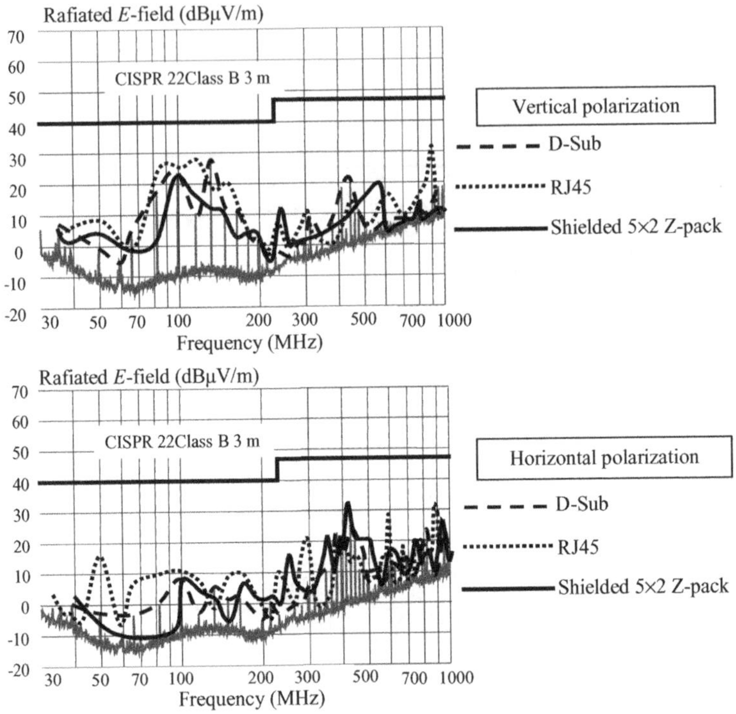

Figure 9.39 Radiated emission measurement profiles with three types of connector and driver RS422. The profiles are the envelopes of the maximum value of the harmonics. The harmonics are shown for a shielded 5 × 2 Z-pack only

- Radiated emission measurements confirm a direct relationship with conducted emission measurements.
- Radiated emission measurements with the antenna in vertical polarization give higher values than those with the antenna in horizontal polarization in the low-frequency range, although the cable is parallel to the metallic floor of the chamber. This can be explained by considering the *common-mode* current path in Figure 9.33.

As the RS422 device provides the worst emission profile, it was used to carry out radiated emission measurements with an SFTP cable and the three types of shielded connector. The results are shown in Figure 9.39. Observe that the emission profiles are now well below the regulation limit, and the tested connectors offer comparable EMC performances. To understand the mechanisms that govern these types of emission, the measurements were reproduced by simulations with RS422 devices.

(ii) Simulations

A SPICE-oriented circuit model of the set-up was developed and verified experimentally in order to investigate the radiated emission mechanism and to predict the emission levels. The circuit model includes:

- a macromodel of the driver;
- a transmission-line model of the connectors and cable;
- a lumped model of the termination.

The same circuit can allow analysis in the time and frequency domains for signal integrity and emission investigation. The circuit simulations to estimate radiated emissions were developed in accordance with the following assumptions:

- The RS422 driver, UTP cable, and SFTP cable with a 5×2 pin Z-pack were used to validate the model experimentally, as the other two connectors have the same EMC performance.
- A simple model for the differential driver was used with two voltage sources having as excitation a trapezoidal waveform with skew and different rise/fall times.
- As the cable is 35 cm high from the reference ground plane, the TL model is valid with good approximation up to 300 MHz, where the wavelength is 3 m.
- The cables were modeled with a cascade of five lossless transmission lines (TLs) of equal length and having as reference ground plane the metallic floor of the semi-anechoic chamber. The distributed model outlined in *Section 6.4* for two coupled lines was used for each section of the cable.
- The per-unit-length inductance and capacitance matrices of the UTP/SFTP cables were numerically calculated in order to take into account the dielectric of the insulation surrounding the wires.
- The link between the internal part of the shielded cable where the *common-mode* current $\hat{I}_{CM}$ flows and the external part of the shielded cable where the antenna current $\hat{I}_{an}$ responsible for the radiated emission flows was modeled by the transfer impedance of the cable $\hat{Z}_{tc}$ and of the connectors $\hat{Z}_{tg}$.
- The value of the inductance $L_{tc} = 0.1$ nH/m at the frequency $f = 100$ MHz was extrapolated from the measurements performed by the triaxial method on a SFTP sample of 30 cm length [42].
- The transfer impedance of the shielded connectors is inductive. For the Z-pack connector this impedance was measured by other researchers [41], resulting in an impedance of 0.11 Ω at 300 MHz. As, at this frequency, the SFTP cable has an estimated impedance of 0.19 Ω calculated by Equation (9.37) with $L_{tc} = 0.1$ nH/m, and neglecting the DC term, it is possible to assert that the emission from the cable dominates over the emission from the connectors, especially at high frequencies.
- The boxes were simulated as perfect short circuits for the SFTP case.
- The radiated field was calculated by dividing the cable into five segments and calculating the *common-mode* current $\hat{I}_{CM}$ in each segment.
- The radiated fields were computed at a distance of 3 m, considering the horizontal and vertical paths of the *common-mode* current and the image effect of the metallic floor of the semi-anechoic chamber and using the closed-form equations of *Appendix D* for both E-field polarizations.

Details about the circuit model of the full structure are given by Caniggia and Santi [42]. Here, for brevity, the results of simulations are shown and comments are provided.

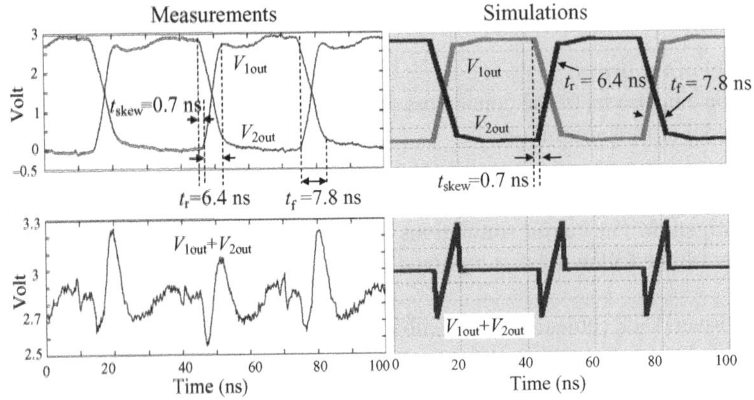

Figure 9.40 Differential-mode and common-mode voltages in UTP cables: measured and simulated waveforms of single-ended signals V_{1out} and V_{2out} at the driver output, and the corresponding common-mode voltage $V_{1out} + V_{2out}$

As a first validation of the circuit model for *common-mode* current calculation, the single-ended voltages V_{1out} and V_{2out} at the driver output and their sum with the UTP cable attached were compared with measurements, as shown in Figure 9.40. The measured and simulated waveforms use the same vertical scale. Note that the measured skew is 0.7 ns (i.e. the difference between the complementary rise and fall time parameters divided by 2). This skew is responsible for *common-mode* emission, and it was used in the simulations.

Measurements (envelope) and circuit simulations (harmonics) of the radiated fields for horizontal polarization of the antenna are shown in Figure 9.41a. The simulations were performed by harmonics because in the frequency domain the two voltage sources simulating the differential driver output were modeled by the Fourier series of a periodic trapezoidal waveform with different rise and fall times [1], that is, with $t_r = 6.4$ ns and $t_f = 7.8$ ns, as shown in Figure 9.40, in order to reproduce the skew. The results confirm that, in the low-frequency range of 30–300 MHz, there is a difference of about 40 dB between UTP and SFTP cables owing to the action of the transfer impedance parameter of the SFTP cable.

Measurements (envelope) and circuit simulations (harmonics) of the radiated fields for vertical polarization of the antenna are shown in Figure 9.41b. Simulated results confirm that the emissions with the antenna in vertical polarization are higher than those with the antenna in horizontal polarization owing to the path of the *common-mode* current $\hat{I}_{CM}$ returning to the source, as shown in Figure 9.33.

(iii) Conclusions

In conclusion of this experiment, the following remarks can be made:

- The main source of *common-mode* current and emission with a digital differential transmission using UTP/SFTP Cat.5e cables is the imbalance between the rise and fall times of the differential outputs of the driver.
- The driver for the RS422 standard produces more significant *common-mode* current than the LVDS and LVPECL drivers which have the drawback of less *common-mode* immunity.

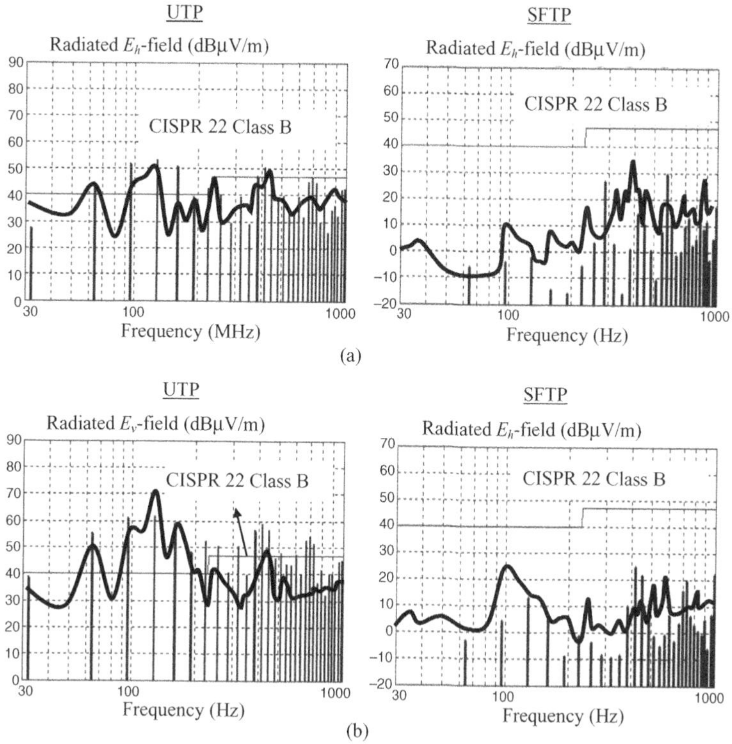

Figure 9.41 Measured (profile) and computed (harmonics) radiated E-fields for UTP and SFTP cables: (a) in horizontal polarization; (b) in vertical polarization

- Shielded connectors such as a 5×2 pin Z-pack, RJ45, and 9 pin D-Sub provide the same EMC performance as with SFTP Cat.5e cables.
- The *common-mode* emission mechanism reproduced by a circuit model approach suitable for SPICE has shown a satisfactory accuracy by comparison with measurements, in spite of the simplifications used for the radiated field computation.
- Although the cable that links the two boxes is in the horizontal position, the radiated emission with the receiving antenna in vertical polarization dominates in the low-frequency range owing to the difference in phase and path of the *common-mode* currents on the boxes. This was also verified by numerical simulation, as reported by Caniggia and Santi [42].
- Measurements and simulations show that, to comply with the CISPR22 Class B limit, shielded cables and connectors are required with RS422 differential drivers and receivers. With unshielded cables, EMI filters such as transformers, chokes, and capacitors should be used at the driver output, as will be discussed in *Section 9.8* and *Section 10.3*.

9.8 Emission from a Complex System

The main goal of a system regarding radiated emission is to be compliant with the limits required by the standards. To this end, a specific design strategy must be adopted. In general, for standing floor equipment, the layout required for measurement is that sketched in Figure 9.42a. The rack contains PCBs, cables, and power supplies. The external cables are positioned to maximize the emissions. In a bottom-up design procedure, it is very important to take care of the following:

(a) The choice of the logic families to be employed or, more generally, the technology TTL, CMOS, ECL, LVDS, etc. Bandwidth, currents, and transition edges must be chosen strictly to satisfy the technical requirements. For instance, devices with a very high-speed edge should not be used when not required by timing specifications.
(b) The PCB layout should have a sufficient number of ground planes. In this way it is easier to minimize the length of the interconnects and the associated loop area.
(c) Shielded racks are often necessary for complex systems composed of several PCBs and cables, although multilayer PCBs are used.
(d) Screened and/or filtered cables must be used for the interconnect between racks.
(e) All screened cables leaving the shielded structure must be tied with a 360° contact to the rack.

With shielded racks, two main sources of emission should be distinguished:

- cables (UTP, SFTP, coaxial, triaxial, power supply, etc.);
- apertures (holes, narrow slots, etc.).

These two issues will be investigated in this section. Simple models to predict the emission from shielded cables and from apertures are presented and validated by comparison with measurements.

Example 9.12: Measurements of Emission from a Shielded Rack Equipped with PCBs and Power Supply
To investigate how apertures and cables act as sources of emission, an experimental shielded rack equipped as shown in Figure 9.42 was placed in a semi-anechoic chamber for 3 m measurements. The *Equipment Under Test* (EUT) consists of DC/DC converter for PCB power supply, switching digital devices in the PCB, and outgoing cables for signals and power supply. Power supply for DC/DC converters enters into the rack by filtered cords at the top of the rack.

Clamped and unclamped cable installations were considered:

- Unclamped cable means that the shield of the outgoing cable passes through the rack without touching it.
- Clamped cable means that the shield of the outgoing cable passes by a good contact with the rack at 360°.

This investigation shows that radiation from cables (i.e. *common-mode* emission) dominates at low frequencies according to the law stating that, after the first breakpoint frequency of the trapezoidal signal spectrum, the emission profile is constant and decreases at a rate of −20 dB/dec after the second breakpoint frequency. Besides, the radiation from the circuits within the racks dominates at high frequencies according to the law stating that, after

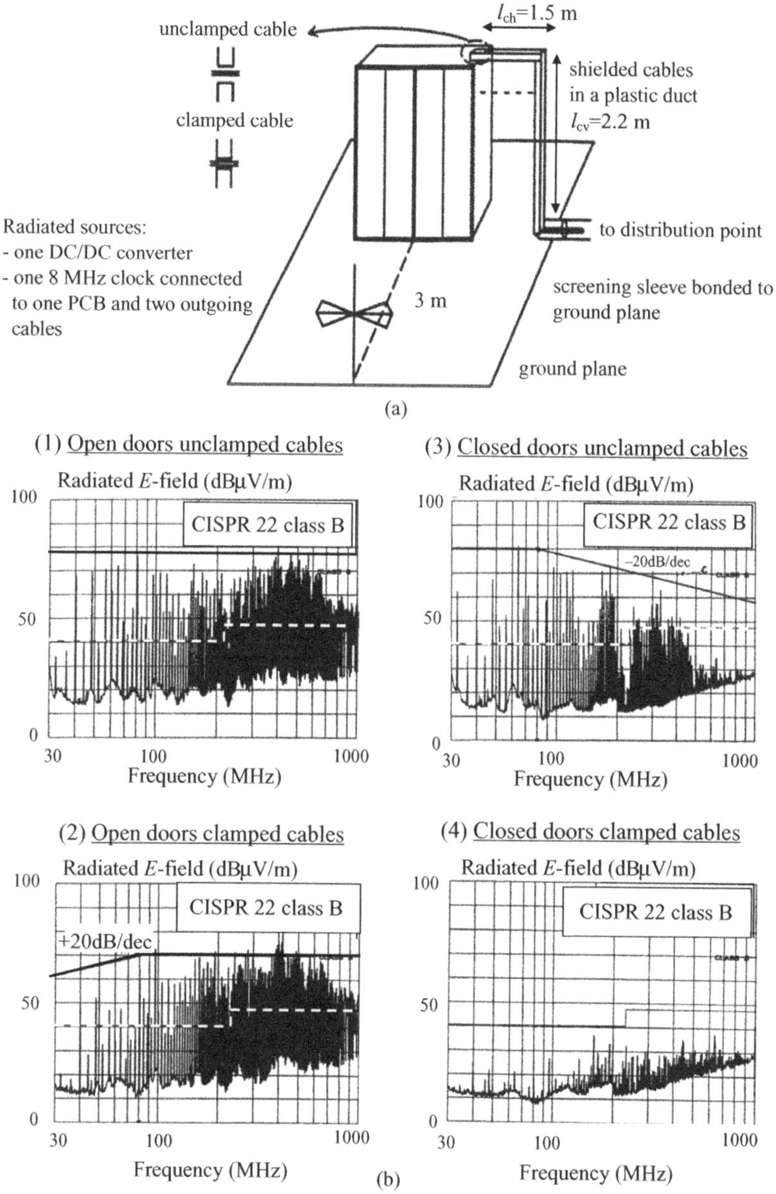

Figure 9.42 Radiated emission at system level: (a) measurement set-up; (b) radiated emission measurements at 3 m with the antenna in horizontal polarization for different conditions: (1) open doors and unclamped cables; (2) open doors and clamped cables; (3) closed doors and unclamped cables; (4) closed doors and clamped cables

the first breakpoint frequency, the emission profile increases at a rate of +20 dB/dec and after the second breakpoint frequency is constant at its maximum value, as demonstrated in *Section 9.2*. In this experiment, the PCB was equipped with TTL devices having an average switching time $t_r = 4$ ns and therefore a second breakpoint frequency of $1/\pi t_r = 80$ MHz.

The measured radiated E-fields are shown in Figure 9.42 for the following different conditions of the system:

1. *Open doors and unclamped cables.* The emission profile is substantially flat in the entire frequency range 30–1000 MHz. The harmonic emissions produced by the 8 MHz clock are about 40 dB above the CISPR 22 Class B limit for 3 m measurements.
2. *Open doors and clamped cables.* The emission profile is that typical of *differential-mode* emission, characterized by an increase at a rate of +20 dB/dec up to the second breakpoint frequency of 80 MHz, and remaining constant above this frequency. This means that the emissions are due to the circuits.
3. *Closed doors and unclamped cables.* The emission profile is that typical of *common-mode* emission, characterized by a constant value up to the second breakpoint frequency of 80 MHz, and decreasing above this frequency with a slope of −20 dB/dec. This means that the emissions are due to the external cables.
4. *Closed doors and clamped cables.* The emissions are reduced drastically in the entire frequency range and are below the CISPR 22 limit for Class B equipment. In this case, the measurement is about the contribution of the current flowing on the external part of the cable shield and of the small apertures in the shielded rack.

To obtain the results of point 2, it is very important to connect the shield of the cable to the rack with very good contact at 360° and not to use wire or a similar practice such as the one usually indicated as *pigtail*. The apertures for cooling or visual inspection should have a maximum dimension much smaller than the minimum wavelength corresponding to the maximum frequency of interest, which for CISPR 22 is 1 GHz, and therefore $\lambda_{min} = 30$ cm.

In the following, models for predicting the radiated emission from cables and apertures will be outlined and experimentally verified.

9.8.1 Emission Model of Coaxial Cables

In *Section 9.7*, a circuit model described by Caniggia and Santi [42] for computing the antenna current on the external part of the shield of an SFTP cable was used and experimentally verified. In this section, closed-form expressions are provided and validated by measurements for coaxial cables.

Radiation from coaxial cables is due to the signal current inside the cable $\hat{I}_c$ and to the per-unit-length transfer impedance $\hat{Z}_t$ of the cable (see Figure 9.43a). These quantities generate distributed voltage sources $d\hat{V}_{dis}(f,x) = \hat{Z}_t(f)\hat{I}_c(f,x)dx$ on the external part of the cable shield, as schematically shown in Figure 9.43b, that produce the circulation on the shield of a current indicated as antenna current $\hat{I}_a(f,x)$, as it is responsible for the radiation from the cable. The antenna current can be calculated by simple transmission-line closed-form expressions under the assumptions of neglecting cable losses and of weak coupling between the internal and external parts of the cable [39, 43]. Note that both assumptions are conservative, as a worst-case emission is obtained from them.

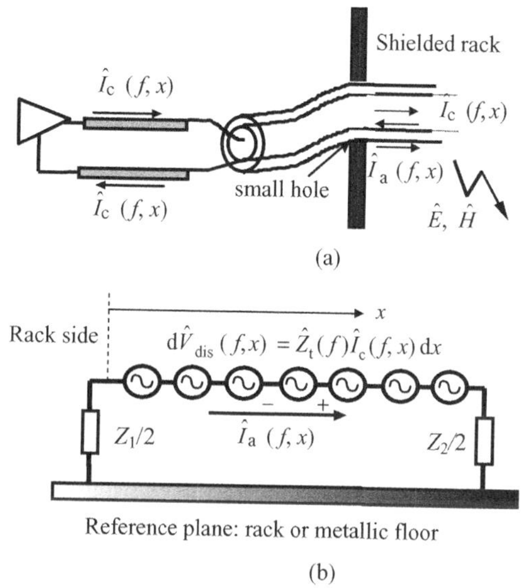

Figure 9.43 Calculation of radiated fields from a signal cable outgoing from a shielded rack: (a) schematic representation of the structure; (b) equivalent circuit based on the TL model for the part of the cable outside the rack

The antenna current $\hat{I}_a(f, x)$ is then given by [43]

$$\begin{aligned}\hat{I}_a(f, x) = \hat{K}_1 \int_0^x 2\hat{Z}_t(f)\hat{I}_c(f, \xi)(Z_0 \cos(\beta\xi) + jZ_1 \sin(\beta x))\, d\xi \\ +\hat{K}_2 \int_x^l 2\hat{Z}_t(f)\hat{I}_c(f, \xi)(Z_0 \cos(\beta(l - \xi)) + jZ_2 \sin(\beta(l - \xi)))\, d\xi\end{aligned} \tag{9.38}$$

$$\hat{K}_1 = \frac{Z_0 \cos(\beta(l - x)) + jZ_2 \sin(\beta(l - x))}{Z_0 \hat{D}} \tag{9.39a}$$

$$\hat{K}_2 = \frac{Z_0 \cos(\beta x) + jZ_1 \sin(\beta x)}{Z_0 \hat{D}} \tag{9.39b}$$

$$\hat{D} = (Z_0 Z_1 + Z_0 Z_2)\cos(\beta l) + j\left(Z_0^2 + Z_1 Z_2\right)\sin(\beta l) \tag{9.39c}$$

$$Z_0 = 120 \ln(2h/a) \tag{9.39d}$$

$$\hat{I}_c(f, \xi) = \frac{\hat{V}_S(f)}{Z_S + Z_L} e^{-j\beta_c \xi} \tag{9.40}$$

where x is the coordinate along the cable starting from the rack, l is the length of the line, a is the radius of the shield of the cable, Z_0 is twice the characteristic impedance of the shield-ground TL (i.e. round wire of radius a placed at height h above a ground plane), $\hat{Z}_t(\omega)$ is the transfer impedance of the cable (which can be modelled by the approximate expression $\hat{Z}_t = R_t + j\omega L_t$), $\hat{I}_c(f, \xi)$ is the signal current within the cable in matched condition (with

$\hat{V}_S(f)$ being the source voltage, Z_S the source impedance, and Z_L the load impedance equal to the characteristic impedance of the internal path of the cable), $Z_1/2$ and $Z_2/2$ are respectively the termination impedances of the cable shield at the rack side and at the other end with respect to the ground metallic floor (the factor 1/2 is used because image theory is applied to the structure of a wire above a ground plane [43]), $\beta_c = 2\pi/\lambda_c$ is the phase constant of the TL internal to the cable, $\beta = 2\pi/\lambda$ is the phase constant in free space, $\lambda = c/f$ is the wavelength in meters, $c = 1/(\mu_0\varepsilon_r\varepsilon_0)^{0.5}$ is the velocity of propagation, and f is the frequency in Hertz.

If the cable layout has horizontal and vertical paths of length l_{ch} and l_{cv} respectively, as in the set-up shown in Figure 9.42, in a first approximation the characteristic impedance is assumed to be equal for the two paths. Setting the distance between the cable and the reference metallic plane (i.e. rack or ground floor) equal to $(l_{ch}l_{cv})^{0.5}$, Equation (9.39d) becomes $Z_0 = 120\ln(2\sqrt{l_{ch}l_{cv}}/a)$. Moreover, in this case the cable length in Equations (9.38) and (9.39) is the total length of the cable outside the rack $l = l_{ch} + l_{cv}$.

Note that the integrals in Equation (9.38) may be solved using the symbolic calculation of commercial mathematical programs such as MathCad. More details regarding this calculation procedure are reported by Caniggia *et al.* [35] and Smith [43].

A typical structure of a coaxial cable with a braided shield is shown on the left in Figure 9.44. Note that the braid is characterized by very small holes, responsible for magnetic and electric field leakage at high frequencies. Typical measured values of transfer impedance $\hat{Z}_t(\omega)$ as a function of the number of braids are shown in Figure 9.44 [44]. The dotted

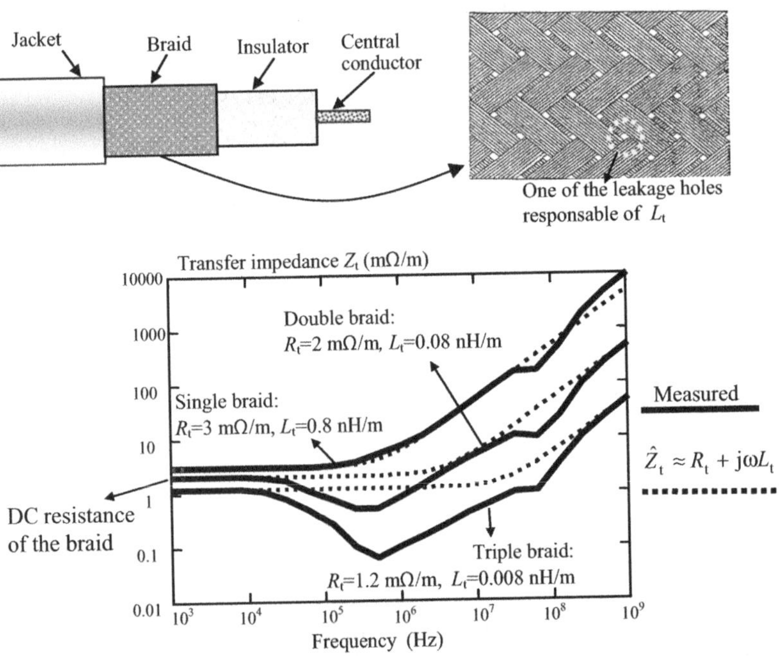

Figure 9.44 Typical structure of a coaxial cable and typical Z_t values for coaxial cables with different layers of braids

lines are an extrapolation for obtaining R_t and L_t parameters for the approximate expression $\hat{Z}_t = R_t + j\omega L_t$. Note that the transfer impedance of the triple braid up to 1 MHz is very similar to a solid shield having $\hat{Z}_t(\omega)$ tending to zero for higher frequencies. Other considerations and examples of transfer impedances can be found in reference [45]. As stated in *Section 9.7*, the transfer admittance due to the capacitive leakage can be neglected for this type of prediction because the shielded cable is usually well connected to ground at least at one end, where the active circuits are located, to guarantee the required EMC performance.

Example 9.13: Calculations and Measurements of Radiated Emission from a Coaxial Cable
To verify the influence of the transfer impedance on the radiated emission, the following structure was considered: a coaxial cable of 1.15 m length in a vertical position, fed by an 8 MHz oscillator driving a digital gate with 50 Ω output resistance, as shown in Figure 9.45. The circuit was put within a shielded box installed on the ground plane of the chamber, in order to avoid the contribution of the circuits to the emission measurements. The cable was terminated with its characteristic impedance.

The radiated emissions obtained with two types of coaxial cable outgoing from the shielded box are shown in Figure 9.45. The cables considered were the RG58 cable with one braided shield and the RG214 cable with two braided shields. The radiated emissions are due to the parameter $\hat{Z}_t(\omega)$ only because an optimal bonding of the shielded cable to the enclosure by

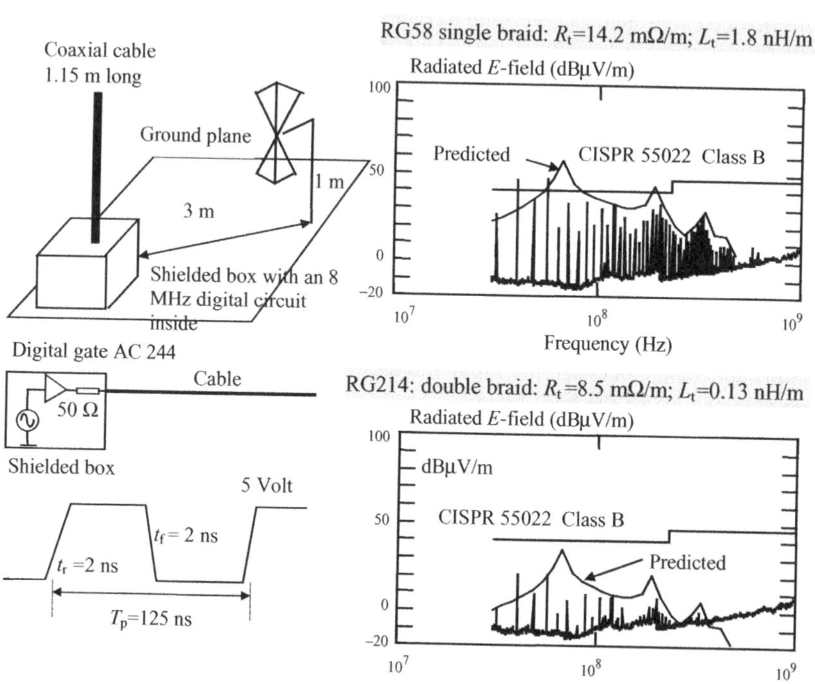

Figure 9.45 Radiated emission from coaxial cables: set-up and equivalent circuit (on the left); measured (harmonics) and computed (envelope) *E*-field for two types of coaxial cable (on the right)

a 360° contact was provided. As expected, emission from a double-braid cable is less than the emission from a single-braid cable. Although $\hat{Z}_t(\omega)$ was modelled by the simplified form (9.37), very good agreement between experimental (harmonics) and analytical (profile) results can be observed. The values of transfer impedances used for computations are indicated in Figure 9.45.

The radiated field was calculated by the procedure of *Section 9.8.1*, adopting the following values: $l = 1.15$ m, $l_{ch} = 1.15$ m, $l_{cv} = 0$, $Z_1 = 0$, $Z_2 = 10^6\ \Omega$, $Z_S = Z_L = 50\ \Omega$. The characteristic impedance Z_0 given by Equation (9.39d) was calculated assuming that the cable–reference plane distance was equal to the cable length $h = l_{ch}$. The average antenna current was calculated by dividing the cable into 10 segments. Image theory for a vertical antenna was applied to calculate the radiated fields (see the long wire in Table D.1 of *Appendix D*, where the metallic floor of the shielded room for measurements is located at $\xi = 0$). According to image theory, the currents above and below the reference plane have the same direction.

Example 9.14: Measurements of Radiation from UTP Cable Filtered with a Common-mode Choke

In *Section 9.7* it was demonstrated by experimental data that an UTP cable is not suitable for complying with the emission limits required by the EMC standards without the aid of some fixes. When the high speed is not a requirement for the transmission of a digital signal by cables, an unshielded cable such as UTP with a *common-mode* filter placed near to the connector of the PCB is one of the most effective methods for reducing *common-mode* currents. This type of filter, referred to as the *common-mode choke* [1], can be effective in blocking *common-mode* currents. In order to provide high impedance to *common-mode* currents, the wire of the filter must be wound around the high-permeability core such that the fluxes due to the two *common-mode* currents add in the core, whereas the fluxes due to the two *differential-mode* currents subtract in the core. To be effective, the choke must ideally have equal self and mutual inductances, i.e. $L = M$. Referring to *Section 3.3*, where the concepts of *differential-mode* and *common-mode* inductances were introduced, this means that the *common-mode* current flowing in each wire finds a high impedance $j\omega(L + M)$, while the *differential-mode* or signal current finds an impedance $j\omega(L - M)=0$. Ideal behavior does not occur because there is a parasitic capacitance between the input and the output terminals that shunts the effective inductance of each wire of the EMI filter.

To verify the efficiency of a choke in the real world, the experiment illustrated in Figure 9.46a was performed. A PCB was built in order to transmit a differential signal to an UTP cable matched with a 100 Ω resistor. The two parallel traces had a *differential-mode* characteristic impedance of 116 Ω and a *common-mode* characteristic impedance of 172 Ω.The signal was an 8 MHz clock provided by an oscillator that drove an RS422 device (34C87). The devices were powered by a DC battery placed very close to the PCB to avoid extra *common-mode* current. The equivalent circuit of the EMI filter used is shown in Figure 9.46a. It consists of a choke plus two capacitors, indicated as $C = 120$ pF, with the task of diverting to ground the *common-mode* currents. These two capacitors are used in practice to avoid the effect of the parasitic shunt capacitances of the choke, which would allow the high-frequency *common-mode* currents to bypass the inductances. The choke was a PLM250H10 of Murata. The cable was realized with two twisted wires in order to simulate an UTP commercial cable with

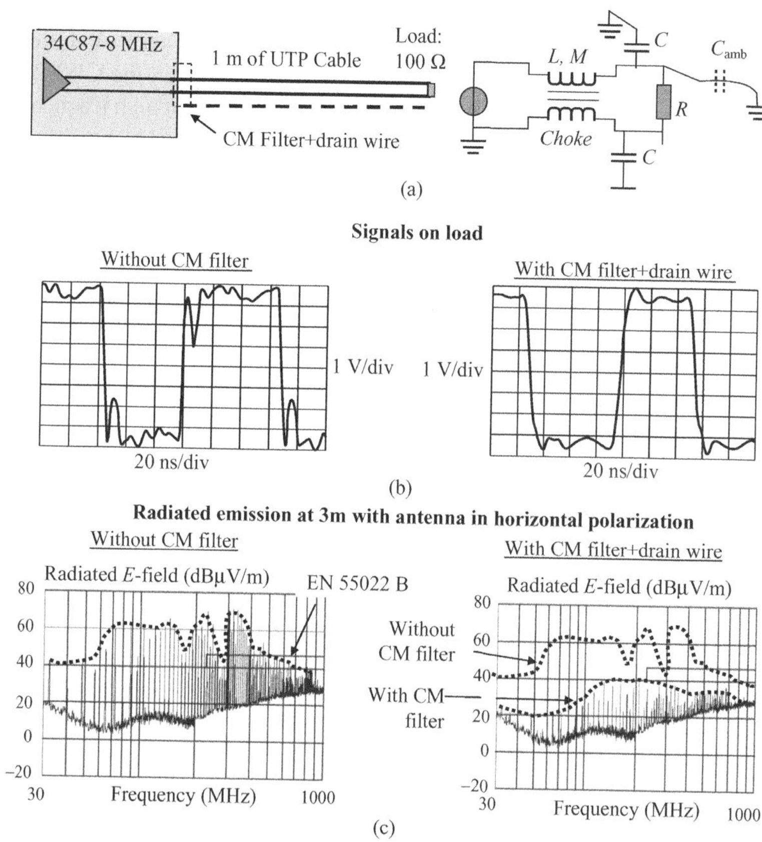

Figure 9.46 Radiated emission from an UTP cable driven by a differential driver 34C87: (a) structure under test and equivalent circuit of the filter; (b) measured signals on the load; (c) measured *E*-field at 3 m with and without a common-mode filter

worst-case EMC performance, and an additional wire connected to the ground of the PCB, in order to simulate the drain wire present in actual UTP cables.

The differential signals on load with and without an EMI filter are shown in Figure 9.46b. Without an EMI filter the signal integrity is not good, as unwanted negative reflections due to the mismatch between PCB and cable are present on the signal. Note that higher rise and fall times and therefore fewer reflections were measured with an EMI filter. The drawback is that the solution of an EMI filter must always be verified in terms of transmission speed, as the switching edges are increased by the filter. If this fix is acceptable, the benefit offered in reducing the radiated emission is evident in the graphs of Figure 9.46c.

For radiated field measurements, the antenna in horizontal polarization was located at 3 m from the PCB with the same height of 1 m from the reference ground of the chamber. The PCB and cable were in the horizontal position. Note that the emission profile is around 60 dBμV/m without filter and drain wire, and diminishes just below the EN 55022 Class B (CISPR 22) limit, especially at higher frequencies, with filter and drain wire. The benefit provided by the drain wire in reducing the emission was verified by comparing the emission profile without and with drain wire in the absence of an EMI filter. It was observed that the emissions remained high at frequencies below 200 MHz and became lower than the CISPR 22 limit above 200 MHz. This can be justified considering that the drain wire, at high frequencies, acts as a return conductor for the *common-mode* current produced by the unbalance of the driver, thereby having a cancellation effect. Other measurements showed that added capacitors of 120 pF to the choke are necessary to have emission below the CISPR 22 limit and to improve the signal integrity. This balances the negative effect of the parasitic shunt capacitances of the choke. For higher clock frequencies, as shown in *Section 9.7*, drivers with less unbalance and faster rise and fall times, as offered by LVDS and LVPECL, should be used. Consider that, by using a commercial UTP Cat.5e cable with good wire symmetry, a better emission profile could be obtained, as there is less conversion of *differential-mode* currents into *common-mode* currents. In *Chapter* 12 it will be shown that the use of a choke plus pulse transformer and capacitors to ground with LVDS devices also offers great benefit in terms of immunity.

9.8.2 Low-Frequency Model of an Aperture

Among the main weak points of a shielded structure are the apertures, such as slots and holes, that are generally present owing to the need to maintain the temperature of the apparatus in the required range for good functionality. To achieve this task, apertures must be numerous, nearby, and of small dimensions compared with the wavelength corresponding to the maximum frequency of interest. For this reason, the *Small-Aperture* (SA) model will be proposed in this section to calculate radiation from apertures. The model is the well-known *equivalent dipole model* which makes it possible to calculate a set of two dipoles, electric and magnetic, to replace the aperture as radiating elements. The method is illustrated in Figure 9.47 [36, 46]. The first picture on the left shows the original problem, which consists in calculating the field at point P owing to an incident field on the left of the metal wall with an aperture. It can be shown that the problem can be solved by short-circuiting the aperture and considering the dipole function of the short-circuited tangential $\hat{H}_{\mathrm{t}}^{\mathrm{sc}}$ and normal $\hat{E}_{\mathrm{n}}^{\mathrm{sc}}$ fields on the aperture. These fields depend on the incident fields $\hat{\boldsymbol{H}}^{\mathrm{i}}$ and $\hat{\boldsymbol{E}}^{\mathrm{i}}$ according to the polarization, as indicated in Figure 9.47. Hence, the fields at point P can be calculated as radiation due to the value of electric $\hat{\boldsymbol{p}}_{\mathrm{e}}$ and magnetic $\hat{\boldsymbol{p}}_{\mathrm{m}}$ dipole moments when image theory is applied. Calculation of these dipoles implies knowledge of the electromagnetic field incident on the aperture and the geometric characteristics of the aperture.

The electric dipole moment $\hat{\boldsymbol{p}}_{\mathrm{e}}$ is perpendicular to the aperture, and the magnetic dipole moment $\hat{\boldsymbol{p}}_{\mathrm{m}}$ is tangential to the aperture plane. The following equations hold [36]:

$$\hat{\boldsymbol{p}}_{\mathrm{e}} = 2\alpha_{\mathrm{e}}\varepsilon\hat{\boldsymbol{E}}_{\mathrm{n}}^{\mathrm{sc}} \tag{9.41a}$$

$$\hat{\boldsymbol{p}}_{\mathrm{m}} = -2\alpha_{\mathrm{m}}\hat{\boldsymbol{H}}_{\mathrm{t}}^{\mathrm{sc}} \tag{9.41b}$$

where α_{e} and α_{m} are the electrical and magnetic polarizability of the apertures.

These are the only existing components of the fields. Values of the polarizability are available in the literature for different aperture shapes [36, 46]. For some typical apertures encountered in practice such as rectangular slots and round holes, see Table D.3 in *Appendix D*.

Once the two dipoles are known, the electromagnetic field at point P can be calculated considering that the problem is dual: the electric dipole moment is $\hat{\boldsymbol{p}}_{\mathrm{e}} = \hat{p}_{\mathrm{e}}z = (1/\mathrm{j}\omega)\hat{I}\ \mathrm{d}lz$,

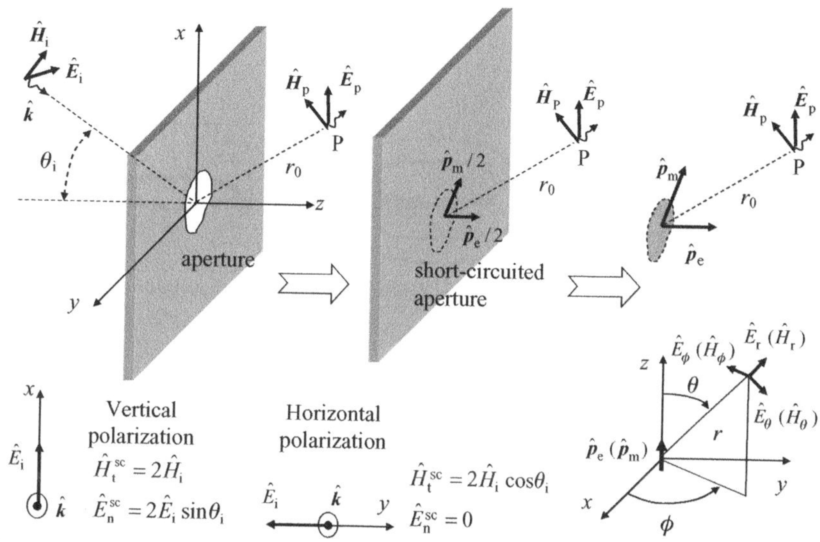

Figure 9.47 Low-frequency approximation by dipole moments (aperture polarizabilities). In the notation, 't' stands for tangential, 'n' for normal, and 'sc' for short-circuited

where $\hat{I}$ is the current of an element of infinitesimal length $\mathrm{d}l$, located at the origin of the coordinate system; the magnetic dipole moment is $\hat{\boldsymbol{p}}_{\mathrm{m}} = \hat{p}_{\mathrm{m}} z = \hat{I}\,\mathrm{d}Az$, where $\hat{I}$ is the current around a loop of infinitesimal area $\mathrm{d}A$, located at the origin of the local coordinate system and perpendicular to the xy plane of Figure 9.47. For the electric dipole, the radiated field in the local coordinate system of Figure 9.47 is given by [1]

$$\hat{H}_\phi = \frac{\mathrm{j}\omega \hat{p}_{\mathrm{e}}}{4\pi}\beta^2 \sin\theta \left(\frac{\mathrm{j}}{\beta r} + \frac{1}{\beta^2 r^2}\right) \mathrm{e}^{-\mathrm{j}\beta r} \tag{9.42a}$$

$$\hat{E}_{\mathrm{r}} = 2\frac{\mathrm{j}\omega \hat{p}_{\mathrm{e}}}{4\pi}\eta\beta^2 \cos\theta \left(\frac{1}{\beta^2 r^2} - \frac{\mathrm{j}}{\beta^3 r^3}\right) \mathrm{e}^{-\mathrm{j}\beta r} \tag{9.42b}$$

$$\hat{E}_\theta = \frac{\mathrm{j}\omega \hat{p}_{\mathrm{e}}}{4\pi}\eta\beta^2 \sin\theta \left(\frac{\mathrm{j}}{\beta r} + \frac{1}{\beta^2 r^2} - \frac{\mathrm{j}}{\beta^3 r^3}\right) \mathrm{e}^{-\mathrm{j}\beta r} \tag{9.42c}$$

$$\hat{H}_{\mathrm{r}} = \hat{H}_\theta = \hat{E}_\phi = 0 \tag{9.42d}$$

and for the magnetic dipole [1]

$$\hat{E}_\phi = -\mathrm{j}\frac{\omega\mu \hat{p}_{\mathrm{m}}}{4\pi}\beta^2 \sin\theta \left(\frac{\mathrm{j}}{\beta r} + \frac{1}{\beta^2 r^2}\right) \mathrm{e}^{-\mathrm{j}\beta r} \tag{9.43a}$$

$$\hat{H}_{\mathrm{r}} = \mathrm{j}2\frac{\omega\mu \hat{p}_{\mathrm{m}}}{4\pi}\frac{\beta^2}{\eta} \cos\theta \left(\frac{1}{\beta^2 r^2} - \frac{\mathrm{j}}{\beta^3 r^3}\right) \mathrm{e}^{-\mathrm{j}\beta r} \tag{9.43b}$$

$$\hat{H}_\theta = \mathrm{j}\frac{\omega\mu \hat{p}_{\mathrm{m}}}{4\pi}\frac{\beta^2}{\eta} \sin\theta \left(\frac{\mathrm{j}}{\beta r} + \frac{1}{\beta^2 r^2} - \frac{\mathrm{j}}{\beta^3 r^3}\right) \mathrm{e}^{-\mathrm{j}\beta r} \tag{9.43c}$$

$$\hat{E}_{\mathrm{r}} = \hat{E}_\theta = \hat{H}_\phi = 0 \tag{9.43d}$$

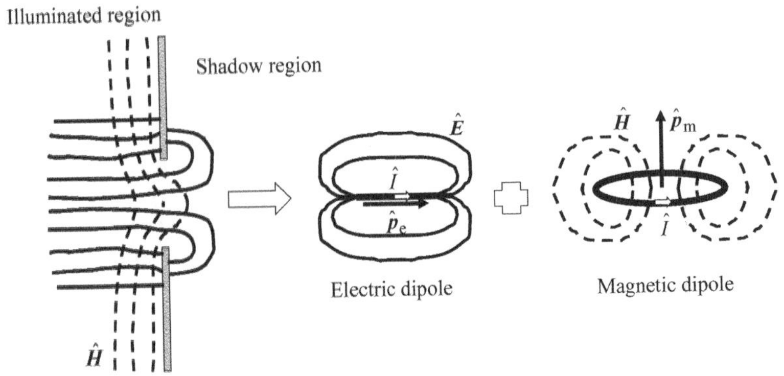

Figure 9.48 Static electric and magnetic fields in an aperture and the equivalent dipole polarization

Note that Equations (9.43) are obtained from Equations (9.42) by the following conversions using the dual relationship between electric and magnetic sources [36]:

- E-field converts to H-field.
- H-field converts to E-field with a changed sign.
- $j\omega\hat{p}_e = \hat{I}\ dl$ converts to $j\omega\mu\hat{p}_m = j\omega\mu\hat{I}\ dA$.
- Electric current converts to magnetic current.
- μ converts to ε.
- ε converts to μ.
- $\eta = (\mu/\varepsilon)^{0.5} = 377\ \Omega$ converts to $1/\eta$.

Recall that, in the case of aperture problems, the computation of the fields produced by the two dipoles requires a change of coordinate system for one of the two dipoles.

An intuitive method for justifying the physical existence of dipoles in the aperture is to consider the static field lines as illustrated in Figure 9.48. Note that the electric dipoles reproduce the paths of the electric field lines on the right of the aperture. On the other hand, the magnetic dipoles reproduce the paths of the magnetic field lines.

Example 9.15: Calculations and Measurements of Radiation from P-Test Boards within a Shielded Rack with a Rectangular Aperture

The small aperture model presented in the previous section is applied here to a practical case. Consider the configuration depicted in Figure 9.49a, where a PCB is inserted within a shielded box aligned with a rectangular aperture of 7.5 cm × 30 cm size. At 1 GHz the wavelength $\lambda = 300/f_{\mathrm{MHz}}$ is exactly the maximum dimension of the aperture l, and therefore the aperture should be electrically small up to $f_{\mathrm{MHz}} = 300/(l/10) = 300/(0.3/10) = 100$ MHz. However, it will be shown that in practice the method can be extended up to the frequency where the maximum dimension of the aperture is $l = \lambda/2$, and hence for a frequency up to 500 MHz, with acceptable accuracy from an engineering viewpoint.

To calculate the emission from PCBs on account of apertures, the *Shielding Effectiveness* (SE) concept is applied. The shielding effectiveness of a shield is defined as the ratio between the magnitudes of the field at point P in the absence of a shield $E_p = |\hat{E}_p|$ and the field at the

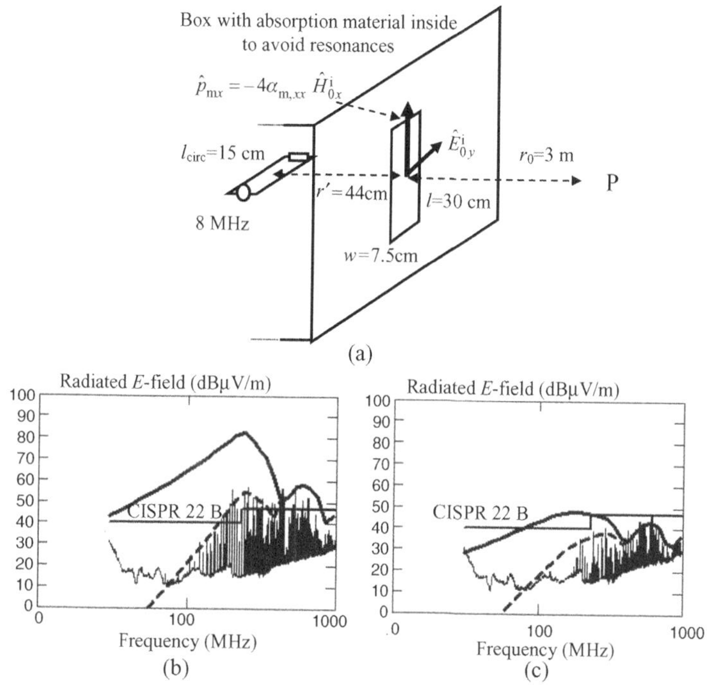

Figure 9.49 Radiated emission from a rectangular aperture: (a) schematic set-up; (b) radiated *E*-field from a PCB with parallel wires; (c) radiated *E*-field from a PCB with a wire above a ground plane. Calculated results without a shield (solid line) and with a shield (dashed line), and measured radiated field (harmonics)

same point with a shield present $E_{\mathrm{p\,sh}} = |\hat{E}_{\mathrm{p\,sh}}|$:

$$\mathrm{SE_{dB}} = 20\log\left(\frac{E_{\mathrm{p}}}{E_{\mathrm{p\,sh}}}\right) \tag{9.44}$$

Therefore, the field in the presence of a shield (in dB) can be obtained as

$$E_{\mathrm{p\,sh,dB}} = E_{\mathrm{p,dB}} - \mathrm{SE_{dB}} \tag{9.45}$$

The field radiated from the aperture was predicted by Equation (9.45), where the radiation from P-test boards when a shield is not present is calculated as described in *Example 9.2*, and SE is calculated by applying the small aperture model described in this section. Once the dipoles on the aperture are known, it is possible to calculate the shielding effectiveness.

To find a simple expression for the shielding effectiveness, an incident plane wave field traveling in a negative z direction with the electric field oriented along the y axis $\hat{E}^{i}_{0y}$ (i.e. horizontal polarization) was considered. In this case the electric dipole vanished (i.e. $\hat{E}^{i}_{0z} = 0$), and the magnetic dipole was generated by the magnetic component of the incident field along the x axis $\hat{H}^{i}_{0x} = \hat{E}^{i}_{0y}/\eta$. Moreover, the observation point P was chosen to be aligned with the centre of the aperture. In this situation the electric field magnitude $E_{\mathrm{p\,sh}}$ (electric field with a shield present) oriented along the y axis is obtained by Equation (9.43a) and, considering that $\omega\mu/\beta = \eta$, is given by

$$E_{\mathrm{p\,sh}} = \left| \eta \frac{\beta^2}{4\pi} p_{\mathrm{m}} \left(\frac{1}{r_0} - \frac{\mathrm{j}}{\beta r_0^2} \right) \right| \approx \frac{\eta\beta^2}{4\pi r_0} 4\alpha_{\mathrm{m},xx} \frac{E^{\mathrm{i}}_{0y}}{Z_{\mathrm{w}}} \tag{9.46}$$

where $E^{\mathrm{i}}_{0y} = |\hat{E}^{\mathrm{i}}_{0y}|$, and Z_{w} is the wave impedance at the aperture, defined as the ratio between the orthogonal incident electric and magnetic fields (i.e. $Z_{\mathrm{w}} = E^{\mathrm{i}}_{0y}/H^{\mathrm{i}}_{0x}$).

Assuming an attenuation inversely proportional to the distance, the field E_{p} at point P in the absence of a shield is

$$E_{\mathrm{p}} = A E^{\mathrm{i}}_{0y} \tag{9.47}$$

where $A = r'/(r_0 + r')$. Then, the shielding effectiveness SE (in dB) for $r_0 \geq 3$ m is

$$\mathrm{SE_{dB}} \cong 20 \log \left(\frac{A\pi r_0}{\beta^2 \alpha_{\mathrm{m},xx}} \frac{Z_{\mathrm{w}}}{\eta} \right) \tag{9.48}$$

In this formula it is important to note the dependence of SE on the wave impedance Z_{w}. This impedance in far-field condition is equal to $\eta = 377\ \Omega$, while in the near field it depends on the type of source [18]. If the source produces *common-mode* emission, such as in the case of a PCB with two parallel wires, as considered in *Section 9.2*, the source can be modeled as a long dipole and the wave impedance is $Z_{\mathrm{w}} \geq 377\ \Omega$. If the source produces *differential-mode* emission, such as in the case of a PCB with a wire above a ground plane, the source can be modeled as a loop loaded with the terminal resistance of the circuit and the wave impedance is $Z_{\mathrm{w}} \leq 377\ \Omega$.

In order to reproduce this situation, some measurements were performed. The test PCBs with two parallel wires and a wire above a ground plane (see *Section 9.2*) were put within a box of dimensions $b_1 = 0.86$ m, $b_2 = 1.5$ m, and $b_3 = 0.6$ m, centered at 1.2 m from the floor of the chamber. The receiving antenna was placed at 1.2 above the floor. The circuit was 44 cm distant from the slot oriented in order to excite $\hat{E}_y$ only on the rectangular aperture. The walls within the box were covered with anechoic material in order to reduce, as much as possible, the resonances of the structure. The dimensions of the aperture were 30 cm × 7.5 cm, large enough to have the correct amplitude level for the measurements. For this type of slot, $\alpha_{\mathrm{m},xx}$ was assumed to be given by the formula of the equivalent ellipse (see Table D3 in *Appendix D*):

$$\alpha_{\mathrm{m},xx} = \frac{\pi}{24} \frac{l^3 e^2}{K(e) - E(e)} \tag{9.49}$$

where

$$e = \sqrt{1 - \left(\frac{w}{l}\right)^2} \tag{9.50}$$

and K and E are the complete elliptic integrals of the first and second kind.

The results of the simulations are shown in Figures 9.49b and c, and a good correspondence with the measurements can be observed. For the PCB with two parallel wires, $Z_w = 377\ \Omega$ was used. For the PCB with one wire above a ground plane, it is difficult to compute the wave impedance. An approximate method adopted here consists in using the load of the circuit (i.e. $Z_w = 50\ \Omega$ for the configuration considered). Observe that there is good agreement until $\lambda/l > 2$ instead of the theoretical value $\lambda/l > 10$. Other measurements were carried out with apertures of various dimensions and different kinds of source, finding good agreement between measurements and calculations. It is interesting to note that, if the well-known very simplified formula of a slot provided by Ott [47] is used, the shielding effectiveness is given by $SE_{dB} = 20\log(\lambda/(2l))$, and an underestimation of about 20 dB is obtained.

In all these calculations, the loss of SE owing to the resonance phenomena within the shielded box was not taken into account. To this end, a simple method was developed by Robinson *et al.* [48], where a shielded box with its rectangular aperture is modeled as a transmission line, and the SE is computed considering the first resonance of the box. It can be shown that the dipole method and the transmission-line method provide very close results for electrically small apertures. Consider that, in a real shielded box with many PCBs, owing to the loading effect of the PCBs, the resonance frequencies are shifted towards very high frequencies, and hence the dipole method can be suitably used for a first grade of prediction.

For a rectangular metal box of dimension $l \times w \times h$ (in meters), the natural resonance frequencies of the $TE_{m,n,p}$ modes are given by [18]

$$f_{\text{res}} = 150\sqrt{\left(\frac{m}{l}\right)^2 + \left(\frac{n}{w}\right)^2 + \left(\frac{p}{h}\right)^2} \quad \text{(in MHz)} \tag{9.51}$$

where m, n, and p are integer numbers that can take any value, but no more than one at a time equal to zero. At these frequencies, an empty box could exhibit fields on the aperture so high that the SE could be negative with an apparent gain. Fortunately, the box of actual equipment is never empty but filled with PCBs, components, and cables which behave as many elements with losses.

Finally, consider that arrays of similar adjacent holes with edge-to-edge separation of less than the minimum aperture dimension leak as only one hole. Currents in each rib are equal and opposite, and magnetic moments are mutually canceling [18]. Therefore, Equation (9.45) can be used for estimating the shielding effectiveness of arrays of this type.

9.9 Radiation Diagrams

As recommended by the EMC authorities throughout the world, such as the FCC and CISPR, the testing of equipment should be performed at an *Open Area Test Site* (OATS) or in a semi-anechoic chamber that reproduces the same conditions as an OATS. Both sites are characterized by a ground plane, and the *Equipment Under Test* (EUT) is installed on a rotating turntable in order to measure by an antenna with variable height over the ground plane (1–4 m) the radiated E-field in horizontal and vertical positions for several angles. The spectrum analyzer records the maximum E-field at each frequency. Most electronic systems incorporate a large number of fast digital devices, assembled in functionally interactive circuits, and packaged in separate units interconnected by cables: PCBs, shelves, boxes, racks. The problem is to know about the synergy of ICs, PCBs, and cables seen as sources of emission within a system. The problem is further complicated by the near-field and ground-plane effects of the site of measurements. In this case, to predict the directions of maximum emission and their polarization, the models presented in the previous sections are not appropriate, and full-wave numerical simulations are required. In this way, additional information can be obtained on the radiated field origin and distribution in space. The EUT radiated emission profile is determined by superimposition of emissions from individual components and subassemblies, as well as by their interaction. Thus, the EUT can be modeled by a set of elementary sources. To account for elementary source interaction, it is assumed that each generator in the system can feed one or several radiators, and the radiators can be mutually coupled.

The purpose of this section is to provide the necessary information to build up and verify models of simple PCBs with an attached cable in order to create the bases for other models that reproduce the radiated emission patterns of more complicated structures.

Correct modeling for an elementary source of emission such as long dipoles, circular or square loops, helixes, patches, boxed-in slots, etc., is very important because, when their dimensions exceed the wavelength, the pattern becomes extremely sensitive to the frequency. In real life the ideal pattern of these elementary sources is distorted by other factors such as cable and interconnect layout irregularities, proximity effects, and finite ground planes. The effects of the radiation pattern change with the frequency variations for the same size of radiator, as will be illustrated in *Example 9.17*. It might be expected that a radiating pattern of actual EMC test systems with complex cabling would be very complex. This is illustrated by radiation patterns of several EMC test systems, configured with different degrees of complexity [49].

Example 9.16: Measurement of Radiated Emission Patterns from Telecommunication Equipment

This example deals with the measurement of radiated emission patterns from commercial telecommunication switching equipment with shielded racks and screened cables according to the set-up shown in Figure 9.50a. All the patterns were measured in a semi-anechoic chamber test site at 3 m separation distance from the EUT. The receiving biconical antenna was mounted on a mast, and the EUT was installed on a turntable. The field intensity was measured in dBμV/m. A radiation pattern for the case of a shielded cable attached to a test PCB and with the shield in full contact with a rack is shown in Figure 9.50b. The maximum deviation $\Delta E = E_{max} - E_{min}$ is measured for eight positions of rotation angle φ. Note that ΔE is around 10 dB, with a peak of 20 dB.

The radiation pattern for the case of a test circuit within a shielded box with an aperture is given in Figure 9.50c. Note that ΔE is again around 10 dB, with a peak of 20 dB near 200 MHz.

Radiated emission measurement for the case of commercial telecommunication switching equipment with shielded racks and screened cables is shown in Figure 9.50d. The radiation pattern at 100 MHz and the maximum deviation are shown for frequencies up to 300 MHz. Observe that, for a real complex system, most points are below 10 dB of deviation, and only very few are above. Therefore, it seems that for a system there is less directivity than in a simple structure such as a cable or aperture excited by a test board. The pattern shape of a complex system is much smoother than a simple circuit such as formed by a single clock line of two parallel wires driven by a digital device battery powered as considered in *Section 9.2*. This can be explained by treating the system as a 3D radiator of large size. This

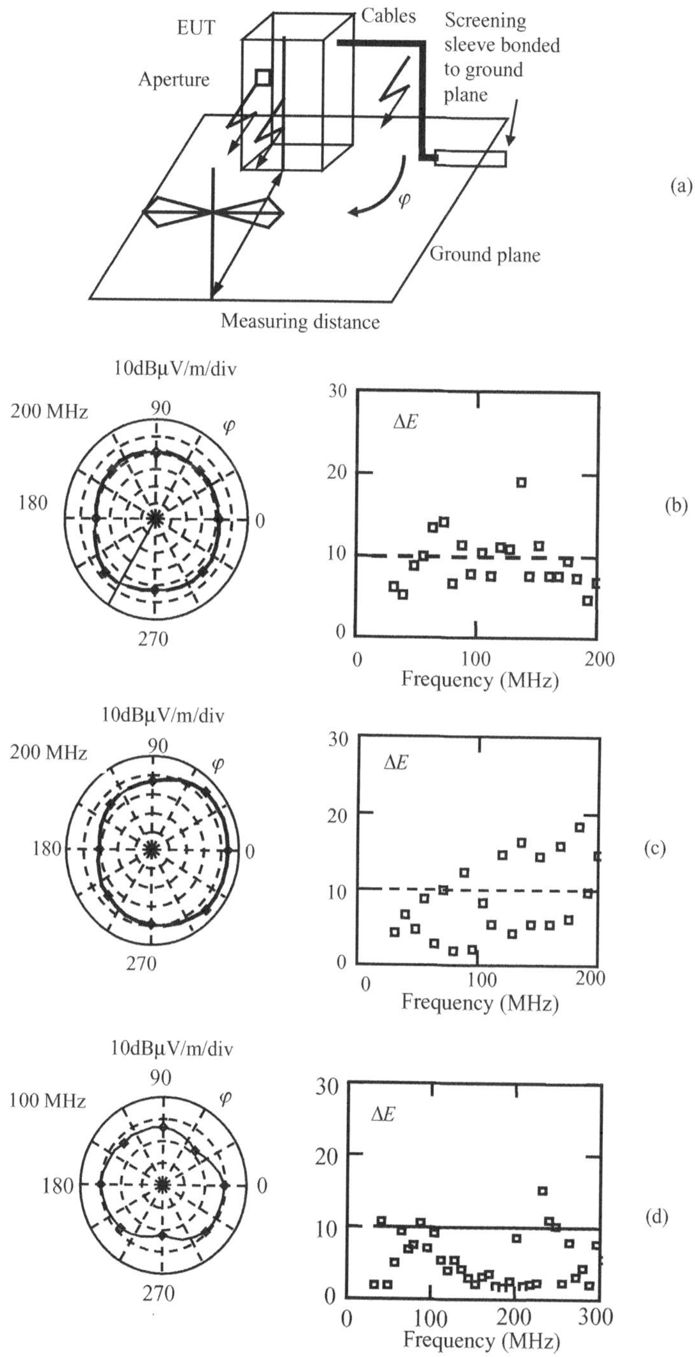

Figure 9.50 Radiated field pattern and deviation $\Delta E = E_{max} - E_{min}$: (a) radiated emission set-up with angle of rotation φ; (b) shielded cable leaving a shielded rack; (c) slot in a rack; (d) complex system formed by two shielded racks of a commercial switching exchange operating in normal condition

fact has important implications for modeling an electronic system by numerical codes in order to interpret correctly the results considering not only the contribution of the circuits but also all the metallic parts used for assembling the various functionalites which act as antennas (see *Section 11.3*). In conclusion, the following remarks can be made:

- For simple structures, deviations ΔE are around 10 dB, with peaks of less than 20 dB.
- For complex structures, most deviations ΔE are below 10 dB, and only very few points are above.
- For a complex system there is less directivity than for simple structures such as one PCB, one cable, or one aperture for testing.
- This means that computation for the expected maximum field direction can be used for estimating the radiation in other directions and frequencies.
- Numerical codes and transmission-line and antenna models can help in this process.

Example 9.17: Computation of Radiation from a Wire above a Finite Ground Plane with an Attached Cable

To demonstrate how to use numerical codes to model a simple PCB for predicting radiated emission patterns, two different full-wave numerical tools were used: the first tool, NEC-Win, is based on the *Method of Moment* (MOM) in the frequency domain [50] and was developed by Lawrence Livermore Laboratory (CA) for antenna modeling; the second tool, MWS, is based on the *Finite Integration Technique* (FIT) [51]. The NEC-Win tool allows fast simulation with some limitation in describing the structure under test, while MWS runs simulations in the time domain with great potentiality but requires much more time for performing the computation. When users without any great experience intend to begin modeling by these tools, it is suggested they begin with simple PCB structures and verify whether the results are those expected by theory. Another way is to make comparisons with measurements or with different simulation approaches.

The simplest PCB to consider is a microstrip-like structure, such as a wire above a ground plane, in order to simulate an actual microstrip trace as often met in a multilayer PCB. The test PCB considered is shown in Figure 9.51a. Since NEC-WIN does not enable structures with a dielectric to be simulated, the PCB considered was chosen without a dielectric substrate. This does not matter for our task of comparison, as the presence of a dielectric has the practical function of shifting the resonance frequencies of the structure under simulation. The simulated board has an 18×30 cm ground plane, and the signal line is a wire of 18 cm length centered 1 cm above the ground plane. The signal wire is fed at one end by a 1 V voltage generator and is terminated at the other end with a 100 Ω load. The I/O cable is simulated by a 0.6 m long wire. The dielectric of the PCB is air, and therefore $\varepsilon_r = 1$. Another important limitation of NEC-Win is that the structure under test can be simulated by a grid of wires or patches for a closed surface. Here, the first feature was chosen, and the ground plane was simulated with a grid of wires reinforced just under the signal wire, considering that at high frequencies the current density concentrates in that area (see *Section 10.2*). When using wires, there are some rules to follow in order to obtain the most accurate model. The most important rules pertain to the wire radius and segmentation. The segment length Δ depends upon the wavelength λ, so that $\Delta < 0.1\lambda$. For long wires with no abrupt changes, such as a cable attached to a PCB, a longer segment length might be acceptable. Shorter segments of 0.05λ or less may be needed to model critical regions. For wire radius a, the following rule to increase the validity of the

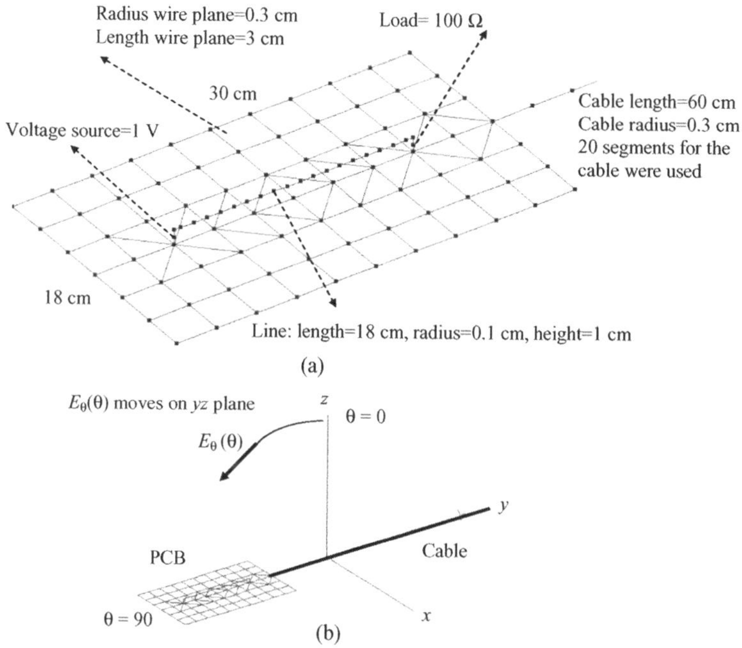

Figure 9.51 Test board: (a) view of the test board simulated by NEC-Win with wires; (b) radiation pattern angle

approximations should be applied: $2\pi a/\lambda \ll 1$. Furthermore, for accurate computations, the ratio between the segment length and the wire radius should be greater than 2 (i.e. $\Delta/a > 2$). The PCB structure of Figure 9.51a was modeled following these rules. It is important to note that sources are placed in the middle of a segment. Therefore, in order to place a source in the middle of an antenna, the wire should have an odd number of segments. In the test PCB considered, one segment was used for the voltage source. The field E_θ was computed as a function of the elevation angle θ shown in Figure 9.51b. This angle was chosen because the electric field should have its maximum values in the plane orthogonal to the ground plane, considering the orientation of the PCB.

The same structure was simulated by MWS which enables a finite ground plane to be modeled with its conductivity and without using wires. MOM is a surface current technique. FIT is a volume-based solution of time-domain Maxwell's equations in integral form. The volume includes the structure under test and some added space computed automatically by MWS. A cut view of the PCB as simulated by MWS is shown in Figure 9.52. For the PCB without cable, a mesh of 29 988 cells was used. With cable, the number of cells increases to 57 528. Since the simulations are performed in the time domain, and the FFT is used to obtain frequency-domain solutions, a Gaussian waveform having a flat spectrum of 1 V in the range 0–1000 MHz was adopted for the source. Once the interval of frequencies is defined, the mesh is automatically generated. Observe that the mesh is finer at the edge of the plane and in the

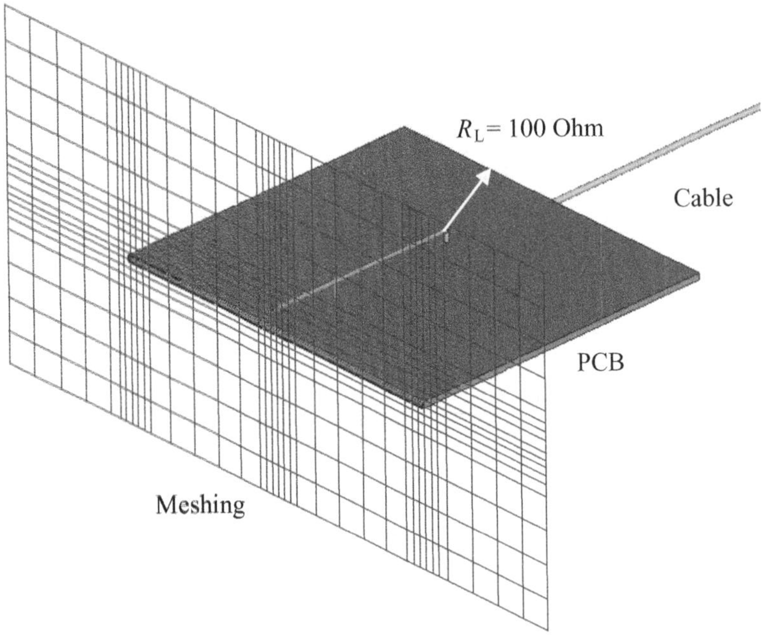

Figure 9.52 Cut view with meshing of the test board as simulated by MWS

zone under the signal wire. Apart from the default meshing offered by MWS, which in many cases is appropriate, a more suitable mesh can be obtained by the user.

Comparisons of the radiated emission patterns obtained by the two codes at four frequencies are shown in Figure 9.53 for a PCB only and for a PCB with an attached cable. The electric field is given in linear scale and not in dB in order to evidence the lobes that arise at high frequencies. Observe how the radiation pattern changes, especially at low frequencies where the *common-mode* emission from the cable dominates, as shown in *Section 9.2*.

Although the NEC-Win tool uses few wires instead of the numerous cells used by MWS, a very good agreement can be observed between MOM and FIT results. The great advantage of NEC-Win over MWS is the speed of computation: some seconds with a Pentium PC 4 having a clock frequency of 3.2 GHz, as opposed to several minutes with the MWS code. Of course, the great drawback of modeling a structure by wires is when more complicated structures such as a PCB with an irregular ground plane within a shielded box with an aperture for an outgoing I/O cable need to be simulated. *Section 10.3.4* will show how to use MWS to set design rules for such complex structures.

At this point it would be interesting to compare the results obtained by these numerical methods with those obtained by the TL models outlined in *Section 9.2* for *differential mode* in PCBs and in *Section 9.6* for *common mode* in the case of cables attached to PCBs. The model for the *current-driven* mechanism is used with its equivalent circuit as shown in Figure 9.54. The ground inductance $L_{gnd} = 3.99$ nH was obtained by the closed-form expression of Table A.2 in *Appendix A*, adopting a ground plane thickness $t = 3$ mm. The current $\hat{I}_{sig}$ was computed as the average current along the signal wire divided into six segments of equal length.

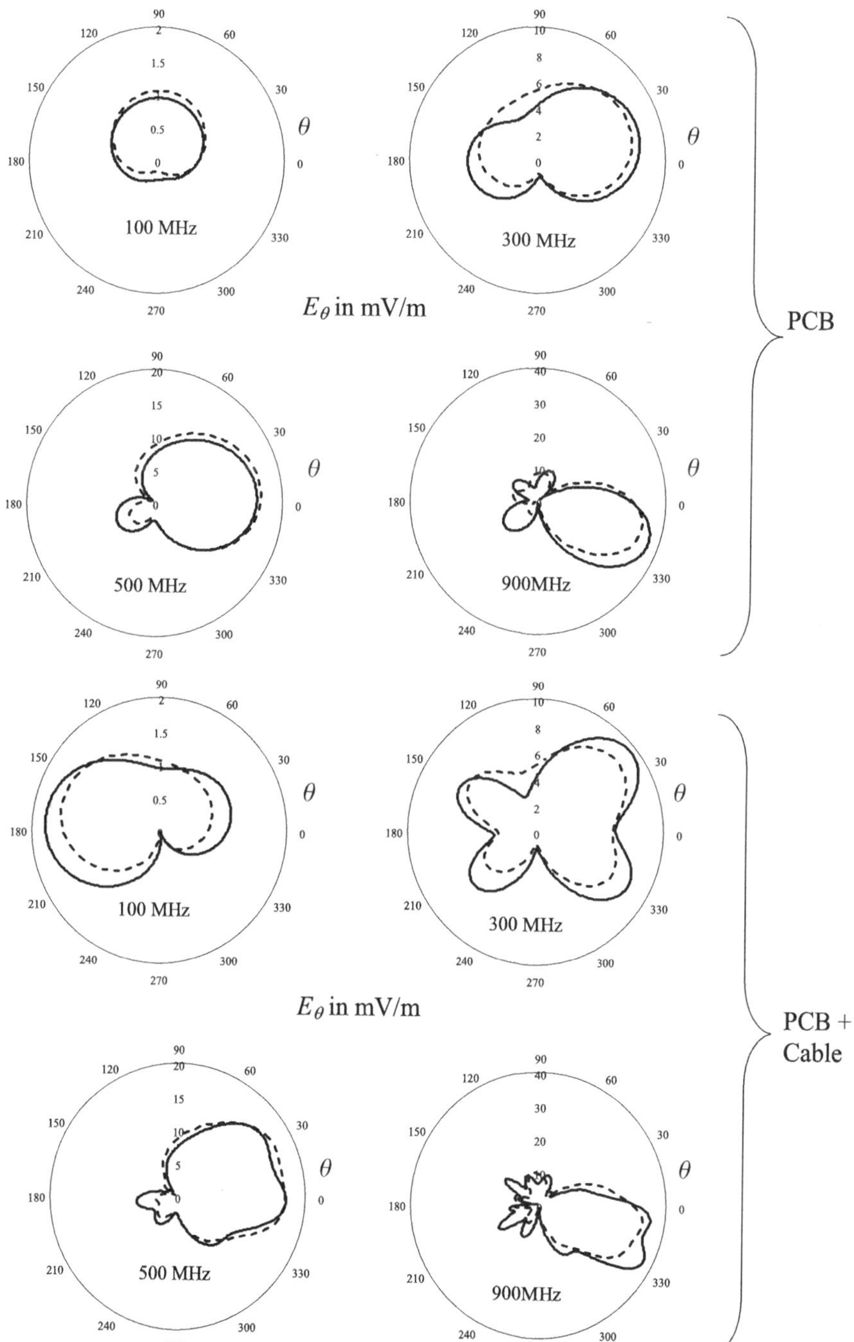

Figure 9.53 Computed radiation pattern by NEC (solid line) and MWS (dashed line) of the test PCB without attached cable (PCB) and with attached cable (PCB + Cable)

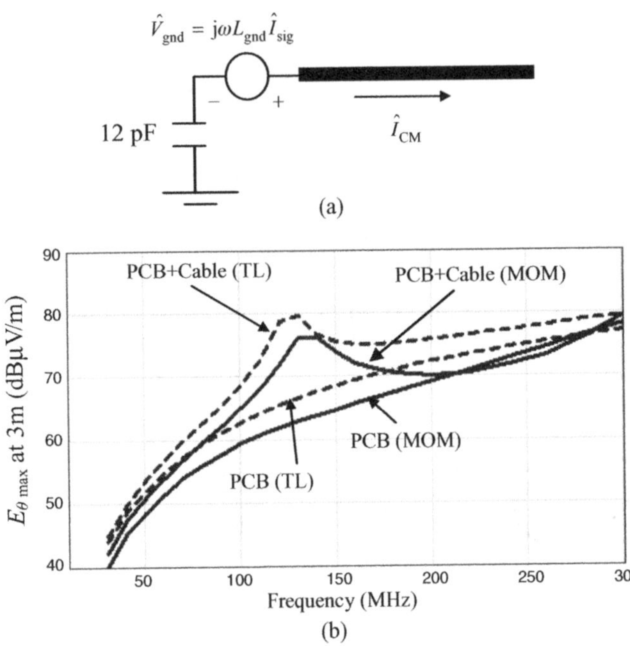

Figure 9.54 Analytical and numerical model comparison of a PCB with an attached cable: (a) equivalent circuit for radiated field computation with the TL model; (b) computed maximum field values by MOM and the TL model

The current on each segment was calculated by transmission-line equations for a wire over a ground plane having $Z_0 = 180\ \Omega$, terminated with $100\ \Omega$, and excited by a 1 V voltage source (see *Section 9.2*). Once the voltage drop $\hat{V}_{gnd} = j\omega L_{gnd}\hat{I}_{sig}$ was known, the current along the cable was computed by the TL model, assuming a line with characteristic impedance $Z_{0cable} = 60\ \ln(2l_{cab}/r_{cab}) = 360\ \Omega$. The cable was also divided into six segments to calculate its average currents. Then, the radiated field at 3 m was calculated by the closed-form expression of *Appendix D* for the relative structure.

The results of this calculation and comparison with the MOM method are shown in Figure 9.54. The maximum $\hat{E}_\theta$ values for rotation angle θ obtained by MOM are reported for each frequency. The field calculations by the TL approach were obtained adopting an elevation angle $\theta = 0$. The slight differences in values (see Figure 9.54) are due to the fact that the numerical code simulates a finite ground plane while the TL model assumes an infinite ground plane. In fact, it will be shown in *Subsection 10.3.4* that, when the numerical method models an infinite ground plane, the results are coincident.

This final result demonstrates that, when interest is not focused on the radiation pattern calculation, the TL approach is appropriate for finding maximum values. Observe the resonance peak at about 130 MHz, as evidenced by the two methods when the PCB has an attached cable.

To conclude this section, the following observations can be made:

- Numerical codes are the most appropriate tools for predicting radiated emission patterns from a complex digital system formed by elementary sources such as long dipoles, round or square loops, simple PCBs with parallel wires, or a wire above a ground plane, because they interact as an array of antennas.

- Although elementary sources present lobes in different directions depending on frequency, a complex system has a much smoother radiation pattern, as all sources act as a 3D array of antennas, including the metallic parts of the structure with their apertures.
- This simplification of the emission pattern makes it possible to use TL models to predict maximum emission fields with acceptable accuracy for structures such as simple PCBs (finite ground plane) with an attached cable.
- Numerical methods such as MOM, FEM, and FIT should be used to analyze complex PCBs in order to develop design rules.
- With MOM, FEM, and FIT methods, the PCB structures should have dimensions in agreement with the guidelines for wire modeling (MOM based on the original NEC2) or meshing (FEM, FIT) in order to avoid numerical errors.
- MOM based on the original NEC2 is much faster than FEM and FIT by using the wire-grid technique for ground or image planes, but it has the drawback of the impossibility of modeling actual PCBs with dielectric substrates inserted in shielded boxes.

9.10 Points to Remember and Design Rules for Radiated Emission

This section addresses important points to remember when dealing with radiated emission and design guidelines, summarizing what has previously been presented and discussed. This is not an exhaustive list – the intention is to consider the most important fixes to be implemented during design and development phases of systems that must meet the international standard regulations. A detailed treatment of this important aspect can be found in textbooks written for this purpose [17, 18], where several examples of realization for actual PCBs are provided. A merit of this book is that the benefits in minimizing radiated emissions that are offered by the different fixes can be quantified by simple models that have been validated experimentally, or by comparison with other approaches.

(i) General Comments

- The radiated emission is mainly due to the clocks in the system. To estimate the emission profile, the magnitude spectrum envelope of a trapezoidal waveform, representing signals, and of a triangular waveform, representing switching currents, can be used (*Section 9.1*).
- The envelope of the radiated field depends on current amplitude, clock frequency, and rise and fall times. Current must be minimized, and edge time should be appropriate for timing purposes to avoid using faster devices when it is not required (*Section 9.1*).
- Radiated emission is due to two types of current: (1) the *differential-mode* current, which represents the signal current and can be easily controlled; (2) the *common-mode* current, which represents a displacement current between the PCB and the nearby metallic object, does not transport information, and is difficult or almost impossible to control (*Section 9.2*).
- Continuous power/ground planes as return paths for signal currents or image planes placed beneath double-side boards are good ways to avoid uncontrolled *common-mode* emission. The emission generated by *common-mode* current in PCBs is very difficult to predict, and an estimation can be made only by simple antenna models (*Section 9.2*).
- *Common-mode* radiation usually dominates in the low-frequency range (30–300 MHz), and *differential-mode* dominates at high frequency (>300 MHz). This fact helps at the

diagnostic stage, as with a system built with multilayer PCBs the *common-mode* radiation is due to cables and *differential-mode* radiation to circuits (*Section 9.2* and *Section 9.8*).

- In PCBs with power/ground planes, *differential-mode* emission dominates, as the ground plane for image theory makes the circuit symmetric with respect to the reference plane. Considering the signal current and its return path, the *differential-mode* current can be easily predicted by using transmission-line models (*Section 9.2*).

(ii) Comments and Design Rules on Traces

- The use of traces with controlled characteristic impedance, such as microstrips or striplines, is preferable (*Section 9.3*). In this manner, reduction in the radiated field can be accomplished by reducing the current loop size and by using appropriate terminations such as parallel termination for oscillation reduction on the signal waveforms or series termination for matching at the source end and for reducing the current in line (*Section 9.2*).
- Slots or gaps in power/ground planes crossed by signal traces increase dramatically the effective inductance associated with the return path, and therefore strong *common-mode* emission is generated. This should be avoided if stitching capacitors are not used across the gap, to ensure a return path to the signal current very close to the trace (*Section 9.3*, *Section 10.2*, and *Section 12.2*).
- Every large metallic part of the PCB, such as an image plane or headsinks, should be connected to the ground plane of the PCB at several points with electrically short separation (*Section 9.3*).
- Striplines should be used for critical traces because they generate much lower emissions (about 20 dB) than microstrips (*Section 9.3*).
- Critical traces must not be placed near the edge of the PCB, to avoid *common-mode* emissions, as the finite ground plane makes the circuit no longer symmetric with respect to the ground plane.

(iii) Comments and Design Rules on Integrated Circuits

- In multilayer PCBs with controlled characteristic impedance of the traces, the associated loop area becomes small compared with the sum of the loops associated with the ICs, so that the ICs, especially if mounted with sockets, can be the main source of emission owing to the large switching current flowing in the loop area associated with each device and its package (*Section 9.4* and *Section 9.5*).
- Fast switching devices should be used only if this is necessary for timing requirements.
- The IC package influences the radiation emission. Therefore, it is important to avoid using sockets and non-controlled connections of the die with the traces. Surface-mount devices (SMDs) and surface-mount technology (SMT) allow the best possible loop reduction, as they exhibit significant reduction in component dimensions (*Section 9.4* and *Section 12.2*).
- Devices with a controlled ground bounce phenomenon (i.e. ICs with a large number of pins for power and ground functions) should be used. A good choice is to use ICs with the ground and power pins placed in the center of the package rather than the traditional end-pin location. This minimizes the loop area and the effects of the ground/power bounce phenomenon (*Section 8.3*).
- Decoupling capacitors are very important when a cable is attached to the PCB. In this case the best solution is to use buried capacitance technology which increases significantly the

capacitance between the power and ground planes and therefore the self-filtering action of the PCB (*Section 9.6*).

- The decoupling capacitors must be located in order to minimize their associated connection inductance to the power/ground planes, and they must be chosen in a quantity and value suitable for shifting the resonance frequencies of the PCB out of the frequency range of interest (*Section 8.1* and *Section 8.2*).
- It is strongly recommended that the IEC 61967-1–6 technical documents be consulted, as they provide useful information on measurements of conducted and radiated electromagnetic disturbances from integrated circuits.
- It is also recommended that the IEC 62433-2 technical document be consulted. The objective of this standard is to propose a model for predicting conducted emissions at the chip or multichip and PCB level and for power integrity analysis, consult [52] for an example of modeling a complex IC.

(iv) Comments and Design Rules on Cables

- The radiated emission from cables attached to PCBs is due to the fact that the structure becomes a monopole or a dipole antenna fed by a voltage source that represents noises in the ground and power distribution network (*Section 9.6*).
- The antenna voltage source is the result of two phenomena: the *current-driven* mechanism, which concerns the signal current and line inductance associated with the signal line and its return, and the *voltage-driven* mechanism, which concerns the signal voltage and the parasitic capacitance of the PCB and its environment (see *Section 9.6*).
- The use of ground planes for PCBs is recommended because the effective inductance associated with the return path, responsible for the *current-driven* mechanism, is much lower than that of a strip used as ground-return conductor (*Section 9.6*).
- The use of decoupling capacitors with very low connection to the power/ground planes is recommended, as their associated inductance is responsible for the *voltage-driven* mechanism. Buried capacitance technology, which increases the intrinsic capacitance between power and ground planes, is the best solution (*Section 9.6*).
- Differential drivers with low skew between the two complementary outputs should be chosen in order to have low values of *common-mode* current generated by the driver (*Section 9.7*).
- With unshielded twisted-pair (UTP) cables, transformers should be used with a *common-mode choke* in series, located at the output of the differential driver to stop the *common-mode* current, and shunt capacitors should be added to divert the *common-mode* current to ground (*Section 9.8* and *Section 10.3*).
- With shielded twisted-pair (STP) cables, cables with double-braid shields connected at 360° to the chassis or shielded rack should be used (*Section 9.7*).
- Shielded connectors such as Z-Pack, D-SUB 9, and RJ45 are appropriate with STP cables.
- Do not use pigtails to connect the shield of the cable with the ground of the PCB. A large strip of metal could be appropriate (*Section 10.3*).

(v) Comments and Design Rules on Shielded Racks

- The emissions are due to apertures for cooling and ongoing cables if all metallic parts are bonded with care in order to ensure electrical continuity at very low impedance (*Section 9.8*).

- Apertures with maximum dimensions electrically short should be used to control and predict emissions by accurate and simple closed-form expressions (*Section 9.8*).
- Emission from apertures are affected by resonance frequencies within the shielded box which can be considered as a resonant cavity excited by the internal circuits. This must be taken into account when performing predictions, especially when the box is considered empty (*Section 9.8*).
- STP or coaxial cables with low transfer-characteristic impedance should be used: Z_t = 10 mΩ/m up to 100 MHz is usually appropriate. The cable shield should be connected to the rack at 360° (*Section 9.8*).
- When using UTP cables, the PCB should be connected to a chassis by stitches, and EMI filters should be located very close to the cable connector (*Section 10.3*).

(vi) Comments on the Radiation Pattern

- Radiation pattern prediction performed by a 3D full-wave numerical code is very useful for determining the direction of maximum emission at each frequency, as required by the EMC standards (*Section 9.9*).
- The radiation pattern of a simple PCB with an attached cable presents lobes that change direction with frequency (*Section 9.9*).
- Fortunately, maximum deviation from the maximum and minimum fields is around 10 dB for complex systems in the frequency range of interest. This fact enables closed-form expressions to be used to predict maximum field emissions in the same direction for each frequency for each elementary radiating source that forms the system (for instance, ICs, traces, cables, apertures, etc.) with an acceptable accuracy considering the uncertainty of the emission measurements (*Section 9.9* and *Section 11.3*).

References

[1] Paul, C., *'Introduction to Electromagnetic Compatibility'*, John Wiley & Sons, Inc., Hoboken, NJ, 2006.
[2] Balanis, C., *'Antenna Theory'*, 2nd edition, John Wiley & Sons, Inc., New York, NY, 1997.
[3] Hardin, K., Paul, C., and Naishadham, K., 'Direct Prediction of Common-mode Currents', *IEEE International Symposium on Electromagnetic Compatibility*, Cherry Hill, NJ, 12–16 August 1991, 67–71.
[4] Gonshor, D., 'Wire Radiation Impact on System-level EMC', *IEEE International Symposium on Electromagnetic Compatibility*, Washington, DC, 21–23 August 1990, 667–670.
[5] Paul, C. and Bush, D., 'Radiated Emissions from Common-mode Currents', *IEEE International Symposium on Electromagnetic Compatibility*, Atlanta, GA, August 1987, 197–203.
[6] Paul, C., 'A Comparison of the Contributions of Common-mode and Differential-mode Currents in Radiated Emissions', *IEEE Trans. on Electromagnetic Compatibility*, **31**(2), May 1989, 189–193.
[7] Ramo, S., Whinnery, J., and Van Duzer, T., *'Fields and Waves in Communication Electronics'*, 3rd edition, John Wiley and Sons, Inc., New York, NY, 1994.
[8] German, R., Ott, H., and Paul, C., 'Effect of Image Plane on Printed Circuit Board Radiation', *IEEE International Symposium on Electromagnetic Compatibility*, Washington, DC, 21–23 August 1990, 284–291.
[9] Caniggia, S., Catello, L., and Costa, V., 'On the Computation of Electromagnetic Field Components from a Transmission Line: Theory and Application', *IEEE International Symposium on Electromagnetic Compatibility*, Washington, DC, 21–23 August 1990, 651–656.
[10] Schelkunoff, S. and Friis, H., *'Antennas: Theory and Practice'*, John Wiley & Sons, Inc., New York, NY, 1952.
[11] Hsu, T., 'The Validity of Using Image Plane Theory to Predict Circuit Board Radiation', *IEEE International Symposium on Electromagnetic Compatibility*, Cherry Hill, NJ, 12–16 August 1991, 58–60.
[12] Kami, Y., 'Mechanism of Electromagnetic Radiation from a Transmission Line', *IEICE Trans. on Communications*, **E75-B**(3), March 1992, 115–123.

[13] Simpson, J., Goulette, R., and Costache, G., 'Radiation from Microstrip Transmission Lines', *IEEE International Symposium on Electromagnetic Compatibility*, Seattle, WA, 2–4 August 1988, 340–343.

[14] Khan, R. and Costache, G., 'Finite Element Method Applied to Modeling Crosstalk Problems on Printed Circuit Boards', *IEEE Trans. on Electromagnetic Compatibility*, **31**(1), February 1989, 5–15.

[15] Kayano, Y., Tanaka, M., Drewniak, J., and Inoue, H., 'Common-mode Current due to a Trace near a PCB Edge and its Suppression by a Guard Band', *IEEE Trans. on Electromagnetic Compatibility*, **46**(1), February 2004, 46–53.

[16] Wang, H., Ji, Y., Hubing, T., Drewniak, J., Van Doren, T., and DuBroff, R., 'Experimental and Numerical Study of Radiation from Microstrip Bends', *IEEE International Symposium on Electromagnetic Compatibility*, Washington, DC, 21–25 August 2000, 739–741.

[17] Montrose, M., *'Printed Circuit Board Design Technique for EMC Compliance'*, 2nd edition, IEEE, Inc., New York, NY, 2000.

[18] Mardiguian, M., *'Controlling Radiated Emissions by Design'*, 2nd edition, Kluwer Academic Publishers, Norwell, MA, 2001.

[19] Caniggia, S., Catello, L., Costa, V., Marano, G., O'Riordan, N., and Viticci, L., 'Controlling Radiated Emission from a Physically Large Telecommunication System', *IEEE Trans. on Electromagnetic Compatibility*, **38**(3), August 1996, 290–299.

[20] Daijavad, S., Pence, W., Rubin, B., Heeb, H., Ponnapalli, S., and Ruehli, A., 'Methodology for Evaluating Practical EMI Design Guidelines Using EM Analysis Programs', *IEEE International Symposium on Electromagnetic Compatibility*, Anaheim, CA, 17–21 August 1992, 30–34.

[21] Leone, M., 'Design Expression for the Trace-to-Edge Common-mode Inductance of a Printed Circuit Board', *IEEE Trans. on Electromagnetic Compatibility*, **43**(4), November 2001, 667–671.

[22] Slattery, K., Muccioli, J., and North, T., 'Measuring the Radiated Emission from a Family of Microprocessors using a 1 GHz TEM Cell', *IEEE Trans. on Electromagnetic Compatibility*, **41**(2), May 1999, 146–152.

[23] Robinson, M., Benson, T., Christopoulos, C., Dawson, J., Ganley, M., Marvin, A., Porter, S., Thomas, D., and Turner, J., 'Effect of Logic Family on Radiated Emission from Digital Circuits', *IEEE Trans. on Electromagnetic Compatibility*, **40**(3), August 1998, 288–293.

[24] Criel, S., Bonjean, F., De Smedt, R., Moerloose, J., Martens, L., Olyslager, F., and De Zutter, D., 'Accurate Characterization of Some Radiative EMC Phenomena at Chip Level, Using Dedicated EMC-testchips', *IEEE International Symposium on Electromagnetic Compatibility*, Denver, CO, 22–28 August 1998, 734–738.

[25] Wada, O., Kosaka, M., Oka, H., Koga, R., and Sano, H., 'Power-sum Estimation of Electromagnetic Noise Radiated from High-speed CMOS Printed Circuit Boards', *IEICE Trans. on Communications*, **E75-B**(3), March 1992, 165–173.

[26] Caniggia, S., Costa, V., and Vitucci, L., 'Investigation of EMI on Multilayer Circuit Boards: Radiated Emission', *IEEE Int. Symp. on Electromagnetic Compatibility*, Santa Clara, CA, 19–23 August 1996, 316–321.

[27] Caniggia, S., Costa, V., Meani, C., and Preatoni, R., 'Printed Circuit Post-layout Analysis: from Signal Integrity to Radiated Emission', *IEEE International Symposium on Electromagnetic Compatibility*, Denver, CO, 22–28 August 1998, 975–980.

[28] Hockanson, D., Drewniak, J., Hubing, T., Van Doren, T., Sha, F., and Wilhelm, M., 'Investigation of Fundamental EMI Source Mechanism Driving Common-mode Radiation from Printed Circuit Boards with Attached Cables', *IEEE Trans. on Electromagnetic Compatibility*, **38**(4), November 1996, 557–566.

[29] Hubing, T. and Kaufman, F., 'Modeling the Electromagnetic Radiation from Electrically Small Table-top Products', *IEEE Trans. on Electromagnetic Compatibility*, **31**(1), February 1989, 74–84.

[30] Shim, H. and Hubing, T., 'Model for Estimating Radiated Emissions from a Printed Circuit Board with Attached Cables due to Voltage-driven Sources', *IEEE Trans. on Electromagnetic Compatibility*, **47**(4), November 2005, 899–907.

[31] Leferink, F. and van Doom, M., 'Inductance of Printed Circuit Board Ground Planes', *IEEE International Symposium on Electromagnetic Compatibility*, Dallas, TX, 9–13 August 1993, 327–329.

[32] Hockanson, D., Drewniak, J., Hubing, T., Van Doren, T., Sha, F., Lam, C., and Rubin, L., 'Quantifying EMI Resulting from Finite-impedance Reference Planes', *IEEE Trans. on Electromagnetic Compatibility*, **39**(4), November 1997, 266–297.

[33] Holloway, C. and Kuester, E., 'Net and Partial Inductance of a Microstrip Ground Plane', *IEEE Trans. on Electromagnetic Compatibility*, **40**(1), November 1998, 33–46.

[34] Bergervoet, J., 'EMC Measurements and Models Connecting the System Level with the Module Level', *Philips Journal Research*, **48**(1–2), 1994, 63–81.

[35] Caniggia, S., O'Riordan, N., and Vitucci, L., 'A Simple Method to Predict Radiated Emission from Shielded Cables from a Physically Large Digital System', *International Symposium on Electromagnetic Compatibility – EMC ROMA'94*, Rome, Italy, 13–16 September 1994, 273–278.

[36] Tesche, F., Ianoz, M., and Karlsson, T., *'EMC Analysis Methods and Computational Models'*, John Wiley & Sons, Inc., New York, NY, 1997.

[37] Vance, E., *'Coupling to Shielded Cables'*, John Wiley & Sons, Inc., New York, NY, 1978.

[38] Tsaliovich, A., *'Electromagnetic Shielding Handbook for Wired and Wireless EMC Applications'*, Kluwer Academic Publishers, Norwell, MA, 1999.

[39] Caniggia, S. and Maradei, F., 'Circuit Macromodels for the Analysis of the Conducted and Radiated Immunity of Shielded Cables', *IEEE Trans. on Electromagnetic Compatibility*, **46**(4), November 2004, 881–886.

[40] Vanlandschoot B. and Martens, L., 'New Method for Measuring Transfer Impedance and Transfer Admittance of Shields Using a Triaxial Cell', *IEEE Trans. on Electromagnetic Compatibility*, **39**(2), May 1997, 180–185.

[41] Martens, L., Madou, A., Konè, L., Demoulin, B., Sjoberg, P., Anton, A., Van Koetsem, J., Hoffman, H., and Schricker, U., 'Comparison of Test Methods for the Characterization of Shielding of Board-to-Backplane and Board-to-Cable Connectors', *IEEE Trans. on Electromagnetic Compatibility*, **42**(4), November 2000, 427–440.

[42] Caniggia, S. and Santi, P., 'Common-mode Radiated Emissions from UTP/STP Cables with Differential High-speed Driver/Receivers', *IEEE International Symposium on Electromagnetic Compatibility*, Boston, MA, 18–22 August 2003, 564–569.

[43] Smith, A.A., *'Coupling of External Electromagnetic Fields to Transmission Lines'*, 2nd edition, Interference Control Technologies, New York, NY, 1987.

[44] Hoeft, L., *'Measurement of Surface Transfer Impedance of Cables and Connectors'*, EMC Expo 1986, Washington, DC.

[45] Goedbloed, J.J., *'Electromagnetic Compatibility'*, Prentice Hall International (UK) Limited, Deventer, The Netherlands, 1992.

[46] Butler, C., Rahmat-Samii, Y., and Mittra, R., 'Electromagnetic Penetration through Apertures in Conducting Surface', *IEEE Trans. on Electromagnetic Compatibility*, **20**(1), February 1978, 82–93.

[47] Ott, H., *'Noise Reduction Techniques in Electronic Systems'*, 2nd edition, AT&T Bell Laboratories, John Wiley & Sons, Inc., New York, NY, 1988.

[48] Robinson, M., Benson, T., Christopoulos, C., Dawson, J., Ganley, M., Marvin, A., Porter, S., and Thomas, D., 'Analytical Formulation for the Shielding Effectiveness of Enclosures with Apertures', *IEEE Trans. on Electromagnetic Compatibility*, **40**(3), August 1998, 240–248.

[49] Tsaliovich, A., 'Modeling Electronic System Radiated Field Patterns', *Internation Zürich Symposium and Technical Exhibition on Electromagnetic Compatibility*, Zürich, Switzerland, 1989, pp. 323–328.

[50] www.nittany-scientific.com.

[51] www.cst.com.

[52] Labussière-Dorgan, C., Bendhia, S., Sicard, E., Tao, J., Queresma, H.J., Lochot, C., and Vrignon, B., 'Modeling the Electromagnetic Emission of a Microcontroller Using a Simple Model', *IEEE Trans. on Electromagnetic Compatibility*, **50**(1), February 2008, 22–34.

10

Grounding in PCBs

In this chapter the concept of ground as reference for signals and power distribution is introduced. Some ground strategies to avoid *common-mode* coupling among circuits are outlined. An example of grounding for high-power, analog, and digital circuits in the same PCB is given.

Techniques for distributing power and ground planes in multilayer PCBs are considered. It is shown that the return current of a trace is not uniformly distributed in the return plane but is concentrated near to the trace. Some design rules such as splits, moats, and stitches with power and ground planes are qualitatively discussed. *Crosstalk* and *common-ground noise* at connector level are investigated. The concept of transfer impedance of a connector is introduced. Grounding solutions to mitigate radiated emissions when an I/O cable is attached to a PCB are compared experimentally and by numerical simulations. Numerical simulations quantify fixes such as a split in a ground plane, *common-mode* EMI filters, and PCBs in a shielded box with stitch connections.

Three test boards having power and ground planes are used to investigate the problem of partitioning the power plane distribution. Measurements, circuit simulations, and numerical simulations are compared for the first test board with a split in the power plane. The effects of a bridge or ferrite bead across the moat with capacitors are studied by numerical simulations in the case of the second test board having an island in the power supply. The third test board is used to introduce a device behavioral modeling technique for 3D interconnects in order to predict both noise on power distribution and crosstalk on signal lines.

10.1 Common-Mode Coupling

Grounding is a very critical point in designing an electronic system. Very often a non-expert designer makes the mistake of coping with the signal distribution current path in a PCB without considering its return path. The consequence is that the signal and its return current could form a large loop producing strong radiation or a dangerous interfering area. Another important aspect to consider is that the conductor or plane devoted to return current could be common to several signals. This means that a voltage drop on the ground conductor produced by the return currents and the effective inductance associated with the ground path could

Signal Integrity and Radiated Emission of High-Speed Digital Systems Spartaco Caniggia and Francescaromana Maradei

interfere with the circuits and could cause strong *common-mode* current on cables eventually attached to the PCB. All this is indicated as *common-mode coupling*, and the significance of ground must be clarified.

10.1.1 What is Ground?

When talking about the ground concept, the following points should be kept in mind:

- Ideal ground is considered to be a zero potential region with zero resistance and zero impedance at all frequencies.
- This is just not the case in practical high-speed design.
- All metal has some amount of resistance, and the current flowing through a conductor in a loop creates inductance.
- This means that the metal ground plane/wire/bar/etc. has a voltage drop across it.
- This voltage drop is responsible for interferences.

In the following sections of this chapter, grounding effects for signal integrity and radiated emission will be investigated in detail.

10.1.2 Ground Loop Coupling and Transfer Impedance

A general situation in transmitting a signal from PCB 1 to PCB 2 is shown in Figure 10.1 [1], where the two PCBs are placed inside enclosures or chassis box 1 and box 2. A similar representation could be used for the connection between two devices sharing the same return path, which could be a wire, a trace, or a plane. In this case, the A–B path in Figure 10.1 corresponds to the ground path (indicated as 'gnd path'). Common ground impedance

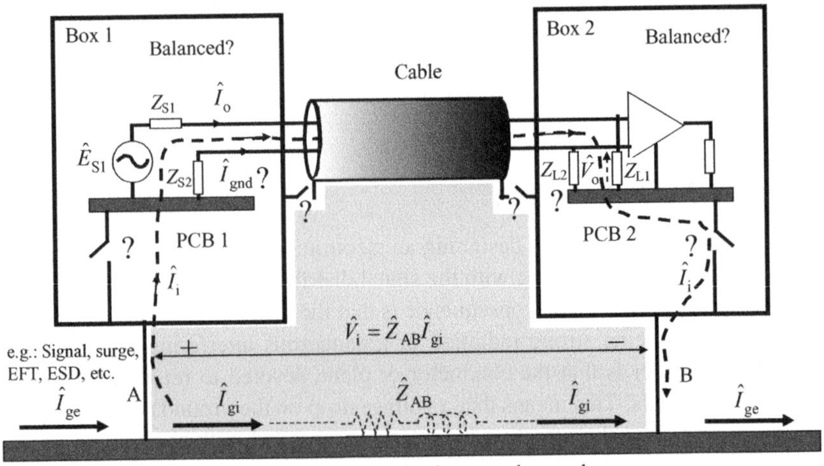

Figure 10.1 Grounding problems with the interconnect between two PCBs

$\hat{Z}_{AB}$ converts ground currents $\hat{I}_{gi}$ to *common-mode* interfering voltage $\hat{V}_i$. The ground current $\hat{I}_{gi}$ could be produced by signal or external disturbances such as *Electrostatic Discharge* (ESD), *Electrical Fast Transient* (EFT), surge, etc. The impedance $\hat{Z}_{AB}$ may correspond either to a metallic connection between two items of equipment or a return path of two or more logic devices. The impedance $\hat{Z}_{AB}$ is normally the series of a resistance and an inductance, both functions of frequency. The external ground current $\hat{I}_{ge}$ is almost equal to the internal ground current $\hat{I}_{gi}$ (i.e. $\hat{I}_{ge} \approx \hat{I}_{gi}$), so that the interfering current $\hat{I}_i$ flowing towards the interconnect is much smaller than the ground current (see Figure 10.1). In spite of this, the current $\hat{I}_i$ cannot be neglected, as it can cause interference with possible malfunctioning. Once the current $\hat{I}_i$ arrives on PCB 1, it splits into $\hat{I}_o$ flowing through the signal line and $\hat{I}_{gnd}$ flowing through the return wires (i.e. $\hat{I}_i = \hat{I}_o + \hat{I}_{gnd}$). It should be pointed out that the currents $\hat{I}_i$ and $\hat{I}_o$ shown in Figure 10.1 are those produced by the external disturbance when $\hat{E}_{S1} = 0$. The noise voltage $\hat{V}_o$ is the induced disturbance on the receiver, produced by the voltage drop $\hat{V}_i = \hat{Z}_{AB}\hat{I}_{gi}$ which generates the noise current $\hat{I}_i$.

For a single-ended interconnect, $Z_{S2} = Z_{L2} = 0$ in Figure 10.1, and the second line acts as return wire for the signal current. Therefore, a noise current $\hat{I}_{gnd}$ produced by the voltage drop $\hat{V}_i$ flows on the return signal conductor between PCB 1 and PCB 2 because its impedance is much lower than the signal line impedance. However, the value of the impedance associated with the return conductor, denoted by $\hat{Z}_{gnd}$, could be sufficiently large at high frequencies to cause malfunctioning at the receiver owing to the voltage drop $\hat{V}_{gnd} = \hat{Z}_{gnd}\hat{I}_{gnd}$. Generally, $\hat{Z}_{gnd}$ is represented by an effective partial inductance L_{gnd} associated with the return conductor, and hence $\hat{Z}_{gnd} \approx j\omega L_{gnd}$, as shown in *Section 3.2*. In fact, for a single-ended connection, the noise voltage that appears at the interconnect output (i.e. the input of the receiver) is $\hat{V}_o = \hat{V}_{gnd}Z_{L1}/(Z_{L1} + Z_{S1})$, and, as $Z_{L1} \gg Z_{S1}$, $\hat{V}_o \approx \hat{V}_{gnd}$. This means that almost all the noise sums to the signal voltage on the load Z_{L1} generated by $\hat{E}_{S1}$. This mechanism will be explained later using appropriate equivalent circuits and introducing the concept of transfer impedance.

To create an obstacle to the flow of the current $\hat{I}_{gnd}$, a common practice is to interrupt the path by isolating one of the PCBs or making it float from the reference ground (i.e. the chassis). If this is a usual solution at low frequencies, it is not recommended at high frequencies when the structure has dimensions comparable with the minimum wavelength of interest, as high-voltage disturbing peaks are generated at resonance frequencies of the structure. When it is necessary to preserve the isolation at low frequencies, a capacitor can be used to close the current path in order to avoid resonance problems [2].

To improve considerably the immunity of the system, *differential-mode* signaling is used, as will be discussed in *Chapter 12*. In this case, the Thévenin equivalent circuit of the driver has low values of $Z_{S1} = Z_{S2}$ and a second voltage source $\hat{E}_{S2} = -\hat{E}_{S1}$ in series with Z_{S2} in order to excite a *differential-mode* signal. The interconnect is terminated with a load consisting of two impedances that satisfy the following conditions: $Z_{L1} = Z_{L2}$ for balancing, and $Z_{L1} + Z_{L2} = Z_0$, where Z_0 is the nominal *differential-mode* characteristic impedance of the interconnect (see *Section 6.2*). In this case, the interfering current $\hat{I}_i$ produced by the voltage drop $\hat{V}_i$ is a *common-mode* current for the link of the two conductors between the driver and the receiver. Hence, $\hat{V}_{od}$ is the voltage drop between line 1 and line 2 at the input of the differential receiver, produced by the conversion of the *common-mode* voltage $\hat{V}_o$ to the *differential-mode* noise due to possible asymmetries in the interconnects including source and load. For a perfectly balanced interconnect, $\hat{V}_{od} = 0$. A differential receiver has an intrinsic rejection to the *common-mode* disturbance up to a voltage level characteristic of each type of device. The *common-mode* noise voltage in this context is the voltage drop on impedances Z_{L1} and Z_{L2} produced by $\hat{V}_i$.

If the *common-mode* rejection is not sufficient for the environment where the connection is placed, or the link does not guarantee a sufficient level of balancing necessary to avoid dangerous *common-mode* to *differential-mode* noise conversion, the solution of a shielded cable is the most appropriate. In fact, *common-mode* filters might not be suitable, as they stretch the signal edges and slow the signal speed (see *Example 9.14*). In the case of shielded cables, such as coaxial, *shielded foil twisted-pair* (SFTP), and twinax cables, the classical question regarding the connection of the cable shield has a certain answer: the cable shield must be connected at both ends to the chassis with a very low impedance connection, especially at high frequency [2]. In fact, with the cable shield well connected to the chassis, in the high-frequency region where the skin effect becomes significant, the current $\hat{I}_{gnd}$ produced by $\hat{V}_i$ flows on the external part of the shield for both single-ended (coaxial cable) and differential cables. Only a very small quantity of disturbing current flows within the cable owing to the distributed voltage sources $\hat{Z}_t\hat{I}_{gnd}\Delta x$ on the internal side of the cable, where $\hat{Z}_t$ is the shield transfer impedance and Δx is an infinitesimal portion of the cable (see *Section 9.8.1*).

These considerations for immunity can be reversed for emission. Indeed, if the current $\hat{I}_o$ is produced by the circuits, the current $\hat{I}_g$ assumes the significance of an antenna current that flows between the structure and the environment. As discussed in *Section 9.7*, the current $\hat{I}_g$ could produce high values of radiated fields. The link between $\hat{I}_o$ and $\hat{I}_g$ is always the transfer impedance $\hat{Z}_t$. The same fixes as used for immunity are required to mitigate the radiated missions. These cases were thoroughly investigated and discussed in *Chapter 9*, with numerous examples.

The concept of transfer impedance is not used only for shielded cables by the EMC community, but it assumes a larger significance, and it is also used to characterize *common-mode* ground impedance between a loop with high current that produces unwanted current in a second loop and the second loop itself. The two loops are characterized by the fact that they share a segment of conductor with a low effective impedance [3]. Consider a simplified case of Figure 10.1 with a single-ended transmission and with the two PCBs directly connected to the reference ground. This case is shown in Figure 10.2a, where the current $\hat{I}_{gnd}$ on the return conductor mentioned above is denoted by $\hat{I}_{sh}$. The subscript 'sh' is adopted because the return conductor very often is the shield of a coaxial cable. The disturbing voltage $\hat{V}_i$ can also be a voltage produced by an external H-field interfering in the loop formed by the return conductor of the single-ended link and the reference ground (see the gray area in Figure 10.1). As $\hat{I}_o \ll \hat{I}_{sh}$, $\hat{I}_{sh} \approx \hat{I}_i$, where $\hat{I}_i$ is the current provided by the voltage generator $\hat{V}_i$.

The resistances R_{gndS} and R_{gndL} are associated respectively with the 'reference ground–PCB 1' path and 'reference ground–PCB 2' path. At high frequencies these resistances have the meaning of impedance. For now, for the sake of simplicity, the interconnect is considered to be electrically short. The voltage $\hat{V}_i$ acts as a potential EMI source pushing current around the path indicated by the dashed line in Figure 10.1. The induced noise voltage at the interconnect output is the voltage $\hat{V}_o$ occurring at the input of the receiver. To quantify the interference, the *Ground Loop Coupling* (*GLC*) parameter is defined as

$$GLC = 20\log\left(\left|\hat{V}_o/\hat{V}_i\right|\right) \tag{10.1}$$

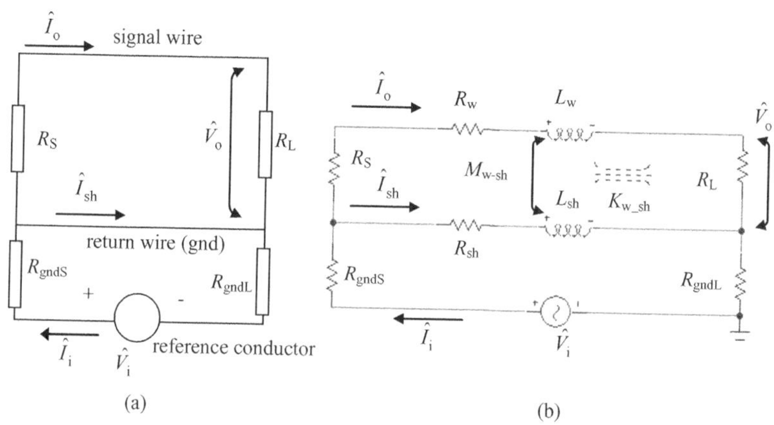

Figure 10.2 Grounding problems in PCBs: (a) simplified structure of Figure 10.1; (b) equivalent circuit

Considering EMC practice, the parameter *GLC* can be mitigated using ground strategy, choosing balanced signal transmission with shielded cables, and adopting techniques for opening 'ground loops' such as isolation transformers, *common-mode chokes*, opto-isolators, and buffer amplifiers [1].

When the line between source and load is electrically short, the equivalent circuit of Figure 10.2b can be used to compute *GLC*, or equivalently $\hat{V}_o$, setting $\hat{V}_i = 1$ V in the entire frequency range of interest. In this circuit, the capacitive effects are neglected because the resistances of the circuit, R_S and R_L, normally have low values for digital signaling, and therefore inductive effects prevail. In the circuit of Figure 10.2b, the inductances L_w of the signal wire and L_{sh} of the return wire are the self partial inductances, and $M_{w\text{-}sh}$ is the mutual partial inductance between the two conductors. The partial inductance concept is here applied to model the structure of Figure 10.2a by the equivalent circuit of Figure 3.12b, where at each of the three conductors an inductive effect is assigned and the coupling among the wires are modeled by controlled sources. The equivalent circuit of Figure 10.2b is derived by assuming that the coupling between the signal/return and reference can be neglected and the self partial inductance effect of the reference conductor is included in V_i. The coupling factor $K_{w\text{-}sh} = M_{w\text{-}sh}/(L_w L_{sh})^{1/2}$ is normally used to take into account mutual inductance coupling in SPICE-like circuit simulators. The parameters R_w and R_{sh} are the resistances of the signal line conductors. The resistances R_{gndS} and R_{gndL} are associated with the connections to the ground and therefore have very low values. The parameters R_S and R_L are the source and load resistances respectively. By applying loop equations to the circuit of Figure 10.2b, the new equivalent circuit shown in Figure 10.3 can be obtained. Each dependent current source captures the current $\hat{I}_i$ and the current $\hat{I}_o$ to force it respectively into the series impedance R_t and L_t. The dependent voltage sources $\hat{Z}_t\hat{I}_i$ and $\hat{Z}_t\hat{I}_o$, with $\hat{Z}_t = R_t + j\omega L_t$, determine the current in the respective loops.

Using SPICE, it is possible to verify that the two circuits provide the same results. However, the second circuit offers the great advantage that the grade of coupling is described by one parameter, the transfer impedance $\hat{Z}_t$, defined as

$$\hat{Z}_t(\omega) = R_t + j\omega L_t = R_{sh} + j\omega(L_{sh} - M_{w\text{-}sh}) \quad (10.2)$$

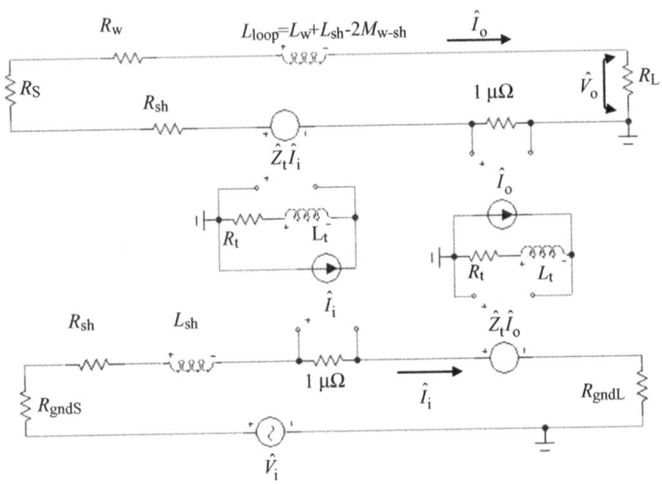

Figure 10.3 Equivalent circuit of Figure 10.2a with the transfer impedance concept Z_t

From an EMI point of view, the transfer impedance $\hat{Z}_t$ should be minimized. The parameter $\hat{Z}_t$ depends on R_t, equal to the DC value of the return conductor at low frequencies, and the difference between the self partial inductance L_{sh} and the mutual partial inductance $M_{w\text{-}sh}$. As discussed in *Chapter 7*, the transfer resistance R_t is a function of frequency when skin and proximity effects become significant.

As usually $\hat{I}_i \gg \hat{I}_o$, the contribution of the dependent voltage source $\hat{Z}_t\hat{I}_o$ in Figure 10.3 can be neglected. A method for measuring or computing numerically the transfer impedance $\hat{Z}_t$ in a low-frequency approximation consists in forcing a current $\hat{I}_i$ in the ground conductor and closing the signal loop with a very high load R_L so that the current $\hat{I}_o \approx 0$. In this manner

$$\hat{Z}_t = \left.\frac{\hat{V}_o}{\hat{I}_i}\right|_{\hat{I}_o=0} \qquad (10.3)$$

The equivalent circuit of Figure 10.3 can also be used for emission computation, as presented in *Section 9.6*, where the *current-driven* mechanism was introduced. In this case the upper loop is the signal interconnect and the bottom loop is the PCB–cable–environment. The parameter L_t also assumes the meaning of the effective partial inductance L_{gnd} associated with the return conductor of the signal line, as discussed at the beginning of this section.

If the bottom loop of Figure 10.2 is open, it is possible to derive the circuit shown in Figure 10.4 to calculate the parameter L_t or L_{gnd}. The shape of the signal and return conductors strongly influences the value of $\hat{Z}_t$, and closed-form expressions are not always available. In this case, the signal and return conductors can be simulated by a number of parallel filaments having self and mutual partial inductances, as reported in Table A.1 of *Appendix A* for an isolated wire and for two parallel wires. When the signal conductor is a trace, a filament can be used for it, and a certain number of filaments for the return conductor, depending on its shape. With reference to this equivalent structure based on filaments, the following

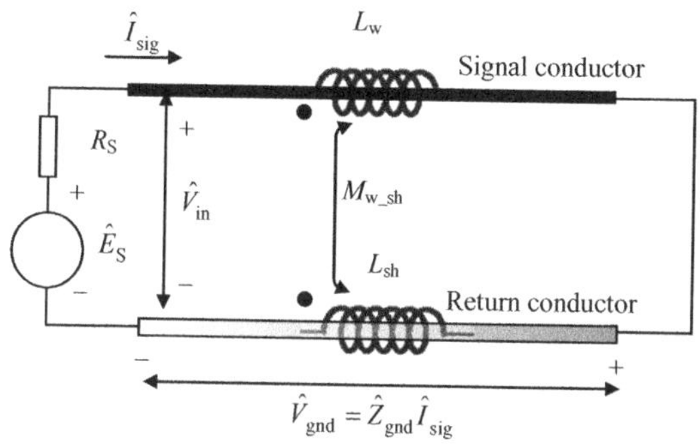

Figure 10.4 Equivalent circuit used to calculate transfer inductance $L_t = L_{gnd}$

relations hold:

$$L_{loop} = \frac{1}{\omega}\text{Im}\left[\frac{\hat{V}_{in}}{\hat{I}_{sig}}\right] \tag{10.4a}$$

$$L_{sh} = \frac{1}{\omega}\text{Im}\left[\left.\frac{\hat{V}_{gnd}}{\hat{I}_{sig}}\right|_{M_{w-sh}=0}\right] \tag{10.4b}$$

$$L_{gnd} = L_t = \frac{1}{\omega}\text{Im}\left[\frac{\hat{V}_{gnd}}{\hat{I}_{sig}}\right] \tag{10.4c}$$

where L_{loop} is the loop inductance of the interconnect, L_{sh} is the self partial inductance of the return conductor, computed by setting all the mutual inductances between the filament of the trace and the filaments of the return conductor to zero, and L_{gnd} is the effective partial inductance associated with the return conductor and also assumes the meaning of transfer inductance L_t.

As illustrated in *Appendix E*, the parameters in Equations (10.4) can be calculated by SPICE or the nodal method. The following example explains how this approach can be used to quantify the EMC performance of a given structure.

Example 10.1: Computation of Transfer Impedance for Several Interconnection Structures Consider the structures shown in Figure 10.5, characterized by a trace and its return conductor with different shapes:

(a) a trace having another trace of equal size to the return conductor: a trace–trace (tt) structure;
(b) a trace sandwiched between two return traces of equal size: a trace–trace–trace (ttt) structure;
(c) a trace over a finite ground plane: a microstrip-like (tp) structure;
(d) a trace over a finite ground plane that also extends in the vertical direction: a conduit-like (tc) structure;
(e) a trace between two finite ground planes: a stripline-like (ptp) structure;

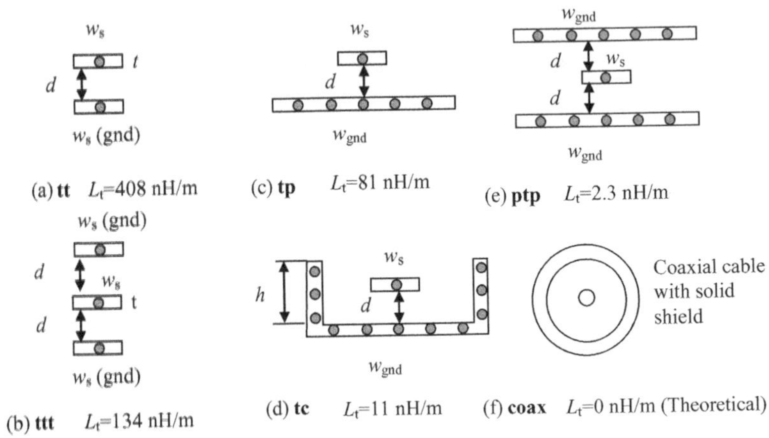

Figure 10.5 Transfer inductance L_t of several structures with signal (s) and return (gnd) conductors of different shape ($t = 0.1$ mm, $w_s = 0.25$ mm, $w_{gnd} = 10w_s$, $d = 0.5$ mm, $h = 6w_s$)

(f) a round wire enclosed within a tubular conductor: a coaxial-like (coax) structure, used as reference for computation (in theory, it should have $L_t = 0$; in practice, as explained in *Section 9.8.1*, $L_t > 0$, but in any case much less than the transfer inductance associated with the other structures considered).

If the reference conductor in Figure 10.1 is assumed to be sufficiently far away from the signal conductor and its return, the concept of partial inductance can be used to compute L_t by SPICE. The number of filaments used for computations as round wires is indicated in gray in Figure 10.5. The parameters computed by Equations (10.4) are shown in Table 10.1.

Details of the calculation of these parameters by the nodal approach or by SPICE are given in *Appendix E*. Note that the closed-form expression provided in Table A.2 of *Appendix A* for a flat conductor over a finite-return ground plane gives $L_{gnd} = 75.1$ nH/m, which is close to the value of 80.63 nH/m reported in Table 10.1 for the tp structure.

The parameter *GLC* defined by Equation (10.1) was calculated by the equivalent circuit of Figure 10.3, adopting the inductances of Table 10.1. The results obtained, setting $\hat{V}_i = 1$ V, $R_S = 10\ \Omega$, $R_L = 100\ \Omega$, and all other resistances equal to 1 mΩ for a length of 1 m, are shown in Figure 10.6. The maximum frequency is 10 MHz in order to have the line electrically short. As, in practice, a coaxial cable with a solid shield is not used, the result shown in the graph concerns an RG214 coaxial cable with a double-braided shield, which is very often

Table 10.1 Computed per-unit-length (p.u.l.) inductances for the structures of Figure 10.2. All the values are in nH/m. The value of L_t for coaxial cable comes from *Section 9.8.1*

nH/m	(a) tt	(b) ttt	(c) tp	(d) tc	(e) ptp	(f) coax (RG214)
L_{loop}	815.4	542	562.8	499.7	441.2	250
L_{sh}	1830	1557.1	1409	1300.9	1316	998.3
$L_{gnd}=L_t$	407.7	134.6	80.63	9.5	2.292	0.13

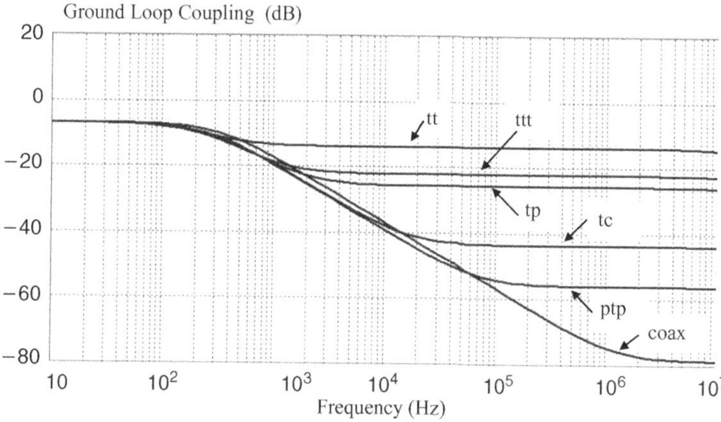

Figure 10.6 Voltages on the load (i.e. *ground loop coupling – GLC*) for the structures of Figure 10.5, computed by the equivalent circuit of Figure 10.3

used for high-speed data transmission. The coaxial cable values shown in Table 10.1 were obtained assuming $L_t = 0.13$ nH/m, $Z_0 = 50\ \Omega$, $t_{pd} = 5$ ns/m, and radius $r_w = 5$ mm, and using the equation for partial inductance of a round wire of radius r_w reported in Table A.1 of *Appendix A*. From Figure 10.6 it can be noted that the best performance is with stripline-like and coaxial cable structures. With more filaments, better values can be obtained for the stripline-like structure. For the other structures, the values of the transfer inductance L_t are practically stable. Note that, increasing the frequency, the smaller the inductance L_t or L_{gnd}, the lower is the *GLC* parameter. The values of Table 10.1 could also be used to predict the radiated emission field by the circuit model of the *current-driven* mechanism that occurs when a cable is attached to a PCB (see *Section 9.6*). If the transfer impedance refers to an electrically long shielded cable, the equivalent circuit of Figure 10.3 becomes that described in *Section 9.7.2*. The concept of transfer impedance will be used again in *Section 10.3* for investigating the EMI effects in connectors for PCBs.

10.1.3 Grounding Strategy

Grounding strategy is the most effective way to obtain a low *GLC* parameter. There are two basic grounding strategies that are used in PCBs and system design [4–8]:

- single-point ground reference;
- multipoint ground reference.

The configurations of several possible solutions are shown in Figure 10.7. The indicated currents are those flowing through the ground connections with their associated impedances. By *system* we mean the connection between two devices in a PCB as well as between two PCBs or racks.

A multipoint (MP) ground reference is used at frequencies higher than 100 kHz to minimize ground loop impedance. For PCBs with high-speed digital devices, the reference ground

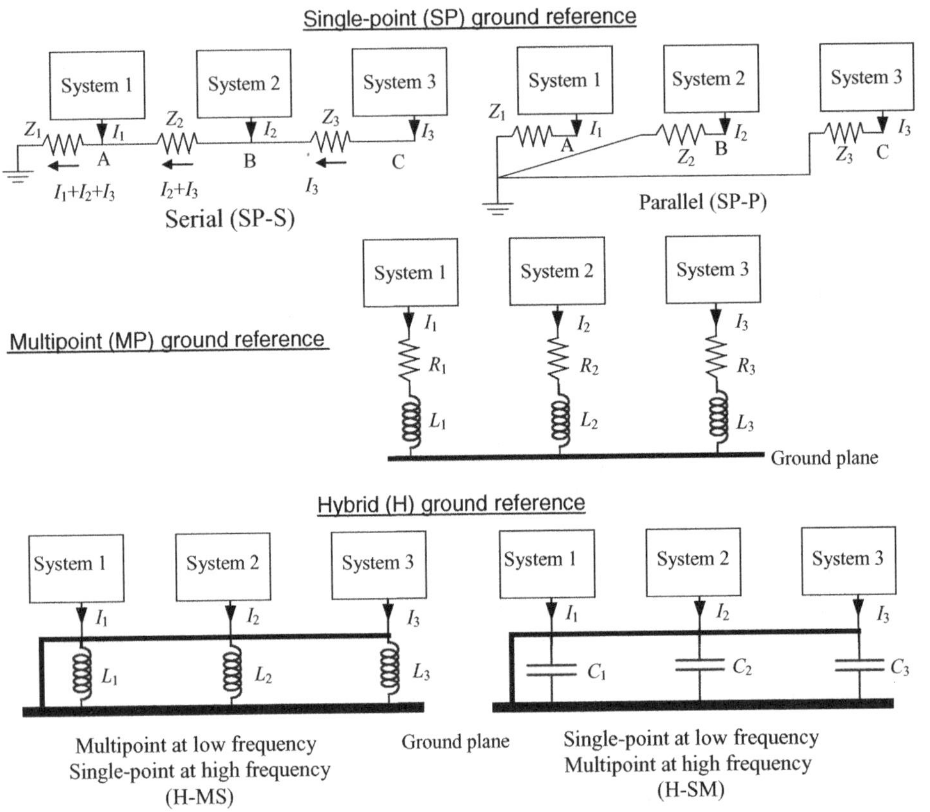

Figure 10.7 Grounding strategies: single-point ground reference (serial and parallel); multipoint ground reference; hybrid ground reference (multipoint at low frequencies and single-point at high frequencies, and vice versa)

conductor is usually a ground plane owing to its lower associated inductance, as shown in Figure 10.7. For systems it is generally a grid of conductors in order to realize a reference ground at quasi-constant potential with the function of a Faraday cage. The connections between each circuit or system and the ground plane should be kept as short as possible to minimize their impedance. At high frequencies, the length of these ground leads must be kept to a small fraction of the minimum wavelength of interest. This solution offers the great advantage that restrictions for interconnection paths are not generally required regarding the *GLC* parameter.

A hybrid (H) ground is a solution in which the ground reference must have different behaviors at different frequencies. A hybrid ground connection with inductances acts as a multipoint ground reference at low frequencies and a single-point ground reference at high frequencies (H-MS). The opposite occurs when capacitances are used (H-SM).

An example of a multipoint reference ground at low frequencies is when a ferrite bead is used to connect two separate ground planes in a PCB. An example of a multipoint reference ground at high frequencies is when a capacitor is used to connect analog and digital grounds in a telephone line interface card with an analog–digital device, or when a capacitor is used

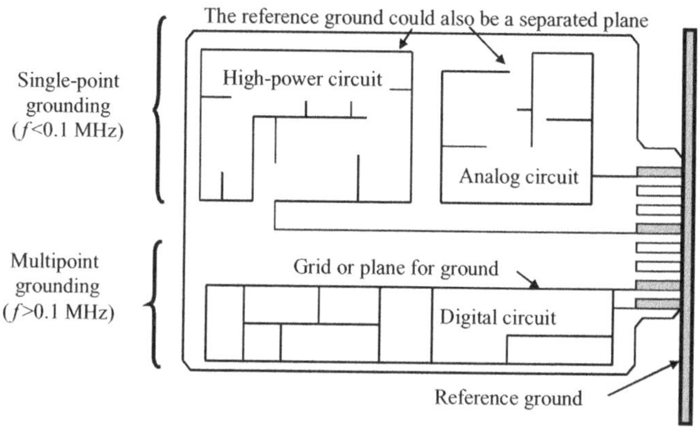

Figure 10.8 Grounding in a PCB with high-power, analog, and digital circuits

to connect one end of a shielded cable to the ground, and the other end is short-circuited to the ground.

A single-point ground reference is often used in a PCB with different types of circuit such as high-power, analog, and digital circuits. The different ground circuits should be tied together, usually at a single point, as shown in Figure 10.8, where the reference ground is usually the chassis. When analog circuits are located in a multilayer PCB with digital circuits, an isolated area is created for the analog circuits, splitting the power and ground planes and connecting the analog ground to the digital ground at one point. The purpose is to prevent the digital return current from modulating the analog return current. This problem will be discussed in *Section 10.2.2*.

The three types of circuit can be characterized in the following way:

1. Analog signal circuits:
   - narrow band;
   - no signal regeneration;
   - gain devices;
   - low signal level (μV, mV).
2. Digital signal circuits:
   - broadband;
   - moderate signal regeneration;
   - no gain;
   - moderate signal level (V).
3. 'Noisy' (control and power circuits):
   - broadband;
   - little or no signal parameter control;
   - extremely high signal levels (up to kV).

10.2 Ground and Power Distribution in a Multilayer PCB

In this section, the return current path problem concerning the digital signal in a PCB is investigated. At first, by using numerical 2D computations, it is shown that the return current density in microstrip and stripline structures follows the position of the signal conductor and minimum emission can be obtained with the conductor far away from the edges of the board. In a multilayer PCB, where the signal routing could involve several layers, the return current could flow through uncontrolled paths that ensure minimum impedance. The appropriate use of decoupling and stitching capacitors to mitigate EMI problems that could arise is discussed. Fixes such as moats/barriers and stitches in reducing *common-mode* emissions are also considered.

10.2.1 Return Path for the Signal

Many EMI problems associated with high-speed devices and their interconnects are due to improper design of the return current path of the signal current [9]. PCB designers must always consider the path made by the signal current to return to its source, that is, the driver device. These EMI problems can be avoided by considering the following rules for the return path of a digital signal:

- Each digital signal needs a return current path.
- The return path could be another trace for a low-speed circuit.
- For a high-speed circuit, a metallic ground plane is required.
- Care should be taken to ensure that this path is not interrupted and does not pick up noise from other circuits.
- In multilayer PCBs, two structures can be used: microstripline and stripline configurations.
- Owing to the low impedance of the decoupling capacitors between the power and ground planes at medium–high frequencies, the reference return path could also be the power plane.
- The decoupling capacitor is an obstacle for the high-frequency components of the return current because the associated inductance makes the equivalent impedance of the capacitor too high.

Example 10.2: Current Distribution and Radiation Pattern for Microstrip- and Stripline-like Structures

To investigate how the return current is distributed on the return ground plane, consider a microstrip-like structure of a wire above a ground plane, as depicted in Figure 10.9, with finite dimensions in the xy plane and infinite dimensions along the z axis. The field radiated by a line source of constant current $\hat{I}_z$, in the absence of a ground plane, and at a generic distance ρ_{m} when the observation point is in the middle of the ground plane (see point P in Figure 10.9) is [10]

$$\hat{E}_z^{\mathrm{d}}(\rho_{\mathrm{m}}) = -\frac{\beta^2 \hat{I}_z}{4\omega\varepsilon} \hat{H}_0^{(2)}(\beta\rho_{\mathrm{m}}) \tag{10.5}$$

where $\hat{H}_0^{(2)}(\beta\rho_{\mathrm{m}})$ is the Hankel function of the second kind of zero order and β is the phase constant. Part of the field given by Equation (10.5) is directed towards the ground plane, and it induces on it a linear current density $\hat{J}_{\mathrm{Sm}}(x)$. If the ground plane is subdivided into N segments, each of width $\Delta = w_{\mathrm{m}}/N$, each element can be associated with a current $\hat{J}_{\mathrm{Sm}}(x)\Delta$ that produces an electric field component indicated as a reflected or scattered field. If this

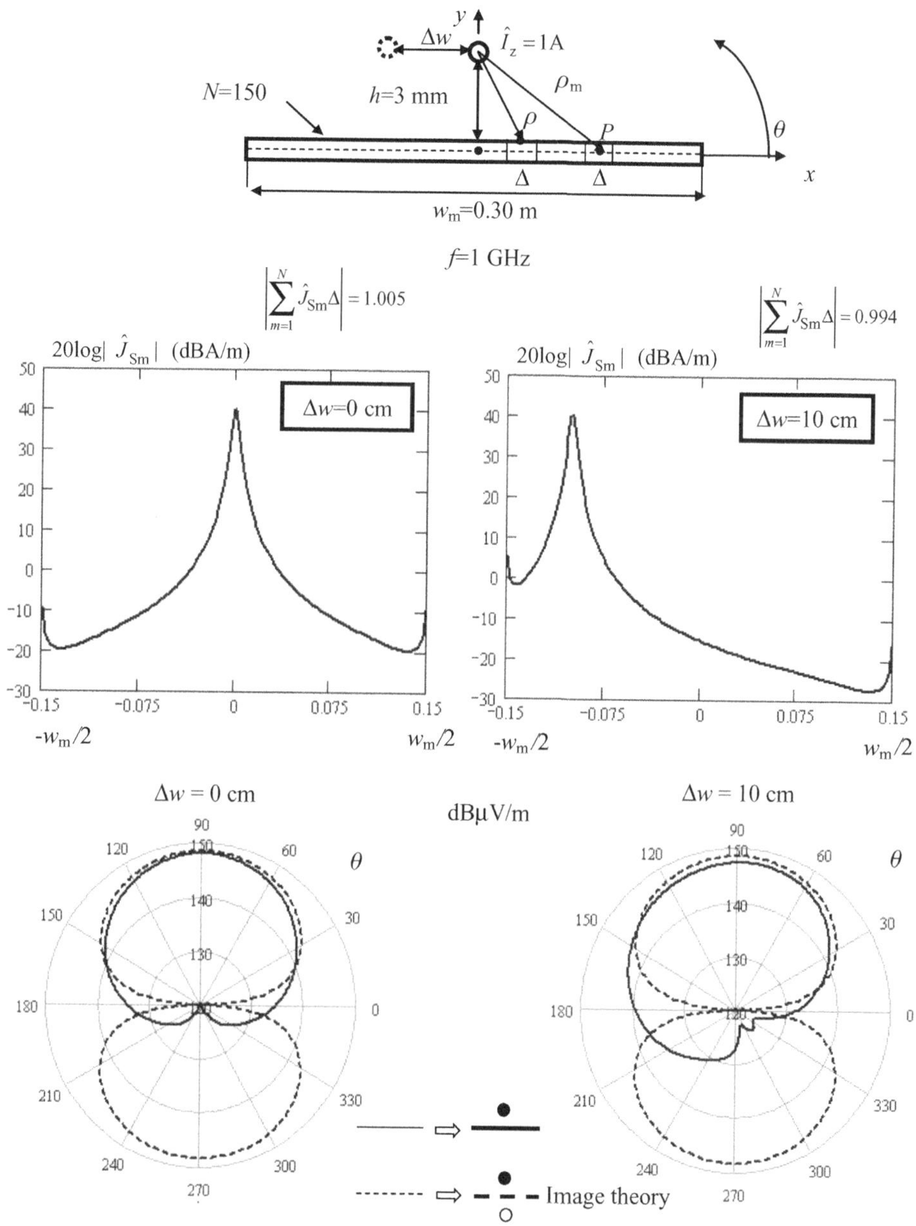

Figure 10.9 Microstrip-like structure: current distribution on the finite ground plane with and without offset Δw at a frequency of 1 GHz, and radiated emission patterns computed at 3 m

current is assumed to be located on the upper external surface of the ground plane, the total scattered field can be written according to Equation (10.5) as

$$\hat{E}_z^{\mathrm{s}}(\rho_{\mathrm{m}}) = -\frac{\beta^2}{4\omega\varepsilon}\sum_{n=1}^{N}\hat{H}_0^{(2)}(\beta\,|\rho_{\mathrm{m}} - \rho(x_n)|)\hat{J}_{\mathrm{Sm}}(x_n)\Delta \tag{10.6}$$

The total field at any observation point, produced by both signal line and ground currents, will be the sum of the direct and scattered fields given by Equations (10.5) and (10.6) respectively. However, to determine the scattered component, it is necessary to know the induced current density $\hat{J}_{\mathrm{Sm}}(x_n)$. This can be accomplished by choosing the observation point on the ground plane itself, and imposing that, for any of these observation points, at a distance ρ_{m}, the total tangential field vanishes:

$$\hat{E}_z^{\mathrm{t}}(\rho_{\mathrm{m}}) = \hat{E}_z^{\mathrm{d}}(\rho_{\mathrm{m}}) + \hat{E}_z^{\mathrm{s}}(\rho_{\mathrm{m}}) = 0 \tag{10.7}$$

Substituting Equations (10.5) and (10.6) into Equation (10.7), and for $N \to \infty$, yields the electric field integral equation (EFIE) for the line source above the strip, and it can be used to find the current density $\hat{J}_{\mathrm{Sm}}(x)$ on the basis of a unit current $\hat{I}_z$. This equation can be solved by using a 2D *Method of Moment* (MOM) as described by Balanis [10]. The results of this computation are shown in Figure 10.9 for a wire centered above the ground plane and for a wire with an offset.

Note that the return current flows directly beneath the signal line. The computed radiation patterns at a distance of 3 m (see the solid line in Figure 10.9) are also shown, using Equation (10.6), as the contribution of the current on the wire and the currents flowing through each of the 150 sections used to model the ground plane. The dashed line represents the radiation pattern of a wire above an infinite ground plane, computed by applying image theory and the expression given in Table D.1 of *Appendix D* for an infinitesimal dipole setting $\hat{I}(\xi) = 1$ A and $\Delta\xi = 1$ m. Of course, for this last structure, only the fields with an angle θ between 0 and 180° are significant. Note that the maximum fields in the direction orthogonal to the plane computed by the two methods are equal.

By applying the same procedure for a stripline-like structure, the results shown in Figure 10.10 are obtained. Observe that there is more dense current beneath and above the signal conductor. As shown in Figure 10.10, for a symmetric stripline the signal return current uses both planes equally. The radiated emission patterns are also shown in Figure 10.10. Note that the maximum electric field is about 40 dB lower than the maximum field of the microstrip-like structure. This fact confirms the experimental results reported in *Section 9.3*, where the emissions from microstrip and stripline PCBs are compared. For both structures, when there is an offset, the radiated field increases in the direction of the shift [11]. By simulations, it can be shown that the maximum emission is obtained when the wire approaches the edge. Therefore, to reduce radiated emission, clock traces must be far away from the edges.

The results of Figures 10.9 and 10.10 can also be obtained by a different approach based on the concept of partial inductances. The wire can be considered as a filament, and the ground plane as N filaments in parallel, connected together at both ends, as shown in Figure 3.13. Taking into account all the self and mutual partial inductances, the current density in the

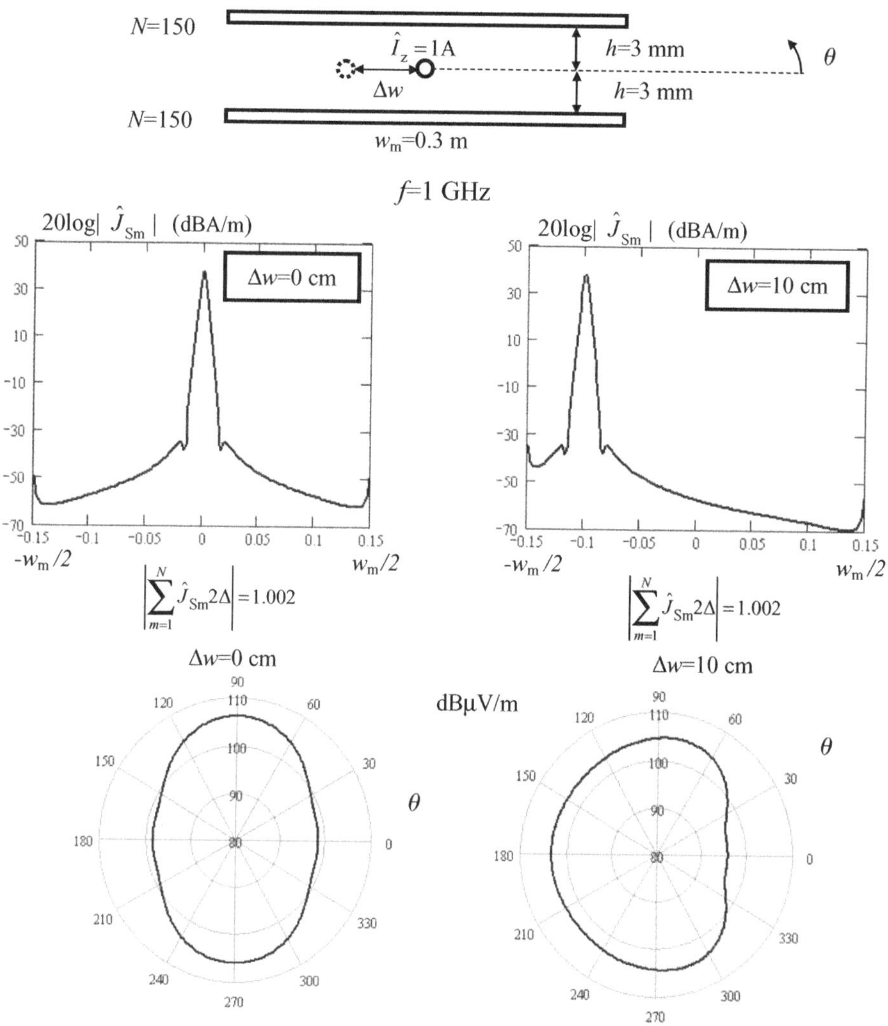

Figure 10.10 Stripline-like structure: current distribution on the finite ground plane with and without offset Δw at a frequency of 1 GHz, and radiated emission patterns computed at 3 m

ground plane is reconstructed as shown in *Appendix E*. If a frequency of 1 MHz is considered instead of 1 GHz, the current distributions obtained by the two methods also practically coincide near the edges of the ground planes.

The configurations adopted for this investigation are obviously an idealization of what occurs in actual PCBs. However, as long as no discontinuity in the reference plane exists, such as a break in the plane, a via transition, etc., the return current remains closely coupled to the signal current, with a distribution of current density as shown in Figures 10.9 and 10.10, and an effective transmission line is created.

In a multilayer PCB with several power and ground planes, the majority of the signal current returns on the reference plane closest to the signal line. This means that the signal reference plane could either be a power or a ground plane. This is due to the fact that the decoupling capacitors connect the power and the ground planes together at high frequencies for a limited frequency range. In fact, the intrinsic inductance of the capacitor (i.e. the inductance of vias, pads, and traces connected to the capacitor) limits the high-frequency performance of the capacitor. This results in a non-zero impedance between the power and ground planes. How to minimize these undesirable parasitic effects associated with the decoupling capacitors has been described in *Section 8.1*. Other fixes that make it possible to avoid the interruption of the desired path for the return current will be described in the following sections and concern:

- a signal trace changing the reference plane (see *Section 10.2.3*);
- a split in the reference plane (see *Section 10.2.4*);
- a signal going through a connector between two different circuit boards (see *Section 10.3.1*).

10.2.2 Power (PWR) and Ground (GND) Layer Planning and Topology

The following rules should be considered for power and ground layer planning:

- Design the power and ground layers first.
- Always use power and ground layers in pairs in order to increase the capacitance between layers and reduce the power distribution network impedance.
- Sprinkle many ground vias to connect the ground planes by low impedances in order to allow low inductance and layer jumping of return current.
- Bypass capacitors help in reducing the impedance between power and ground planes up to about 100 MHz. However, they exhibit an impedance that is too high for the high-frequency components of the return currents. This means that, over a certain frequency, the return current uses uncontrolled paths in the form of displacement current.
- Prefer GND–Signal–GND to GND–Signal–PWR layers for stripline transmission lines for high-speed signal routing, as GND–Signal–PWR requires bypassing by capacitors. The drawback is that more than four layers are required to make the PWR plane a second return path for the signal current.

The following rules should be considered for power and ground topology:

- There should be no overlapping area of *Analog* GND (AGND) and *Digital* GND (DGND) in order to reduce parasitic capacitance between the areas.
- A voltage source supplying several different circuits on a PCB can be distributed separately to the different circuits, with dedicated traces or planes tied together at the edge connector by decoupling inductors.

10.2.3 Trace Changing Reference Plane

In a multilayer PCB it is very common to change the layer in routing a signal trace [9]. This is due to the fact that, to prevent crosstalk, traces in adjacent layers are routed in the orthogonal

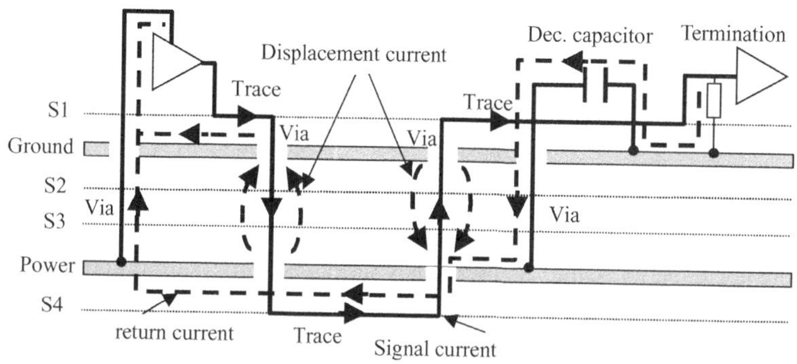

Figure 10.11 Signal and return current path in a multiplayer PCB.

direction. The path of the return current must also change layer to be near the signal trace, as discussed in *Section 10.2.1*.

Consider a six-layer PCB as depicted in Figure 10.11, where four layers are reserved for routing traces and indicated as S1, S2, S3, and S4. The signal trace changes from layer S1 to layer S4, and returns after a certain length to layer S1. The assumption is made that the driver switches from low to high voltage level. The path of the signal current is indicated by the solid line, with arrows to highlight the direction of the current flow. There are two reference planes: the ground plane for the signal path in layer S1, and the power plane for the signal path in layer S4. For the return current, indicated by the dashed line in Figure 10.11, there are two parallel paths for crossing the bottom reference layer to the top reference layer. For low-frequency components of the current, the path is through a decoupling capacitor, which is very often positioned near to the via for signal, to make an easier path for its return current. For higher frequencies, the path is through the displacement current of the interplane capacitance. The preferred path for the current depends on which path has the lower impedance at a given frequency, considering also that, owing to the skin effect, the current flows on the internal or external surface of the reference planes. Since the decoupling capacitor is physically located on one side of the PCB, the intrinsic impedance of the capacitor and the inductance of traces and vias connecting the capacitor to the power and ground planes affect the return current path, as discussed in *Section 8.1*. If the total impedance associated with the decoupling capacitor is so high as to oblige the return current to follow the path of the displacement current, this current causes a voltage drop between the power and ground planes responsible for EMI in the PCB and radiated emissions. Therefore, the designer should always verify the impedance associated with the decoupling capacitor in order to make it useful. If the reference plane must be changed, then the decoupling capacitor should be used close to the via where the layers are changed. As discussed in *Section 3.2.8* and *Section 8.1*, the signal and capacitor vias should be as close as possible and have the associated current in opposite directions in order to increase the mutual partial inductance, which, subtracted from the self partial inductance, produces a low effective inductance associated with each via.

When changing layer, a routing strategy should be applied. For instance, the example of Figure 10.11 is not a good practice for changing layer, even when changing from layer S1 to layer S3. A good routing strategy should be to change from layer S1 to layer S2, and vice versa. Return currents can flow from one side of the plane to the other side of the same plane through the antipad opening of the via hole, and no voltage drop is created between the power and ground planes.

10.2.4 Split Power Plane

Power planes with different voltages, e.g. 3.3/5 V planes, are usually divided by a split while the ground plane can be the same. As previously discussed, the return current tends to make a path close to the signal trace, and it does not matter if the reference plane is a ground or a power supply, as they are short-circuited by the decoupling capacitors at the frequency where their associated inductance yields very low impedance. When a trace runs across a split between two different DC power voltages, the return current cannot flow across the split, which acts as an obstacle, the path of the return current becomes uncontrolled, and an undesirable large loop of emission could be created.

To avoid this problem, stitching capacitors connected between two power islands and positioned close to the location where the trace crosses the split are often used to provide a return current path across the split. An example of a stitching capacitor across a split between two power planes is shown in Figure 10.12. A switching of the driver from low to high voltage level is considered. Other connections where the currents do not flow are omitted for simplicity of representation. In this case the signal and the return current follow the path indicated by arrows in Figure 10.12. This solution is particularly effective when the ground plane also has a split. The decoupling capacitor acts as a low-impedance connection between the power and ground plane. To be effective in providing a suitable return path for the signal current, the stitching capacitor, as well as the decoupling capacitor, must have its intrinsic inductance, and the additional inductance associated with their connections, minimized so that the total impedance at the frequencies of interest is very low. The actual capacitor has a capacitive behavior up to the resonance frequency owing to the series impedance formed by its capacitance and its associated total inductance. After this frequency, its behavior becomes inductive. The choice of stitching capacitors should be done considering that, up to the ninth harmonic of the clock, the impedance offered by the stitching capacitor should be no more than some ohms. Usually, capacitors of 1–10 nF should be appropriate. Archambeault [9] shows that the emission of a PCB with an ideal stitching capacitor could be up to 20 dB lower than that from a PCB without a stitching capacitor. For a real capacitor, where its total inductance is accounted

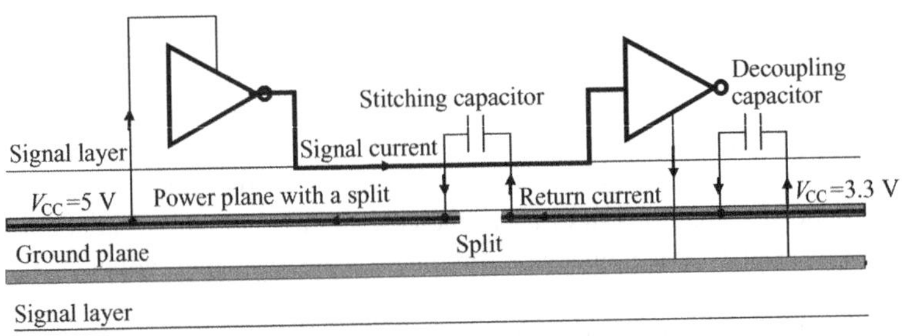

Figure 10.12 Illustration of the signal and return current paths when a stitching capacitor is applied to a power plane with a split

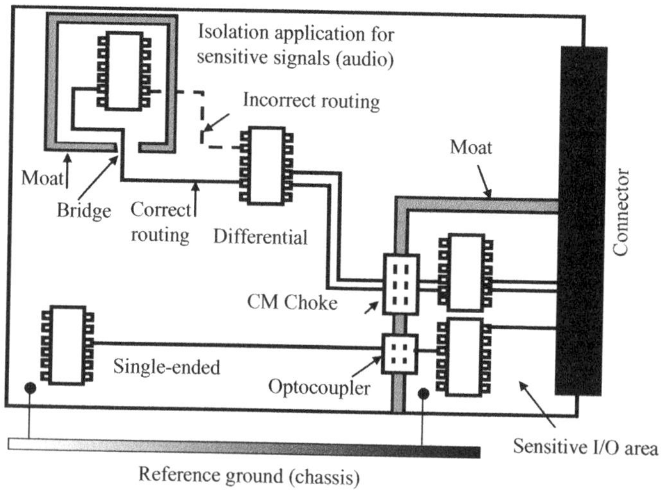

Figure 10.13 Isolation applied to a PCB with sensitive signals and I/O circuitry

for, the 20 dB of difference remains at low frequencies but vanishes at frequencies approaching 1 GHz. Using more stitching capacitors, the EMC performance is slightly improved. A quantification of the stitching benefits and drawbacks in a vast range of frequencies will be discussed in *Section 10.4* by using simulations and measurements.

10.2.5 Moats/Barriers and Bridges

Moats and barriers are techniques often used for limiting *Ground Loop Coupling (GLC)* in power and ground planes [12, 13]. Moats consist in removing a narrow strip of metal in the ground plane in order to isolate the area with sensible devices. The bridge is a small copper land across the gap to permit a return path without obstacle to the signal current. An example is shown in Figure 10.13, where a protected zone for sensitive analog circuitry with a bridge and an isolated zone for low frequency (some kHz) I/O devices can be distinguished. In this way, the analog devices are protected from the spread of the digital signal return currents which could interfere with the return path of the analog devices, and the *common-mode* current produced by the switching of the digital devices does not flow in the I/O cables. Note that, when a complete isolation is required, single-ended signals are not allowed to cross the moat, and in this case a decoupling such as an optocoupler device must be used. For differential signaling, the decoupling must be implemented by using a *common-mode choke* filter with transformer in order to maintain isolation and, thereby, to stop the *common-mode* currents.

When these fixes are realized, the following considerations should be taken into account:

- Moating can be used to isolate any circuit, e.g. oscillators.
- The power and ground planes are divided by moats into areas where different circuits are allocated.
- A bridge is allowed between these areas, with the task of ensuring an equal potential at DC and to provide an obstacle to the flow of the noise generated by one of the two areas.

- The 'bridge' acts as a *common-mode choke* on the power and ground conductors.
- Traces should never cross between different PCB areas, except over the bridge.
- For mixed-signal devices (analog and digital), the better solution is to consult the application notes prepared by the device manufacturer for layout and grounding. In general, the mixed-signal device should be placed across the moat between the two analog (AGND) and digital (DGND) grounds which could be isolated or connected at one point through a 0 Ω resistor according to the final validation of the EMI performance of the PCB.

In *Section 10.3* it will be shown that these practices could be useful under certain decoupling conditions and only for some ranges of frequency by means of numerical simulations.

10.2.6 Stitches

Stitches are conductive connections from the PCB through a supporting member such as a mechanical standoff to the chassis itself [13, 14]. This configuration is repeated both in and around the periphery of the circuitry to control the eddy current flow.

Without stitches, the total *common-mode* voltage across the board $\hat{V}_{CM}$ will drive leads coming off the board at opposite ends, such as I/O and power cables attached to the PCB (see Figure 10.14). This structure acts like a dipole antenna. The voltage drop $\hat{V}_{CMi}$ is caused by the signal return current that flows on the transfer impedance or partial effective inductance associated with segment i of the ground plane, in this case $i = 1, 2$ (see *current-driven* mechanism in *Section 9.6.2*).

Stitching the board to the metal chassis in at least two places causes a circulating current to flow through the board chassis and the stitches. This reduces the action of the voltage drops $\hat{V}_{CM}$ in producing *common-mode* currents on the attached cables by shorting out the dipole antenna formed by the leads. The drawback is that the circulating current between the PCB and the chassis causes a slot antenna to be formed. For this reason, enough stitches must be used to keep the size of this slot small compared with the wavelength λ corresponding to the frequencies generated on the board. The spacing between chassis ground connections should not exceed λ/20. High-frequency components should also be grouped together, with several close stitches enclosing the components. At each stitch, a *common-mode* 'null' is created for

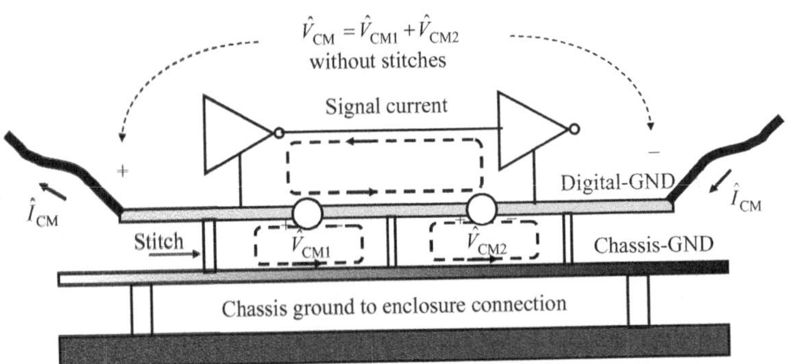

Figure 10.14 Illustration of a PCB with stitches

the resonance frequencies. A bypass capacitor in the range 470 pF–4.7 nF should be connected between power supply and ground on the PCB at each stitch. These bypass capacitors should also be used when solid connection cannot be made between the digital ground (DGND) used as return conductor for the signal current and the chassis. The choice of the value depends on the maximum frequency of the desired isolation. The chassis–DGND structure must also be connected to the metallic enclosure case when present. The benefit of this solution will be quantified in *Section 10.3.4* by using numerical simulations.

10.3 Grounding at PCB Connectors

This section focuses the investigation on PCB connectors by using an analytical as well as a circuit approach to show the importance of a good pin assignment in reducing *common-mode* coupling and improving signal integrity. The effectiveness of grounding practice at the connector level and the use of fixes such as *common-mode* filters, ground splitting, and stitches in reducing emissions from a cable attached to a PCB inserted in a shielded box is also shown by measurements and numerical simulations.

10.3.1 Ground Noise and Transfer Impedance

When designing connectors in a PCB, the following points should be considered [9, 15, 16]:

- The connector is one of the most critical devices in signal integrity and EMI problems.
- Connector pin assignments are very important in designing a PCB.
- Care should be taken as to how to ground the nearby area of the connector to the chassis, considering unshielded and shielded cables.

How the inductance of power and ground pins can cause common impedance coupling is shown schematically in Figure 10.15. The two circuits characterized by *loop A* (dash-dotted

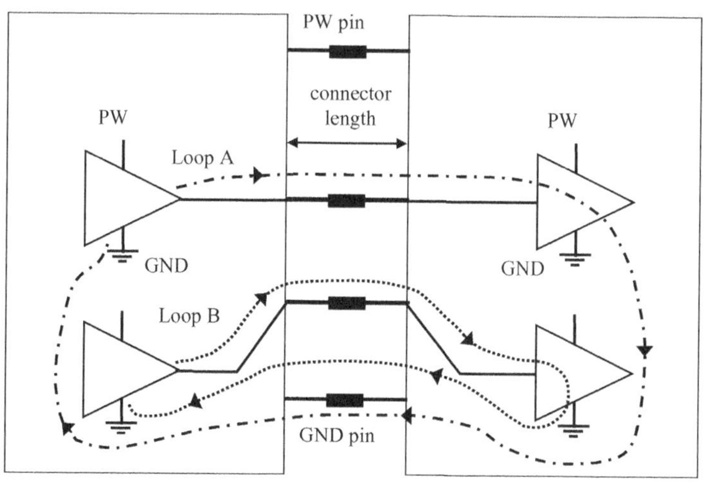

Figure 10.15 Illustration of discontinuities and return currents at connector level

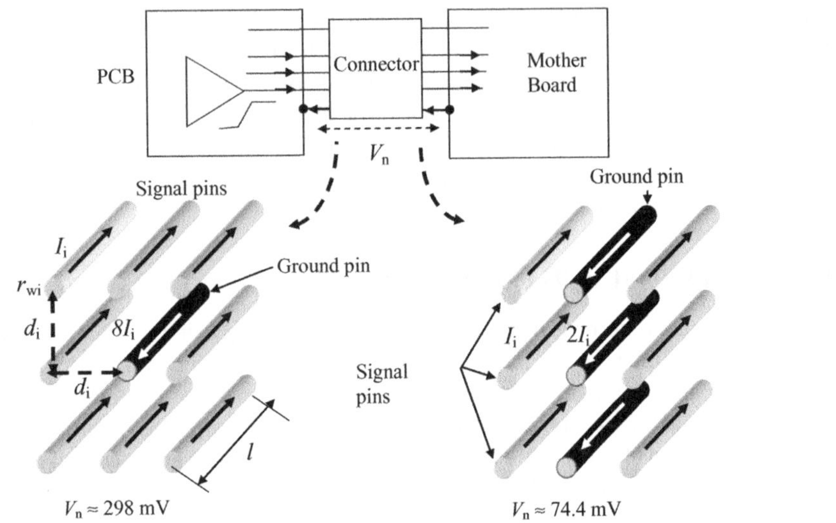

Figure 10.16 Currents and voltage noise in a connector with one ground pin (on the left) and three ground pins (on the right). The ground pin is in black and the signal pins are in light-gray

line) and *loop B* (dotted line) share the same ground pin. Therefore, return current paths are very important. To compute the *common-mode* noise, or, in other words, the voltage drop on the ground pin, the most appropriate approach is that based on the application of the partial inductance concept. In the connector area, infinity can be considered as reference for flux computation, and therefore the partial inductances associated with each pin can be computed (see *Section 3.6.2*). In a first approximation, the capacitance effects can be neglected as long as the connector is electrically short and all the critical lines such as clocks are terminated on their characteristic impedance. When this hypothesis does not hold, TL models with coupled lines as described in *Section 6.4* must be applied to simulate the connector.

The problem of *common-ground noise* at the connector level of a digital system is illustrated by the example shown in Figure 10.16. A driver on a PCB sends a signal to a receiver located on another board through a motherboard. Consider that the connector between the PCB with the driver and the motherboard has the structure shown on the left of Figure 10.16 and has only one ground pin. If there are eight simultaneously switching drivers that inject a current I_i into the respective signal pin, in the ground pin there is a return current equal to $8I_i$. Assuming for simplicity that the distance between signal pin i and the ground pin is d_i, equal for each pin, the voltage drop V_n, which represents the ground noise, can be calculated by (see *Appendix A*):

$$L_{p,i} = \frac{\mu_0}{2\pi} l \left[\ln\left(\frac{2l}{r_{wi}}\right) - 1 \right] \tag{10.8a}$$

$$M_{p,ij} = \frac{\mu_0}{2\pi} l \left[\ln\left(\frac{l}{d_{ij}} + \sqrt{1 + \frac{l^2}{d_{ij}^2}}\right) - \sqrt{1 + \frac{d_{ij}^2}{l^2}} + \frac{d_{ij}}{l} \right] \tag{10.8b}$$

$$V_n = L_{p,0} \sum_{i=1}^{n} \frac{dI_i}{dt} - \sum_{i=1}^{n} M_{p,i0} \frac{dI_i}{dt} \approx \left(L_{p,0} - M_{p,i0}\right) \sum_{i=1}^{n} \frac{dI_i}{dt} \tag{10.8c}$$

$$L_t = L_{p,0} - M_p \tag{10.8d}$$

where l is the length of the pins (equal for the ground and signal pins), $L_{p,i}$ is the self partial inductance of signal pin i, $L_{p,0}$ is the self partial inductance of the ground pin, equal to $L_{p,i}$ with pins of the same geometry, $M_{p,i0} \approx M_p$ is the mutual partial inductance between signal pin i and the ground pin 0, equal to the mutual inductance M_p, which represents the coupling of a generic pin i with the ground pin, because the separation was assumed to be equal for each pin, and Z_t is the transfer impedance of the connector, which depends on $(L_{p,0} - M_p)$ (ideally, this difference should be zero in order to avoid ground noise, or, in other words, the signal pin should be very close to the ground pin).

Therefore, the dominant effect of connectors is accounted for by a series lumped inductance model. As stated by Equation (10.8c), V_n is the algebraic sum of two terms:

- *The common-ground noise* effect dominated by the self partial inductance of the ground pin $L_{p,0}$;
- The *crosstalk* effect dominated by the mutual partial inductance $M_{p,i0}$.

Example 10.3: Calculation of the Noise in a Connector with an ECL Device

As a numerical example, consider ECL switching devices with $dI/dt = 20$ mA/ns and a connector characterized by the geometrical parameters shown in Figure 10.16, where $r_{wi} = 0.25$ mm, $d_i \approx 2.5$ mm, $l = 5$ mm, and $\mu_0 = 4\pi 10^{-7}$ H/m, and therefore $L_{p,0} = 2.69$ nH, $M_p = 0.826$ nH, and $L_t = 1.86$ nH. Introducing these values into Equation (10.8c) yields $V_n \approx 298$ mV, which is a noise that is too high for the immunity of an ECL receiver. To lower the noise V_n, the number of ground pins must be increased. For instance, with the pin assignment of the connector shown on the right in Figure 10.16, the noise is reduced at $V_n \approx 74.4$ mV.

Another noise reduction can be achieved by acting on the connector structure in order to minimize its transfer impedance Z_t. This can be done by the following practices:

- The pin connectors should be 'short'.
- The number of power and ground pins should be larger than the number of signal pins.
- Signal pins should be close to their current return.

Recall that the number of power and ground pins assumes the same importance because the return current chooses the path of minor impedance, as discussed in *Section 10.2*. Reducing Z_t is very important because V_n can also be a source of *common-mode* current on cables attached to the PCB, and therefore a source of emission.

Example 10.4: Calculation of Noise in a Connector using SPICE

In order to verify by a circuit approach that the simple calculation outlined above provides appropriate results, the connector structure shown in Figure 10.17a is considered, adopting the same parameters as the previous example. A pin for a quiet line at low level is added to compute induced noise. The simulations were performed by the Microcap simulator based on

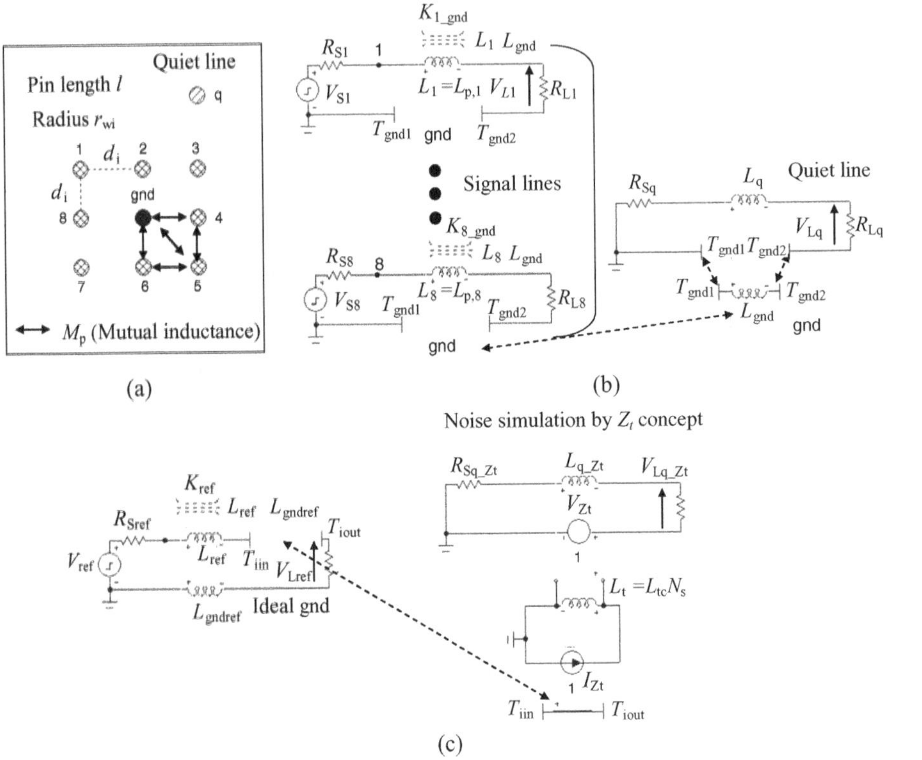

Figure 10.17 Connector with one ground pin and eight signal pins: (a) structure; (b) equivalent circuits of switching signals and of the quiet line sharing the same ground pin; (c) reference circuit for signal current computation and equivalent circuit for noise computation using the transfer impedance concept for the connector

SPICE [17] under the following assumptions:

- There are eight simultaneous switching devices.
- There is magnetic coupling between adjacent signal wires only.
- There is magnetic coupling between signal and ground wire.
- The quiet line shares the same ground pin without magnetic coupling with other signal wires.
- The magnetic coupling between two signal pins or between signal and ground pins is taken into account by the coupling factor $K = M_p/L_p$.
- Capacitances are neglected, as the length of the pin connector is electrically short compared with the minimum wavelength of interest and the lines are matched.
- Each line is terminated with a $R_{Li} = 50\ \Omega$ load, with $i = 1,8$.

The equivalent circuits of the eight drivers are shown in Figure 10.17b. The drivers are simulated with a voltage source having a step voltage of 1 V and a rise time $t_r = 1$ ns. The output resistance of i-th driver with $i = 1,8$ is $R_{Si} = 1\ \text{m}\Omega$ in series with the self and mutual inductances of the pin connector. The i-th circuit is completed with $R_{Li} = 50\ \Omega$ and by the self and mutual inductances of the ground pin. Therefore, in each line there is a current step

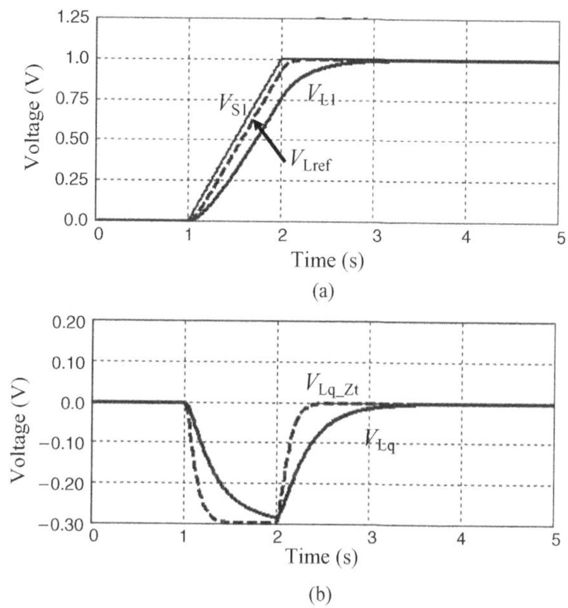

Figure 10.18 Simulated waveforms for eight simultaneous switchings: (a) signals; (b) noises

of 20 mA and a rise time $t_r = 1$ ns. The coupling factor K is indicated by the symbol with three parallel dotted lines, as provided by Microcap. The factor K between two adjacent signal pins is not indicated in Figure 10.17b. The pin in common is represented by the circuit in Figure 10.17b, where it is possible to distinguish the equivalent circuit of the quiet line and the common inductance L_{gnd} with tie circuit connectors T_{gnd1} and T_{gnd2}. The tie symbol in Figure 10.17 denotes direct connection among the parts of the circuit having the same symbol. For instance, all the circuits in Figure 10.17b have the same return in common.

The ideal reference equivalent circuit with one ground pin for each signal is shown in Figure 10.17c. 'Ideal' means no interference from the other circuits by the common ground pin. The tie circuit connectors T_{iin} and T_{iout} collect the signal current of the reference circuit and assign it to the dependent current source I_{Zt}. The current I_{Zt} injects the signal current onto the inductance L_{tc} of the connector transfer impedance Z_{tc} multiplied by the number of signals simultaneously switching N_s. The voltage drop across the inductance L_t is transferred to the dependent voltage source V_{Zt} to compute the induced noise into the quiet line.

Simulated signals and induced noises are shown in Figures 10.18a and 10.18b respectively. Note that the signal V_{L1} on signal line 1 and the induced noises V_{Lq} in the quiet line are affected by the voltage drop in the common ground pin. On the other hand, signal V_{Lref} on the load of an ideal connector with a ground pin for each signal has the same waveform as the input signal V_{S1} with delay. Note that V_{L1} has more rise time and extra delay by comparison with the ideal signal V_{Lref}. Moreover, the induced noise V_{Lq_Zt} computed by the circuit in Figure 10.7c and the noise V_{Lq} computed by the circuit in Figure 9.17b have different shapes. However, the maximum peaks are equal to the value calculated analytically of about 300 mV, as calculated in *Example 10.3*.

In the case of one ground pin for two signals, as depicted in Figure 10.19a, a significant reduction in coupled noise occurs, 75 mV instead of 300 mV, as given by the analytical

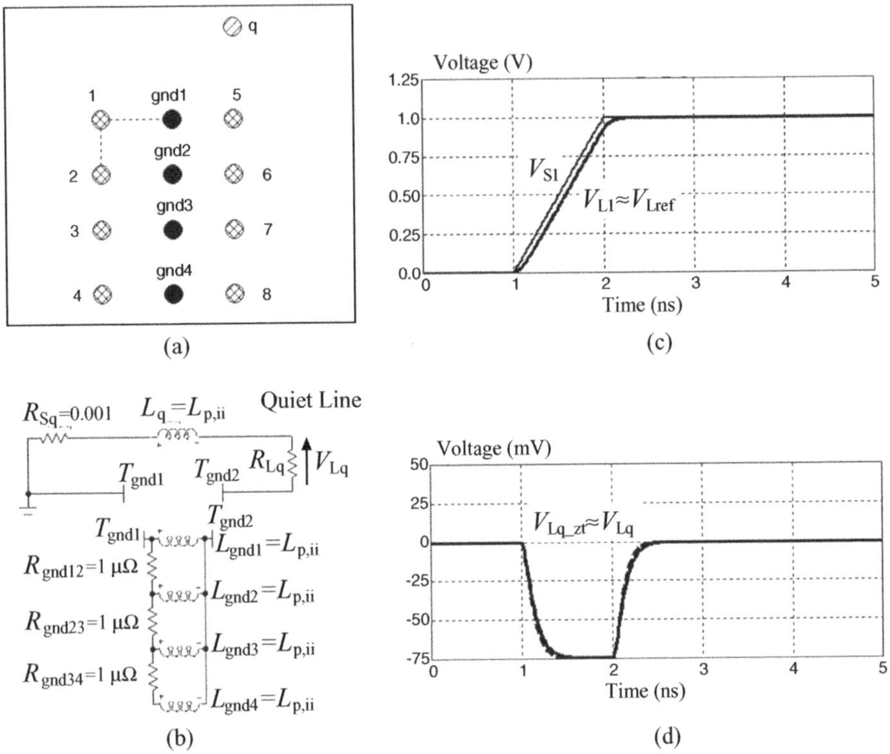

Figure 10.19 Connector with four ground pins and eight signal pins: (a) structure; (b) equivalent circuit of quiet line with ground pins; (c) simulated signal waveforms; (d) simulated noise waveforms

calculation of *Example 10.3*, and signal integrity is very close to the ideal situation as shown in Figures 10.19c and 10.19d. The equivalent circuit of the ground pins from connector ties T_{gnd1} and T_{gnd2} is shown in Figure 10.19b. For accurate results it is important to take into account the magnetic coupling between adjacent pins, in other words, between inductances L_{gnd1} and L_{gnd2}, L_{gnd2} and L_{gnd3}, and L_{gnd3} and L_{gnd4}. For simplicity, the factor K between adjacent pins is not shown in Figure 10.19b. These circuit simulations confirm a reduction by a factor of 4 for the induced noise into the quiet line, as calculated analytically in *Example 10.3*.

Example 10.5: Measurements of Transfer Impedance for Three Test Connectors

Two solutions for lowering the transfer impedance Z_t in the case of board-to-backplane connectors, such as those used in telecommunication and computer systems (also known as Hard Metric 2 mm connectors), are shown in Figure 10.20 [18]. For comparison purposes, the pin assignments shown in Figure 10.20 were chosen to measure the connector transfer impedance Z_t. The reference structure is without a shield. The extra ground pins, indicated with an 'X', act as a low-performance shield. The connector realizes a 90° connection between the component board and the backplane. This means that the pins do not have the same length. 'S' stands for single-ended excitation, and the signal pins were chosen in two opposite corners in order to maximize the difference in length.

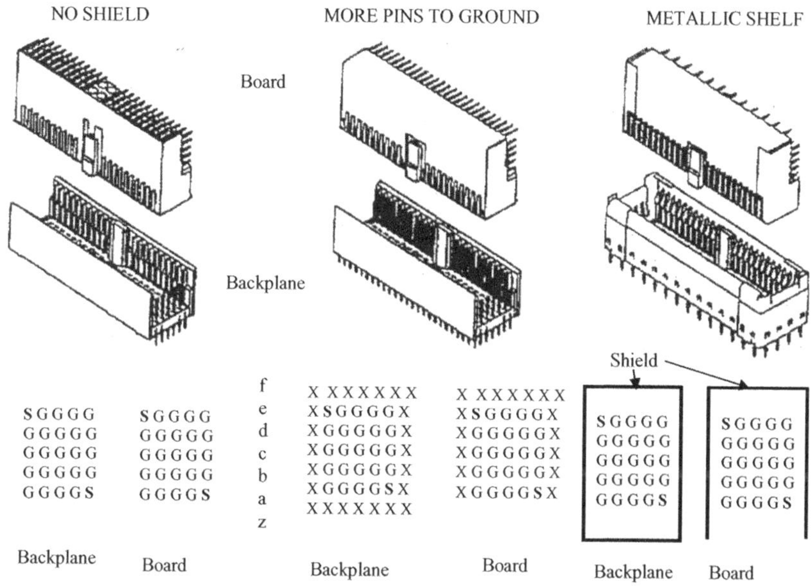

Figure 10.20 Shielding configurations for connector 2 mm grid pins: S = Signal; G = ground; X = extra ground pins. Based on reference [18]

The transfer impedance shown in Figure 10.21b was measured by the triaxial cell method (see Figure 10.21a) which measures the voltage drop $\hat{Z}_t\hat{I}_s$, where $\hat{I}_s$ is the known signal current across the connector [18]. The entire set-up, including the test board, the connector under test, and the terminations of the signal lines in the inner conductor of the triaxial cell, was built and assembled with care in order to measure the voltage drop $\hat{Z}_t\hat{I}_s$ only. For instance, the signals on the test board were traveling on stripline traces of 50 Ω impedance to avoid radiated emissions. The non-excited pins were short-circuited to grounds in the component and backplane boards. At the back side of the backplane, the signal pins were terminated with 50 Ω loads. A metal hood was soldered over the pins that protruded from the backplane, together with termination resistors to avoid radiation from these pins and loads.

From the measured transfer impedance of Figure 10.21b it can be seen that:

- Z_t has an inductive behavior, as expected, and L_t values can be extrapolated.
- Connectors with more pins to ground and metallic shelf have the same behavior up to 1 GHz.
- A metallic shelf connector has better performance than more pins to ground connector above 1 GHz.
- With shielded connectors there is an improvement in Z_t of about 30 dB.
- This 30 dB attenuation is consistent with the results reported in *Section 9.7* for UTP and SFTP cables.

10.3.2 Pin Assignment

Some pin assignment solutions are shown in Figure 10.22. It is obvious that solution 4 is expensive and difficult to realize. A trade-off solution in terms of costs and performance is

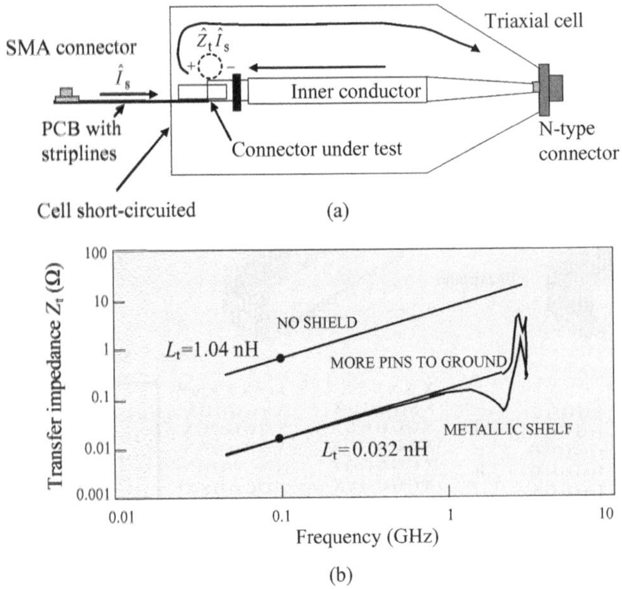

Figure 10.21 Transfer impedance Z_t for a single-ended board-to-backplane connector: (a) measurement set-up in a triaxial cell; (b) measurement results for three types of connector. Based on reference [18]

often required. The pin assignment solutions of case 2 and 3 may be suitable. In any case, note that the same number of pins should be assigned for power and ground, considering that the power plane also acts as return conductor for the signal current.

Example 10.6: Measurements of Radiation from a Test Cable with a Pigtail

As was shown in *Chapter 9*, other important sources of emission are the I/O cables attached to a PCB. This experiment will point out the importance of a good grounding practice at connector level to mitigate radiated emission from cables.

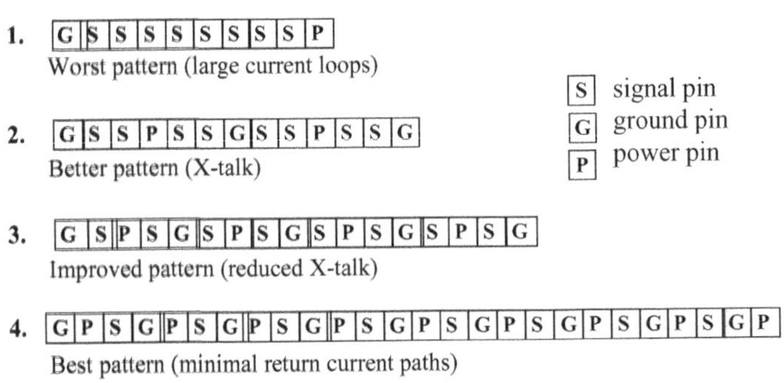

Figure 10.22 Example of connector pin assignments

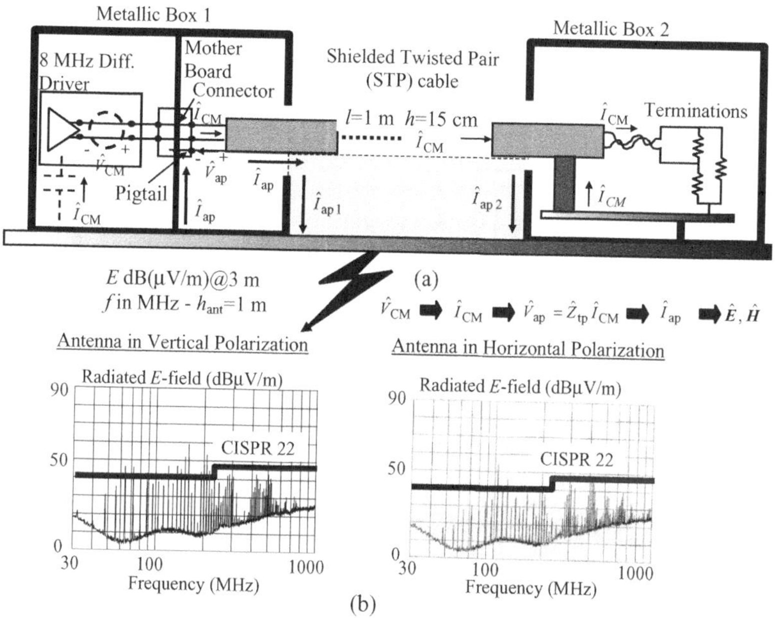

Figure 10.23 Radiated emission from a shielded twisted-pair (STP) cable with connection of the shield to the PCB ground by a pigtail: (a) measurement set-up; (b) measurement results

The set-up used to measure the radiated emission at a distance of 3 m from a *Shielded Twisted-Pair* (STP) cable is shown in Figure 10.23. The antenna is placed at $h_{\mathrm{ant}} = 1$ m above the floor. The shield of the STP cable positioned at $h = 15$ cm above the reference plane is connected to the ground of a PCB by a wire or 'pigtail'. As was explained in *Section 9.7*, the unbalance inside the differential driver of the test board generates the *common-mode* current $\hat{I}_{\mathrm{CM}}$. Assuming that the cable transfer impedance is very low, the antenna current $\hat{I}_{\mathrm{ap}}$ responsible for the radiated emissions is caused by the voltage drop $\hat{V}_{\mathrm{ap}}$ on the pigtail, given by the product of the connector transfer impedance $\hat{Z}_{\mathrm{tp}}$ due to the effective inductance associated with the pigtail and the *common-mode* current $\hat{I}_{\mathrm{CM}}$. With the set-up shown in Figure 10.23, the emissions are about 10 dB over the standard limit of CISPR 22 for Class B equipment.

To lower the radiated emission of about 20 dB, it is necessary to connect the cable shield to the test PCB ground with a very low impedance connection, as done on the termination side with a large strip of metal. The new radiated emission is shown in Figure 10.24. This figure also shows the radiation mechanism: the current $\hat{I}_{\mathrm{CM}}$ produces a voltage drop $\hat{V}_{\mathrm{a}} = \hat{Z}_{\mathrm{t}}\hat{I}_{\mathrm{CM}}$ that is much smaller than the voltage drop $\hat{V}_{\mathrm{ap}}$ with the pigtail because $\hat{Z}_{\mathrm{t}} \ll \hat{Z}_{\mathrm{tp}}$. Hence, the antenna current $\hat{I}_{\mathrm{a}}$ is also much lower than the antenna current $\hat{I}_{\mathrm{ap}}$ with the pigtail, and as a consequence there is a strong reduction in radiated emission. Once again the importance of a good connection to ground of the cable shield in order to have a low radiated emission profile is demonstrated. This good connection can be realized by a 360° contact [3], p. 198 or by a conductive bracket which diverts the *common-mode* current to a conductive strip well connected to the chassis by screws [3], p. 199.

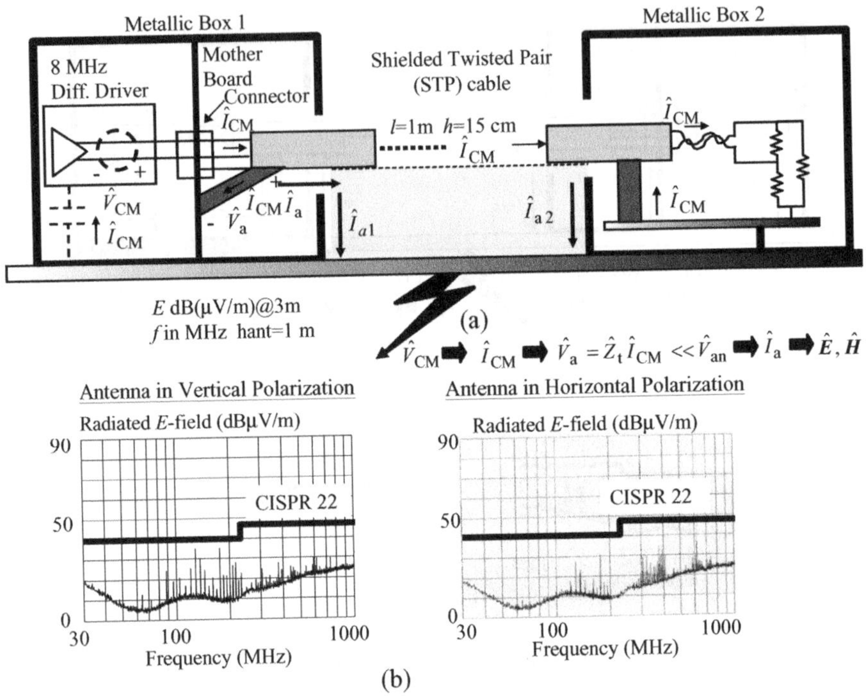

Figure 10.24 Radiated emission from a shielded twisted cable with connection of the shield to the PCB ground by a conductive strap: (a) measurement set-up; (b) measurement results

10.3.3 Grounding a PCB to a Chassis

U*sually*, a PCB is not left floating but is connected to a nearby metallic enclosure. It should be considered that:

- Most PCBs are within a shielded box.
- Unshielded cables for relatively low-frequency I/O connections (of the order of MHz) require EMI filters.
- Shielded cable must be connected to the shielded box with a very low impedance up to 1 GHz.
- The metallic walls of the box (i.e. the chassis) are the reference ground for the entire system composed of PCBs and cables.
- The most common cause of external emissions in typical products is unwanted *common-mode* voltages from unintentional signals on the I/O cables and wires relative to the chassis.
- The chassis and cables act as antennas owing to the voltage drop between the chassis and the PCB where the cable is connected.
- It is this voltage drop that must be mitigated.

In the following, grounding at connector level will be investigated by using numerical simulations.

10.3.4 Techniques to Limit Emission from Cables

The first technique to mitigate emission from I/O cables is to provide a low-impedance connection between the reference planes of PCBs and the chassis [9]. The main reason for this fix is the need to locate I/O devices and the EMI filter on the PCB area near to the PCB connector. EMI filters instead of shielded cable are used because many I/O signals have relatively low frequency (of the order of MHz). The task of an EMI filter is to divert the chassis or/and stop unwanted or *common-mode* currents that otherwise would return to the source by the cable and the environment. EMI filters for this application are normally low-pass filters. The filter must be designed to pass the significant harmonics of the current signal (seventh–ninth), and to stop other unwanted currents. Assigning a nominal input impedance of 100 Ω to the cable at all frequencies of interest (this is a trade-off between the real impedance, which changes according to position, and the need to have a reference value for filter designing), the series filter components (inductors or ferrite beads) must be selected with an impedance much greater than 100 Ω, while the parallel filter components (capacitors) must be selected with an impedance much less than 100 Ω. A factor of 100 is usually chosen. This means that the series impedance of the filter should have a value of about 10 kΩ, and the parallel impedance a value of about 1 Ω [9].

In *Section 9.8.1* it was shown that, to make the filter action of a *common-mode choke* really effective, a couple of capacitors, diverting the *common-mode* current to the quiet area well connected to the chassis, must be added. These capacitors have the function to compensate for the bypass action of the parasitic capacitances associated with the inductances of the *common-mode choke*. The most suitable EMI filter for single-ended connection (i.e. the shunt series configuration) is shown in Figure 10.25. This is a good solution for minimizing radiated emission, as an unwanted current on the trace traveling towards the I/O connector sees first the low impedance offered by the capacitor to return to the source, and then the high impedance

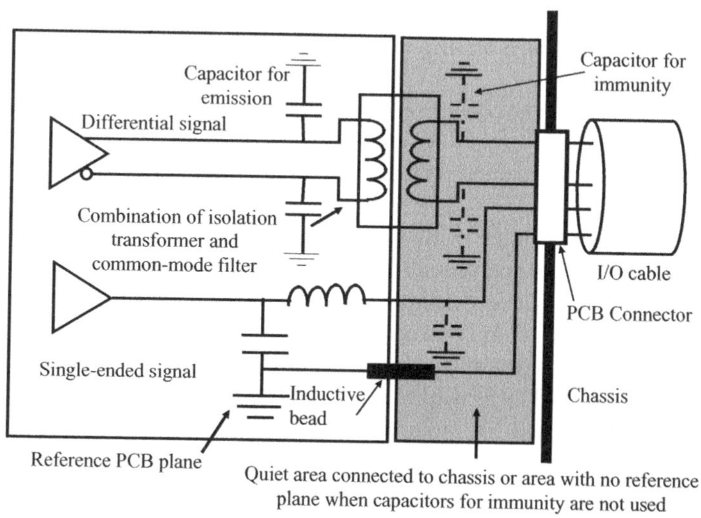

Figure 10.25 Illustration of a technique for limiting the radiated emission from cables attached to a PCB, based on the use of EMI filters and ground partitioning

offered by the inductor. The inductive bead on ground wire of the I/O cable is necessary to stop the unwanted current generated by the voltage noises on the ground of the PCB and to ensure a return current path for the I/O signal at DC and low frequencies. If the capacitor is shifted to the right of the inductor, the circuit acts as EMI filter for susceptibility. The better solution for emission and immunity is to place a capacitor also on the right of the inductor (indicated by the dashed line) in order to form a π-type filter.

To avoid undesirable parasitic elements, which can compromise the filter action, grounding strategy is required. The strategy consists in creating a quiet area near the connector or an area with no reference plane if the capacitors for immunity are not used. Other fixes consist in providing stitches (metallic connections between PCB ground and chassis) or using shielded connectors. When using stitches, the spacing between chassis ground connections should not exceed $\lambda/20$, as discussed in *Section 10.2.6*. The quiet area can only be removed in the case of differential signal transmission, and without capacitors for immunity. In fact, if the ground reference plane is present, RF noise currents on the ground can couple onto the differential signal traces and effectively bypass the *common-mode* filter. All power planes should also be removed from this area of the PCB for the same reason. Recall that, in the case of differential signaling, in spite of the effort of minimizing the voltage drops in the digital ground, the differential driver itself produces *common-mode* currents on cables owing to the unbalanced output stage, as shown in *Section 9.7*. Therefore, *common-mode* filters are always required in the case of unshielded cables.

After this discussion, indications and methodologies will be provided on how to quantify these fixes and others by means of full-wave electromagnetic codes, considering the non-ideal behavior of the *common-mode choke*, ferrite bead, and capacitors.

Example 10.7: Numerical Simulations of a Test Board with an Attached Cable

The aim of this example is to validate by numerical simulations the practical design rules for grounding in PCBs that have been previously discussed. The code used is *MicroWave Studio* (MWS), based on the *Finite Integration Technique* [19], and the test board has the structure shown in Figure 10.26.

The following points should be considered:

- The test board is a structure composed of a round wire of 6 cm length above a finite ground plane of 6 cm $\times$ 10 cm size.
- The geometrical parameters of the PCBs were chosen to speed up the calculation.
- The signal source is a current source with an internal impedance $R_S = 50\ \Omega$.
- Computations were performed in the time domain by using a Gaussian waveform for the source that has a flat spectrum in the frequency range 0–1000 MHz of value $2/\sqrt{50}$ A at each frequency. This value is determined by the code, as the S-parameter option was chosen.
- *Fast Fourier Transform* (FFT) was used to obtain radiated fields in the frequency domain.
- An E-field probe is positioned 3 m from the board, with polarization parallel to the signal line.
- Radiated emission up to 1 GHz from a board with an attached cable is investigated, considering several technical solutions to mitigate emissions.
- For some basic structures, emission profiles calculated by MWS are compared with analytical values coming from simple transmission line and antenna models outlined in *Chapter 9*.

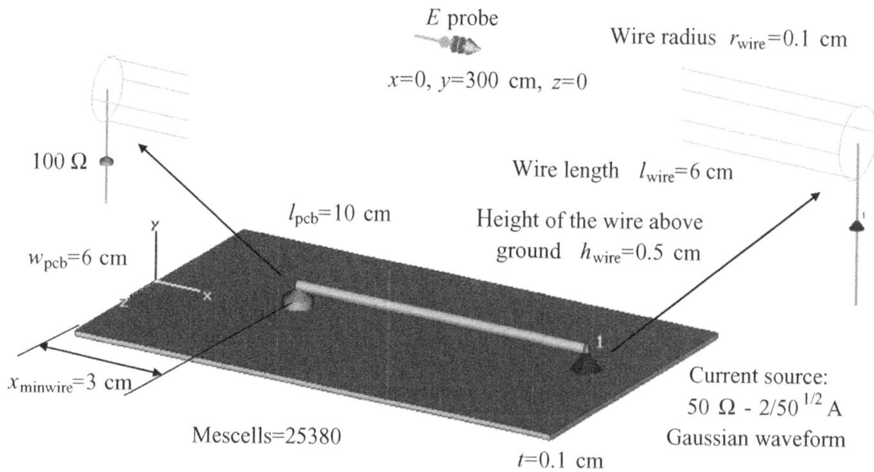

Figure 10.26 Test board used for numerical simulation

- Other structures such as a PCB with an attached cable, a ground plane with a split, a PCB inserted in a shielded box, and a PCB with a *common-mode* EMI filter are studied.
- *Common-mode* filters are simulated by a capacitance of 0.1 pF located at the point where the cable is attached to the PCB.

For basic structures such as a PCB with and without an attached cable, analytical computations were performed and compared with the numerical results. The procedures outlined in *Section 9.2* for *differential-mode* emission and in *Section 9.6* for *common-mode* emission by the *current-driven* mechanism were applied with the following assumptions:

- The analytical procedure considers the ground plane of the PCB as infinite, and the radiated field is computed by using the image method.
- The wire over the ground plane has a characteristic impedance $Z_{0,\text{wire}} = 60\ \ln(2h_{\text{wire}}/r_{\text{wire}}) = 138\ \Omega$.
- The PCB with an attached cable does not have a nearby reference plane (see *Section 9.6* for more details). Therefore, a PCB–cable capacitance of 2.5 pF (estimated) was used, and the characteristic impedance of the cable $Z_{0,\text{cable}} = 60\ \ln(2l_{\text{cable}}/r_{\text{cable}}) = 425\ \Omega$, with $\mathrm{l}_{\text{cable}} = 60$ cm and $r_{\text{cable}} = r_{\text{wire}}$. In this calculation it was assumed that the cable is at a distance lcable from a hypothetical reference plane.
- The partial inductance associated with the ground plane of the PCB is $L_{\text{gnd}} = (\mu_0 l_{\text{wire}}/2\pi)\ln(h_{\text{wire}}\pi/w_{\text{pcb}} + 1) = 2.79$ nH (see Table A.2 of Appendix A).
- To compute the *common-mode* current on the cable, the TL associated with the cable should be considered lossy with, $\sigma_{air}(\omega) = \omega\varepsilon_0/10$, in order to avoid very high peaks of resonance (see *Example 9.8*).
- The capacitance associated with the split was estimated to be 0.3 pF.

Radiated field results computed numerically and analytically are shown in Figure 10.27. It is interesting to note that there is a perfect agreement of results between the analytical and

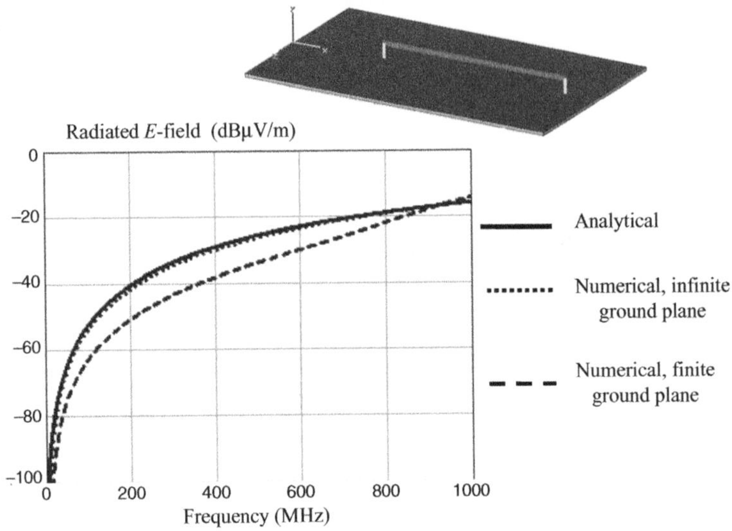

Figure 10.27 Computed radiated E-field at 3 m from the PCB with a wire over a finite and infinite ground plane

numerical procedures when the reference plane is defined as infinite. With a finite reference plane, the emission is slightly lower than with an infinite reference plane.

The computed *common-mode* current and radiated field from the PCB with an attached cable are shown in Figure 10.28. The cable has a length of 60 cm, a radius of 0.1 cm, and a height from the ground plane of 0.5 cm, and it touches the ground plane at 1 cm from the edge with a resistance of 1 mΩ. It can be observed that there is good agreement between the two methods of computation: the resonant points of the *common-mode* current are reproduced, and the radiation due to the cable is dominant in the low-frequency range.

The computed *common-mode* current and radiated field from the PCB with an attached cable and a split are shown in Figure 10.29. The emission from the wire does not change, while the emission from the cable decreases, as the split acts as a partial barrier to the *common-mode* current which decreases by approximately one-third. There is again good agreement between the currents, while the E-field computed numerically is slightly lower than the E-field computed analytically owing to the finite dimension of the ground plane. To reproduce the same resonant frequencies by the analytical method, the length of the cable must be enhanced to $l_{\mathrm{cableeq}} = 64$ cm to take into account that in the case of a split the radiating cable structure consists of the cable and the portion of plane between the connector and the split.

To verify the effect of a ferrite bead inserted between the two edges of the split in the middle of the PCB, the structure shown in Figure 10.30a was simulated. The equivalent circuit of the ferrite bead is shown in Figure 10.30a, and its impedance versus frequency is shown in Figure 10.30b. Ferrite beads are largely used as filters. They behave like an inductor because their impedance increases with frequency up to frequencies where the parasitic capacitance becomes dominant. They usually have a low Q-factor, so the loss they provide to the filter has a wide bandwidth. An example is given in Figure 10.30b, where the impedance of the ferrite

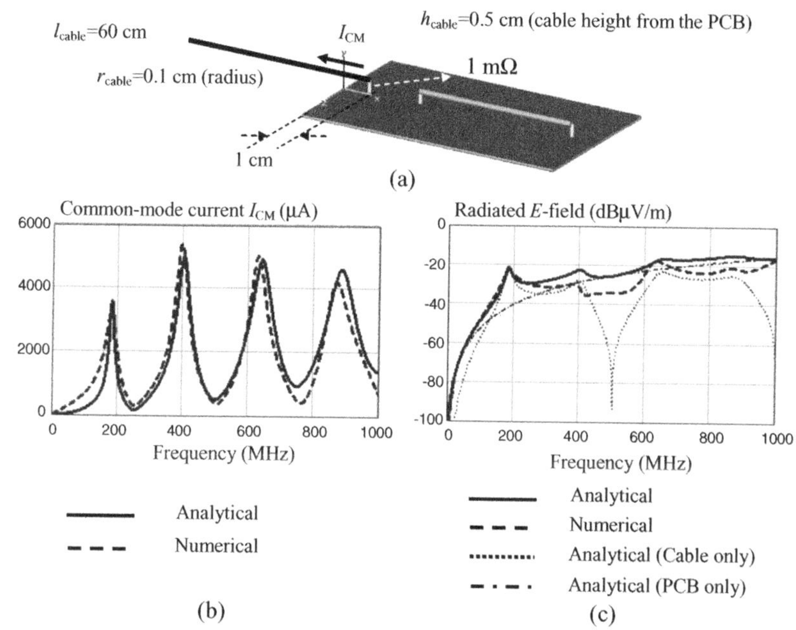

Figure 10.28 PCB with an attached cable: (a) PCB structure; (b) computed common-mode current I_{CM}; (c) computed radiated fields

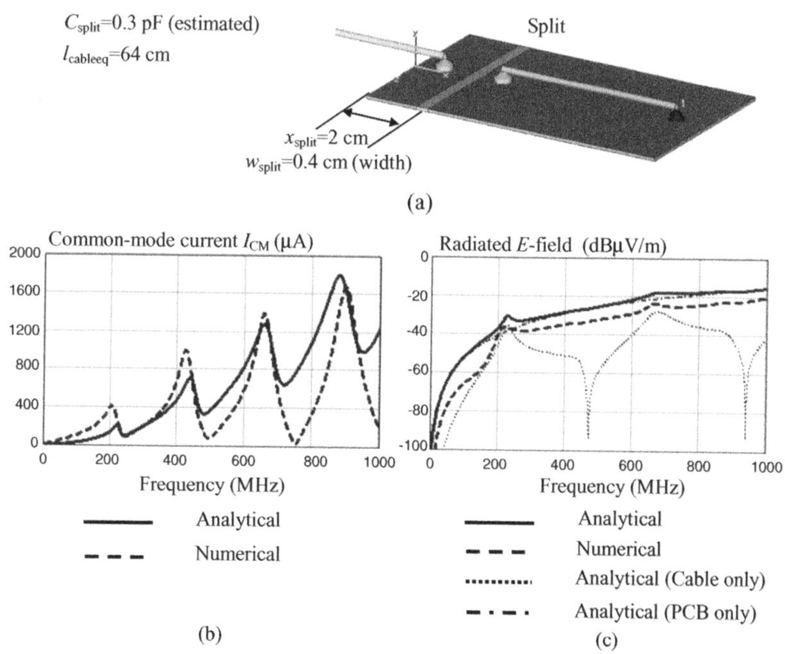

Figure 10.29 PCB with an attached cable and a split on the ground plane between wire and connector area: (a) PCB structure; (b) computed common-mode current ICM; (c) computed radiated fields

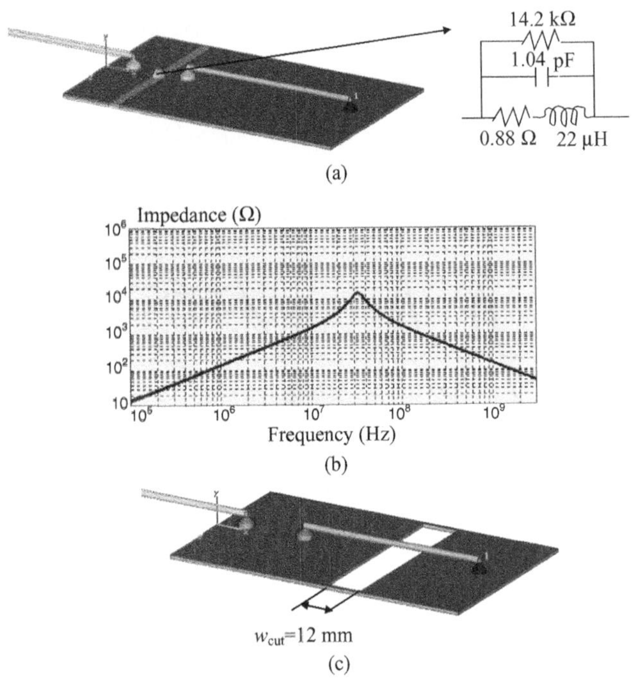

Figure 10.30 (a) Configuration of the PCB with a ferrite bead used for simulations and the equivalent circuit; (b) impedance of ferrite bead versus frequency; (c) PCB with a cut

bead has its maximum value of 14.2 kΩ at about 30 MHz. A cut in the ground plane is also considered, for comparison, according to the configuration shown in Figure 10.30c.

Computed radiated fields for four PCB structures are shown in Figure 10.31: (1) a PCB and cable without any split (used as reference); (2) a PCB and cable with a split; (3) a PCB and cable with a split and ferrite; (4) a PCB and cable with a cut. As expected, the worst-case

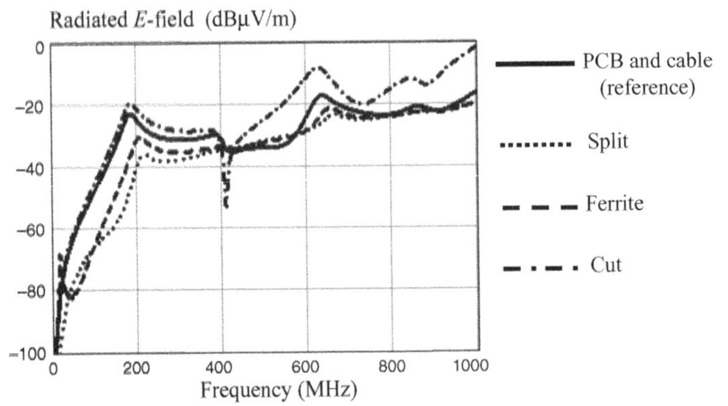

Figure 10.31 Computed radiated field comparisons for four structures of PCB

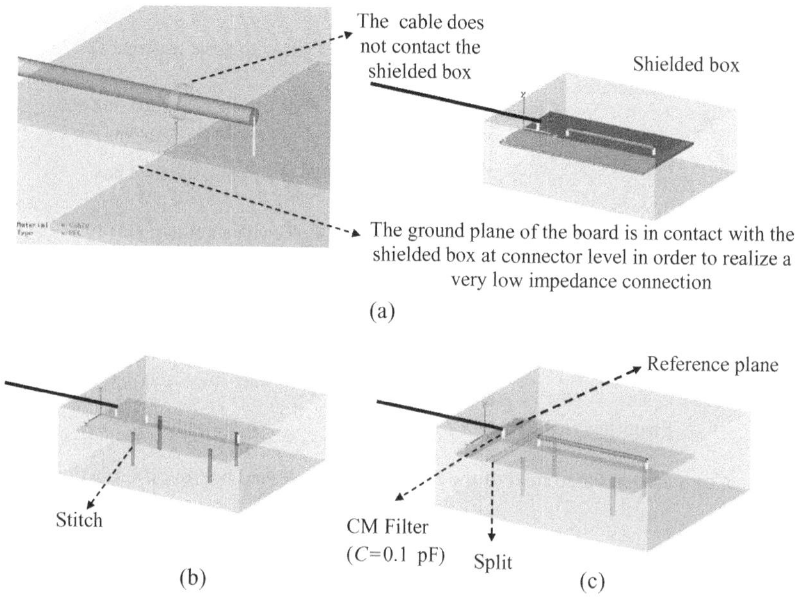

Figure 10.32 PCB with an attached cable within a shielded box (10 cm × 12 cm × 10 cm): (a) detail of the cable outgoing from the shielded box; (b) PCB with stitches; (c) PCB with a split, a common-mode (CM) filter, and a reference plane

radiated emission occurs when the ground plane has a cut in the middle area of the signal wire, especially in the high-frequency range. The best-case radiated emission occurs with the presence of a split located between the signal and the cable area. If a ferrite bead is used to give a path for the return signal current, the EMC performance is slightly reduced, as shown by the dashed curve. Note that a split with a ferrite bead is effective up to 400 MHz, where the emission from the cable dominates. In *Section 12.2.4*, the closed-form expression for calculating the inductance associated with the cut or slot, which adversely affects the current-driven mechanism of emission when the cable is attached to the PCB, is provided and verified numerically.

Many apparatuses use a shielded container or rack to improve EMC performance. Therefore, consider the same test board within a shielded box with different conditions of grounding, as shown in Figure 10.32. The reference plane is the portion of ground plane area between the wall of the shielded box and the point where the conductor representing the cable touches the ground plane. As stated before, this metallic part can be removed or not, according to the need to increase the decoupling between the cable and the currents on the ground plane, and considering whether the I/O connection is single-ended or differential.

Several simulations were performed with different solutions for grounding, and the main results are shown in Figure 10.33. The presence of a *common-mode* filter was simulated with a capacitance that bypasses the inductors at high frequencies. Looking at the results, it can be noted that, with the PCB within a shielded rack and with a cable attached, the emission is drastically reduced at low frequencies but has resonance peaks due to the metallic walls of the box. The reduction at low frequencies can be explained, as the ground plane of the PCB

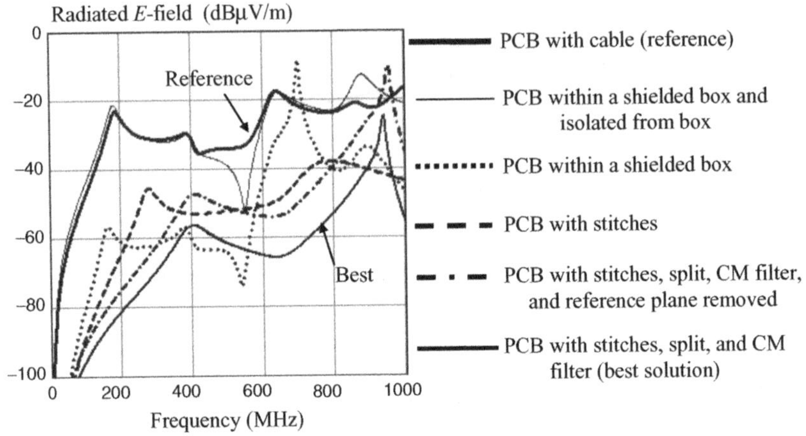

Figure 10.33 Computed radiated field of the PCB within a shielded box in different configurations

is well connected to the wall of the box where the cable exits. With the PCB isolated, there is no emission reduction up to 400 MHz.

To eliminate resonance peaks of the *E*-field, stitches are the most appropriate solution. However, an increase in emission between 200 and 600 MHz occurs. To improve the performance at low frequencies, a *common-mode choke* and a split between the *common-mode choke* and the circuit should be introduced. Therefore, as shown in Figure 10.33, the best-case radiated emission for a PCB within a shielded box occurs when stitches, a *common-mode* filter, a split and a reference plane are present. It was verified that removing the metallic area between the connector and the *common-mode* filter location is effective in reducing emission when it is far from the connector. Better results can be obtained by increasing the number of stitches in order to attenuate and shift resonance peaks above 1 GHz.

The proposed investigation leads to the following conclusions:

- Analytical and numerical calculated values for basic structures are in good agreement.
- Maximum emissions occur with the reference PCB and an attached cable.
- A large cut in the ground plane under the signal wire has a significant effect on radiated emission in the high-frequency range.
- A split in the ground plane between the signal line and the attached cable reduces emission by about 15 dB in the low-frequency range 30–300 MHz.
- A smaller reduction, about 10 dB, can be obtained if a ferrite bead is used through the split.
- No emission reduction is obtained when the board is within a shielded box, the cable goes out through a small aperture without contact, and the PCB is isolated with respect to the box.
- Strong emission reduction can be obtained when the board is within a shielded box and the PCB ground is well connected to the box from the side where the cable exits.
- Stitches are effective in reducing emissions below 200 MHz and to avoid peaks of resonance in the high-frequency range.

- Removing the metallic area between the connector and the *common-mode* filter location is effective in reducing emission when the *common-mode* filter is far from the connector.
- To reduce radiated emission from unshielded cable, the best solution is to use a *common-mode* filter, stitches, and a split in the ground plane.

10.4 Partitioning and Modeling

Section 10.2.4 introduced the concept of a split in the power plane in order to isolate two areas that power the ICs with different DC voltages. In *Section 10.2* it was also mentioned that the return signal current chooses the path with minor impedance. Therefore, the power plane can act as return conductor when the trace is physically close to it, and good filtering must be provided to the power and ground planes to have a very low impedance between these two planes. To facilitate the flow of the return current between the power areas with different DC voltages, stitching capacitors are used to form a very low AC impedance.

Another need that often arises in a digital PCB is to isolate some particular devices placed in a certain area of the PCB from the interference coming from other ICs in the same PCB, or vice versa [20, 21]. The high-frequency noise generated by the digital devices can propagate throughout the entire power bus, creating significant signal integrity and electromagnetic interference problems. In addition to techniques based on decoupling capacitors to mitigate the noise, the isolation technique using power-plane segmentation or partitioning can also be an effective method for minimizing noise propagation. A power island can be employed to provide power to fast-switching or noisy IC devices, and, if these devices share a common power supply with the rest of the circuit, a conductive bridge could be employed to connect the power island to the larger power area. However, the low-frequency performance of this bridge is poor, and ferrite beads are used, as they have a frequency-dependent impedance and are therefore more suitable for isolation.

In this section, two types of segmentation will be investigated by measurements and circuit and numerical simulations: a split in the power planes and a power island.

Example 10.8: S-parameter Measurements and Simulations of a Test Board with a Split
To show the accuracy of numerical methods for studying segmentation in power planes, the results obtained by two commercial codes are compared with measurements and circuit simulations. This experiment can be used as a benchmark by the reader intending to begin with these types of simulation. The numerical codes used for the simulations were:

- *MicroWave Studio* (MWS) of CST [19], which is based on the *Finite Integration Technique* (FIT), performs analysis in the time domain, and transforms the results into the frequency domain by *Fast Fourier Transform* (FFT);
- HFSS of ANSOFT [22], which is based on the *Finite Element Method* (FEM) and performs analysis in the frequency domain.

Both software tools can compute S-parameters which can be compared with measurements carried out by a *Vector Network Analyzer* (VNA). The VNA will be introduced in detail in *Section 11.2*.

The impedance seen by the port acting as a source of noise and the transfer function between the exciting port and another port placed at any other point of the PCB can be derived by

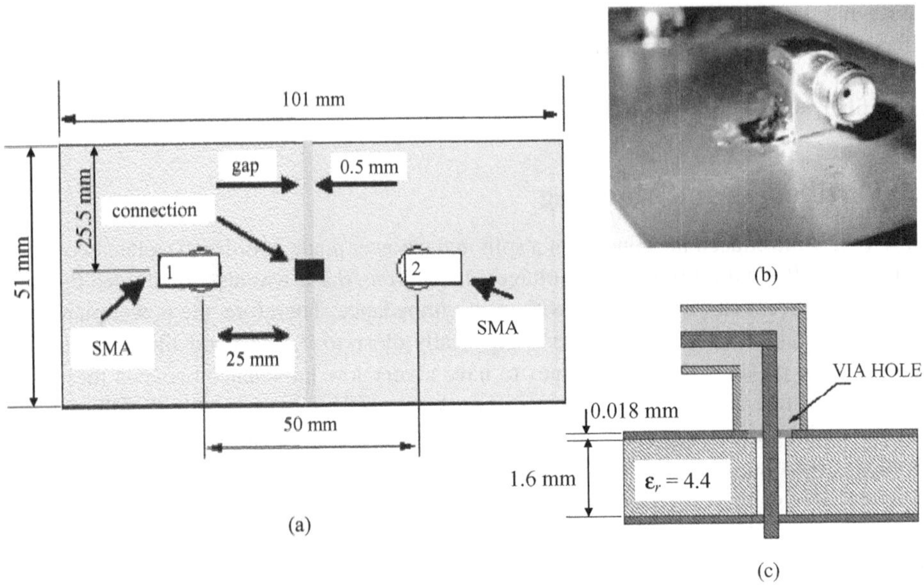

Figure 10.34 Test board with a gap in the power plane: (a) PCB layout; (b) SMA connector attached to the PCB board; (c) SMA connector cut view

measurement or calculation of S_{11} and S_{21} respectively. According to Equation (8.7), the closer S_{11} is to 1, the higher is the power distribution network impedance magnitude Z_{PDN}. On the other hand, the closer S_{11} is to -1, the lower is Z_{PDN}, and the lower the value of S_{21}, the better is the decoupling between the two points.

Two kinds of simple test boards were realized to analyze the effects of gapped planes [23]: the first consists of two solid copper planes (indicated as type A), and the second consists of one plane of type-A board with a split in the middle (indicated as type B). The topology of the test board of type B is shown in Figure 10.34a. The SMA connector used to connect the PCB to the input and output ports of the VNA is shown in Figure 10.34b, and its cross-section, as simulated by MWS and HFFS codes, is shown in Figure 10.34c. For an accurate comparison with measurements at high frequencies, it is very important to excite the PCB by a waveguide port, as will be shown in *Section 11.2*. Both types of board measured 101 mm × 51 mm × 1.6 mm. The dielectric substrate was an FR-4 material with $\varepsilon_r = 4.4$. The type-B board was analyzed with different kinds of connection between the copper islands:

- open connection (no connection between the two islands) – indicated as B1 board;
- connection by a thin wire to simulate a bridge – indicated as B2 board;
- connection by a 10 nF stitching capacitor – indicated as B3 board;
- connection by a small 22 μH ferrite bead – indicated as B4 board.

The meshes used with the two numerical codes are shown in Figure 10.35. Tetrahedral and cubic cells were used by HFSS and MWS respectively. Figure 10.35 also shows a zoom of the wire that connects the two power planes in HFSS simulation.

The test board was also simulated by an equivalent circuit obtained by applying the segmentation approach: the structure was segmented into cells whose dimensions were smaller

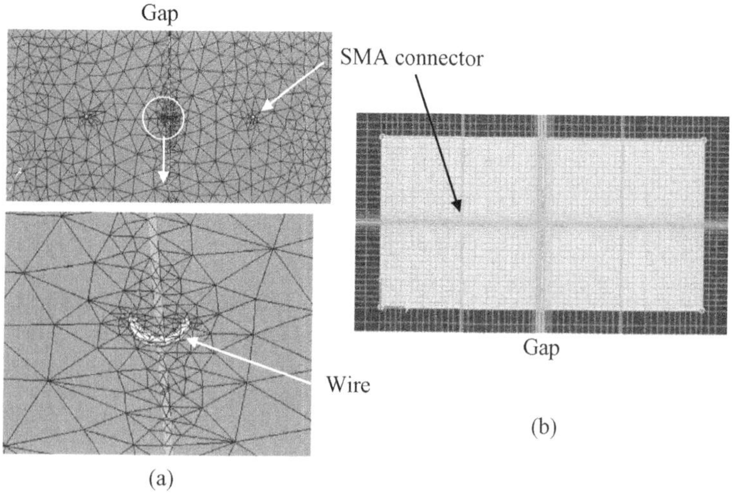

Figure 10.35 Meshing of the test board: (a) HFSS; (b) MWS

than the minimum wavelength λ_{min} and equal to λ_{min}/k, with k (number of cells per wavelength) set by the user. The SMA connectors at the cable input and output were modeled by using equivalent electrical circuits.

Two kinds of cell characterize the structure, as shown in Figure 10.36: cells of type 1 are those with the conductor at both the bottom and the top; cells of type 2 are those with the conductor located only at the bottom and occur at the gap.

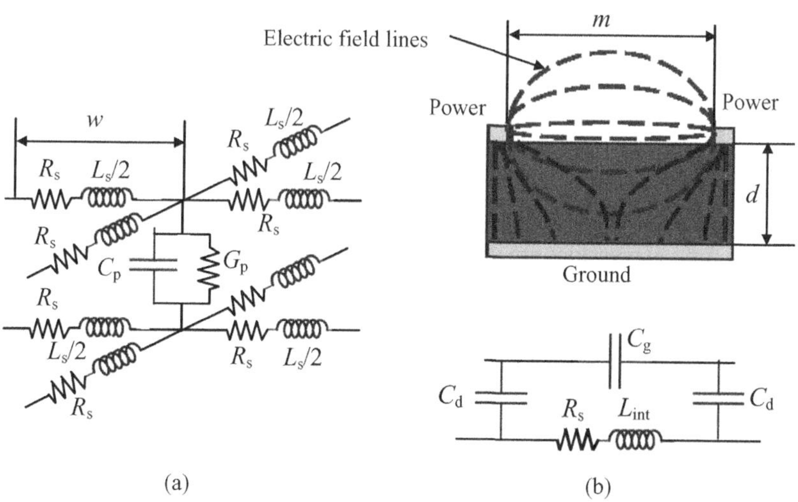

Figure 10.36 Equivalent circuits of one cell of (a) power and ground (type 1) and (b) gap area (type 2)

The electrical parameters of a type-1 cell are given by (see reference [24] and *Appendix C*)

$$R_s \approx 1/(\sigma t) + \sqrt{\mu_0 \pi f/\sigma} \tag{10.9a}$$

$$C_p = \varepsilon_0 \varepsilon_r w^2/d \tag{10.9b}$$

$$G_p = 2\pi f C_p \tan\delta \tag{10.9c}$$

$$L_s = \mu_0 d \tag{10.9d}$$

where t is the copper thickness, w is the cell dimension, d is the dielectric thickness, ε_0 and μ_0 are the vacuum permittivity and permeability, σ is the copper conductivity, ε_r is the relative dielectric constant, and tanδ is the dielectric loss tangent.

The electrical parameters of a type-2 cell are given by [25]

$$L_{\text{int}} = \frac{1}{\omega}\sqrt{\mu_0 \pi f/\sigma} \tag{10.10a}$$

$$R_s \approx 1/(\sigma t) + \sqrt{\mu_0 \pi f/\sigma} \tag{10.10b}$$

$$C_d = \frac{w}{2}\left(\frac{m}{w}\right)^{m_e} e^{K_e}\left(\frac{\varepsilon_r}{9.6}\right)^{0.9} \tag{10.10c}$$

$$C_g = \frac{w}{2}\left(\frac{m}{w}\right)^{m_o} e^{K_o}\left(\frac{\varepsilon_r}{9.6}\right)^{0.8} - \frac{C_d}{2} \tag{10.10d}$$

$$m_o = (w/d)(0.619\log(w/d) - 0.3853) \tag{10.10e}$$

$$K_o = 4.26 - 1.453\log(w/d) \tag{10.10f}$$

$$m_e = \begin{cases} 0.8675 & \text{for } \dfrac{1}{10} \le \dfrac{m}{w} \le \dfrac{3}{10} \\ \dfrac{1.565}{(w/d)^{0.16}} - 1 & \text{for } \dfrac{3}{10} \le \dfrac{m}{w} \le 1 \end{cases} \tag{10.10g}$$

$$K_e = \begin{cases} 2.043(w/d)^{0.12} & \text{for } \dfrac{1}{10} \le \dfrac{m}{w} \le \dfrac{3}{10} \\ 1.97 - \dfrac{0.03}{(w/d)} & \text{for } \dfrac{3}{10} \le \dfrac{m}{w} \le 1 \end{cases} \tag{10.10h}$$

where m is the width of the gap or split.

These closed-form expressions are valid for $0.5 \le w/d \le 2$ and $2.5 \le \varepsilon_r \le 15$.

These equivalent circuits can be implemented either into SPICE or into MatLab/MathCad codes by using the *Nodal Method* (NM) as described in *Appendix E*.

The components for isolation were simulated by lumped elements, as shown in Figure 10.37 for the capacitor. The capacitor has a minimum impedance of 0.2 Ω at 25 MHz. The inductor bead model is shown in Figure 10.30. The ferrite bead has a maximum impedance of 14.2 Ω at about 30 MHz. It has been verified that the computed impedances are practically the same as the measured values.

A comparison of the values of parameter S_{21} obtained by measurement, numerical simulations (HFSS, MWS), and a SPICE-like simulator with continuous power and ground planes

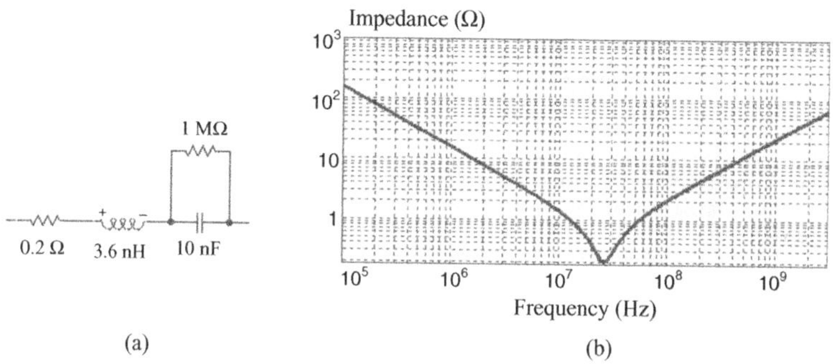

Figure 10.37 Capacitor used for decoupling: (a) equivalent circuit model; (b) capacitor impedance

is shown in Figure 10.38. A very good agreement can be observed between measurements and simulations up to 2.5 GHz. Above this frequency there are some slight differences, but the points of resonance are verified. MWS and HFSS are in excellent agreement in the entire frequency range.

The comparisons for type-B test boards with different connections are shown in Figure 10.39. The case of a type-B1 board is not so useful in real-world applications; nevertheless, it is a useful exercise to test the models and to see the frequency behavior of the split. It was verified that, with an increase in the width of the split from 0.5 mm to 1 mm, parameter S_{21} decreases by about 4 dB in the entire frequency range 0–1 GHz. Once the models are validated, other topologies and geometries of the PCBs can be simulated, as reported by Costa *et al.* [23].

From the results obtained, the following observations can be made:

- Comparing measurements with simulations, a very good agreement can be observed up to 1000 MHz; above this frequency the simulations slightly overestimate the parameter S_{21}.
- Comparing measurements with the circuit model results, a good agreement is obtained up to 1.6 GHz. The main reason why, above this frequency, the simulation values fail seems to be the wire model and its interactions with the board.

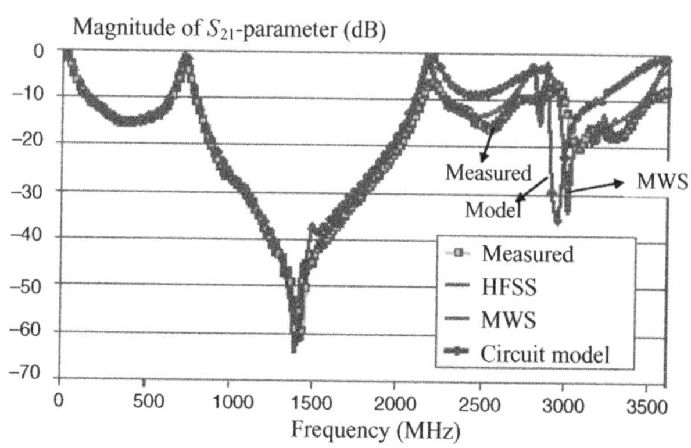

Figure 10.38 Scattering parameter S_{21} in the case of a test board of type A with continuous planes

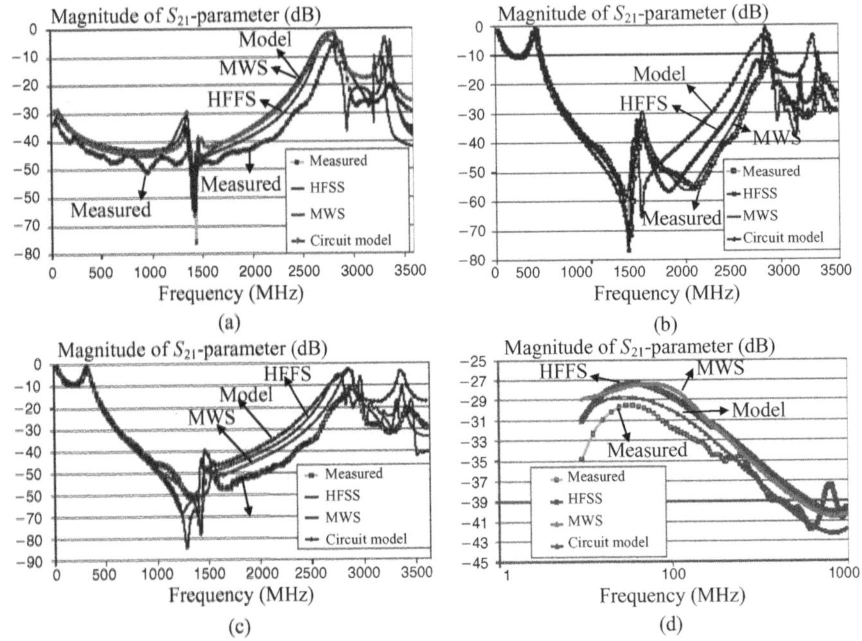

Figure 10.39 Magnitude of the scattering parameter S_{21} for various connections on the gap: (a) open (type B1); (b) thin wire (type B2); (c) capacitive (type B3); (d) inductive for frequency range 10–1000 MHz (type B4)

- Suitable isolation can be obtained at very low frequencies by using a stitching capacitor.
- Suitable isolation can be obtained at high frequencies by using a ferrite bead inductor.
- The circuit approach is a powerful tool for modeling the power-plane partitioning in the case of simple boards. The advantage consists in the possibility of using a SPICE model for the IC, as shown in *Section 8.2*.
- The numerical approach is a powerful tool for modeling the power-plane partitioning for more complex structures in order to achieve significant noise isolation between different power areas.

In the next example, the possibility will be discussed of simulating more complex boards by SPICE using the vector fitting technique, which makes it possible to extract an equivalent circuit between two points of interest in the PCB in order to perform simulations directly in the time domain, as done for a lossy line in Section 7.2.

Example 10.9: S-parameter Measurements and Simulations of a Test Board with an Island
In this example, MWS code is used to predict parameter S_{21} for a power distribution with an island to separate or protect a particular area of a PCB, as discussed in *Section 10.2.4*. The PCB considered has the same structure as the one reported Cui *et al.* [21].

The topology of the PCB under investigation is shown in Figure 10.40. Current sources of impedance 50 Ω were placed at port 1 and port 2. The code computed the parameter S_{21} as a transfer function between the two ports. Port 1 acted as the source of noise, and the goal for

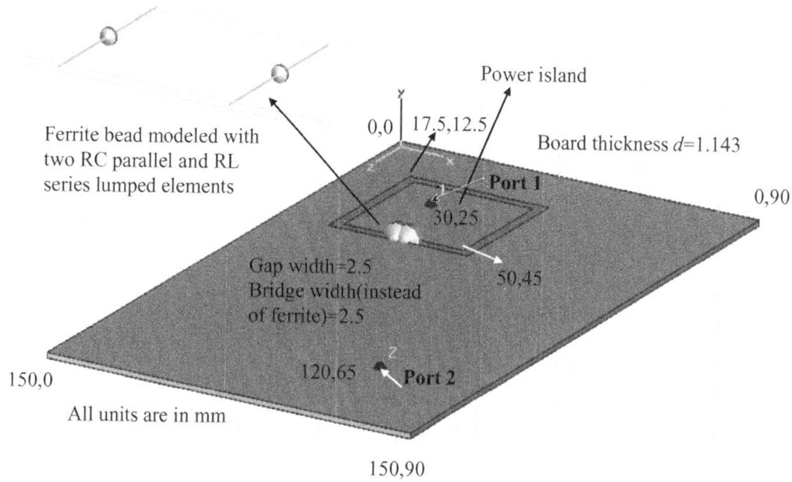

Figure 10.40 Geometry of the test board with an island on the power plane

an effective isolation was to have S_{21} as low as possible at all frequencies of interest, in this case 0–3 GHz. The dielectric substrate was characterized by $\varepsilon_r = 4.5$, dielectric loss tangent $\tan\delta = 0.02$, and thickness $d = 1.143$ mm. The metallic plane had a thickness $t = 0.1$ mm.

Simulations were performed considering the following cases:

- continuous power and ground planes (reference);
- isolated power island;
- power island with a bridge of 2.5 mm width;
- power island connected with a ferrite bead of 14.2 kΩ impedance at 30 MHz.

The magnitudes of parameter S_{21} computed by MWS are shown in Figure 10.41 for three PCB configurations. The curves are very close to those of reference [21], considering that in

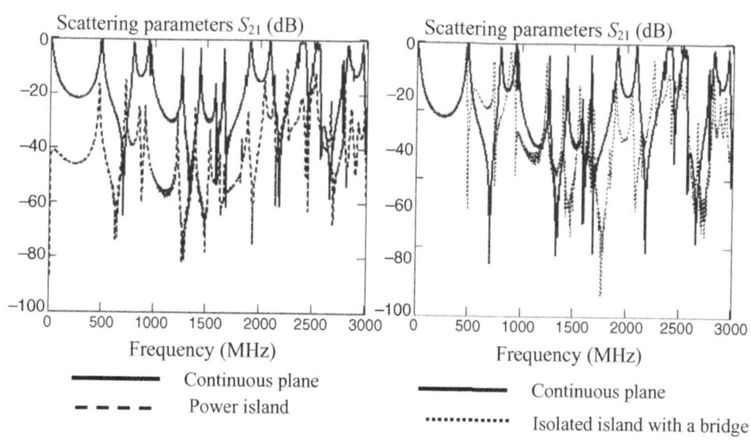

Figure 10.41 Magnitude of parameter S_{21} computed by MWS for three power plane cases

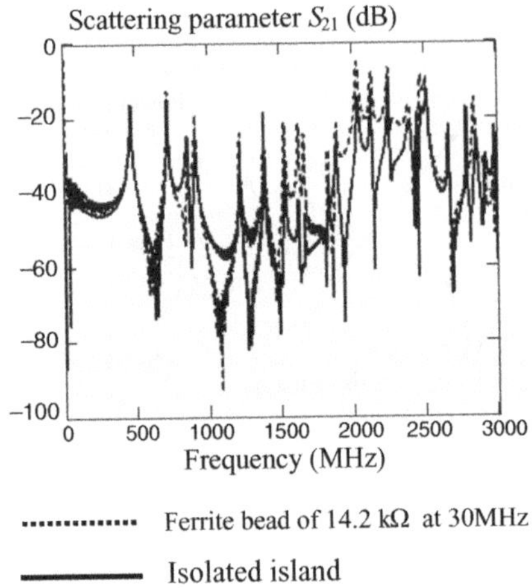

Figure 10.42 Magnitude of parameter S_{21} computed by MWS with a ferrite bead instead of a bridge

MWS the ports were simulated by a simple ideal current source with 50 Ω impedance. As shown in the previous section, SMA connectors should be modeled by MWS to reproduce the measured values obtained by the VNA. In any case, it is confirmed that a metallic bridge through the island has very little effect in decoupling power planes, as it is effective only in some short ranges of frequencies. It is important to implement fixes that maintain isolation. Looking at the results reported in Figure 10.42, it seems that the ferrite bead is a good practical solution when it is necessary to ensure that power plane areas are at the same DC voltage. In fact, little difference in parameter S_{21} can be noted between ferrite and the isolated island case which is effective in the entire frequency range.

10.4.1 Modeling the Power Distribution with a Driver for Simulations

In this section, essential information is provided regarding the last frontier in modeling digital devices inserted in a 3D structure such as a PCB [26]. For instance, consider the structure of two planes as depicted in Figure 10.43a: one plane is used as ground and the other for distributing power to an IC device. The task is to simulate by SPICE the voltage on the capacitive load at the output of the driver, considering the power distribution of the two planes without using a grid of cells as done in *Example 10.8*.

The process is based on the following steps (see Figure 10.43b):

(a) Electromagnetic characterization of the PCB structure by measured or calculated S-parameters between the ports of interest, in this case port 1 and port 2 in Figure 10.43a.
(b) Extraction of a linear macromodel for the PCB structure, starting from S-parameters by using the Vector Fitting (VF) technique [27] to obtain an equivalent circuit between the two ports in a similar way to *Section 7.2* for lossy lines.
(c) Create a behavioral model instead of a transistor-level model for the non-linear driver [28].
(d) Circuit simulation of the full structure in the time domain.

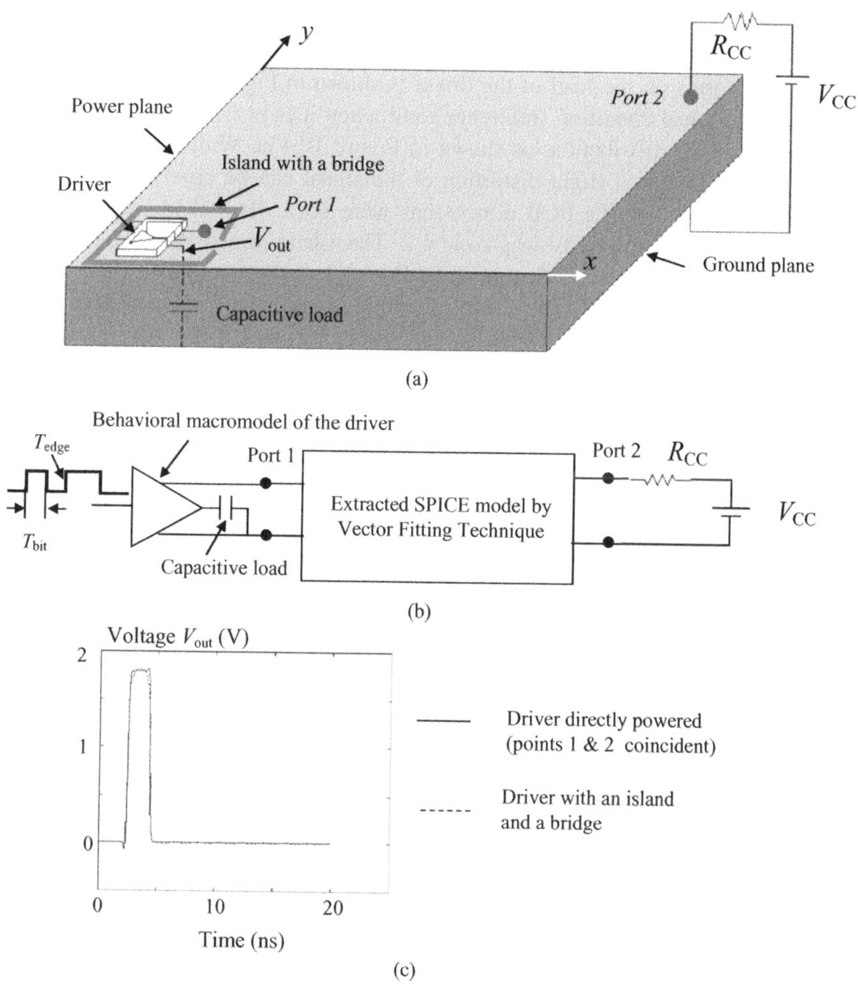

Figure 10.43 Test board with a switching digital device: (a) topology; (b) equivalent circuit to compute load voltage; (c) simulated load voltage

The characterization between the two ports should be done by using:

1. Frequency-domain full-wave simulations (MOM, FEM, etc.).
2. Time-domain full-wave simulation (FIT, FDTD) + FFT post-processing.
3. Direct *Vector Network Analyzer* (VNA) measurement.

The digital device should be simulated by:

- A behavioral model that reproduces the static and dynamic input/output characteristics and switching noise [29].
- Adopting as excitation a random bit pattern of '...11010...'.
- Using digital signal characteristics such as the time bit T_{bit} and edge times t_r and t_f.

Once these models are created, the load voltage, ΔI-*noise* voltage on power supply, and crosstalk can be simulated with minimum effort directly in the time domain. An example of simulated waveforms on the load of the driver is shown in Figure 10.43c for the case of a driver powered in ideal condition (reference) and when it is placed inside an island with a bridge in power plane distribution, as shown in Figure 10.43a. With the island surrounding the driver and a bridge, a slight distortion of the signal can be observed. The example comes from Genovese [29]. The PCB dimensions were $16 \times 10 \times 0.14$ cm, and the substrate relative dielectric permittivity was $\varepsilon_r = 4.2$. The island dimensions were 3×3 cm, and the cut and the bridge were 2.5 mm wide. Port 1 had the coordinate (4, 3) cm, and port 2 the coordinate (14, 9) cm. The driver, an Mplog-type behavioral model suitable for simulating accurately the I/O characteristics and switching noise [28], was powered with $V_{CC} = 1.8$ V and $R_{CC} = 1\ \Omega$, and had $T_{bit} = 2$ ns, $t_{edge} = 0.2$ ns, and a capacitive load of 1 pF. Details about this example, theory, and other examples are given by Genovese [29].

10.5 Points to Remember and Design Rules for Grounding in PCBs

In this last section of the chapter devoted to grounding in PCBs, our intention is twofold: to highlight some important concepts regarding the grounding item; to provide a list of design rules for ensuring a correct functionality of the PCB with an attached cable, considering also the presence of the chassis and discontinuities in power and ground planes.

(i) General Comments

- By the term 'ground', an engineer mainly means two things: (a) the metallic parts of an interconnect devoted to the flow of the return signal or power supply current to the origin; (b) the metallic parts used as voltage reference for the system (i.e. chassis), where intentional currents do not flow to avoid unwanted voltage drops, and with the ultimate task of earthing all metallic objects.
- Very often a ground conductor, wire, or plane used as a signal return path is shared by several signals and could be crossed by disturbing currents externally produced (ESD, fast transients, etc.). As a consequence of these facts there is a voltage drop due to the impedance associated with the ground and the unwanted crossing current. This coupling mechanism on PCBs is indicated as *common impedance coupling*. The *Ground Loop Coupling* (GLC) parameter, defined in *Section 10.1.2*, characterizes the effects on the circuits of this phenomenon. GLC is directly dependent on the transfer impedance parameter Z_t related to signal integrity and radiated emission effects.
- The transfer impedance Z_t must be minimized by using large solid conductors as the return current path, such as one or two ground planes having the structure of a microstrip or stripline respectively (see *Section 10.1.2*).
- A grounding strategy based on single-point or multipoint connection must be considered in order to minimize the *GLC* parameter. For instance, low-frequency sensitive analog circuits must have their own ground area, separated from the high-speed digital circuit ground realized by one or more ground planes. All the different grounds must then be connected together at a reference ground such as the chassis (see *Section 10.1.3*).
- Inductors and capacitors can be used for a hybrid grounding. Inductors make it possible to have a single-point connection of each type of system at high frequency; capacitors make

it possible to have a multipoint connection to a reference ground of several single-point ground parts at high frequencies in order to avoid dangerous resonance frequencies when the dimensions of the whole grounding structure are comparable with the minimum wavelength of interest (see *Section 10.1.3*).

- The techniques for mitigating *Simultaneously Switching Noise* (SSN) are: (1) decoupling capacitors located over the entire PCB (see *Chapter 8)*; (2) splitting planes to block the propagation of unwanted electromagnetic waves (see *Section 10.4*); (3) power islands (see *Section 10.4*); (4) shorting vias; (5) novel electromagnetic band gap (EBG) structures [30].

(ii) Comments on the Return Current

- Solid copper power and ground planes are an excellent solution for the return current path in a multilayer PCB because they minimize ΔI*-noise* (see *Section 8.1*) and radiated emissions (see *Chapter 9*).
- It is necessary to distinguish between a DC signal return path, which coincides with the nominal ground plane layers, and an AC return path, which can be the ground layer or the power. At PCB common working frequencies, these two planes are practically short-circuited by the distributed interplane capacitance, which provides an alternative path for displacement return current, and by the decoupling capacitors in the frequency range where they act as lumped capacitance (see *Section 8.2.1*).
- To enhance the effectiveness of decoupling capacitors in frequency, the parasitic equivalent series inductance and the effective inductance associated with the connections must be minimized (see *Section 8.2.1*).
- In microstrip and stripline structures, most of the return current of a high-speed digital connection crowds beneath the trace. The peak current density lies directly under the trace, while falling off sharply away from the trace. To avoid increasing the radiated emission, critical traces must be far away from the PCB edges (see *Section 10.2.1*).
- The AC return signal current flows along a path that minimizes the associated loop impedance formed with the signal trace, or, in other words, the high-frequency currents follow the paths exhibiting the lowest impedance. Therefore, the return path could be a ground or a power plane (see *Section 10.2.3*).
- Power and ground plane layers must be placed as close as possible in order to maximize their interplane capacitance. This distributed capacitance, together with the action of the decoupling capacitors, allows the signal current to use either the ground plane or the power plane as its return path, depending on which of these planes is geometrically closer to the trace (see *Section 10.2.2*).
- The need for a trace of changing layer is a critical point, as the signal via could not have another nearby via, such as the connection of the decoupling capacitor to the ground or power plane, as return current. Therefore, frequent changes of layer should be avoided and, when required, adjacent signal layers with a ground plane interposed for routing critical traces should be used (see *Section 10.2.3*).
- Recall that, above about 100 MHz, the intrinsic inductance of the decoupling capacitors and those associated with the connections to the ground and power planes make the impedance between the two planes too high for the high-frequency components of the return currents. This means that the return currents can flow along uncontrolled paths as displacement currents, causing noise between the two planes. Therefore, the solution to be used is

ground–signal–ground, as offered by the stripline structure for routing critical traces such as those for clocks (see *Section 10.2.3*).

- Stitching capacitors should be used between the split of two power planes to ensure continuity to the return current path when the AC signal return path is a gapped power plane for powering the devices with different voltages (5 V, 3.3 V, etc.). Recall that this is a frequency-limited solution owing to the effective inductance associated with the capacitor and its connection leads (see *Section 10.2.4*).
- The installation of an SMT stitching capacitor must be done with care in order to minimize the effective inductance associated with the mounting pad and connection leads such as traces (to be avoided) and vias (see *Section 3.2.8* and *Section 8.1*).
- The return current should not encounter obstacles such as isolation splits or gaps in either the power or the ground planes in order to avoid generation of *common-mode* currents (see *Section 9.3*) and inductive signal integrity effects. The inductive effects are much less trouble for a pair of differential signal traces (see *Section 12.2.4*).
- Traces that cross the gaps in split power planes or slots in ground planes, for example a motherboard connector with high-density holes and excessive clearance (antipad), are affected by more crosstalk, as shown in *Section 6.5.3*.
- Use buried vias to reduce the number of pads and antipads in the reference layers (see *Section 12.2.5*).

(iii) Comments on the Partitioning Technique

- Moats or barriers are techniques to isolate some area of the PCB with sensitive or analog circuitry from the interference of other noisy circuits such as high-speed digital circuits or low-frequency I/O devices (of the order of kHz) to prevent excessive radiated emission from cables (see Section 10.2.5).
- Partitioning is achieved by totally removing the copper from all layers and planes on the PCB with the use of an intentional gap, typically of 1.27 mm minimum, between the two zones.
- As very often the devices on the protected area must communicate with the other devices in the same PCB, to preserve isolation, optocoupler devices for single-ended signaling and *common-mode* chokes with transformers for differential signaling are required. These components must be placed across the moat (see *Section 10.2.5*).
- For safety reasons, each area must be connected with a very low impedance to a reference ground such as the chassis.
- When a complete isolation cannot be realized, ferrite-bead inductors should be used to ensure a return path across the moat for the low-frequency signals between the two separated planes (see *Example 10.8*). This fix can reduce the *common-mode* emission from the PCB by up to 20 dB in the range 30–1000 MHz [9].
- The use of a bridge that connects the two areas with a narrow strip of copper should be avoided because it is effective as a *common-mode* barrier only for a very small interval of frequencies, and, for other higher-frequency regions where resonances occur, it could make the isolation worst (see *Example 10.9*).
- Mixed analog and digital signal devices have their own design rules for grounding, provided by the manufacturer as application notes (see *Section 10.2.5*).

- Simulations or measurements of S-parameters are the most appropriate procedures for assessing the effectiveness of several techniques used for AC isolation in PCBs. In this way, the resonance frequencies of the structure are also found (*see Section 10.4*).
- A new technique, known as the planar *Electromagnetic Bandgap* (EBG) structure, can be used for switching noise mitigation and for isolation in mixed signal boards. This structure consists of a two-layer power distribution system, with one of the layers, patterned in a periodic fashion, forming a high-impedance surface to prevent the propagation of electromagnetic waves over some frequency range. For signal integrity preservation, differential signaling is recommended. For more details, see reference [30].

(iv) Comments on Edge Connectors

- The connection between a motherboard and boards with the ICs by edge connectors is a critical point for ensuring signal integrity and performance against EMI problems. The parameter that quantifies the EMI performance of a connector is the connector transfer impedance Z_t (see *Section 10.3*).
- A suitable number of pins for ground and power must be chosen to have an acceptable value of transfer impedance. The best solution should be that each signal pin has a ground or a power adjacent pin as the return path. Normally, a compromise using 2:1 is acceptable for a suitable value of Z_t (see *Example 10.4*).
- To minimize Z_t, a metallic shelf or more pins to ground around the connector should be used (see *Example 10.5*).
- For differential connectors, two columns of adjacent signal pins must have at both sides a column of ground pins (see *Section 12.1.2*).

(v) Comments on I/O Cables

- Cables attached to a PCB are among the main sources of emissions. To cope with this problem, it is very important to implement appropriate grounding techniques in the area of the I/O cables, and using the chassis as reference ground (see *Section 10.3.4*).
- For high-speed interface devices, the partitioning technique is ineffective. Shielded cables or EMI filters for unshielded cables must be used with appropriate grounding techniques to a reference quiet area or chassis (see *Section 10.3.3*).
- When the shield of a cable must be connected to the ground of the PCB, the connection must be characterized by a very low impedance, and a solution such as pigtails must not be used (see *Example 10.6*).
- For unshielded cables, EMI filters such as inductors and capacitors should be used for *common-mode* decoupling of the signal and return conductors. For differential transmission, *common-mode chokes*, transformers, and capacitors to ground should be used (see *Section 10.3.4*).
- An isolated area between the interface devices and the connector should be reserved in order to divert the *common-mode* currents to the chassis. This area can be removed in the case of differential signaling only (see *Section 10.3.4*).
- Using a shielded box is the most effective solution for lowering the emission significantly, especially at high frequencies (see *Section 10.3.4*).
- A shielded box is ineffective in reducing radiated emission if the ground of a PCB with an attached cable is not connected to the chassis in the I/O connector zone (see *Section 10.3.4*).

- Connection of the PCB to the chassis by stitches is an effective way to divert *common-mode* current to the chassis instead of the cables. Stitches must be positioned at regular intervals and with a distance less than the minimum wavelength at the frequency of interest (see *Section 10.3.4*).
- Maximum reduction in emission can be obtained by a combination of fixes consisting in using stitches, *common-mode* filters, and an I/O quiet area realized by a split (see *Section 10.3.4*).

References

[1] Mardiguian, M., *'EMI Control Methodology and Procedures'*, Vol. 8, 1989, Interference Control Technologies, Gainsville, VA, 1989.

[2] Caniggia, S. and Maradei, F., 'SPICE-like Models for the Analysis of the Conducted and Radiated Immunity of Shielded Cables', *IEEE Trans. on Electromagnetic Compatibility*, **46**(4), November 2004, 606–616.

[3] Goedbloed, J., *'Electromagnetic Compatibility'*, Prentice Hall, Deventer, The Netherlands, 1990.

[4] Ott, H., *' Noise Reduction Techniques in Electronic Systems'*, 2nd edition, AT&T Bell Laboratories, John Wiley & Sons, Inc., New York, NY, 1988.

[5] *'Information Technology – Cabling Installation. Part 2: Installation Planning and Practices Inside Buildings'*, EN 50174-2.

[6] *'Bonding Configuration and Earthing Inside a Telecommunication Building'*, K.27, ITU-T.

[7] *'Earthing and Bonding of Telecommunication Equipment in Telecommunication Centres'*, ETS 300 253.

[8] *'Installation and Mitigation Guidelines – Section 2: Earthing and Cabling'*, IEC 61000-5-2.

[9] Archambeault, B., *'PCB Design for Real-World EMI Control'*, Kluwer Academic Publishers, Norwell, MA, 2002.

[10] Balanis, C., *' Advanced Engineering Electromagnetics'*, John Wiley & Sons, Inc., New York, NY, 1989.

[11] Celozzi, S. and Orlandi, A., 'Numerical Modeling of Finite Size PCB for Radiated EMI Prediction', in *Proceedings of 11th International Zürich Symposium on Electromagnetic Compatibility*, Zürich, Switzerland, 7–9 March 1995, 33–38.

[12] Montrose, M., *'EMC and the Printed Circuit Board: Design, Theory, and Layout Made Simple'*, IEEE, Inc., New York, NY, 1999,.

[13] Montrose, M., *'Printed Circuit Board Design Technique for EMC Compliance'*, 2nd edition, IEEE, Inc., New York, NY, 2000.

[14] Heirman, D., Crosby, S., and Smith, D., 'International Mandatory Products Standards', *AT&T Technical Journal*, March/April 1993, 25–40.

[15] Rainal, A., 'Computing inductive noise of chip packages', *AT&T Bell Laboratories*, **63**(1), January 1984, 177–195.

[16] Ye, X., Drewniak, J., Nadolny, J., Hockanson, D., 'High-performance inter-PCB connectors: analysis of EMI characteristics', *IEEE Trans. on Electromagnetic Compatibility*, **44**(1), February 2002, 165–174.

[17] www.spectrum-soft.com.

[18] Martens, L., Madou, A., Konè, L., Demoulin, B., Sjoberg, P., Anton, A., Van Koetsem, J., Hoffman, H., and Schricker, U., 'Comparison of Test Methods for the Characterization of Shielding of Board-to-Backplane and Board-to-Cable Connectors', *IEEE Trans. on Electromagnetic Compatibility*, **42**(4), November 2000, 427–440.

[19] www.cst.com.

[20] Fan, J., Drewniak, J., Shi, H., and Knighten, J., 'DC Power-bus Modeling and Design with a Mixed-potential Integral-equation Formulation and Circuit Extraction', *IEEE Trans. on Electromagnetic Compatibility*, **43**(4), November 2001, 426–436.

[21] Cui, W., Fan, J., Ren, Y., Shi, H., Drewniak, J., and DuBroff, R., 'DC Power-bus Noise Isolation with Power-plane Segmentation', *IEEE Trans. on Electromagnetic Compatibility*, **45**(2), May 2003, 436–443.

[22] www.ansoft.com.

[23] Costa, V., Gualdoni, M., and Politi, M., 'Analysis of Power and Ground Islands in Printed Circuit Boards', *IEEE International Symposium on Electromagnetic Compatibility*, S. Clara, CA, 2004, 39–44.

[24] Kim, J.-H. and Swaminathan, M., 'Modeling of Irregular Shaped Power Distribution Planes Using Transmission Matrix Method', *IEEE Trans. on Advanced Packaging*, **24**(3), August 2001, 334–346.
[25] Gupta, K.C., Garg, R., and Bahl, I.J., *'Microstrip Lines and Slot Lines'*, Artech House Inc., Dedham, MA, 1979.
[26] Grivet-Talocia, S., Stievano, I.S., Canavero, F.G., and Maio, I.A., 'A Systematic Procedure for the Macromodeling of Complex Interconnects and Packages', *International Symposium on Electromagnetic Compatibility – EMC Europe 2004*, Eindhoven, The Netherlands, 2004.
[27] Grivet-Talocia, S., 'Passivity Enforcement via Perturbation of Hamiltonian Matrices', *IEEE Trans. on Circuit and Systems*, **51**, September 2004, 1755–1769.
[28] Stievano, I.S., Canavero, F.G., and Maio, I.A., 'Parametric Macromodels of Digital I/O Ports', *IEEE Trans. on Advanced Packaging*, **25**(2), May 2002, 255–264.
[29] Genovese, E., '*Noise Propagation on Power Bus Planes: Circuit and Electromagnetic Simulations*', Thesis at Politecnico of Torino (in Italian), Prof. F. Canavero, 2004.
[30] Qin, J., Ramahi, O.M., and Granatstein, V., 'Novel Planar Electromagnetic Bandgap Structures for Mitigation of Switching Noise and EMI Reduction in High-Speed Circuits', *IEEE Trans. on Electromagnetic Compatibility*, **49**(3), August 2007, 661–669.

11

Measurement and Modeling

Signal Integrity (SI) is a primary concern for system functionality and *Electromagnetic Compatibility* (EMC) compliance, and allows a product to achieve the required certification to be legally sold. *Printed Circuit Boards* (PCBs) are the physical structures used for the mechanical support of transmission lines and connect components. Signal integrity is based on how efficiently the information propagates through a transmission line, and can be verified by measurements in the time and frequency domains. A direct relationship exists between time- and frequency-domain measurements, and is provided by Fast Fourier Transform (FFT) and Inverse FFT (IFFT), respectively.

Two key instruments useful for extracting circuit parameters regarding interconnects and discontinuities occurring in high-speed digital systems are presented. The first is the *Time-Domain Reflectometer* (TDR) which performs measurements in the time domain, and the second is the *Vector Network Analyzer* (VNA) which performs measurements in the frequency domain. The advantage of performing time-domain measurements is that the effects of a discontinuity along the interconnect on signal integrity can be directly observed, and models of the parasitic elements can be developed for the switching time of interest. The advantage of performing frequency-domain measurements is that resonance effects at some frequencies can be determined and appropriate models for components can be developed.

The theory of the TDR instrument, based on the 'closed-loop radar' principle, is introduced. The main characteristics of this technique are investigated by using SPICE models that reproduce the set-up for measurements: the TDR, the coaxial cable and its tip connector, and the *Device Under Test* (DUT). It is shown how to measure transmission-line parameters for traces in PCBs and lumped parameters associated with discontinuities, such as small inductances, capacitances, vias, bends, etc. Errors in using the TDR are also discussed.

The definition of scattering S-parameters for a two-port network and correlation with the impedance Z-parameters is briefly outlined. A VNA measures the amplitude and phase of S-parameters in the frequency domain. The procedure for calibration and error correction of a VNA is presented. The VNA measurement set-up can be modeled by a SPICE equivalent circuit and by a full-wave numerical code such as MicroWave Studio (MWS). Both these prediction tools are suitable for calculating S-parameters which can be compared with those obtained by VNA measurements. A discussion is proposed in *Section 11.2* to highlight the importance of the parasitic parameters of the ports used to excite the MWS model in order to establish

Signal Integrity and Radiated Emission of High-Speed Digital Systems Spartaco Caniggia and Francescaromana Maradei

a consistent comparison of the measured and calculated results. Examples are presented concerning the extraction from S-parameters of circuit parameters for a lossy line, a short coaxial connector, and a via. Possible errors in performing simulations are also discussed.

The last section of this chapter addresses the problem of uncertainty in the measured data when radiated emission measurements are carried out in EMC laboratories certified according to the standard for model validation. It is shown with a test carried out in two different EMC labs that the differences between computed and measured radiated fields from a shielded rack with an attached cable are close to the verified ± 5 dB of uncertainty due to site, instruments, and cables. It is also shown that, for a consistent comparison, the test site (the presence or absence of a ground plane, the presence of metallic objects, etc.) and the set-up for measurements (antenna and source position, cables connected to the source, etc.) must be exactly reproduced. A loop fed by a coaxial cable connecting a shielded oscillator is used as a test to discuss all these items.

11.1 Time-Domain Reflectometer (TDR)

The Time-Domain Reflectometer (TDR) is a technique used in signal integrity to measure impedance characteristics, delays, and step responses of traces in PCBs and cables. It is also used to measure discontinuities along interconnects in order to determine a lumped-circuit model to be associated with each discontinuity.

11.1.1 TDR as a 'Closed-Loop Radar'

The TDR employs a system known as the 'closed-loop radar' [1–5]. A voltage step propagates down the transmission line connecting the oscilloscope and the *Device Under Test* (DUT). The incident step and the reflected voltage waves, which are algebraically summed, are monitored by the oscilloscope at a particular point on the line. The step generator has a 50 Ω output impedance, and the interconnection between the high-speed oscilloscope and the DUT is realized by a coaxial cable having a 50 Ω characteristic impedance. Therefore, only the reflections from the DUT are present. The schematic representation of the set-up used by the TDR to characterize a DUT is shown in Figure 11.1.

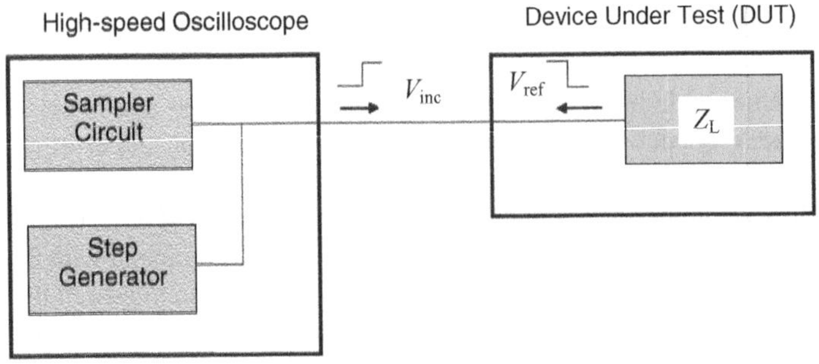

Figure 11.1 Schematic representation of the set-up used by the TDR to characterize a *Device Under Test* (DUT)

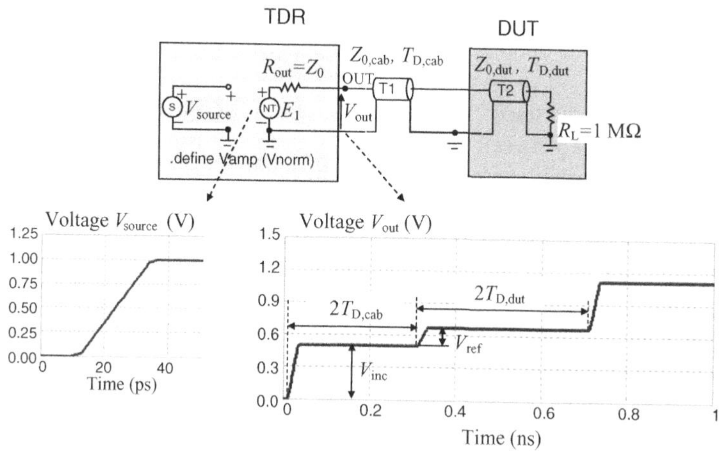

Figure 11.2 SPICE equivalent circuit of the TDR set-up and simulated waveforms when the DUT is a trace terminated with an open load

The advantage of the TDR over frequency-domain measurements is the ability to extract electrical data relevant to digital systems, represented by time-domain signals. By TDR measurements it is possible to extract the nominal impedance, the delay time, and a possible discontinuity of an interconnect. Examples of measurements performed by the TDR can be found in the references. In this section, SPICE simulations of the equivalent circuit that models the usual TDR measurement set-up will be used to highlight the TDR performance.

Example 11.1: Simulation of TDR Measurements Assuming a Trace in a PCB as the DUT
The basic TDR theory is applied to the example shown in Figure 11.2, where the TDR and the DUT, which in this case is a trace, are simulated by SPICE. The TDR is modeled by two voltage sources. One, V_{source}, is an independent voltage source with the voltage assigned in table form to build the actual waveform with a smooth rise time $t_r = 25$ ps and an amplitude of 1 V. The other, E_1, is a voltage-controlled voltage source depending on V_{source}, with the control parameter V_{norm} able to change the amplitude, if required. The DUT considered is a trace of characteristic impedance $Z_{0,dut} = 100\ \Omega$ and delay time $T_{D,dut} = 200$ ps. The parameters used in the simulation were: $V_{norm} = 1$ V; $R_{out} = Z_0 = 50\ \Omega$ (source impedance); $Z_{0,cab} = 50\ \Omega$, $T_{D,cab} = 150$ ps. Lossless TL models are used.

At time $t = 0$, an incident voltage step V_{inc} of 0.5 V amplitude is launched onto the 50 Ω coaxial cable owing to the partitioning effect between the source and the characteristic impedance of the coaxial cable. After a time $2T_{D,cab}$, a reflected step $V_{ref} = (V_{out} - V_{inc})$ due to the mismatch of the trace sums to $V_{inc} = (v(E_1)/2)$, as shown in Figure 11.2, for a time $2T_{D,dut}$. The waveform at the TDR output (point OUT in Figure 11.2) changes at the arrival of the wave reflected from the line T2 (i.e. the DUT), which is practically open because it is terminated with a very high resistance ($R_L = 1$ MΩ). It is important to point out that the line parameters $Z_{0,dut}$ and $T_{D,dut}$ of DUT can be obtained from inspection of voltage at the output of the TDR. In particular: $Z_{0,dut} = Z_0(1 + \rho_r)/(1 - \rho_r)$, with the reflection coefficient $\rho_r = V_{ref}/V_{inc}$. In this case: $V_{inc} = 0.5$ V, $V_{ref} = 0.17$ V, then $\rho_r = 0.34$, then $Z_{0,dut}$ (measured) = 100 Ω, as expected.

Consider that the TDR gives the possibility of reading directly the reflection coefficient ρ_r. When the DUT is a resistance of 50 Ω, $\rho_r = 0$; when it is an open circuit, $\rho_r = 1$; when it is a short circuit, $\rho_r = -1$.

The delay time $T_{D,dut}$ of the trace can be read from the TDR waveform as twice (incident and reflected wave) the time between the first and the second mismatch, as shown in Figure 11.2. However, considering the 'aberration' error, which will be explained later, the simplest way to derive the propagation delay time is to use the TDR and measure the delay between two identical test structures of different length. The time delay is determined by subtracting the delays between the two structures. The instant to consider is when the reflected pulse begins to rise owing to the open-end termination.

TDR measurements are suitable for establishing the effect of high- and low-impedance discontinuities occurring along the trace. Consider, as the DUT, a trace with a characteristic impedance $Z_{0,t} = 100$ Ω and a delay time $T_{D,t} = 1$ ns, with the following two discontinuities:

- discontinuity 1 – a narrow short trace having a characteristic impedance $Z_{0,d1} = 150$ Ω and a delay time $T_{D,d1} = 50$ ps, positioned after one-third of the main trace;
- discontinuity 2 – a large short trace having a characteristic impedance $Z_{0,d2} = 67$ Ω and a delay time $T_{D,d2} = 50$ ps, positioned after two-thirds of the main trace.

The TDR response obtained by the SPICE model is shown in Figure 11.3. It can be noted that the narrow trace has an inductive effect and the large trace has a capacitive effect. To support this statement, further simulations were performed in which the two discontinuities were modeled by a lumped inductance and by a lumped capacitance respectively, instead of using TLs. In particular, discontinuity #1 was modeled by a lumped inductance of value $L = Z_0 T_D = 150 \times 0.05$ ns $= 7.5$ nH, while discontinuity 2 was modeled by a lumped capacitance of value $C = T_D/Z_0 = 50$ ps/67 $= 0.75$ pF. The results of this second simulation are shown by the dotted lines in Figure 11.3.

11.1.2 TDR Resolution and Aberrations

TDR resolution depends on the system rise time [5]. Two narrowly spaced discontinuities may be indistinguishable to the measurement instrument if they are separated by less than half the system rise time $T_{r\text{-system}}$.

The system rise time is characterized by the fall or rise time of the reflected edge from an ideal short or open load at the probe tip. TDR resolution can be calculated by the following equation:

$$\text{TDR}_{\text{resolution}} \geq T_{r\text{-system}}/2 \tag{11.1a}$$

$$T_{r\text{-system}} = \sqrt{(T_{r\text{-step generator}})^2 + (T_{r\text{-sampler}})^2 + (T_{r\text{-probe}})^2} \tag{11.1b}$$

where $T_{r\text{-step generator}}$, $T_{r\text{-sampler}}$ and $T_{r\text{-probe}}$ are rise times of the step generator, sampler, and probe. When using the TDR, the user should take into account the rise/fall time of the actual digital system. In fact, the fast rise time of the TDR also shows small discontinuities that could be masked by the slower edge rate of actual signals. Therefore, in practical cases, it is better to use a TDR-like generator with a slower rise time.

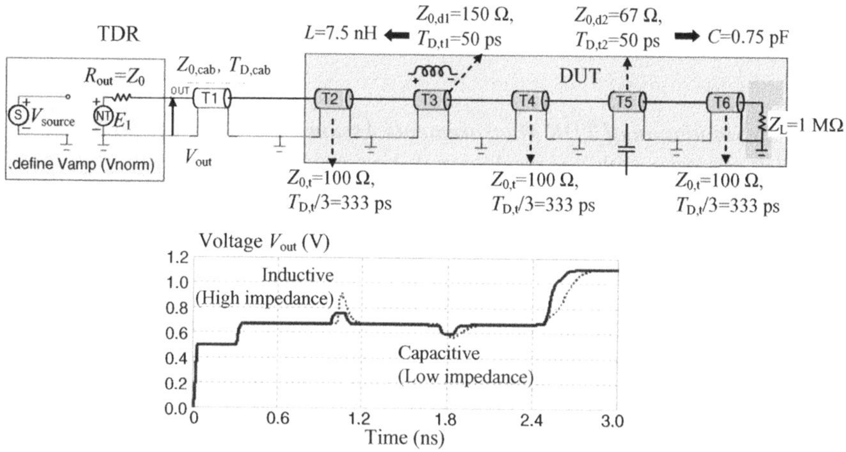

Figure 11.3 SPICE equivalent circuit of the TDR and simulated waveforms in the case of a DUT given by a trace with discontinuities modelled as a TL (solid line) or modelled as L or C lumped elements (dotted line)

Aberrations, e.g. ringing (see Figure 11.4), that occur prior to the main incident step can be particularly troublesome because they arrive at a discontinuity and begin generating reflections before the main step arrives [5]. These reflections reduce resolution by obscuring closely spaced discontinuities. Aberrations that occur after the incident step will cause corresponding aberrations in the reflections. They will be difficult to distinguish from the reflections due to the DUT discontinuities.

Many factors contribute to the accuracy of a TDR measurement. These include the TDR system's step response, interconnect reflections, DUT losses, step amplitude accuracy, and the accuracy of the reference impedance used in the measurements. All these factors have to be kept under control.

When the incident step is very fast (20–40 ps), the spectrum extends up to 20 GHz. The echo technique reveals at a glance the position and the nature (resistive, inductive, or capacitive) of each discontinuity along the line in terms of bumps and dips. A bump

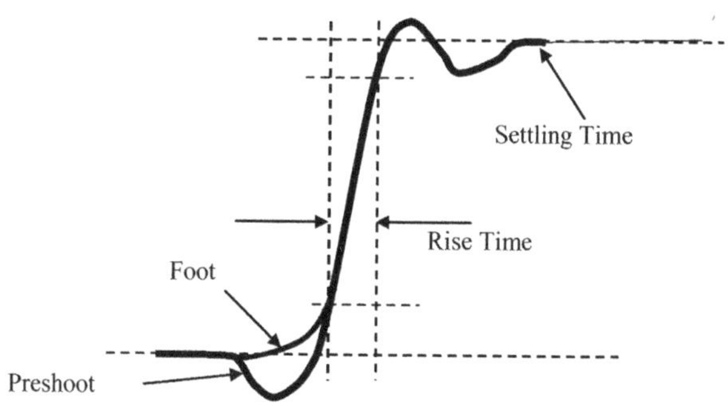

Figure 11.4 Illustration of aberration phenomena

indicates a higher-impedance event (e.g. open or reduction in line width). A dip indicates a lower-impedance event (e.g. short or increase in line width).

Example 11.2: Simulation of TDR Measurements Assuming a Via as the DUT
The same PCB structure with a via, which will be considered in *Example 11.4*, is used to illustrate the echo technique. The electrical parameters of Figure 11.5 are: $R_{out} = Z_0 = 50\ \Omega$ (source impedance), $Z_{0,cab} = 50\ \Omega$, $T_{D,cab} = 150$ ps, $Z_{0,con} = 50\ \Omega$, $T_{D,con} = \sqrt{L_{con} C_{con}} = 22$ ps ('con' = connector), $Z_{0,t} = 36.3\ \Omega$, $T_{D,t} = 80$ ps ('t' = trace), $L_{via} = 0.776$ nH, $C_{via} = 0.472$ pF.

Two TDR waveforms are compared in Figure 11.5. The solid line was obtained under the assumption of an ideal connector between the coaxial cable and the DUT, simulated with a lossless TL having a characteristic impedance of 50 Ω. The dotted-line waveform is obtained by simulating a non-ideal connector with a loop inductance $L_{con} = 1$ nH and a capacitance $C_{con} = 0.5$ pF for an impedance of 44.7 Ω.

Note that a non-ideal connector introduces 'aberration', and the via effect is significantly different from the case of an ideal connector. The choice of a suitable probe connection between the TDR and the DUT is therefore very important. Any time the probe is not exactly 50 Ω, an impedance discontinuity will occur in the set-up. This discontinuity will cause reflections at the probe, decreasing the measurement accuracy. This is the main factor affecting the accuracy. Hand-held probes should be avoided and used for a quick check only. The error is due to the ground connection mechanism of the probe, which introduces loop inductance. SMA connectors soldered to the board ensure good repeatability. The drawback is due to its capacitance, which slows down the edge rate of the TDR and reduces the resolution. Controlled 50 Ω impedance microprobes are the most accurate probe type owing to their very low parasitic parameters such as capacitance and loop inductance.

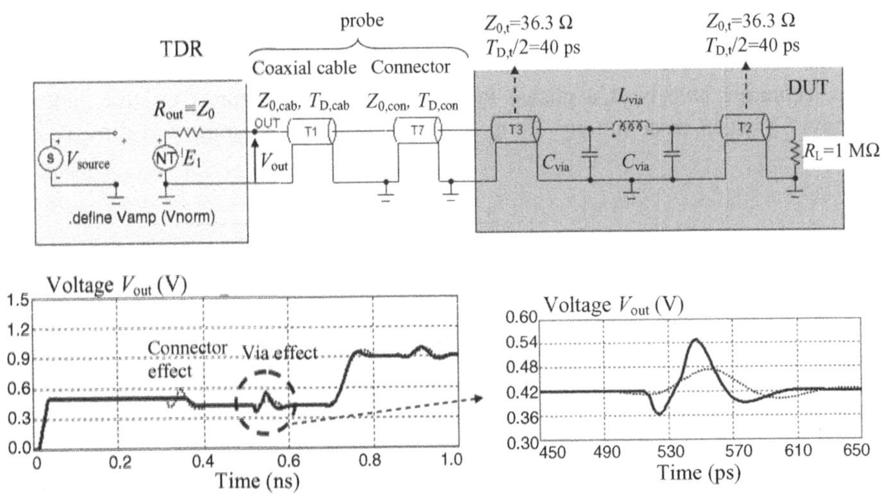

Figure 11.5 Equivalent circuit and simulated waveforms of a coaxial connector and via effects along a trace by the TDR: ideal coaxial cable connector (solid line); actual coaxial connector (dotted line)

11.1.3 TDR and Lossy Lines

TDR measurements can also be useful for characterizing lossy lines. As an example, the measured TDR waveforms for two different trace structures are shown in Figure 11.6. The first has the structure of a microstrip, and the second the structure of a stripline. Both traces have the same length $l = 70$ cm, the same strip width $w = 0.16$ mm, and the same nominal characteristic impedance $Z_{0,t} = 50\ \Omega$. From Figure 11.6c it can be seen that the microstrip has losses greater than those occurring in striplines.

TDR measurements on traces can be very useful for determining:

- the coefficient K_p, defined in *Section 7.1.3*, related to the proximity effect between the trace and its return ground plane (for microstrip) and ground planes (for stripline) respectively;
- the coefficient θ_0, defined in *Section 7.1.4*, related to the dielectric loss tangent provided at frequency f_0.

The coefficients K_p and θ_0 for lossy lines can be determined by fitting the waveforms obtained by measurements with the simulated waveforms obtained as the step response of a lossy line

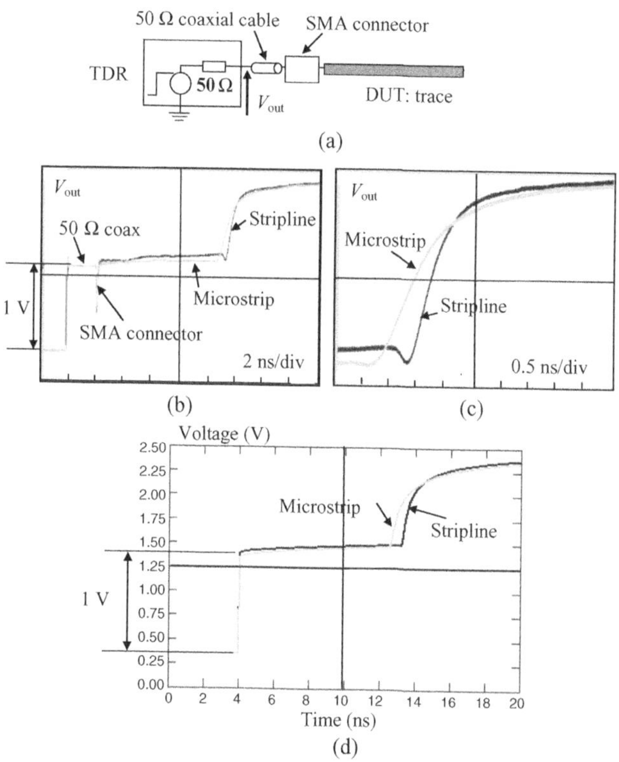

Figure 11.6 Microstrip and stripline of length $l = 70$ cm, width $w = 0.160$ mm, and nominal characteristic impedance $Z_0 = 50\ \Omega$: (a) TDR measurement set-up with open line; (b) measured output voltage normalized to 1 V; (c) detail of the rise times for both traces; (d) simulated waveforms in the absence of a coaxial cable and connector (courtesy of Dr Vittorio Ricchiuti, Technolabs, Italy)

by using the equivalent circuit presented in *Section 7.1.5.2* and shown in Figure 7.14b. The closed-form expressions of the model make it possible to perform calculations in the frequency domain, and the waveforms in the time domain are obtained by the inverse Fourier transform (IFT). The equivalent circuit must simulate the set-up in Figure 11.6a. In a first approximation, the coaxial cable and SMA connector are assumed to have losses much lower than those occurring in the trace, so that only the TDR and the trace are simulated. The initial point for loss evaluation starts at SMA connector level in Figures 11.6b and 11.6c because the 50 Ω coaxial cable used for measurements can be practically considered as a lossless line. The computed waveforms are shown in Figure 11.6d and were obtained using the following parameters:

- Microstrip: $w = 0.160$ mm (trace width), $t_w = 35$ μm (trace thickness), $l = 70$ cm (trace length), $Z_0 = 50\ \Omega$ (characteristic impedance), $t_{pd} = 6$ ns/m (per-unit-length propagation delay time), $K_p = 5$ (proximity-effect coefficient), $\delta_0 = 0.0086$ (loss tangent).
- Stripline: $w = 0.160$ mm (trace width), $t_w = 18$ μm (trace thickness), $l = 70$ cm (trace length), $Z_0 = 55\ \Omega$ (measured characteristic impedance), $t_{pd} = 6.5$ ns/m (per-unit-length propagation delay time), $K_p = 4$ (proximity-effect coefficient), $\delta_0 = 0.0086$ (loss tangent).

Good agreement with the measurements can be observed. Note that the nominal impedance of the stripline was assumed to be 55 Ω according to the TDR measured data, and the propagation delay of the stripline was set higher than that of the microstrip because the trace in the stripline structure is completely surrounded by the dielectric and therefore the signal propagates with less speed.

11.1.4 Differential TDR

The differential TDR is very useful for measuring even-mode Z_{0e} and odd-mode Z_{0o} characteristic impedances in coupled lines used for differential transmission. Even and odd propagation modes were defined in *Section 6.2*. According to the definition, the TDR measurement of these two impedances should be performed as follows:

- Z_{0o} by using two complementary outputs of the instrument and observing one line while the other line is driven by the other output;
- Z_{0e} by using two equal outputs and observing one line while the other line is driven by the other output.

Once these impedances are known, with their associated delays, the equivalent circuit based on even and odd modes of *Section 6.2.2* can be used for signal integrity and EMI predictions. Recall that the *differential-mode* characteristic impedance is $Z_{0DM} = 2Z_{0o}$, and the *common-mode* impedance is $Z_{0CM} = Z_{0e}$. This last impedance can also be measured by using only one output of the TDR and connecting the two lines at both ends. In this case the TDR measures the impedance $Z_{0e}/2$. With a differential TDR it is possible to measure losses for *differential-* and *common-mode* propagation, and the model based on even and odd modes can be extended to lossy lines, as discussed in *Section 7.2.2*. For other information and examples, see reference [6]. An example of using differential TDR for signal integrity and crosstalk characterization is provided by *Example 12.3* in *Chapter* 12.

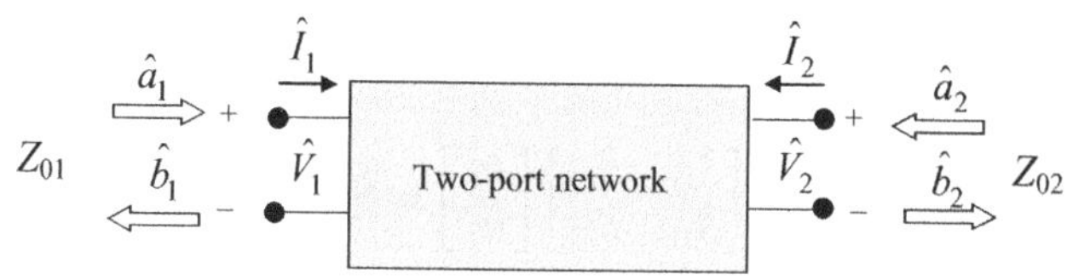

Figure 11.7 Incident and reflected waves of a two-port network

11.2 Vector Network Analyzer (VNA)

Network analyzers are instruments used to analyze the properties of electrical networks, especially those properties associated with the reflection and transmission of electrical signals known as scattering parameters (S-parameters) [7, 8]. Network analyzers are used mostly at high frequencies and can operate in the range from 9 kHz to 100 GHz. In the microwave region it is more convenient and direct to analyze the properties of a two-port network in terms of waves. The two independent quantities required for each waveguide terminal are incident and reflected waves replacing the voltage and current, as shown in Figure 11.7. The two main categories of network analyzers are the *Scalar Network Analyzer* (SNA) and the *Vector Network Analyzer* (VNA). The SNA measures amplitude properties only, while the VNA, which is most used, measures both amplitude and phase properties. An SNA is functionally identical to a spectrum analyzer in combination with a tracking generator.

Examples of using S-parameters for circuit parameter extraction in *Signal Integrity* are provided in this section. Other examples involving S-parameters were proposed in *Section 7.2* for lossy-line characterization, and are proposed in *Appendix C* for PCB characterization as a resonance cavity.

11.2.1 Scattering Parameter Definition

Suppose that incident and reflected voltage waves at port 1 are given in magnitude and phase by $\hat{V}_{1+}$ and $\hat{V}_{1-}$ respectively. Similarly, looking at port 2, incident and reflected waves are $\hat{V}_{2+}$ and $\hat{V}_{2-}$. It is common to normalize incident and reflected waves as follows:

$$\hat{a}_n = \frac{\hat{V}_{n+}}{\sqrt{Z_{0n}}} \tag{11.2a}$$

$$\hat{b}_n = \frac{\hat{V}_{n-}}{\sqrt{Z_{0n}}} \tag{11.2b}$$

where $\hat{a}_n$ is the normalized incident wave, $\hat{b}_n$ is the normalized reflected wave and $n = 1, 2$. The relationships of these parameters with the voltages and currents at the two ports are:

$$\hat{V}_n = (\hat{V}_{n+} + \hat{V}_{n-}) = \sqrt{Z_{0n}}(\hat{a}_n + \hat{b}_n) \tag{11.3a}$$

$$\hat{I}_n = \frac{1}{Z_{0n}}(\hat{V}_{n+} - \hat{V}_{n-}) = \frac{1}{\sqrt{Z_{0n}}}(\hat{a}_n - \hat{b}_n) \tag{11.3b}$$

S-parameters are defined as

$$\begin{bmatrix} \hat{b}_1 \\ \hat{b}_2 \end{bmatrix} = \begin{bmatrix} \hat{S}_{11} & \hat{S}_{12} \\ \hat{S}_{21} & \hat{S}_{22} \end{bmatrix} \begin{bmatrix} \hat{a}_1 \\ \hat{a}_2 \end{bmatrix} = \hat{\mathbf{S}} \begin{bmatrix} \hat{a}_1 \\ \hat{a}_2 \end{bmatrix} \tag{11.4}$$

The advantages of using S-parameters are:

- VNAs are commercially available and cover the range from 300 kHz to 100 GHz.
- General-purpose electromagnetic simulators can usually produce their results in the form of S-parameters.
- S-parameters can easily be converted into other multiport network representation such as Z, Y and ABCD parameters [9].

If the Z-parameter representation

$$\begin{bmatrix} \hat{V}_1 \\ \hat{V}_2 \end{bmatrix} = \begin{bmatrix} \hat{Z}_{11} & \hat{Z}_{12} \\ \hat{Z}_{21} & \hat{Z}_{22} \end{bmatrix} \begin{bmatrix} \hat{I}_1 \\ \hat{I}_2 \end{bmatrix} = \hat{\mathbf{Z}} \begin{bmatrix} \hat{I}_1 \\ \hat{I}_2 \end{bmatrix} \tag{11.5}$$

is considered, it can be shown that the impedance matrix $\hat{\mathbf{Z}}$ is related to the scattering matrix $\hat{\mathbf{S}}$ by

$$\hat{\mathbf{Z}} = Z_{\text{ref}}(\mathbf{I} + \hat{\mathbf{S}})(\mathbf{I} - \hat{\mathbf{S}})^{-1} \tag{11.6}$$

where $\mathbf{I}$ is the identity matrix and Z_{ref} is the reference impedance (usually the nominal characteristic impedance Z_0 of the transmission line driving the ports). An application of Equation (11.6) to build an equivalent circuit for a lossy twisted-pair cable by the *Vector Fitting* (VF) technique was illustrated in *Section 7.2*. Another frequent application is the characterization of multilayer boards as resonant cavities, as done in *Appendix* **C**, where the impedance values are obtained from the S-parameters measured by VNA or computed by numerical simulations.

The physical interpretation of the S-parameters is illustrated in Figure 11.8. It can be noted that $\hat{S}_{11}$ has the meaning of a reflection coefficient at port 1, and $\hat{S}_{21}$ has the meaning of a transfer function when $a_2 = 0$, or port 2 is matched (no reflections), according to the following

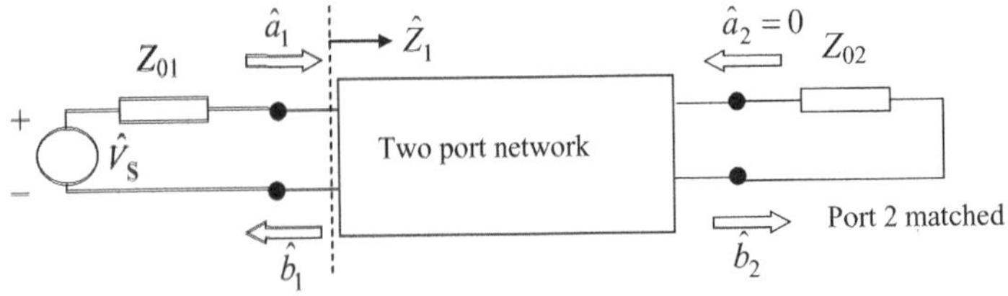

Figure 11.8 Physical interpretation of S-parameters

equations:

$$\hat{S}_{11} = \left.\frac{\hat{b}_1}{\hat{a}_1}\right|_{\hat{a}_2=0} \tag{11.7a}$$

$$\hat{S}_{21} = \left.\frac{\hat{b}_2}{\hat{a}_1}\right|_{\hat{a}_2=0} \tag{11.7b}$$

From Equations (11.3), the incident and reflected waves are given by

$$\hat{a}_1 = \frac{\hat{V}_1 + \hat{I}_1 Z_{01}}{2\sqrt{Z_{01}}} \tag{11.8a}$$

$$\hat{a}_2 = \frac{\hat{V}_2 + \hat{I}_2 Z_{02}}{2\sqrt{Z_{02}}} \tag{11.8b}$$

$$\hat{b}_1 = \frac{\hat{V}_1 - \hat{I}_1 Z_{01}}{2\sqrt{Z_{01}}} \tag{11.8c}$$

$$\hat{b}_2 = \frac{\hat{V}_2 - \hat{I}_2 Z_{02}}{2\sqrt{Z_{02}}} \tag{11.8d}$$

Introducing Equations (11.8) into Equations (11.7) yields

$$\hat{S}_{11} = \frac{\hat{V}_1 - Z_{01}\hat{I}_1}{\hat{V}_1 + Z_{01}\hat{I}_1} = \frac{\hat{V}_1/\hat{I}_1 - Z_{01}}{\hat{V}_1/\hat{I}_1 + Z_{01}} = \frac{\hat{Z}_1 - Z_{01}}{\hat{Z}_1 + Z_{01}} = \rho_1 \tag{11.9}$$

Therefore, $\hat{S}_{11}$ has the meaning of a reflection coefficient at port 1 according to the definition given in *Section 5.2*. Similar operations for $\hat{S}_{21}$ yield

$$\hat{S}_{21} = \frac{\hat{V}_2 - Z_{02}\hat{I}_2}{\hat{V}_1 + Z_{01}\hat{I}_1} = 2\frac{\hat{V}_2}{\hat{V}_S} \tag{11.10}$$

and it is shown that $\hat{S}_{21}$ has the meaning of a transfer function.

When performing measurements by VNA, usually $Z_{01} = Z_{02} = Z_0 = 50\ \Omega$. In fact, one port of the VNA having an output impedance Z_0 is used to source the two-port network at port 1 by a coaxial cable of characteristic impedance Z_0, while the other VNA port with input impedance Z_0 is connected to port 2 by a coaxial cable of characteristic impedance Z_0. Thus, the S-parameter measurement provides directly the reflection coefficient at port 1 and the transfer function of the two-port network loaded at both ends by the reference impedance Z_0. VNA measurements are widely used to characterize lossy lines, to determine the power distribution impedance in multilayer PCBs seen by a switching digital device, and to investigate how the noise produced by the device propagates along the PCB between the point where the switching device is placed and another chosen point on the PCB. Examples can be found in *Section 7.2* and *Section 10.4*.

11.2.2 VNA Calibration

At this point, a brief overview of the utilization of the VNA will be given. The reader wishing to have more information is invited to consult references [10–12]. This is useful for introducing some parasitic effect of the instrument, which must be taken into account when dealing with modeling of the VNA set-up as illustrated immediately below.

When using VNA, two operations are necessary: *S*-parameter calibration and error correction.

Calibration is important to determine the systematic artifacts of the measurement system by measuring a number of known standards such as thru lines, transmission lines, loads, shorts, and opens. Each standard has its associated calibration algorithm. The error model consists of error boxes into which all the non-idealities of the instrument can be lumped. These non-idealities concern the network analyzer, test set, cables, adapters, fixtures, etc. The parameters corresponding to these boxes, found during calibration, can be mathematically removed.

The purpose of the calibration is to quantify each systematic error term through measurement of standards. There are two types of calibration:

- coaxial environment: Short-Open-Load-Thru (SOLT) calibration.
- non-coaxial environment: Thru-Reflect-Line (TRL) calibration.

(i) SOLT Calibration

SOLT calibration is the most common calibration selection, and usually works well for coaxial systems. It removes the effects of connectors and coaxial cables connecting the VNA to the DUT, shifting the reference plane to the tips of the coaxial cables, see Figure 11.9. The systematic errors are determined from the difference between the measured and known responses of the standards. In a conventional SOLT full two-port calibration, three known coaxial impedance standards (short, open, load) and a single coaxial transmission standard (thru) are required. These standards are provided by the VNA manufacturer. Once characterized, the error can be mathematically related by solving a signal flow graph. The 12-term error model includes all the significant systematic effects for the measurement of a two-port device.

The reference plane is the position (relative to each port) where 0 phase is defined. Reference planes coincide with the calibration planes at the ends of the measurement probes. Sometimes there may be a test fixture between the coaxial calibration planes and the DUT. To remove the effects of this fixture, TRL calibration can be used [12]. This is the process of mathematically subtracting networks from the measured results (see Figure 11.9). If one can treat the fixture as a simple 50 Ω transmission line, the problem reduces to that of shifting the reference plane: *port extension*. This assumes that launch parasitics are small enough and, usually, that the frequencies involved are not too high: a frequency-dependent phase shift is all that is required.

Shorts (or opens if the capacitance is small enough) can be placed at the actual DUT plane, and an 'auto reference plane' extension function can be used to calculate the distance required to put the reference plane exactly where the artifact was placed.

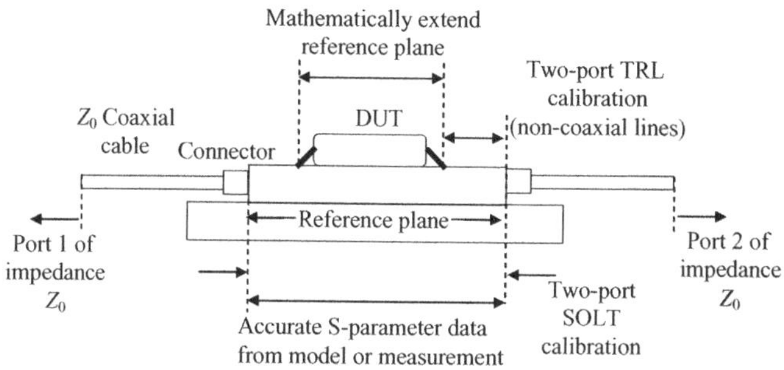

Figure 11.9 Illustration of calibration and port extension operations

The fixture in the port extension procedure is considered to be an ideal matched transmission line (no mismatch, no attenuation) that generates only a phase shift (time delay).

(ii) TRL Calibration

TRL calibration is used for network measurements in non-coaxial media [12]. The problem is how to separate the effects of the transmission medium (in which the device is embedded for testing) from the device characteristics see Figure 11.9. Consider that SOLT calibration standards can be difficult – if not impossible – to build in many non-coaxial measurement applications. The advantages of TRL calibration are: the calibration kit is easy to build; the TRL method is easy to apply in dispersive media such as microstrip, stripline, and waveguide structures; it achieves the highest accuracy. The main disadvantages of TRL calibration consist in a limited field of application, as it can be used for two-port calibration only and it cannot work at low frequencies because the length of the lines would be too long. Moreover, a dedicated test board has to be built.

TRL refers to the following three basic steps in the calibration process:

1. THRU – connect port 1 and port 2 directly or with a short length of transmission line (the same impedance as that feeding the DUT).
2. REFLECT – connect identical one-port high reflection coefficient devices at each port.
3. LINE – insert a short length of transmission line between port 1 and port 2 (different line lengths are required for THRU and LINE).

11.2.3 Extraction of Equivalent Circuits by S-Parameter Simulations

Network analyzer is very useful in extracting components and line parameters, as *S*-parameters can be compared with those obtained by numerical tools such as *MicroWave Studio* (MWS) based on the *Finite Integration Technique* (FIT) [13]. Because in this book many examples are provided to solve signal integrity and radiated emission problems by using MWS, some important aspects will be outlined in using this code to extract circuit parameters. The information given may also be useful when codes based on similar techniques such as FDTD are used. The source used by MWS to calculate *S*-parameters is shown in Figure 11.10 (left). It is a current source of impedance Z_0 that causes a dissipation of 1 W on a load of impedance Z_0. The equivalent SPICE model in terms of voltage source is shown in

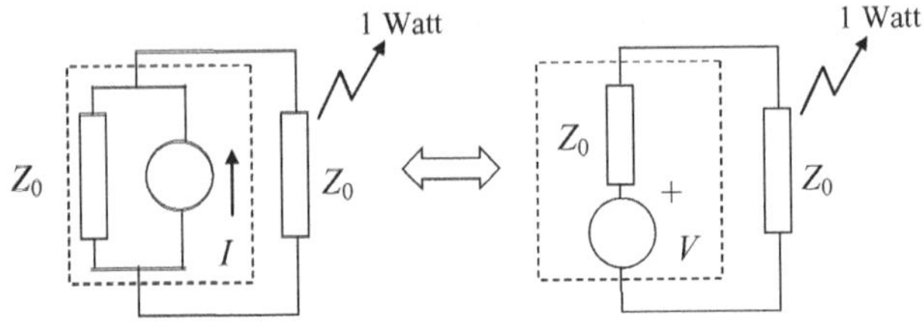

Figure 11.10 Equivalent circuits used by MWS (left) and SPICE (right) when the computation of S-parameters is performed

Figure 11.10 (right). According to the MWS definition of source, the ideal current and voltage sources of Figure 11.10 assume the magnitude values $I = 2/\sqrt{Z_0}$ and $V = 2\sqrt{Z_0}$ respectively.

The SPICE model suitable for simulating two-port network measurements is shown in Figure 11.11, where V_{in}, V_{out}, I_{in}, and I_{out} are voltage and current sources set to zero, which are used as probes for currents and voltages at port 1 and port 2. When these voltages and currents are solved by the circuit simulator as $V_1 = V(I_{\text{in}})$, $V_2 = V(I_{\text{out}})$, $I_1 = I(V_{\text{in}})$, and $I_2 = I(V_{\text{out}})$, S-parameters can be calculated by using Equations (11.7) and (11.8). In the following, this circuit model will be used for some practical cases.

When the DUT is a transmission line of length l with known p.u.l. parameters as a function of frequency, the model of Figure 11.12 can be used, where the dependent voltage sources in the dashed rectangle are used to reproduce Equation (11.5). For a uniform transmission line $\hat{Z}_{11} = \hat{Z}_{22}$ and $\hat{Z}_{12} = \hat{Z}_{21}$, and these frequency-dependent impedances (see Figure 7.1) are given by [14]

$$\hat{Z}_{11}(\omega) = -\sqrt{\frac{\hat{Z}_{\text{puls}}(\omega)}{\hat{Y}_{\text{puls}}(\omega)}}\,\frac{\cosh\left(\sqrt{\hat{Z}_{\text{puls}}(\omega)\hat{Y}_{\text{puls}}(\omega)}\,l\right)}{\sinh\left(\sqrt{\hat{Z}_{\text{puls}}(\omega)\hat{Y}_{\text{puls}}(\omega)}\,l\right)} \tag{11.11a}$$

$$\hat{Z}_{12}(\omega) = \sqrt{\frac{\hat{Z}_{\text{puls}}(\omega)}{\hat{Y}_{\text{puls}}(\omega)}}\,\frac{1}{\sinh\left(\sqrt{\hat{Z}_{\text{puls}}(\omega)\hat{Y}_{\text{puls}}(\omega)}\,l\right)} \tag{11.11b}$$

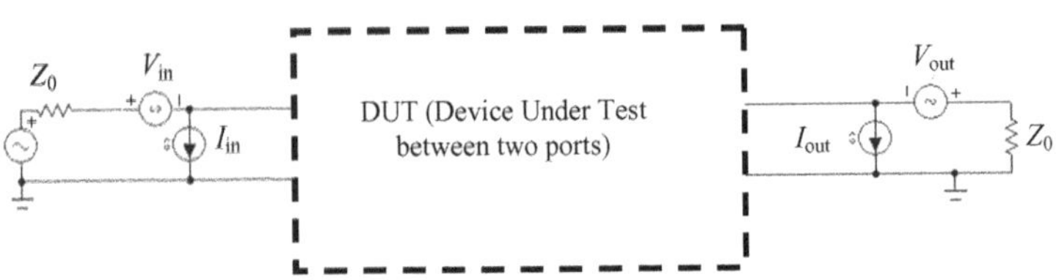

Figure 11.11 Equivalent circuit for S-parameter calculation in the frequency domain

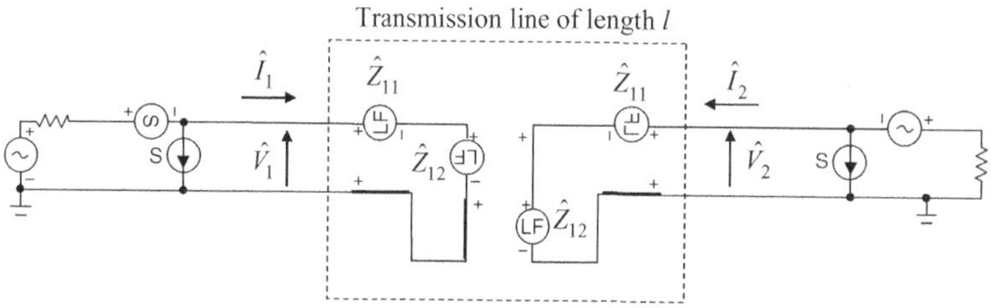

Figure 11.12 Equivalent circuit for S-parameter calculation in the frequency domain when the DUT is a transmission line with p.u.l. parameters described by known expressions

where

$$\hat{Z}_{\text{puls}}(\omega) = \hat{Z}_{\text{i}}(\omega) + \text{j}\omega L_0 \quad (11.12\text{a})$$

$$\hat{Y}_{\text{puls}}(\omega) = 1/R_{\text{d}}(\omega) + \text{j}\omega C_0 \quad (11.12\text{b})$$

An application of this circuit model to calculate S-parameters of a coaxial cable can be found in *Section 7.2*. In this case $\hat{Z}_{\text{i}}(\omega)$ is the sum of the inner wire impedance and of the shield internal impedance; $1/R_{\text{d}}$ is neglected, and L_0 and C_0 are the nominal external inductance of the wire–shield structure and the nominal capacitance between the inner wire and the shield respectively.

MWS performs the calculation in the time domain, and frequency-domain S-parameters are obtained by FFT. The excitation signal used by MWS is a Gaussian pulse narrow enough to have a flat spectrum up to the maximum frequency of interest. For the purpose of comparing time-domain MWS with SPICE results, the SPICE circuit can be used to perform transient analysis. In this case, the circuit is excited with the same Gaussian pulse used in MWS simulation by means of the circuit shown in Figure 11.13. The Gaussian waveform is assigned in table form to V_{source}. V_{s} is a dependent voltage source that raises the amplitude of the Gaussian pulse at value $2\sqrt{Z_0}$ according to Figure 11.10 in order to compute the actual voltages and currents of the structure.

Example 11.3: Transmission-line Parameter Extraction of a Simple Connector
Consider the simple configuration shown in Figure 11.14, consisting of:

- Two waveguide ports or two discrete ports for excitation.
- Two equal coaxial cables with the task of connecting the sources to the DUT.

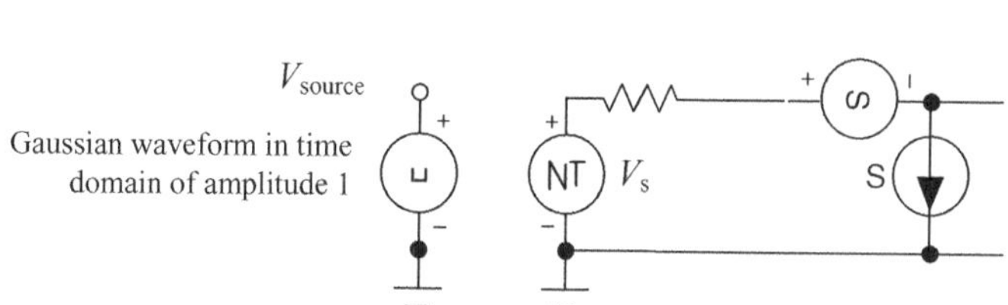

Figure 11.13 Equivalent circuit of the source for time-domain simulation

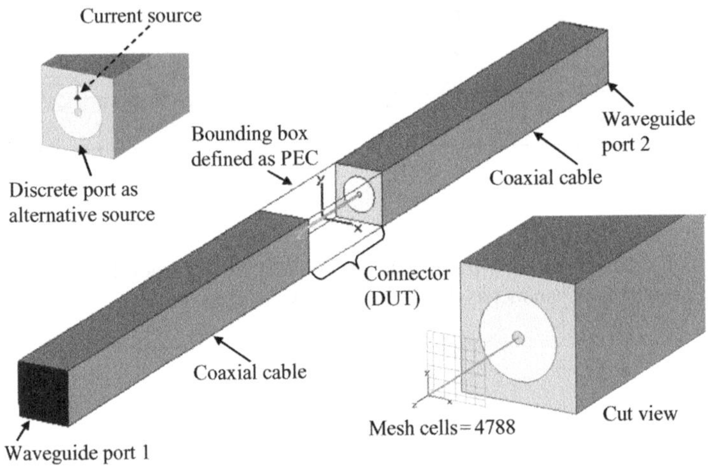

Figure 11.14 Two coaxial cables and a connector drawn by MWS

- The DUT is a connector represented by a thin wire connecting the inner conductor of the two coaxial cables and having as the return path for the current the relative bounding box declared as a *Perfect Electric Conductor* (PEC).

The waveguide port is an ideal port used to excite the structure and generates electromagnetic modes that match perfectly with the coaxial cable characteristic impedance. The discrete port, often used for practical reasons, has the same excitation function, but it has an important drawback that should be considered in interpreting *S*-parameter data: it has an associated parasitic inductance element that must be taken into account in extracting the DUT circuit parameters. The purpose of this example is to show that:

- Parasitic elements associated with discrete ports can affect the results.
- Normalization of *S*-parameters to an impedance different from that of the coaxial cable characteristic impedance can make difficult the extraction of circuit parameters of the DUT.

As the lines are considered lossless, the coaxial cable consists of a cylindrical solid object defined as a PEC, and the shield is formed by PEC material filling the space between the internal shield and the bounding in order to limit the zone of computation. The dimension of the coaxial cable is characterized by the parameters: radius of the internal wire $r_c = 1$ mm, radius of the internal shield $r_{si} = 6.66$ mm, relative dielectric constant of the cable $\varepsilon_r = 2.3$, cable length $l_{coax} = 17$ cm. With these parameters, the theoretical characteristic impedance may be analytically computed as

$$Z_{0,coax} = \frac{60}{\sqrt{\varepsilon_r}} \ln(r_{si}/r_c) = 75\ \Omega \tag{11.13}$$

This should be the nominal characteristic impedance. The actual impedance is computed by MWS and the result is 72 Ω. This is the value that should be assigned to the ports and coaxial

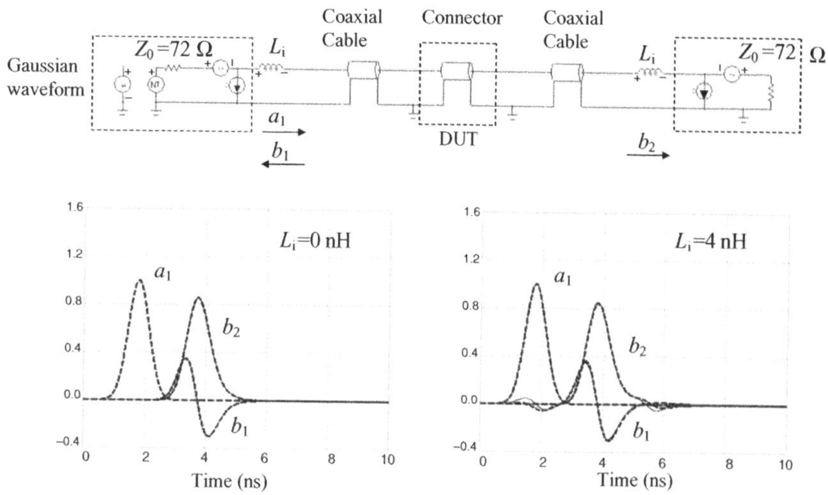

Figure 11.15 Equivalent circuit and computed waveforms in the time domain for the port–coaxial–DUT–coaxial–port structure: SPICE (solid line); MWS (dashed line)

cables when performing the simulations by SPICE for the purpose of establishing a consistent comparison between SPICE and MWS results. The delay time of the coaxial cable can be analytically calculated as

$$T_{\mathrm{D,coax}} = \sqrt{\varepsilon_0\varepsilon_r\mu_0} l_{\mathrm{coax}} = 860\,\mathrm{ps} \tag{11.14}$$

The goal of this example is to extract the characteristic impedance and the delay time of the connector represented as a transmission line. The delay time of the connector can be calculated considering its length, $l_{\mathrm{con}} = 55$ mm, and the fact that the wire is surrounded by air. Therefore, the connector delay time is $T_{\mathrm{D,con}} = \sqrt{\varepsilon_0\mu_0} l_{\mathrm{con}} = 183.5$ ps. The value of the characteristic impedance may be calculated by fitting the results obtained by SPICE with those given by MWS.

The result of this procedure is shown in Figure 11.15, where the incident and reflected waveforms in the time domain, computed by MWS (dashed line), are compared with the SPICE simulations (solid line). The Gaussian waveform assigned in table form to the source is the same as that determined by MWS to have a flat spectrum in the frequency range 0–1000 MHz. The waveforms in the left-hand figure are those obtained by MWS using waveguide ports. In this case the parasitic inductance L_i=0, and the perfect superimposition of the waveforms provided by SPICE is obtained when the characteristic impedance of the connector is $Z_{0,\mathrm{con}} = 230\ \Omega$. Looking at the MWS waveforms in the right-hand plot obtained by using discrete ports, some slight distortion can be observed on the reflected waveforms b_1 and b_2. This distortions can be approximately reproduced by SPICE, assuming that $L_i = 4$ nH. This is consistent with the self partial inductance equation for a thin wire given in *Appendix A* as

$$L_{\mathrm{pw}} = \frac{\mu_0}{2\pi} l_{\mathrm{w}} \left[\ln\left(\frac{2l_{\mathrm{w}}}{r_{\mathrm{w}}}\right) - 1\right] \tag{11.15}$$

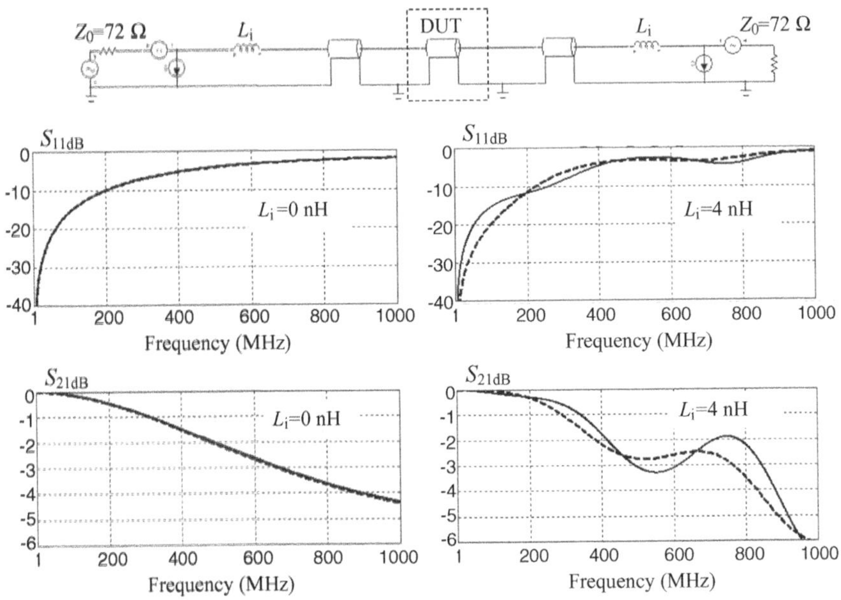

Figure 11.16 Equivalent circuit and simulated S-parameters in the frequency domain for the DUT structure coaxial–connector–coaxial: SPICE (solid line); MWS (dashed line)

By representing the current source in MWS with a wire of radius $r_{\mathrm{w}} = 0.1$ mm and length $l_{\mathrm{w}} = r_{\mathrm{si}} - r_{\mathrm{c}} = 5.66$ mm, the self partial inductance (11.15) is $L_{\mathrm{pw}} = 4.2$ nH.

The distortion produced by L_{i} is more evident if MWS computations and SPICE simulations of S-parameters are performed in the frequency domain, as shown in Figure 11.16. If the analysis is extended up to 6 GHz, the dangerous effect of the parasitic element associated with the discrete ports becomes more evident, as shown in Figure 11.17.

The results of normalizing S-parameters to a fixed impedance different from the characteristic impedance of the coaxial cables connecting the ports with the DUT are shown in Figure 11.18. In this case it becomes difficult to distinguish reflections owing to the mismatch between port impedance and coaxial cable from those produced by the DUT. This is the case, for example, when measurements by VNA are carried out with coaxial cables with different characteristic impedances with respect to the port impedances of the instruments.

Example 11.4: Lumped-circuit Element Extraction of a Simple Via

This example describes how to extract lumped-circuit elements of a simple via from numerical computation. The PCB under investigation has the dimensions 20 × 20 mm and is modeled by MWS as shown in Figure 11.19. The microstrip has width $w_{\mathrm{ms}} = 3.6$ mm, thickness $t_{\mathrm{ms}} =$ 0.03 mm, height $h_{\mathrm{ms}} = 1.6$ mm, and length $l_{\mathrm{ms}} = 10.8$ mm. Considering that the relative dielectric constant of the substrate is $\varepsilon_{\mathrm{r}} = 9$, the nominal characteristic impedance of the microstrip can be analytically computed as (see *Appendix B*)

$$Z_{0,\mathrm{ms}} = \frac{60}{\sqrt{\varepsilon_{\mathrm{re}}}} \ln\left(\frac{5.98 h_{\mathrm{ms}}}{0.8 w_{\mathrm{ms}} + t_{\mathrm{ms}}}\right) = 32.1\,\Omega \tag{11.16}$$

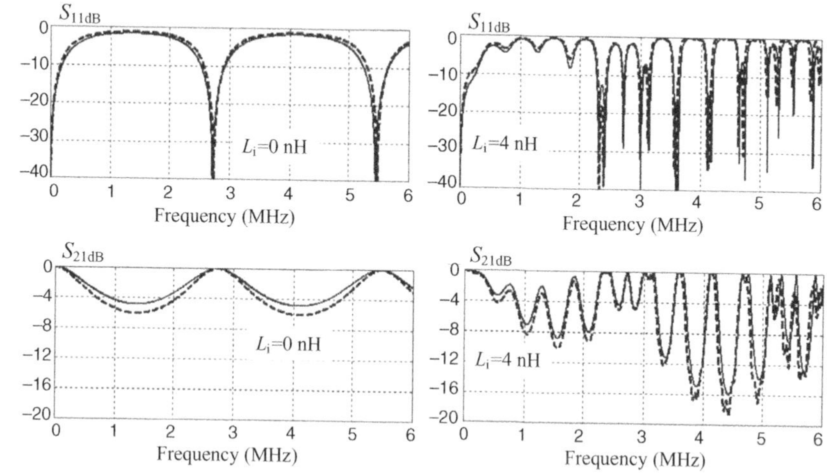

Figure 11.17 Equivalent circuit and simulated *S*-parameters in the frequency domain for the DUT structure coaxial–connector–coaxial when the analysis is extended to 6 GHz: SPICE (solid line); MWS (dashed line)

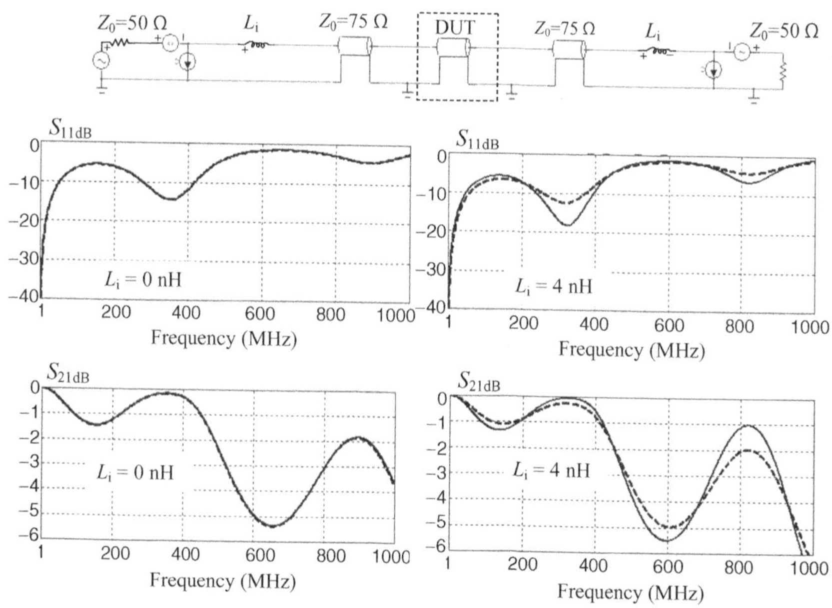

Figure 11.18 Equivalent circuit and simulated *S*-parameters in the frequency domain for the DUT structure coaxial–connector–coaxial when the *S*-parameters are normalized to a 50 Ω impedance: SPICE (solid line); MWS (dashed line)

where ε_{re} is the effective dielectric constant given by $\varepsilon_{re} = 0.475\varepsilon_r + 0.67$. The microstrip delay time is calculated as

$$T_{D,ms} = \sqrt{\mu_0 \varepsilon_0 \varepsilon_{re}} l_{ms} = 80.1\,\text{ps} \tag{11.17}$$

The via may be approximately simulated by a Π-type model. The via has a barrel radius $r_{via} = 0.7$ mm, a barrel length $l_{via} = 2h_{ms}$, and a pad radius $r_{pad} = 1.8$ mm. With these values, the self partial inductance (see *Appendix A*) and the capacitance [15] of the via are estimated as

$$L_{pvia} = \frac{\mu_0}{2\pi} l_{via} \left[\ln\left(\frac{2l_{via}}{r_{via}} \right) - 1 \right] = 0.776\,\text{nH} \tag{11.18}$$

$$C_{via} = \frac{1.41 \times 10^{-12} \varepsilon_r (39.37 h_{ms}) r_{via}}{r_{pad} - r_{via}} = 0.509\,\text{pF} \tag{11.19}$$

where $C_{via} = 2\,C_{pad}$ is the pad capacitance. In the simulations it was used $C_{via} = 0.472$ pF. In this case, also, the results obtained by waveguide and discrete ports are compared. The waveguide ports were applied as shown in Figure 11.19, and the bonding box for computation was defined as perfect magnetic material ($H_t = 0$). The reason why it is possible to use the option $H_t = 0$ is explained in *Appendix C*. The two discrete ports are located in the middle of the microstrip and connect the line to the ground plane. The MWS code has provided a characteristic impedance of the microstrip of 36.2 Ω, and this is the value that was used in the SPICE simulation instead of the nominal value of 32.1 Ω calculated analytically by

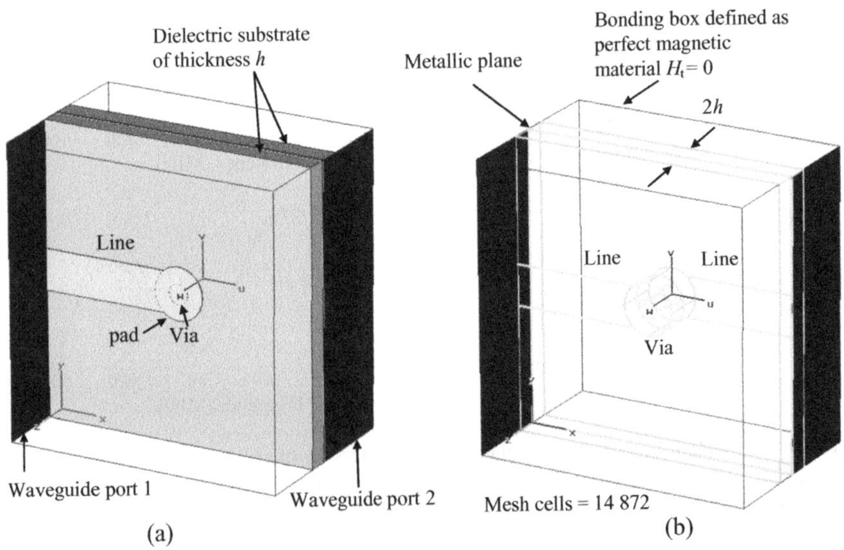

Figure 11.19 PCB with a via: (a) PCB structure; (b) the same PCB with a view of the via in detail

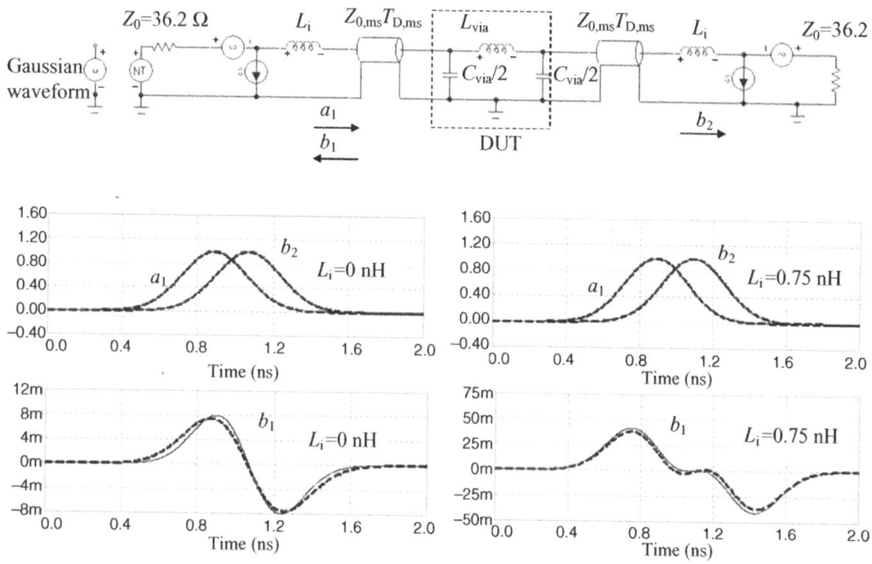

Figure 11.20 Equivalent circuit and computed waveforms in the time domain for the port–microstrip–DUT–microstrip–port structure: SPICE (solid line); MWS (dashed line)

Equation (11.16). The partial inductance associated with each discrete port is estimated by analogy with the previous example by using Equation (11.15). Considering that the port has a length of 1.52 mm and a wire radius of 0.1 mm, it is estimated that $L_{\text{pport}} = 0.734$ nH. Assigning $L_i = 0.75$ nH to the ports in the SPICE simulations, the waveforms shown in Figure 11.20 are obtained. The Gaussian waveform was defined in MWS to have a flat spectrum in the frequency range 0–2 GHz. Observe that, in this example also, the parasitic inductance associated with the discrete ports introduces a distortion in the reflected wave b_1. The results reported in Figure 11.20 show that the Π-type model works very well in the frequency range considered. This is also confirmed by the S-parameters in the frequency domain, as shown in Figure 11.21. A very good agreement can be observed between S-parameters computed by MWS and the simulated parameters produced by SPICE.

If the range of analysis is extended up to 6 GHz, the Π-type model is not perfectly suitable anymore, as significant differences can be observed between MWS and SPICE in the b_1 reflected waveform with both excitations, as shown in Figure 11.22. The Π-type model keeps its validity because the incident waveform a_1, which becomes the transmitted waveform b_2, arrives at the load practically unchanged. Consider also that the length of the via is still electrically short, as, at 6 GHz, $\lambda/10 = 300/(6000 \times 10) = 5$ mm $> l_{\text{via}} = 3.2$ mm.

11.2.4 Conclusions Concerning VNA Measurements and Simulations

To conclude this section, the following observations can be made:

- The VNA instrument performs measurements of S-parameters of two-port networks, making an automatic correction for the variation introduced by the parasitic circuit element associated with the test fixture by techniques such as calibration and de-embedding procedures.

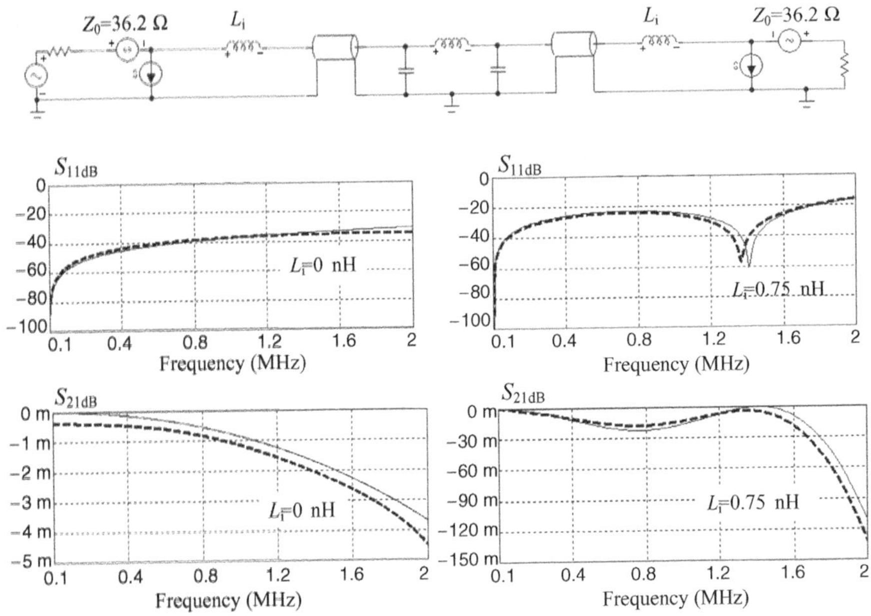

Figure 11.21 Equivalent circuit and computed *S*-parameters in the frequency domain for the port–microstrip–DUT–microstrip–port structure: SPICE (solid line); MWS (dashed line)

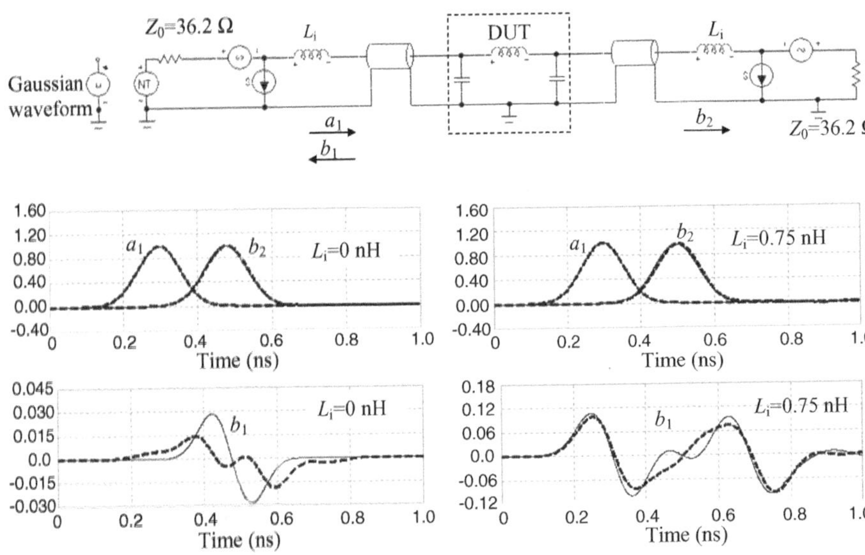

Figure 11.22 Equivalent circuit and computed waveforms in the time domain for the port–microstrip–DUT–microstrip–port structure when the range of analysis is extended to 6 GHz: SPICE (solid line); MWS (dashed line)

- When used to extract circuit parameters by simulating the function of a VNA, numerical codes do not have an automatic correction procedure, and the use of waveguide ports is required, when possible, to avoid errors caused by parasitic elements associated with the discrete ports.
- Analysis in the time domain is more suitable, as it is possible to compare reflected and incident waves computed by the full-wave numerical code and by SPICE simulation of the DUT, which can be implemented in the form of transmission lines or lumped elements.
- Normalization of the ports to a fixed impedance different from the characteristic impedance of the coaxial cables used to connect the ports to the DUT is not convenient, as the mismatch at port level hides the contribution of the reflections due to the DUT device.
- This condition of matching at port level is particularly important for extracting losses from lossy lines, as outlined in *Section 7.2*.

11.3 Prediction Model Validation by Radiated Emission Measurements

When a prediction model is built up with the goal of simulating *Radiated Emission* (RE) from structures such as PCBs, cables, or shielded boxes, it is very important to perform validation of the model by measurements or comparison with other models based on different approaches. If measurements are used for validation, the people in charge of carrying out the test must be aware of the errors that the test site and instruments can introduce for a correct comparison with the computed results. This aspect is known as the measurement uncertainty and is important when preparing new EMC standards, as briefly discussed in *Section 1.2*. Here, the problem of model validation is considered in more detail with some examples.

11.3.1 Uncertainty of the EMC Lab for Radiated Field Measurements and Numerical Simulations

The most common validation of prediction models is to compare results with measurements, taking the measurement accuracy into account [16]. In fact, a critical problem is the repeatability of measurements in a test laboratory. The uncertainty of the measurements is due to the measurement equipment, antenna factors, measuring-site reflection errors and cables. The agreement between the modeled data and the measurements can be no better than the test laboratory's uncertainty.

Example 11.5: Radiated Emission Measurements from a Rack in Two Different Semi-anechoic Rooms

An example of uncertainty investigation regarding two EMI/EMC laboratories will be described and discussed [17], under the following assumptions:

- Measurements were carried out in two different *Semi-Anechoic Rooms* (SARs) that fulfilled the normalized site attenuation requirements (± 4 dB) in the frequency range 30–1000 MHz.
- The first SAR has the dimensions 7 m $\times$ 5 m $\times$ 6 m, allows measurements at a distance of 3 m, and is lined with hybrid absorbers (this is referred to as the Italtel Lab).
- The second SAR has the dimensions 22 m $\times$ 4 m $\times$ 9 m and allows measurements at distances of 3 and 10 m. It is lined with absorber of length 2.5 m in order to fulfill the

requirement of a site attenuation within ±4 dB to reproduce the EMC performance of an ideal open-area test site in the whole frequency range 30–1000 MHz (see CISPR 16-1-4) (this is referred to as the TI Lab).

- The radiating system was an experimental shielded rack having the dimensions 0.98 m × 0.67 m × 2.2 m, connected to an output cable represented by a wire of 1 mm diameter and fed by an RG214 coaxial cable with the shield connected at 360° in contact with the rack. At the other end, the coaxial cable was connected to a tracking generator.
- Generator level was characterized by $P_{out} = -15$ dBmW (voltage source = 79 mV).
- A broadband antenna (BiLog antenna) with a height scan between 1 and 4 m and a step of 0.5 m, and in horizontal and vertical polarization, was used, and the maximum E-field was recorded according to the CISPR 22 standard. The same antenna was used for both labs.
- The system was placed on a turntable with 0–360° rotation, with a step of 30°.

A photograph of the set-up is shown in Figure 11.23. Details concerning the rack raised 10 cm above the ground floor are shown in Figure 11.24. The difference ΔE between the radiated E-field measured in the Italtel Lab at a distance of 3 m and the fields measured in the TI Lab at distances of 3 and 10 m, as required by the standards, is shown in Figure 11.25.

Figure 11.23 An experiment to assess the importance of the uncertainty of a test site for model validation by radiated emission measurement: photograph of the source (a shielded rack with a cable) and measuring antenna

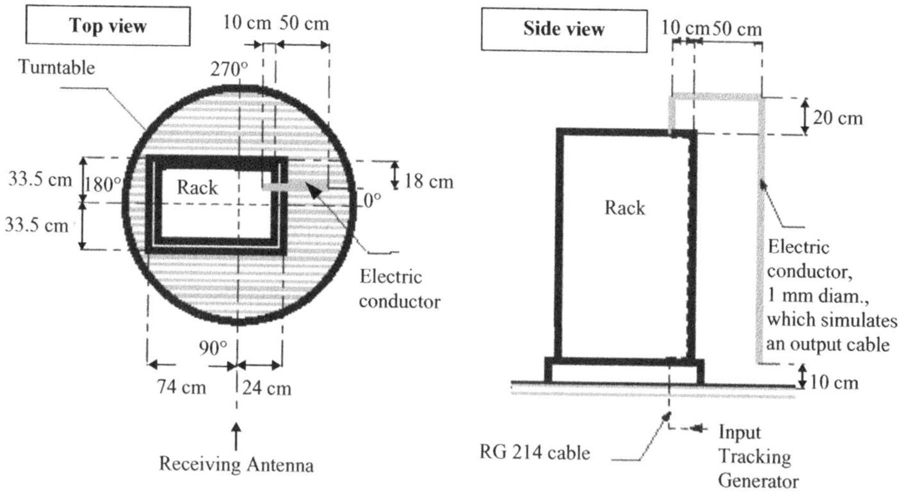

Figure 11.24 Physical dimensions of the radiating source within an SAR: a shielded rack with an attached cable and fed by a tracking generator

The results indicate that:

- Between 3 and 10 m, the expected difference should be 10 dB ± 5 dB. The 10 dB is due to the fact that in the *far-field* zone the field decreases by about 10 dB, moving from 3 to 10 m, and the ±5 dB is due to theoretical measurement uncertainty.
- At 3 m the expected difference should be ±5 dB.

These requirements are confirmed by measurements.

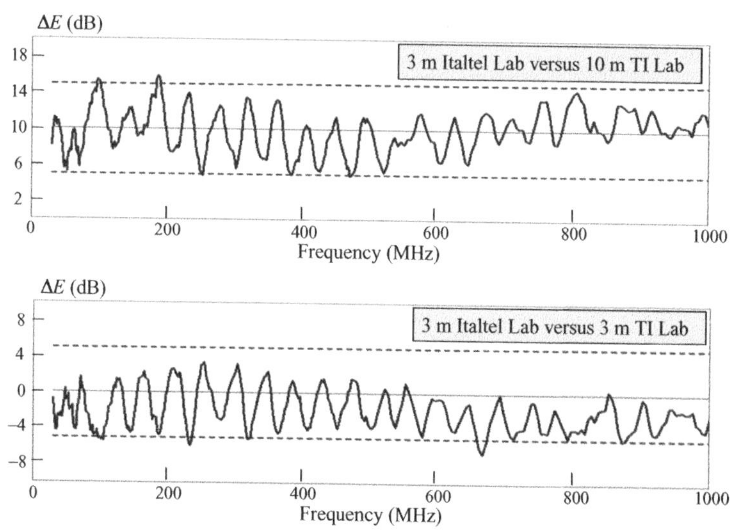

Figure 11.25 Measured difference in electric field ΔE of Italtel Lab and TI Lab

In order to verify whether the results obtained by prediction models fall in this range of uncertainty, the simulations were performed as follows:

- The structure made of the shielded rack and the attached cable was simulated by two different numerical codes: NEC based on the method of moments in the frequency domain [18], and MWS based on the *Finite Integration Technique* (FIT) in the time domain, with conversion of results in the frequency domain by FFT [13].
- The shielded rack was modeled with 10 cm segments by NEC and 234 432 mesh cells by MWS.
- The radiation pattern at 10 m was calculated at several frequencies and compared with the measurements. To appreciate the comparison better, a linear scale in mV/m is used to emphasize the main lobes. An example of these simulations is shown in Figure 11.26.
- The NEC simulation was performed in the low-frequency range (up to 300 MHz), including both the rack and the receiving antenna in the model.

Once the prediction model was validated, the receiving antenna was modeled by NEC and emissions at 3 m and 10 m were computed using the same procedure as adopted for measurements, i.e. by recording the maximum E-field for all rotation angles of the rack and for all positions of the receiving antenna. The comparison between measured and calculated values is shown in Figure 11.27. Although the number of calculated values is much smaller

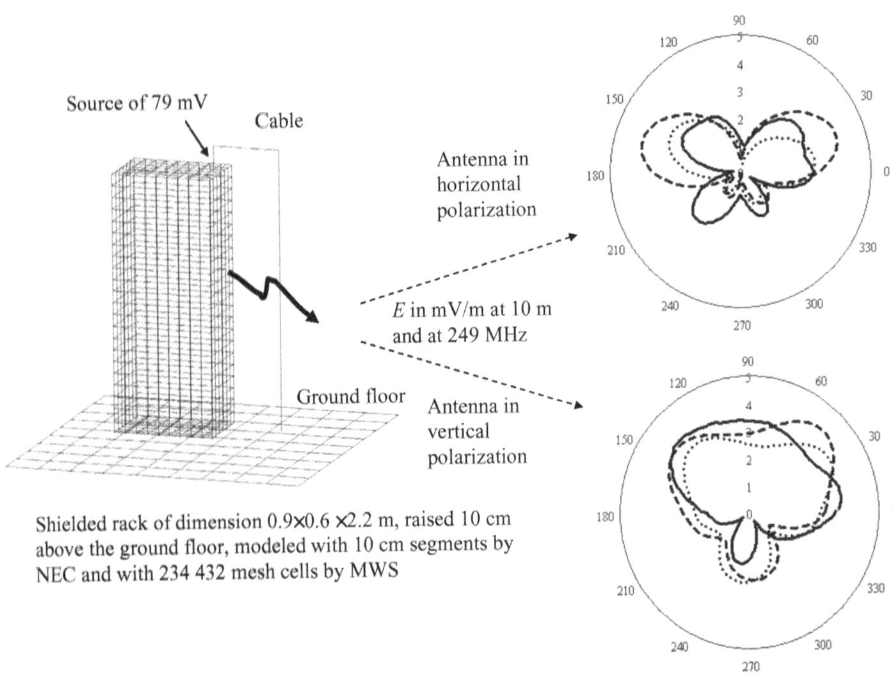

Figure 11.26 Radiation patterns of the E-field: comparison among measurements (solid line), NEC simulation (dotted line), and MWS simulation (dashed line)

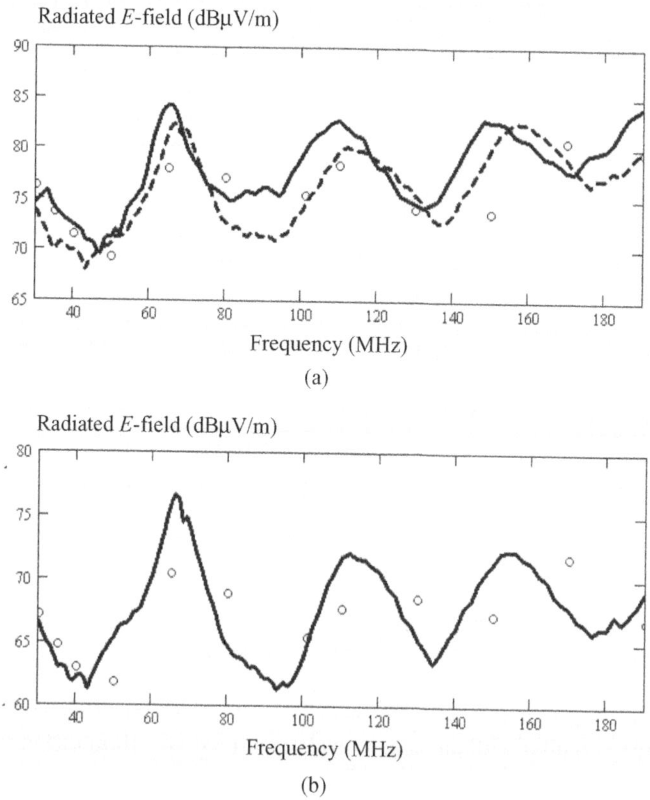

Figure 11.27 Comparison between measured and computed radiated E-fields: (a) measurement at 3 m in Italtel Lab (solid line), measurement at 3 m in TI Lab (dashed line), and computed values by NEC (circle); (b) measurement at 10 m in TI Lab (solid line), computed values by NEC (circle)

than the number measured, it can be noted that the shapes of the curves are similar. Even the simulations confirm a difference of about 10 dB between emissions at 3 and 10 m.

In order to compare the measurements carried out at the two sites at different distances, an uncertainty budget was assessed according to standards ISO [19], CISPR 16-4-1 [20], and CISPR 16-4-2 [21]. Combined uncertainty was calculated on the basis of the uncertainty associated with each element in the measurement system and using the law of uncertainty propagation. To characterize the measurements, the expanded uncertainty was used with a coverage factor $k = 2$ (level of confidence 95 %). An example of the uncertainty contributions calculated for the Italtel Lab is shown in Table 11.1. The standard uncertainty $u(x_i)$ in decibels and the sensitivity coefficient c_i will be evaluated for the estimate x_i of each quantity. The combined standard uncertainty $u_c(y)$ of estimate y of the measurand will be calculated as [21]

$$u_c(y) = \sqrt{\sum_i c_i^2 u^2(x_i)} \tag{11.20}$$

Table 11.1 Measurement uncertainty: u_d denotes declared in the norm; u denotes evaluated uncertainty

		Antenna at 3 m			
		Dipole antenna		Bilog antenna	
		Uncertainty (dB)			
Contribution	Probability distribution	u_d	u	u_d	u
Antenna factor calibration	Normal ($k = 2$)	±0.1	±0.05	±1.2	±0.6
Cables loss calibration	Normal ($k = 2$)	±0.2	±0.1	±0.2	±0.1
Receiver	Rectangular	±1.5	±0.87	±1.5	±0.87
Site imperfections	Rectangular	±4	±2.31	±4	±2.31
Combined uncertainty u_c	Normal		±2.47		±2.54
Expanded uncertainty U_{lab}	Normal ($k = 2$)		±4.94		±5.08

The expanded measurement instrumentation uncertainty U_{lab} for a test laboratory will be calculated as

$$U_{lab} = 2u_c(y) \tag{11.21}$$

and will be stated in the test report. For the examined case, $c_i = 1$.

To conclude this test, the following observations can be made:

- The results show that radiated emission measurements at a 3 m distance, corrected by a 10 dB factor, may conflict with the actual levels measured at a 10 m distance, particularly in the lower-frequency range, as discussed in *Section 1.2*.
- An evaluation of the measurement uncertainty compared with the different results obtained in the two laboratories shows that an estimated uncertainty value of 5 dB is in good agreement with the actual results.
- Simulations were made to confirm that the difference between computed data and measurements is within 5 dB for the majority of frequencies of interest, confirming the validity of the models used.

11.3.2 Modeling the Radiating Source

Once the test set-up and environment are characterized in terms of uncertainty, one of the most common errors is that some essential radiating part of the source is neglected in building up the prediction model [16]. This is the case, for example, when a PCB is fed with a coaxial cable starting from the floor of the chamber [22]. The emission contribution of the cable can be removed by placing ferrite along the cable or by feeding the PCB by a shielded local battery-powered oscillator, placed very closed to the PCB ground plane. To compare computed data and measurements correctly, it is important to take into account that the *extra* conductor is part of the physical measurement test, as shown by the following example.

Example 11.6: Modeling a Radiating Loop in a Semi-anechoic Room

As a test for validating the radiation source modeling, an 8 MHz clock signal produced by a shielded ACMOS digital gate with an output series resistor $R_{source} = 50\ \Omega$, feeding a 28 cm diameter loop with a coaxial cable, was considered. The simulation of the configuration was performed by the full-wave tool MWS, considering that the current source feeding the loop delivers 1 W to an output resistance equal to the internal resistance of the current source (see Figure 11.10). Computations were performed at a distance of 3 m from the loop, and the reflecting plane of the semi-anechoic chamber was taken into account using the condition $E_t = 0$. By this test it is shown that, although the current source is placed on the loop, the coaxial cable and shielded box are parts of the radiating source and must be taken into account for correct predictions and consistent comparison between simulation and measurements.

This experiment is interesting because, to characterize a shielded box or rack, a loop source is very often used to determine the shielding effectiveness of the structure by measuring the difference between the radiated field with the loop standing alone and then within the shielded structure [23]. The emission profile produced by the loop is usually a good representation of the emission profile of circuits and cables.

This test is useful for showing that, for a correct prediction of the measured radiated field, it is necessary to simulate all the parts of the radiating source, especially the part that brings the signal through the coaxial cable to the large radiating conductor, which in this case is given by the loop.

The considered set-up and the MWS model are shown in Figure 11.28. For a correct prediction of the measured data it is important to model:

- the whole space of the SAR;
- the metallic floor;
- the location of the antenna with respect to the floor and its orientation;
- the location of the radiating source with respect to the floor and all the metallic and dielectric parts.

Observe that the loop in the prediction model is fed by a current source of 50 Ω applied very close to the coaxial cable attached to a shielded box. The electric field generated by the

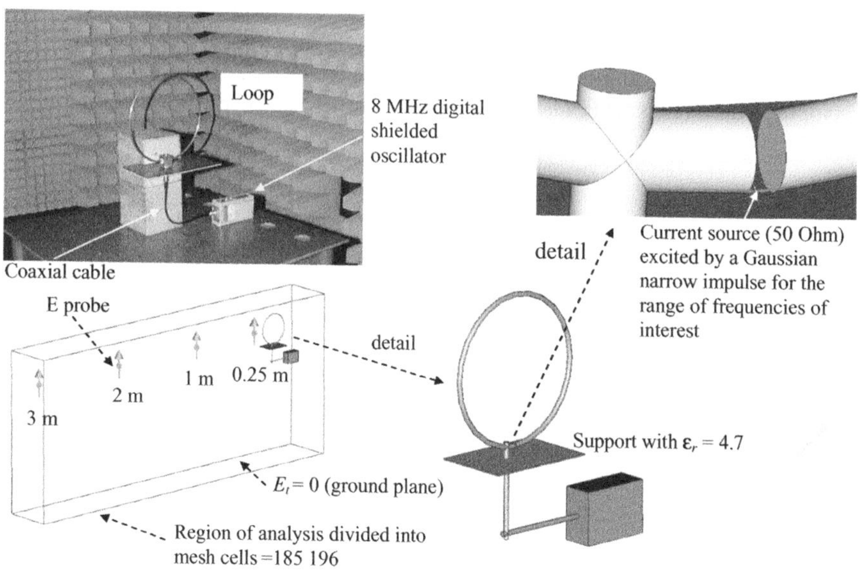

Figure 11.28 Loop in a semi-anechoic room: photograph of the set-up and MWS model with details

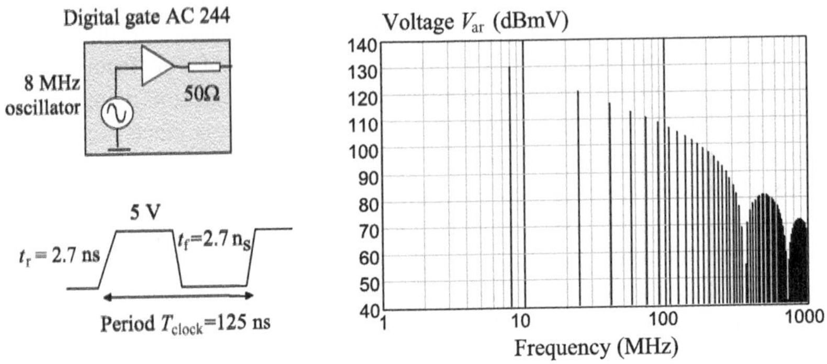

Figure 11.29 Time and frequency harmonics representation of the digital signal used to feed the loop

loop is computed by using electric probes for the position and orientation of interest, as shown in Figure 11.28.

MWS provides the *current source–electric probe* transfer function. To establish a consistent comparison with measurements, the predicted results E_{an} to be compared are given by

$$E_{an} = V_{ar} E_y \frac{1}{2\sqrt{R_{source}}} \tag{11.22}$$

where V_{ar} is the spectrum of the 8 MHz trapezoidal waveform (see Figure 11.29), and E_y is the field computed by the code as the response of a Gaussian normalized impulse determined for the frequency range of interest. The last term is a correction factor that takes into account that the source delivers 1 W over a resistance $R_{source} = 50\ \Omega$ (see Figure11.10). As the spectrum analyzer provides root-mean-square (rms) values, 3 dB must be subtracted from the computed field E_{an}.

The results of this computation are shown in Figure 11.30, where a good agreement between computed harmonics and measured emission profile can be observed. The difference

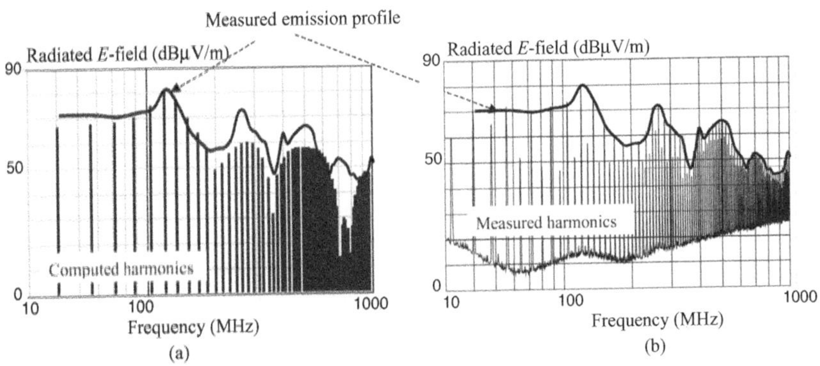

Figure 11.30 Vertical electric field at 3 m: (a) comparison between calculated harmonics E_{an} and measured emission profile; (b) measured harmonics and emission profile

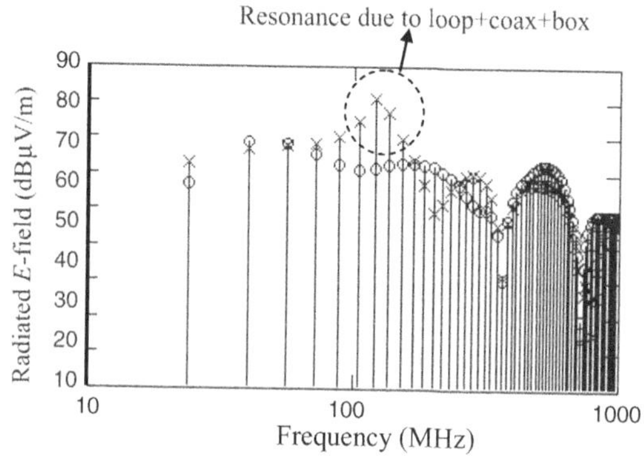

Figure 11.31 Computed radiated field modeling different structures: loop–coaxial–box (maximum values indicated by crosses) and loop alone (maximum values indicated by circles)

is less than 10 dB in the full range of frequency. This is a good result because, as previously shown, 5 dB can be assigned to the uncertainty of the SAR and related instrumentation, and another 5 dB to the uncertainty of the source in terms of devices, geometry, and material. The even components in the computed harmonics are absent because the spectrum is calculated adopting a 50 % duty cycle. On the other hand, the measurement shows both even and odd harmonics because the actual duty cycle is not perfectly 50 %. This also explains why the dips in harmonics at about 360 MHz and at 700–800 MHz are not present in the measured results.

Figure 11.31 shows the error that would occur if the loop were modeled without the coaxial cable and shielded box objects. The point of resonance at about 120 MHz, also present in the measurements, would be completely neglected.

11.3.3 Conclusion Concerning Validation of the Numerical Prediction Model for Radiated Emission by Comparison with Measurements

To end this section, the following conclusions can be drawn:

- When an EMC laboratory for radiated measurements is qualified according to the requirements of the standard, a difference of about ±5 dB should be expected between computed and measured values owing to the uncertainty of the test site and instruments.
- All the metallic parts of the source and the cables eventually attached, if any, must be simulated. This is especially true when the radiating source is fed by a cable. A further ±5 dB of uncertainty, approximately, can be assigned to the source.
- When there is a difference of less than about ±10 dB in the entire frequency range, with the exclusion of a very few separated frequency points, considering that the excitation in the simulation is ideal, the comparison between predicted and measured radiated emission profiles can be considered good.
- For a correct comparison, the simulation must reproduce the exact location of the radiating source and antenna, the metallic floor of the open site or of the shielded room, if present, and the orientation of the antenna used for measurements.

- The best way to perform measurements with a test PCB is to excite the traces by a digital oscillator powered by a battery and placed in a shielded box. The cable connection between the PCB and the oscillator should be as short as possible, and the shield should be very close to the ground plane and connected to it at a minimum of four points.

References

[1] Blood, W., *'MECL System Design Handbook'*, 4th edition, Motorola Semiconductor Products, Inc., Phoenix, AZ, 1983.

[2] *'Time Domain Reflectometry Theory'*, AN 1304-2, Agilent Technologies, 2002. Available online: www.agilent.com.

[3] *'Evaluating Microstrip with Time Domain Reflectometry'*, AN 1304-1, Agilent Technologies, 2000. Available online: www.agilent.com.

[4] *'TDR and S-parameters Measurements – How Much Performance do you* Need', AN, Tektronix, 2006. Available online: www.tektronix.com/tdr.

[5] *'TDR Impedance Measurements: a Foundation for Signal Integrity'*, AN, Tektronix, 2005. Available online: www.tektronix.com/oscilloscopes.

[6] Caniggia, S., 'EMC Design of Digital System Using Macromodelling Procedures for Integrated Circuits and their Interconnections', *5th Symposium and Technical Exhibition on Electromagnetic Compatibility*, Zürich, Switzerland, March 1983, 465–470.

[7] *'Understanding the Fundamental Principles of Vector Network Analysis'*, AN 1287-1, Agilent Technologies, 2000. Available online: www.agilent.com.

[8] Ramo, S., Whinnery, J., and Van Duzer, T., *'Fields and Waves in Communication Electronics'*, 3rd edition, John Wiley & Sons, Inc., New York, NY, 1994.

[9] *'EMC IC Modeling – Part 1: General Modeling Framework'*, IEC 62433-1, Ed. 1.0, Current Document 47A/785/CD, 2008-02-15.

[10] *'In-fixture Microstrip Device Measurements Using TRL Calibration'*, Product Note PN 8720-2, Agilent Technologies, 2000. Available online: www.agilent.com.

[11] *'Applying Error Correction to Network Analyzer Measurements'*, AN 1287-3, Agilent Technologies, 2002. Available online: www.agilent.com.

[12] *'Embedding/De-embedding'*, AN, Anritsu, 2002. Available online: www.us.anritsu.com

[13] www.cst.com.

[14] D'Amore, M., *'Compatibilità Elettromagnetica'* (in Italian), Siderea, Rome, Italy, 2003.

[15] Johnson, H. and Graham, M., *'High-Speed Signal Digital Design: a Handbook of Black Magic'*, Prentice Hall PTR, Upper Saddle River, NJ, 1993.

[16] Archambeault, B., Brench, C., and Ramahi, O., *'EMI/EMC Computational Modeling Handbook'*, 2nd edition, Kluwer Academic Publishers, Norwell, MA, 2001.

[17] Caniggia, S., Ballesio, F., Borsero, M., Giunta, M., and Marano, G., 'Comparison of Large System Radiated Emission at 3 m and 10 m: Analysis, Simulation and Measurement Uncertainty', *International Symposium on Electromagnetic Compatibility – EMC EUROPE 2002*, Sorrento, Italy, September 2002, 579–584.

[18] www.nittany-scientific.com.

[19] *'Guide to Expression of Uncertainty in Measurements* (GUM)', ISO Guide, 1995. Available online: www.iso.org.

[20] *'Specification for Radio Disturbance and Immunity Measuring Apparatus and Methods – Uncertainties, Statistics and Limit Modelling – Uncertainties in Standardized EMC Tests'*, CISPR 16-4-1, Ed. 1.1, 02-2005.

[21] *'Specification for Radio Disturbance and Immunity Measuring Apparatus and Methods – Uncertainties, Statistics and Limit Modelling – Uncertainty in EMC Measurements'*, CISPR 16-4-2, Ed. 1, 11-2003.

[22] Hubing, T. and Kaufman, F., 'Modeling the Electromagnetic Radiation from Electrically Small Table-top Products', *IEEE Trans. on Electromagnetic Compatibility*, **31**(1), February 1989, 74–84.

[23] Caniggia, S. and Santi, P., 'Shielding Effectiveness of an Enclosure for Digital Switching Equipment: FIT Modelling and Measurements', *International Symposium on Electromagnetic Compatibility – EMC EUROPE 2002*, Sorrento, Italy, September 2002, 913–918.

12

Differential Signaling and Discontinuity Modeling in PCBs

The differential transmission technique is the best way to ensure high-speed data functionality and immunity to the system for signaling at PCB and cable level. The advantages of using *differential-mode* (DM) transmission versus single-ended transmission are presented in this last chapter. Techniques for implementing differential signal transmission in a system are outlined, referring to the *Advanced Telecommunications Computing Architecture* (ATCA) standard. This standard is adopted by many companies involved in the development of very high-speed systems for telecommunications. LVDS is one of the most popular devices for *differential-mode* transmission. For this reason, LVDS characteristics and performances are shown in comparison with other families. The results obtained in terms of *signal integrity* (SI) and induced noise from an experimental set-up that uses test boards with LVDS drivers and receivers are presented and discussed. It is also shown that, by adding pulse transformers at driver and receiver locations, the immunity of the LVDS to a *common-mode* (CM) noise voltage can rise over 40 V instead of the specified ±1 V. Crosstalk in differential signaling with overhead and coplanar traces is investigated by SPICE simulations. Moreover, some design rules are provided for trace routing. An example of the realization of a motherboard according to the ATCA standard is presented, and its performance in terms of crosstalk and data rate transmission are verified by measurements.

The chapter ends by considering how to model discontinuities occurring in PCB interconnects and IC packages in order to extract equivalent circuit models to be used in SPICE simulations. Equivalent circuits of bends, vias, connectors and ground slots are presented. Some investigations and model validations are performed by 3D numerical codes. It is shown that differential transmission is appropriate when a signal must cross a gap in a ground plane without significant deterioration in the *signal integrity* and EMI performance. Finally, package types of connection for ICs are presented and discussed.

Signal Integrity and Radiated Emission of High-Speed Digital Systems Spartaco Caniggia and Francescaromana Maradei

12.1 Differential Signal Transmission

In the past, differential signal transmission was mainly used to transmit signals between PCB racks or apparatus by using twisted-pair cables as an economical solution capable of improving the immunity of the interconnects. Differential transmissions were implemented in PCBs by ECL devices for EMI and speed reasons. However, the price to pay was high power dissipation due to the nature of ECL devices operating with the transistor in the linear region, thereby sinking a large amount of current from the power supply in static conditions as well. With the improvements in CMOS technology in terms of transistor size and speed, and with the possibility of powering the devices with lower voltages, differential transmission makes it possible to operate over 1 GHz with appropriate *signal integrity* and EMI performance.

The basic concepts of a differential signaling transmission are treated in several textbooks [1–5] and *Application Notes* (ANs) prepared by the companies in charge of the design, development and selling of devices [6–14]. This chapter begins by considering the advantages offered by the differential technique, and some experimental and simulated results will be discussed to set design rules for enhancing receiver immunity, trace routing and line terminations in a PCB.

12.1.1 Single-Ended Versus Differential Signal Transmission

A differential signal provides maximum noise immunity. This is because any noise coupled with a pair of closed–parallel conductors generally appears equally on both conductors, so that this noise has the form of a *common-mode* noise. Whereas the receiver responds only to a voltage difference across the lines, in a twisted-pair cable the crosstalk noise can be ignored in many practical cases, as it is picked up equally by each of the two lines. This holds true up to the *common-mode* noise rejection limit of the receiver [5]. In the following, this feature will be discussed in more detail.

Three kinds of signal transmission are shown in Figure 12.1: single-ended, unbalanced and balanced (differential). In this figure, the following notation is adopted:

- V_s is the signal at the driver (i.e. source) output.
- V_r is the signal at the receiver input affected by noise.
- V_i is an interfering disturbance on the signal conductor owing to crosstalk or radiated fields.
- V_n is a noise on the ground conductor owing to interfering currents such as the return current of other signals, electrostatic discharge, surge, etc.
- gnd is the reference voltage point for the system.
- Δ before a symbol of a *common-mode* voltage means the fraction of that voltage that converts to a *differential-mode* voltage owing to asymmetries in the interconnect structure.

Single-ended Transmission

The structure consists of two conductors: one is used for transporting signal current and the other is used as the return path, as shown in Figure 12.1a. The ground conductor could be in common with other signal conductors. The voltage V_r is the algebraic sum of the signal voltage V_s, with delay when the line is matched, and the noise voltages V_i and V_n caused by external interferences. This means that the total disturbance ($\pm V_i \pm V_n$) affects the receiver directly. For this reason, the immunity is poor.

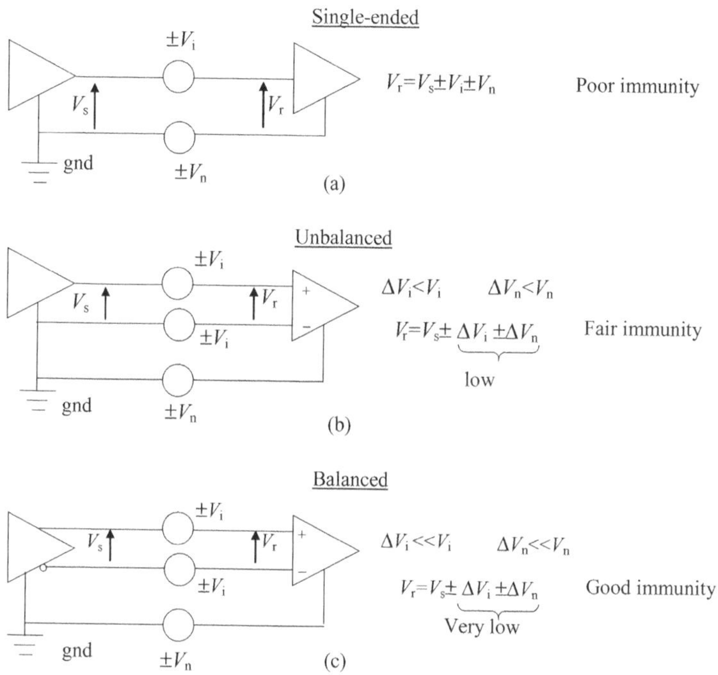

Figure 12.1 Signaling in the presence of external noise: (a) single-ended; (b) unbalanced; (c) balanced (differential transmission)

Unbalanced Transmission

The structure consists of three conductors: one conductor is used for transporting signal current, a second conductor, symmetric to the first, is used as the return path, and a third is used as reference ground where the ground pins of the driver and receiver are connected. The driver is a single-ended device and the receiver is differential, as shown in Figure 12.1b. This means that the receiver inputs recognize the voltage difference across the two ends of the signal conductors. The differential signal voltage V_r at the receiver is the algebraic sum of the signal voltage V_s and the voltages ΔV_i and ΔV_n of the disturbances V_i and V_n respectively. These disturbances come from the conversion of *common mode* to *differential mode* and are mainly caused by the unbalanced structure of the driver. This improves the immunity of the interconnect with respect to the single-ended solution, but it is not suitable for high-speed interconnects.

Balanced Transmission

The structure is similar to the previous one, the difference being that the driver is also differential. The driver has two outputs switching with opposite polarity and equal output impedance in order to excite a *differential mode* according to the definition given in *Section 6.2*. In this case, if the structure is perfectly symmetric, V_r should be equal to V_s with the delay of the interconnect. This does not occur in practice because of the slight non-symmetry of the driver, conductors, and receiver with respect to the reference ground. *Common-mode* to *differential-mode* conversion always exists, but the fractions ΔV_i and ΔV_n, which sum algebraically to V_s, are very small.

An ideal receiver should be able to reject *common-mode* disturbances of any values. In practice this is not the case, and a receiver is characterized by a parameter, the *common-mode* rejection, that defines the ability of the receiver to work properly up to a defined amount of *common-mode* noise. This parameter varies from a few to several volts, depending on the speed of the device. For example, a receiver of series RS422 has immunity to *common-mode* disturbance of ± 7 V. LVDS devices are faster but offer less immunity (± 1 V).

A differential signal is transmitted on a dual-signal path, and the two signals are driven as a complementary pair, with one signal being the logic inverse of the other. In this case, the signal quality can be measured by the technique described in reference [15].

As shown in Figure 12.1c, the differential signal involves a differential transmitter, a differential interconnect, and a differential receiver. The ground potential difference between the transmitter and the receiver is modeled as a noise voltage source and can have both DC and AC components. High-speed data links use differential signaling at much higher frequencies where noise problems tend to be more severe, even for relatively short connections.

The advantages of differential signaling in high-speed data transmission include the following:

- higher *common-mode* noise rejection;
- increased noise immunity;
- reduced crosstalk;
- reduced ground noise;
- reduced EMI;
- a better eye diagram than a single-ended signal, with a non-solid reference plane of the PCB.

Differential signaling is able to reject *common-mode* noise from ground potential variations between transmitter and receiver, and other injected noises that are common to both signal paths. The transmitted differential signal is processed at the receiver as the voltage difference between the two lines. By taking the difference between the two complementary signals, the differential receiver also produces twice the signal swing of a single-ended signal for improved noise immunity.

The balanced nature of differential signals in general leads also to a more constant switching current than that of single-ended circuits. The signal currents in differential drivers tend to be steered between the two outputs as the signal polarity switches, which results in a more constant load current compared with the load current spikes commonly seen with single-ended drivers. Reduced load current transients should also result in a reduction in that portion of ground noise that is caused by current spikes passing through the ground lead inductance. The improved noise immunity and high sensitivity of differential receivers also allows the use of reduced logic swings with differential signaling. This smaller signal swing and the balanced field distribution associated with differential signaling generate less EMI.

The drawbacks of using differential signaling are:

- the increased layout complexity;
- the need for balanced signals and interconnects.

The common use of point-to-point connections rather than a shared bus structure results in separate paths for transmitted and received signals, which effectively doubles the number of signal pairs required in a high-speed serial link. Consider also that imbalance in differential signaling paths leads to the generation of *common-mode* currents and reduced *common-mode* rejection at the receiver. An experimental investigation regarding the generation of *common-mode* current by differential drivers such as RS422, LVDS, and PECL and its effects in producing radiated fields is reported in *Section 9.7*, considering the EMI performance of UTP and SFTP cables and their connectors.

12.1.2 Differential Interconnect with Traces in PCBs and the ATCA Standard

When considering the transmission characteristics for a differential signaling interconnect between two boards, the complete end-to-end path of the connection must be taken into account. End-to-end signal skew and propagation delay time must be properly understood to ensure interoperability between boards and backplanes. These requirements are fundamental to the *Advanced Telecommunications Computing Architecture* (ATCA) standard proposed by a consortium of companies for the development of very high-speed systems for telecommunications [16]. ATCA is the largest specification effort in the history of the PCI *Industrial Computer Manufacturers' Group* (PICMG), with more than 100 companies participating. The official specification designation is PICMG 3.*x*. Here, AdvancedTCA^{TM} is targeted to requirements for the next generation of 'carrier-grade' communications equipment. This series of specifications incorporates the latest trends in high-speed interconnect technologies, next-generation processors, and improved *Reliability*, *Availability* and *Serviceability* (RAS). In this section, the main characteristics and performance will be highlighted.

A typical point-to-point differential signal connection is defined by three main components, as shown in Figure 12.2: the trace routed across the backplane, pair connectors at each end, and the trace routing on the board between the connector and the transmitter/receiver.

Equalization is a high-pass filtering technique applied to the signal interconnect to compensate for the increase in interconnect attenuation with frequency. The simplest implementation of equalization is to use an AC-coupling capacitor C_e in series with the pair of conductors in order to form, with the termination resistor R_t, a high-pass filter as shown in Figure 12.2. As the termination resistor is set by the differential characteristic impedance of the interconnect, the value of the AC-coupling capacitor should be selected to match the desired equalization response [15].

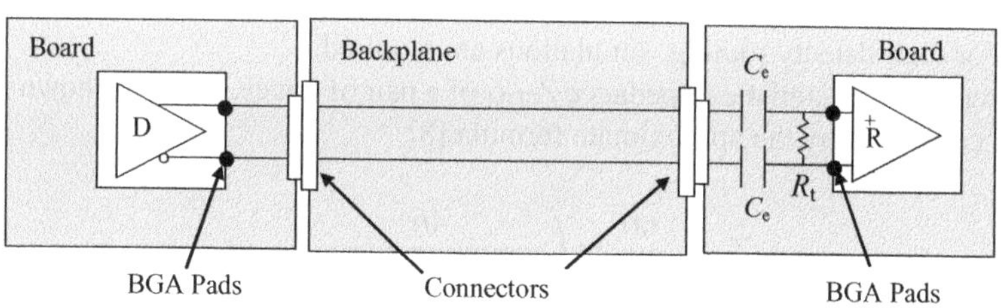

Figure 12.2 Typical point-to-point differential connection with *Ball Grid Arrays* (BGAs)

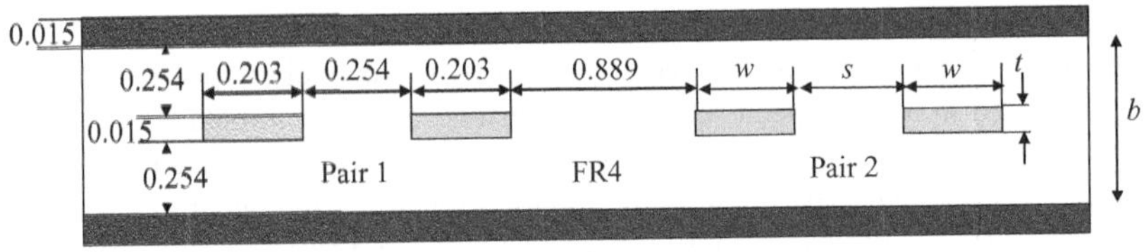

Figure 12.3 Suggested differential interconnect structure in a PCB. All dimensions are in mm

If vias must be used in the signal routing path, via design should try to minimize the associated inductance and capacitance, and these parasitic parameters should be matched in both traces of a differential pair. Other examples of routing parasitic parameters are the parasitic input capacitances present at the input pin of a serial data receiver, and package parasitic parameters present in a circuit-board-mounted serial data connector.

An electrical cable for the transmission of high-speed serial data is often shielded, and the differential signal path is typically implemented with twisted pairs. Twisted pairs are used for differential signal impedance control and to minimize crosstalk between the transmit and receive signaling pairs.

High-speed serial data interconnects on circuit boards also require the use of a differential signal routing technique. Differential signal interconnects are routed as coupled transmission lines. Generally, there is close spacing between circuit board traces and board ground planes, so that the degree of coupling between differential lines on a circuit board is much less than that for twisted pairs, particularly for edge-coupled lines. This limited coupling between circuit differential pairs means that it is not uncommon to see differential pairs routed separately on circuit boards as two uncoupled transmission lines.

Coupled lines can be routed in several different ways, depending on the layout requirements. Edge-coupled lines, where the traces are routed side by side on the same circuit board layer, are commonly placed on the outer layers as microstriplines, although they can be embedded as inner-layer striplines. Broadside-coupled lines should generally be routed only on inner layers as striplines in order to provide a symmetrical structure.

An example of an edge-coupled backplane single stripline with FR4 dielectric, taken from reference [16], is shown in Figure 12.3. The coupling between traces within a pair in this structure is significant. Therefore, maintaining the typical 0.254 mm spacing throughout the length of the pair is very important to keep the 100 Ω differential impedance constant. The distance between any two pairs must be no smaller than 0.889 mm edge to edge, as shown in Figure 12.3. This limits crosstalk to lower than 0.2 % (near-end coupling coefficient, NEXT), as will be shown by simulations in the next section devoted to LVDS devices. The 0.889 mm spacing derives from the relative dielectric constant of the FR4 material, as outlined in *Chapter 6*. For high-density routing, simulations are required.

The differential characteristic impedance Z_{ODM} of a pair of traces, like that shown in Figure 12.3, can be calculated by the approximate formula [8]

$$Z_0 = \frac{60}{\sqrt{\varepsilon_r}} \ln\left(\frac{4b}{0.67\pi(0.8w + t)}\right) \tag{12.1a}$$

$$Z_{\mathrm{ODM}} = 2Z_0\left(1 - 0.347\ \mathrm{e}^{-2.9s/b}\right) \tag{12.1b}$$

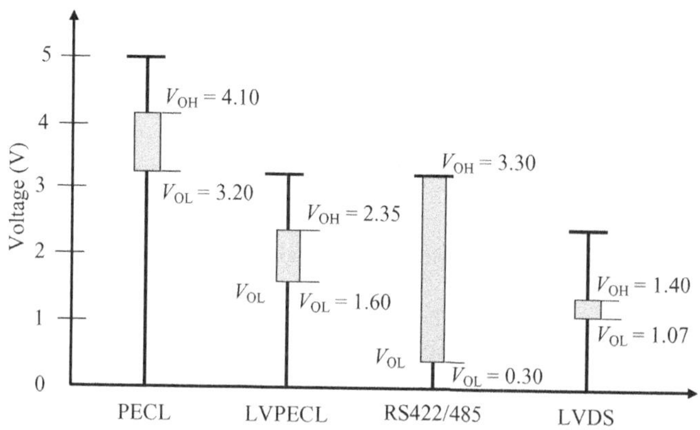

Figure 12.4 Output level comparison of four differential family devices

The parameters b, w, and s are shown in Figure 12.3. Applying Equations (12.1) to the board of Figure 12.3, and assuming that $\varepsilon_r = 4$, it is found that $Z_0 = 51.7\ \Omega$ (nominal characteristic impedance of an isolated trace) and $Z_{0DM} = 94.6\ \Omega$. Other studies regarding differential trace routing in PCBs can be found elsewhere [17–19].

12.1.3 Differential Devices: Signal Level Comparison

Signal level comparison for typical differential driver/receiver devices is shown in Figure 12.4 [10], [14]. The most interesting device is LVDS. The LVDS standard was created to address applications in the data communications, telecommunications, server, peripheral, and computer markets where high-speed data transfer is necessary. LVDS offers low-cost, high-speed, low-power solution by comparison with the standards of the past. LVDS is defined in the TIA/EIA-644 standard [20]. It is a low-voltage, low-power, differential technology used primarily for point-to-point and multidrop cable driving applications. The standard was developed under the *Data Transmission Interface* subcommittee TR30.2. It specifies a maximum data rate of 655 Mbps, although some of today's applications are pushing well above this specification for a serial data stream.

Compared with other differential cable driving standards such as PECL, LVPECL, and RS422/RS485, LVDS has the lowest differential swing, with a typical single-ended voltage swing of 350 mV and with a typical offset voltage of 1.25 V above ground. The differential swing is therefore 700 mV. Examples of data transmission with RS422, LVPECL, and LVDS are shown in *Section 9.7.2* with UTP and SFTP cables.

12.1.4 Differential Signal Distribution and Terminations

Termination of LVDS is necessary at the receiver input to generate the output differential voltage [9]. The TIA/EIA-644 specification [20] stipulates an internal termination resistor value of between 90 and 132 Ω. Termination of LVDS is much easier than that of most other technologies such as ECL and PECL (see Figure 12.5, where Z_{0DM} is the differential characteristic impedance of the interconnect). In a point-to-point system configuration, the termination

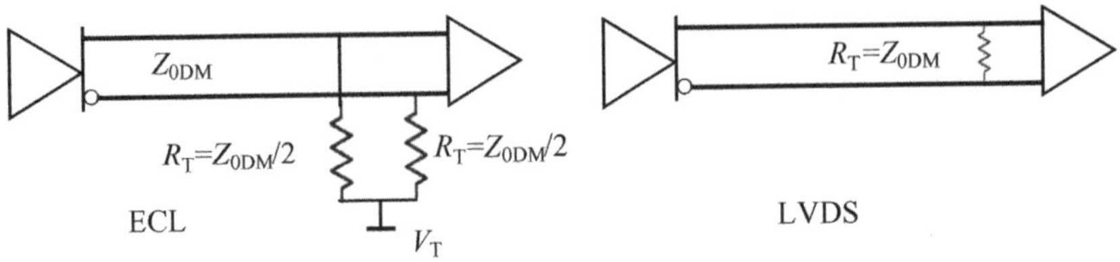

Figure 12.5 Example of differential signal terminations

resistor should be placed within 2 cm of the receiver. For a multidrop configuration, the termination resistor should also be located within 2 cm of the last receiver. To avoid reflections, it is essential for the impedance of all cables, connectors, buses, and termination resistors to be closely matched. The majority of twisted-pair cables are designed to have a characteristic impedance of about 100 Ω, so a 100 Ω termination resistor is recommended to avoid transmission-line mismatches, which will result in reflections and other discontinuities.

LVDS can also be used in a bus or multidrop structure typically found in backplanes [23], as well as in box-to-box applications, providing that the media transmission distance is short. In a typical multidrop system, the termination resistor must be located at the receiver positioned at the far end of the bus (see Figure 12.6).

Although, as defined in the RS644 standard, LVDS does not have the dynamic current drive to support a multipoint bus system, there is a high-drive LVDS available, which has a higher drive compared with the 3.5 mA drive of a standard LVDS. In a multipoint system (see Figure 12.6), the driver can be located at any point along the bus. For this reason, much like the multidrop center-driven bus previously discussed, a termination resistor is required at each end of the bus. This also means that the driver sees the two resistors in parallel and must supply twice the current to the bus. An 11 mA dynamic drive is provided on the high-drive version of LVDS to address a multipoint configuration. The standard described in reference [16] recommends the use of *Multipoint* LVDS (MLVDS). MLVDS is specifically designed to improve performance in bused designs. Compared with standard *Bused* LVDS (BLVDS), MLVDS has controlled edge rates, a slightly larger signal swing, and tighter thresholds. The simulations indicated that frequencies of up to 100 MHz were feasible with the MLVDS implementation. The TIA/EIA-899 standard [21] on the electrical characteristics of multipoint low-voltage differential signaling (MLVDS) provides a description of the driver requirements. The simulations of MLVDS bused clocks indicated significant performance improvement, with a differential impedance of 130 Ω instead of 100 Ω. This helps to increase the effective impedance of the clock bus when heavily loaded.

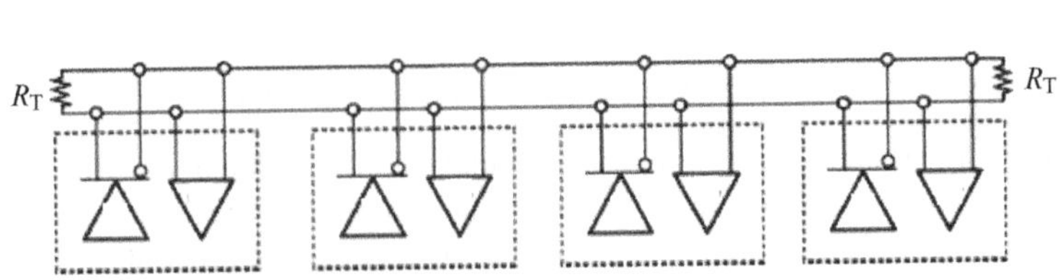

Figure 12.6 Multipoint configuration with a differential driver/receiver

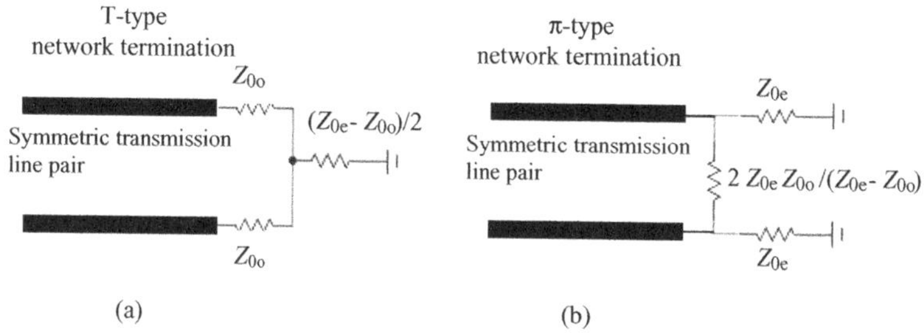

Figure 12.7 Optimal termination for a symmetric transmission-line pair: (a) T-type network termination; (b) π-type network termination

For a symmetric transmission pair, both even and odd modes of propagation should have a termination equal to the characteristic impedances Z_{0e} and Z_{0o}, respectively, to avoid any type of reflection. This can be accomplished by either a π-type network [1] or by a T-type network, as shown in Figure 12.7.

However, as previously discussed, differential pairs are not typically terminated with the networks in Figure 12.7. Three common termination schemes used in practice are shown in Figure 12.8. The bridge solution (Figure 12.8a) provides termination without reflections for *differential mode*, while the *even* or *common mode* is completely reflected. Examples of signaling with this solution are given in *Section 9.7.2*. This means that *common-mode* disturbances such as external interfering fields or ground loop coupling noises (see *Section 10.1*) can cause *common-mode* oscillations on account of the fact that the *even* or *common mode* is not properly terminated. This should not be a problem as long as there is no mode conversion and the differential receiver has sufficient *common-mode* rejection. The single-ended termination (Figure 12.8b) offers the advantage of matching the odd mode and partially the even mode. The drawback is that an extra resistor is required. A mix solution with a bridge at the receiver location and a single-ended termination at the differential driver offers a matched condition for the *odd mode* at both ends, while the driver damps the *even mode*, mitigating the *common-mode* noise produced by external interferences. The AC termination (Figure 12.8c) has the same advantages as the single-ended termination without increasing static power dissipation. The drawback is that a capacitor is required as a third element.

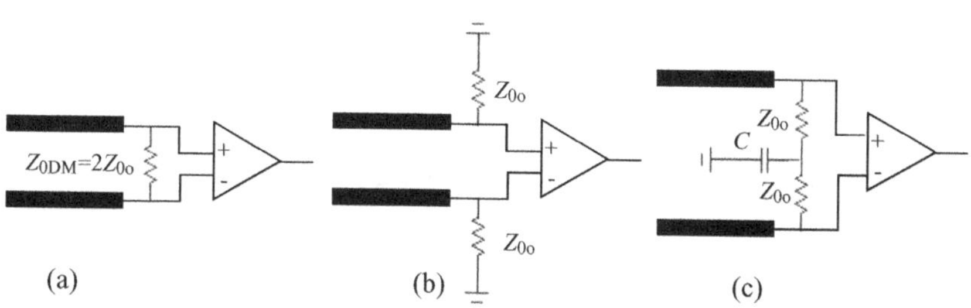

Figure 12.8 Termination schemes used in practice for a differential transmission-line pair: (a) bridge; (b) single-ended; (c) AC

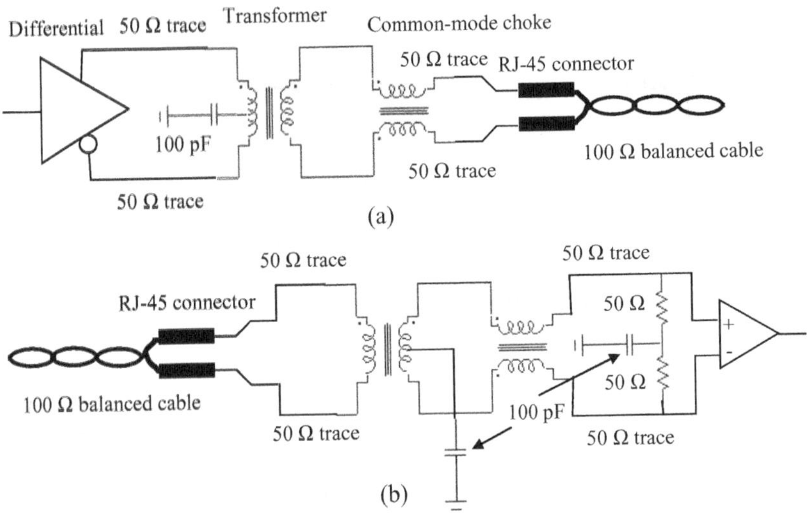

Figure 12.9 Solutions to enhance EMI performance of a differential link communication by an unshielded twisted pair: (a) EMI filters at driver location to mitigate emission; (b) EMI filters at receiver location to increase common-mode rejection

Differential transmission is often used to link PCBs sharing different racks or equipment as a lower-cost solution with respect to coaxial cables. For economic reasons, the cable is often an unshielded twisted pair (UTP) of categories 3, 5, 5e, 6, and 7, depending on the speed of the transmitted signal [2]. The main problem with these types of cable is that, although cables 5e, 6, and 7 are well balanced, the radiated fields produced by the *common-mode* current, generated by the non-balanced condition of the switching edges within the differential driver, dominate over the radiated fields produced by the *differential mode*, as explained in *Section 9.7.2*, where the performances of RS422, LVDS, and LVPECL drivers are compared. It is shown that, without EMI filters, the emission profile can be higher than the emission limit imposed by the CISPR 22 standard for 3 m Class B equipment. An EMI filter useful for mitigating emission is the one indicated in Figure 12.9a, where the transformer followed by a *common-mode choke* has the task of stopping the *common-mode* current, and the capacitor connected to the central point of the transformer has the task of diverting the *common-mode* current to the ground of the PCB. Reducing the *common-mode* current leads to minimization of cable emissions. The efficiency of this type of EMI filter is shown in *Section 10.3.4*, where several grounding solutions are also compared using a 3D numerical code. To increase significantly the immunity of the receiver to *common-mode* noises, the same EMI filter solution should be implemented at the receiver location as indicated in Figure 12.9b. The capacitors with the task of diverting the interfering *common-mode* current should be connected to a PCB clean ground connected to the chassis with a very low impedance. How these filters act in improving the *common-mode* rejection of the receiver will be explored in *Example 12.1* by measurements. Note that in Figure 12.9 the traces used to connect the components with the cable connectors must have an impedance of 50 Ω in order to have a differential impedance of 100 Ω, which matches the 100 Ω characteristic impedance of the UTP cables.

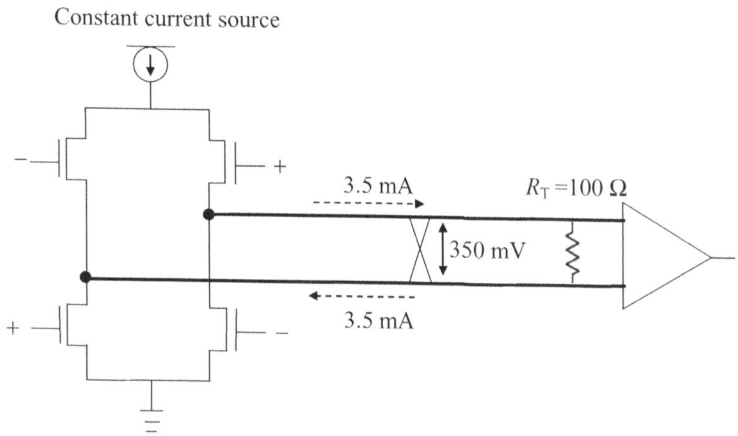

Figure 12.10 Output equivalent circuit of an LVDS device driving a line terminated with a resistance

Moreover, the traces must be symmetrically positioned with respect to all nearby grounded objects.

12.1.5 LVDS Devices

LVDS features a low-swing differential constant current source configuration that supports fast switching speeds and low power consumption [12]. The configuration is shown in Figure 12.10. The output equivalent circuit of the LVDS device is a current source with high impedance that provides termination resistance R_T with a typical 3.5 mA current for one logic state and a typical −3.5 mA current for the other logic state. This means that the receiver sees a differential signal of $2 \times 350 = 700$ mV.

Differential signaling also offers *common-mode* rejection. The receiver ignores any noise that is coupled equally with the differential signals and considers only the difference between the two signals. The receiver has a *common-mode* voltage in the range 0.25–2.25 V. LVDS receivers will operate with as much as a ±1 V ground shift between driver and receiver, as shown in Figure 12.11.

Failsafe is a feature offered in LVDS that will help system reliability by preventing errors. Failsafe guarantees that the outputs are in a known state (high or low) when the receiver inputs are under certain fault conditions. Without the failsafe feature, any external noise above receiver thresholds could trigger the output to an unknown state. With the failsafe feature, the receiver outputs will always be in a known state as long as the inputs are not receiving a valid signal.

Example 12.1: LVDS Signal Integrity and Common-mode Rejection Investigation by Measurements

Two test boards with LVDS devices were built in order to check signal integrity in serial link and immunity to *common-mode* disturbances. To carry out the test, commercial devices satisfying the LVDS standard were chosen according to the set-up shown in Figure 12.12,

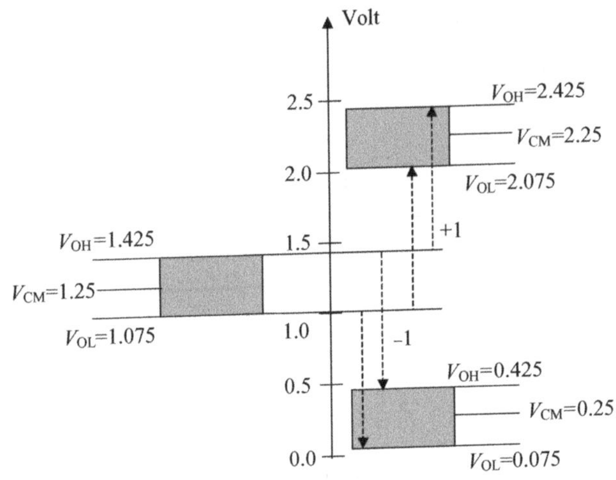

Figure 12.11 LVDS common-mode noise range

where:

- DS90LV31A is one buffer with TTL-compatible input and differential LVDS output, characterized by a clock frequency $f_{clock} = 200$ MHz and an NRZ data sequence.
- DS90LV32A is one buffer with a differential LVDS input and a TTL-compatible output.
- The cable is a 100 Ω UTP of Cat-5e with the drain wire connected to the ground board (GND_TX, GND_RX) and length $l_{cable} = 10$ m.
- Traces on the board have a differential characteristic impedance of 100 Ω.
- The connectors are of type Z-pack.
- The termination resistance $R_T = 100$ Ω.

With this structure, the ground noise between the two boards should be less than 1 V. To allow a greater noise level than 1 V, decoupling capacitors and impulse transformers with *common-mode chokes* should be used.

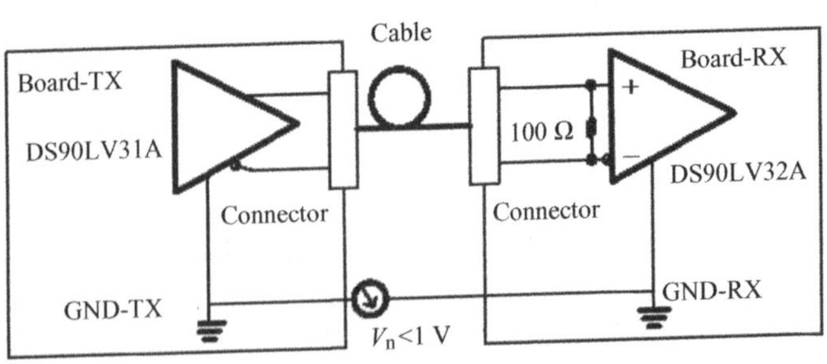

Figure 12.12 Test set-up for common-mode rejection in an LVDS serial link

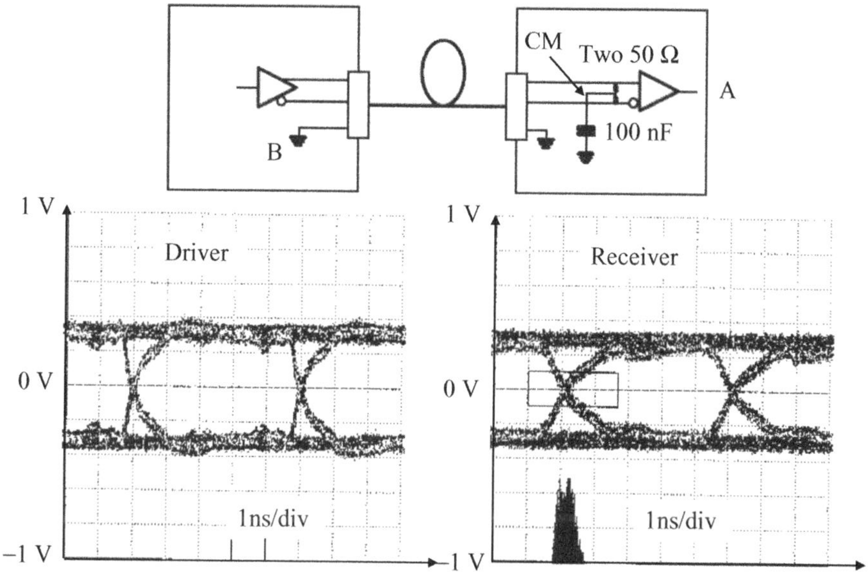

Figure 12.13 Measured eye diagrams for the basic link

Signal Integrity

The signal integrity was investigated by performing eye diagram measurements on the basic configuration in Figure 12.13 and on four other configurations shown in Figure 12.14. The jitter value was calculated between the thresholds ±100 mV and given in histogram form (just below the eye diagram of Figure 12.13). Measurements on a direct link using a 100 nF capacitor to provide a partial match condition for *common-mode* disturbances are shown in Figure 12.13. In this case, the differential signal is matched with $2 \times 50 = 100\ \Omega$, and the *common-mode* disturbances are partially matched by 50 Ω termination resistors connected to ground by the 100 nF decoupling capacitor.

Looking at the other configurations shown in Figure 12.14, the following comments can be made. The decoupling action of the series capacitors has the purpose of increasing the basic value of ±1 V as immunity to *common-mode* noise. In this case, it is necessary to provide a polarization voltage of 1.2 V at the receiver by a resistive net. The central point of the transformer is connected to ground by a 100 nF capacitor to divert to ground the *common-mode* noise. The choke has the task of stopping the *common-mode* noise passing through the parasitic capacitances between the two coils of the transformer.

In all the cases investigated, it was verified that the signal at the receiver is characterized by an open eye diagram, symmetric with respect to the 1.2 V voltage, and with a clear crossing between the differential decision thresholds of ±100 mV. The jitter values of the different solutions are summarized in Table 12.1.

Common-mode Immunity

For the purpose of investigating how EMI filters such as decoupling capacitors, transformers and *common-mode chokes* enhance the immunity of receivers to *common-mode* disturbances,

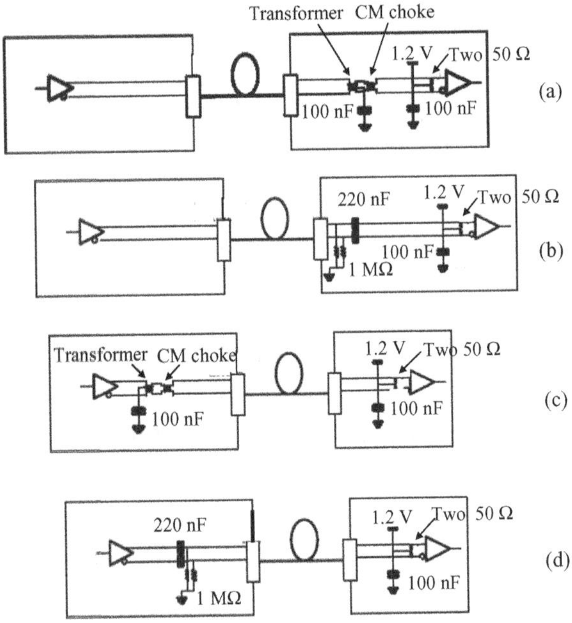

Figure 12.14 Configurations: (a) transformer and CM choke at receiver end; (b) capacitors at receiver end; (c) transformer and CM choke at driver end; (d) capacitors at driver end

the set-up shown in Figure 12.15 was realized. The measurements were carried out under the following conditions:

- clock signal;
- noise amplitude 0–50 V;
- noise pulse duration 80 μs;
- trigger given to the oscilloscope by the noise generator;
- all measurements have ground 2 of the receiver as the reference point.

When the noise is applied, the isolation transformer acts like the wire drain of the cable. For this reason, during this type of test, the wire drain of the cable was disconnected from the board ground.

Table 12.1 Jitter of LVDS links

Type of link	Jitter in ns	% of time bit
Basic (direct)	1.16	23
Capacitors at driver end	1.44	29
Capacitors at receiver end	1.04	21
Transformer and CM choke at driver end	1.28	25
Transformer and CM choke at receiver end	1.42	28

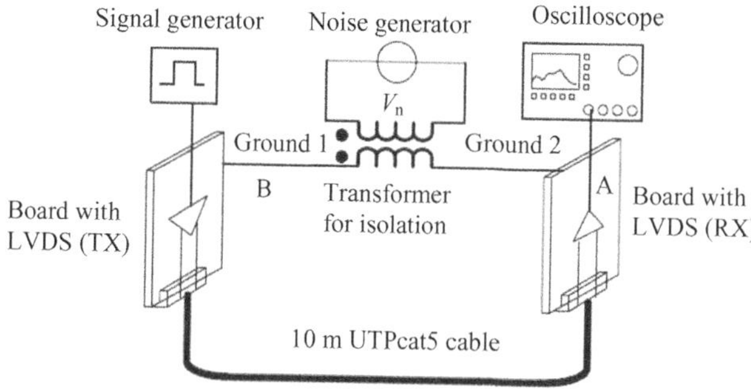

Figure 12.15 Set-up for immunity noise measurements between two grounds of the LVDS link

The waveform of the impulse noise depends on the type of load applied to the source. An example of noise without load is shown in Figure 12.16. The duration of the pulse was always set to 80 μs.

Without the drain wire connected to both board grounds, the *common-mode* noise at point CM of Figure 12.13 follows the induced noise at point B of Figure 12.15 between the two grounds (same waveforms). There is a loss of data for a voltage noise of the generator of 2 V.

With 220 nF decoupling capacitors at the receiver end, the noise generator is connected to a circuit with high impedance. The noise at point CM is lower than the noise at point B. Data failure occurs with low values of the noise generator, i.e. with a voltage at point B of about 4.6 V, with a slight improvement in immunity, of about 2.5 V. Loss of data also occurs with decoupling capacitors at the driver end with less immunity. The results obtained are summarized in Table 12.2.

With a transformer and a *common-mode choke* at the receiver or driver end, no data failure was recorded, even in the case of a 40 V voltage noise assigned to the source noise generator. In this case, noise at point CM is also very low for high values of voltages at point B.

The immunity to *common-mode* noise voltages of LVDS devices was checked, considering the basic link of Figure 12.13 without the presence of the drain wire. The signal voltage V_A at

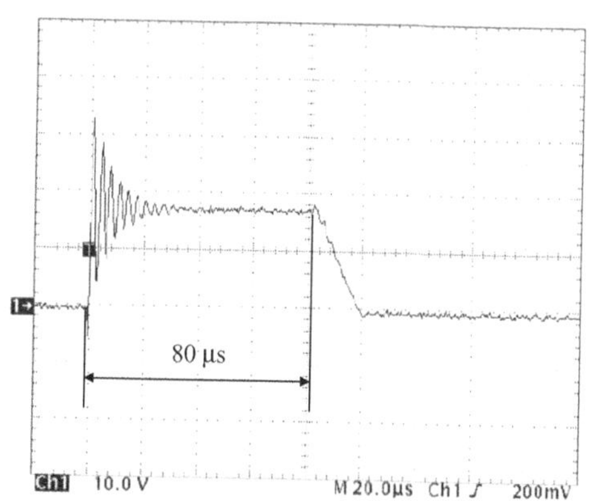

Figure 12.16 Output voltage of the noise generator without load

Table 12.2 Maximum peak noises that cause data failure with decoupling capacitors in the LVDS data link

LVDS with decoupling capacitors	Capacitors at driver end (maximum peak)	Capacitors at receiver end (maximum peak)
Noise at point CM	2.7	3.8
Noise at point B	3.44	4.64

the receiver output at point A and the applied noise voltage V_B at point B increased gradually from 0 V to 15 V where monitored, and the results are shown in Figure 12.17. In particular, in graphs 1 to 9 the following situations can be observed:

1. Regular signaling without noise.
2. Signal with one single error due to noise.
3. Signal with several errors.
4. The driver returns to transmit when the noise ceases.

5-8. There is an advanced change in the device characteristics with increasing applied voltage noise. The impedance seen by the noise source decreases, as can be seen from the deformation of the impulse noise from a trapezoidal shape to a form with steps.

9 The driver stops transmitting. The data transmission is correct again only after switching the power supply off and on in sequence, and in the absence of noise. This behavior of the driver was verified with a noise voltage of up to 40 V.

To check the breakdown of LVDS devices in the static condition, a DC voltage source was connected directly without an isolation transformer to the two ground boards in Figure 12.15. The drain wire of the cable was not connected to the two grounds. Three drivers and receivers were used for the test. The noise voltage V_{nDC} was slowly varied from 0 to 9.5 V.

Breakdown did not occur suddenly but after the following stages:

- $V_{nDC} = 0$–2.4 V, the link worked correctly.
- 2.4 V $< V_{nDC} <$ 8.5 V, data transmission was interrupted, but, by decreasing the offset voltage, it was corrected again;
- $V_{nDC} = 8.5$–9.5 V, the device broke down (the drivers twice and the receivers once).

After these experiments, it can be concluded that:

- No failures on the data stream were observed with the drain wire connected, which highlights the importance of this connection for receiver immunity.
- With a direct link, without drain wire, data failures were observed at a noise voltage of about 2 V. *Common-mode* voltage at the receiver has the same waveform of the impulsive noise applied between the two PCB grounds.
- With decoupling capacitors of 220 nF and without drain wire, the noise peak of 3.5–4 V can cause data failure. The series capacitor is useful when the offset voltage remains constant. *Common-mode* voltage at the receiver has a smaller peak value than the applied impulse noise.

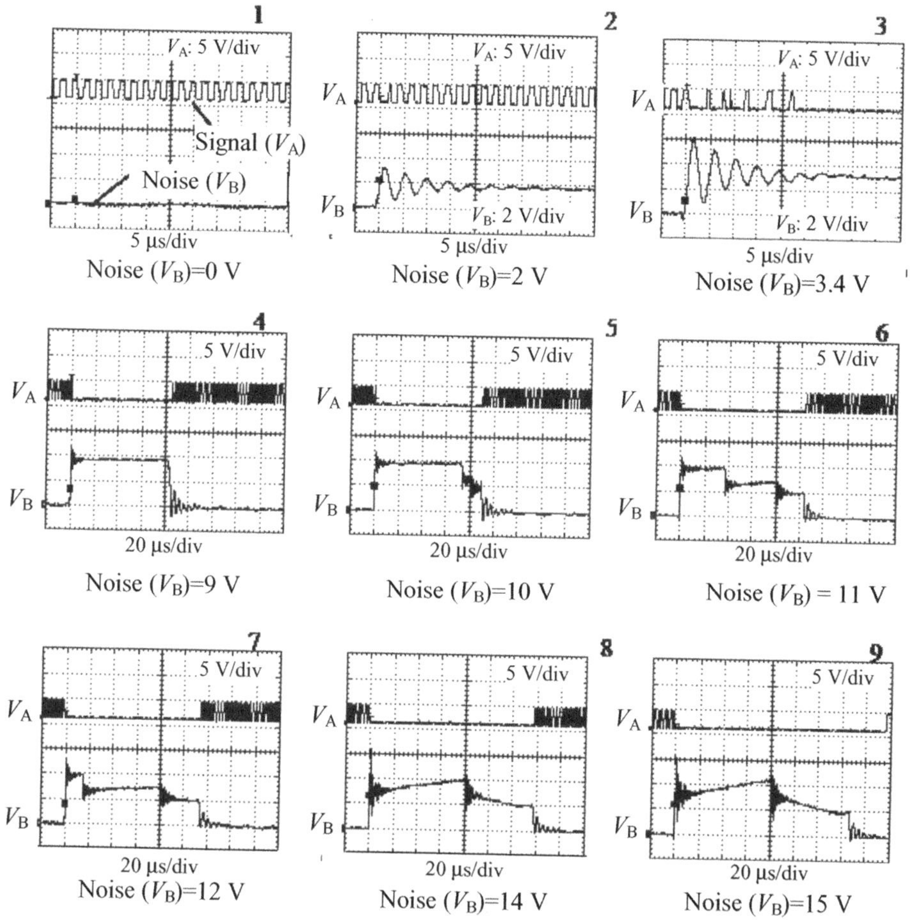

Figure 12.17 Signal interruption during a noise impulse with the basic link without drain wire: 1, signal without noise; 2, one bit error; 3, several bit errors; 4, transmission becomes regular when the noise application ends; 5–8, the devices change their characteristics with increasing noise level; 9, the driver stops working and transmits again after the return of power supply

- With a transformer and *common-mode choke*, and without drain wire, no data failures were observed with high-noise voltage up to 40 V.
- With static noise, breakdown of the LVDS devices can occur with a noise voltage of 8 V.
- Jitter increases slightly when a transformer and *common-mode choke* are used.

As recommendations for maximum noise immunity when using *Unshielded Twisted Pairs* (UTPs):

- Links with a transformer and *common-mode choke* should be used.
- The drain wire of the UTP cable must be connected to both grounds at both ends.

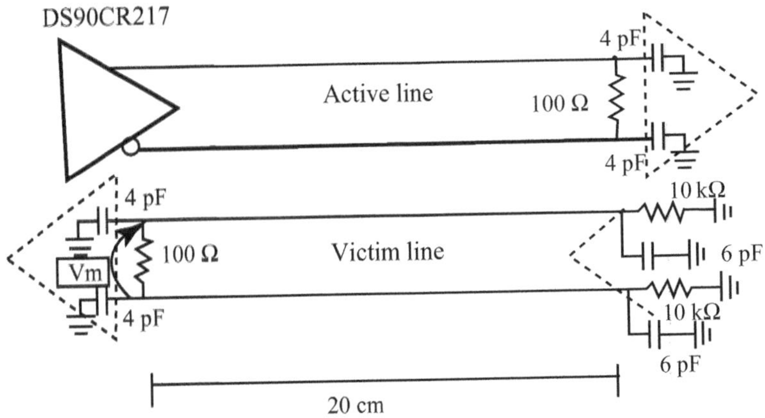

Figure 12.18 Two coupled pairs of differential lines with LVDS equivalent circuits for crosstalk investigations

Example 12.2: LVDS Crosstalk Investigation by Simulations

Performing simulations is the best way to set design rules for routing differential traces in PCBs. In the following, the spacing between two pairs of traces is investigated, considering their relative position.

The circuit model analyzed by the HSPICE simulator was carried out by simulating the lines with the distributed models presented in *Section 6.4*, and computing the per-unit-length line parameter with a field solver. DC and skin-effect losses in traces were accounted for. Dielectric losses were neglected, as the interest was focused on the maximum-level crosstalk at the near end (NEXT). The length of the lines was chosen to permit crosstalk to rise to its maximum value, and the orientations of the driver and receiver were chosen to simulate the worst case. The simulated PCB structure is typical of an actual design.

The adopted topology in the case of two differential lines with LVDS devices is shown in Figure 12.18. The driver in the active line was simulated by a macromodel of the DS90CR217 device, starting with the IBIS model provided by the manufacturer (The National Semiconductor). The rise and fall times of the driver $t_r = t_f = 700$ ps were computed between 10 and 90 % of the 700 mV swing. In the victim line, the driver is represented with a high impedance (10 kΩ between the two traces and 6 pF connected to ground) for the purpose of avoiding results dependent too much on a particular device and for better reproducibility. For the same reason, the receiver was simulated by two 4 pF capacitors connected to ground. The length of the lines was 20 cm and the differential traces were matched at the receiver end by a 100 Ω resistance.

Simulations were performed by considering several locations of the traces, as shown in Figure 12.19a. In the base configuration indicated by '0', the couples of traces for differential signaling had a width $w = 0.2$ mm, a thickness $t = 0.018$ mm, and a separation $s = 0.2$ mm. Adopting a relative dielectric constant $\varepsilon_r = 4.3$, the characteristic differential impedance is about 100 Ω. The simulated waveforms of the near-end crosstalk are shown in Figure 12.19b for a driver switching frequency of 33 MHz. The results refer to structures in which a couple of traces are kept at the same location while the other is horizontally shifted with a step equal to the width of the trace. Observe that the waveforms 0 and 1 in

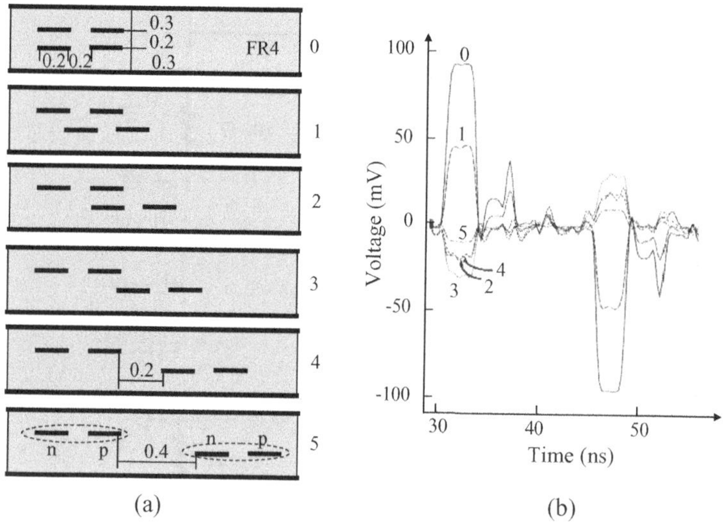

Figure 12.19 Crosstalk with LVDS: (a) PCB structures with dimension in mm; (b) Vm simulated waveforms by SPICE (see Figure 12.18)

Figure 12.19b have opposite sign to the each other. This is due to the fact that, for configurations 0 and 1, conductor p of the first pair of traces is mainly coupled with conductor p of the second pair of traces. For the other structures, the coupling between conductor p and conductor n is dominant. However, this fact is not so important, as the purpose is to reduce the crosstalk magnitude.

Another case of practical interest is the crosstalk between coplanar differential pairs with LVDS devices, as shown in Figure 12.20. There are two active lines disturbing a victim line in the middle. The receiver of the victim line is located at the same end as the drivers. Simulations with the receiver at the opposite end did not provide significant differences.

The basic structure is shown in Figure 12.21a, where the separation parameter s was varied. The results of the simulations are shown in Figure 12.21b. It was verified that the presence of a second active line doubles the crosstalk. Considering this fact, the case $s = 0.2$ mm provides a maximum crosstalk similar to case 4 in Figure 12.19a. For $s = 0.6$ mm, and considering one driver, the crosstalk is $0.5 \times 10/700 \times 100 = 0.7$ %. Making $s = 0.9$ mm as in Figure 12.3, the target of 0.2 % is obtained.

Other simulations of coplanar traces with different logic families led to the rules for minimum spacing shown in Table 12.3. The values reported ensure that crosstalk is a reasonable part of the total immunity of the receiver. Structure with parallel overhead traces in two adjacent layers must be absolutely avoided owing to the high level of crosstalk. Other information for signal integrity interconnect can be found in reference [24].

Example 12.3: Characterization of a Backplane for High-speed Application up to 3.125 GHz by Measurements

In this example, the experimental characterization of a backplane to verify the compliance with the AdvancedTCA requirements will be described. The backplane should meet the following specifications:

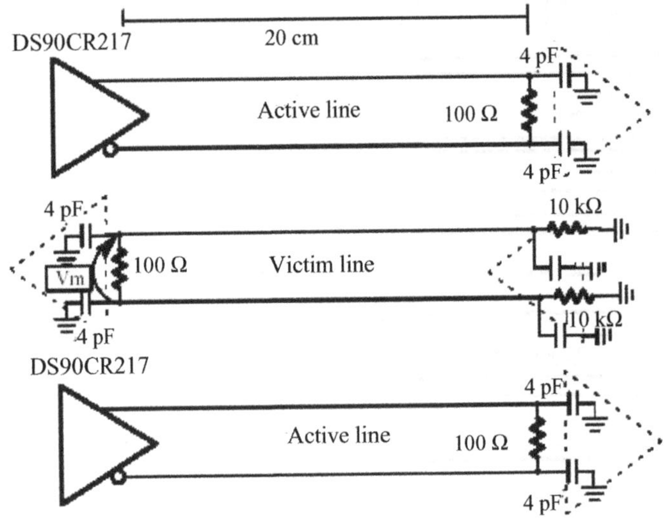

Figure 12.20 Three coupled pairs of differential lines with LVDS equivalent circuits for crosstalk investigations

- Maximum data rate 3.125 GHz.
- The differential impedance of the backplane serial links for the high-speed interfaces shall be 100 Ω ± 10 %.
- The backplane signal lengths, within the differential pair, shall be matched better than 3.4 ps (0.5 mm separation between the two ground planes and FR4 at $\varepsilon_r = 4$).

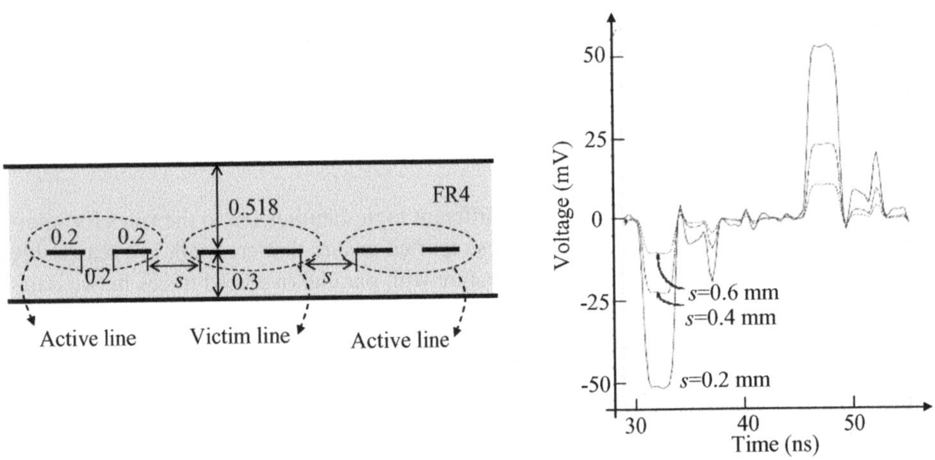

Figure 12.21 Crosstalk with two active LVDS devices and one victim line: (a) PCB structures with dimensions in mm; (b) Vm simulated waveforms by SPICE (see Figure 12.20)

Table 12.3 Minimum spacing between two coupled lines as a function of the technology

Active driver	Victim receiver	Minimum spacing between two-coupled lines
LVT (Single-ended)	LVDS (Differential)	1 mm
LVDS (Differential)	LVDS (Differential)	0.4 mm
RS422 (Differential)	RS422 (Differential)	0.4 mm
RS422 (Differential)	LVDS (Differential)	1 mm
LVT (Single-ended)	RS422 (Differential)	0.8 mm

- The transmitting and receiving pairs within a fabric channel routed across the backplane shall have a delay time in matched condition of 17 ps or less (0.254 mm trace separation in FR4 at $\varepsilon_r = 4$).
- Crosstalk lower than 0.2 % (near-end coupling coefficient NEXT) for a pair of traces with a separation of 0.889 mm (edge to edge).

The physical stack-up of a 14-layer motherboard is shown in Figure 12.22a. The notation GND means a layer dedicated to ground, POWER means an area of the plane dedicated to the power supply, and HF_i means layer i dedicated to traces, with $i = 1,2,\ldots,6$. The geometries of a pair of traces for differential transmission are chosen according to the ATCA requirements (see Figure 12.22b) in order to have an odd characteristic impedance $Z_{0o} = 50\ \Omega$ and therefore a differential characteristic impedance $Z_{0DM} = 2Z_{0o} = 100\ \Omega$.

Three types of measurement were carried out:

- TDR measurements to verify the required characteristic impedance;
- TDR measurements to verify the crosstalk;
- oscilloscope measurements to verify the jitter.

For all the measurements, a commercial test board, referred to as the *test blade*, was used to connect the motherboard with the instruments, as shown in Figure 12.23a. The test blade is an off-the-shelf ATCA solution designed by NESA and F9 Systems [24] for measurements up to and including 10 Gbps. It is particularly suitable to monitor the active transmission and received signals across an ATCA backplane to assure signal quality, to measure impedance and coupled *Near End Crosstalk* (NEXT), to measure skew and propagation delay. SMA connectors assure the connection between the traces and the instruments, as shown in Figure 12.23a.

To verify the odd impedance Z_{0o} of each trace of a differential interconnect, it is necessary to perform differential TDR measurements in order to excite both traces at the same time with complementary steps of amplitude 250 mV and a rise time of 25 ps, as shown in Figure 12.23b. The results of differential TDR measurements are shown in Figure 12.23c, where it can be noted that, for both traces in the test blade and in the backplane, the required odd impedance $Z_{0o} = 50\ \Omega$ is verified, as the reflected waveforms for the two boards are very low considering the 50 Ω reference waveform of the TDR. Some inevitable slight distortions can be observed owing to the SMA and daughter–motherboard connectors. The delay time of the 20 cm trace can be deduced by examining the time interval between the blade connector and

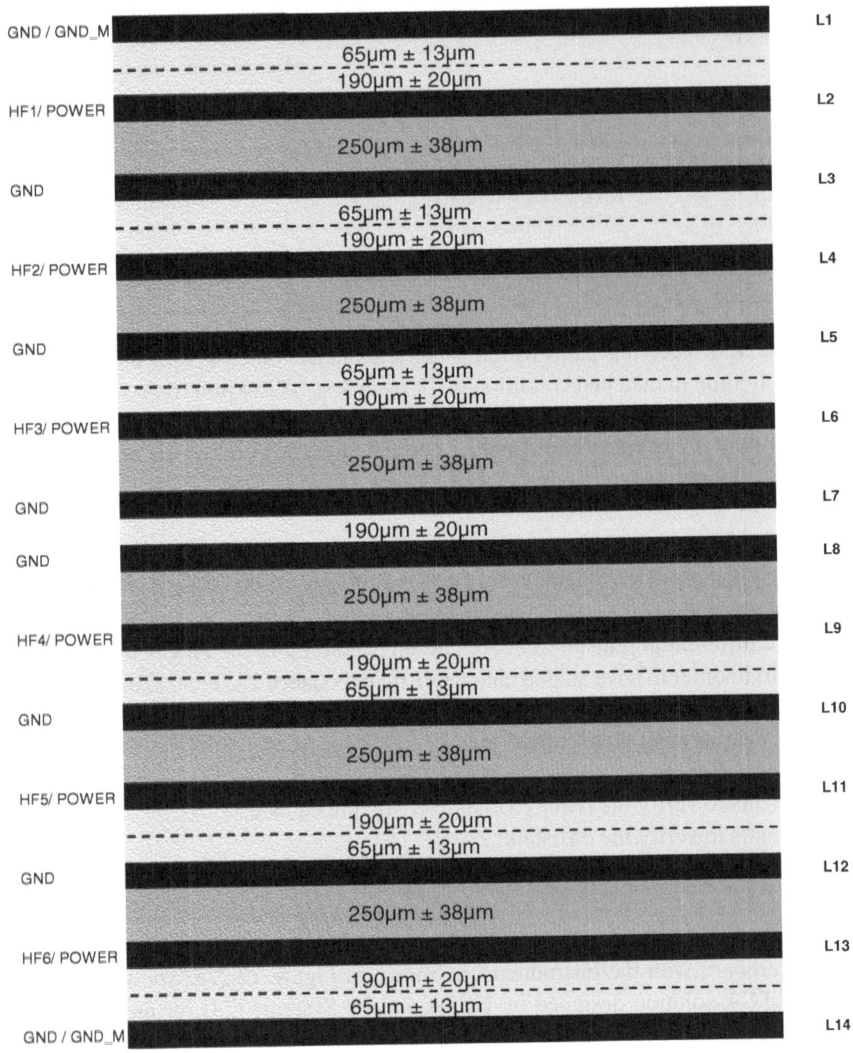

(a)

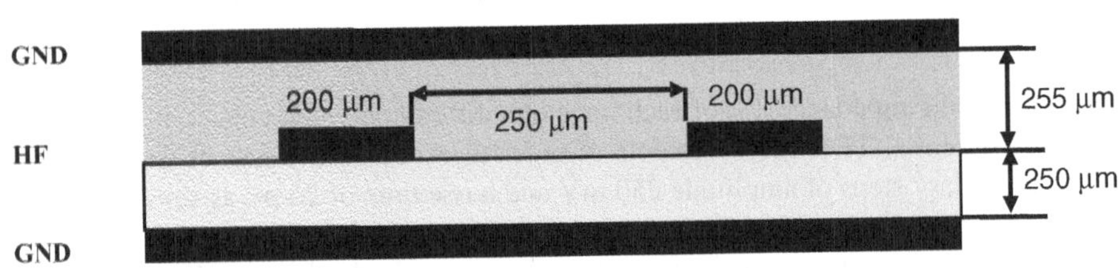

(b)

Figure 12.22 Example of a 14-layer board in FR4, thickness $t = 3.472$ mm $\pm$ 10 %: (a) stack-up; (b) details of differential traces for a differential characteristic impedance $Z_{0DM} = 100\ \Omega$ (courtesy of Dr Vittorio Ricchiuti, Technolabs, Italy)

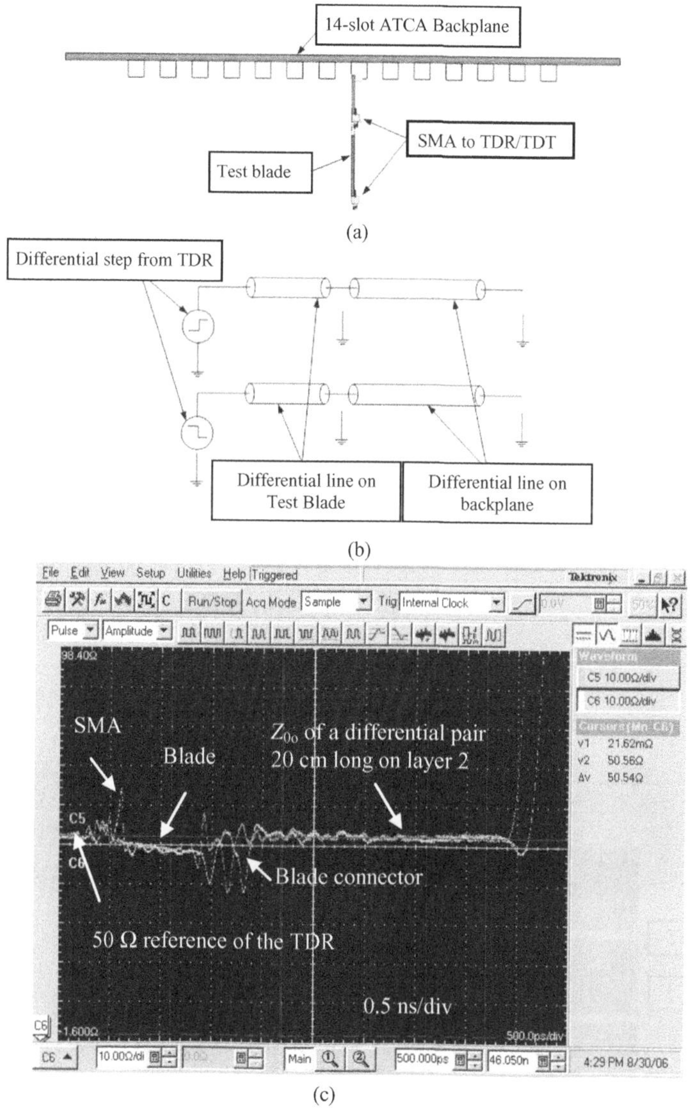

Figure 12.23 TDR measurements: (a) view of test board and backplane; (b) schematic of TDR and traces; (c) measured TDR waveforms showing that each trace for differential transmission has an odd characteristic impedance $Z_{0o} = 50\ \Omega$ (courtesy of Dr Vittorio Ricchiuti, Technolabs, Italy)

the moment when the waveform rises to high values, as the incident step sees an open line. In fact, this time is twice the time required by the TDR step to go from the connector to the open end of the trace. This delay being about 2.5 ns, the per-unit-length delay time of the trace is $t_{pd} = (2.5/2)/0.2 = 6.25$ ns/m.

Owing to its complementary outputs, the differential TDR is also suitable for crosstalk measurements, as shown in Figure 12.24a. The equivalent circuit is shown in Figure 12.24b.

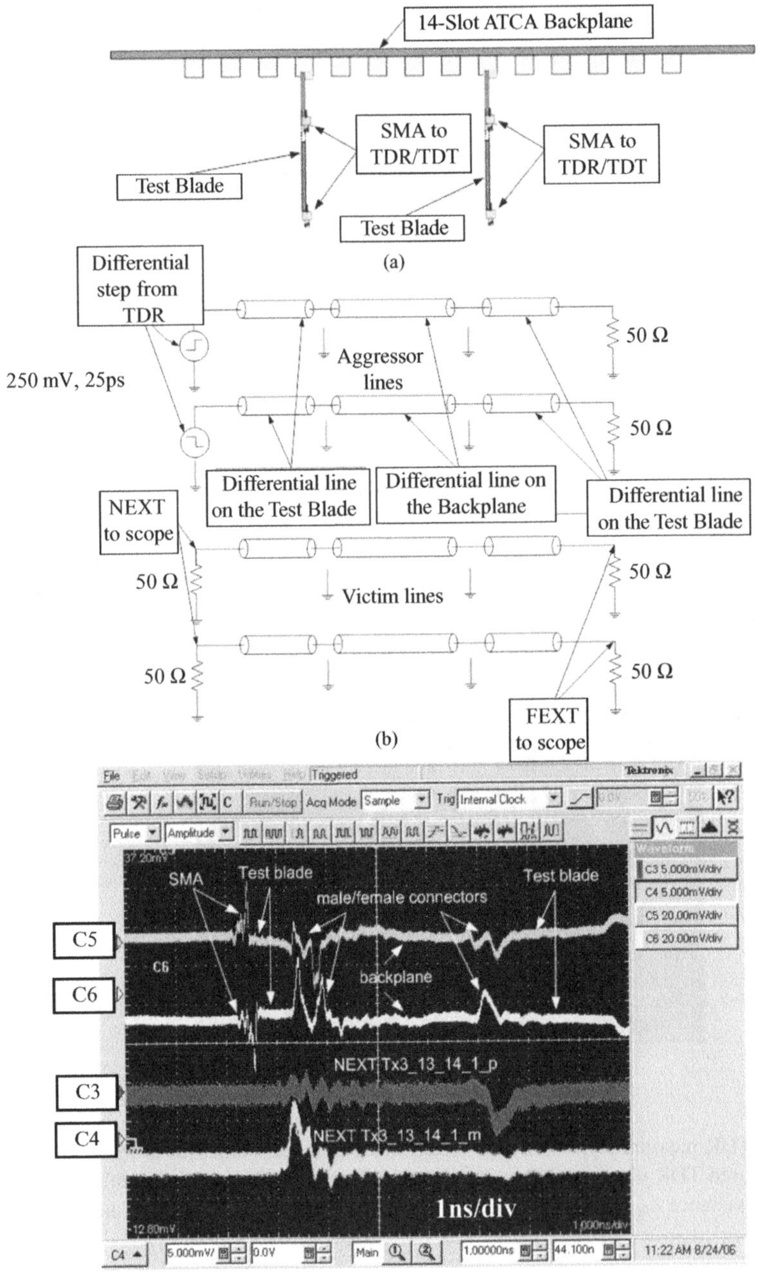

(c)

Figure 12.24 Crosstalk measurements: (a) view of test boards and backplane; (b) schematic of TDR and coupled traces; (c) measured waveforms showing distortions on aggressor lines owing to SMA and male/female connectors and near-end (NEXT) crosstalk on victim line (courtesy of Dr Vittorio Ricchiuti, Technolabs, Italy)

Of course, the two test boards have isolated pairs of traces for differential signaling. Therefore, the single-ended NEXT waveforms of Figure 12.24c concern the traces on the backplane only. As each trace is matched by its characteristic impedance, according to crosstalk theory, discussed in *Section 6.1*, NEXT should have a width equal to twice the delay time of the trace of 26 cm length in the backplane (i.e. $2 \times 6.25 \times 0.26 = 3.25$ ns), and a rise/fall time of 25 ps. With the scale adopted for waveforms *c3* and *c4* (i.e. 5 mV/div and 1 ns/div, as shown in Figure 12.24c), the crosstalk waveforms are flat, and therefore the requirement of <0.2 % crosstalk is satisfied. The peaks of about 5 mV and of less than 1 ns width are due to a slight mismatch at the connector, as evidenced in waveforms *c5* and *c6* of the aggressive line.

Eye diagram measurements are shown in Figure 12.25. In this case, one test blade is used to launch signals generated by pattern generators, and the other board is used to monitor the waveforms at the end of the pair of differential lines (see Figure 12.25a). The location of the differential line under test in the motherboard is shown in Figure 12.25b. Measured eye diagrams are shown in Figure 12.25c. Observe that the peak-to-peak jitter $t_{cs}/t_{ui} \times 100 = 40/320 \times 100 = 12.5$ % is suitable for this type of frequency and application, as the eye is evidently open.

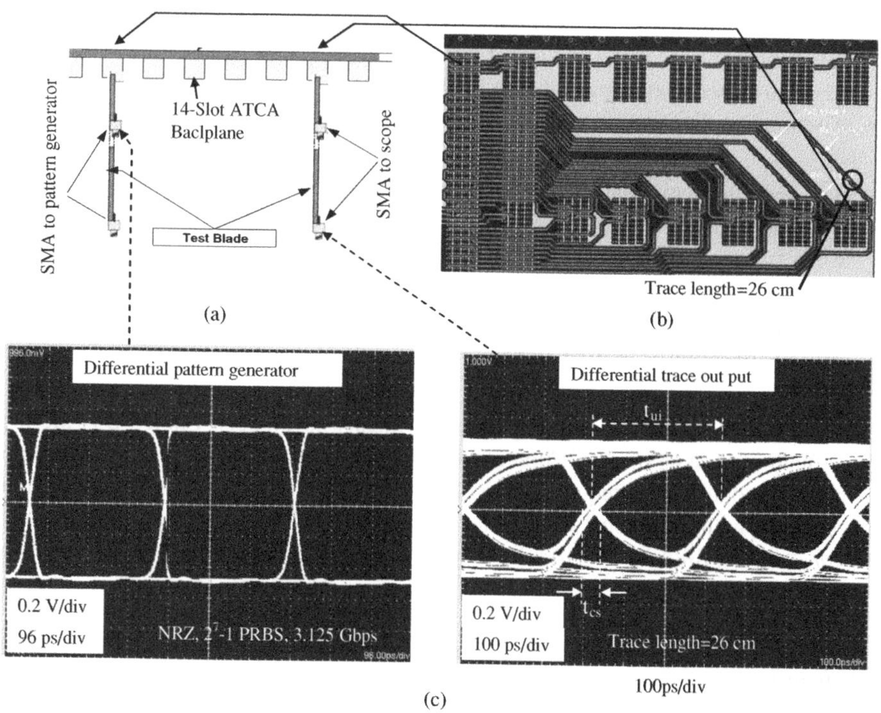

Figure 12.25 Eye diagram measurements: (a) view of test boards and backplane; (b) backplane layout; (c) measured waveforms at pattern generator location and at the output of the test blade (courtesy of Dr Vittorio Ricchiuti, Technolabs, Italy)

12.2 Modeling Packages and Interconnect Discontinuities in PCBs

In this last section of the book, the problem of how to extract equivalent circuits to simulate discontinuities in PCBs with SPICE is considered. These models can be used for simulating reflections and for computing extra delays along an interconnect with several discontinuities. Very useful tools for extracting the fundamental parameters for the model are those based on full-wave numerical solutions of field equations. Some examples are provided. The section ends with a brief discussion concerning the types of package used for digital devices in order to minimize the parasitic effects of their connections to the PCB.

12.2.1 Multilayer Boards

Discontinuities in a multilayer PCB such as bends, vias, connectors, IC packages, ground gaps, etc., should be simulated with an appropriate equivalent circuit to investigate their effects on signal integrity and how they generate EMI [25]–[28]. Every layout discontinuity must be properly accounted for, performing the simulations of the full interconnect including suitable macromodels of drivers and receivers, especially for high-speed differential transmission [29, 30]. Of particular importance are the discontinuities affecting current return paths of the signals. They have inductive effects, and the consequences are:

- to slow the edge rate by filtering out high-frequency components;
- signal integrity problems at the receiver if the current divergence path is long;
- to increase the current loop area and then EMI;
- to increase the coupling coefficient between signals.

Bends, vias, and connectors have complicated structures, and three approaches can be used for modeling their effects:

1. Transmission-line models.
2. Lumped-circuit elements.
3. Full-wave analysis and scattering responses.

Choice 1 is difficult to realize, which practically means no models. Choice 2 is the most useful for qualitative assessments. Choice 3 should be applied when an accurate equivalent circuit of the discontinuity is required for SPICE simulations. The equivalent circuit should reproduce the S-parameters computed by the full-wave code [31]–[33] or measured directly by a *Vector Network Analyzer* (VNA). Some examples of these procedures will be provided in this section.

12.2.2 Bends

Bends in a real PCB are necessary owing to the high density and consequently routing constraints regarding traces. An example is shown in Figure 12.26a. A lumped model for a bend is a shunt capacitor, as shown in Figure 12.26b. It has been demonstrated through correlation with empirical measurements that a simple lumped-capacitance model is adequate for most

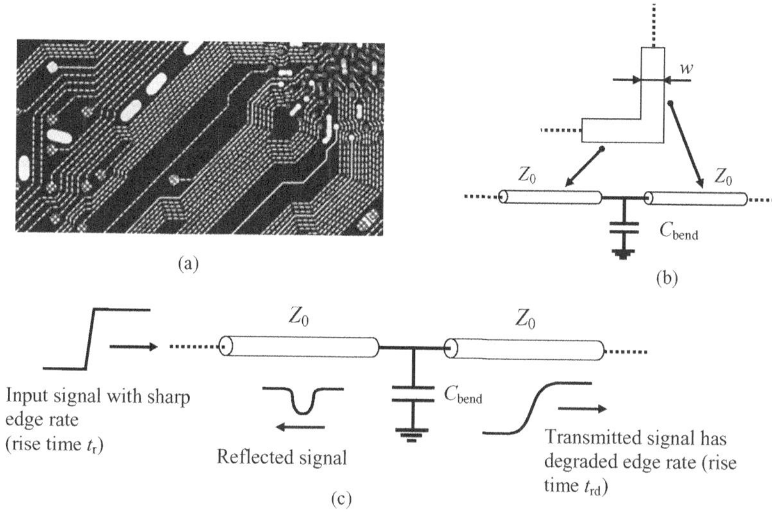

Figure 12.26 Bends: (a) example in a PCB; (b) equivalent circuit; (c) rise time degradation

systems [34]. The value of the bend capacitance C_{bend} is

$$C_{\text{bend}} = Cw \quad \text{for } 90^\circ \text{ bends} \tag{12.2a}$$

$$C_{\text{bend}} \ll Cw \quad \text{for } 45^\circ \text{ bends} \tag{12.2b}$$

where $C = t_{\text{pd}}/Z_0$ is the per-unit-length line capacitance, w is the width of the strip, and t_{pd} and Z_0 are respectively the per-unit-length delay time and the characteristic impedance of the trace forming the bend. For a 90° bend the capacitance value is equal to that of a transmission-line segment equal to its width or, in other words, it is the capacitance of the square portion of trace area that joins the two vertical traces forming the bend. Although this capacitance is small, it can cause signal integrity problems if several bends are present along the trace. A good solution for mitigating bend effects significantly is to chamfer the edge by 45°, as shown in Figure 12.26a. The capacitance of a 45° bend is significantly smaller than that of a 90° bend. Figure 12.26c illustrates how the signal rise time is changed by the bend according to

$$t_{\text{rd}} = \sqrt{t_{\text{bend}}^2 + t_{\text{r}}^2} \approx \sqrt{(2.2Z_0C_{\text{bend}}/2)^2 + t_{\text{r}}^2} \tag{12.3}$$

where t_{bend} takes into account the effect of the shunt capacitance C_{bend}.

For a more accurate assessment of signal integrity regarding bends, a 3D simulation is the most appropriate tool, especially when other phenomena such as coupling between the same trace occurs, as will be shown with the serpentine example in the following section.

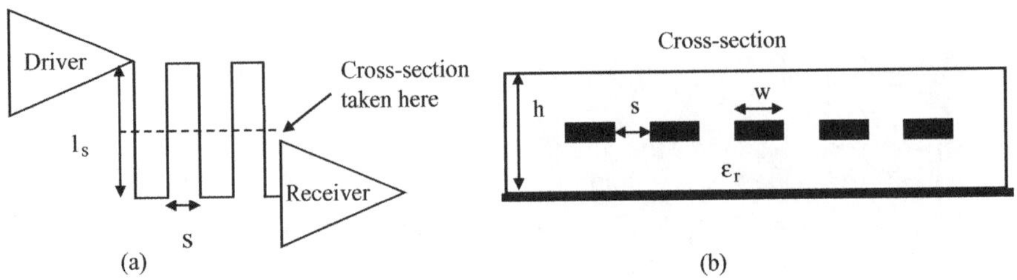

Figure 12.27 Serpentine: (a) routing; (b) cross-sectional view

12.2.3 Serpentines

Usually, in actual PCBs, it is not possible to route a trace in a perfectly straight line in order to have a well-controlled characteristic impedance and delay time. Board constraints such as geometries, high density layout, and timing requirements force the traces to be routed in serpentine patterns, as illustrated in Figure 12.27a. For timing reasons, serpentine traces are also often used to delay the data with respect to the clock in order to enhance the hold time, or to equalize trace lengths.

Care must be taken in routing the trace, as parallel sections could be coupled, causing signal integrity problems and EMI effects. The design rules to follow can be summarized as follows [34]:

- A spacing s greater than 3–4 times the substrate height h should be used.
- The length l_s of parallel sections should be minimized.
- Serpentines should be avoided in the case of clock signals.

12.2.4 Ground Slot

In actual PCBs it is not rare for a trace with the structure of a microstrip to have to cross a gap in the return ground plane, as shown in Figure 12.28. A slot in the ground adds inductance to a trace passing perpendicularly over the slot, creating signal integrity and EMI problems

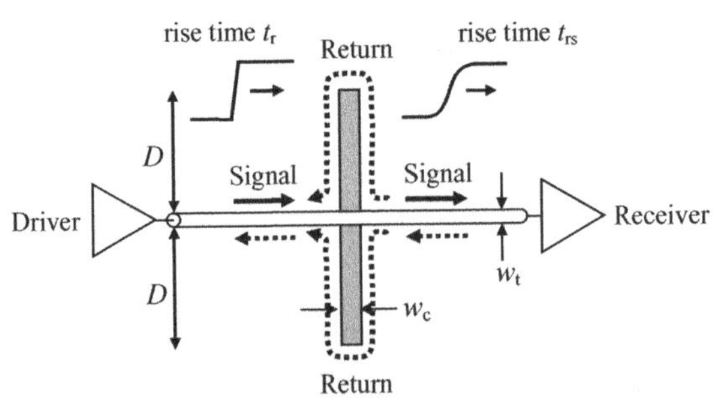

Figure 12.28 Current paths in a slotted ground plane and signal rise time degradation

[35, 36]. As shown in Figure 12.28, the return current from the driver cannot flow directly under the trace but diverts around the ends of the ground slot. Only a little portion of the return signal current flows through the gap capacitance. The diverted current flows on a large loop and greatly increases the inductance of the signal path. The effective inductance associated with the gap can be considered in series with the trace, and it can be estimated approximately using the expression of a loop formed by two flat parallel conductors with center-to-center distance d, and having width w and length l. Assuming that the return current around the gap flows onto two parallel conductors of width $w = 3w_t$ and length $l = D$, and spaced by $d = w_c + 3w_t$, the inductance associated with the gap according to Table A2 of *Appendix A* is

$$L_{slot} = \frac{1}{2} 2 \frac{\mu_o}{2\pi} D \left[\ln \left(\frac{3w_t + w_c}{3w_t + t_t} \right) + \frac{3}{2} \right] \quad \text{in H} \tag{12.4}$$

where t_t is the thickness of the trace. In Equation (12.4), the factor 1/2 is added to take into account that the slot causes the effect of two loop inductances in parallel, and factor 2 takes into account that we are interested in calculation of the inductance of one of the loops which is twice the effective inductance associated with one branch of the loop. In this calculation, for simplicity, the edge effects are neglected.

Example 12.4: Numerical Simulation of the Distortion Introduced on Signals by a Slot in the Ground Plane

To validate the slot inductance expression (12.4), numerical computations were performed for the test PCB shown in Figure 12.29a by using MWS [22]. The test PCB has the dimensions 20 mm × 100 mm × 0.77 mm and a thickness of the dielectric layer of 0.7 mm, with $\varepsilon_r = 4.4$, $w_t = 0.35$ mm, $t_t = 0.035$ mm, trace length $l = 100$ mm, a gap created in the middle of the PCB of size $D = 7.5$ mm, and $w_c = 5$ mm (see Figure 12.28 for the notation of different parameters). The source was an ideal voltage source with a trapezoidal waveform of 1 V amplitude, rise and fall times $t_r = t_f = 0.1$ ns, pulse width $t_{pw} = 5$ ns, and $t_{tot} = 10$ ns. The trace was loaded with its characteristic impedance $Z_0 = 90.36\ \Omega$. Introducing these values in Equation (12.4) yields $L_{slot} = 4.82$ nH. This is a high value that produces significant distortion on the signal line and increases *common-mode* emission, especially when a cable is attached to the PCB; recall that a typical effective inductance associated with a ground plane is less than 1 nH for actual PCBs having the traces close to its reference plane (see *Chapter 9*).

It is interesting to look at the surface current distribution around the slot, computed by MWS at 1 GHz and shown in Figure 12.29b. Note that the assumption adopted in Equation (12.4) that most of the return current concentrates in a space equal to $3w_t$ is realistic.

The comparison between waveforms obtained by SPICE and MWS is shown in Figure 12.30, where a good agreement can be observed for two kinds of slot having different width w_c. These results confirm the validity of Equation (12.4). For a gap of 5 mm there is a negative reflection of 25 % of the signal. This signal distortion cannot be tolerated in a high-speed digital system, and a trace crossing a gap must be absolutely avoided. If routing a trace across a gap cannot be avoided, a stitching capacitor placed across the gap and close to the

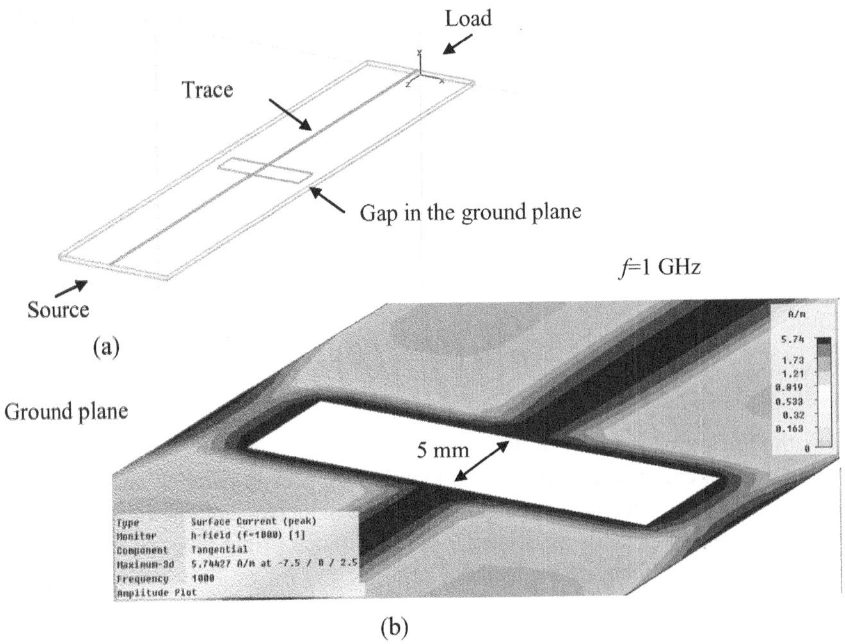

Figure 12.29 PCB with a gap in the ground plane: (a) PCB structure modelled by MWS; (b) tangential surface current plot in proximity of the gap at 1 GHz

signal trace can be a solution to overcome this problem. However, this fix cannot always be realized.

A good solution is to use differential transmission. Simulations by MWS were performed for the PCB structure shown in Figure 12.31a, where a second trace was added to the PCB of Figure 12.29a to form a differential line. The separation between the two traces (edge to edge) was $s = 0.65$ mm. Two ideal voltage sources having complementary waveforms were used to excite the traces with equal but opposite currents. The load was a resistance between the end of the pair traces with value $Z_{0DM} = 200\ \Omega$. In fact, to have a perfect matching, the load should be equal to twice the odd characteristic impedance of two symmetric coupled lines. Note that, with a differential transmission, the negative reflection is a very low percentage of the signal. This confirms an assertion made in *Section 10.3.4*: to increase the efficiency of a *common-mode choke* used as an EMI filter, the reference area between the device and the connector could be removed in the case of differential transmission. This fix lowers the parasitic capacitances, which could create an alternative path between the traces before and after the filter. The best performance of differential transmission over single-ended transmission is seen when the reference plane for return current is not solid. This is also confirmed by the investigation reported in reference [44] on PCBs with the power plane having a planar *Electromagnetic BandGap* (EBG) structure for mitigating simultaneous switching noise propagation. In reference [44] it is shown that an eye diagram with differential transmission is significantly better than that obtained with single-ended transmission for EBG structures.

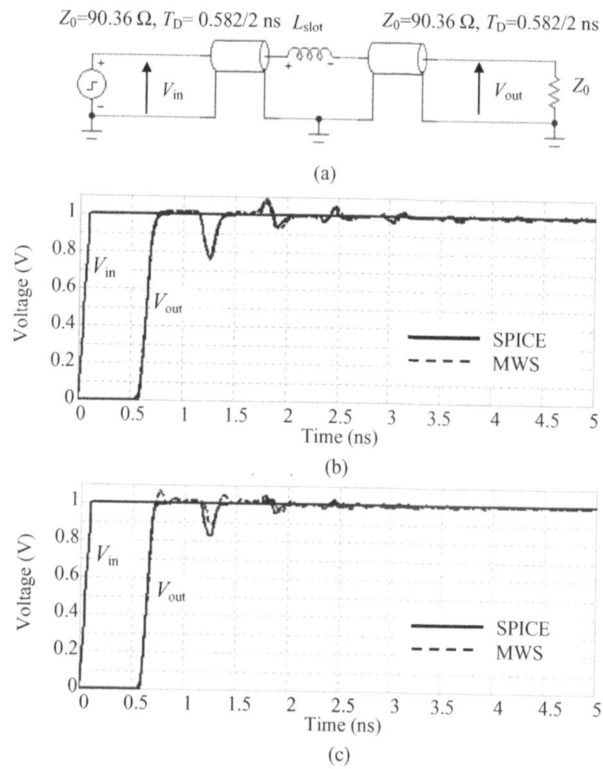

Figure 12.30 Simulated waveforms when a ground plane has a gap: (a) equivalent circuit for SPICE; (b) results with a gap having $w_c = 5$ mm; (c) results with a gap having $w_c = 1$ mm

To conclude this section, for single-ended connections the degradation of the signal rise time can be computed as

$$t_{rs} = \sqrt{t_{slot}^2 + t_r^2} \approx \sqrt{(2.2(L_{slot}/2Z_0))^2 + t_r^2} \tag{12.5}$$

12.2.5 Vias

A via is the way to connect two traces belonging to different layers of a PCB or to connect components to traces [34]. Figure 12.32a shows an example of a via connecting a trace on layer 1 with a trace on layer 2. The via consists of the barrel, the pad, and the antipad, as depicted in Figure 12.32b. The barrel is a conductive material that fills the hole to permit an electrical connection between layers, the pad is used to connect the barrel to the trace component, and the antipad is a clearance hole between the pad and the metal plane on a layer to which connection is not required. The via could be of the through-hole type because it is made by drilling a hole through the board. Other vias are blind, buried, and microvias (see

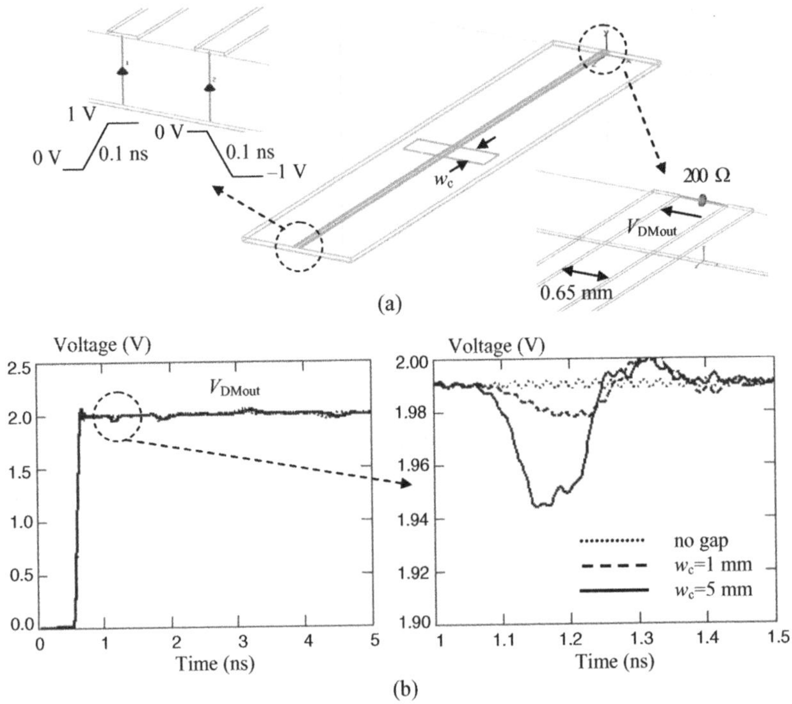

Figure 12.31 Simulated waveforms when a ground plane has a gap and the signal transmission is differential: (a) MWS model of the structure; (b) voltage V_{DMout} on the load without a gap, with a gap having $w_c = 5$ mm and $w_c = 1$ mm

Figure 12.32a for examples). When the maximum dimension of a via is much less than the minimum wavelength of interest, or, in other words, is electrically short, the via can be represented by the π-net circuit of Figure 12.32c. The capacitors represent the via pad capacitance on layers 1 and 2. The series inductance represents the barrel. The capacitance [37] and the inductance (see the isolated wire in Table A.1 of *Appendix A*) can be estimated by the following closed-form expressions:

$$C_{pad} \approx 0.71 \times 10^{-12} \varepsilon_r 39.37t \frac{d_1}{d_2 - d_1} \quad \text{in F} \tag{12.6a}$$

$$L_{barrel} \approx \frac{\mu_0}{2\pi} h \left[\ln(4h/d_1) - 1\right] \quad \text{in H} \tag{12.6b}$$

where d_1 is the via barrel diameter, d_2 is the via pad diameter, t is the distance between the pad and the nearest reference plane, h is the via length, $\mu_0 = 4\pi 10^{-7}$ H/m, and the length must be in meters.

However, the best way to extract an equivalent circuit of a via is to calculate S-parameters by using a 3D numerical code and compare the results with an equivalent circuit implemented in SPICE. This procedure is explained by *Example 11.4* in *Section 11.2.3*, where the validation

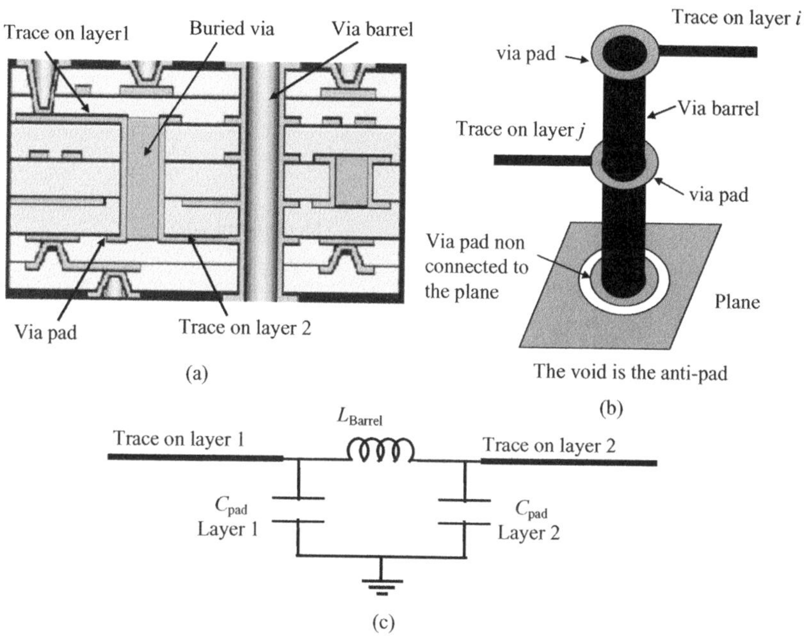

Figure 12.32 Via: (a) cross-view of a via in a PCB; (b) 3D view; (c) equivalent circuit

limit of the π-model is also discussed. More accurate procedures for characterizing vias can be found elsewhere [38–40]. It is important to point out that the inductance computed by Equation (12.6b) refers to an isolated round wire, and it is this value that mainly deteriorates the signal integrity. The effective inductance associated with the via can be lowered if the return signal current flows in a via close to the signal via, as explained in *Section 10.2.3*. In fact, the mutual inductance between the vias has a positive action, as it must be subtracted from the self-inductance of the signal via to calculate the effective inductance associated with the via. This can be accomplished by taking care to place a decoupling capacitor in a manner such that its via, connecting the capacitor to the ground or power plane of interest and acting as signal return current, must be close to the signal via. If this does not occur, because the signal current follows a return path with minimum impedance, the return current has two possibilities: one is the nearest via available for the return current, usually causing a large loop of current at low frequencies; the other is the displacement current between the planes crossed by the signal via, and this occurs when the frequencies are so high that the local capacitance between the plane allows the flow of the return current. Of course, since this last situation cannot be controlled, it should be avoided in order to minimize signal integrity and EMI problems.

12.2.6 Connectors

Connectors are components that are used to connect one printed circuit board to another. An example of a connector is shown in Figure 12.33. Note that the conductors of the connector

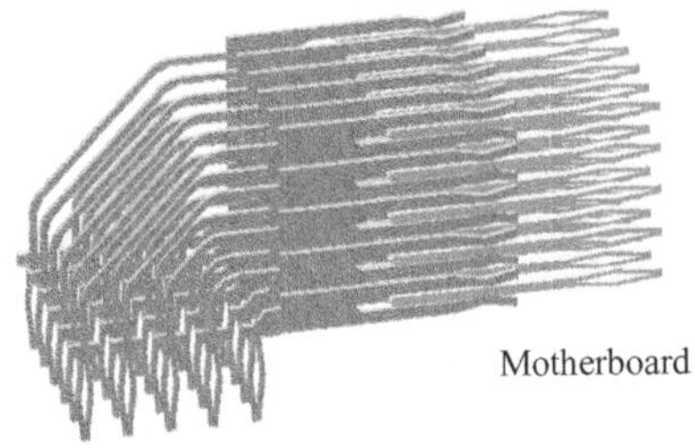

Figure 12.33 Example of a connector for PCBs

have a complex geometry; they do not follow a straight path, and only those that share the same row have equal length. The ground pins are generally longer than the others for safety reasons. In fact, during the insertion of the PCB onto the powered system, it is essential that the ground pins of the PCB contact the ground system first to avoid damage. This makes modeling of the connector extremely hard without using measurements or 3D field solvers to extract a connector equivalent circuit. Good examples of connector modeling are given elsewhere [41, 42]. All the self and mutual inductances and capacitance parameters should be considered, as there are two problems in the connector: crosstalk among the pins and voltage drops on the ground and power pins owing to the return of the signal currents. In fact, the inductance associated with power and ground pins produces common-impedance coupling, as the return current of the signals shares the same ground and power pin.

This section presents an approximate approach for investigating connector problems, based on the fact that the capacitance effect can be neglected when the maximum dimension of the connector is electrically short. This can be done because the main effect of the connector capacitances is to slow down the system edge rate [34]. For this reason, the discussion will be focused on the inductance effects.

Consider the simple connector structure depicted in Figure 12.34a. The dominant effect of connectors is accounted for by a series-lumped self partial inductance given by the simplified expression (see Table A1 of *Appendix A* for an isolated wire)

$$L_{\text{con}} \approx \frac{\mu_0}{2\pi} l\,(\ln(2l/r) - 1) \quad \text{for } r \ll l \text{ and } f \to \infty \tag{12.7a}$$

and a mutual partial inductance between two pins given by the simplified expression (see Table A1 of *Appendix A* for two parallel wires with opposite currents)

$$M_{\text{con}} \approx \frac{\mu_0}{2\pi} l\,(\ln(2l/s) - 1) \quad \text{for } s \ll l \tag{12.7b}$$

where l is the length of the pin, r is the average radius of the pin, and s is the spacing between two pins. For s comparable with l, M_{con} is given by the more complete closed-form expression of Table A.1 regarding two parallel wires.

These expressions can be used for segments of parallel pins (see Figure 12.34a and its equivalent circuit shown in Figure 12.34b). The shape of the connector pin is not relevant for

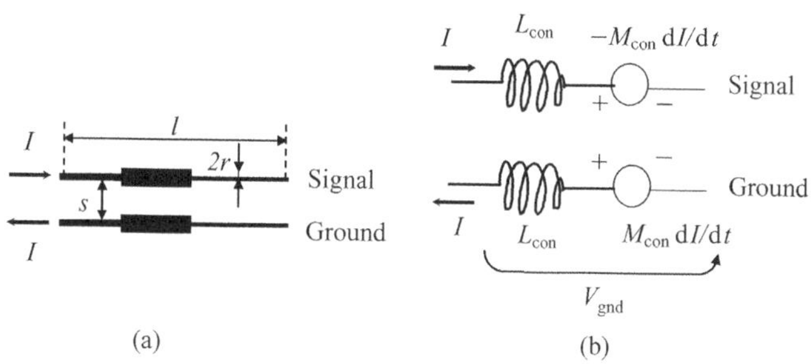

Figure 12.34 Connector: (a) two coupled pins; (b) equivalent circuit

modeling, and therefore an average pin radius can be adopted for calculations. The voltage drop on the ground pin caused by other $n - 1$ signal pins with current I, having the same self inductance and approximately the same mutual inductance, is

$$V_{\mathrm{gnd}}(t) \approx n(L_{\mathrm{con}} - M_{\mathrm{con}})\frac{\mathrm{d}I}{\mathrm{d}t} \tag{12.8}$$

The higher the coupling between signal and ground/power pins, the higher is the mutual inductance. As the self inductance of a pin is independent of the spacing between two coupled pins, a lower voltage drop V_{gnd} on the ground/power pins occurs. Therefore, this simple expression for V_{gnd} makes it possible to establish some fundamental design rules:

- The number of power/ground pins should be larger than the number of signal pins, which in practice means $n = 1$ in Equation (12.8).
- Signal pins should be close to their current return pins, which means high mutual inductance.
- Power and ground pins should be adjacent to maximize associated mutual inductance.
- Pin connectors should be 'short', as pin inductances depend on length l.

In *Section 10.3*, examples of ground noise calculation and SPICE simulations to explore EMI effects and deterioration in the signal integrity caused by *common-mode* coupling in the connector have been provided. Pin assignment for connectors with several pins has also been discussed.

Once all the self and mutual partial inductances are known by Equations (12.7), calculation of the voltage drop along each signal pin, useful for calculating crosstalk, can be done following the procedure outlined in *Section 3.2.6*, which links the concept of loop inductances with the concept of partial inductances.

In the case of differential transmission and the connector of Figure 12.33, it is very important to choose as pair pins two adjacent pins in the same row in order to avoid skew.

12.2.7 IC Package

Connections of chips to boards are also points of discontinuity [34]. Figure 12.35 shows two examples. Holders provide mechanical, thermal, and electrical connections of chips to boards.

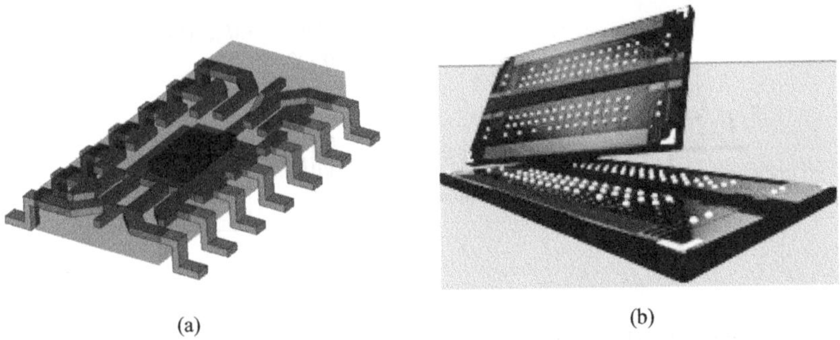

Figure 12.35 IC package: (a) lead frame directly soldered on the PCB; (b) *Pin Grid Arrays* (PGAs) or *Ball Grid Arrays* (BGAs) stick out of the bottom of the package (as in the flip chip attachment) to a socket or to the board

They can be classified according to:

- type of die attachment;
- type of package connection;
- type of package attachment to board.

Some considerations about packages can be summarized as follows:

- Connection on packages can be routed via either controlled or non-controlled impedance traces.
- A controlled impedance package looks like a miniature multilayer PCB (it can accommodate more than one die).
- Non-controlled impedances are mostly used with wire bond attachment, where wires are directly connected to the lead frame (this means worse electrical performance).

Die attachment can be performed by wire bond or by the flip chip technique, as shown in Figure 12.36. A wire bond consists of a chip, mounted with pads on top, connected to the package by thin wires (~25 μm). The characteristics of the wire bond are:

- good mechanical and thermal connection;
- simple routing;
- large (and variable from wire to wire) series inductance;
- difficult to model.

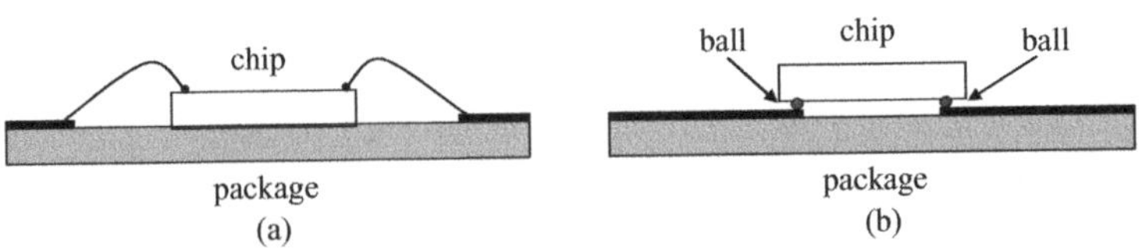

Figure 12.36 Type of die attachment: (a) wire bond; (b) flip chip

Table 12.4 Inductances and minimum spacing between two coupled lines as functions of the technology

	Wire bond	Flip chip
Series inductance (nH)	1–5	0.1
Minimum pitch (μm)	100–150	50–80

A flip chip consists of a chip mounted upside-down with pads on the bottom, which are connected to the package by solder balls. The characteristics of the flip chip are:

- worse thermal connection;
- finer pad pitch;
- smaller and predictable series inductance.

Typical electrical and geometrical parameters of the wire bond and flip chip are shown in Table 12.4.

Designers must be aware of package effects, which can be summarized as follows:

- Wire bond packages with lead frame attachment have the worst electrical performance, causing large series inductance and mutual inductances between wires, slow down edge rates, cause crosstalk, and exacerbate *Simultaneous Switching Noise* (SNN) effects.
- Wire bond packages with controlled impedance interconnects mainly cause a slower edge rate.
- Flip chip packages are mandatory for very high-speed applications.
- Parasitic circuit elements depend on the package shape and vary from pin to pin.
- Square packages minimize the mismatch of the parasitic elements of the pins.

When using IBIS models, make sure that the parasitic elements of the package are included. Further details about SPICE models of integrated circuits for EMI behavioral simulation can be found in the IEC 62433-2 standard [43]. The objective of this standard is to propose a model for describing the conducted emissions of an integrated circuit at the chip or multichip and PCB level and for power integrity analysis.

References

[1] Young, B., *'Digital Signal Integrity: Modeling and Simulation with Interconnects and Packages'*, Prentice Hall PTR, Upper Saddle River, NJ, 2001.

[2] Johnson, H. and Graham, M., *'High-Speed Signal Propagation: Advanced Black Magic'*, Pearson Education, Inc., Prentice-Hall PTR, Upper Saddle River, NJ, 2003.

[3] Thierauf, S.C., *'High-Speed Circuit Board Signal Integrity'*, Artech House, Inc., Norwood, MA, 2004.

[4] Montrose, M., *'EMC and the Printed Circuit Board: Design, Theory, and Layout Made Simple'*, IEEE, Inc., New York, NY, 1999.

[5] Blood, W., *'MECL System Design Handbook'*, 4th Edition, Motorola Semiconductor Products, Inc., Phoenix, AZ, 1983.

[6] *'Designing with PECL'*, AN-1406, Motorola Semiconductor. Available online: www.freescale.com.

[7] *'High-Speed Layout Design Guidelines'*, AN-2536D, Motorola Semiconductor. Available online: www.freescale.com.

[8] *'LVDS Owner's Manual'*, National Semiconductor, Spring 1997. Available online: www.national.com.

[9] *'A Comparison of Differential Termination Techniques'*, AN-903, National Semiconductor. Available online: www.national.com.

[10] *'Summary of Well Known Interface Standards'*, AN-216, National Semiconductor. Available online: www.national.com.

[11] *'Transmission Line Drivers and Receivers for TIA/EIA Standards RS-422 and RS-423'*, AN-214, National Semiconductor. Available online: www.national.com.

[12] *'LVDS Description and Family Characteristics'*, MS500530, Fairchild Semiconductor. Available online: www.fairchildsemi.com.

[13] *'LVDS Reduces EMI'*, AN-5020, Fairchild Semiconductor. Available online: www.fairchildsemi.com.

[14] *'LVDS Fundamentals'*, AN-5017, Fairchild Semiconductor. Available online: www.fairchildsemi.com.

[15] High-speed Differential Data Signalling and Measurements, AN, Tektronix, 2003. Available online: www.tektronix.com.

[16] AdvancedTCA, *'Base Specification'*, PICMG_3_0_R2.0. Available at: www.picmg.org.

[17] Leone M. and Navratil V., 'On the External Inductive Coupling of Differential Signaling on Printed Circuit Boards', *IEEE Trans. on Electromagnetic Compatibility*, **46**(1), February 2004, 54–61.

[18] Kam, D., Lee, H., and Kim, J., 'Twisted Differential Line Structure on High-speed Printed Circuit Boards to Reduce Crosstalk and Radiated Emission', *IEEE Trans. on Advanced Packaging*, **27**(4), November 2004, 590–596.

[19] Shiue, G.-H., Guo, W.-D., Lin, C.-M., and Wu, R.-B., 'Noise Reduction Using Compensation Capacitance for Bend Discontinuity of Differential Transmission Lines', *IEEE Trans. on Advanced Packaging*, **29**(3), August 2006, 560–569.

[20] *'Electrical Characteristics of Low Voltage Differential Signalling (LVDS) Interface Circuits'*, TIA/EIA-644, 1 February 2001.

[21] *'Electrical Characteristics of Multipoint-Low-Voltage Differential Signalling (M-LVDS) Interface Circuits for Multipoint Data Interface'*, TIA/EIA-899, 2 March 2000.

[22] www.cst.com.

[23] *'GTLP: An Interface Technology for Bus and Backplane Applications''*, AN-1065, Fairchild Semiconductor. Available online: www.fairchildsemi.com.

[24] www.nesa.com and www.f9-systems.com.

[25] Fan, J., Drewniak, J., and Knighten, J., 'Lumped-circuit Model Extraction for Vias in Multilayer Substrates', *IEEE Trans. on Electromagnetic Compatibility*, **45**(2), May 2003, 272–280.

[26] McGibney, E. and Barrett, J., 'An Overview of Electrical Characterization Techniques and Theory for IC Packages and Interconnects', *IEEE Trans. on Advanced Packaging*, **29**(1), February 2006, 131–139.

[27] Wang, C.-L. and Wu, R.-B., 'Modeling and Design for Electrical Performance of Wideband Flip-chip Transition', *IEEE Trans. on Advanced Packaging*, **26**(4), November 2003, 385–391.

[28] Eo, Y., Eisenstadt, W., Jin, W., Choi, J., and Shim J., 'A Compact Multilayer IC Package Model for Efficient Simulation, Analysis, and Design of High-performance VLSI Circuits', *IEEE Trans. on Advanced Packaging*, **26**(4), November 2003, 392–401.

[29] Stievano, I., Maio, I., Canavero, F., and Siviero, C., 'Parametric Macromodels of Differential Drivers and Receivers', *IEEE Trans. on Advanced Packaging*, **28**(2), May 2005, 189–196.

[30] Stievano, I., Maio, I., Canavero, F., and Siviero, C., 'Reliable Eye-diagram Analysis of Data Links via Device Macromodels', *IEEE Trans. on Advanced Packaging*, **29**(1), February 2006, 31–38.

[31] Araneo, R. and Celozzi, S., 'A General Procedure for the Extraction of Lumped Equivalent Circuits from Full-wave Electromagnetic Simulations of Interconnect Discontinuities', *Proceedings of 15th International Zürich Symposium on Electromagnetic Compatibility*, Zürich, Switzerland, February 2003, 419–424.

[32] Araneo, R. and Celozzi, S., 'Extraction of Equivalent Lumped Circuits of Discontinuities Using the Finite-difference Time-domain Method', *IEEE 2002 International Symposium on Electromagnetic Compatibility*, Minneapolis, MN, August 2002, 119–122.

[33] Araneo, R. and Maradei, F., 'Passive Equivalent Circuits of Complex Discontinuities: an Improved Extraction Technique', *IEEE 2005 International Symposium on Electromagnetic Compatibility*, Chicago, IL, August 2005, 700–704.

[34] Hall, S., Hall, G., and McCall, J., *'High-Speed Digital System Design – a Handbook of Interconnect Theory and Design Practice'*, John Wiley & Sons, Inc., New York, NY, 2000.

[35] Chen, J., Shi, W., Norman, A., and Ilavarasan, P., 'Electrical Impact of High-speed Bus Crossing Plane Split', *IEEE Symposium on Electromagnetic Compatibility*, Minneapolis, MN, August 2002, 861–865.

[36] Kim, J., Lee, H., and Kim, J., 'Effects on Signal Integrity and Radiated Emission by Split Reference Plane on High-speed Multilayer Printed Circuit Boards', *IEEE Trans. on Advanced Packaging*, **28**(4), November 2005, 724–735.

[37] Johnson, H. and Graham, M., *'High-Speed Signal Digital design: a Handbook of Black Magic'*, Prentice Hall PTR, Upper Saddle River, NJ, 1993.

[38] Cui, W., Ye, X., Archambeault, B., White, D., Li, M., and Drewniak J., 'EMI Resulting from Signal via Transitions through the DC Power Bus', *IEEE Symposium on Electromagnetic Compatibility*, Washington, DC, August 2000, 821–826.

[39] Park, J., Kim, H., Jeong, Y., Kim, J., Pak, J., Kam, D., and Kim, J., 'Modeling and Measurements of Simultaneous Switching Noise Coupling Trough Signal via Transaction', *IEEE Trans. on Advanced Packaging*, **29**(3), August 2006, 548–559.

[40] Parker, J., 'Via Coupling within Parallel Rectangular Planes', *IEEE Trans. on Electromagnetic Compatibility*, **39**(1), February 1997, 17–23.

[41] Ye, X., Drewniak, J., Nadolny, J., and Hockanson, D., 'High-performance Inter-PCB Connectors: Analysis of EMI Characteristics', *IEEE Trans. on Electromagnetic Compatibility*, **44**(1), February 2002, 165–174.

[42] Katzier, H. and Reischl, R., 'SPICE-models for High-pincount Board Connectors', *IEEE Trans. on Components, Packaging, and Manufacturing Technology – Part B*, **19**(1), February 1996, 3–6.

[43] IEC 62433-2: *'Models of Integrated Circuits for EMI Behavioral Simulation. Part 2: ICEMCE, ICEM Conducted Emission Model'*.

[44] Qin, J., Ramahi, O., and Granatstein, V., 'Novel Planar Electromagnetic Bandgap Structures for Mitigation of Switching Noise and EMI Reduction in High-speed Circuits', *IEEE Trans. on Electromagnetic Compatibility*, **49**(3), August 2007, 661–669.

Appendix A

Formulae for Partial Inductance Calculation

This appendix provides closed-form expressions for calculating partial inductances for round wires and busbars or strips useful for modeling connectors, vias, traces and planes in PCBs. The concept of effective inductance L_e associated with one conductor is used here to compute the voltage drop $V(t) = \mathrm{L_e}\, \mathrm{d}I(t)/\mathrm{d}t$ on the conductor that is caused by the current $I(t)$. For two conductors with currents flowing in the opposite direction (i.e. series connection), the overall inductance is $L_{et} = L_{e1} + L_{e2}$. For two conductors with currents flowing in the same direction (i.e. parallel connection) the overall inductance is $L_{et} = L_{e1}L_{e2}/(L_{e1} + L_{e2})$.

A.1 Round Wires

A collection of formulae for round wire structures such as pin connectors or vias in PCBs, is shown in Table A.1. It is worth making the following observations:

- *Isolated wire*. This is the situation that occurs when the current I on the conductor returns through another far away conductor and, therefore, the contribution of the mutual inductance can be neglected. The associated inductance is the self partial inductance L_p function of the length l and wire radius r_w [1]. The low-frequency expression includes the internal wire inductance $L_{int} = \mu_0/8\pi$, and the resulting inductance is higher than the inductance at high frequency when the skin effect becomes dominant.
- *Two parallel wires*. The mutual partial inductance is calculated as mutual inductance between two filaments and provides exact values [1].
- *Two parallel wires with currents in opposite directions*. This is a favorable situation for the effective inductance associated with each conductor (i.e. signal and current return path), as the mutual partial inductance M_p must be algebraically subtracted from the self partial inductance L_p and the overall inductance of the two conductors is $L_{et} = 2(L_p - M_p)$.
- *Two parallel wires with currents in the same direction*. This is the case of a structure that consists of a return conductor far away from two parallel signal conductors where currents flow in the same direction. When the two currents are equal, the total effective inductance

Signal Integrity and Radiated Emission of High-Speed Digital Systems Spartaco Caniggia and Francescaromana Maradei

Table A.1 Inductance formulae for round-wire conductors

Configuration	Formulae All dimensions in meters, inductance in Henry $\mu_0 = 4\pi \times 10^{-7}$ A/m
Isolated wire 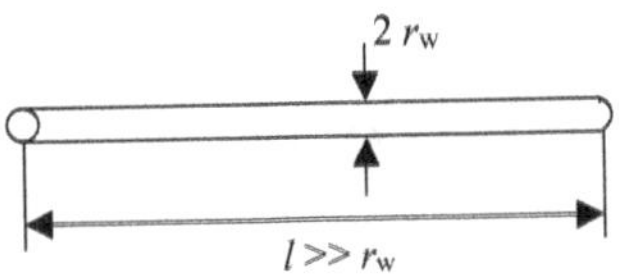	Self partial inductance $L_p = \frac{\mu_0}{2\pi} l \left[\ln\left(\frac{2l}{r_w}\right) - \frac{3}{4} \right]$ low frequency (LF) $L_p = \frac{\mu_0}{2\pi} l \left[\ln\left(\frac{2l}{r_w}\right) - 1 \right]$ high frequency (HF)
Two parallel wires 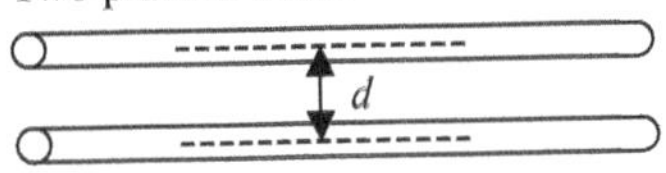	Mutual partial inductance $M_p = \frac{\mu_0}{2\pi} l \left[\ln\left(\frac{1}{d} + \sqrt{1 + \frac{l^2}{d^2}}\right) - \sqrt{1 + \frac{d^2}{l^2}} + \frac{d}{l} \right]$ $M_p \approx \frac{\mu_0}{2\pi} l \left[\ln\left(\frac{2l}{d}\right) - 1 \right]$ for $d \ll 1$
Two parallel wires with currents in opposite directions 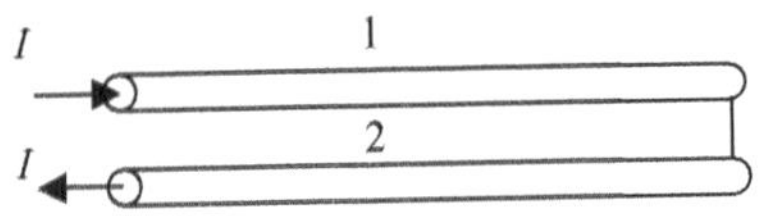	Effective inductance associated with one wire: $L_{e1} = L_{p1} - M_p \qquad L_{e2} = L_{p2} - M_p$ when the wires have the same radius: $L_e = L_p - M_p$ $L_e \approx \frac{\mu_0}{2\pi} l \left(\ln\left(\frac{d}{r_w}\right) + \frac{1}{4} \right)$ (LF)
Two parallel wires with currents in same direction 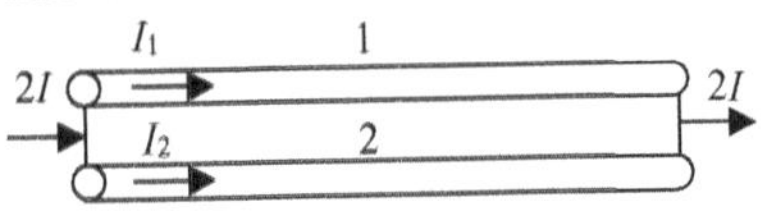	Effective inductance associated with one wire: $L_{e1} = \frac{L_{p1}L_{p2} - M_p^2}{L_{p2} - M_p} \qquad L_{e2} = \frac{L_{p1}L_{p2} - M_p^2}{L_{p1} - M_p}$ If the wires have the same radius, $I_1 = I_2 = I$: $L_e = L_p + M_p$ $L_e \approx \frac{\mu_0}{2\pi} l \cdot 2 \left(\ln\left(\frac{2l}{\sqrt{r_w d}}\right) - \frac{7}{8} \right)$ (LF)
Wire above a large return ground plane 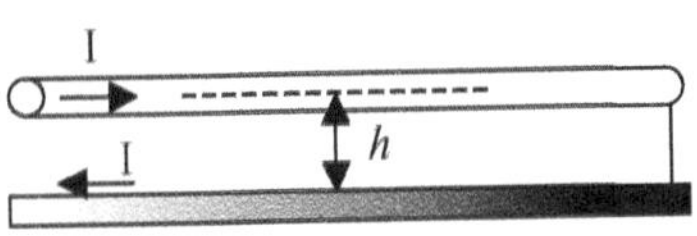	Effective inductance associated with the wire only: $L_e = L_p - M_p(d = 2h)$ $= \frac{\mu_0}{2\pi} l \left(\ln\left(\frac{2h}{r_w}\right) + \frac{1}{4} \right)$ (LF) The ground is replaced by an image parallel wire with opposite current I and distance $2h$ from the wire. For HF the term $^1/_4$ can be omitted.

associated with the two conductors is $L_{et} = 1/2(L_p + M_p)$, which is lower than the self partial inductance of an isolated conductor L_p, as the mutual partial inductance between the two conductor $M_p < L_p$.

- *Wire over a large return ground plane.* For a wire above a ground plane with dimensions much larger than the height h, the method of image can be applied to calculate the effective inductance associated with the wire. The plane is replaced by a conductor parallel to the signal conductor, with equal current flowing in the opposite direction, and having a distance from the other twice the distance between the signal conductor and the ground plane. The result is the simple expression in Table A.1. This is also the overall inductance of the signal–ground loop, as the effective inductance associated with a large ground plane can be neglected (see the expression of the effective inductance $L_{e\ gnd}$ in Table A.2).

A.2 Busbars

A collection of formulae for busbar structures such as traces in PCBs, is shown in Table A.2. It is interesting to note the following:

- *Isolated busbar*. As in the case of round wire, this is the situation when the current I on the busbar, i.e. a PCB trace, returns through another conductor far away so that the contribution of the mutual inductance is negligible. The exact formula for the self partial inductance of a trace of width w, thickness t, and length l is reported elsewhere [2, 3]. As this expression is quite complicated, a simplified formula that works well for practical cases of interest is shown in Table A.2. For instance, it can be shown that, for $0 < t < 1$ mm, $w = 1$ mm, and $l \geq w$, the simple formula provides the same results as the exact formula.
- *Two parallel busbars*. In the case of parallel, rectangular cross-section conductors (bars), the exact expression of mutual partial inductance is again given elsewhere [2, 3]. If the bars are not too close, then a reasonable approximation is to treat them as filaments and use the formula of Table A.1. Another more accurate method is to divide the cross-section of each bar into sub-bars, treat each of them as a filament, then use the filament approximation of Table A.1 to characterize each sub-bar, and finally sum the results as indicated in Table A.2. It can be shown that the three methods give the same results for practical situations. For instance, for a parallel busbar with $w = 0.25$ mm and $t = 0.1$ mm, and assuming that $K_x = M_x = 2$ and $K_y = M_y = 5$, the results are practically coincident for $d \geq 0.25$ mm in configuration (a) and for $d \geq 0.35$ in configuration (b). For both cases, $l \geq 1$ mm. In conclusion, the filament expression works as well as the expression with the summations for most cases of interest, except when the bar separation is of the order of the bar thickness.
- *Parallel busbars with differential currents*. The effective inductance to be associated with each conductor can be calculated as the difference between the self and mutual inductance by using the formulae given in Table A.2.
- *A bar (trace) above a finite ground plane*. This is a very important practical case because, once the effective inductance associated with the ground of a PCB is known, it is possible to calculate the radiated field from a PCB with an attached cable. Numerous examples are provided in *Chapter 9*. The formula given in Table A.2, taken from reference [4], works well, as demonstrated experimentally in *Chapter 9*. Other formulae can be found elsewhere [5, 6]. A general, accurate method, useful for various ground plane cross-sections, consists in computing the voltage drop along the ground plane that is caused by current I. The ground plane is divided into sub-bars, or filaments, and self and mutual partial inductances regarding all the conductors are accounted for. An example of application of this method for some structures of practical interest is outlined in *Section 10.1*.
- The onset frequency between the low- and high-frequency regions is the frequency where the skin effect becomes significant (see *Section 7.1*).

Table A.2 Inductance formulae for busbar conductors

Configuration	Formulae for low frequencies All dimensions in meters, inductance in Henry $\mu_0 = 4\pi \times 10^{-7}$ A/m
Isolated busbar 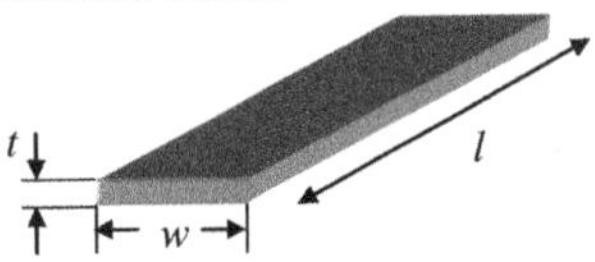 	$L_p = \frac{\mu_0}{2\pi} l \left[\ln\left(\frac{2l}{w+t}\right) + \frac{1}{2} + \frac{2}{9}\left(\frac{w+t}{l}\right) \right]$
Two parallel busbars 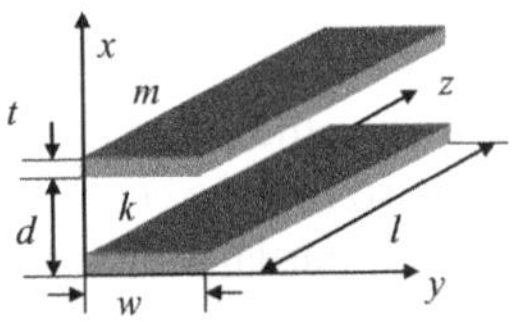 (a) 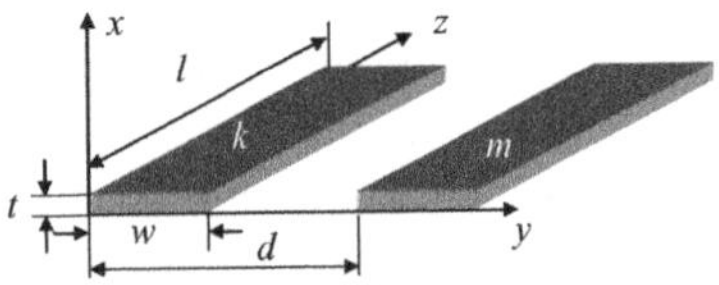 (b)	Accurate partial mutual inductance calculation considering proximity effect: $M_p = \frac{1}{(K_x K_y)(M_x M_y)} \sum_{k_1=1}^{K_x} \sum_{k_2=1}^{K_y} \sum_{m_1=1}^{M_x} \sum_{m_2=1}^{M_y} M_{\mathrm{pf}_{k_1 k_2 m_1 m_2}}$ where bar k is divided into $K_x\ K_y$ sub-bars or filaments and bar m is divided into $M_x M_y$ sub-bars or filaments. $M_{\mathrm{Pf}_{k_1 k_2 m_1 m_2}}$ is the mutual partial inductance between the filament on the bar k characterized by k_1 and k_2 and filaments on the bar m characterized by m_1 and m_2. For busbars that are not too close: $M_p = \frac{\mu_0}{2\pi} l \left[\ln\left(\frac{1}{d} + \sqrt{1 + \frac{l^2}{d^2}}\right) - \sqrt{1 + \frac{d^2}{l^2}} + \frac{d}{l} \right]$
Busbar and adjacent return bus 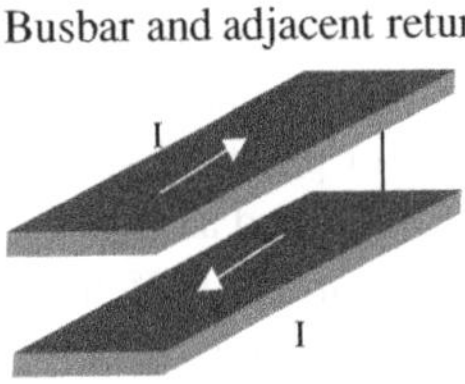 	Effective inductance associated with one wire: $L_e = L_p - M_p$ $L_e \approx \frac{\mu_0}{2\pi} l \left[\ln\left(\frac{d}{w+t}\right) + \frac{3}{2} \right]$ (LF) and $w,\ d \ll l$
Flat conductor and adjacent return-path conductor 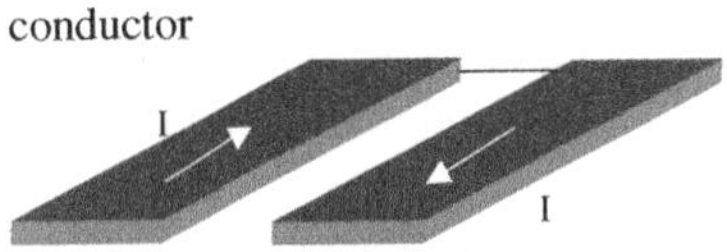 	Effective inductance associated with one wire: $L_e = L_p - M_p$ $L_e \approx \frac{\mu_0}{2\pi} l \left[\ln\left(\frac{d}{w+t}\right) + \frac{3}{2} \right]$ (LF) and $w,\ d \ll l$
Flat conductor over a finite return ground plane 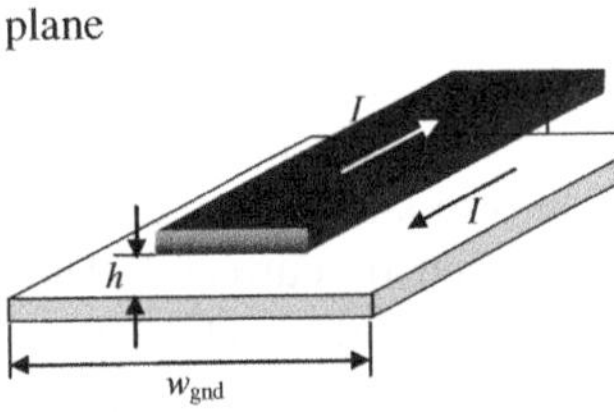 t = thickness of the conductor	Equivalent inductance associated with bar and finite ground plane thickness t: $L_{\mathrm{e\,bar}} \approx L_p - M_p(d = 2h) = \frac{\mu_0}{2\pi} l \left[\ln\left(\frac{2h}{w+t}\right) + \frac{3}{2} \right]$ (LF) and $w,\ 2h \ll l$ $L_{\mathrm{e\,gnd}} \approx \frac{\mu}{2\pi} l \ln \left(\frac{w_{\mathrm{gnd}} + t + \pi\left(h - \frac{t}{2}\right)}{w_{\mathrm{gnd}} + \left(1 + \frac{\pi}{2}\right) t} \right)$ $L_{\mathrm{e\,gnd}} \approx \frac{\mu}{2\pi} l \ln \left(\frac{\pi h}{w_{\mathrm{gnd}}} + 1 \right)$ for $t = 0$

Table A.3 Table of sample inductance calculations

Effective partial inductance L_e (nH) associated with 12 inch (30.48 cm) wire long with 0.01 inch (0.254 mm) diameter at low frequency

	Single wire	Parallel currents in same direction	Parallel opposite currents	Wire having a plane for return current path
Seperation (inch)	Isolated	$d = 0.02$	$d = 0.02$	$h = 0.02$
Inductance (nH)	471	842	100	142
Seperation (inch)	Isolated	$d = 0.2$	$d = 0.2$	$h = 0.2$
Inductance (nH)	471	702	240	282

Effective partial inductance L_e (nH) associated with 12 inch busbar of 0.01 inch width, 0.003 inch thickness, center-to-center seperation d and height above a ground plane h, at low frequency.

	Single bus	Parallel opposite currents (vertical)	Parallel opposite currents (horizontal)	Bus over a ground plane
Seperation (inch)	Isolated	$d = 0.02$	$d = 0.02$	$h = 0.02$
Inductance (nH)	489	118	118	160

A.3 Examples of Application of the Inductance Formulae

An example of application of the inductance formulae is given in Table A.3. The numerical values refer to the effective inductance L_e associated with each conductor when not isolated. The total inductance of the loop formed by the two conductors is $2L_e$ when the conductors have equal and opposite currents and $L_e/2$ when the conductors have equal current. Observe that, for conductors with opposite currents, the effective inductance $L_e = L_p - M_p$ decreases when the conductors are closer, as the self partial inductance L_p remains the same while the mutual partial inductance M_p increases. For conductors with currents in the same direction, to have a low value of $L_e = L_p + M_p$, the mutual partial inductance M_p must be minimized by increasing the separation between the two conductors. On the other hand, to have high values of L_e to stop the *common-mode* currents, M_p must be maximized by increasing the magnetic flux coupled between the two conductors, as done with choke EMI filters. In the case of two parallel busbars, the inductance does not change with the reciprocal position of the bars when the bars have the same center-to-center separation. For comparison purposes, the same structures were chosen as those considered in reference [7], where the dimensions are expressed in inches (1 inch = 2.54 cm).

References

[1] Grover, F.W., *'Inductance Calculations'*, Dover, New York, NY, 1946.

[2] Ruehli, A.E., 'Inductance Calculations in a Complex Integrated Circuit Environment', *IBM Journal of Research and Development*, **16**, September 1972, 470–481.

[3] Hoer, C. and Love, C., 'Exact Inductance Equations for Rectangular Conductors with Applications to More Complicated Geometries', *J. Research Natl. Bar. Stand. – C. Eng. Instrum.*, **69C**, 1965, 127–137.

[4] Leferink, F., 'Inductance Calculations: Methods and Equations', *IEEE International Symposium on Electromagnetic Compatibility*, Atlanta, GA, 14–18 August 1995, 16–22.

[5] Hockanson, D., Drewniak, J., Hubing, T., Van Doren, T., Sha, F., Lam, C.-W., and Rubin, L., 'Quantifying EMI Resulting from Finite-impedance Reference Planes', *IEEE Trans. on Electromagetic Compatibility*, **39**(4), November 1997, 286–297.

[6] Holloway, C. and Kuester, E., 'Net and partial inductance of a microstrip ground plane', *IEEE Trans. on Electromagnetic Compatibility*, **40**(1), February 1998, 33–46.

[7] Rostek, P., 'Avoid Wiring–Inductance Problems', *Electronic Design*, **25**, 6 December 1974.

Appendix B

Characteristic Impedance, Delay Time, and Attenuation of Microstrips and Striplines

This appendix provides closed-form expressions for calculating the characteristic impedance, delay time, and attenuation of traces having microstrip and stripline structures. A procedure for computing analytically the proximity-effect parameter K_p as defined in *Chapter 7* is also outlined. Some results are compared with those found in the literature by using field-solver software.

B.1 Microstrip

Microstrip has the structure shown in Figure B.1 and is characterized by fields propagating in two different dielectrics: air and substrate with relative permittivity ε_r. This is particularly true if the trace is not covered by a soldermask to prevent corrosion. Usually, an effective dielectric constant, ε_{re}, is used for electric parameter calculations, which is given by

$$\varepsilon_{re} = \begin{cases} \varepsilon_{re1} & \text{for } w/h < 2 \\ \varepsilon_{re2} & \text{for } w/h \geq 2 \end{cases} \tag{B.1}$$

where

$$\varepsilon_{re1} = 0.475\varepsilon_r + 0.67 \tag{B.2a}$$

$$\varepsilon_{re2} = \frac{\varepsilon_r + 1}{2} + \left(\frac{\varepsilon_r - 1}{2\sqrt{1 + 10h/w}}\right) \tag{B.2b}$$

The per-unit-length (p.u.l.) propagation delay time t_{pd} is then given by

$$t_{pd} = \sqrt{\mu_0 \varepsilon_0 \varepsilon_{re}} \text{ ns/m} \tag{B.3}$$

where $\mu_0 = 4\pi \times 10^{-7}$ H/m and $\varepsilon_0 = 8.854 \times 10^{-12}$ F/m.

Signal Integrity and Radiated Emission of High-Speed Digital Systems Spartaco Caniggia and Francescaromana Maradei

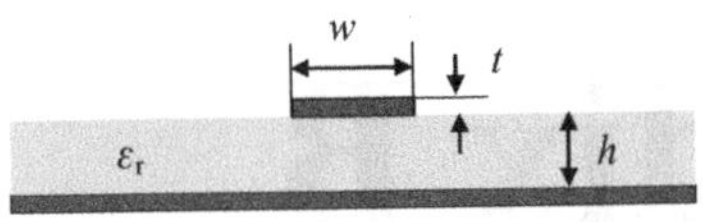

Figure B.1 Microstrip structure

The microstrip characteristic impedance $Z_{0,\mathrm{ms}}$ is given in closed form by [1]

$$Z_{0,\mathrm{ms}} = \begin{cases} Z_{0,\mathrm{ms1}} & \text{for } w/h < 2 \\ Z_{0,\mathrm{ms2}} & \text{for } w/h \geq 2 \end{cases} \tag{B.4}$$

where

$$Z_{0,\mathrm{ms1}} = \frac{60}{\sqrt{\varepsilon_{\mathrm{re1}}}} \ln\left(\frac{5.98\,h}{0.8w + t}\right) \tag{B.5a}$$

$$Z_{0,\mathrm{ms2}} = \frac{120\pi}{\sqrt{\varepsilon_{\mathrm{re2}}}\left(\dfrac{w}{h} + 1.393 + 0.667\ \ln\left(\dfrac{w}{h} + 1.444\right)\right)} \tag{B.5b}$$

The microstrip attenuation $\alpha_{\mathrm{t,ms}}$ can be calculated by the closed-form expression [1]

$$\alpha_{\mathrm{t,ms}} = \alpha_{\mathrm{d,ms}} + \alpha_{\mathrm{c,ms}} \tag{B.6}$$

where $\alpha_{\mathrm{d,ms}}$ and $\alpha_{\mathrm{c,ms}}$ are the attenuation due to losses in dielectric and conductor media, respectively, considering the return path, also, as defined in *Chapter 7*. These parameters are given by

$$\alpha_{\mathrm{d,ms}} = 2.318 f \sqrt{\varepsilon_{\mathrm{re}}} \tan\,\delta \tag{B.7}$$

$$\alpha_{\mathrm{c,ms}} = \frac{11.411\sqrt{f}}{h Z_{0,\mathrm{ms}}}\left(1 - \left(\frac{w_{\mathrm{p}}}{4h}\right)^2\right)\left(1 + \frac{h}{w_{\mathrm{p}}} + \frac{h}{\pi w_{\mathrm{p}}}\left(\ln\left(\frac{2h}{t}\right) - \frac{t}{h}\right)\right) \tag{B.8}$$

$$w_{\mathrm{p}} = w + \frac{t}{\pi}\left(\ln\left(\frac{2h}{t}\right) + 1\right) \tag{B.9}$$

where $\tan\,\delta$ is the dielectric loss tangent. The results of this calculation are in dB/inch, and the frequency must be assigned in GHz. The parameters w, t, and h must be set in mils. The proposed closed-form expressions give results in good agreement (less than 4 %) with those obtained by a field-solver code. Usually the trace has the structure of an embedded microstrip. The soldermask chemistry and final thickness is a 0.6–0.8 mil thick coating over the copper with $\varepsilon_{\mathrm{r}} = 3.1$–$3.3$ and loss tangent $\tan\,\delta \approx 0.02$. In this case there are three dielectrics involved, and the prediction of the effective dielectric permittivity $\varepsilon_{\mathrm{re}}$ and of the characteristic impedance Z_0 change slightly. However, from an engineering viewpoint, the proposed formulation can still be used without losing significant accuracy, as will be demonstrated later with an example.

The accuracy of α_d is better than 1 %, and α_c is suitable for w/h ranging between 0.159 and 2 or for microstrip on FR4 with Z_0 from roughly 50 Ω and 100 Ω.

B.2 Stripline

Stripline is a conductor immersed in a dielectric and sandwiched between two return planes. The structure is symmetric when the trace is centered in the dielectric, as shown in Figure B.2, with $h = b - t/2$. An offset stripline is a structure with the trace closer to one plane. In contrast to microstrip, the field lines are confined into the dielectric, and therefore the p.u.l. propagation delay time depends on the relative permittivity ε_r and is given by

$$t_{pd} = \sqrt{\mu_0 \varepsilon_0 \varepsilon_r} \tag{B.10}$$

The stripline characteristic impedance $Z_{0,sl}$ can be calculated by the closed-form expressions [1]

$$Z_{0,sl} = \begin{cases} Z_{0,sl1} & \text{for } w/(b-t) \geq 0.35 \\ Z_{0,sl2} & \text{for } w/(b-t) < 0.35 \end{cases} \tag{B.11}$$

where

$$Z_{0,sl1} = \frac{94.15}{\sqrt{\varepsilon_r}\left(\dfrac{w}{b}k + \dfrac{C_f}{8.854\varepsilon_r}\right)} \tag{B.12a}$$

$$Z_{0,sl2} = \frac{60}{\sqrt{\varepsilon_r}} \ln\left(\frac{4b}{\pi d}\right) \tag{B.12b}$$

$$k = \frac{1}{1 - t/b} \tag{B.12c}$$

$$C_f = \frac{8.854\varepsilon_r}{\pi}\left(2k\ln(k+1) - (k-1)\ln\left(k^2 - 1\right)\right) \tag{B.12d}$$

$$d = \frac{w}{2}\left(1 + \frac{t\left(1 + \ln(4\pi w/t) + 0.51\pi\,(t/w)^2\right)}{\pi w}\right) \tag{B.12e}$$

The expressions for the case $w/(b - t) \geq 0.35$ are valid for traces no thicker than 25 % of the plate spacing. For a 1 oz trace, this means $b \geq 5.6$ mils, which is generally met in practical PCB design. The trace thickness is rated in plating weight, typically reported in ounces. A 1

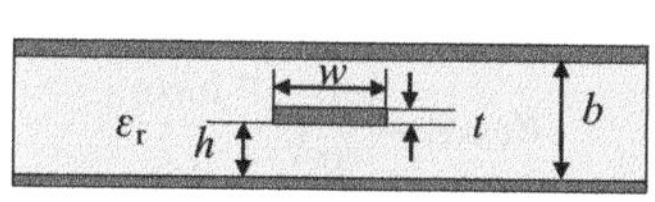

Figure B.2 Stripline structure

oz plating corresponds to a thickness of 34.8 μm. The thickness scales in proportion to plating weight [2].

The proposed closed-form expressions provide results with discrepancies lower than 2 % with respect to field-solver software solutions for a wide range of impedances and trace widths.

The stripline attenuation $\alpha_{\mathrm{t,sl}}$ can be calculated as the sum of attenuation due to dielectric $\alpha_{\mathrm{d,sl}}$ and conductor $\alpha_{\mathrm{c,sl}}$ by the following closed-form expression:

$$\alpha_{\mathrm{t,sl}} = \alpha_{\mathrm{d,sl}} + \alpha_{\mathrm{c,sl}} \tag{B.13}$$

where

$$\alpha_{\mathrm{d,sl}} = 2.318 f \sqrt{\varepsilon_{\mathrm{r}}} \tan\,\delta \tag{B.14}$$

$$\alpha_{\mathrm{c,sl}} = \frac{2.02 \times 10^{-3} \varepsilon_{\mathrm{r}} Z_{0,\mathrm{sl}} \sqrt{f}}{b} \left(k + \frac{2w/b}{(1 - t/b)^2} + \frac{1}{\pi} \frac{1 + t/b}{(1 - t/b)^2} \ln \left(\frac{k+1}{k-1} \right) \right) \tag{B.15}$$

$$k = \frac{1}{1 - t/b} \tag{B.16}$$

The results of this calculation are in dB/inch, and the frequency must be assigned in GHz. The parameters w, t, and b must be set in mils. The accuracy of α_{d} is better than 1 %. Considering α_{c}, the expression should be valid only when $w/(b - t) \geq 0.35$, i.e. for a wide trace having impedance below about 65 Ω. However, the expression provides acceptable accuracy for a higher-impedance trace.

B.3 Trace Attenuation and the Proximity-Effect Parameter

The closed-form expressions previously given were used to calculate the attenuation for stripline and microstrip traces with a characteristic impedance of 50 Ω, and the results are shown in Figure B.3. It can be noted that, above 1 GHz, the dielectric losses dominate, and this fact makes the accuracy of the formula used to calculate α_{c} less important.

The proposed closed-form expressions can be very useful for calculating the proximity-effect parameter K_{p} as an alternative to field-solver software. The proximity factor takes into account the additional resistance due to redistribution of current on both the signal conductor and the reference planes, as defined in *Section 7.1*. The procedure consists of the following steps:

1. The real part of the skin-effect impedance $R_{0\mathrm{Skin}}$ at a particular frequency of interest f_0 is calculated by combining Equations (7.5) and (7.19) and adopting a proximity factor $K_{\mathrm{p}} = 1$:

$$R_{0\mathrm{Skin}} = \frac{1}{p} \sqrt{\frac{\pi f_0 \mu_0}{\sigma}} \tag{B.17}$$

where $p = 2(t + w)$ is the perimeter of the trace, and σ is the conductivity of the trace.

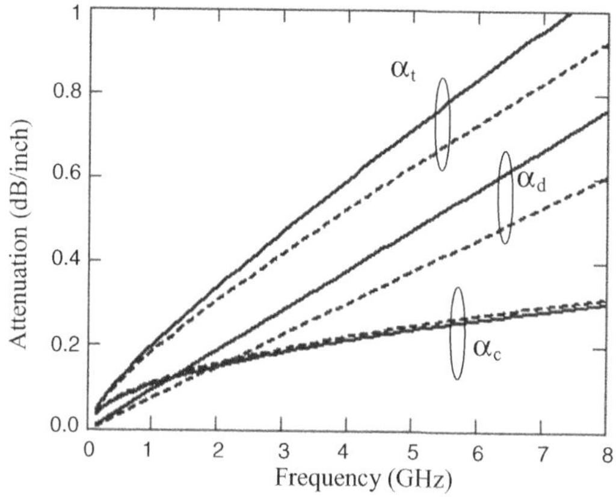

Figure B.3 Attenuations in a 50 Ω stripline (solid line) and a 50 Ω microstrip (dashed line). Trace parameters: $\varepsilon_r = 4.25$, loss tangent $\tan\delta = 0.02$, $w = 5$ mils, $t = 0.65$ mils, $b = 12.85$ mils (stripline), $h = 3.0$ mils (microstrip)

2. The characteristic impedance $Z_{0,i}$ is calculated by Equation (B.4) with $i =$ 'ms' for a microstrip or by Equation (B.11) with $i =$ 'sl' for a stripline trace.
3. The attenuation due to the skin effect only at frequency f_0 and indicated here as $\alpha_{0\text{Skin},i}$ (with $i =$ 'ms' or $i =$ 'sl') is calculated in dB, considering that 1 neper $= 8.686$ dB and 1 inch $= 0.0254$ m (see Equations (7.32) and (7.39)). Therefore

$$\alpha_{0\text{Skin},i} = \frac{8.686}{2} \frac{R_{0Skin}}{Z_{0,i}} 0.024 \tag{B.18}$$

4. The proximity factor $K_{\text{p0},i}$ at frequency f_0 is computed as the ratio of the attenuation $\alpha_{\text{c},i}(f_0)$ at frequency f_0, computed by Equation (B.8) for $i =$ 'ms' or by Equation (B.15) for $i =$ 'sl', and the attenuation $\alpha_{0\text{Skin},i}$ for the skin effect only. Remember that Equations (B.8) and (B.15) take into account the skin effect and the proximity effect between the trace and its return path:

$$K_{\text{p0},i} = \frac{\alpha_{\text{c},i}(f_0)}{\alpha_{0\text{Skin},i}} \tag{B.19}$$

5. The total resistance of the trace $R_{0,i}$ which incorporates the factor $K_{\text{p0},i}$ at frequency f_0 is calculated as the product of the skin-effect resistance $R_{0\text{Skin}}$ and the proximity-effect parameter $K_{\text{p0},i}$ calculated by Equation (B.19):

$$R_{0,i} = R_{0\text{Skin}} K_{\text{p0},i} \tag{B.20}$$

To check the accuracy of this procedure, the values reported in Table 5.1 of reference [2] were used for comparison at a frequency $f_0 = 1$ GHz and shown in round brackets in Table B.1. The resistance in [2] was calculated by a method-of-moments magnetic field simulator, and the authors estimate the accuracy of the data generated by this simulator at approximately ±2 %.

Table B.1 Proximity-effect coefficient K_{p0} and AC resistance R_0 including K_{p0} for single-ended microstrips and striplines computed for $f_0 = 1$ GHz. The values in brackets come from Table 5.1 of reference [2]

h (mils)	w (mils)	b (mils)	Z_0 (Ω)	$\alpha_c(f_0)$ (dB/inch)	K_{p0}	R_0 Ω/m
Microstrip $\alpha_d(f_0)$=0.076 dB/inch						
9	16	na	48.8 (50)	0.038 (0.040)	1.824 (1.928)	16.9 (18.1)
6	10.2	na	48.6 (50)	0.056 (0.058)	1.766 (1.851)	24.6 (26.1)
9	11.1	na	60.7 (60)	0.042 (0.042)	1.782 (1.754)	23.0 (23.0)
6	6.9	na	60.5 (60)	0.061 (0.061)	1.711 (1.675)	33.4 (33.0)
9	7.8	na	71.6 (70)	0.045 (0.045)	1.668 (1.609)	29.5 (28.7)
6	4.8	na	70.8 (70)	0.065 (0.064)	1.600 (1.542)	41.9 (40.8)
Stripline $\alpha_d(f_0)$=0.096 dB/inch						
7	5.8	15	49.64 (50)	0.094 (0.088)	1.707 (1.586)	42.3 (39.9)
10	8	20	49.94 (50)	0.072 (0.068)	1.768 (1.637)	32.7 (30.8)
15	12.7	30	49.74 (50)	0.050 (0.046)	1.865 (1.729)	22.3 (21.0)

Note: $\varepsilon_r = 4.3$ at 1 GHz, dielectric loss tangent $tan\delta = 0.02$. The microstrip examples assume copper traces of 1 oz (1.3 mils) thickness (including plating) with $\sigma = 5.98 \times 10^7$ S/m plus a conformal coating (soldermask) consisting of a 12.7 μm layer having a dielectric constant of 3.3. The stripline examples assume copper trace of 1/2-oz (0.65 mils) thickness with no plating. All parameters are calculated for a frequency $f_0 = 1$ GHz.

The results of the proposed analytical procedure are summarized in Table B.1 for some trace structures. The data of Table 5.1, in brackets, are those reported in reference [2]. The comparison shows that there is a very good agreement, although the microstrips are with soldermask, and the analytical procedure of this appendix refers to bare microstrip structures. For the striplines there is a slight overestimation. Note that, at $f = f_0$, the attenuation due to the dielectric $\alpha_d(f_0)$ is higher than the attenuation due to the conductor $\alpha_c(f_0)$ right from 1 GHz. This makes the inaccuracy introduced by the analytical calculation of the attenuation α_c less critical. On the other hand, the attenuation α_d has an accuracy lower than 1 %.

References

[1] Thierauf, S.C., '*High-speed Circuit Board Signal Integrity*', Artech House, Inc., Norwood, MA, 2004.

[2] Johnson, H. and Graham, M., '*High-speed Signal Propagation: Advanced Black Magic*', Prentice Hall PTR, Upper Saddle River, NJ, 2003.

Appendix C

Computation of Resonances in the Power Distribution Network of a PCB

This appendix describes three methods for computing resonances in the *Power Distribution Network* (PDN) of PCBs, which consists of two parallel metallic plates separated by a dielectric substrate, also indicated as power bus. The structure of the PCB used for simulations is the same as that presented in *Section 8.2*. The first method is based on the cavity model, the second is based on a SPICE equivalent circuit, and the third is based on a 3D model for full-wave numerical electromagnetic codes. The computed results for three points on the PCB are compared for the case of a bare PCB and a PCB populated by decoupling capacitors. It is shown that the three methods provide the same results from a practical viewpoint.

C.1 Cavity Model

The cavity model developed for a microstrip patch antenna can be applied to a PDN having a rectangular power-bus structure, as demonstrated in reference [1]. In this very useful paper, rich in references, it is also demonstrated that the method can be extended to PCBs with arbitrary shape by applying the segmentation method, and a mathematical simplification is given. Herein, the basic mathematical expressions are provided.

The cavity method is based on the following assumptions:

(a) The close proximity between the two ground planes suggests that the E-field has only the z component and the H-field has only the xy components in the region bounded by the power and ground planes.
(b) The field in the aforementioned region is independent of the z coordinate for all frequencies of interest.
(c) The electric current in the plane must have no component normal to the edge at any point on the edge, implying a negligible tangential component of the H-field along the edge.

Signal Integrity and Radiated Emission of High-Speed Digital Systems Spartaco Caniggia and Francescaromana Maradei

In fact, if electric sources $\vec{J}_S$ are present along the interface between a perfect conductor and a dielectric medium, the following equation holds: $\vec{n} \times \vec{H} = \vec{J}_S$. This means that the condition of considering the tangential magnetic field at the edge to be zero, $H_t = 0$, can be used.

The region between the two planes can therefore be treated as a cavity bounded by magnetic walls along the edges and by electric walls from above and below. This is also an important factor when building a model for numerical codes in order to speed up the computation. In fact, the condition $H_t = 0$ at the edge can be used instead of adding open field space.

For a rectangular PCB like that shown in Figure C.1, the impedance matrix $\hat{Z}_{ij}$ between two generic ports i and j placed at coordinates (x_i, y_i) and (x_j, y_j), respectively, and having an electrically small rectangular cross-section of size (P_{x0}, P_{y0}), can be obtained analytically by the expressions

$$\hat{Z}_{ij}(f) = \sum_m \sum_n \frac{j2\pi f \mu_0 w_z N_{mni} N_{mnj}}{w_x w_y \left(k_{mn}^2 - \hat{k}_p^2\right)} \tag{C.1}$$

where

$$N_{mni} = c_m c_n \cos\left(\frac{m\pi x_i}{w_x}\right) \cos\left(\frac{n\pi y_i}{w_y}\right) N_{mx} N_{ny} \tag{C.2}$$

$$N_{mx} = \begin{cases} 1 & \text{if} \quad m = 0 \\ \dfrac{\sin\left(\dfrac{m\pi P_{x0}}{2w_x}\right)}{\dfrac{m\pi P_{x0}}{2w_x}} & \text{if} \quad m \neq 0 \end{cases} \tag{C.3a}$$

$$N_{ny} = \begin{cases} 1 & \text{if} \quad n = 0 \\ \dfrac{\sin\left(\dfrac{n\pi P_{y0}}{2w_y}\right)}{\dfrac{n\pi P_{y0}}{2w_y}} & \text{if} \quad n \neq 0 \end{cases} \tag{C.3b}$$

$$k_{mn} = \sqrt{\left(\frac{m\pi}{w_x}\right)^2 + \left(\frac{n\pi}{w_y}\right)^2} \tag{C.4}$$

$$\hat{k}_p = 2\pi f \sqrt{\varepsilon_0 \varepsilon_r \mu_0} \left(1 - j\frac{\tan\delta + r(f)/w_z}{2}\right) \tag{C.5}$$

$$c_m, c_n = \begin{cases} 1 & \text{for} \quad m, n = 0 \\ \sqrt{2} & \text{for} \quad m, n > 0 \end{cases} \tag{C.6}$$

$$r(f) = \sqrt{1/(\pi f \mu_0 \sigma)} \tag{C.7}$$

N_{mnj} has the same expression of N_{mnj} with j instead of i. In Equation (C.5), $\tan\delta$ is the dielectric loss tangent, and $r(f)$ is the skin depth at frequency f.

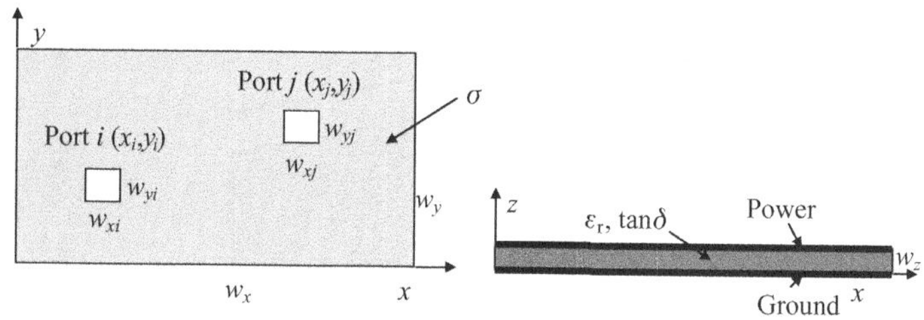

Figure C.1 Configuration of a rectangular PCB

These expressions provide the driving point impedance at port *i*, and the transfer impedance between ports *i* and *j*. Inspecting the various terms in (C.1)–(C.7), the following observations can be made:

- The term N_{mni} describes the wave physics associated with the cavity geometry.
- The terms in the double summation of Equation (C.1) are the propagating modes in the cavity and have modal resonant frequencies given by

$$f_{\mathrm{res}}(m,n) = \frac{1}{2\pi\sqrt{\varepsilon_0\varepsilon_{\mathrm{r}}\mu_0}}\sqrt{\left(\frac{m\pi}{w_x}\right)^2 + \left(\frac{n\pi}{w_y}\right)^2} \tag{C.8}$$

- The term $\hat{k}_{\mathrm{p}}$ accounts for the dielectric and skin-effect conductor losses.
- The double infinite summations must be truncated in practice. A trade-off between accuracy and time computation could be $m = n = 20$, as will be shown later.
- The computed impedance is valid for bare boards.

Impedance plots versus frequency are often used to evaluate the behavior of a power bus. In general, $\hat{Z}_{ij}$ is defined as

$$\hat{Z}_{ij} = \left.\frac{\hat{V}_i}{\hat{I}_j}\right|_{\hat{I}_k=0 \text{ for } k\neq j} \tag{C.9}$$

where $\hat{V}_i$ is the voltage at a location on the power bus, labeled as port *i*, and $\hat{I}_j$ is the current at a location on the power bus, labeled as port *j*, and all other ports are open circuits, i.e. $\hat{I}_k = 0$ for $k \neq j$.

Therefore, $\hat{Z}_{11}$ provides an indication of the voltage created by the injection of a noise current. On the other hand, $\hat{Z}_{21}$ indicates the noise transmission from a noise source to anywhere on the board. $\hat{Z}_{21}$ is very useful for circuit susceptibility and radiated emission studies. Although $\hat{Z}_{21}$ is a complex quantity, often, for these types of study, it is just the magnitude that is considered.

For a PCB with three decoupling capacitors, as shown in Figure C.2 (the same as in *Section 8.2*), where the noise source is placed at port P1, the structure can be characterized by a five-port network: 1 and 2 are the observation ports, while 3, 4, and 5 are the ports associated with

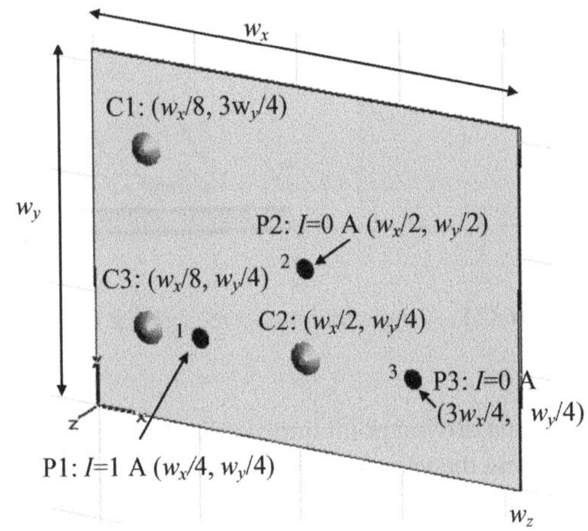

Figure C.2 PCB structure with two parallel planes modelled by MWS. Locations of ports (P) and decoupling capacitors (C) are shown

the decoupling capacitors. The **Z** matrix of the network can be written as

$$\begin{bmatrix} \hat{\mathbf{V}}_{\mathrm{S}} \\ \hat{\mathbf{V}}_{\mathrm{C}} \end{bmatrix} = \begin{bmatrix} \hat{\mathbf{Z}}_{\mathrm{SS}} & \hat{\mathbf{Z}}_{\mathrm{SC}} \\ \hat{\mathbf{Z}}_{\mathrm{CS}} & \hat{\mathbf{Z}}_{\mathrm{CC}} \end{bmatrix} \begin{bmatrix} \hat{\mathbf{I}}_{\mathrm{S}} \\ \hat{\mathbf{I}}_{\mathrm{C}} \end{bmatrix} \tag{C.10}$$

where $\hat{\mathbf{V}}_{\mathrm{S}} = [\,\hat{V}_1\ \hat{V}_2\,]^{\mathrm{T}}$ is the vector containing the voltages at the observation ports, and $\hat{\mathbf{V}}_{\mathrm{C}} = [\hat{V}_3\ \ \hat{V}_4\ \ \hat{V}_5]^{\mathrm{T}}$ is the vector containing the voltages of the three decoupling capacitor ports. Note that Figure C.2 shows the model simulated by MWS, and the observation ports can be P1 and P2 or P1 and P3. For the ports concerning the decoupling capacitors, the currents are related to the voltages as

$$\hat{\mathbf{V}}_{\mathrm{C}} = -\hat{\mathbf{Z}}_{\mathrm{L}}\hat{\mathbf{I}}_{\mathrm{C}} \tag{C.11}$$

where $\hat{\mathbf{Z}}_{\mathrm{L}}$ is a diagonal matrix with the diagonal elements $\hat{Z}_{\mathrm{L}ii} = R_i + \mathrm{j}\omega L_i + 1/(\mathrm{j}\omega C_i)$, $i = 1, 2, 3$. For this example (see Figure 8.12): $R_1 = 0.05\ \Omega$, $L_1 = 5$ nH, $C_1 = 1\ \mu$F; $R_2 = 0.05\ \Omega$, $L_2 = 1$ nH, $C_2 = 10$ nF; $R_3 = 0.05\ \Omega$, $L_3 = 8$ nH, $C_3 = 10$ nF. Inserting Equation (C.11) into Equation (C.10), and solving for $\hat{\mathbf{V}}_{\mathrm{S}}$, yields

$$\hat{\mathbf{V}}_{\mathrm{S}} = \left[\hat{\mathbf{Z}}_{\mathrm{SS}} - \hat{\mathbf{Z}}_{\mathrm{SC}}\left(\hat{\mathbf{Z}}_{\mathrm{L}} + \hat{\mathbf{Z}}_{\mathrm{CC}}\right)^{-1}\hat{\mathbf{Z}}_{\mathrm{CS}}\right]\hat{\mathbf{I}}_{\mathrm{S}} = \hat{\mathbf{Z}}_{\mathrm{SStotal}}\hat{\mathbf{I}}_{\mathrm{S}} \tag{C.12}$$

where $\hat{\mathbf{Z}}_{\mathrm{SStotal}}$ is the total impedance matrix of the two observation ports when the decoupling capacitors are accounted for.

For the example considered, the voltages at the two observation points, computed by Equation (C.12), have the following matrix form:

$$\hat{\mathbf{V}}_S = \begin{bmatrix} \hat{V}_1 \\ \hat{V}_2 \end{bmatrix} = \left(\begin{bmatrix} \hat{Z}_{11}(f) & \hat{Z}_{12}(f) \\ \hat{Z}_{21}(f) & \hat{Z}_{22}(f) \end{bmatrix} - \begin{bmatrix} \hat{Z}_{13}(f) & \hat{Z}_{14}(f) & \hat{Z}_{15}(f) \\ \hat{Z}_{23}(f) & \hat{Z}_{24}(f) & \hat{Z}_{25}(f) \end{bmatrix} \cdot \right.$$
$$\left. \left(\begin{bmatrix} \hat{Z}_{L11}(f) & 0 & 0 \\ 0 & \hat{Z}_{L22}(f) & 0 \\ 0 & 0 & \hat{Z}_{L33}(f) \end{bmatrix} + \begin{bmatrix} \hat{Z}_{33}(f) & \hat{Z}_{34}(f) & \hat{Z}_{35}(f) \\ \hat{Z}_{34}(f) & \hat{Z}_{44}(f) & \hat{Z}_{45}(f) \\ \hat{Z}_{35}(f) & \hat{Z}_{45}(f) & \hat{Z}_{55}(f) \end{bmatrix} \right)^{-1} \cdot \begin{bmatrix} \hat{Z}_{13}(f) & \hat{Z}_{23}(f) \\ \hat{Z}_{14}(f) & \hat{Z}_{24}(f) \\ \hat{Z}_{15}(f) & \hat{Z}_{25}(f) \end{bmatrix} \right) \cdot \begin{bmatrix} 1 \\ 0 \end{bmatrix} \tag{C.13}$$

where the port impedances $\hat{Z}_{ij}(f)$ with $i, j = 1, 2, \ldots, 5$ are calculated by the cavity model expression (C.1). With this formulation, two computations are required: port 1 with $\hat{I} = 1$ and port 2 with $\hat{I} = 0$; port 1 with $\hat{I} = 1$ and port 3 with $\hat{I} = 0$. The application of this formulation to a PCB with more than three decoupling capacitors is straightforward, considering that the numeration for the capacitors starts from 3.

C.2 Spice Model

When the quasi-static approximation can be used (i.e. when the dielectric separation d is much less than both the metal dimensions and the wavelength λ), each power and ground plane pair can be divided into unit cells of size $w \ll \lambda$ and a lumped-element model can be assigned to each cell [2]. A cell consists of an equivalent circuit with R, L, C, and G_d components, as shown in Figure C.3.

Using the equation for a parallel plate [3], the equivalent circuit parameters of a unit cell are given by

$$C = \varepsilon_0 \varepsilon_r \frac{w^2}{d} \tag{C.14a}$$

$$L = \mu_0 d \tag{C.14b}$$

$$R = R_{dc} + R_{ac} \tag{C.14c}$$

$$G_d = \omega C \tan \delta \tag{C.14d}$$

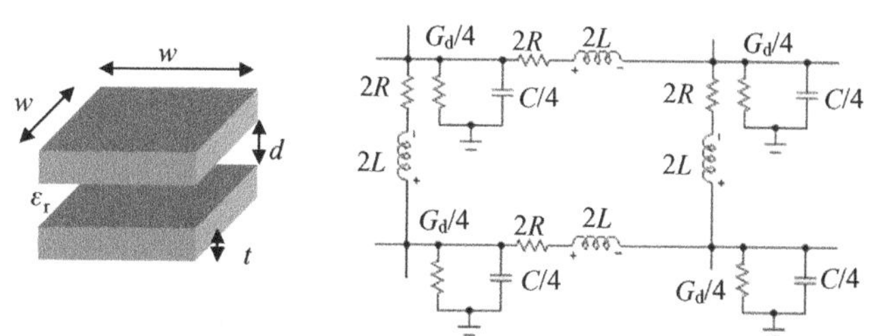

Figure C.3 Unit cell of power and ground planes and its equivalent circuit

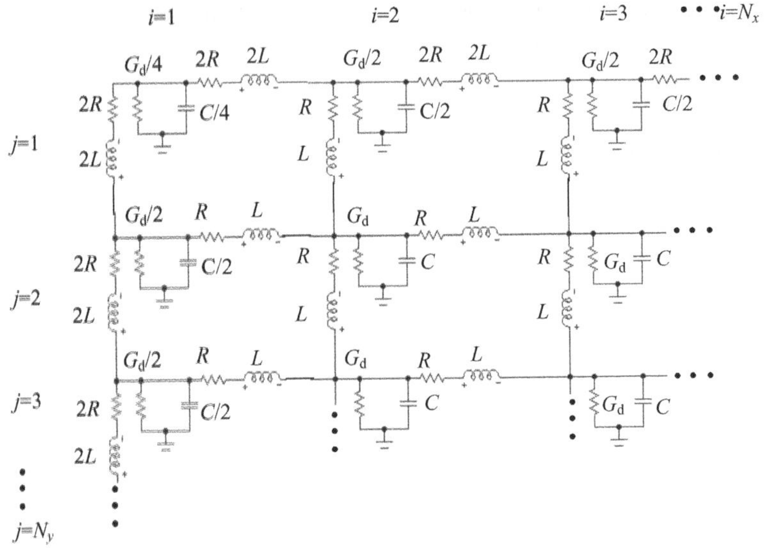

Figure C.4 Zoom of the upper left side of the board equivalent circuit

where C is computed by the parallel-plate capacitor expression, L is computed by using the property $\varepsilon_0\varepsilon_r\mu_0 w^2 = LC$, σ_c is the conductivity of the metal, $R_{dc} = 2/(\sigma_c t)$ is the resistance of the power and ground planes for a steady DC current, $R_{ac} = (1 + \mathrm{j})\sqrt{\pi f \mu_0/\sigma_c}$ accounts for the skin effect on both conductors, and G_d represents the dielectric loss in the substrate (see *Section 7.1* for these last two quantities).

Using a distributed network of $RLCG_d$ elements, each rectangular power/ground plane pair can be divided into $N_x \times N_y$ unit cells. The upper left side of the equivalent circuit is shown in Figure C.4.

C.3 Numerical Model

The full-wave numerical simulation of the PCB structure shown in Figure C.2 was performed by the software tool *MicroWave Studio* (MWS) based on the finite integration technique [4]. The condition $H_t = 0$ was used at x and y direction edges and $E_t = 0$ to simulate the planes, according to the explanation provided in *Section C.1*. This makes it possible to have a non-excessive number of mesh cells, and the simulation runs in a few minutes. The impedances $\hat{Z}_{11}$, $\hat{Z}_{12}$, and $\hat{Z}_{13}$ were calculated by placing a current source of 1 A at position 1 in Figure C.2 and computing the voltages at positions 1 (current source with $I = 1$ A), 2 (current source with $I = 0$) and 3 (current source with $I = 0$).

C.4 Results of the Simulations

The results of simulations performed with the three methods are shown in Figure C.5. Ports and decoupling capacitors have the locations indicated in Figure C.2. The following parameters were used for the cavity model: $w_x = 20.8$ cm, $w_y = 15.6$ cm, $w_z = 1.5$ mm,

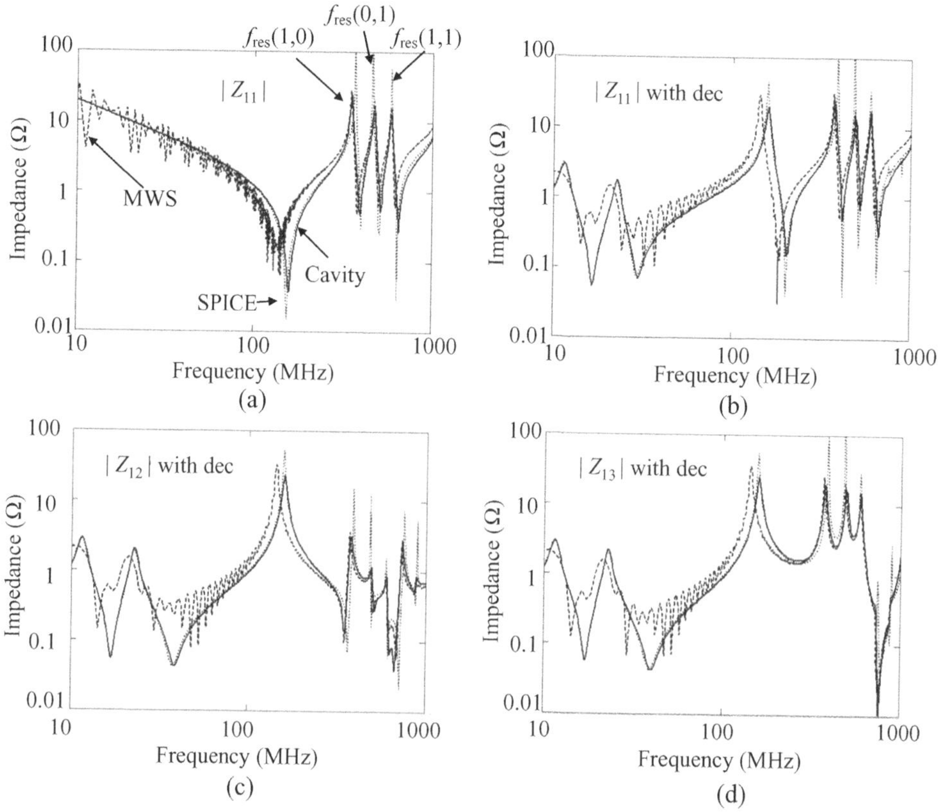

Figure C.5 Computed impedances: (a) at port 1 with bare board; (b) at port 1 with decoupling capacitors; (c) between ports 1 and 2 with decoupling capacitors; (d) between ports 1 and 3 with decoupling capacitors. Cavity model (solid line), SPICE (dotted line), MWS (dashed line)

$\varepsilon_r = 4.25$, $\varepsilon_0 = 8.854 \times 10^{-12}$ F/m, $\mu_0 = 4\pi 10^{-7}$ H/m, $\sigma_c = 5.9 \times 10^7$ S/m, $\tan\delta = 0.02$, $m_{max} = n_{max} = 20$, $P_{x0} = P_{y0} = 1$ mm. The parameters of the decoupling capacitors have the values reported in *Section 8.2*. The circuit parameters were computed by Equations (C.14), adopting $N_x = 20$ and $N_y = 16$, according to the notation of Figure C.2. As the computations were performed up to a frequency of 1 GHz, seeing as $w \ll \lambda_m = 300/f_{MHz} = 30$ cm, the condition of an electrically short unit cell is verified.

Figure C.5 shows a good agreement among the curves obtained by the three methods. MWS presents fast oscillations in the low-frequency range because the simulation in the time domain is stopped at a time when the points of resonance are clearly visible in order to save computation time. Recall that MWS provides results in the frequency domain by FFT. When the cavity structure is highly resonant, as in this case, a long simulation time is necessary to have a low level of voltage oscillations at the required port and therefore stable impedance values at low frequencies. When the interest is mainly focused on the values of the resonance frequencies, the computation can be stopped at a convenient time, referring to the time and frequency results which are continuously displayed during the simulations. The three resonance frequencies with the bare boards correspond to the first modal resonance frequencies: $f_{\text{res}}(1, 0) = 350$ MHz, $f_{\text{res}}(0, 1) = 467$ MHz, $f_{\text{res}}(1, 1) = 583$ MHz.

References

[1] Wang, C., Mao, J., Selli, G., Luan, S., Zhang, L., Fan, J., Pommerenke, D., DuBroff, R., and Drewniak, J., 'An Efficient Approach for Power Delivery Network Design with Closed-form Expressions for Parasitic Interconnect Inductances', *IEEE Trans. on Advanced Packaging*, **29** (2), May 2006, 320–334.

[2] Kim, J.-H. and Swaminathan, M., 'Modeling of Multilayer Power Distribution Planes Using Transmission Matrix Method', *IEEE Trans. on Advanced Packaging*, **25**(2), May 2002 189–199.

[3] Pozar, M., ' *Microwave Engineering*', 2nd edition, John Wiley & Sons, Inc., New York, NY, 1998, Chapter 4.

[4] www.cst.com.

Appendix D

Formulae for Simple Radiating Structures

This appendix provides closed-form expressions for calculating radiated fields from wire structures, as often found in PCBs with attached cables. A method for considering also the metallic reference plane of a semi-anechoic chamber for measurements is given. A low-frequency approximation method for computing radiated fields from an aperture is introduced and explained for the case of a rectangular aperture.

D.1 Wire Structures

Formulae for calculating the radiated E-field in the *Far-Field* (FF) region when the currents on the wires are known are shown in Table D.1. It is important to make the following observations:

- *Infinitesimal dipole.* In general, the EM fields radiated by an infinitesimal dipole have components in spherical coordinates that depend on $1/r$, $1/r^2$, and $1/r^3$ [1]. However, as the interest is focused on E-field prediction at a distance $r \geq 3$ m, where the observation point is normally in the far-field region, the expression for the E-field can be simplified and becomes dependent on $1/r$ only, as shown in Table D.1. Note that, in the far field, E and H are mutually orthogonal and $E/H = 377\ \Omega$.
- *Electrically short differential-mode structure.* As a first consequence of using the far-field approximation for an infinitesimal dipole, the simple expression in Table D.1 is obtained for the case of two parallel electrically short wires with opposite currents. This is a practical case of signal current and its return path. Note that the E-field depends on f^2.
- *Electrically short common-mode structure.* When the two parallel wires have the same current in amplitude and sign, the expression for E depends on f. For example, this is the case for a PCB where the ground plane is absent and the signal current flowing on a wire returns through another parallel wire. It is a very dangerous case, because *common-mode* currents of some μA can produce a radiated field above the limit of emission required by the standards.

Signal Integrity and Radiated Emission of High-Speed Digital Systems Spartaco Caniggia and Francescaromana Maradei

Table D.1 Formulae for calculating the radiated E-field from wire structures

<table>
<tr><th>Configuration</th><th>Formulae
All dimensions in meters, Field in V/m</th></tr>
<tr><td>

Infinitesimal dipole ($\Delta\xi < \lambda/20$)

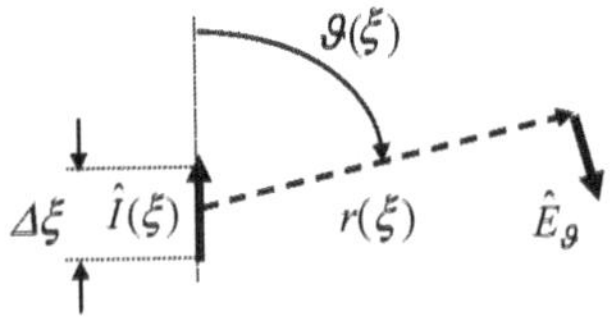

</td><td>

Radiated E-field in far-field region (i.e., $\beta r \gg 1$):

$$\hat{E}_\vartheta(r(\xi)) = \frac{j377\beta(\omega)\hat{I}(\xi)\Delta\xi}{4\pi}\frac{e^{-j\beta(\omega)r(\xi)}}{r(\xi)}\sin\vartheta(\xi)$$

where: $\beta(\omega) = 2\pi/\lambda = \omega/300$, and the wavelength in meters is given by: $\lambda_m = 300/f_{MHz}$ with the frequency in MHz.

</td></tr>
<tr><td>

Electrically short differential-mode structure

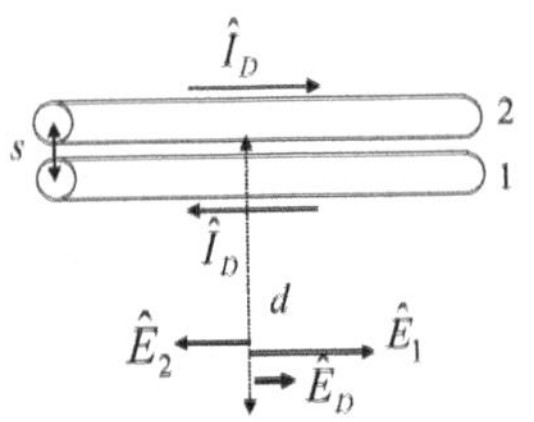

</td><td>

Maximum *differential-mode* E-field in far-field region valid for $l < \lambda/20$:

$$E_{D\max} = |\hat{E}_D| = 1.316 \times 10^{-14}\left|\hat{I}_D\right| f^2 l \frac{s}{d}$$

where l is the length of the conductors.

</td></tr>
<tr><td>

Electrically short common-mode structure

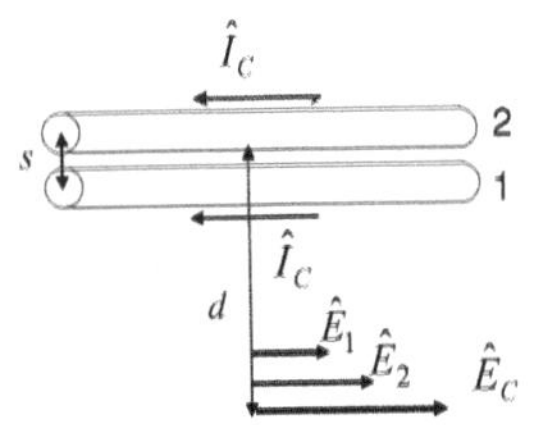

</td><td>

Maximum common-mode E-field in far-field region valid for $l < \lambda/20$:

$$E_{C\max} = |\hat{E}_C| = 1.257 \cdot 10^{-6}\left|\hat{I}_C\right| f l \frac{1}{d}$$

where l is the length of the conductors.

</td></tr>
<tr><td>

Long wire

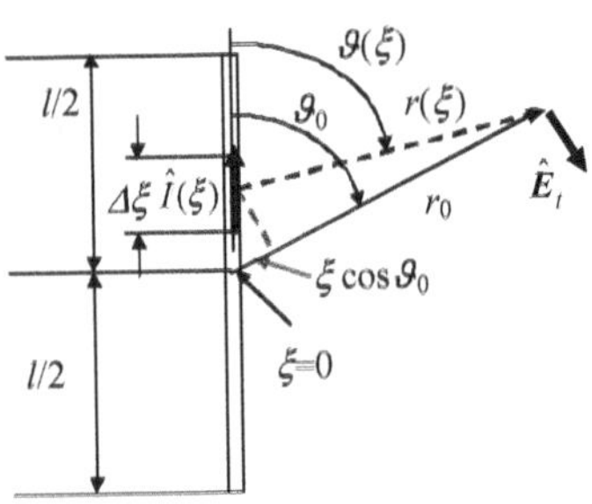

</td><td>

Far-field approximations:

$$\hat{E}_t = \sum_{i=1}^{n}\hat{E}_\theta(r(\xi_i))$$

$r(\xi) \approx r_0 - \xi\cos\vartheta_0$ for phase terms
$r(\xi) \approx r_0$ for amplitude terms
n = number of segment $\Delta\xi$.

When $\vartheta_0 = 90^\circ$ the vertical E-field is given by:

$$\hat{E}_t = \frac{j377\beta(\omega)\hat{I}_t l}{4\pi}\frac{e^{-j\beta(\omega)r_0}}{r_0}$$

with $\hat{I}_t = \sum_{i=1}^{n}\hat{I}(\xi_i)/n$.

</td></tr>
</table>

Table D.2 Formulae for differential-mode and common-mode emissions produced by wire structures in the presence of a metallic reference plane

Configuration	Formulae All dimensions in meters, Field in V/m
DM radiated field with a reference metallic plane 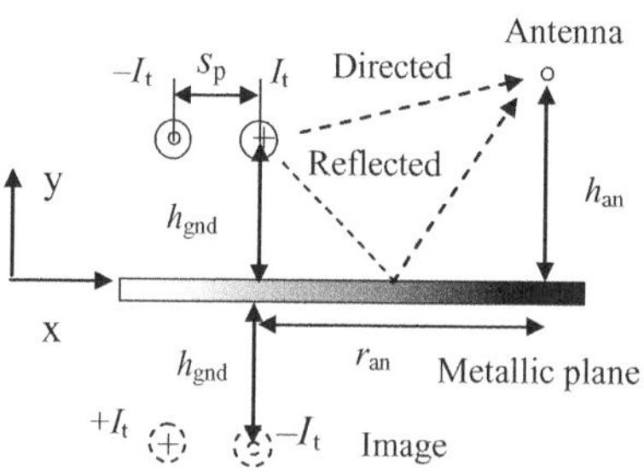 	$\mu = 4\pi \times 10^{-7}$A/m, $\varepsilon = 8.854 \times 10^{-12}$F/m, $c = 1/(\mu\varepsilon)^{1/2}$, $\lambda = c/f$, $\beta = 2\pi/\lambda$ $\eta = (\mu/\varepsilon)^{1/2} = 377\Omega$ $r(x, y) = \sqrt{x^2 + y^2}$ *E*-field *z*-component: $\hat{E}(\omega, x, y) = \frac{j\eta\beta(\omega)\hat{I}_t(\omega)l}{4\pi} \frac{e^{-j\beta(\omega)r(x,y)}}{r(x, y)}$ $\hat{E}_d(\omega) = \hat{E}(\omega, r_{an}, h_{an} - h_{gnd}) - \hat{E}(\omega, r_{an} + s_p, h_{an} - h_{gnd})$ $\hat{E}_i(\omega) = \hat{E}(\omega, r_{an} + s_p, h_{an} + h_{gnd}) - \hat{E}(\omega, r_{an}, h_{an} + h_{gnd})$ $\hat{E}_t(\omega) = \hat{E}_d(\omega) + \hat{E}_i(\omega)$
CM radiated field with a reference metallic plane 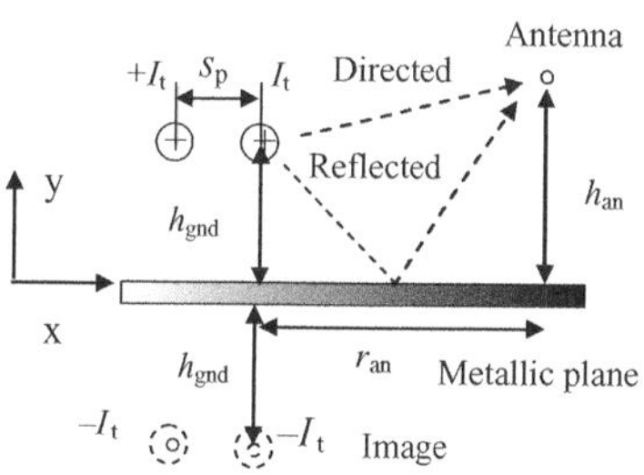 	$\hat{E}_d(\omega) = \hat{E}(\omega, r_{an}, h_{an} - h_{gnd}) + \hat{E}(\omega, r_{an} + s_p, h_{an} - h_{gnd})$ $\hat{E}_i(\omega) = -\hat{E}(\omega, r_{an} + s_p, h_{an} + h_{gnd}) - \hat{E}(\omega, r_{an}, h_{an} + h_{gnd})$ $E_t(\omega) = E_d(\omega) + E_i(\omega)$

- *Long wire*. This is a particular case of a more general structure in which, once the current distribution is known, the field can be computed by segmenting the structure into several electrically short dipoles. The resulting field at the observation point is determined as the sum of all the individual dipole contributions. The number of segments is dictated by the maximum frequency of interest. Note that, for a long wire and angle $\vartheta_0 = 90°$, the expression to be used is the same as for an infinitesimal dipole in which the distance $r(\xi)$ is replaced by r_0, that is, the distance between the central point of the antenna and the observation point. The average current $\hat{I}_t$ is assigned to the generic current $\hat{I}(\xi)$.

D.2 Wires and Ground Planes

Very often the calculation of the radiated field from a PCB with an attached cable is required when a reference ground plane is present. This is the case when the models are to reproduce radiated emission measurements performed in a semi-anechoic chamber for EMC limit

Table D.3 Formulae for calculating the radiation from apertures in a shielded box

<table>
<tr><th>Configuration</th><th>Formulae
All dimensions in meters, Field in V/m</th></tr>
<tr><td>

Small radiating loop

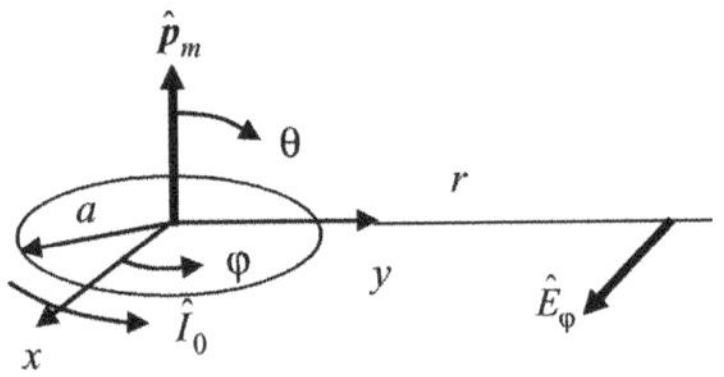

</td><td>

$$\hat{E}_\varphi = \eta \frac{(\beta a)^2 \hat{I}_0 \sin\theta}{4r} \left[1 + \frac{1}{\mathrm{j}\beta r}\right] \mathrm{e}^{-\mathrm{j}\beta r}$$

or

$$\hat{E}_\varphi = \eta \frac{\beta^2 \hat{p}_\mathrm{m} \sin\theta}{4\pi} \left[\frac{1}{r} + \frac{1}{\mathrm{j}\beta r^2}\right] \mathrm{e}^{-\mathrm{j}\beta r}$$

A magnetic dipole of moment $\hat{p}_\mathrm{m} = \pi a^2 \hat{I}_0$ is equivalent to a small electric loop of radius a and constant electric current $\hat{I}_0$.

</td></tr>
<tr><td>

Shielded box with a rectangular aperture

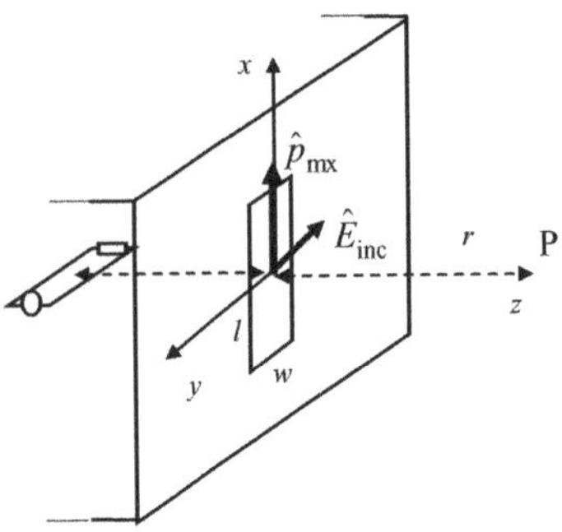

The radiating source (a circuit), the centre of the aperture, and the point of observation P are aligned. The dimensions w and l are electrically short, in other words, $w, l < \lambda/20$, were λ is the wavelenth at maximum frequency of interest.

</td><td>

The far-field E_φ at point P is due to a magnetic dipole of moment $\hat{\boldsymbol{p}}_\mathrm{m} = \hat{p}_{\mathrm{mx}} \boldsymbol{x}$ located in the aperture:

$$\hat{p}_{\mathrm{mx}} = -4\alpha_{\mathrm{m,xx}} \hat{H}_{\mathrm{inc}} = -4\alpha_{\mathrm{m,xx}} \frac{\hat{E}_{\mathrm{inc}}}{\eta}$$

$$\alpha_{\mathrm{m,xx}} = \frac{\pi}{24} \frac{l^3 \cdot \mathrm{e}^3}{K(\mathrm{e}) - E(\mathrm{e})}, \; \eta = 377\Omega$$

with

$$\mathrm{e} = \sqrt{1 - \left(\frac{w}{l}\right)^2}$$

K and E are the complete elliptic integrals of the first and second kind.
$\hat{E}_{\mathrm{inc}}$ is the field on the aperture produced by the radiating source.
For a narrow ellipse or slit ($w \ll l$):

$$\alpha_{\mathrm{m,xx}} = \frac{\pi}{24} \frac{l^3}{\ln(4l/w) - 1}$$

For a circle of diameter d: $\alpha_{\mathrm{m,xx}} = d^3/6$

</td></tr>
</table>

compliance. *Differential-mode* emission occurs in a case of two parallel wires or in case of a wire above a PCB ground plane. For a radiating structure like this, the image method can be applied twice: once for the PCB structure and once for the PCB and metallic floor of the semi-anechoic chamber. The method consists in replacing the ground plane with an image conductor parallel to the signal conductor at a distance twice that between the signal conductor and the ground plane. This operation results in the two cases depicted in Table D.2. For both cases the expression for long wire must be applied 4 times, considering the directions of the currents and the distance of the signal and image wires from the observation point (usually the location of the antenna for measurements).

D.3 Emission from Apertures

Especially for complex digital equipment consisting of several PCBs connected by motherboards and cables, the solution of shielding the equipment is adopted to mitigate radiated emission. Unfortunately, apertures are necessary for cooling, and the apertures become sources of emission excited by the fields produced by the internal circuits. Generally, the computation of these fields is a complicated process and full-wave codes are required. However, when the maximum dimension of the aperture is electrically short, or less than the wavelength at the maximum frequency of interest, the analytical method illustrated in Table D.3 can be adopted [2]. The starting point is that an electrically small loop with constant current $\hat{I}_0$ radiates as a small magnetic dipole of moment $\hat{\boldsymbol{p}}_m$. It can be shown that an electrically short aperture radiates as an electric (perpendicular to the aperture) and two magnetic dipoles (orthogonal to and on the plane of the aperture). The dipole moment is determined by the geometrical dimensions, as discussed in *Section 9.8.2*. The structure with a rectangular aperture shown in Table D.3 is a particular case where the circuit within the shielded box produces in the aperture an electric field oriented as the y axis and a magnetic field oriented as the x axis. In this situation, the magnetic dipole moment $\hat{p}_{mx}$ only is present. The radiated field from the aperture in the far field can be calculated using the expression for the small loop, where $\hat{p}_{mx}$ is directly proportional to the coefficient $\alpha_{m,xx}$ which depends on the minimum w and maximum l dimension of the aperture.

For generic orientation of the EM incident fields on the aperture, magnetic dipole moments $\hat{p}_{mx}$ and $\hat{p}_{my}$ along the x and y axes, respectively, and the electrical dipole moment $\hat{p}_{ez}$ along the z axis are present. The link with the incident fields on the aperture are coefficients tabulated according to the type of aperture [2].

References

1. Balanis, C., '*Antenna Theory: Analysis and Design*', 2nd edition, John Wiley & Sons, Inc., New York, NY, 1997.
2. Tesche, F., Ianoz, M., and Karlsson, T., '*EMC Analysis Methods and Computational Models*', John Wiley & Sons, Inc., New York, NY, 1997.

Appendix E

The Nodal Method to Calculate the Partial Inductance of Finite Ground Planes

In this appendix the nodal method for analysis in the frequency domain is introduced and applied to compute the effective partial inductance L_{gnd} associated with the return path of two typical trace structures in PCBs: a microstrip and a stripline with finite ground planes. This inductance is very important for *Ground Loop Coupling* (GLC) calculation (see *Section 10.1*) and radiated emission prediction (see *Chapter 9*). The method is validated by comparing the results with those obtained by SPICE and by the method of moments.

E.1 Nodal Method Equations

Consider a network with $(N + 1)$ nodes (0 reference node and $h = 1, 2, \ldots, N$ remaining nodes) and B branches, each represented as in Figure E.1. Currents and voltages at each branch can be computed by the following equations in matrix form [1]:

$$\hat{\mathbf{E}} = \mathbf{A}\left[\left(\mathbf{A}^{t}\hat{\mathbf{Y}}\mathbf{A}\right)^{-1}\mathbf{A}^{t}(\hat{\mathbf{I}}_{s} - \hat{\mathbf{Y}}\hat{\mathbf{E}}_{s})\right] \tag{E.1}$$

$$\hat{\mathbf{V}} = \hat{\mathbf{E}} + \hat{\mathbf{E}}_{s} \tag{E.2}$$

$$\hat{\mathbf{I}} = \hat{\mathbf{Y}}\hat{\mathbf{E}} - \left(\hat{\mathbf{I}}_{s} - \hat{\mathbf{Y}}\hat{\mathbf{E}}_{s}\right) \tag{E.3}$$

$$\hat{\mathbf{J}} = \hat{\mathbf{Y}}\hat{\mathbf{V}} \tag{E.4}$$

where:

- $\hat{\mathbf{E}}_{s} = [\hat{E}_{s1}, \ldots, \hat{E}_{sr}, \ldots, \hat{E}_{sB}]^{T}$ and $\hat{\mathbf{I}}_{s} = [\hat{I}_{s1}, \ldots, \hat{I}_{sr}, \ldots, \hat{I}_{sB}]^{T}$ are the vectors containing respectively the independent voltage and current sources of the B branches.
- $\mathbf{A}$ is the reduced incident matrix of B rows and N columns. The generic element A_{hk} assumes the following values: 1 if node h is an initial node of branch r, -1 if node h is a final node of branch r, and 0 otherwise.

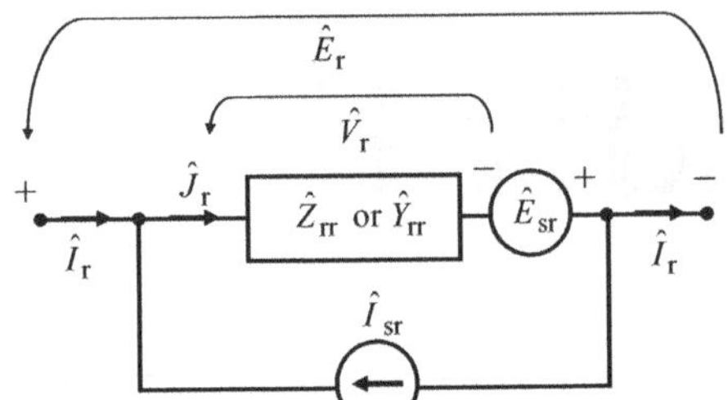

Figure E.1 Generic equivalent circuit of a network branch

- $\hat{\mathbf{E}} = [\hat{E}_1, \ldots, \hat{E}_r, \ldots, \hat{E}_B]^T$ and $\hat{\mathbf{I}} = [\hat{I}_1, \ldots, \hat{I}_r, \ldots, \hat{I}_B]^T$ are vectors containing the voltages and currents of the B branches respectively.
- $\hat{\mathbf{V}} = [\hat{V}_1, \ldots, \hat{V}_r, \ldots, \hat{V}_B]^T$ and $\hat{\mathbf{J}} = [\hat{J}_1, \ldots, \hat{J}_r, \ldots, \hat{J}_B]^T$ are vectors containing the voltages and currents on the impedance $\hat{Z}_{rr} = 1/\hat{Y}_{rr}$ of the B branches respectively.
- $\hat{\mathbf{Y}} = 1/\hat{\mathbf{Z}}$ is the squared matrix of dimension $B \times B$. $\hat{\mathbf{Y}}$ is the admittance matrix and $\hat{\mathbf{Z}}$ is the impedance matrix. $\hat{\mathbf{Y}}$ and $\hat{\mathbf{Z}}$ are diagonal matrices if there is no coupling between the branches. When the coupling is due to voltage-dependent current sources, some or all of the off-diagonal elements have a non-zero value equal to the value of the coupling parameter between h and k branches. For instance, in the case of mutual inductance between the branches, $\hat{Z}_{hh} = \mathrm{j}\omega L_{hh}$ and $\hat{Z}_{hk} = \mathrm{j}\omega L_{hk}$, with $h \neq k$, where L_{hk} is the mutual inductance between the hth and kth branches. When the coupling is due to current-dependent voltage sources, it is the coupling parameter that must be directly added to the off-diagonal elements of $\hat{\mathbf{Y}}$.

E.2 Nodal Method Applied to Compute the Partial Inductance Associated with a Finite Ground Plane

The *Nodal Method* (NM) can be very useful in signal integrity investigations when the nodal matrix contains many elements. The following example treats the computation of the partial inductance L_{gnd} associated with a finite ground plane of a structure composed of a plane and a conductor above the ground plane, as shown in Figure E.2. This is one of the cases presented in *Section 10.1* and solved by the nodal network approach.

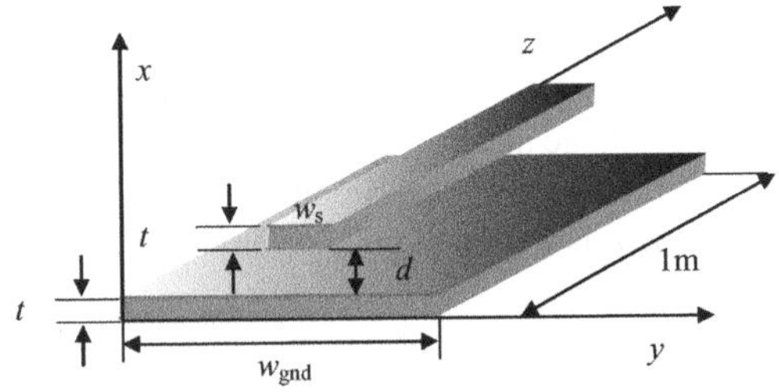

Figure E.2 A trace above a finite ground plane

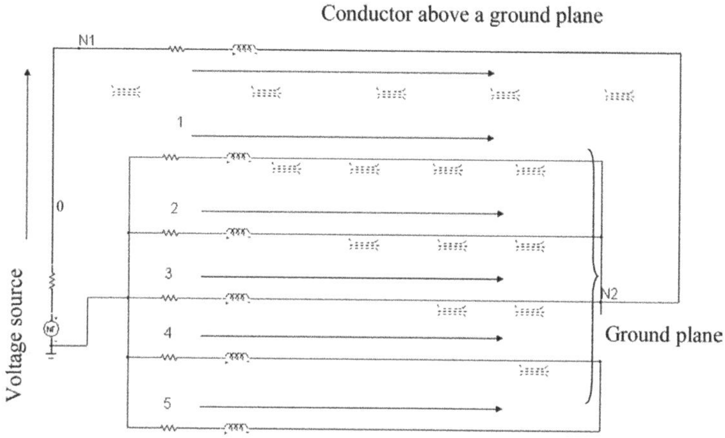

Figure E.3 Equivalent circuit for the trace above a ground plane. The ground is divided into five sub-bars or filaments. All mutual inductances are accounted for

The geometrical dimensions of the structure under investigation (see Figure E.2) are: $w_s = 0.25$ mm, $w_{gnd} = 2.5$ mm, $d = 0.5$ mm, $t = 0.1$ mm. Both conductors are 1 m long. The equivalent circuit of the structure is shown in Figure E.3, where the ground plane is divided into five filaments. Therefore, there are seven branches in total: index 0 for the source, indices 1–5 for the filaments of the ground plane, and index 6 for the conductor above the ground plane.

With the orientation of the currents as indicated in Figure E.3, the matrices for computations are:

$$\mathbf{A} = \begin{bmatrix} -1 & 0 \\ 0 & -1 \\ 0 & -1 \\ 0 & -1 \\ 0 & -1 \\ 0 & -1 \\ 1 & -1 \end{bmatrix} \qquad \hat{\mathbf{E}}_s = \begin{bmatrix} 1 \\ 0 \\ 0 \\ 0 \\ 0 \\ 0 \\ 0 \end{bmatrix} \qquad \hat{\mathbf{I}}_s = \begin{bmatrix} 0 \\ 0 \\ 0 \\ 0 \\ 0 \\ 0 \\ 0 \end{bmatrix}$$

$$\hat{\mathbf{Z}} = j\omega\mathbf{L} = j\omega \begin{bmatrix} 0 & 0 & 0 & 0 & 0 & 0 & 0 \\ 0 & 1722 & 1459 & 1320 & 1239 & 1182 & 1290 \\ 0 & 1459 & 1722 & 1459 & 1320 & 1239 & 1370 \\ 0 & 1320 & 1459 & 1722 & 1459 & 1320 & 1422 \\ 0 & 1239 & 1320 & 1459 & 1722 & 1459 & 1370 \\ 0 & 1182 & 1239 & 1320 & 1459 & 1722 & 1290 \\ 0 & 1290 & 1370 & 1422 & 1370 & 1290 & 1830 \end{bmatrix} \cdot 10^{-9}$$

The values in matrix **L** are calculated by the expressions for the self partial inductance of a rectangular-cross-section conductor and the mutual inductance of two filaments, as reported in Table A.2 of *Appendix A*, and are expressed in H/m.

As stated in *Appendix A*, L_{gnd} is the immaginary part of the ratio between the voltage drop on the ground plane that is caused by a current $\hat{I}$ flowing on the conductor above the ground and the term $\omega\hat{I}$. Therefore, using the elements of matrices $\hat{\mathbf{E}}$ and $\hat{\mathbf{I}}$:

$$L_{gnd} = \frac{1}{\omega}\mathrm{Im}\left[\frac{\hat{E}_1(\omega_0)}{\hat{I}_0(\omega_0)}\right] \tag{E.5}$$

where $\hat{E}_1(\omega_0)$ is the second element of the voltage vector $\hat{\mathbf{E}}$, $\hat{I}_0(\omega_0)$ is the first element of the current vector $\hat{\mathbf{I}}$, and ω_0 is the angular frequency used for calculation. In the case considered, $f_0 = 10$ MHz. As, for simplicity, the resistances are not considered in the calculation of matrix $\hat{\mathbf{Z}}$, L_{gnd} is not dependent on the frequency chosen for calculation. The result of the *Nodal Method* is $L_{gnd\text{-}NM} = 80.632$ nH, as against the SPICE result of $L_{gnd\text{-}SPICE} = 80.630$ nH. With the approximate closed-form expression of Table A.2, $L_{gnd\text{- Analytical}} = 75.1$ nH.

The advantage of NM over the SPICE method consists in the fact that NM can be easily implemented in mathematical software such as MathCad or Matlab, and the number of sub-bars used to decompose the ground plane can be expanded to higher values without great effort. For instance, consider the same microstrip structure as in Figure 10.9 of *Section 10.2*, with $w_s = 2$ mm, $w_{gnd} = 30$ cm, $d = 3$ mm, and $t = 0.1$ mm for 0.3 m of conductor length. Choosing 150 filaments for the ground plane, sub-bars of width 2 mm are obtained. With these values, $L_{gnd\text{-}NM} = 1.26$ nH and $L_{gnd\text{-}Analytical} = 1.16$ nH. As validation of the NM, the curve of the distribution of the current density along the ground plane is very close to the distribution obtained applying the *Method of Moment* (MOM) (compare Figure E.4a and Figure 10.9). The quantity $\hat{I}_0(\omega_0)$ is the current on the trace, and $\hat{I}_h(\omega_0)/\Delta$ is the density of current associated with the hth filament of the ground plane, where $\Delta = w_m/150$, computed at the angular frequency ω_0.

The same computation was performed for the stripline structure of Figure 10.10. The result shown in Figure E.4b should be compared with the curve of Figure 10.10 for $\Delta w = 0$. Although the two approaches are completely different (the NM method applies the partial inductance concept, whereas the MOM method applies incident and scattered fields), the results are in very good agreement, except for some slightly different values at the edge of

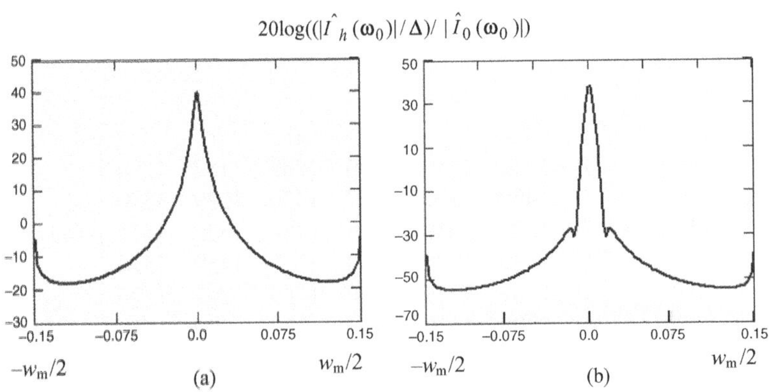

Figure E.4 Current density distribution computed by the *Nodal Method*, normalized to the total current injected into the ground plane: (a) microstrip structure; (b) stripline structure

the planes, as the computation with MOM (see *Section 10.2.1*) was performed at 1 GHz. Performing computation at 1 MHz by MOM, it can be shown that there is also perfect agreement of the curves on the edges.

References

[1] Adby, P.R., '*Applied Circuit Theory: Matrix and Computer Methods*', Ellis Horwood Limited, distributed by John Wiley & Sons, Ltd, Chichester, UK, 1980.

Appendix F

Files on the Web

In this last appendix the reader can find a list of files available on the web concerning computations by MathCad (file.MCD) or simulations by MicroCap9 (file.CIR) reported in the book.

F.1 Program Files of Chapter 1

- *CMOS_TL_DIODE_L_CMOS.cir*. The model computes reflected waveforms in an interconnect point-to-point structure where driver and receiver are a CMOS gate formed by three MOS inverters. It is shown, with the option stepping of parameter *pL*, how the signals change according to the presence of the parasitic inductances. Since macros are used for X1 and X2 models, check the path of the required file on your computer. For more details, see *Section 1.3* of the textbook, where the other examples can be easily reproduced with this model. Run transient analysis.

F.2 Program Files of Chapter 2

- *AC244_TR_5pin_TL.cir*. The program calculates the waveforms in a point-to-point interconnect structure by using the IBIS model of the CMOS device AC244. The user can choose a min, typ, and max characteristic of the driver and receiver. For more details concerning the I/O characteristics of AC244, see *Section 2.4*. Run transient analysis. Check the path of the IBIS model on your computer.

F.3 Program Files of Chapter 5

- *TERMINATION_LH.cir*. The program computes the reflections in a point-to-point interconnect owing to low-to-high switching by using the exact lossless line model outlined in *Section 5.2*. Driver and receiver are represented by Thévenin equivalent circuits. The computed waveforms in several termination conditions (unmatched, series, Thévenin, parallel *RC*) can be compared with those shown in Figure 5.27.

- *TERMINATION_HL.cir*. The program computes the reflections in a point-to-point interconnect owing to high-to-low switching by using the exact lossless line model outlined in *Section 5.2*. Driver and receiver are represented by Thévenin equivalent circuits. The computed waveforms in several termination conditions (unmatched, series, Thévenin, parallel *RC*) are obtained.
- *Lossless_line_term_LH.mcd*. The program computes the reflections in a point-to-point interconnect by using the exact lossless line model outlined in *Section 5.2*. Driver and receiver are represented by Thévenin equivalent circuits. The computed waveforms in several termination conditions (unmatched, series, Thévenin, parallel) can be compared with those obtained by SPICE and shown in Figure 5.27.
- *Lossless_line_term_HL.mcd*. The program computes the reflections in a point-to-point interconnect by using the exact lossless line model outlined in *Section 5.2*. Driver and receiver are represented by Thévenin equivalent circuits. The computed waveforms in several termination conditions (unmatched, series, Thévenin, parallel) are derived.

F.4 Program Files of Chapter 6

- *Even&odd mode model.cir*. The program computes the signal and crosstalk waveforms of two coupled lines by using the exact model based on even and odd modes outlined in *Section 6.2*.
- *F00_BOOK_IN_OUT_DC.cir*. The program computes the I/O static characteristics of the F00 device used in the simulation of two and five coupled lines presented in *Chapter* 6. For more details concerning the model parameters, see *Section 6.3*. Run DC analysis.
- *AC00_BOOK_IN_OUT_DC.cir*. The program computes the I/O static characteristics of the AC00 device used in the simulation of five coupled lines presented in *Section 6.4*. For more details concerning the model parameters, see *Section 6.4*. Run DC analysis.
- *Distributed5L_F00_mac.cir*. The program computes the signal and crosstalk waveforms of five coupled lines with F00 digital devices. Details about the device and the coupled-line models with experimental validations can be found in *Section 6.4*. Run transient analysis. As macros are used, check the correct path on your computer.
- *Distributed5L_AC00_package_mac.cir*. The program computes the signal and crosstalk waveforms of five coupled lines with AC00 digital devices. Details about the device and the coupled-line models with experimental validations can be found in *Section 6.4*. Run transient analysis. Since macros are used, check the correct path on your computer.
- *LC_Xtalk_F00_AC00_5lines.mcd*. The program computes the parameters required by the coupled-line model outline in Section 6.4 starting from the per-unit-line matrices of capacitance and inductance. The output of the program is the input for the file: *Distributed5L.mac* in the directory "component_library" of MicroCap.

F.5 Program Files of Chapter 7

- *S_LOSSYTL_ANALYTICAL_10GHz.CIR*. The program computes S_{11} and S_{21} parameters of a coaxial cable by using closed-form expressions for the per-unit-length line impedance and admittance. The VNA has an output and input impedance of 50 Ω. The coaxial cable has a characteristic impedance $Z_0 = 49.94$ Ω. The line is modeled by Laplace sources. The same circuit can be used for any DUT expressed by an equivalent circuit. For more details, see *Section 7.2* and *Section 11.2*. Run AC analysis.

- *S11_S12_trace.mcd*. The program computes the voltages at source and load when the lossy interconnect (a trace in this case) is matched at both ends. The computation is performed in the frequency domain, and the voltages in the time domain are obtained by IFFT. The waveforms obtained can be used as input for the SPICE model, based on the S-parameters outlined in *Section 7.2*. With a simple modification, the program can compute voltages with different sources and loads.
- *Alpha_stripline_Kp.mcd*. The program computes the characteristic impedance and attenuation of a symmetric ($h = b/2$) stripline, considering both skin and proximity effects, and dielectric losses. The coefficient K_p for proximity-effect prediction is computed as the ratio between the attenuation with both skin and proximity effects and the attenuation with the skin effect only. For more details, see *Section 7.1* and *Appendix B*.
- *Alpha_microstrip_Kp.mcd*. The program computes the characteristic impedance and attenuation of a microstrip, considering both skin with proximity effects, and dielectric losses. The coefficient K_p for proximity-effect prediction is computed as the ratio between the attenuation with both skin and proximity effects and the attenuation with the skin effect only. For more details, see *Section 7.1* and *Appendix B*.
- *Twisted75m_eye.mcd*. This program, written in MathCad language, computes the response of a 75 m twisted-pair cable when the line is sourced by a NRZ sequence of bits. The lossy-line model outlined in *Section 7.2*, based on convolution integrals, is used. The required inputs are the S-parameters of the cable in the time domain, measured or computed, and the NRZ sequence for the voltage source. The outputs are the voltages at source and load, obtained by lossless and lossy-line models and the eye diagram on the load. The reader can learn how to implement:
  – the lossy-line model based on S-parameters in the time domain;
  – the convolution integral to solve the lossy line;
  – the expression to obtain the eye diagram.
  Remark: the computation with the line models must be performed on the variation in the voltages and currents in line and not on the total values: DC plus variations.

F.6 Program Files of Chapter 8

- *CMOS.cir*. This circuit implements three MOS inverters in series. It is used in file 'CMOS_2_TL_3chip_mac.cir' for bounce investigation, and in file 'CMOS_TL_DIODE_L_CMOS.cir' of *Chapter* 1 for reflection investigation. View transient and DC analysis for the appropriate simulations.
- *CMOS_IN_OUT_DC_3gates.cir*. The model computes the DC I/O characteristics of a CMOS inverter used in file 'CMOS_2_TL_3chip_mac.cir' for bounce investigation. Run DC analysis.
- *CMOS_2_TL_3chip_mac.cir*. The program simulates the ground and power bounce of a CMOS IC as a result of the inductances associated with the lead and package conductors of the IC. Two gates switch simultaneously, and a third gate is set quiet at low- and high-output level by using the stepping option. For more details, see *Section 8.3.2*. Run transient analysis. Since macros are used for the gates, check the correct path of the required file for X1, X2, X3 models on your computer.
- *Power-Bus_port_matrix.mcd*. This program computes the impedance Z_{11} at port 1 between two parallel rectangular planes separated by a dielectric support of thickness w_z, the

transfer impedance Z_{12} between ports 1 and 2, or Z_{13} between ports 1 and 3 by using the cavity model outlined in *Appendix C*. The excitation is a current source of 1 A amplitude located at port 1. The computation can be performed with the bare board or with three distributed decoupling capacitors. The extension of the program to more capacitors is straightforward.

F.7 Program Files of Chapter 9

- *TRAP.mcd*. The program computes the spectrum of a trapezoidal waveform representing a clock signal. Three expressions are compared: envelope, analytical, and harmonic representation. For more information, see *Section 9.1*.
- *D_INOISE.mcd*. The program computes the spectrum of three waveforms with period T_p representing impulsive noises in a PCB: triangular, Gaussian, and damped oscillation. The signal parameters are defined in *Section 9.1.3*.
- *EMISCM.mcd*. The program computes the *common-mode* current and the consequent radiated field in the far field of a PCB formed by two parallel wires, driven by a digital signal, and terminated on a resistive load. The PCB is isolated and has as reference plane the metallic floor of an open field site. The equivalent circuit of the dipole consists of lumped elements such as the inductance associated with the wires, the capacitance between wires, and the radiation resistance. The dipole is fed by half the voltage source of the interconnect, considering that the *common-mode* current is generated by the asymmetric position of the source with respect to the line. For more details, see *Section 9.2*.
- *EMISDM.mcd*. This program computes the *differential-mode* current and the consequent radiated field in the far field of a PCB formed by a wire above a ground plane, driven by a digital signal, and terminated on a resistive load. The PCB is isolated and has as reference plane the metallic floor of an open field site. Applying image theory, the equivalent circuit of the PBC is a line of two parallel wires spaced at twice the height of the wire from the plane, driven by a doubled voltage source, and terminated with a doubled load. This program can also be used for a microstrip-line structure once the characteristic and propagation delay are known and extended to more complex sources and loads. For more details, see *Section 9.2*.
- *COUTRACK_2wires.mcd*. The program calculates the radiated field in the far-field zone, generated by a cable attached to a PCB and outgoing from a shielded rack. The PCB consists of two parallel wires driven by a digital signal. The *common-mode* current responsible for the radiation is produced by the voltage drop on the return wire as the product of the *differential-mode* current of the circuit and the associated partial inductance L_{gnd} of the return wire. It is assumed that the rack is the reference plane for the vertical path of the cable and the metal floor of the open site is the reference plane for the horizontal path of the cable. For both polarizations, image theory is applied. The phase difference between the direct and image field is accounted for along the distance from the coordinate origin and the observation point. The model is valid up to the frequency where the distance of the cable from the reference plane is electrically short, in this example about 300 MHz. For more details concerning the source of the circuit and the circuit itself, see 'EMISCM.mcd' and *Section 9.6*.
- *COUTRACK_1wire_1plane.mcd*. The program calculates the radiated field in the far field, generated by a cable attached to a PCB and outgoing from a shielded rack. The PCB consists of a wire above a ground plane driven by a digital signal. The *common-mode* current

responsible for the radiation is produced by the voltage drop on the return plane as the product of the *differential-mode* current of the circuit and the associated partial inductance L_{gnd} of the return plane. It is assumed that the rack is the reference plane for the vertical path of the cable and the metal floor of the open site is the reference plane for the horizontal path of the cable. For both polarizations, image theory is applied. The phase difference between the direct and image field is accounted for along the distance from the coordinate origin and the observation point. The model is valid up to the frequency where the distance of the cable from the reference plane is electrically short, in this example about 300 MHz. For more details concerning the source of the circuit and the circuit itself, see 'EMISDM.mcd' and *Section 9.6*.

- *EMSHCM.mcd*. The program computes the radiated field in the far field of a PCB formed by two parallel wires within a shielded box with a rectangular aperture. The low-frequency model of an aperture, described in *Section 9.8.4*, is used. For details about the PCB and set-up, see the file EMISCM.MCD.
- *EMISHDM.mcd*. The program computes the radiated field in the far field of a PCB formed by a wire above a ground plane within a shielded box with a rectangular aperture. The low-frequency model of an aperture, described in *Section 9.8.4*, is used. For details about the PCB and set-up, see the file EMISDM.mcd.

F.8 Program Files of Chapter 10

- *DEFZT_V_K_full.cir*. The program computes the disturbance on the load R_L produced by a voltage source V_i for two return-wire diameter values: 1 mm and 10 mm. Two equivalent circuits are used: one based on mutual inductance and the other based on the transfer impedance concept. It is shown that the two equivalent circuits provide the same results. For more details, see *Section 10.1.2*. Run AC analysis.
- *wrplane.mcd*. The program computes the current density in a finite ground plane produced by an impressed current I_z of frequency f flowing in a wire above the plane. The structure has infinite dimension along the z axis. The method of moment, as outlined in *Section 10.2*, is used. The radiated E-field is also computed at a distance $r = 3$ m, and the consequent radiation pattern is compared with that obtained by using an equivalent short dipole of length l.
- *stripsim.mcd*. The program computes the current density in the two ground planes of a stripline structure produced by an impressed current I_z of frequency f flowing in a wire placed between the two planes. The structure has infinite dimension along the z axis. The method of moment, as outlined in *Section 10.2*, is used. The radiated E-field is also computed at a distance $r = 3$ m.
- *trace_plane_circ_MC.mcd*. The program computes the current density in the plane of a microstrip structure by using the concept of partial inductance. The plane is divided into NM_y filaments. The nodal method outlined in *Appendix E* is used.
- *trace_plane_plane_circ_MC.mcd*. The program computes the current density in the two planes of a stripline structure by using the concept of partial inductance. Each plane is divided into NM_y filaments. The nodal method outlined in *Appendix E* is used.

responsible for the radiation is produced by the voltage drop on the return plane as the cond [illegible]

[illegible]

Index

Printed and bound by CPI Group (UK) Ltd, Croydon, CR0 4YY
16/09/2026
14963389-0006